AF821960

Instrumentation Measurement and Analysis

Third Edition

About the Authors

B C Nakra is presently Professor Eminence, Mechanical and Automobile Engineering Department at the Institute of Technology and Management Gurgaon, Haryana. He did his PhD from Imperial College of Science and Technology, London, and started his academic career at IIT Kharagpur, followed by long service at IIT Delhi during which he worked as Professor and Head, Mechanical Engineering Department; Head, Instrument Design and Development Centre; Head, ITMME Centre and held BHEL and RRM Chairs and several other positions. He has been involved in teaching and research in Vibration Engineering, System Dynamics, Instrumentation, Automatic Controls, Mechatronics and Engineering Design for over four decades.

K K Chaudhry is presently Professor, Mechanical and Automobile Engineering Department at the Institute of Technology and Management, Gurgaon, Haryana. Prior to joining this department, he was Professor in the department of Applied Mechanics of IIT Delhi. During his service at IIT Delhi, he had brief tenures of visiting assignments to Imperial College, London; University of Technology, Baghdad; and Department of Medical Sciences, University of Paris VII, Paris. He has been involved in teaching, research and industrial consultancy for more than four decades in the areas of Biomechanics, Fluid Mechanics, Instrumentation, Environmental Engineering, Wind Engineering and Industrial Aerodynamics.

Instrumentation Measurement and Analysis

Third Edition

B C Nakra
Professor Eminence
Department of Mechanical and Automobile Engineering
Institute of Technology and Management
Gurgaon, Haryana

K K Chaudhry
Professor
Department of Mechanical and Automobile Engineering
Institute of Technology and Management
Gurgaon, Haryana

Tata McGraw Hill Education Private Limited
NEW DELHI

McGraw-Hill Offices
New Delhi New York St Louis San Francisco Auckland Bogotá Caracas
Kuala Lumpur Lisbon London Madrid Mexico City Milan Montreal
San Juan Santiago Singapore Sydney Tokyo Toronto

Tata McGraw Hill

Published by the Tata McGraw Hill Education Private Limited,
7 West Patel Nagar, New Delhi 110 008.

ISBN (13): 978-0-07-068108-8
ISBN (10): 0-07-068108-2

Managing Director: *Ajay Shukla*

Head—Higher Education Publishing: *Vibha Mahajan*
Manager—Sponsoring: *Shalini Jha*
Assoc. Sponsoring Editor: *Suman Sen*
Development Editor: *Devshree Lohchab*
Executive—Editorial Services: *Sohini Mukherjee*
Sr Manager—Production: *PL Pandita*

General Manager: Marketing—Higher Education: *Michael J Cruz*
Senior Product Manager: SEM & Tech Ed: *Biju Ganesan*

General Manager—Production: *Rajender P Ghansela*
Asst General Manager—Production: *B L Dogra*

Typeset at The Composers, 260, C.A. Apt., Paschim Vihar, New Delhi 110 063

Cover: Rashtriya

RAXCCRQFDDQQQ

The McGraw-Hill Companies

Contents

Preface

We have always felt the need for a suitable textbook on instrumentation encompassing the three main features, viz., instrumentation principles, measurement techniques and data analysis, presented in a form that is lucid and easily comprehensible to students. Currently, both students and teachers have been experiencing difficulty in finding these three aspects highlighted in a single textbook. In fact, the syllabi of most courses on instrumentation/experimental methods for various undergraduate and postgraduate disciplines comprise all the three aspects. Keeping in view the above-mentioned requirements, we have endeavoured to bring out the present textbook based on our wide and long-standing experience of teaching and research in this interdisciplinary field of instrumentation.

The first edition of *Instrumentation, Measurement and Analysis* was published in 1985 and the second edition was published in 2004. There have been several reprints subsequently every year. In view of the area being truly interdisciplinary and several developments in the area taking place, a need for revision was felt by a number of institutes of Science, Engineering and Technology. Comments were invited and received by the publisher from several reputed teachers in the area. These were carefully looked into by the authors and the third edition is based on the above suggestions from various reviewers.

The third edition includes the following new features:

- Two new chapters namely, **Dimensional Metrology** and **Electrical Measurements**
- A new section on **Virtual Instruments**
- Revision of the chapter on **Condition Monitoring and Signature Analysis** and inclusion of sections on **Material Defect Monitoring** and **Acoustic Emission Monitoring** after deletion of the chapter on **Non-Destructive Testing**
- Revision of the chapter on **Motion Measurements** with due emphasis on **Vibration Measurements**
- Revision of the chapter on **Introduction to Instruments and Their Representation**
- Addition of new problems in a number of chapters
- Addition of an appendix on **Derivation of Solution for Step Response of Second-Order System Response**
- Deletion of the chapter on Application of **Digital Computers on Experimental Data Analysis**

We have divided the book into three main parts: ***Part I*** deals with the general treatment of instruments and their characteristics, without referring to a particular measurement situation, and contains chapters 1 to 6. **Chapter 1** gives an introduction to instruments and their representation. **Chapters 2** and **3** discuss the static performance characteristics and dynamic characteristics of instruments respectively. **Chapters 4** and **5**, on the other hand, discuss transducer elements and intermediate elements respectively. The last chapter of Part I, **Chapter 6**, describes the various types of indicating, recording and display elements.

Part II gives the details of measurement of actual physical variables referring to Part I whenever necessary. In addition, this section incorporates signal and system applications as well as miscellaneous measurements including process instruments, biomedical devices and environmental air-pollution measuring systems. This part contains chapters 7 to 20.

Chapter 7 describes motion and vibration measurements, while **Chapter 8** deals with dimensional metrology. Balances, hydraulic and pneumatic load cells and elastic force devices are explained in **Chapter 9** on force measurement. **Chapter 10,** on torque and power measurement, describes different types of dynamometers and their calibration. Moderate, high and low pressure measurement is dealt with in **Chapter 11** on pressure measurement, while **Chapter 12** which is on temperature measurement explains different types of temperature scales, and electrical, non-electrical and radiation methods of measuring temperature. **Chapter 13** on flow measurement discusses the various types of flow meters and measuring methods.

Characteristics of sound, loudness, microphones and sound-measuring systems are explained in **Chapter 14** on acoustics measurement. **Chapter 15** on signal and systems analysis deals with analog and digital filters, and system analysis by harmonic and transient testing. Condition monitoring and signature analysis applications are discussed in **Chapter 16**. **Chapter 17** describes the miscellaneous instruments in industrial, biomedical and environmental applications. Computer-aided measurements, fiber optic transducers, microsensors, smart sensors and the like are discussed in **Chapter 18** on recent developments in instrumentation and measurement. Control engineering applications are explained in detail in **Chapter 19**, while **Chapter 20** is on electrical measurements. Different types of electrical measuring instruments, measurement of resistance, inductance, capacitance, voltage, current, magnetic flux, waveform generation, frequency and phase are described in this chapter.

Lastly, ***Part III*** discusses statistical analysis of data with emphasis on computer applications in data analysis. This part contains chapters 21 to 23. **Chapter 21** describes basic statistical concepts like types of measured quantities, central tendency of data, measures of deviation, evaluation of mean and standard deviation of the mean. **Chapter 22** is on normal distribution and discusses Gaussian distribution, normal distribution, central limit theorem and important tests like the significance test and chi-square test. Finally, **Chapter 23** is on graphical representation and curve fitting of data and explains equations of approximating curves, least squares equations and such other topics.

Besides the 23 chapters, this book also has five appendices. **Appendix A-1** is on fundamental and derived quantities in international system of units. **Appendix A-2** deals with the derivation of solution for step response of second-order systems. **Appendix A-3** explains the auto-correlation functions of a random signal. **Appendix A-4** describes the principal strain and stress relations. **Appendix A-5** discusses the statistical properties of a pair of random signals. A **Bibliography** is also provided at the end of the book, which has a list of reference material for further study.

The website of the book can be accessed at http://www.mhhe.com/nakra/ima3 and contains the following material:

For Instructors

- PowerPoint slides
- Solution Manual

For Students

- Chapter on *Non-Destructive Testing (NDT)*
- Chapter on *Applications of Digital Computers in Experimental Data Analysis*
- Web links for additional reading
- Interactive Objective Questions

We have attempted to incorporate the following notable features in the text:

- Interdisciplinary treatment in selecting the contents of the book by incorporating applications from various engineering and applied science disciplines
- Discussions of latest developments including digital computer applications in instrumentation, measurements and analysis
- Discussion of the measurement principles, constructional features, advantages, limitations, etc., of various possible instruments for a particular measurement situation
- Current applications in the area of condition monitoring and signature analysis of machines, in process measurements, biomedical and environmental air-pollution measurement applications
- Emphasis on measurement standards and calibration methods which are essential features of any measurement programme
- Inclusion of a sufficient number of solved examples followed by review questions including objective-type questions within each chapter
- Simple and lucid treatment of statistical analysis of data
- Inclusion of a fairly large number of pertinent and functional figures, relevant tables wherever necessary in various chapters as well as a bibliography at the end
- An introduction to the various systems of units in use
- Suitability to the practising engineers in industry as the text not only emphasises the fundamentals but also gives practical details in the various aspects of instrumentation including the latest advances in this area

We wish to acknowledge our thanks to the following for permitting the use of figures/tables in the present text:

- M/s VDI—Verlag GmbH, Dusseldorf for the use of *Vibration Criterion Chart* (Fig. 15.12) from their publication VDI-Guideline 2056 (1957)
- M/s Butterworth and Co. (Publishers) Ltd., Kent, UK, for the use of *Vibration Criterion* (Fig. 15.13) from their publication *Tribology Handbook* (1973)
- M/s Biometrika Trustees C/o, Imperial College, London, UK, for the use of normal distribution tables (Tables 16.1 and 16.2) and chi-square table (Table 16.3) from their publication *Biometrika Tables for Statisticians*, vol. 1, third edn. (1966)

We would also like to acknowledge the various reviewers who took out time to review the book. Their names are given below.

Surekha Bhanot	*BITS Pilani, Rajasthan*
R K Srivastav	*MNIT Allahabad, Uttar Pradesh*
Zachariah C Alex	*Vellore Institute of Technology, Vellore, Tamil Nadu*
K Arun Kumar	*Easwari Engineering College, Anna University, Chennai, Tamil Nadu*
S Sharmila	*P R Engineering College, Anna University, Thanjavur, Tamil Nadu*
T Sindhuja	*Dr Sivanthi Aditanar College of Engineering,Tiruchendur, Tamil Nadu*
K Udhayakumar	*College of Engineering Anna University, Chennai, Tamil Nadu*
Monojit Mitra	*Bengal Engineering and Science University, Howrah, West Bengal*
M K Paswan	*National Institute of Technology, Jamshedpur, Jharkhand*
P Chattapadhya	*Techno India College of Technology, Howrah, West Bengal*
T Roy	*Dr B C Roy Engineering College, Durgapur, West Bengal*

Maheshappa *Reva Institute of Technology, Belgaum, Karnataka*
Manjunath *University Visveswaraya College of Engineering, Bangalore, Karnataka*

We are grateful to many of our colleagues and numerous students at the Indian Institute of Technology Delhi who have contributed by way of their constructive and useful discussions. Last but not the least, we also owe our gratitude to the Director, Co-ordinator, Quality Improvement Programme; Co-ordinator, Curriculum Development Cell of Electrical Engineering Department of IIT Delhi for sponsoring the writing of this textbook and the National Book Trust of India for subsidising its publication for the benefit of the readers.

B C Nakra
K K Chaudhry

Visual Walkthrough

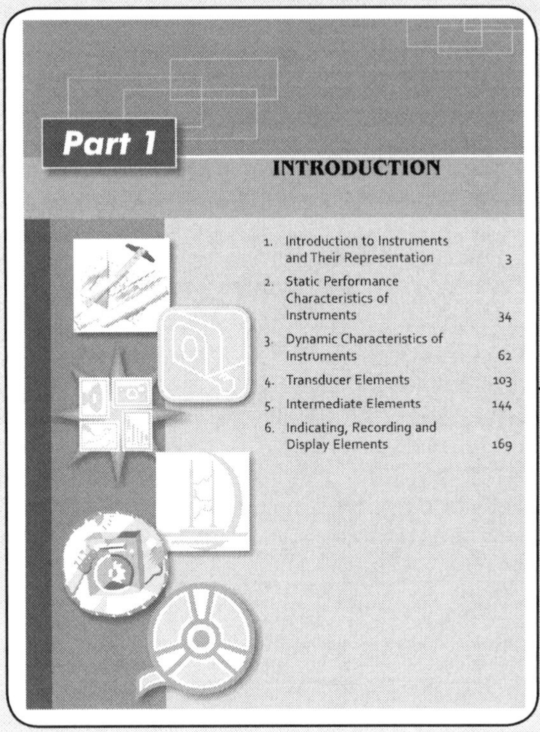

Part 1

INTRODUCTION

Presentation of Text

The various chapters of the book have been sub-divided in three parts. They are

1. Part 1—Introduction
2. Part 2—Measurement, Methods and Applications
3. Part 3—Data Analysis

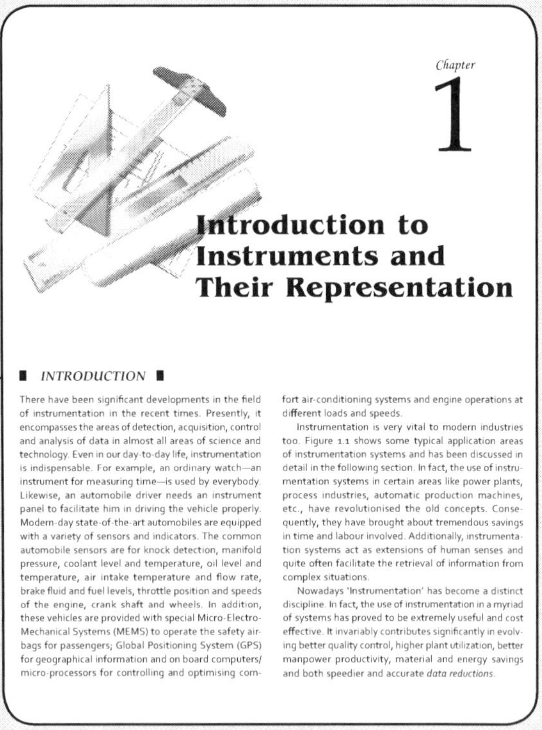

Chapter 1

Introduction to Instruments and Their Representation

■ *INTRODUCTION* ■

There have been significant developments in the field of instrumentation in the recent times. Presently, it encompasses the areas of detection, acquisition, control and analysis of data in almost all areas of science and technology. Even in our day-to-day life, instrumentation is indispensable. For example, an ordinary watch—an instrument for measuring time—is used by everybody. Likewise, an automobile driver needs an instrument panel to facilitate him in driving the vehicle properly. Modern-day state-of-the-art automobiles are equipped with a variety of sensors and indicators. The common automobile sensors are for knock detection, manifold pressure, coolant level and temperature, oil level and temperature, air intake temperature and flow rate, brake fluid and fuel levels, throttle position and speeds of the engine, crank shaft and wheels. In addition, these vehicles are provided with special Micro-Electro-Mechanical Systems (MEMS) to operate the safety air-bags for passengers; Global Positioning System (GPS) for geographical information and on board computers/micro-processors for controlling and optimising comfort air-conditioning systems and engine operations at different loads and speeds.

Instrumentation is very vital to modern industries too. Figure 1.1 shows some typical application areas of instrumentation systems and has been discussed in detail in the following section. In fact, the use of instrumentation systems in certain areas like power plants, process industries, automatic production machines, etc., have revolutionised the old concepts. Consequently, they have brought about tremendous savings in time and labour involved. Additionally, instrumentation systems act as extensions of human senses and quite often facilitate the retrieval of information from complex situations.

Nowadays 'Instrumentation' has become a distinct discipline. In fact, the use of instrumentation in a myriad of systems has proved to be extremely useful and cost effective. It invariably contributes significantly in evolving better quality control, higher plant utilization, better manpower productivity, material and energy savings and both speedier and accurate *data reductions*.

Chapter Introduction

Each chapter has a brief introductory paragraph which gives an overview of the background and contents of the chapter.

8.1 ▌ LINEAR DIMENSIONAL GAUGING

Common linear dimensional measurements include measurement of lengths, widths and heights of components. In addition, quite often depths of holes and slots, etc. also need to be measured. In general, the dimensional linear gauging consists of comparing the unknown dimension of components by means of measuring tools which have been previously calibrated with the standard of known traceability. As discussed in Ch. 1, the international standard of length has been defined in terms of universally reproducible wavelength standard and has accuracy of 1 part in 10^8. In this, one metre length corresponds to 1,650,763.763 wavelengths of light emitted by Kr^{86} orange-red lamp. However, the National level dimensional metrology reference standards often adopted by various countries consist of very high quality slip gauges and length bars of hardened steel whose end faces are lapped flat and parallel to within ±10 nm. Therefore, these reference standards have accuracy specifications of the order of ±0.01 μm and are generally employed to check the calibrations of commonly used industrial instruments in dimensional metrology.

The dimensional gauging instruments can be classified as follows:

- Mechanical type
- Electro-mechanical type
- Pneumatic type
- Hydraulic type
- Optical type
- Special instruments like opto-electronic or fibre-optic type.

8.2 ▌ MECHANICAL TYPE OF DIMENSIONAL GAUGING DEVICES

These devices have marked scales and can be conveniently obtained in the required accuracy specifications. It is easier and quicker to use them over the required range of dimensional measurements. However, the wear and tear due to long useage may introduce inaccuracies in measurements. Some commonly used mechanical types of dimensional gauging devices have been discussed.

8.2.1 Rulers and Tapes

Rulers and tapes are most commonly used tools in our day-to-day lives and shop floors. An engineer's steel rule is also termed *scale*, is a low cost and easy to use, length measuring device. It is made up of hardened steel and is generally available to measure dimensions up to 1000 mm i.e., 1 m. The accuracy of readings using the steel rule of 1 mm engravings is generally ± 0.5 mm, i.e., half the distance between millimeter markings using the judgement of interpolation by eye alone. However, improved type of steel rules marked with 0.5 mm engravings are also available. For such rulers the accuracy of measurements is of the order of ±0.25 mm.

For the measurement of larger dimensions up to 3000 mm or more, retractable type of the steel tapes are generally used. The end of the tape is usually provided with a small hook at 90° to the tape length for convenient placement with the wall for dimensional measurements of buildings/rooms, etc. The thickness of this hook is included in the tape and hence no correction or compensation for its thickness is necessary.

Advantages

1. They provide simplest, low cost, easy and quicker way of measuring a wide range of lengths.
2. They are useful shop floor instruments of measuring lengths where high levels of accuracies is not a requirement.

Sections and Sub-sections

Each chapter has been neatly divided into relevant sections and sub-sections so that the text material is presented in a logical progression of concepts and ideas.

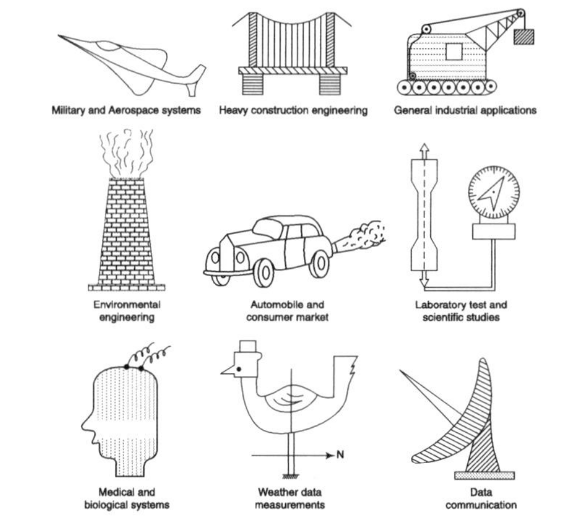

Fig. 1.1 *Typical application areas of instrumentation systems*

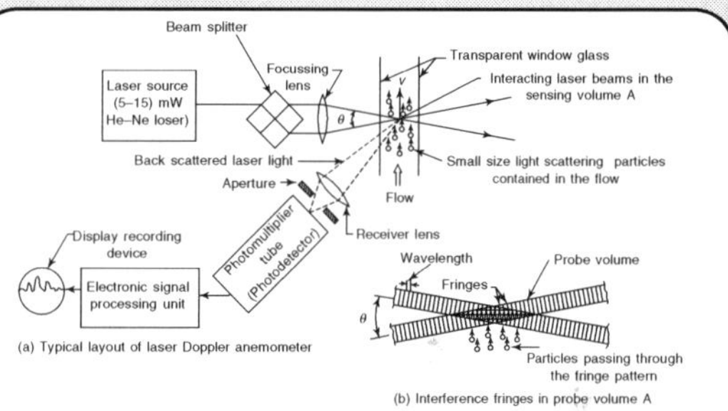

Fig. 13.16 *Details of laser Doppler anemometer in dual beam or fringe mode*

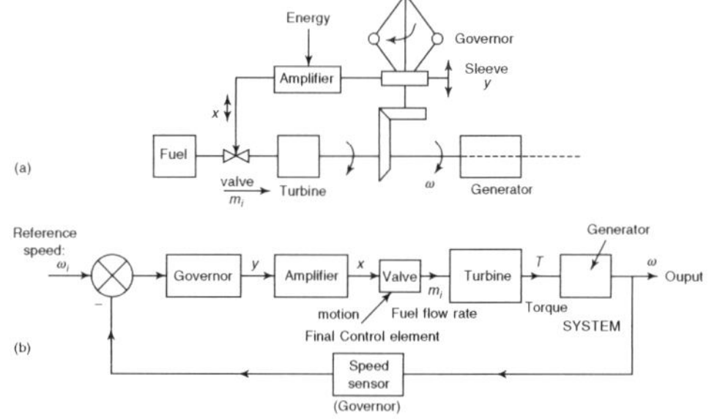

Fig. 19.3 *Representation of a speed control system*

Illustrations

Illustration is an important tool while presenting text material in a clear and lucid manner. Ample number of diagrams/illustrations are provided in each chapter to effectively discuss the concepts of instrumentation principles, measurement techniques and data analysis situations.

Special Topics

Special topics from chapters 15 to 18 have been included which are of interest not only to academicians but also to practising engineers in industry.

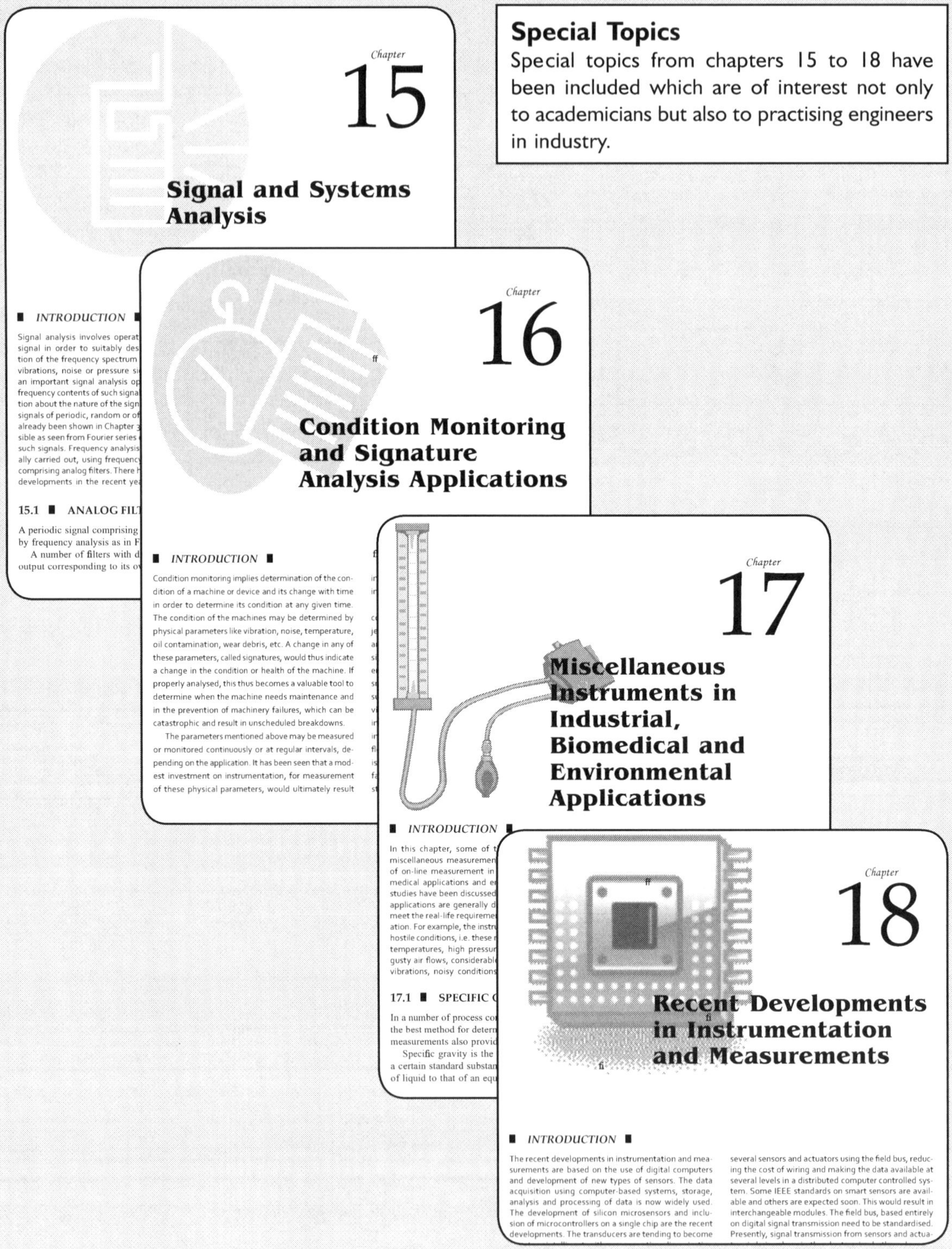

of practical importance. Some useful bridges have been obtained by making two of the four arms of an ac bridges purely resistive. The measurement of unknown capacitance or inductance can be conveniently carried out by standard capacitor/inductances, using the ac bridges with two purely resistive arms.

Problem 20.1 *The impedances of an AC bridge hawing an excitation voltage of 1 kHz are as follows:*
Arm AB with impedance $Z_1 = 100\ \Omega \angle 60°$ (inductive impedance)
Arm AD with impedance $Z_1 = 300\ \Omega \angle 0°$ (purely resistive)
Arm BC with impedance $Z_3 = 50\ \Omega \angle 30°$ (inductive impedance)
and Arm DC with impedance Z_4 = unknown impedance
Determine the R, L or C components of the unknown impedance considering it as series circuit.

Solution For bridge balance, we get,

$$Z_1 Z_4 = Z_2 Z_3$$

Writing the impedances in polar form, we get

$$[100\ \Omega \angle 60°]\ [Z_4] = [300\ \Omega \angle 0°]\ [50\ \Omega \angle 30°].$$

Using Eqs 20.11 and 20.12, we get :

$$Z_4 = (300)\ (50)/(100) = 150\ \Omega \text{ and}$$
$$\theta_4 = 0° + (30°) - (60°) = -30°$$

Therefore, the unknown impedance

$$Z_4 = 150\ \Omega \angle -30°.$$

Further, the negative angle of impedance indicates that Z_4 consists of a series $R - C$ circuit.

Now resistance $R_4 = 150 \cos 30° = 129.9\ \Omega$

and capacitive impedance

$$X_{C_4} = 150° \sin 30° = 75\ \Omega$$
$$= 1/(2\pi \times 1000 \times C_4)$$

or

$$C_4 = 2.12 \times 10^{-6}\ \text{F}$$
$$= 2.12\ \mu\text{F}$$

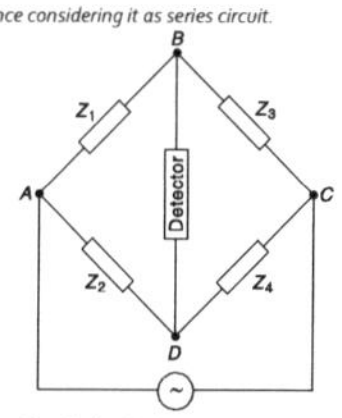

Fig. Prob. 20.1

20.2.2 Measurement of Resistance

Wheatstone-bridge Method A Wheatstone bridge is commonly used for both accuracy and precise measurements of resistance in the range of 1 Ω to 100 kΩ. The ac bridge discussed earlier takes the shape of the Wheatstone bridge if all the arms are purely resistive. The excitation voltage to the bridge may be either ac or dc type. This has been discussed in detail in chapter 4.

Advantages

1. It is a low-cost device and does not require skilled operation.
2. The accuracy of measurement of resistance depends on the accuracy of adjustable, standard resistor which provides null condition for the determination of the unknown resistance. With the use of high-quality standard resistors, accuracies of ± 0.5 % can be achieved.
3. It is used extensively in industrial applications like quality control of resistance wires, determination of resistance of transformers, motor windings, relay coils and solenoids.

Disadvantages

1. It is not possible to measure with reasonable accuracy low values of resistances below 1 Ω, as well as high values of resistances above 100 kΩ.
2. Small errors are caused due to the resistance of connecting wires and contact resistances of the binding posts.

Worked-out Numerical Problems

Sufficient number of worked-out numerical problems have been provided at appropriate places to impart understanding of concepts.

(b) $F = 1 + 2\nu$

(c) $F = 1 + \mu + \dfrac{\frac{d\rho}{\rho}}{\frac{dL}{L}}$

(d) $F = 1 + 2\mu + \dfrac{\left(\frac{d\rho}{\rho}\right)}{\left(\frac{dL}{L}\right)}$

where μ = Poisson's ratio, L = length and ρ = resistivity

(ii) The value of gauge factor for a semiconductor strain gauge used in practice can be approximately
(a) 0.48 (b) 2.05 (c) 3.5 (d) 150

(iii) The most usual value of resistance, suitable for a wire resistance strain gauge is
(a) 12 Ω (b) 50 Ω (c) 120 Ω (d) 2400 Ω

(iv) The calibration of strain gauge bridge circuit is carried out by
(a) heating the active gauge to a known temperature
(b) applying the known voltage across the dummy gauge
(c) applying a known mechanical strain on the active gauge
(d) shunting a known resistance across a dummy gauge

(v) Name the most sensitive type of sensing element for strain measurement
(a) potentiometric transducer (b) wire resistance strain gauges
(c) extensometer (d) semiconductor strain gange

(vi) The most common transducer for shock and vibration measurement is
(a) dial gauge (b) ring type of load cell
(c) LVDT (d) Piezoelectric pick-up

(vii) LVDT, used for displacement measurement is:
(a) an externally power operated transducer
(b) a self generating passive transducer
(c) a capacitive transducer
(d) a digital transducer

(viii) Wheatstrone bridge has got three resistances taken in one direction as 120.3 Ω, 119.2 Ω and 119.2 Ω. The value of the fourth resistance for null balance would be
(a) 120.3 Ω (b) 119.2 Ω (c) 120.0 Ω (d) 118.9 Ω

(ix) LVDT works on the principle of
(a) variable resistance (b) variable self-induction
(c) variable mutual induction (d) variable capacitance

(x) A solar cell is
(a) photo-voltaic transducer (b) photo-emissive transducer
(c) photo-conductive transducer (d) photo-resistive transducer

(xi) Which material out of the following has got the property of generating emf when subjected to mechanical strain
(a) strain gauge material (b) piezo-electric material
(c) steel conductor (d) thermosetting plastics

Objective-Type Questions

Objective-type questions have been included to enable the readers have a clear comprehension of the subject matter. Answers to all the objective-type questions have also been provided.

56 *Instrumentation, Measurement and Analysis*

Therefore, the 'normal' procedure of selecting a particular instrument consists of carefully studying the positive and negative points of each instrument including the prevailing market price and the availability. This combined with mature judgement, intuition and experience helps to arrive at the 'value guided' optimal selection of the instrument for the given application.

Review Questions

2.1 Match the following. Give your answers in the space within the brackets. For example, the answer of part (i) is (6).

(i) Relative error (6)	1. A device whose output is an enlarged reproduction of the essential features of the input wave and which draws power from a source other than the input signal.
(ii) Null type device ( )	2. The act or process of making adjustments or markings on the scale so that the instrument readings conform to an accepted standard.
(iii) Amplifier ( )	3. Measurand generates an opposing effect to maintain zero deflection.
(iv) Drift ( )	4. An action used to convey information.
(v) Attenuator ( )	5. An element which converts the input of energy in a form of an output with different form of energy.
(vi) Signal ( )	6. The ratio of difference between measured value and true value to the true value of the measurand.
(vii) Transducer ( )	7. Maximum distance or angle through which any part of mechanical system may be moved in one direction without causing the motion of the next part.
(viii) Precision ( )	8. Unwanted signal tending to obscure the transducer signal.
(ix) Calibration ( )	9. Gradual departure of the instrument output from the calibrated value.
(x) Resolution ( )	10. A device which causes decrease in amplitude of the signal without causing appreciable distortions in it.
(xi) Noise ( )	11. Smallest increment in measurand that can be detected with certainty by the instrument.
(xii) Backlash ( )	12. The ability of the device to give identical output when repeat measurements are made with the same input signal.

2.2 Indicate if the following statements are true or false. If false, then write the correct statement.

(i) Correctness or exactness in measurements is associated with the accuracy and not with the precision.
(ii) Reproducibility and consistency are expressions that best describe precision in measurements.
(iii) It is not possible to have precise measurements which are not accurate.
(iv) Instrument bias refers to the random errors in the instrument.
(v) An instrument with 1% accuracy is considered better than another with 5% accuracy.
(vi) It is worthwhile to improve the accuracy of the instrument beyond its precision.
(vii) Any measurement is expressed by a numerical value alone.
(viii) Error and uncertainty are synonymous terms.
(ix) To prevent loading of the circuit under test, the input impedance of the voltmeter must be very low.

Review Questions

Each chapter contains a set of review questions which are either design oriented or of numerical type. Solutions of these ptoblems involve the use of application material covered in the chapter. In addition, these are very helpful to the instructors as they can conveniently assign class-work problems, and give home assignments. Answers of the review questions have also been provided.

Bibliography

A relevant list of books and references has been listed for further reading.

Bibliography

Allocca, J.A. and Stuart Allen, *Transducers: Theory and Applications*, Reston Publishing Co., VA-1984.

Barney, G.C., *Intelligent Instrumentation*, Prentice-Hall of India Pvt. Ltd., New Delhi, 1988.

Beckwith, Thomas G., N. Buck Lewis and D. Marangoni Roy, *Mechanical Measurements*, 3rd Ed., Addison-Wesley Reading, Massachussetts, 1982.

Bentley, J.P., *Principles of Measurement Systems*, Pearson Education, New Delhi, 1995.

Bolton, W., *Mechatronics* 3rd Ed., Pearson Education, New Delhi, 2003.

Brignell, J. and White, N., *Intelligent Sensor Systems*, Institute of Physics Publishing, London, Revised Ed., 1996.

Collacott, R.A., *Mechanical Fault Diagnosis and Condition Monitoring*, Chapman and Hall, London, 1977.

Cromwell, Leslie, Weibell, F.J. and Pfeiffer E.A., *Biomedical Instrumentation and Measurements*, 2nd Ed., Prentice-Hall, N.J. 1991.

Dally, J.W. and W.F. Riley, *Experimental Stress Analysis*, 3rd Ed., McGraw-Hill, New York 1991.

Dally, J.W., William, R.F. and McConnell K.G., *Instrumentation for Engineering Measurements*, 2nd Ed., John Wiley and Sons, N.Y. 1993.

Doeblin, E.A. and Manik D.N., *Measurement Systems, Application and Design*, 5th Ed., Tata McGraw Hill Education Private Ltd., 2004.

Figliola, R.S. and Beasley, D.E. *Theory and Design for Mechanical Measurements*, John Wiley and Sons, N.Y. 1991.

Frank, R., *Understanding Smart Sensors*, Artech House Inc., U.S.A., 2000.

Helfrick, A.D. and Cooper, W.D., *Modern Electronic Instrumentation and Measurement Techniques*. Prentice-Hall of India, 1990.

Holman, J.P., *Experimental Methods for Engineers*, 7th Ed., Tata McGraw Hill Education Private Ltd., 2001.

Khandpur, R.S., *Handbook of Biomedical Instrumentation*, Tata McGraw Hill Education Pvt. Ltd., New Delhi, 1987.

Murty, D.V.S., *Transducers and Instrumentation*, 2nd Ed. Prentice-Hall of India Pvt. Ltd., New Delhi, 2008.

Nakra, B.C., *Theory and Applications of Automatic Controls*, New Age International (P) Ltd. New Delhi, 1998.

Nakra, B.C., Yadava, G.S. and Thuestad, L., *Vibration Measurement and Analysis*, National Productivity Council, New Delhi, 1989.

Nottingk, B.E. (Editor), "*Instrumentation Reference Book*, Butterworths, London, 2nd Ed. 1996.

Padmanabhan, T.R., *Industrial Instrumentation—Principles and Design*, Springer-Verlag, London, 2000.

Patranabis, D., *Sensors and Transducers*, Wheeler Publishing, New Delhi, 1997.

Rangan, C.S., G.R. Sarma, and V.S.V. Mani, *Instrumentation—Devices and Systems*, Tata McGraw Hill Education Private Ltd., New Delhi, 1997.

Raj, B, Jayakumar T. and Thavasimuthu M., *Practical Non-destructive Testing*, Narosa Publishing House, N. Delhi, 2nd Ed. 2002.

Part 1

INTRODUCTION

Chapter

1

Introduction to Instruments and Their Representation

INTRODUCTION

There have been significant developments in the field of instrumentation in the recent times. Presently, it encompasses the areas of detection, acquisition, control and analysis of data in almost all areas of science and technology. Even in our day-to-day life, instrumentation is indispensable. For example, an ordinary watch—an instrument for measuring time—is used by everybody. Likewise, an automobile driver needs an instrument panel to facilitate him in driving the vehicle properly. Modern-day state-of-the-art automobiles are equipped with a variety of sensors and indicators. The common automobile sensors are for knock detection, manifold pressure, coolant level and temperature, oil level and temperature, air intake temperature and flow rate, brake fluid and fuel levels, throttle position and speeds of the engine, crank shaft and wheels. In addition, these vehicles are provided with special Micro-Electro-Mechanical Systems (MEMS) to operate the safety air-bags for passengers; Global Positioning System (GPS) for geographical information and on board computers/micro-processors for controlling and optimising comfort air-conditioning systems and engine operations at different loads and speeds.

Instrumentation is very vital to modern industries too. Figure 1.1 shows some typical application areas of instrumentation systems and has been discussed in detail in the following section. In fact, the use of instrumentation systems in certain areas like power plants, process industries, automatic production machines, etc., have revolutionised the old concepts. Consequently, they have brought about tremendous savings in time and labour involved. Additionally, instrumentation systems act as extensions of human senses and quite often facilitate the retrieval of information from complex situations.

Nowadays 'Instrumentation' has become a distinct discipline. In fact, the use of instrumentation in a myriad of systems has proved to be extremely useful and cost effective. It invariably contributes significantly in evolving better quality control, higher plant utilization, better manpower productivity, material and energy savings and both speedier and accurate *data reductions*.

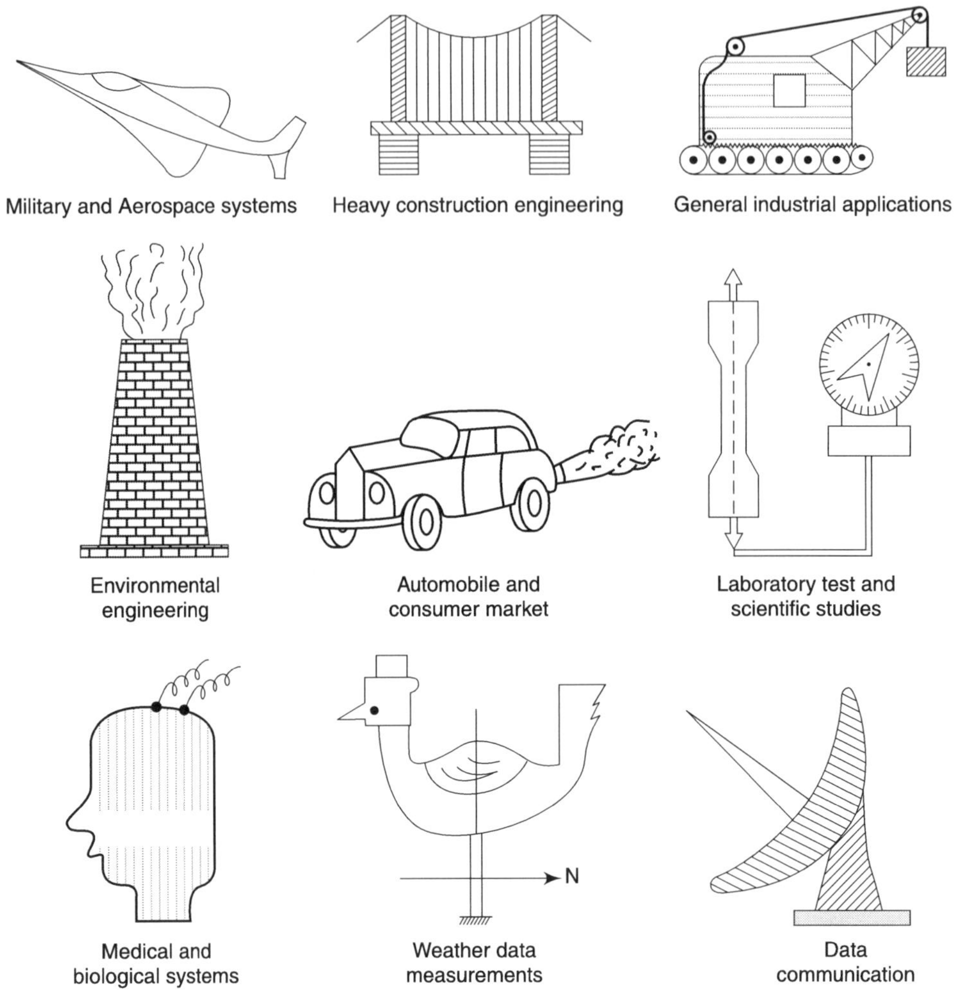

Fig. 1.1 *Typical application areas of instrumentation systems*

1.1 ■ TYPICAL APPLICATIONS OF INSTRUMENT SYSTEMS

The objectives of performing experiments are too numerous to be enumerated. However, certain common motivating factors for carrying out the measurements are as follows:

Measurement of system parameters informations One of the important functions of the instruments is to determine the various parameters/informations of the system or a process. In addition, they present the desired information about the condition of the system in the form of visual indication/registering/recording/monitoring/suitable transmission according to the needs and requirements of the system. In fact, condition-based system of operation is being used very widely these days in a number of situations like the medical care of patients or the maintenance of machines/systems where shut downs are costly/prohibitive, etc.

Control of a certain process or operation Another important application of measuring instruments is in the field of automatic control systems. The measurement system forms an integral part of such systems (Fig. 1.2) which in turn provides deliberate guidance or manipulation to maintain them at a set point or to change it according to a pre-set programme.

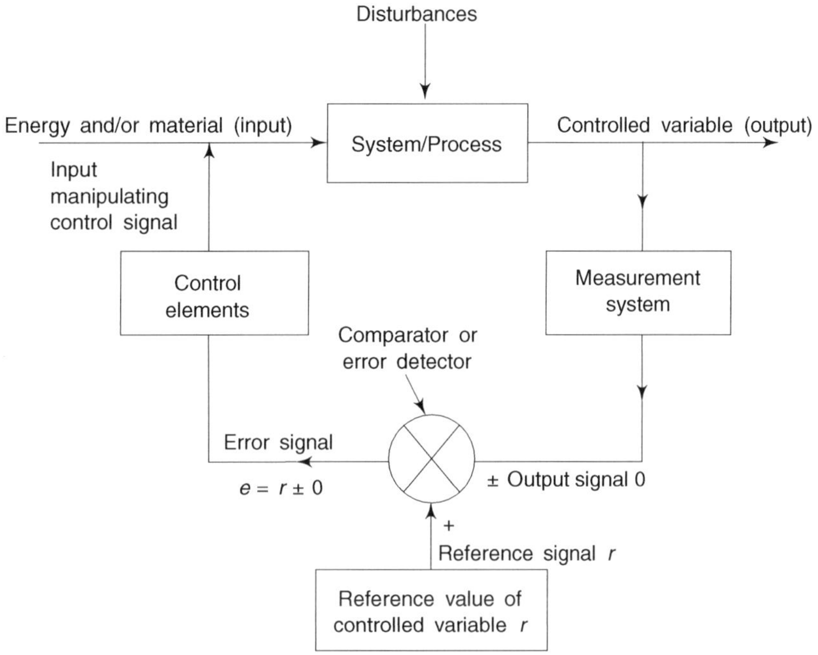

Fig. 1.2 *A typical block diagram of automatic (feedback-type) control system*

The very concept of any control in a system requires the measured discrepancy between the actual and the desired performance. It may be noted that for an accurate control of any physical variable in a process or an operation, it is important to have an accurate measurement system. Further, ***the accuracy of the control system cannot be better than the accuracy of measurement of the control variable***. For example, a thermostat fitted in a domestic refrigerator is a control device for maintaining the temperature in a specified range. Currently, automatic control systems are widely used in process industries like oil refineries, chemical plants, textile mills, etc. for controlling variables like temperature, pressure, humidity, viscosity, flow rate and other relevant parameters. Furthermore, they are also used in modern sophisticated systems like autopilots, automatic landing of aircraft, missile guidance, radar tracking systems, etc.

Simulation of system conditions Sometimes, it may be necessary to simulate experimentally the actual conditions of complex situations for revealing the true behaviour of the system under different governing conditions. Generally, a scale model may be employed for this purpose where the similarity of significant features between the model and the full-scale prototype are preserved. In such cases, analytical tools like dimensional analysis may also be employed to translate the experimental results on the model to the prototype. The lift, drag and other relevant parameters of aerodynamic bodies are usually obtained by testing the models in controlled air streams generated in wind tunnels that simulate the flow conditions experienced by aerodynamic bodies. The information thus obtained is used in the design and development of the prototype.

Experimental design studies The design and development of a new product generally involves trial-and-error procedures which generally involve the use of empirical relations, handbook data, the standard practices mentioned in design codes as well as design equations based on scientific theories and principles. In spite of this, we sometimes have to resort to experimental design studies to supplement design and development work. For example, a design team of experienced aircraft designers put in a number of years of effort to produce a prototype aircraft. The prototype is flown by a test pilot to determine the various performance/operating parameters. The prototype test data is then used to improve further the design calculations and a modified prototype is produced. This is carried on till the desired design performance is achieved. Thus, experimental design studies quite often play an important role in the design and development of the new products/systems.

To perform various manipulations In a number of cases, the instruments are employed to perform operations like signal addition, subtraction, multiplication, division, differentiation, integration, signal linearisation, signal sampling, signal averaging, multi-point correlations, ratio controls, etc. In certain cases, instruments are also used to determine the solution of complex differential equations or other mathematical manipulations. A simple pocket calculator is an example of a mathematical processing instrument, to some extent. Further, the modern large-memory computers are instruments that are capable of varied types of mathematical manipulations.

Testing of materials, maintenance of standards and specifications of products Most countries have standards organisations that specify material standards and product specifications based on extensive tests and measurements. These organisations are meant to protect the interests of consumers. They ensure that the material/products meet the specified requirements so that they function properly and enhance the reliability of the system. For example, an aircraft engine is subjected to extensive endurance tests by the civil aviation authorities as per their specifications, before it is certified to be airworthy.

Verification of physical phenomena/scientific theories Quite often experimental data is generated to verify a certain physical phenomenon. Coulomb postulated that the friction between two dry surfaces is proportional to the normal reaction and is independent of the area of contact. His hypothesis has since been verified experimentally and is now known as Coulomb's law of dry friction. In fact, such examples are numerous. Whenever a scientist or an engineer proposes any hypothesis predicting the system's behaviour, it needs to be checked experimentally to put the same on a sound footing.

In addition, experimental studies play an important role in formulating certain empirical relations where adequate theory does not exist. For example, a number of empirical relations for the friction factor of turbulent flow in pipes (where theoretical basis is inadequate) have been formulated till date by various investigators based on their hypotheses in which numerical constants have been evaluated from experimental data.

Furthermore, experimental studies may be motivated by the hope of developing new theories, discovering new phenomena or checking the validity of a certain hypothesis which may have been developed using some simplifying assumptions.

Quality control in industry It is quite common these days to have continuous quality control tests of mass produced industrial products. This enables to discover defective components that are outright rejected at early stages of production. Consequently, the final assembly of the machine/system is free from defects. This improves the reliability of the product considerably. For example, a boiler plate has to undergo a number of quality control tests before it is put in actual operation. The various tests are:

X-ray examination of the plate for defects like blow holes, cracks, etc.; metallographic examination for metallurgical defects; periodic strength tests of the samples, etc.

1.2 ■ FUNCTIONAL ELEMENTS OF A MEASUREMENT SYSTEM

A generalised 'Measurement System' consists of the following:

1. Basic Functional Elements, and
2. Auxiliary Functional Elements.

Basic Functional Elements are those that form the integral parts of all instruments. These are shown in Fig. 1.3 using *thick lines*. They are the following:

1. *Transducer Element* that senses and converts the desired input to a more convenient and practicable form to be handled by the measurement system.
2. *Signal Conditioning or Intermediate Modifying Element* for manipulating/processing the output of the transducer in a suitable form.
3. *Data Presentation Element* for giving the information about the measurand or measured variable in the quantitative form.

Auxiliary Functional Elements are those which may be incorporated in a particular system depending on the type of requirement, the nature of measurement technique, etc. They are:

1. *Calibration Element* to provide a built-in calibration facility.
2. *External Power Element* to facilitate the working of one or more of the elements like the transducer element, the signal conditioning element, the data processing element or the feedback element.
3. *Feedback Element* to control the variation of the physical quantity that is being measured. In addition, feedback element is provided in the null-seeking potentiometric or Wheatstone bridge devices to make them automatic or self-balancing.
4. *Microprocessor Element* to facilitate the manipulation of data for the purpose of simplifying or accelerating the data interpretation. It is always used in conjunction with analog-to-digital converter which is incorporated in the signal conditioning element.

1.2.1 Some Examples of Identification of Functional Elements in Instruments

Bourdon tube pressure gauge A Bourdon tube pressure gauge is shown in Fig. 1.4(a) along with a block diagram (Fig. 1.4(b)) showing its functional elements. The pressure applied to the hollow oval-shaped bent tube, known as the Bourdon tube, deforms the cross-section of the tube as well as causes a relative motion, proportional to the applied pressure, of the free end of the tube with respect to its fixed end. Thus, this tube acts as a transducer element as it converts the desired input, i.e. pressure into a displacement x at its free end. This displacement is amplified by the combined lever and the gearing arrangement which may be referred to as the signal conditioning elements. Finally, the movement of the pointer attached to the gear on a scale gives an indication of the pressure and thus the pointer and the scale constitute the data presentation elements of the Bourdon tube pressure gauge.

Bourdon pressure gauge with electrical read-out The use of the linear variable differential transducer (LVDT) for sensing the movement of the tip of the Bourdon tube shown in Fig. 1.5(a) improves the performance of the pressure measuring device. The main advantage is that the output of the instrument is electrical and is quite convenient for suitable signal conditioning operations. Further, to achieve other desirable features like linearity, rapidity of response and a small volume displacement, a very stiff and short Bourdon tube is used. The block diagram of this instrument is shown in Fig. 1.5(b). The first block

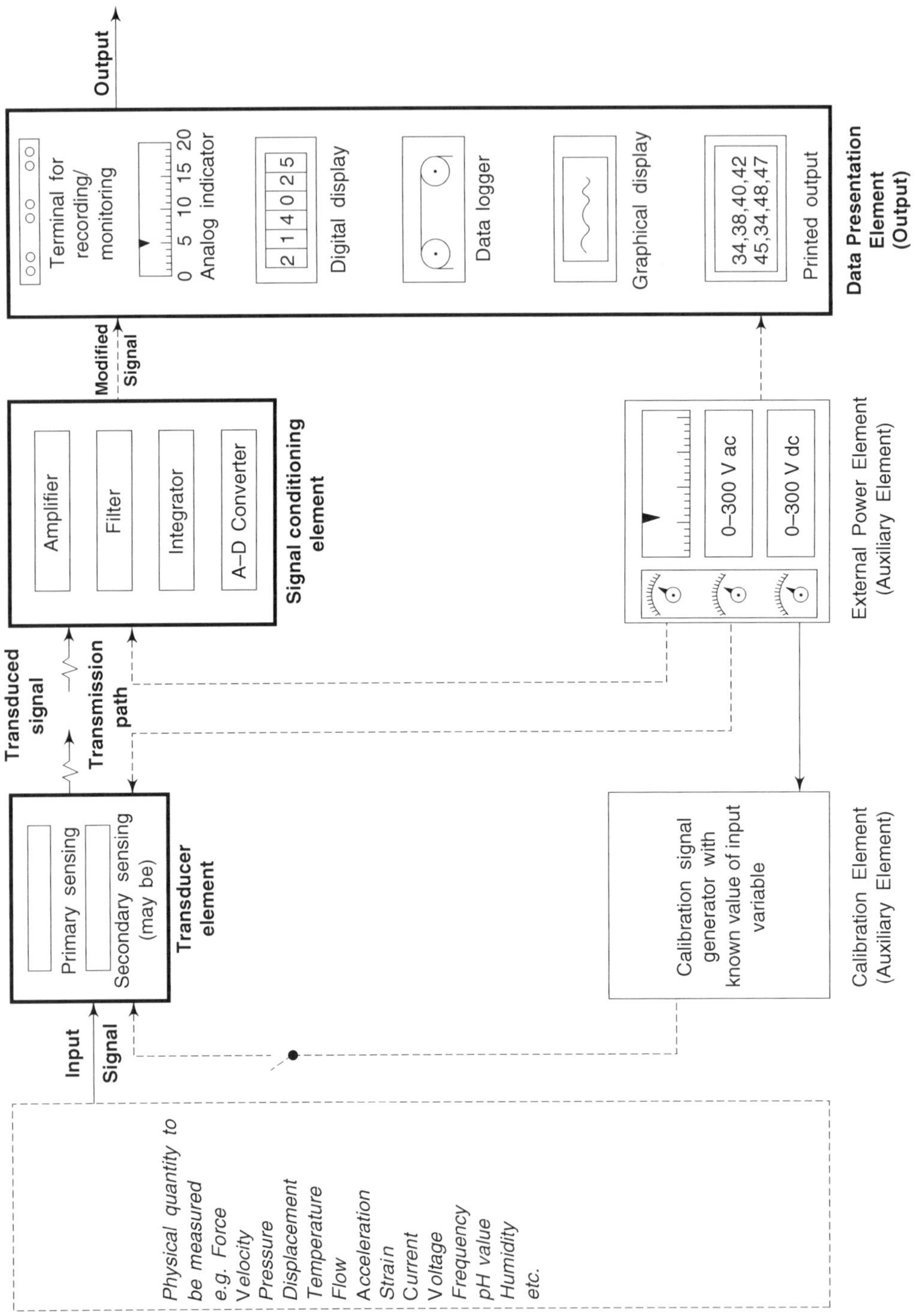

Fig. 1.3 *Basic and auxiliary functional elements of a measurement system*

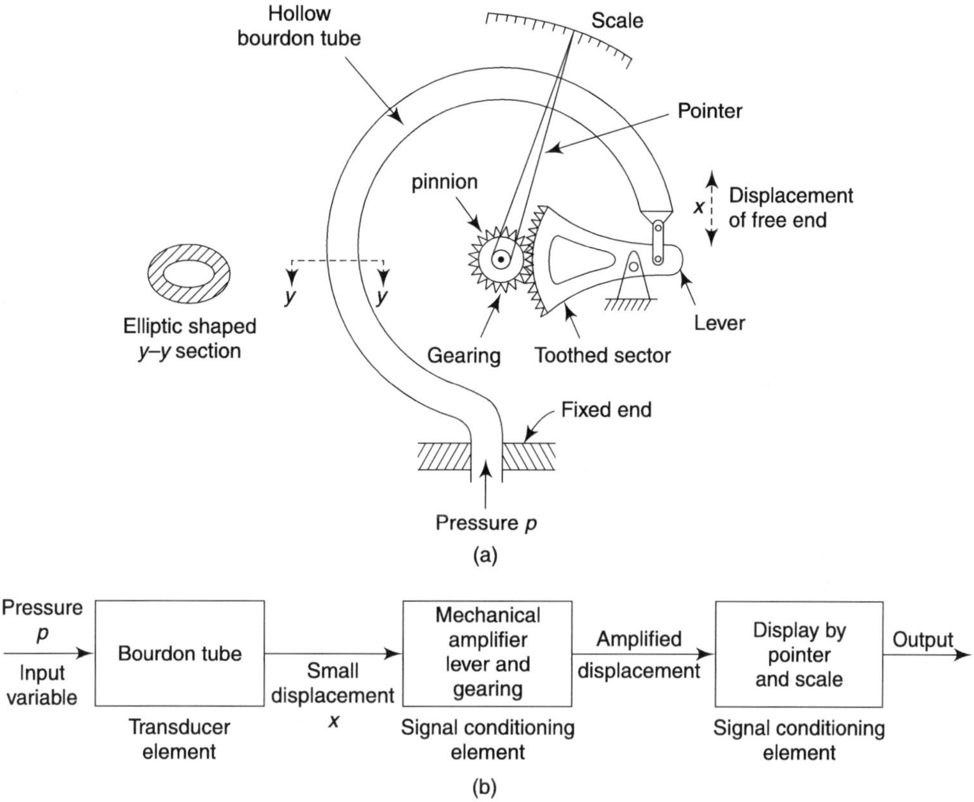

Fig. 1.4 *(a) Bourdon tube pressure gauge (b) Functional elements of the Bourdon tube pressure gauge*

shown in the figure is of the transducer elements. This is because the transduced voltage signal due to the applied pressure is produced by the combined effect of two transducer elements, viz. by the Bourdon tube and the LVDT that may be termed primary and secondary transducer elements, respectively. The output of the transducer elements is processed by the signal conditioning element involving the amplification of the signal and also the filtration of spurious signals present in the transducer signal. Finally, the pressure is indicated in terms of a reading on a suitable analog or digital voltmeter, depending on the form in which the output is desired.

Electrodynamic displacement measuring device For the measurement of linear displacement, a device incorporating the electrodynamic principle is shown in Fig. 1.6(a) along with its block diagram in Fig. 1.6(b). In this device, for measuring the displacement x, a coil wound on a hollow cylinder of non-magnetic material is attached to the moving object. The movement of the coil with respect to a fixed magnet induces a voltage proportional to the rate of change of magnetic flux which in turn is proportional to the velocity of the coil. Thus, the coil and the magnet constitute the transducer element as they produce a voltage signal V_1, proportional to the instantaneous velocity of the object during the course of displacement x of the object. In the signal conditioning element, the transducer signal is suitably amplified and then integrated so that the voltage V_2 is proportional to the displacement. Finally, the output voltage V_2 is indicated on a cathode ray oscilloscope (CRO) which forms the data presentation element of the instrument.

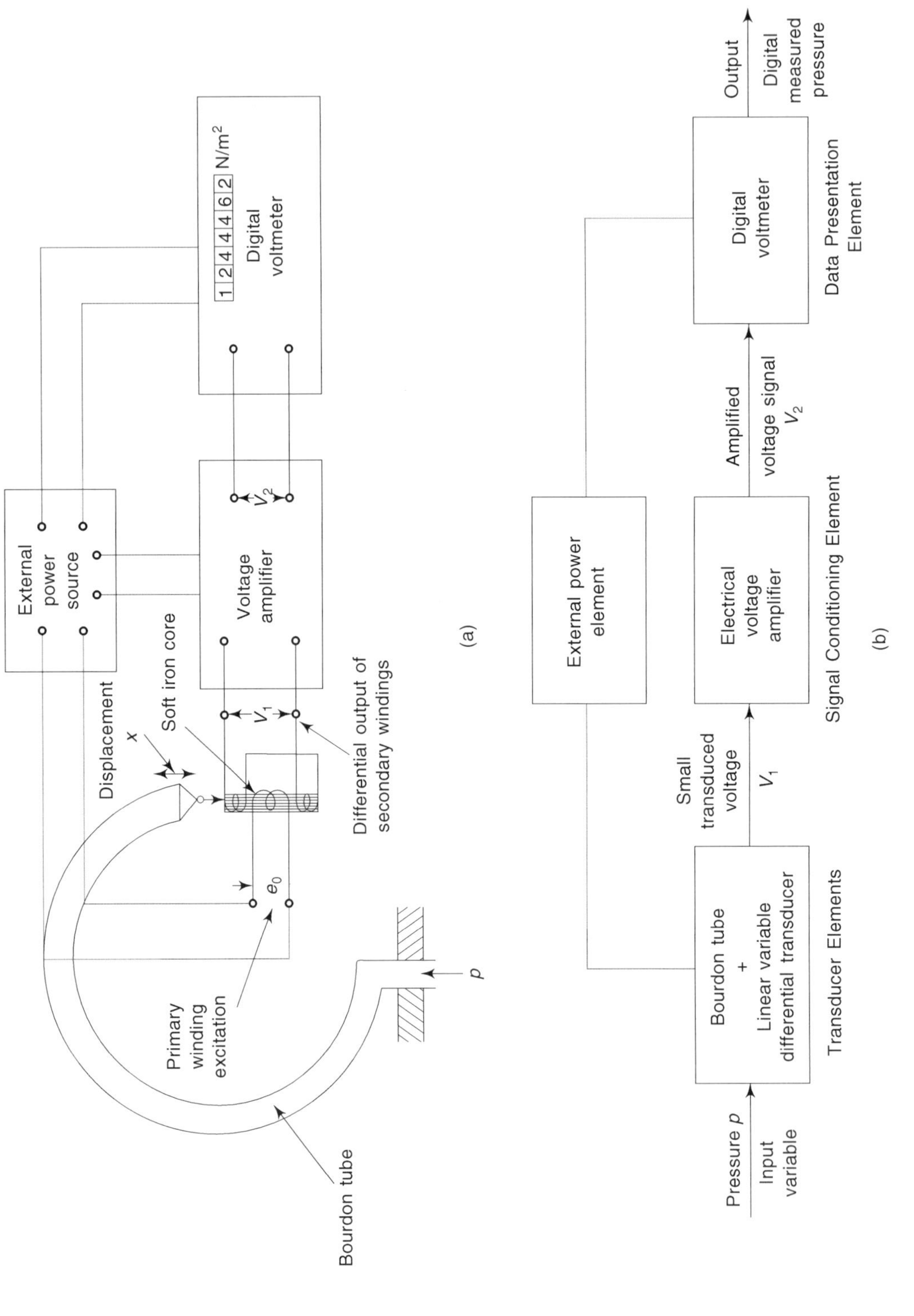

Fig. 1.5 *(a) Bourdon tube pressure gauge with digital read out (b) Functional elements of the digital read out pressure gauge*

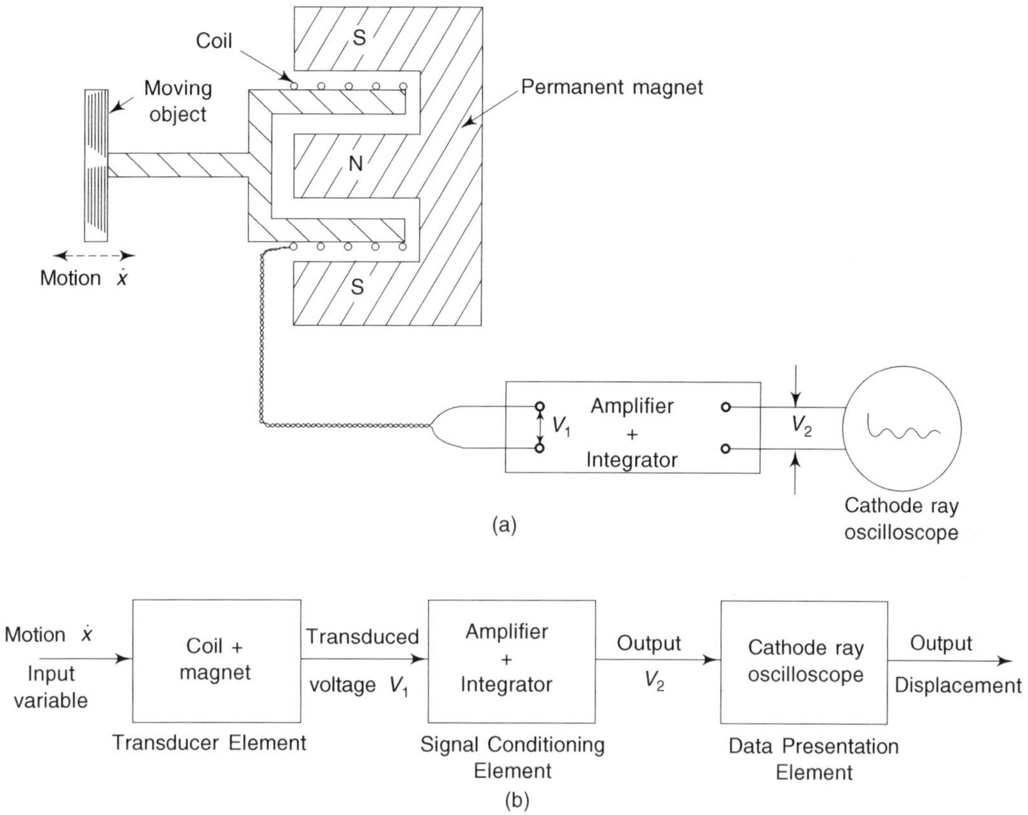

Fig. 1.6 *(a) Electrodynamic type of displacement measuring instrument (b) Functional elements of the Electrodynamic type of displacement measuring instrument*

D'Arsonval Type of Galvanometer For the measurement of either dc voltage V or current I, a permanent magnet moving coil (PMMC) type of galvanometer is shown in Fig. 1.7(a). Herein, the input current I (or current proportional to voltage V) flows in the coil suspended in the magnetic field causes a tangential force on the axial conductors of the coil. This generates a torque T porportional to the input current I. This torque is balanced by means of twin spiral springs, namely, the upper and lower spiral springs. This results in the output θ of the pointer, attached to the shaft, which gives the output indication q_o on the circular scale.

D'Arsonal electromagnetic movement can be conveniently represented in the form of a block diagram shown in Fig. 1.7(b) showing its various functional elements.

- The transducer element converts the input current I in amperes (A) into a torque T (in N.m) with a transfer function K_T in the form of (N.m)/A.
- The signal conditioning element converts the torque T (in N.m) into an angular displacement θ with transfer function K_S in the form of θ/(N.m)
- The data-presentation element converts the angular displacement θ into scale deflection q_0 (in mm) with transfer function K_D in the form of θ/mm.

* For detailed discussion of D'Arsonval galvanometer, refer Ch. 20.

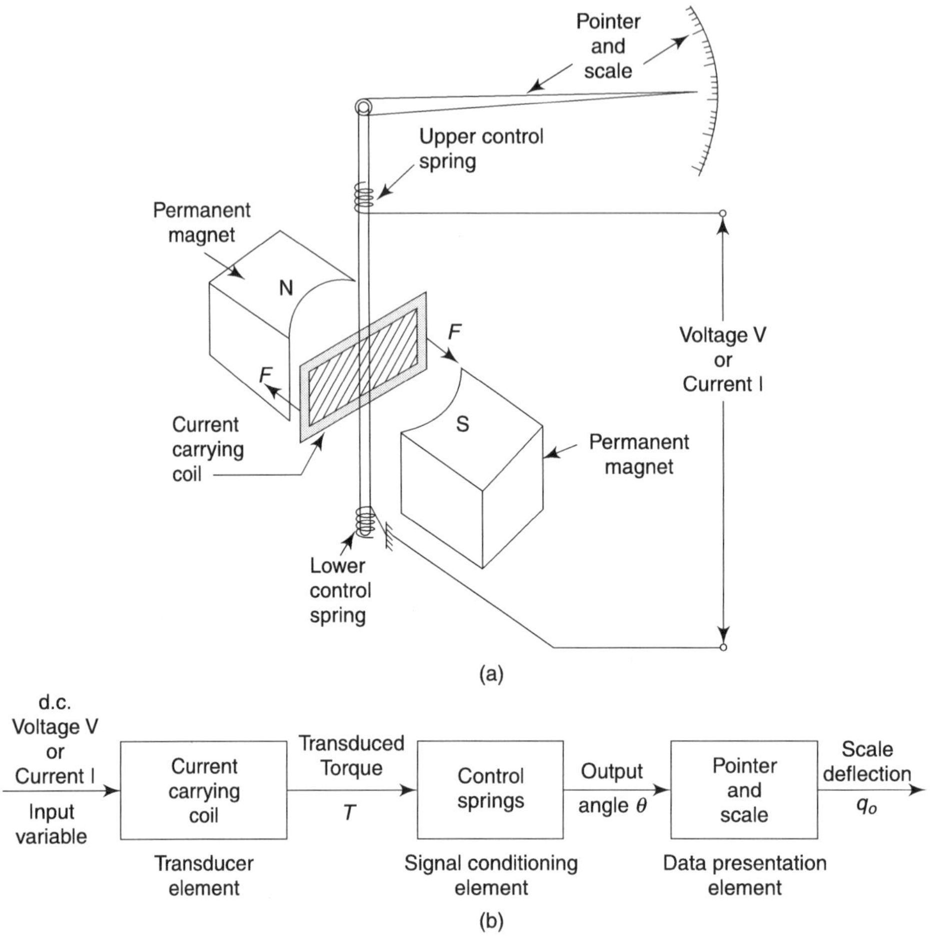

Fig. 1.7 *(a) A Permanent Magnet Moving Coil (PMMC) galvanometer (b) Functional elements of the PMMC galvanometer*

The transfer functions K_T, K_S and K_D can be expressed as follows:

$$\text{Input current } I \text{ (A)} \times K_T = \text{Torque } T \text{(N.m)} \tag{1.1}$$

$$\text{Torque (N} \cdot \text{m)} \times K_S = \text{Angle } \theta \tag{1.2}$$

and

$$\text{Angle } \theta \times K_D = \text{scale deflection } q_o \text{ (mm)} \tag{1.3}$$

From the above equations, we get,

$$\text{Input current } (I) \times (K_T) \times (K_S) \times (K_D) = \text{Scale deflection } q_o \tag{1.4}$$

$$\text{Alternatively, } \frac{dq_o}{dI} = (K_T) \times (K_S) \times (K_D) \text{ mm/A} \tag{1.5}$$

It may be noted that the transfer functions K_T, K_S and K_D are generally constants for steady–state conditions. These values are generally referred as sensitivities or gain or amplifications of the respective functional elements. Further, the overall sensitivity or transfer function of any instrument can be represented as

$$[K]_{\text{overall}} \text{ of the instrument} = (K_T) \times (K_S) \times (K_D) \tag{1.6}$$

Problem 1.1 *An elastic type of pressure-measuring instrument is of diaphragm type. The central deflection of the diaphragm was found to be 0.25 mm of an applied pressure of 10^6 Pa. The output displacement of diaphragm has been fed to an LVDT (linear variable differential transducer) with a built-in amplifier having a sensitivity of 40 V/mm. Finally, the output is displayed on an analog voltmeter which has a radius of scale line as 60 mm and has a voltage range from zero to 10 volts in an arc of 150°. Determine the sensitivity of the given diaphragm gauge in terms of mm/bar (1 bar = 10^5 Pa).*

Solution The block diagram of the pressure-measuring instrument is shown in Fig. 1.8.

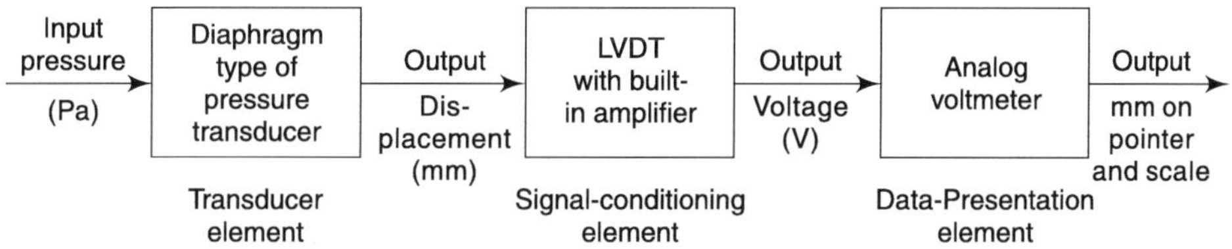

Fig. 1.8 *Block diagram of diaphragm type of electro-mechanical pressure gauge.*

Since the central deflection of the diaphragm gauge is 0.25 mm for an applied pressure of 10^6 Pa; therefore, the sensitivity of the transducer element K_T becomes:

$$K_T = \frac{0.25}{10^6} = 2.5 \times 10^{-7} \text{ mm/Pa.}$$

Further, the output of the transducer is modified by the LVDT system with a built-in amplifier, and the sensitivity K_S of the signal–conditioning system is given as

$$K_S = 40 \text{ V/mm.}$$

Finally the output of the signal–conditioning element is fed to the data presentation element, i.e., an analog voltmeter whose sensitivity K_D can be evaluated is as follows:

$$K_D = \frac{\text{movement of the pointer in mm on the scale}}{\text{voltage range of the voltmeter}}$$

$$= \frac{\left(60 \times 150° \times \dfrac{2\pi \text{ rad}}{360°}\right)}{10} \text{ mm/V} = 15.7 \text{ mm/V}$$

Using Eq. 1.6, the overall sensitivity of the diaphragm pressure gauge becomes

$$(K)_{\text{overall}} \text{ of diaphragm pressure gauge} = (K_T) \times (K_S) \times (K_D)$$

$$= (2.5 \times 10^{-7}) \times (40) \times (15.7) \text{ mm/Pa}$$

$$= 15.7 \times 10^{-5} \text{ mm/Pa}$$

$$= \frac{15.7}{(10^5 \text{ Pa})} = 15.7 \text{ mm/bar}$$

1.3 BRIEF DESCRIPTION OF THE FUNCTIONAL ELEMENTS OF THE INSTRUMENTS

The roles of the various functional elements of the instrument have been explained earlier. The *integrated effect* of all the functional elements results in a useful measurement system. To facilitate mass production, easy maintenance and repairs, the current practice is to have modular type of instruments in which the various functional elements are fabricated in the form of modules or as a combination of certain sub-modules. The brief descriptions of the various functional elements are as follows.

1.3.1 Transducer Element

Normally, a transducer senses the desired input in one physical form and converts it to an output in another physical form. For example, the input variable to the transducer could be pressure, acceleration or temperature and the output of the transducer may be displacement, voltage or resistance change depending on the type of transducer element. Sometimes the dimensional units of the input and output signals may be same. In such cases, the functional element is termed a *transformer*. Some typical examples of transducer elements commonly used in practice are mentioned in Table 1.1.

Table 1.1 ***Typical examples of transducer elements***

S.No.	*Input variable to transducer*	*Output variable of transducer*	*Principle of operation*	*Type of device*
(*1*)	(*2*)	(*3*)	(*4*)	(*5*)
1.	Temperature	Voltage	An emf is generated across the junctions of two dissimilar metals or semiconductors when that junction is heated	Thermocouple or Thermopile
2.	Temperature	Displacement	There is a thermal expansion in volume when the temperature of liquids or liquid metals is raised and this expansion can be shown as displacement of the liquid in the capillary	Liquid in Glass Thermometer
3.	Temperature	Resistance change	Resistance of pure metal wire with positive temperature coefficient varies with temperature	Resistance Thermometer
4.	Temperature	Pressure	The pressure of a gas or vapour varies with the change in temperature	Pressure Thermometer
5.	Displacement	Inductance change	The differential voltage of the two secondary windings varies linearly with the displacement of the magnetic core	Linear Variable Differential Transducer (LVDT)
6.	Displacement	Resistance change	Positioning of a slider varies the resistance in a potentiometer or a bridge circuit	Potentiometric Device
7.	Motion	Voltage	Relative motion of a coil with respect to a magnetic field generates a voltage	Electrodynamic Generator
8.	Flow rate	Pressure	Differential pressure is generated between the main pipe-line and throat of the Venturimeter/Orificemeter	Venturimeter/Orificemeter

(*Contd.*)

(*Contd.*)

(*1*)	(*2*)	(*3*)	(*4*)	(*5*)
9.	Flow velocity	Resistance change	Resistance of a thin wire/film is varied by convective cooling in stream of gas/liquid flows	Hot Wire Anemometer (gas flows). Hot Film Anemometer (liquid flows)
10.	Pressure	Movement of a liquid column	The impressed pressure is balanced by the pressure generated by a column of liquid	Manometer
11.	Pressure	Displacement	The application of pressure causes displacement in elastic elements	Bourdon Gauge
12.	Gas pressure	Resistance change	Resistance of a heating element varies by convective cooling	Pirani Gauge
13.	Force	Displacement	The application of force against a spring changes its length in proportion to the applied force	Spring Balance
14.	Force/torque	Resistance change	The resistance of metallic wire or semiconductor element is changed by elongation or compression due to externally applied stress	Resistance Strain Gauge
15.	Force	Voltage	An emf is generated when external force is applied on certain crystalline materials such as quartz	Piezo-electric Device
16.	Liquid level/thickness	Capacitance change	Variation of the capacitance due to the changes in effective dielectric constant	Dielectric gauge
17.	Speech/music/noise	Capacitance change	Sound pressure varies the capacitance between a fixed plate and a movable diaphragm	Condenser Microphone
18.	Light	Voltage	A voltage is generated in a semiconductor junction when radiant energy stimulates the photoelectric cell	Light Meter/Solar Cell
19.	Light radiations	Current	Secondary electron emission due to incident radiations on the photosensitive cathode causes an electronic current	Photomultiplier tube
20.	Humidity	Resistance change	Resistance of a conductive strip changes with the moisture content	Resistance Hygrometer
21.	Blood flow/any other gas or liquid or two-phase flow	Frequency shift	The difference in the frequency of the incident and reflected beams of ultrasound known as Doppler's frequency shift is proportional to the flow velocity of the fluid	Doppler Frequency Shift Ultrasonic Flow Meter

It may be noted that in certain cases, the transduction of the input signal may take place in two stages or even in the three or more stages namely, primary transduction, secondary transduction, tertiary transduction, etc. For example, in Fig. 1.5(a), the Bourdon tube acts as a primary transducer as it converts the pressure into displacement. The LVDT attached to the free end of the Bourdon tube is the secondary transducer as it converts displacement into electrical voltage. This way the combined effect of primary and secondary transducers coverts the pressure signal into a corresponding voltage signal.

Desirable characteristics of a transducer element: The following points should be borne in mind while selecting a transducer for a particular application are the following:

1. The transducer element should recognise and sense the desired input signal and should be insensitive to other signals present simultaneously in the measurand. For example, a velocity transducer should sense the instantaneous velocity and should be insensitive to the local pressure or temperature.
2. It should not *alter* the event to be measured.
3. The output should preferably be *electrical* to obtain the advantages of modern computing and display devices.
4. It should have good *accuracy.*
5. It should have good *reproducibility* (i.e. precision).
6. It should have *amplitude* linearity.
7. It should have adequate *frequency response* (i.e., good dynamic response).
8. It should not induce phase distortions (i.e. should not induce time lag between the input and output transducer signals).
9. It should be able to withstand hostile environments without damage and should maintain the accuracy within acceptable limits.
10. It should have high signal level and low impedance.
11. It should be easily available, reasonably priced and compact in shape and size (preferably portable).
12. It should have good reliability and ruggedness. In other words, if a transducer gets dropped by chance, it should still be operative.
13. Leads of the transducer should be sturdy and not be easily pulled off.
14. The rating of the transducer should be sufficient and it should not break down.

1.3.2 Signal Conditioning Element

The output of the transducer element is usually too small to operate an indicator or a recorder. Therefore, it is suitably processed and modified in the signal conditioning element so as to obtain the output in the desired form.

The transducer signal may be fed to the signal conditioning element by means of either mechanical linkages (levers, gears, etc.), electrical cables, fluid transmission through liquids or through pneumatic transmission using air. For remote transmission purposes, special devices like radio links or telemetry systems may be employed.

The signal conditioning operations that are carried out on the transduced information may be one or more of the following:

Amplification The term amplification means increasing the amplitude of the signal without affecting its waveform. The reverse phenomenon is termed *attenuation,* i.e. reduction of the signal amplitude while retaining its original waveform. In general, the output of the transducer needs to be amplified in order to operate an indicator or a recorder. Therefore, a suitable amplifying element is incorporated in the signal conditioning element which may be one of the following depending on the type of transducer signal.

1. *Mechanical Amplifying Elements* such as levers, gears or a combination of the two, designed to have a multiplying effect on the input transducer signal.
2. *Hydraulic/Pneumatic Amplifying Elements* employing various types of valves or constrictions, such as venturimeter/orificemeter, to get significant variation in pressure with small variation in the input parameters.
3. *Optical Amplifying Elements* in which lenses, mirrors and combinations of lenses and mirrors or lamp and scale arrangement are employed to convert the small input displacement into an output of sizeable magnitude for a convenient display of the same.
4. *Electrical Amplifying Elements* employing transistor circuits, integrated circuits, etc. for boosting the amplitude of the transducer signal. In such amplifiers we have either of the following:

$$\text{Voltage amplification} = \frac{\text{output voltage}}{\text{input voltage}} = \frac{V_o}{V_i} \tag{1.7}$$

or

$$\text{Current amplification} = \frac{\text{output current}}{\text{input current}} = \frac{I_o}{I_i} \tag{1.8}$$

or

$$\text{Gain} = \frac{\text{output power}}{\text{input power}} = \frac{V_o I_o}{V_i I_i} \tag{1.9}$$

Signal filtration The term signal filtration means the removal of unwanted noise signals that tend to obscure the transducer signal. The signal filtration element could be any of the following depending on the type of situation, nature of signal, etc.

1. *Mechanical Filters* that consist of mechanical elements to protect the transducer element from various interfering extraneous signals. For example, the reference junction of a thermocouple is kept in a thermos flask containing ice. This protects the system from the ambient temperature changes.
2. *Pneumatic Filters* consisting of a small orifice or venturi to filter out fluctuations in a pressure signal.
3. *Electrical Filters* are employed to get rid of stray pick-ups due to electrical and magnetic fields. They may be simple *R–C* circuits or any other suitable electrical filters compatible with the transduced signal.

Other signal conditioning operators Other signal conditioning operators that can be conveniently employed for electrical signals are

1. Signal Compensation/Signal Linearisation.
2. Differentiation/Integration.
3. Analog-to-Digital Conversion.
4. Signal Averaging/Signal Sampling, etc.

1.3.3 Data Presentation Element

This element gathers the output of the signal conditioning element and presents the same to be read or seen by the experimenter. This element should

1. have as fast a response as possible,
2. impose as little drag on the system as possible, and
3. have very small inertia, friction, stiction, etc. (hence using light rays and electron beams is advantageous).

This element may be either of the visual display type, graphic recording type or a magnetic tape. In the visual display type element, devices such as pointer and scale/panel meter, multi-channel CRO, storage CRO, etc. may be employed. The graphic recording type of element gives a permanent record of the input data. The device in this element may be pen recorders using heated stylus, ink recorders on paper charts, optical recording systems such as mirror galvanometer recorders or ultraviolet recorders on special photosensitive paper. Further, a magnetic tape may be used to acquire input data which could be reproduced at a later date for analysis.

In case the output of the signal conditioning element is in digital form, then the same may be displayed visually on a digital display device. Alternatively, it may be suitably recorded either on punched cards, perforated paper tape, magnetic type, typewritten page or a combination of these systems for further processing.

1.4 ■ CLASSIFICATION OF INSTRUMENTS

Instruments may be classified according to their application, mode of operation, manner of energy conversion, nature of output signal and so on. All these classifications usually result in overlapping areas. However, the instruments commonly used in practice may be broadly categorised as follows:

1.4.1 Deflection and Null Types

A deflection type instrument is that in which the physical effect generated by the measuring quantity produces an equivalent opposing effect in some part of the instrument which in turn is closely related to some variable like mechanical displacement or deflection in the instrument. For example, the unknown weight of an object can be easily obtained by the deflection of a spring caused by it on the spring balance as shown in Fig. 1.9. Similarly, in a common Bourdon gauge, the pressure to be measured acts on the C-type spring of the gauge, which deflects and produces an internal spring force to counter balance the force generated by the applied pressure. The deflection of the spring is however magnified by an electrical device like LVDT (Fig. 1.5(a)) or by using suitable lever and gear mechanisms (Fig. 1.4(a)) to be read off from the scale of the instrument.

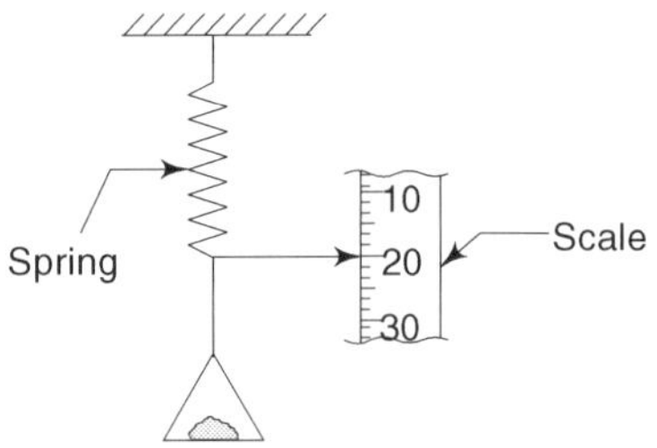

Fig. 1.9 *A typical spring balance—A deflection type weight measuring instrument*

Deflection instruments are simple in construction and operation. In addition, they generally have a good dynamic response. However, the main disadvantage of these instruments are that they interfere with the state of the measured quantity and a small error termed as loading error may be introduced due to this in the measurements.

A null type instrument is the one that is provided with either a manually operated or automatic balancing device that generates an equivalent opposing effect to nullify the physical effect caused by the quantity to be measured. The equivalent null-causing effect in turn provides the measure of the quantity. Consider a simple situation of measuring the mass of an object by means of an equal-arm beam balance shown in Fig. 1.10. An unknown mass, when placed in the pan, causes the beam and pointer to deflect. Masses of known values are placed on the other pan till a balanced or null condition is obtained by means of the pointer. The main advantage of the null-type devices is that they do not interfere with the state of the measured quantity and thus measurements of such instruments are extremely accurate. However, these devices, especially those of the manual type, are quite slow in operation and consequently their dynamic response is quite poor. But, their speed and dynamic response can be improved considerably by using certain feedback type of automatic balancing devices such as instrument servo-mechanisms. Nowadays the instruments of this type are of great importance.

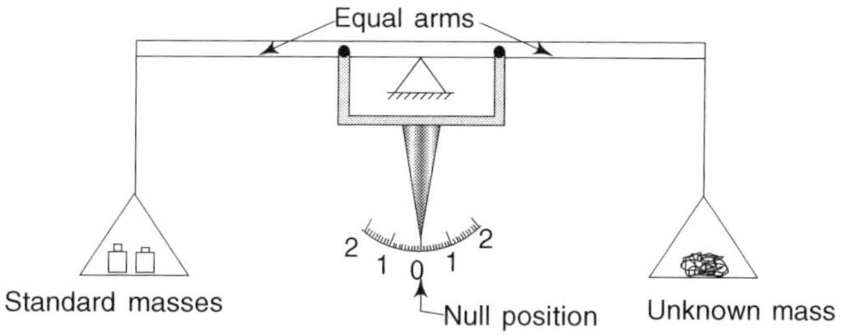

Fig. 1.10 *A schematic diagram of an Equal arm beam balance*

1.4.2 Manually Operated and Automatic Types

Any instrument which requires the services of human operator is a manual type of instrument. The instrument becomes automatic if the manual operation is replaced by an auxiliary device incorporated in the instrument. An automatic instrument is usually preferred because the dynamic response of such an instrument is fast and also its operational cost is considerably lower than that of the corresponding manually operated instrument. A commonly used null-bridge resistance thermometer is shown in Fig. 1.11 which requires manual operation for obtaining the null position. However, the manual operation can be dispensed with by incorporating an automatic self-balancing feedback device known as *instrument servo-mechanism.*

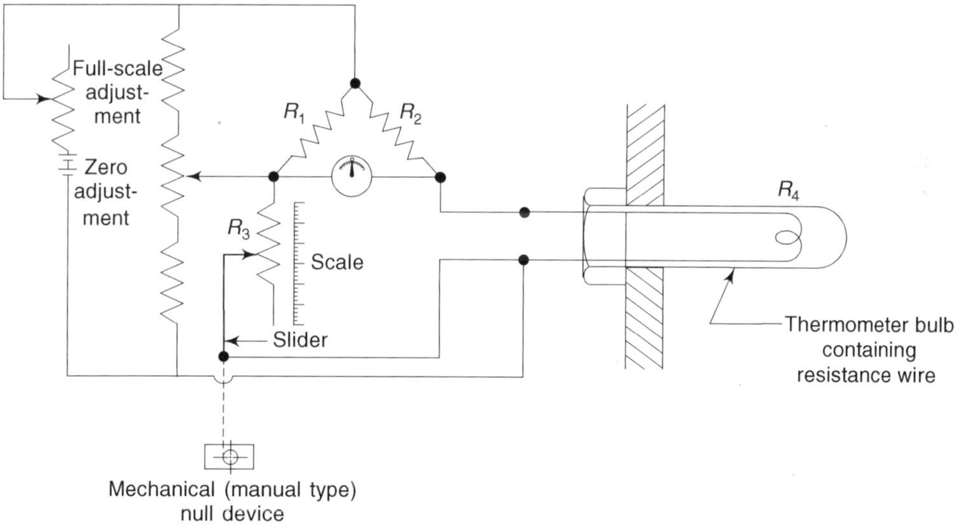

Fig. 1.11 *Manual type null-bridge resistance thermometer*

The block diagram of the automatic self-balancing feedback measuring system is shown in Fig. 1.12. In this device, the amplified output of the error detector actuates the control element (a reversible servo-motor) which in turn causes the movement of an inverse transducer (generally a displacement transducer), the output of which is fed to the error detector after suitable signal condition-

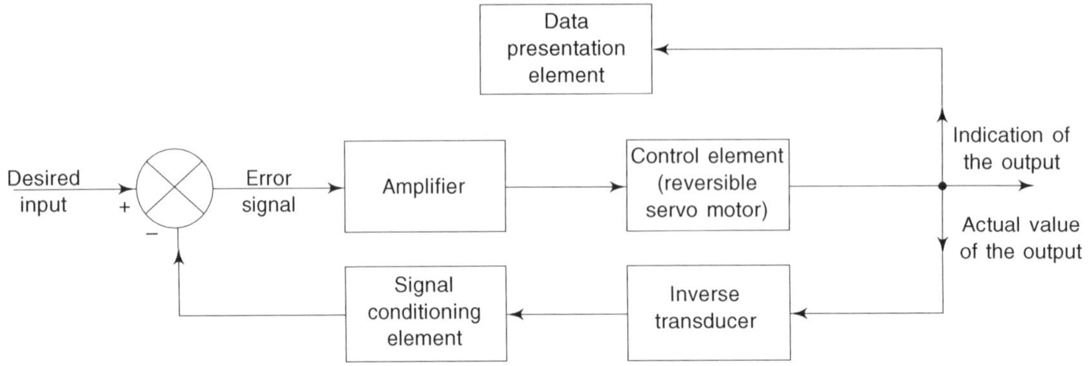

Fig. 1.12 *A typical block diagram of feedback type measurement system*

ing. This way, the feedback loop performs *tracking* of the desired input automatically (i.e. without any human operator) till the error signal vanishes. Further, the device is designed to indicate the value of the desired input on the data presentation element when the error signal becomes zero. In fact, such a device is specially suitable for null-seeking potentiometric or Wheatstone bridge devices, etc.

Figure 1.13 shows a typical automatic null-bridge resistance thermometer in which a mirror type galvanometer (an error detector) is used in conjunction with a photoelectric device. The advantage of this system is that the galvanometer is not subjected to any physical load since it is used to direct light on to a photo cell. The photo cell receives the light due to reflection from the galvanometer mirror whose

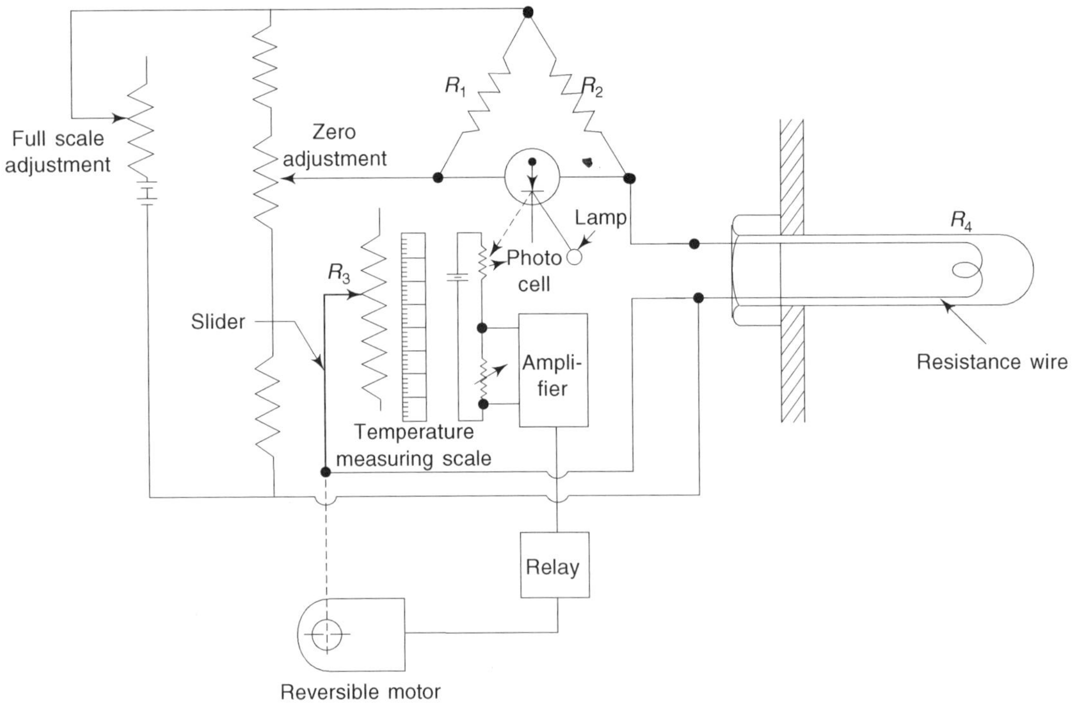

Fig. 1.13 *Automatic type null-bridge resistance thermometer*

angular position is a measure of the unbalanced voltage in the bridge circuit. The photo cell is a part of the input circuit to the amplifier and its resistance controls the input voltage to the amplifier. The amplifier now drives the reversible motor (through a relay switch) which in turn causes the movement of the slider (inverse transducer), the output of which tends to bring the bridge circuit in the null position. When the null condition is reached, the motor would stop running and consequently the movement of the slider on the variable resistance element would also cease at that particular point. Thus, corresponding to this point the temperature can be read off from the calibrated scale of the instrument.

Servo-controlled or self-balancing automatic devices are widely used in industry because they do not require the constant attention of the operator. Further, they have the advantage of giving remote indication and are also suitable for continuous recording. The commonly used devices are: self-balancing recording potentiometer, hot wire anemometer, electro-magnetic flow meter, torque sensor, servo-manometer, capacitance pick-up, servo-controlled accelerometer, etc.

1.4.3 Analog and Digital Types

Analog instruments are those that present the physical variables of interest in the form of continuous or stepless variations with respect to time. These instruments usually consist of simple functional elements. Therefore, the majority of present-day instruments are of analog type as they generally cost less and are easy to maintain and repair.

On the other hand, digital instruments are those in which the physical variables are represented by digital quantities which are discrete and vary in steps. Further, each digital number is a fixed sum of equal steps which is defined by that number. The relationship of the digital outputs with respect to time gives the information about the magnitude and the nature of the input data. The main drawback of such devices is that they are unable to indicate the quantity which is a part of the step value of the instrument. For example, a digital revolution counter cannot indicate, say, 0.65 of a revolution as it measures only in steps of one revolution. However, there are number of distinct advantages of these instruments. The main advantage of the digital representations centres on the on-line use of digital computers for data processing. This has afforded vast possibilities in the areas of computer-assisted decision making, computer-aided design, computer-operated automatic control systems, etc.

Another advantage of digital signals is their noise immunity during transmission. For example, it is easier to recognise the transmitted digital pulse which may be in the form of binary 1 or binary 0 than to distinguish the analog value of voltage say 10.1 or 10.0 or 9.99 V. In other words, it is easier to detect the presence or absence of an electrical pulse (in digital mode) than to discern the precise value of the analog signal in the presence of noise induced along the transmission path.

In addition, several techniques of coding have been developed for digital signals only. Therefore, in order to take advantage of the error detection and error correction capabilities, it is necessary to convert analog data into digital form.

The analog-to-digital conversion is carried out in two steps. In the first step, the analog data is discretised by sampling the data after every time interval t_s known as sampling time. In the second step, the corresponding digital value is assigned a 4-bit binary code so that analog-to-digital conversion becomes compatible with the codes used in the digital computer. A typical analog signal sampling for corresponding digital values is shown in Fig. 1.14.

1.4.4 Self-Generating and Power-Operated Types

In self-generating (or passive) instruments, the energy requirements of the instruments are met entirely from the input signal. For example, an exposure meter of a camera, which is in effect, a photovoltaic cell (shown in Fig. 1.15) is a self-generating (passive) instrument. In this instrument, the incident light

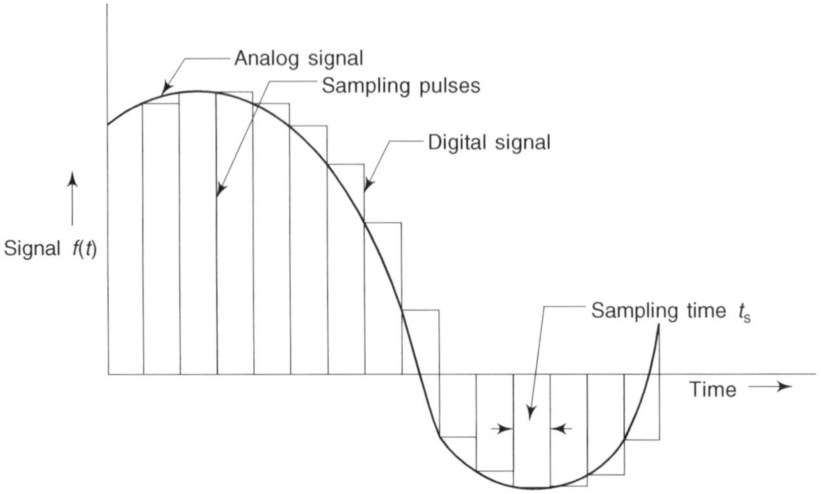

Fig. 1.14 *A typical analog signal sampling for corresponding digital values*

energy whose intensity is being measured, supplies the entire energy for generating the proportional amount of output voltage in the semiconductor junction. Some other common examples of such instruments are: Simple Bourdon gauge (Fig. 1.4(a)) for the measurement of pressure, mercury-in-glass thermometer for the measurement of temperature, Pitot tube for the measurement of fluid flow velocity, a tachogenerator for measurement of rpm, etc.

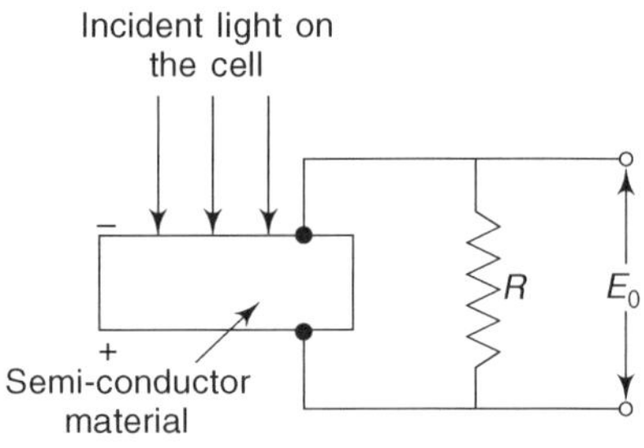

Fig. 1.15 *Schematic diagram of a photo-voltaic cell (self-generating type of instrument)*

On the other hand, power-operated (or active) instruments are those that require some source of auxiliary power such as compressed air, electricity, hydraulic supply, etc. for their operation. In these devices, the input signal supplies only a small portion of the output power. A differential transformer (Fig. 1.5(a)) which is used in the measurement of displacement, force, pressure, etc. is an example of a power-operated instrument. This is because it needs external power to energise its primary as well as two secondary windings. In addition it needs external power in the intermediate element if the output of the differential transformer needs any signal conditioning like amplification, etc.

1.4.5 Contacting and Non-Contacting Types

A contacting type of instrument is one that is kept in the measuring medium itself. A clinical thermometer is an example of such instruments.

On the other hand, there are instruments that are of non-contacting or proximity type. These instruments measure the desired input even though they are not in close contact with the measuring medium. For example, an optical pyrometer monitors the temperature of, say, a blast furnace, but is kept out of contact with the blast furnace. Similarly, a variable reluctance tachometer (Fig. 1.16), which measures the rpm of a rotating body, is also a proximity type of instrument. In this, the toothed rotor made of ferromagnetic material causes variation of flux in the magnetic circuit due to changes in air gap. This in turn induces an emf that is in the form of pulses. The output of the instrument is fed to a frequency counter from which the rpm of the rotor can be determined.

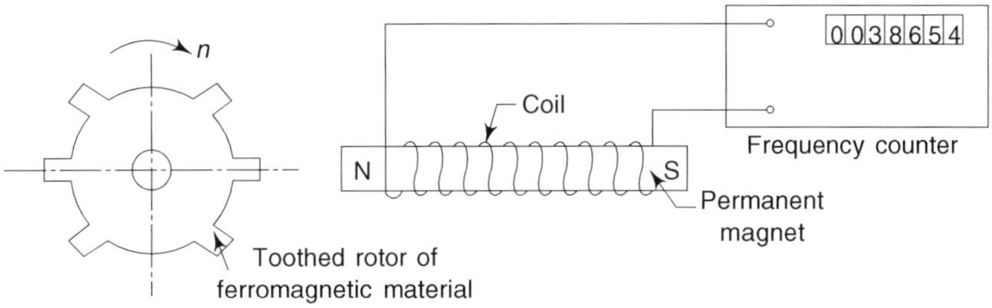

Fig. 1.16 *Schematic diagram of variable reluctance tachometer (proximity type of instrument)*

1.4.6 Dumb and Intelligent Types

A dumb or conventional instrument is that in which the input variable is measured and displayed, but the data is processed by the observer. For example, a Bourdon pressure gauge is termed as a dumb instrument because though it can measure and display a car tyre pressure but the observer has to judge whether the car tyre air inflation pressure is sufficient or not.

Currently, the advent of microprocessors has provided the means of incorporating Artificial Intelligence (AI) to a very large number of instruments. Intelligent or smart instruments process the data in conjunction with microprocessor (μP) or an on-line digital computer to provide assistance in noise reduction, automatic calibration, drift correction, gain adjustments, etc. In addition, they are quite often equipped with diagnostic sub-routines with suitable alarm generation in case of any type of malfunctioning.

An intelligent or smart instrument may include some or all of the following:

1. The output of the transducer in electrical form.
2. The output of the transducer should be in digital form. Otherwise it has to be converted to the digital form by means of anolog-to-digital converter (A–D converter).
3. Interface with the digital computer.
4. Software routines for noise reduction, error estimation, self-calibration, gain adjustment, etc.
5. Software routines for the output driver for suitable digital display or to provide serial ASCII coded output.

It may noted that further details of the intelligent or smart instruments are discussed in Ch. 18.

1.5 ■ MICROPROCESSOR-BASED INSTRUMENTATION

Present-day highly competitive global economy has resulted in the production of best quality products at lowest prices. Currently, highly versatile micro processor chips designed and developed on Very Large Scale Integration (VLSI, i.e. 10^8 circuits/cm^2) have provided very convenient means of adding Artificial Intelligence (AI) into many of our every day facilities. These include automatic teller machines (ATMs) of banks, automatic washing machines, automatic fuel control systems in automobiles, burglar alarms and a large number of modern-day instruments.

In microprocessor (μp)-based instruments, μp forms one of the auxiliary functional elements of the instrument. The word microprocessor has two parts. The first part 'micro' refers to its micro-miniature size/dimensions. Further, the second part signifies its vast potential to perform complex computations at fantastically high speeds, together with pre-programmed logic/software which enhances significantly the capabilities and effectiveness of the instruments. As mentioned in the previous Sub-section 1.4.6, the μp-based instruments are commonly termed as smart or intelligent instruments. Currently, the micro-miniaturisation has resulted in offering these smart instruments of the size of a pocket calculator. Presently, a hand-held oscilloscope weighing less than 500 g can be programmed with soft-key operation to work as voltmeter, analog-to-digital converter or spectrum analyser or a PC note book. Similarly, pocket-size thermocouple sensor of digital type, portable velocity meter, micro-size pressure pick-ups are now common place and are used regularly.

Microprocessor, by itself, is an operational computer. It is incorporated with additional circuits for memory and input/output devices to shape it in the form of a digital computer. A digital computer system essentially has an *Arithmetic and Logic Unit*, control unit, Memory, Input and Output devices, which are shown in Fig. 1.17. Input devices include key board, floppy diskettes, compact Discs (CDs), manually-operated or voice-operated mouse, scanner input, A-D converters or digital transducers in case computers are used for measurement and control applications. Further, the output devices include printers plotters or Visual Display Unit (VDU), etc.

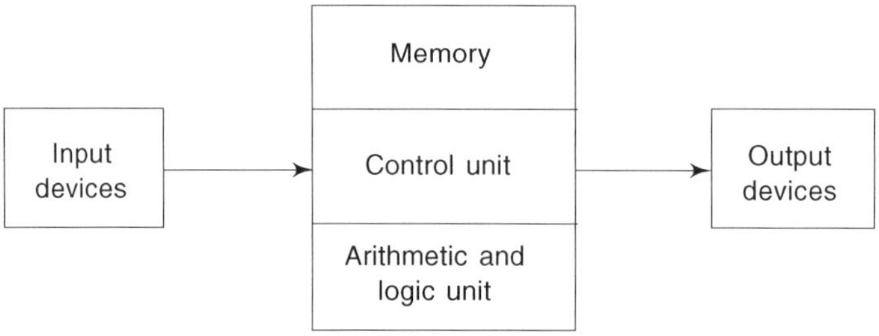

Fig. 1.17 *Digital computer system*

Figure 1.18 shows a typical Digital Computer-based Measurement System. A process or a plant or a system may have to simultaneously measure multiple variables like pressure, temperature, velocity, viscosity, flow rate, etc. A computer-based measurement system has the capability of processing all the inputs and present the data in real time.

A digital computer is fed with a 'sequential list of instructions' termed as a computer program for suitable processing/manipulation of the data. Quite often, Artificial Intelligence (AI) in the form of Artificial Neural Networks (ANNs) or Fuzzy Logic (FL) may be incorporated. With this, the tasks of decision-making in various processes are usually done by the computer itself and not by any human operator.

1.5.1 Advantages and Disadvantages of Computer-Based Instrumentation Systems

The following are the advantages of the computer-based instrumentation systems:

1. They are suitably programmed to automatically carry out the mundane tasks of drift correction, noise reduction/elimination, non-linearity correction, gain adjustments, range and span adjustments, automatic calibration, etc.
2. These instruments have signal conditioning and display which are compact, rugged and reliable. Further, they are capable of performing in tough, industrial, consumer, military, automobile and environmental conditions.
3. Quite often they have built-in diagnostic subroutines, which can detect the fault and automatically correct the same. Alternatively, if it cannot correct the fault, it generates a suitable alarm.
4. The measurement, processing and display of the data of the process variables are done in real time i.e., it is on-line type process information in the desired form and units.
5. Such instruments can be adjusted/programmes with a remote control.
6. They have lower costs, higher accuracy and more flexibility.

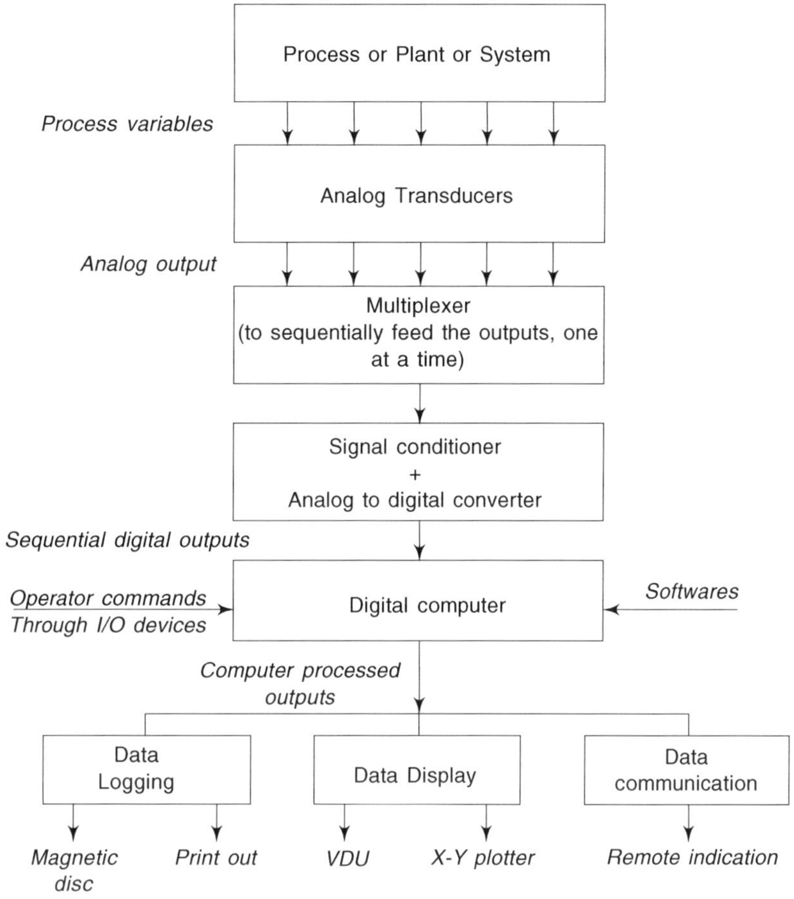

Fig. 1.18 *A typical digital computer based measurement system*

7. Due to the compact size, they are mostly of portable type.
8. They have a low consumption of the order of 40 mW and thus can be battery operated.
9. They do not require skilled operation because of the incorporation of user-friendly subroutines.

However, these instruments have the following disadvantages:

1. They cannot replace the computer programmer/designer of the instruments, i.e., they cannot modify the programmes themselves.
2. These number crunching machines invariably need the processing data in the digital form.
3. The commercial computer softwares become obsolete very fast and periodically updating the same involves extra expenditure.
4. They are prone to virus problems and due to this, they suddenly become sick and inoperative.

1.6 ■ STANDARDS AND CALIBRATION

Basically, measurement is an act of a quantitative comparison between a predefined standard and the unknown magnitude of a physical quantity. In order that the results are meaningful, the following two requirements must be met in the act of measurement:

1. The standard that is used for comparison must be well-established, highly accurate and reproducible; and
2. The measurement devices and the calibration procedures adopted in the act of measurement must have proven reliability.

1.6.1 Standards of Measurements

A standard of measurement is defined as the physical representation of the unit of measurement. A unit of measurement is generally chosen with reference to an arbitrary material standard or to a natural phenomenon that includes physical and atomic constants. For example, the S.I. unit of mass, namely kilogram, was originally defined as the mass of a cubic decimetre of water at its temperature of maximum density, i.e. at 4°C. The material representation of this unit is the International Prototype kilogram which is preserved at the International Bureau of Weights and Measures at Sévrés, France. Further, prior to 1960, the unit of length was the carefully preserved platinum–iridium bar at Sévrés, France. In 1960, this unit was redefined in terms of optical standards, i.e. in terms of the wavelength of the orange-red light of Kr^{86} lamp. The standard metre is now equivalent to 1 650 763.73 wavelengths of Kr^{86} orange-red light. Similarly, the original unit of time was the mean solar second which was defined as 1/86400 of a mean solar day. However, the current internationally recognised unit of universal time is based on a cesium clock. It is now defined in terms of frequency of the cesium transition to its hyperfine state unperturbed by external fields, which occurs at 9 192 631 770 Hz. The use of cesium clock gives an accuracy exceeding 1 μs per day. Therefore, it is considered as the primary time and the frequency standard. In fact, a number of standards have been developed for other units of measurements including standards for fundamental as well as dervied quantities which may include mechanical, electrical, thermal, optical, etc. parameters (Appendix A 1).

There are different types of standards of measurements. They can be classified according to their function and type of application. They are briefly described as follows:

International standards International standards are devices designed and constructed to the specifications of an international forum. They represent the units of measurements of various physical quantities to the highest possible accuracy that is attainable by the use of advanced techniques of production and measurement technology. These standards are maintained by the International Bureau of Weights and Measures at Sévrés, France. For example, the International Prototype kilogram, wavelength of Kr^{86} orange–red lamp and cesium clock are the international standards for mass, length and time, respectively. However, these standards are not available to an ordinary user for purposes of day-to-day comparisons and calibrations.

Primary standards Primary standards are devices maintained by standards organisations/national laboratories in different parts of the world. These devices represent the fundamental and derived quantities and are calibrated independently by absolute measurements. One of the main functions of maintaining primary standards is to calibrate/check and certify secondary reference standards. Like international standards, these standards also are not easily available to an ordinary user of instruments for verification/calibration of working standards.

Secondary standards Secondary standards are basic reference standards employed by industrial measurement laboratories. These are maintained by the concerned laboratory. One of the important functions of an industrial laboratory is the maintenance and periodic calibration of secondary standards against primary standards of the national standards laboratory/organisation. In addition, secondary standards are freely available to the ordinary user of instruments for checking and calibration of working standards.

Working standards These are high-accuracy devices that are commercially available and are duly checked and certified against either the primary or secondary standards. For example, the most widely used industrial working standard of length are the precision gauge blocks made of steel. These gauge blocks have two plane parallel surfaces a specified distance apart, with accuracy tolerances in the 0.25–0.5 micron range (1 micron = 10^{-6} m). Similarly, a standard cell and a standard resistor are the working standards of voltage and resistance, respectively. Working standards are very widely used for calibrating general laboratory instruments, for carrying out comparison measurements or for checking the quality (range of accuracy) of industrial products.

1.6.2 Calibration

Calibration is the act or result of quantitative comparison between a known standard and the output of the measuring system measuring the same quantity. In a way, the process of calibration is in effect the procedure for determining the scale of the measuring system. If the output-input response of the system is linear, then a single-point calibration is sufficient, wherein only a single known standard value of the input is employed. However, if the system response is non-linear, then a set of known standard inputs to the measuring system are employed for calibrating the corresponding outputs of the system. The process of calibration involves the estimation of uncertainty between the values indicated by the measuring instrument and the true value of the input. In fact, calibration procedure is the process of checking the inferior instrument against a superior instrument of known traceability certified by a reputed standards organisation/national laboratory. Herein, the term 'traceability' of a calibrating device refers to its certified accuracy when compared with superior standard of highest possible accuracy. Calibration procedures can be classified as follows:

Primary calibration When a device/system is calibrated against primary standards, the procedure is termed primary calibration. After primary calibration, the device is employed as a secondary calibration device. The standard resistor or standard cell available commercially are examples of primary calibration.

Secondary calibration When a secondary calibration device is used for further calibrating another device of lesser accuracy, then the procedure is termed secondary calibration. Secondary calibration devices are very widely used in general laboratory practice as well as in the industry because they are practical calibration sources. For example, standard cell may be used for calibrating a voltmeter or an ammeter with suitable circuitry.

Direct calibration with known input source Direct calibration with a known input source is in general of the same order of accuracy as primary calibration. Therefore, devices that are calibrated directly are also used as secondary calibration devices. For example, a flow meter such as a turbine flow meter may be directly calibrated by using the primary measurements such as weighing a certain amount of water in a tank and recording the time taken for this quantity of water to flow through the meter. Subsequently, this flow meter may be used for secondary calibration of other flow metering devices such as an orificemeter or a venturimeter.

Indirect calibration Indirect calibration is based on the equivalence of two different devices that can be employed for measuring a certain physical quantity. This can be illustrated by a suitable example, say a turbine flow meter. The requirement of dynamic similarity between two geometrically similar flow meters is obtained through the maintenance of equal Reynold's number, i.e.

$$\frac{D_1 \rho_1 V_1}{\mu_1} = \frac{D_2 \rho_2 V_2}{\mu_2}$$

where the subscripts 1 and 2 refer to the 'standard' and the meter to be calibrated, respectively. For such a condition, the discharge coefficients of the two meters are directly comparable.

Therefore, it is possible to carry out indirect calibration, i.e. to predict the performance of one meter on the basis of an experimental study of another. This way, a small meter may be employed to determine the discharge coefficient of large meters. Also, the discharge coefficient of the meter intended for gas may be determined by carrying out test runs on a liquid provided that similarity through Reynold's numbers is maintained.

Routine calibration Routine calibration is the procedure of periodically checking the accuracy and proper functioning of an instrument with standards that are known to be accurately reproducible. The entire procedure is normally laid down for making various adjustments, checking the scale reading, etc. which conforms to the accepted norms/standards. The following are some of the usual steps taken in the calibration procedure:

1. Visual inspection of the instrument for the obvious physical defects.
2. Checking the instrument for proper installation in accordance with the manufacturer's specifications.
3. Zero setting of all the indicators.
4. Levelling of the devices which require this precaution.
5. Recommended operational tests to detect major defects.
6. The instrument should preferably be calibrated in the ascending as well as descending order of the input values to ensure that errors due to friction/stiction are accounted for.
7. The calibration device should have superior level of traceability of standard as compared to the calibrated device.

Review Questions

1.1 Indicate whether the following statements are true or false. If false, rewrite the correct statement.

(i) Measurements involve the gathering of informations from the physical world and comparison of the same with the agreed standards.

(ii) Accuracy of the control variable can be better than the accuracy of the controlled process variable.

(iii) Instrument engineering is the systematic design of a reliable and cost-effective measuring system which conforms to the requirements of a measurement situation.

(iv) Analog signals are generally more accurate than the corresponding digital signals and therefore are preferred for signal transmission purposes.

(v) International standards have been developed for the fundamental quantities only.

(vi) Working standards are used for primary calibration procedures.

(vii) The term 'Traceability' of a calibrating device refers to its certified accuracy/level of uncertainty when measured against a superior standard by a reputed organisation.

(viii) A clinical thermometer is a self-generating device.

(ix) A null-seeking device can be servo-controlled.

(x) A smart instrument is synonymous to an ideal instrument i.e., with zero error, zero uncertainty and zero least count.

(xi) A transducer can be termed as transformer if the transduced signal has the same units as the measured variable.

(xii) An active transducer is of power-operated type.

(xiii) Null type instruments are comparatively more accurate than the corresponding deflection type instruments.

(xiv) Measurement systems seldom provide exact information about the measurand.

(xv) Bourdon pressure gauge transduces the pressure signal into a large displacement which can be directly read on a measuring scale.

1.2 Fill in the blanks in the following:

(i) Measurements essentially involve the comparison of the magnitudes of unknown variables with pre-determined ________ .

(ii) The term 'Traceability' is related to the ________ of the calibrating device.

(iii) A clinical thermometer has a built-in amplifying element in the form of ________ .

(iv) A thermocouple transduces the temperature signal in the form of ________.

(v) The transducer which converts the sound pressure variations into corresponding changes in capacitance between fixed plate and moveable diaphragm is termed as ________.

(vi) Grocer's equal-arms balance is a ________ type of weight measuring device.

(vii) Microprocessor-based instruments process the data in the ________ form.

(viii) The standard meter corresponds to 1650763.73 wavelengths of ________ light radiations.

(ix) Name four instruments in which transduction principle is change in resistance
(a) ________, (b) ________, (c) ________ and (d) ________.

(x) A portable type of μp-based oscilloscope is a ________ type of instrument.

1.3 Choose the appropriate answer in the following

(i) Power gain in an amplifier is
(a) current gain/voltage gain (b) voltage gain/current gain
(c) input power/output power (d) current gain × voltage gain

(ii) The instrument with null output is
(a) light meter of a camera (b) Bourdon pressure gauge
(c) a platform type of weighing machine (d) a mercury manometer

(iii) The self-generating transducers in the following is
(a) light meter of a camera (b) mercury in glass thermometer
(c) a spring balance (d) a piezo-electric crystal

(iv) The instrument with digital read-out is
(a) an aneroid barometer (b) a milometer/odometer of an automobile
(c) pendulum clock (d) micrometer screw gauge

(v) The proximity type of instrument is a
(a) dial gauge (b) load cell
(c) pyrometer (d) speedometer of a car

(vi) Which of the following is *not* a self-generating type of transducer?
(a) thermocouple (b) LVDT
(c) photo voltaic cell (d) Bourdon tube of a pressure gauge

(vii) The auto-focus and auto-shutter adjustment camera is
(a) a conventional or dumb instrument (b) a self-generating instrument
(c) an analog instrument (d) a smart instrument

(viii) The elastic type of transducer element in the Bourdon pressure gauge is of
(a) circular cross-section (b) square cross-section
(c) rectangular cross-section (d) elliptical cross-section

(ix) The function of the transducer is to
(a) amplify the input signal

(b) average the fluctuating type of input signal
(c) convert the input signal to a form which can be conveniently processed
(d) regulate the signal for a suitable control application.

(x) Self-balancing type of feedback measuring systems are
(a) null type (b) deflection type
(c) self-generating type (d) proximity type
(e) automatic type

1.4 Schematic diagrams of the following measuring devices are given in Fig. 1.19.
(a) diaphragm type liquid level gauge
(b) spring balance with electrical read out
(c) float-operated fuel level gauge used in automobiles
(d) mechanical type of displacement measuring dial gauge.

Indicate the basic and auxiliary functional elements of the devices mentioned above in the form of block diagrams.

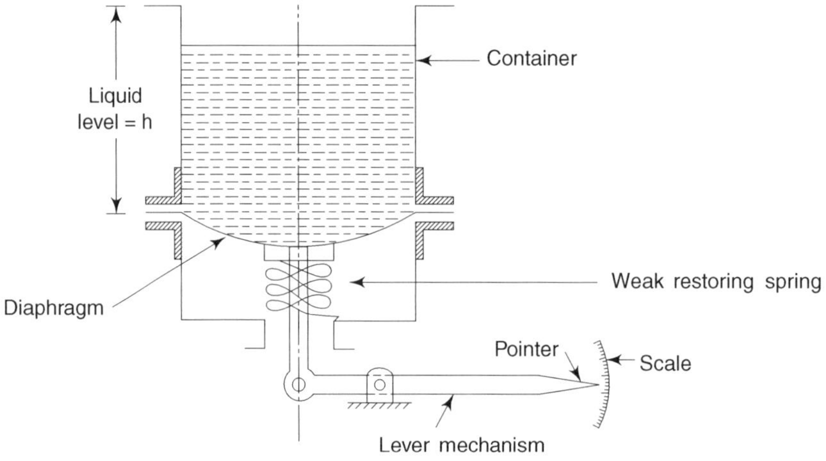

(a)

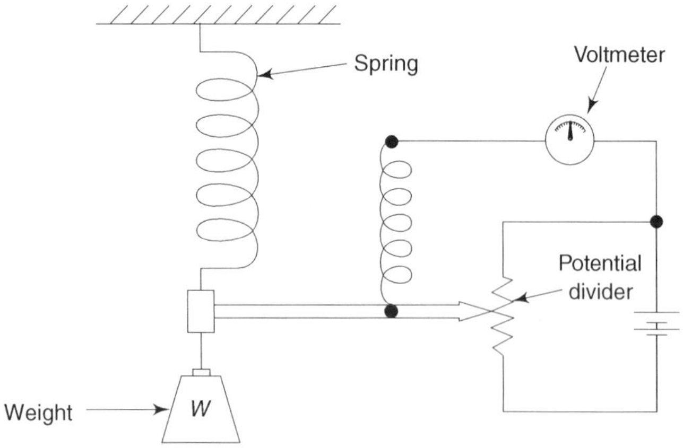

(b)

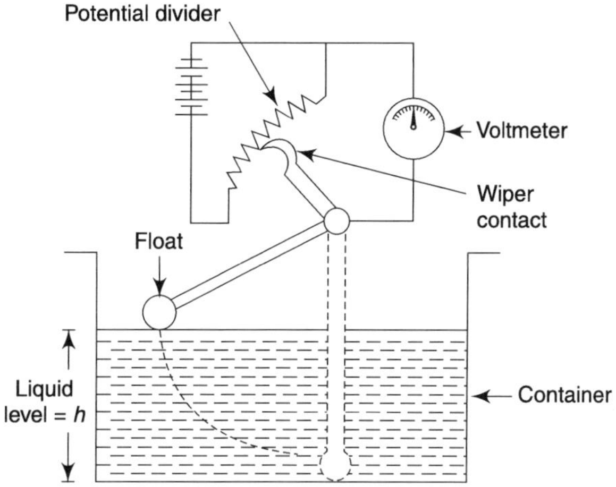

(c)

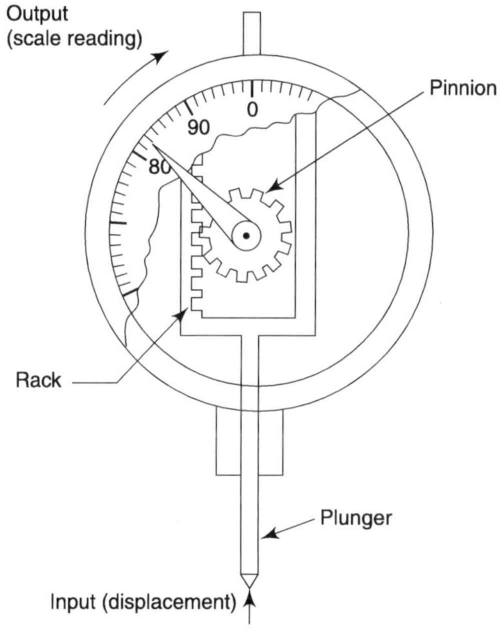

(d)

Fig. 1.19 *(a) Diaphragm type liquid level gauge, (b) Spring balance with electrical read out, (c) Float-operated fuel level gauge used in automobiles, (d) Mechanical type of displacement measuring dial gauge*

1.5 A Bourden pressure gauge having a linear calibration has a 50 mm long pointer. It moves over a circular dial having an arc of 270°. It displays a pressure range of 0 to 15 bar (1 bar = 10^5 Pa). Determine the sensitivity of the Bourden gauge in terms of scale length per bar (i e., mm/bar).

1.6 A thermocouple having a sensitivity of 4.8 mV/°C has been used for the measurement of temperature. Its output is connected to a moving coil millivoltmeter which has a sensitivity of 1°/mV.

If the length of the pointer of the instrument is 30 mm, determine the overall sensitivity of the temperature-sensing system in mm/°C.

1.7 The mean velocity of blood flowing in a blood vessel of 2 mm internal diameter was measured in an experiment by means of a sensor which gave 1 V/(mm/s). The output was connected to a strip chart recorder whose *y*-axis sensitivity was set at 20 mm/V and *x*-axis movement was set at 100 mm/minute. If the length and area of the trace were 200 mm and 2800 mm^2 respectively, determine the velocity, flow rate and total volume of blood that flowed in the said experiment.

1.8 The transducer of an arterial blood pressure instrument consists of a four-element piezo-resistive wheatstone bridge and its sensitivity is 10 μV per volt of excitation per torr of pressure (*Note:* 1 torr = 1 mm Hg). Its output is fed to an amplifier with voltage gain of 100. Determine the output voltage if the bridge is excited by 6 V dc and 120 torr of pressure is applied.

1.9 The overall resistance of a slide wire of an automobile fuel level gauge in Fig. 1.20 is 6 kΩ. If the wiper arm actuated by float is located at 2/3 mark, i.e., R_a = 2 kΩ and R_b = 4 kΩ, determine E_0 if E = 12 V.

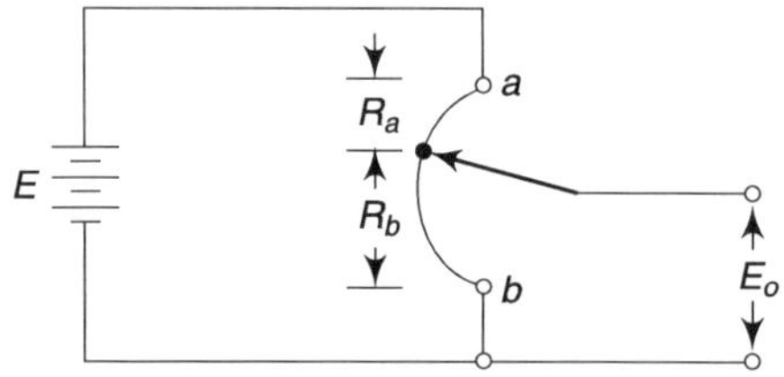

Fig. 1.20 *Variable resistance type of automobile fuel level gauge*

1.10 A diaphragm type of pressure transducer gives a central deflection of 0.2 mm when a pressure of 1.2×10^6 N/m^2 is applied. An electromechanical device, namely, linear variable differential transducer (LVDT) converts the input displacement of pressure transducer into voltage and has a sensitivity of 60 V/mm. Determine the overall sensitivity of the pressure gauge in V/(N/m^2) and determine the unknown pressure when the output voltage of 2.5 V is observed on the meter.

Answers

1.1. (i) T

(ii) F, accuracy of control variable cannot be better than the accuracy of the controlled process variable

(iii) T

(iv) F, analog signals are generally more accurate than the corresponding digital signals; however, for signal transmission purpose, digital signal are preferred.

(v) T

(vi) F, working standards are used for secondary calibrations.

(vii) T

(viii) T

(ix) T

(x) F, a smart instrument is *μp* based but it may not be of an ideal type.

(xi) T

(xii) F, an active instrument is a self-generating type.

(xiii) T

(xiv) T

(xv) F, Bourdon gauge pressure transducer produces micro-displacement which needs amplification.

1.2. (i) standard
(ii) accuracy
(iii) capillary
(iv) voltage
(v) condenser microphone
(vi) null
(vii) digital
(viii) Kr^{86} orange–red
(ix) (a) strain gauge load cell
(b) Pirani gauge
(c) hot wire Anemometer
(d) thermistor
(x) an intelligent/smart

1.3 (i) d
(ii) c
(iii) all
(iv) b
(v) c
(vi) b
(vii) d
(viii) d
(ix) c
(x) a and e

1.5 15.7 mm/bar

1.6 2.5 mm/°C

1.7 4.55 mm/s, 14.3 mm^3/s, 5277 mm^3

1.8 0.72 V

1.9 8 V

1.10 10 V/(10^6 N/m^2), 2.5×10^5 N/m^2

Chapter

2

Static Performance Characteristics of Instruments

INTRODUCTION

The detailed specifications of the functional characteristics of any instrument are termed its performance characteristics. These are in general, indicative of the capabilities and limitations of the instrument for a particular application. Therefore, the knowledge of the performance characteristics is quite important as it enables us to have quantitative estimates of the positive as well as the negative points of various commercially available instruments. Consequently, one can select the optimum type of instrument for the given application.

Instrument performance characteristics can be broadly classified as

1. *Static* characteristics and
2. *Dynamic* characteristics.

In a number of situations, the desired input to the instrument may be constant or varying slowly with respect to time. In these situations, the dynamic characteristics are not important. Therefore, the various static performance parameters like *accuracy, precision, resolution, sensitivity, linearity, hysteresis, drift, overload capacity, impedance loading*, etc. are usually good enough to give meaningful quantitative descriptions of the instrument. Further, we also come across certain situations, where the desired input is not constant but varies rapidly with the time. In such cases, the dynamic characteristics of the instruments should also be known. These are generally represented by the relations between input and output parameters that are governed by the relevant differential equations applicable in the given situation.

Thus, it may be noted that in general, the overall quantitative performance qualities of the instruments are represented by both their static and dynamic characteristics. However, for time-independent signals, only the static characteristics need be considered.

2.1 ■ ERRORS AND UNCERTAINTIES IN PERFORMANCE PARAMETERS

The various static performance parameters of the instruments are obtained by performing certain specified tests depending on the type of instrument, the nature of the application, etc. Some salient static performance parameters are periodically checked by means of a static calibration. This is accomplished by imposing constant values of 'known' inputs and observing the resulting outputs. Quite often, we experience difficulty in obtaining known constant values of the input quantity. Further, we also come across the following difficulties:

1. The change in sensitivity of instruments due to certain perturbations results in influencing all output values, generally equally by a particular quantity. These are caused due to worn out parts, effect of changes in the environment on the equipment or the user, etc.
2. Failure of the instrument to have the same output for repeated applications of any particular value of the input. This effect, i.e. the *scatter* in output values, within a given range, is caused due to *random* variations in the parameter or in the system of measurement.

No measurement can be made with perfect accuracy and precision. Therefore, it is instructive to know the various types of errors and uncertainties that are in general, associated with measurement system. Further, it is also important to know how these errors are propagated. This is because if an error is detected, then it can be eliminated or its effect can be accounted for in the form of a suitable correction. On the other hand, if an error goes unrecognised, then it would make experimental data unreliable.

2.1.1 Types of Errors

Error is defined as the difference between the measured and the true value (as per standard). The different types of errors can be broadly classified as follows.

Systematic or Cumulative Errors Such errors are those that tend to have the same magnitude and sign for a given set of conditions. Because the algebraic sign is the same, they tend to accumulate and hence are known as cumulative errors. Since such errors alter the instrument reading by a fixed magnitude and with same sign from one reading to another, therefore, the error is also commonly termed as *instrument bias*. These types of errors are caused due to the following:

Instrument errors Certain errors are inherent in the instrument systems. These may be caused due to poor design/construction of the instrument. Errors in the divisions of graduated scales, inequality of the balance arms, irregular springs tension, etc., cause such errors. Instrument errors can be avoided by (i) selecting a suitable instrument for a given application, (ii) applying suitable correction after determining the amount of instrument error, and (iii) calibrating the instrument against a suitable standard.

Environmental errors These types of errors are caused due to variation of conditions external to the measuring device, including the conditions in the area surrounding the instrument. Commonly occurring changes in environmental conditions that may affect the instrument characteristics are the effects of changes in temperature, barometric pressure, humidity, wind forces, magnetic or electrostatic fields, etc. For example, change in ambient temperature causes errors due to expansion of the measuring tape. Similarly, buoyant effect of the wind causes errors on weights of a chemical balance.

Loading errors Such errors are caused by the act of measurement on the physical system being tested. Common examples of this type are: (i) introduction of additional resistance in the circuit by the measuring milliammeter which may alter the circuit current by significant amount, (ii) an obstruction type flow meter may partially block or disturb the flow conditions and consequently the flow rate shown by the meter may not be same as before the meter installation, and (iii) introduction of a thermometer alters the thermal capacity of the system and thereby changes the original state of the system which gives rise to loading error in the temperature measurement.

It may be noted that systematic errors can be eliminated or alternatively, a suitable correction can be applied by properly calibrating the instrument. Therefore, instrument calibrations become very necessary as the measurements play a dominant role in any experiment. Further, systematic errors can also be subdivided into static or dynamic errors. Static errors are caused due to limitations of the instruments as well as due to certain shortcomings in the measurement process. For example, a static error is introduced in the micrometer reading when excessive torque is applied on the shaft of the micrometer screw. Dynamic errors are caused in the measurements if the instrument is not responding fast enough to follow the changes in the measured variable.

Accidental or Random Errors These errors are caused due to random variations in the parameter or the system of measurement. Such errors vary in magnitude and may be either positive or negative on the basis of chance alone. Since these errors are in either direction, they tend to compensate one another. Therefore, these errors are also called *chance or compensating type of errors*. The presence of such errors is detected by a lack of consistency in the measured values when the same input is imposed repeatedly on the instrument. In a well-designed experiment very few random errors occur but they become important in a high accuracy work. The following are some of the main contributing factors to random error.

Inconsistencies associated with accurate measurement of small quantities The outputs of the instruments become inconsistent when very accurate measurements are being made. This is because when the instruments are built or adjusted to measure small quantities, the random errors (which are of the order of the measured quantities) become noticeable. For example, if one wishes to measure the weight of a bucket of water to the nearest kilogram, there would be no excuse of one observation differing from another, no matter how many times the measurements are made. However, if one attempts to determine the same weight to the nearest milligram, individual observations are sure to differ unless these are performed with extreme care and without prejudice.

Presence of certain system defects System defects such as large dimensional tolerances in mating parts and the presence of friction contribute to errors that are either positive or negative depending on the direction of motion. The former causes backlash error and the latter causes slackness in the meter bearings. One way of detecting and correcting such errors is to measure the quantity first while increasing and then decreasing the magnitude. This procedure is based on the *method of symmetry.*

Effect of unrestrained and randomly varying parameters Chance errors are also caused due to the effect of certain uncontrolled disturbances which influence the instrument output. Line voltage fluctuations, vibrations of the instrument supports, etc. are common examples of this type. The experimenter should try to reduce the influence of such randomly varying parameters to the minimum. But, in spite of this, there are always certain residual contributions due to these random perturbations.

Miscellaneous Type of Gross Errors There are certain other errors that cannot be strictly classified as either systematic or random as they are partly systematic and partly random. Therefore, such errors are termed *miscellaneous type of gross errors*. This class of errors is mainly caused by the following.

Personal or human errors These are caused due to the limitations in the human senses. For example, one may sometimes consistently read the observed value either high or low and thus introduce systematic errors in the results. While at another time one may record the observed value slightly differently than the actual reading and consequently introduce random error in the data. Therefore, it becomes necessary to exercise extreme care with mature and considered judgement in recording the observations so as to reduce such errors.

Errors due to faulty components/adjustments Sometimes there is a misalignment of moving parts, electrical leakage, poor optics, etc. in the measuring system. These may simultaneously cause, for example, zero shift coupled with zero drift which are systematic and random errors, respectively. In such cases, it becomes necessary to determine the magnitude of the systematic and random errors from the overall gross error. The procedure usually consists of repeating a measurement for a sufficiently large number of times by feeding a 'standard' signal to the instrument. The difference between the mean value of the signal and the standard signal gives the best estimate of systematic error. Further, the estimate of uncertainty which represents the random error in measurement is evaluated from the dispersion of data.

Improper application of the instrument Errors of this type are caused due to the use of instrument in conditions which do not conform to the desired design/operating conditions. For example, extreme vibrations, mechanical shock or pick-up due to electrical noise could introduce so much gross error as to mask the test information. In such cases it becomes necessary to stop the experiment until the disturbing elements causing the *chaotic* type of gross errors are eliminated.

2.1.2 Types of Uncertainties

In any experiment, the exactness of the measured values is affected by deviations caused by the usually associated errors, namely, systematic and random errors. The combined effect of these errors i.e., gross errors can be separated into systematic and random type components. Since systematic errors and the systematic component of gross errors are deterministic in nature, therefore, it is best to eliminate the effect of such errors by introducing a suitable correction factor in the experimental results by a proper calibration procedure. Now, the remaining random errors and the random component of gross errors in general, constitute the chief source of uncertainty in experiments. In fact, the term 'random' error is a misnomer, it should really be termed as '*degree/amount of uncertainty*' which is evaluated from the dispersion/scatter in the data. It is worth mentioning here that even though the magnitude of random type deviations as well as their algebraic sign vary from one observation to another, yet fortunately they follow a definite mathematical law called the normal or Gaussian distribution (Ch. 22). When the various uncertainties are distributed in such a predictable manner, they can be quantified so as to increase the reliability of the experiment.

In general, we come across the two types of uncertainties in measured quantities as discussed here.

External estimate of uncertainty U_E This is usually assessed from the knowledge of the experiment, the limitation of the apparatus, the instrument manufacturer's literature, etc. However, in the absence of the above information, as a thumb rule, the resolution that is the smallest confidently measurable input change is considered as the external estimate of uncertainty. For example, we give a true input of 100 V to a well-calibrated voltmeter which has a resolution of 0.1 V. The expected voltmeter reading would obviously be 100 V. Now, we increase or decrease the input voltage in the range of ±0.1 V, which is equal to the resolution of the instrument. The voltmeter would still indicate 100 V as the instrument is incapable of registering a change up to or below the resolution of the instrument. Therefore, it would be reasonable to say, in this case, that the measured value of 100 V has the external estimate of uncertainty U_E equal to ±0.1 V.

Internal estimate of uncertainty U_I This type of uncertainty is that which is inherent in the data itself and its quantitative assessment is termed as the internal estimate of uncertainty U_I. This occurs because an instrument may indicate a slightly different value every time a given input is fed to the instrument. In other words, we may say that the data is associated with scatter in values. The internal estimate of uncertainty, also termed as internal standard error, is estimated as follows:

1. A reasonably large number of data of a particular measurement is obtained. This data is termed as 'population' and it is assumed to include all possible values.
2. The population is broken up into small portions termed 'samples'. The samples are so selected that they have equitable distribution of all the possible measurements. In other words, the selected values have no bias and this process is known as random sampling.
3. The means and standard deviations of each of the samples are evaluated which are considered estimated true value and uncertainty respectively for the sample.
4. To obtain the best estimate of the mean value, the mean of sample means, which is in effect population mean, is calculated.
5. To obtain the overall estimate of uncertainty, the standard deviation of sample means is calculated (Ch. 21).

Alternatively, this value can also be obtained by calculating the combined estimate of individual uncertainties of the sample values. The value thus obtained is known as the *internal estimate of uncertainty or the internal standard error*. This has been shown to be equal to sample standard deviation divided by square root of the number of samples (Prob. 2.1, Sec. 2.2).

It may be noted that a highly precise instrument with excellent repeatability characteristics is usually associated with quite small value of U_I. But, such instruments are usually costly. On the other hand, a coarse instrument, which may be less expensive, is generally associated with larger value of U_I. This however could be reduced to any desired value by carrying out repeat measurements a sufficiently large number of times.

Generally, *any experiment is considered consistent if the external and internal estimates of uncertainties are of the same order of magnitude.* This criterion ensures minimum overall uncertainty in the experiment. However, if these estimates differ considerably, then the higher of the two is generally taken as uncertainty of the experiment.

2.2 PROPAGATION OF UNCERTAINTIES IN COMPOUND QUANTITIES

In any experiment, a number of different measurements of different quantities may be carried out to determine a certain parameter. For example, if we wish to calculate the magnitude of force, we have to measure the magnitudes of mass and acceleration. Both these measurements would involve uncertainties. Now, our aim is to estimate the uncertainty in the force due to the combined effect of the uncertainties in mass and acceleration. To compute the overall uncertainty due to the combined effect of the uncertainties of different variables, we consider the following equation in most general form:

$$y = f(x_1, x_2, \ldots, x_i, \ldots, x_n) \tag{2.1}$$

where y is a parameter that depends on independent variables $x_1, x_2, \ldots x_i, \ldots, x_n$.
Writing Eq. (2.1) in the differential form we get,

$$dy = \frac{\partial y}{\partial x_1} dx_1 + \frac{\partial y}{\partial x_2} dx_2 + \ldots + \frac{\partial y}{\partial x_i} dx_i + \ldots + \frac{\partial y}{\partial x_n} dx_n \tag{2.2}$$

If the quantities $dy, dx_1, dx_2, \ldots, dx_i, \ldots, dx_n$ are considered to be the uncertainties $U_y, U_{x1}, U_{x2}, \ldots, U_{xi}, \ldots, U_{xn}$ in the quantities $y, x_1, x_2, \ldots, x_i, \ldots, x_n$ respectively, then from Eq. (2.2) we get,

$$U_y = \frac{\partial y}{\partial x_1}(U_{x1}) + \frac{\partial y}{\partial x_2}(U_{x2}) + \ldots + \frac{\partial y}{\partial x_i}(U_{xi}) + \ldots + \frac{\partial y}{\partial x_n}(U_{xn}) \tag{2.3}$$

Now, the maximum value of uncertainty in y, i.e. $U_{y\,max}$ would be obtained when all the uncertainties happen to have the same sign and this would be a worst possible case i.e.,

$$U_{y\,\max} = \left|\frac{\partial y}{\partial x_1}(U_{x1})\right| + \left|\frac{\partial y}{\partial x_2}(U_{x2})\right| + \ldots + \left|\frac{\partial y}{\partial x_n}(U_{xn})\right| \tag{2.4}$$

However, the probability of such an occurrence is very small. Therefore, the more realistic way is to square both sides of Eq. (2.3) which would ensure equal weightage to positive as well as negative values of uncertainties. Therefore, squaring Eq. (2.3) we get,

$$U_y^2 = \left\{\left(\frac{\partial y}{\partial x_1}\right)^2 (U_{x1})^2 + \ldots + \left(\frac{\partial y}{\partial x_i}\right)^2 (U_{xi})^2 + \ldots + \left(\frac{\partial y}{\partial x_n}\right)^2 (U_{xn})^2\right\}$$

$$+ \left\{\left(\frac{\partial y}{\partial x_1}\right)\cdot\left(\frac{\partial y}{\partial x_2}\right)(U_{x1})\cdot(U_{x2}) + \ldots + \left(\frac{\partial y}{\partial x_i}\right)\left(\frac{\partial y}{\partial x_j}\right)(U_{xi})(U_{xj}) + \ldots\right\} \tag{2.5}$$

Because of the random nature of the uncertainties U_{x1}, U_{x2}, . . ., U_{xn}, there is equal probability of these being either positive or negative. Therefore, the term in the second bracket of Eq. (2.5) would tend to be a small quantity which could be neglected.

Therefore, Eq. (2.5) becomes

$$U_y = \pm\left[\left(\frac{\partial y}{\partial x_1}\right)^2 (U_{x1})^2 + \left(\frac{\partial y}{\partial x_2}\right)^2 (U_{x2})^2 + \ldots + \left(\frac{\partial y}{\partial x_i}\right)^2 (U_{xi})^2 + \ldots + \left(\frac{\partial y}{\partial x_n}\right)^2 (U_{xn})^2\right]^{1/2} \tag{2.6}$$

Hence, the overall internal estimate uncertainty due to the combind effect of different variables can be expressed as:

$$(U_I)_{\text{overall}} \pm \left[\sum_{i=1}^{n}\left(\frac{\partial y}{\partial x_i}\right)^2 (U_{xi})^2\right]^{1/2} \tag{2.7}$$

Problem 2.1 *There are n students in a class and each one performs individually a typical simple experiment and determines the mean value of the coefficient of static friction μ between two given surfaces by repeating each measurement m times. If the estimated error of each student is of the order of standard deviation in his data, determine the best estimate of the coefficient of static friction and the internal estimate of the uncertainty on the basis of the results of all the students.*

Solution The best estimate of the coefficient of static friction can be determined by averaging the mean values of all the students so as to obtain the population mean as

$$(\bar{\mu})_{\text{population}} = \frac{\mu_1 + \mu_2 + \mu_3 + \ldots + \mu_n}{n} \tag{2.8}$$

The internal estimate of uncertainty or the internal standard error U_I can be estimated by considering the effect of compounding of errors from a number of variables, i.e.

$$(\bar{\mu})_{\text{population}} = \frac{\mu_1}{n} + \frac{\mu_2}{n} + \frac{\mu_3}{n} + \ldots + \frac{\mu_n}{n} \tag{2.9}$$

Using Eq. (2.7) we get,

$$(U_I)^2_{\text{overall}} = \left(\frac{\partial\bar{\mu}}{\partial\mu_1}\right)^2 \sigma_1^2 + \left(\frac{\partial\bar{\mu}}{\partial\mu_2}\right)^2 \sigma_2^2 + \ldots. + \left(\frac{\partial\bar{\mu}}{\partial\mu_n}\right)^2 \sigma_n^2$$

$$= \left(\frac{1}{n}\right)^2 \sigma_1^2 + \left(\frac{1}{n}\right)^2 \sigma_2^2 + \ldots + \left(\frac{1}{n}\right)^2 \sigma_n^2$$

$$= \frac{\sigma_1^2 + \sigma_2^2 + \ldots + \sigma_n^2}{n^2} \tag{2.10}$$

where $\sigma_1, \sigma_2, \ldots \sigma_n$ are the standard deviations in each student's experiment and represent the estimated error in the same.

Assuming that the estimated error in each case is of the same order, we get,

$$\sigma_1 \simeq \sigma_2 \simeq \sigma_3 \simeq \ldots \simeq \sigma_n \simeq \sigma \tag{2.11}$$

Substituting this in Eq. (2.10) we get,

$$(U_I)^2_{\text{overall}} = \frac{n\sigma^2}{n^2}$$

or

$$(U_I)_{\text{overall}} = \frac{\sigma}{\sqrt{n}} \tag{2.12}$$

In general, the *internal estimate of uncertainty represents the precision error* of the apparatus and can be obtained by dividing the standard deviation of the sample data by the square root of the number of samples. Therefore, for any apparatus, we can write:

$$x = \bar{X} \pm U_I \tag{2.13}$$

where $\bar{X}$ is the population mean and U_I the internal estimate of uncertainty.

Problem 2.2 *It is required to determine the mass of the body shown in the Fig. 2.1. The various dimensions and densities are estimated as follows:*

L = (10.0 ± 0.1) cm

R = (4.00 ± 0.05) cm

ρ_1 = Density of cylindrical portion = (3.50 ± 0.10) g/cm^3

ρ_2 = Density of hemispherical portion = (2.50 ± 0.05) g/cm^3

Calculate the total mass of the body and its overall uncertainty.

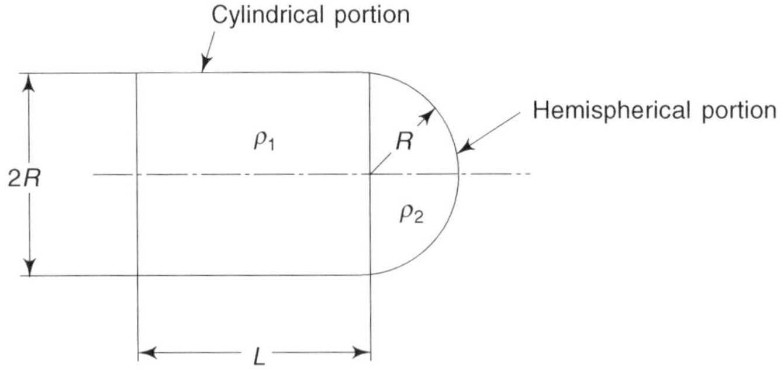

Fig. 2.1 *A typical composite mass consisting of a cylinder and a hemisphere*

Solution Total mass of the body M_T is given by

$$M_T = M_{\text{cylinder}} + M_{\text{hemisphere}}$$
$$= \pi R^2 L \rho_1 + \frac{1}{2}\left(\frac{4}{3}\pi R^3\right)\rho_2 \qquad (2.14)$$

Substituting the mean values of R, L, ρ_1, and ρ_2 in Eq. (2.14) we get,

$$(M_T)_{\text{av}} = \pi(4)^2(10) \times 3.5 + \frac{1}{2}\left(\frac{4}{3}\pi\, 4^3\right) \times 2.5$$
$$= 1759.29 + 335.10$$
$$= 2094.39 \text{ g} \qquad (2.15)$$

Using Eq. (2.7) the overall uncertainty in M_T can be written as:

$$(U)^2_{M_T} = \left(\frac{\partial M_T}{\partial R}\right)^2 U_R^2 + \left(\frac{\partial M_T}{\partial L}\right)^2 U_L^2 + \left(\frac{\partial M_T}{\partial \rho_1}\right)^2 U_{\rho_1}^2 + \left(\frac{\partial M_T}{\partial \rho_2}\right)^2 U_{\rho_2}^2 \qquad (2.16)$$

The various partial derivatives obtained from Eq. (2.14) are

$$\left(\frac{\partial M_T}{\partial R}\right) = 2\pi R L \rho_1 + 2\pi R^2 \rho_2 = 2\pi R[L\rho_1 + R\rho_2]$$

$$\left(\frac{\partial M_T}{\partial L}\right) = \pi R^2 \rho_1$$

$$\left(\frac{\partial M_T}{\partial \rho_1}\right) = \pi R^2 L \qquad (2.17)$$

and

$$\left(\frac{\partial M_T}{\partial \rho_2}\right) = \frac{2}{3}\pi R^3$$

Substituting these values in Eq. (2.16) we get,

$$(U)^2_{M_T} = \{(2\pi)(4)((10 \times 3.5) + (4 \times 2.5))\}^2 (0.05)^2 + (\pi \times 4^2 \times 3.5)^2 (0.1)^2$$
$$+ (\pi \times 4^2 \times 10)^2 (0.1)^2 + \left(\frac{2}{3}\pi\, 4^3\right)^2 (0.05)^2$$
$$= 3080.91 \text{ g}^2$$

$$\therefore \quad (U)_{M_T} = 55.51 \text{ g} \qquad (2.18)$$

Thus, the total mass of the composite body can be expressed as:

$$M_T = 2094.39 \pm 55.51 \text{ g} \qquad (2.19)$$

Alternatively, the uncertainty in the total mass is also sometimes expressed in percentage:

$$\% \text{ Uncertainty in total mass} = \frac{55.51}{2094.39} \times 100$$
$$= \pm 2.65\% \qquad (2.20)$$

Problem 2.3 *The governing equation for the capillary tube viscometer is the well known Hagen–Poiseuelle equation:*

$$Q = \frac{\pi D^4}{128\eta L}\Delta p$$

where *Q is the volume flow rate of the fluid in the capillary tube*
D the diameter of the capillary
η the coefficient of dynamic viscosity of the fluid
L the length of the capillary tube
and *Δp the pressure difference across the two ends of the tube.*

If Q, L, D and Δp are measured with an uncertainty of ± 1%, how accurately is η known? Further, if the uncertainty in the measurement of D is reduced to ± 0.1% by using improved instrumentation, what is the improvement achieved in the uncertainty of η?

Solution From the viscometer equation, the coefficient of dynamic viscosity of the fluid is given by

$$\eta = \frac{\pi D^4}{128 QL}\Delta p \tag{2.21}$$

This shows that η can be written as a function of the variables D, Q, L and Δp as

$$\eta = f(D, Q, L, \Delta p) \tag{2.22}$$

Using Eq. (2.7) the overall uncertainty in η is given by:

$$\{U_I\}^2_{\text{overall}} = \left\{\frac{\partial \eta}{\partial D}\right\}^2 U_D^2 + \left\{\frac{\partial \eta}{\partial Q}\right\}^2 U_Q^2 + \left\{\frac{\partial \eta}{\partial L}\right\}^2 U_L^2 + \left\{\frac{\partial \eta}{\partial (\Delta p)}\right\}^2 U_{\Delta p}^2 \tag{2.23}$$

Differentiating Eq. (2.14) partially with respect to D, Q, L and Δp we get,

$$\frac{\partial \eta}{\partial D} = \frac{\pi D^3}{32 QL}\Delta p$$

$$\frac{\partial \eta}{\partial Q} = -\frac{\pi D^4}{128 Q^2 L}\Delta p \tag{2.24}$$

$$\frac{\partial \eta}{\partial L} = -\frac{\pi D^4}{128 QL^2}\Delta p$$

and

$$\frac{\partial \eta}{\partial (\Delta p)} = \frac{\pi D^4}{128 QL}$$

Substituting the values of the partial derivatives of Eq. (2.24) in Eq. (2.23) and further dividing by η^2 we get from Eq. (2.21),

$$\frac{[U_I]^2_{\text{overall}}}{\eta^2} = \left\{\frac{\frac{\pi D^3}{32 QL}\Delta p}{\frac{\pi D^4}{128 QL}\Delta p}\right\}^2 \cdot U_D^2 + \left\{\frac{\frac{-\pi D^4}{128 Q^2 L}\Delta p}{\frac{\pi D^4}{128 QL}\Delta p}\right\}^2 \cdot U_Q^2 + \left\{\frac{\frac{-\pi D^4}{128 QL^2}\Delta p}{\frac{\pi D^4}{128 QL}\Delta p}\right\}^2 \cdot U_L^2 + \left\{\frac{\frac{\pi D^4}{128 QL}}{\frac{\pi D^4}{128 QL}\Delta p}\right\}^2 \cdot U_{\Delta p}^2$$

$$= 16\left\{\frac{U_D^2}{D^2}\right\} + \left\{\frac{U_Q^2}{Q^2}\right\} + \left\{\frac{U_L^2}{L^2}\right\} + \left\{\frac{U_{\Delta p}^2}{(\Delta p)^2}\right\} \tag{2.25}$$

Substituting the values of proportional uncertainties of D, Q, L and Δp in Eq. (2.25) we get,

$$\frac{[U_{\mathrm{I}}]_{\text{overall}}}{\eta} = \left[16(0.01)^2 + (0.01)^2 + (0.01)^2 + (0.01)^2\right]^{1/2}$$

$$= 0.0436$$

$$= 4.36\% \tag{2.26}$$

Therefore, the overall uncertainty in the measurement of η due to the combined effect of the uncertainties in the measurement of D, Q, L and Δp is ± 4.36%.

It is obvious from Eq. (2.25) that the uncertainty in the measurement of D, which is a fourth order term in Eq. (2.21) has a multiplying factor of 16. Therefore, an improvement in the uncertainty of D would considerably improve the overall uncertainty of η. When the uncertainty in the measured value of D is reduced to ± 0.1%, then the overall uncertainty from Eq. (2.25) becomes:

$$\frac{[U_{\mathrm{I}}]_{\text{overall}}}{\eta} = \left[16(0.001)^2 + (0.1)^2 + (0.01)^2 + (0.01)^2\right]^{1/2}$$

$$= 0.0178$$

$$= 1.78\% \tag{2.27}$$

Therefore, improvement in the overall uncertainty of η becomes

$$= \frac{4.36 - 1.78}{4.36} \times 100$$

$$= 59.2\% \tag{2.28}$$

2.3 ■ STATIC PERFORMANCE PARAMETERS

The definitions and brief descriptions of the various static performance parameters of the instruments are as follows.

2.3.1 Accuracy

Accuracy of a measuring system is defined as the closeness of the instrument output to the true value of the measured quantity (as per standards). However, in usual practice, it is specified as the percentage deviation or inaccuracy of the measurement from the true value. For example, if a chemical balance reads 1 g with an error of 10^{-2}g, the accuracy of the measurement would be specified as 1%.

Accuracy of the instrument mainly depends on the inherent limitations of the instrument as well as on the shortcomings in the measurement process. In fact, these are the major parameters that are responsible for systematic or cumulative errors. In other words, the accuracy of an instrument depends on the various systematic errors involved in the measurement process. For example, the accuracy of a common laboratory micrometer depends on instrument errors like zero error, errors in the pitch of screw, anvil shape, etc. and in the measurement process errors are caused due to temperature variation effect, applied torque, etc.

The accuracy of the instruments (which represents really its inaccuracy) can be specified in either of the following forms:

1. Percentage of true value

$$= \frac{\text{measured value} - \text{true value}}{\text{true value}} \times 100 \tag{2.29}$$

2. Percentage of full-scale deflection

$$= \frac{\text{measured value} - \text{true value}}{\text{maximum scale value}} \times 100 \tag{2.30}$$

It may be noted that *accuracy specification of the instrument as a percentage of full scale deflection (% of fsd) is less accurate than the percentage of true value (% of TV)*. For example, an error ±1% of full-scale deflection of a voltmeter having a range of 1000 V means that a true voltage of 100 V could be read from 90 to 110 V. However, as a percentage of true value, it would be read from 99 to 101 V.

2.3.2 Precision

Precision is defined as the ability of the instrument to reproduce a certain set of readings within a given accuracy. For example, if a particular transducer is subjected to an accurately known input and if the repeated read outs of the instrument lie within say ±1%, then the precision or alternatively the precision error of the instrument would be stated as ±1%. Thus, a highly precise instrument is one that gives the same output information, for a given input information when the reading is repeated a large number of times.

Precision of an instrument is in fact, dependent on the repeatability. The term repeatability can be defined as the ability of the instrument to reproduce a group of measurements of the same measured quantity, made by the same observer, using the same instrument, under the same conditions. As mentioned before, the deviations in repeatability or the inconsistencies in the measured values are caused due to random or accidental errors. Therefore, the precision of the instrument depends on the factors that cause random or accidental errors. As mentioned before, the extent of random errors of alternatively the precision of a given set of measurements can be quantified by performing the statistical analysis (Chs. 21 and 22 of the "Data Analysis-Part III").

Accuracy Versus Precision It may be noted that accuracy represents the degree of correctness of the measured value with respect to the true value. On the other hand, precision represents degree of repeatability of several independent measurements of the desired input at the same reference conditions. As mentioned before, accuracy and precision involved in a measurement are dependent on the systematic and random errors, respectively. Therefore, in any experiment both the quantities have to be evaluated. The former is determined by proper calibration of the instrument and the latter by statistical analysis. However, it is instructive to note that *a precise measurement may not necessarily be accurate and vice versa*. To illustrate this statement we take the example of a person doing shooting practice on a target. He can hit the target with the following possibilities as shown in Fig. 2.2.

1. One possibility is that the person hits all the bullets on the target plate on the outer circle and misses the bull's eye [Fig. 2.2(a)]. This is a case of high precision but poor accuracy.
2. Second possibility is that the bullets are placed as shown in Fig. 2.2(b). In this case, the bullet hits are placed symmetrically with respect to the bull's eye but are not spaced closely. Therefore, this is case of good average accuracy but poor precision.
3. A third possibility is that all the bullets hit the bull's eye and are also spaced closely [Fig. 2.2(c)]. As is clear from the diagram, this is a case of high accuracy and high precision.
4. Lastly, if the bullets hit the target plate in a random manner as shown in Fig. 2.2(d), then this is a case of poor precision as well as poor accuracy.

Based on the above discussion, it may be stated that in any experiment the *accuracy of the observations can be improved but not beyond the precision* of the apparatus.

(a) High precision with poor accuracy (b) Good average accuracy with poor precision (c) High accuracy with high precision (d) Poor accuracy with poor precision

Fig. 2.2 *Illustration of degree of accuracy and precision in a typical target shooting experiment*

2.3.3 Resolution (or Discrimination)

It is defined as the smallest increment in the measured value that can be detected with certainty by the instrument. In other words, it is the degree of fineness with which a measurement can be made. The least count of any instrument is taken as the resolution of the instrument. For example, a ruler with a least count of 1 mm may be used to measure to the nearest 0.5 mm by interpolation. Therefore, its resolution is considered as 0.5 mm. A high resolution instrument is one that can detect smallest possible variation in the input.

2.3.4 Threshold

It is a particular case of resolution. It is defined as the minimum value of input below which no output can be detected. It is instructive to note that resolution refers to the smallest measurable input above the zero value. Both threshold and resolution can either be specified as absolute quantities in terms of input units or as percentage of full scale deflection.

Both threshold and resolution are not zero because of various factors like friction between moving parts, play or looseness in joints (more correctly termed as backlash), inertia of the moving parts, length of the scale, spacing of graduations, size of the pointer, parallax effect, etc.

2.3.5 Static Sensitivity

Static sensitivity (also termed as scale factor or gain) of the instrument is determined from the results of static calibration. This static characteristic is defined as the ratio of the magnitude of response (output signal) to the magnitude of the quantity being measured (input signal), i.e.

$$\text{Static sensitivity,} \quad K = \frac{\text{change of output signal}}{\text{change in input signal}} \tag{2.31}$$

$$= \frac{\Delta q_o}{\Delta q_i} \tag{2.32}$$

where q_o and q_i are the values of the output and input signals, respectively.

In other words, sensitivity is represented by the slope of the input-output curve if the ordinates are represented in actual units. With a linear calibration curve, the sensitivity is constant (Fig. 2.3(a)]. However, if the relationship between the input and output is not linear, the sensitivity varies with the input value and defined as [Fig. 2.3(b)]:

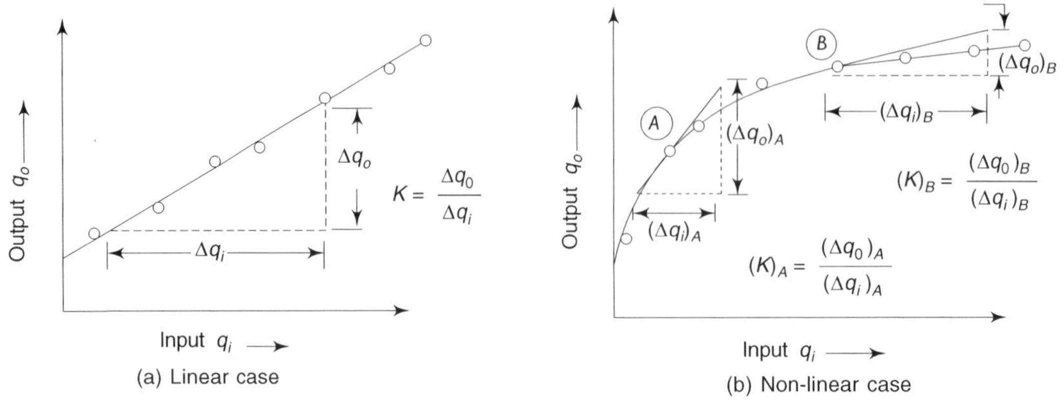

Fig. 2.3 *Static sensitivity of linear and non-linear instruments*

$$\text{Static sensitivity,} \qquad K = \left.\frac{\Delta q_\text{o}}{\Delta q_\text{i}}\right|_{q_\text{i}} \tag{2.33}$$

The sensitivity of a typical linear spring, whose extension is directly proportional to the applied force can be defined as say, 450 N/mm. Similarly, the sensitivity of a non-linear type of copper/constantan thermocouple is found to be maximum at 350 °C and is 60 μV/°C.

It may be noted that in certain applications, the reciprocal of the sensitivity is commonly used. This is termed inverse sensitivity or the deflection factor.

2.3.6 Linearity

A linear indicating scale is one of the most desirable features of any instrument. Therefore, manufacturers of instruments always attempt to design their instruments so that the output is a linear function of the input. However, linearity is never completely achieved and the deviations from the ideal are termed as linearity error. In commercial instruments, the maximum departure from linearity is often specified in one of the following ways.

Independent of the input If the deviations of the output of the instrument from the best fitting straight line (drawn through the calibration points) does not vary with the input [Fig. 2.4(a)], then non-linearity is specified in the terms of higher value of the maximum deviation that occurs on the positive and negative sides of the best fitting or idealised straight line. This value is usually expressed as ± percentage of full-scale deflection.

Proportional to input If the deviations of the output of the instrument from the idealised straight line vary with the input [Fig. 2.4(b)], then non-linearity is specified as a function of the input. In such cases, the maximum deviation points on the positive and negative sides of the idealised straight line are joined with the origin and their slopes are determined. The higher value of the percentage change in slope with respect to the idealised line is usually expressed as ± percentage non-linearity with respect to the magnitude of input values.

Combined independent and proportional to the input In certain cases, the deviations of the output may not vary with the input for a part of the range and may show proportional variation for the rest of the range. In Fig. 2.4(c), a typical calibration curve shows an independent deviation at the lower range and proportional variations at the higher range with respect to the idealised straight line. In this case, the maximum deviations at the lower range is taken care of by specifying say $\pm y\%$ of full scale deflection and by say $\pm x\%$ of the input value.

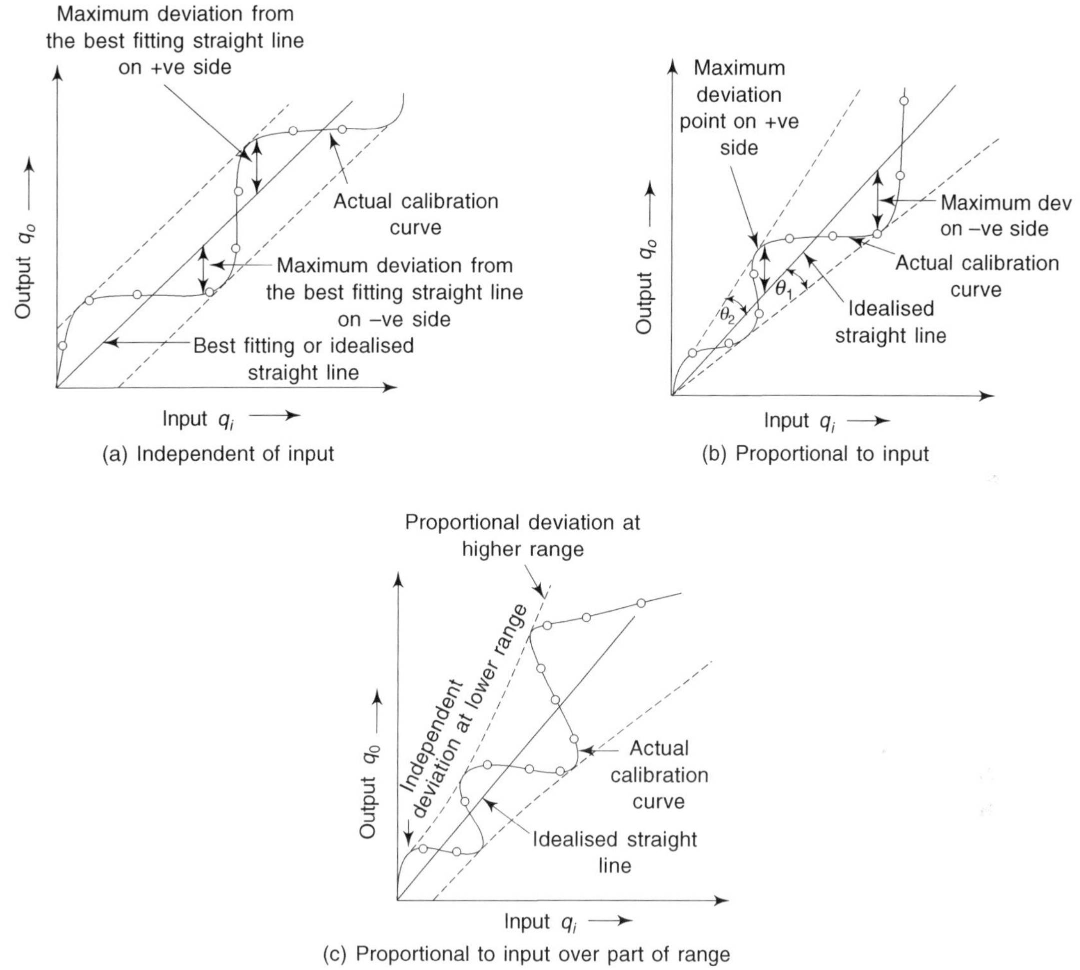

Fig. 2.4 *Typical specifications of non-linearity*

However, it may be noted that in actual practice the non-linearity of a complex type of calibration curve is obtained as say $\pm y\%$ of full-scale deflection and also as $\pm x\%$ of the input value. The non-linearity of the instrument is then stated as $\pm y\%$ of full scale or $\pm x\%$ of the input, whichever is greater. Further, it may be instructive to note that in commercial practice, the term linearity is commonly employed in place of non-linearity. For example, the statement linearity = ±2% of input, means non-linearity = ±2% of input.

In most commercial instruments, linearity is generally implied. In such cases, linearity specifications are equivalent to accuracy specifications and either of them may be specified. However, in instruments that have non-linear characteristics, the read out scale of the instrument is not uniform and is calibrated accordingly.

2.3.7 Range and Span

The range of the instrument is specified by the lower and upper limits in which it is designed to operate for measuring, indicating or recording the measured variable. The algebraic difference between the upper and lower range values is termed as the span of the instrument. The range of the instrument can either be *unidirectional* (e.g., 0–100°C) or *bidirectional* (e.g., –10 to 100 °C) or it can be expanded type (e.g., 80–100°C) or zero suppressed (e.g., 5–40°C).

The *over-range* (or overload capacity) of the instrument is the maximum value of measurand that can be applied to the instrument without causing a perceptible change in its operating characteristics. Further, the *recovery* time of the instrument is the amount of time elapsed after the removal of the overload conditions before it performs again within the specified tolerances.

2.3.8 Hysteresis

It is defined as the magnitude of error caused in the output for a given value of input, when this value is approached from opposite directions, i.e. from ascending order and then descending order. This is caused by backlash, elastic deformations, magnetic characteristics, but is mainly caused due to frictional effects. Whenever, there is solid contact between dry surfaces, stiction (due to Coulomb's friction) comes into play. It is defined as the force or torque necessary to initiate the motion of the instrument. After stiction, dynamic friction comes into play and the output-input characteristics of the instrument takes the shape of a closed curve known as the hysteresis loop shown in Fig. 2.5(a). Further the shape of this loop changes if hydrodynamic or viscous friction is present in the instrument system. In this case, the magnitude of the frictional force depends on the magnitude of the rate of change of input. In other words, the greater the rate of change of input, greater are the deviations in the friction values in the hysteresis loop [Fig. 2.5(b)]. However, if the rate of change of input goes to zero, the magnitude of the viscous friction also approaches zero, i.e. for steady state inputs, there is no error caused due to viscous friction. However, it causes a lag that needs to be compensated.

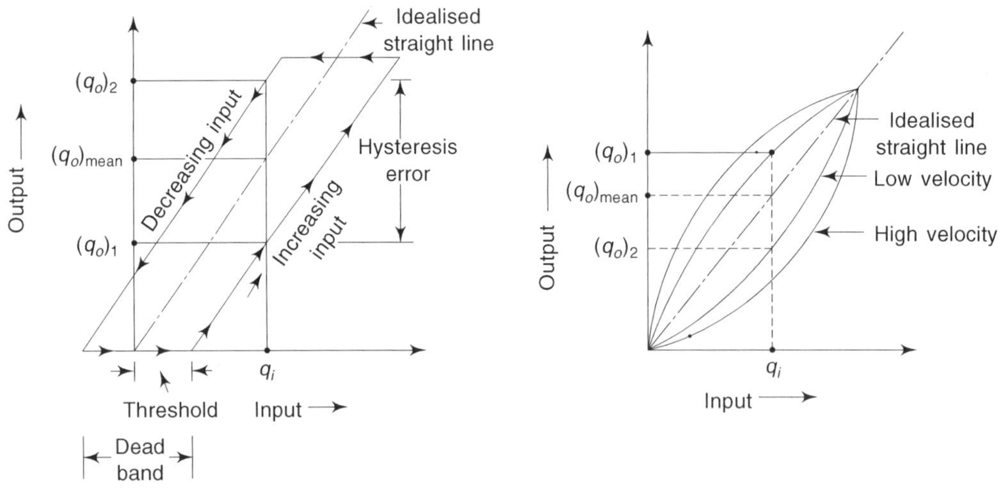

(a) Hysteresis loop due to Coulomb's friction (b) Hysteresis loop due to viscous friction

Fig. 2.5 *Typical output-input curves showing hysteresis effects*

Hysteresis effects are best eliminated by taking the observations both for ascending and descending values of input and then taking the arithmetic mean. For example, in Fig. 2.5(a) and (b), for a value of input q_i, the output in ascending order is $(q_o)_1$ and in descending order is $(q_o)_2$. Then the mean value is:

$$(q_o)_{mean} = \frac{(q_o)_1 + (q_o)_2}{2} \tag{2.34}$$

As is clear from Figs. 2.5(a) and (b), this value is more or less the value obtained from the idealised straight line.

2.3.9 Dead Band

It is defined as the largest change of the measurand to which the instrument does not respond. For example, in the output-input curve with hysteresis effect due to Coulomb's friction, the extent of the dead band is shown in Fig. 2.5(a). In such a case, it is approximately twice the threshold value.

2.3.10 Backlash

It is defined as the maximum distance or angle through which any part of the mechanical system may be moved in one direction without causing motion of the next part. The output-input characteristics of an instrument system with backlash error is similar to hysteresis loop due to Coulomb's friction shown in Fig. 2.5(a). Backlash error can be minimised if the components are made to very close tolerances.

2.3.11 Drift

It is defined as the variation of output for a given input caused due to change in the sensitivity of the instrument due to certain interfering inputs like temperature changes, component instabilities, etc. For example, a strain gauge bridge output without a compensating dummy strain gauge may indicate the changed output-input characteristics as shown in Fig. 2.6 if the ambient temperature changes after the

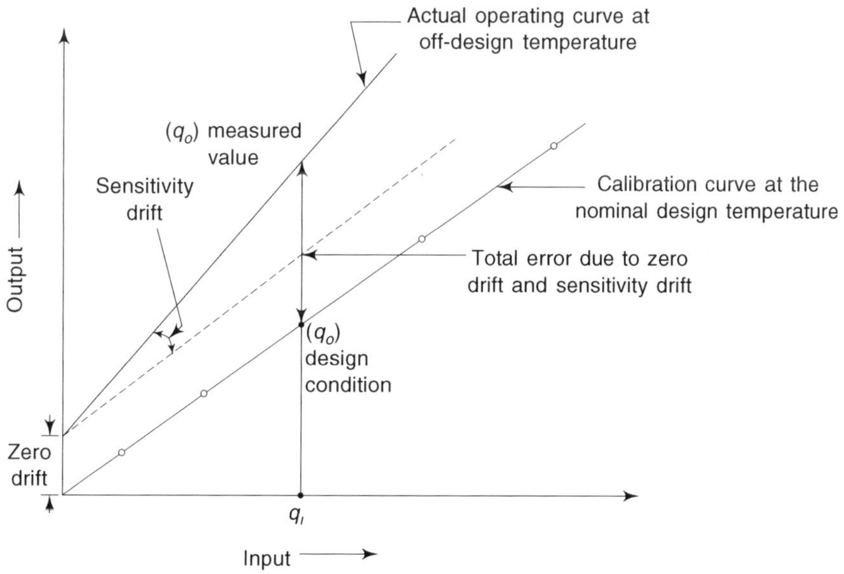

Fig. 2.6 *Zero and sensitivity drift effects on instrument characteristics*

calibration of the instrument. An ambient temperature change causes the change in resistance of the strain gauge. This results in the bridge circuit becoming unbalanced and as a consequence of this, there is an output at zero strain which corresponds to zero drift of the instrument. Further, this resistance change also affects the sensitivity or the scale factor of the instrument which results in the change of slope of the output-input curve (Fig. 2.6). Therefore, the total error involved due to drift is the combined effect of the zero drift as well as the sensitivity drift.

It has been observed that mostly electronic instruments, particularly those involving electron tubes, usually drift considerably for 15 minutes or so after the instrument is switched on. Therefore, it is advisable to switch on the instrument at least half an hour before the commencement of the experiment which ensures that electronic components reach a steady state temperature during this time. This way, the error due to the drift can be minimised. Further, the choice of the components of instruments should be of high quality so that they maintain the desired characteristics during the period of operation.

Problem 2.4 *A load cell calibrated at a temperature of 20°C has the following output/input characteristics:*

Load in kN	*0*	*0.4*	*0.8*	*0.12*	*0.16*	*0.20*
Deflection of meter in mm	*0*	*10*	*20*	*30*	*40*	*50*

When it is used in an environment of 40°C, its characteristics change to the following:

Load in kN	*0*	*0.4*	*0.8*	*0.12*	*0.16*	*0.20*
Deflection of meter in mm	*3*	*14*	*25*	*36*	*47*	*58*

(a) Determine (i) zero drift, (ii) sensitivity drift, and (iii) sensitivity drift per °C change in ambient temperature

(b) If 0.5 mm of scale division can be read with a fair degree of certainity, then determine the resolution of the instrument in both cases, i.e., at 20° C and 40° C.

Solution

(a) (i) As is obvious from the output/input characteristics of the load cell at 40°C, the zero drift which represents no load deflection and is considered the instrument bias or systematic error is found to be 3 mm. By linear interpolation, it represents a systematic error of 0.109 kN at 40°C output-input characteristics.

(ii) At 20°C, the load cell follows a linear output/input characteristics

$$\therefore \quad \text{static sensitivity K at 20°C} = \frac{10}{0.4} = 25 \text{ mm/kN}$$

Further at 40°C also, the output/input characteristics is also linear

$$\therefore \quad \text{static sensitivity K at 40°C} = \frac{11}{0.4} = 27.5 \text{ mm/kN}$$

Hence the sensitivity drift of the instrument

$$= 27.5 - 25 = 2.5 \text{ mm/kN}$$

(iii) Sensitivity drift/°C $= \dfrac{2.5}{20} = 0.125$ (mm/kN)/°C

(b) Inverse sensitivity $\dfrac{1}{K}$ or deflection factor at 20°C

$$= \frac{1}{25} \text{ kN/mm} = 0.04 \text{ kN/mm}$$

∴ resolution of load cell corresponding to 0.5 mm meter scale reading = 0.5 mm × 0.04 kN/mm = 0.02 kN = 20 N

Further, inverse sensitivity, i.e., $\frac{1}{K}$ at 40°C

$$= \frac{1}{27.5} \text{ kN/mm} = 0.036 \text{ kN/mm}$$

∴ resolution of load cell corresponding to 0.5 mm of meter scale reading = 0.5 mm × 0.036 kN/mm = 0.018 kN = 18 N

2.4 ■ IMPEDANCE LOADING AND MATCHING

Any measuring instrument with an input signal source would extract some energy, thereby changing the value of the measured variable. This implies that the input signal suffers a change by virtue of the fact that it is being measured. This effect is termed *loading*. The loading error in the measurements can never be zero, but it must be made as small as possible. It is for this reason that the thermocouple bead is made as small as possible, so that its thermal capacity is small and consequently it does not change the temperature of the system it is sensing. Similarly, a well-calibrated voltmeter (which has high internal resistance) may give a misleading voltage reading when connected across two points in a high resistance circuit due to loading effect [Prob. 2.4(a)]. The same voltmeter may give a more dependable reading when connected across a low resistance circuit with negligible loading error [Prob. 2.4(b)]. Problem 2.4 illustrates the loading effect caused by a high internal resistance voltmeter connected in parallel to a typical resistive circuit. However, the term loading when applied to a general electrical circuit is termed *impedance* loading. Therefore, proper care must be taken to avoid impedance loading or mismatching between electrical devices so as to avoid voltage distortions.

Consider an example of a transducer element (Fig. 2.7) with internal impedance Z_i and say E is the voltage developed in the transducer which is an open circuit voltage. When the transducer is connected in series with a recorder of input impedance Z, the voltage recorded across points A and B would be

$$E_{AB} = \frac{EZ}{Z + Z_i}, \tag{2.35}$$

where $\frac{E}{Z + Z_i}$, is the current in the circuit.

For no impedance loading (the ideal case)

$$E_{AB} = E \tag{2.36}$$

This can be achieved if the value of Z is infinite.

In practice, Z should be $\gg Z_i$, i.e. the larger the value of Z, the more closely does the terminal voltage approach the internal voltage. In fact, this principle is employed in the measurement of voltages using vacuum tube voltmeter (VTVM) which has a high internal resistance of the order of 10 MΩ. Because of this, it does not alter the voltage of the input source, i.e. the impedance loading of the instrument is very small.

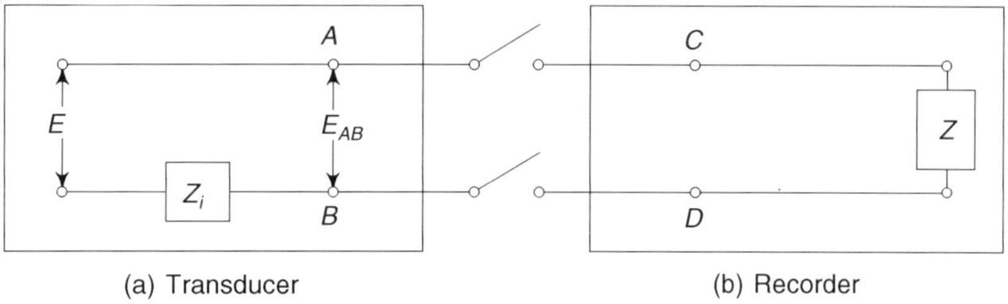

Fig. 2.7 *Loading effect of recorder on the transducer element*

In certain situations, we wish to deliver power from the device (with internal impedance Z_i) to the external load (with impedance Z). A common example of this is a power amplifer connected in series with an electrodynamic vibrator (Fig. 2.8).

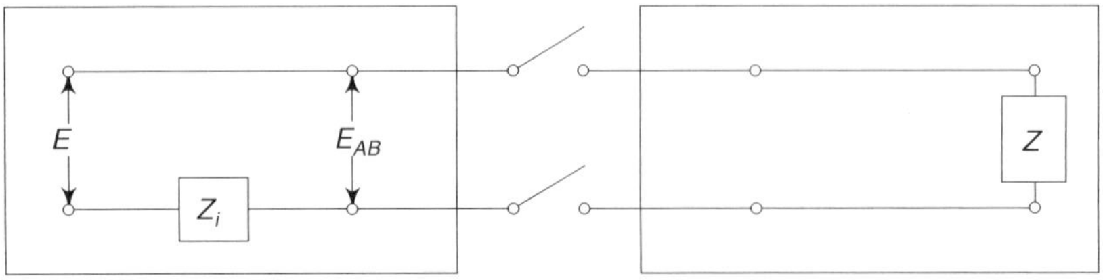

(a) Power amplifier (b) Electrodynamic vibrator

Fig. 2.8 *Impedance matching for maximum power*

Now, the power delivered by the power amplifier to the electrodynamic vibrator is given by:

$$P = \frac{E_{AB}^2}{Z} \tag{2.37}$$

Substituting the value of E_{AB} from Eq. (2.35) we get,

$$P = \frac{E^2}{Z} \cdot \left\{ \frac{Z}{Z + Z_i} \right\}^2 \tag{2.38}$$

For maximum power transmission,

$$\frac{\partial P}{\partial Z} = 0 \tag{2.39}$$

which gives $Z = Z_i$ (2.40)

Thus maximum power may be transmitted from a device if the impedance of the external load just matches its internal impedance. This is the essential principle of impedance matching in electrical circuits.

Problem 2.5 (a) *A voltmeter with internal resistance of 200 kΩ is connected across an unknown resistance. It reads 250 V and the milliammeter (with very small internal resistance) connected in series with the same resistance reads* 10 mA. Determine the apparent resistance, actual resistance and the loading error due to the loading effect of the voltmeter.

(b) If the same voltmeter and milliammeter when connected in another unkown resistance read 100 V and 2 A, respectively, determine the loading error in this case.

Solution

(a) Total resistance in the circuit

$$R_T = \frac{V_T}{I_T} = \frac{250}{10 \times 10^{-3}} = 25\ \text{k}\Omega \tag{2.41}$$

Neglecting the resistance of the milliammeter, the apparent value of unknown resistor,

$$R_{\text{app}} = 25\ \text{k}\Omega \tag{2.42}$$

Since the voltmeter is in parallel with the unknown resistance, therefore,

$$\frac{1}{R_T} = \frac{1}{R_{\text{act}}} + \frac{1}{R_V} \tag{2.43}$$

where R_{act} is the actual resistance of the resistor and R_V the resistance of the voltmeter

or
$$R_{\text{act}} = \frac{R_T R_V}{R_V - R_T}$$

$$= \frac{(25)(200)}{(200-25)}$$

$$= 28.56 \text{ k}\Omega \tag{2.44}$$

Therefore,
$$\% \text{ loading error} = \frac{R_{\text{act}} - R_{\text{app}}}{R_{\text{act}}} \times 100$$

$$= \frac{28.56 - 25}{28.56} \times 100$$

$$= 12.46\% \tag{2.45}$$

(b) In the second case,

$$R_T = \frac{100}{2} = 50 \ \Omega \tag{2.46}$$

Neglecting milliammeter resistance

$$R_{\text{app}} = 50 \ \Omega \tag{2.47}$$

$$R_v = 200 \text{ k}\Omega$$

$$R_{\text{act}} = \frac{50 \times 200}{199.95} = 50.0125 \ \Omega \tag{2.48}$$

$$\% \text{ loading error} = \frac{50.0125 - 50}{50.0125} \times 100$$

$$= 0.025\% \tag{2.49}$$

2.5 ■ SPECIFICATIONS OF INSTRUMENT STATIC CHARACTERISTICS

For the measurement of a physical variable, generally a number of alternatives may be available. For example, pressure in a given system may be measured by means of a simple mechanical type Bourdon gauge or with a diaphragm gauge with electrical read out. Similarly, the velocity of flow may be determined by a simple Pitot-static tube or by means of a hot wire anemometer, depending on the requirement of the particular situation. It may be noted that for a particular application, the types of instruments available normally would differ quite drastically in the scientific principle on which they are based, the mechanism through which this principle is applied and the way in which the result is displayed. Therefore, it is imperative to carefully study the specifications of the different instruments, weigh the pros and cons with regards to their capabilities and limitations and then exercise a judicious choice in the selection of the instrument for a given application.

To illustrate how some of the static characteristics are generally specified in a commercially available instrument, let us take a concrete, commonly occurring situation. For example, the pressure in a system can be measured by means of a simple mechanical Bourdon gauge. However, electrical read out is generally preferable, so we may decide to choose a diaphragm gauge. The deformation of the diaphragm due to applied pressure could be monitored electrically either by a linear variable differential transducer (LVDT) or by the change in the capacitance of a plate with respect to another rigidly fixed parallel plate or by monitoring the variation in the stress (or strain) at the centre of the diaphragm. Thus, in this case,

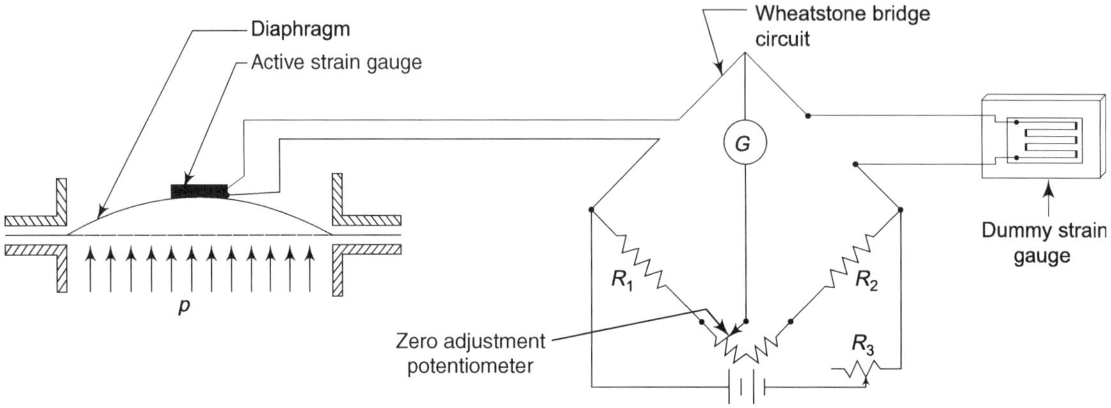

Fig. 2.9 *A typical diaphragm pressure gauge using electrical resistance strain gauges*

we have three choices of the associated electrical circuitry of the diaphragm gauge. The logical step now is to consider the merits and demerits of each type of the associated instrumentation. An experienced experimentalist may think that in LVDT, the mass of the core would result in inertia effects and thus may not be suitable for dynamic measurements. Further, the main drawback of the capacitance type gauge is its non-linear characteristics. Therefore, the strain gauge type of associated instrumentation may be considered quite suitable when used in conjunction with the diaphragm gauge. Then, the specifications of this type of measurements system available commercially may be typically as follows:

1. Maximum bridge excitation = 20 V dc
2. Recommended bridge excitation = 12 V dc
3. Pressure range = 200 kPa
4. Bridge output = 0.1 mV/V/kPa
5. Accuracy = ±0.5% of FS (Full Scale)
6. Repeatability = ±0.2% of FS
7. Linearity and hysteresis = ±0.4% of FS
8. Thermal zero shift = less than 0.02% FS/°C
9. Thermal sensitivity shift = less than 0.02%/°C
10. Bridge resistance = 350 Ω
11. Minimum required impedance of output for indicating/recording unit to be used = 2 kΩ
12. Overload capacity = 250% of FS

Now, let us weigh some further pros and cons of the above-mentioned specifications:

1. When the excitation is 12 V, the static sensitivity = 1.2 mV/kPa
2. Full scale output at this recommended excitation $= 1.2 \times 200 = 240$ mV
3. Accuracy $= \pm \dfrac{(200)(0.5)}{100} = \pm 1.0$ kPa
4. Maximum deviation in linearity and hysteresis $= \pm \dfrac{0.4 \times 240}{100} = \pm 0.96$ mV

5. Overload capacity = up to 500 kPa
6. The specification of impedance greater than 2 kΩ in the voltage measuring device at the output means that in such a case the impedance loading effects would be within the accuracy of the instrument.

2.6 ▮ SELECTION OF THE INSTRUMENT

The selection of any instrument out of those available depends on the performance characteristics of each instrument *vis-à-vis* its cost. In general, the selection procedure seeks to maximise the 'pay-off ratio' or the 'transfer function' of the investment which is the ratio:

$$\frac{\text{Value of useful information}}{\text{Necessary total cost}} \tag{2.50}$$

The various considerations involved in the selection of the instrument include the following from the 'value viewpoint'.

Instrument's qualities, value guided

1. Accuracy and precision characteristics including other specifications like sensitivity, linearity, hysteresis, zero and sensitivity drift, dead band, etc.
2. Nature and type of data available, i.e. whether analog, digital, continuous or sampled.
3. Nature and type of read out, i.e. whether indicating or recording type, etc.
4. Nature of further data computations, if required.
5. Signal-to-noise characteristics of the transducer and the system fidelity especially when extensive data transmission or translation is involved.
6. Dynamic response characteristics if input signal is time-dependent.
7. Susceptibility to environmental disturbances.

Convenience aspects, value judged

1. Suitability for the given application, i.e. whether for laboratory use, field use or both.
2. Adaptability to different sizes of inputs, i.e. scale expansion, range changes, etc.
3. Ease in calibration, when needed.
4. Simplicity and ease of instrument behaviour diagnosis.
5. Material durability and non-fouling design.
6. Fool-proof assembly.
7. Maintenance, repair, local representation and steady delivery.
8. Ready self-indication or check determination in case of instrument malfunction.
9. Safety in use.
10. Proper shape, appealing appearance and necessary protective envelope.

Costs, initial and cumulative total

1. Initial cost of instrument procurement, installation including the various attachments and accessories.
2. Maintenance, repair, recalibration, etc.
3. Running cost.
4. Expected life span considering the 'salvage' value of components which may be used in other similar instruments as interchangeable modules.

It is worth mentioning here that the establishment of true costs and 'values' of an instrument is an ideal which is usually and understandably approached rather than attained. In practice, it is found that a number of instruments may be marketed by different commercial firms for a particular measurement.

Therefore, the 'normal' procedure of selecting a particular instrument consists of carefully studying the positive and negative points of each instrument including the prevailing market price and the availability. This combined with mature judgement, intuition and experience helps to arrive at the 'value guided' optimal selection of the instrument for the given application.

Review Questions

2.1 Match the following. Give your answers in the space within the brackets. For example, the answer of part (i) is (6).

(i) Relative error (6)	1. A device whose output is an enlarged reproduction of the essential features of the input wave and which draws power from a source other than the input signal.
(ii) Null type device ( )	2. The act or process of making adjustments or markings on the scale so that the instrument readings conform to an accepted standard.
(iii) Amplifier ( )	3. Measurand generates an opposing effect to maintain zero deflection.
(iv) Drift ( )	4. An action used to convey information.
(v) Attenuator ( )	5. An element which converts the input of energy in a form of an output with different form of energy.
(vi) Signal ( )	6. The ratio of difference between measured value and true value to the true value of the measurand.
(vii) Transducer ( )	7. Maximum distance or angle through which any part of mechanical system may be moved in one direction without causing the motion of the next part.
(viii) Precision ( )	8. Unwanted signal tending to obscure the transducer signal.
(ix) Calibration ( )	9. Gradual departure of the instrument output from the calibrated value.
(x) Resolution ( )	10. A device which causes decrease in amplitude of the signal without causing appreciable distortions in it.
(xi) Noise ( )	11. Smallest increment in measurand that can be detected with certainty by the instrument.
(xii) Backlash ( )	12. The ability of the device to give identical output when repeat measurements are made with the same input signal.

2.2 Indicate if the following statements are true or false. If false, then write the correct statement.

(i) Correctness or exactness in measurements is associated with the accuracy and not with the precision.
(ii) Reproducibility and consistency are expressions that best describe precision in measurements.
(iii) It is not possible to have precise measurements which are not accurate.
(iv) Instrument bias refers to the random errors in the instrument.
(v) An instrument with 1% accuracy is considered better than another with 5% accuracy.
(vi) It is worthwhile to improve the accuracy of the instrument beyond its precision.
(vii) Any measurement is expressed by a numerical value alone.
(viii) Error and uncertainty are synonymous terms.
(ix) To prevent loading of the circuit under test, the input impedance of the voltmeter must be very low.

(x) For an experiment to be consistent, the internal estimate of uncertainty should be minimum possible.

2.3 Fill in the blanks in the following:

(i) The ratio of the change in output to the change in input is the ______________ of the instrument.

(ii) A spurious signal that modifies the output of the instrument is termed as ______________ .

(iii) Differences in the measured value because of change in direction of the measurand is termed as ______________ .

(iv) External estimate of uncertainty depends on the ______________ of the instrument.

(v) Theoretically, the random errors in any series of measurements follow the law of ______________.

(vi) In a certain factory all the clocks are controlled by electric impulses, so that they indicate the correct time exactly to a second. This type of time system could be described as __________ .

(vii) If $q = \alpha x^a y^b$, then the expression for best estimate of uncertainty in q is ______________.

(viii) Calibration of the instruments removes/accounts for ______________ type of errors.

(ix) Maximum amount of power may be drawn from the device when the internal impedance of the device ______________ the impedance of external load.

(x) Errors caused by the act of measurement on the physical system being tested are termed as ______________ .

2.4 Choose the correct statement from the following:

(i) An ac millivoltmeter has a range of 0 – 1000 mV and its accuracy is ±0.5% of fsd (full-scale deflection). If the input voltage of the instrument is 400 mV, the output of the instrument would be

(a) 402 mV (b) 398 mV
(c) between 398 and 402 mV (d) between 395 and 405 mV

(ii) The smallest change in the value of input variable being measured, that will cause a change in the output signal of the instrument is termed as

(a) hysteresis (b) drfit (c) resolution (d) threshold

(iii) The error which is repetitive in nature is

(a) observational error (personal) (b) environmental error
(c) random error (d) systematic error

(iv) Threshold of the instrument is defined as the

(a) ratio of the output of the instrument to the corresponding input signal
(b) drift of the output of the instrument due to ageing of components
(c) smallest input measurable change (non-zero value)
(d) smallest measurable input signal which can be detected by the instrument.

(v) Zero error of a micrometer is

(a) random error (b) interference error (c) systematic error (d) loading error

(vi) The external estimate of uncertainty in the experiment:

(a) can be reduced by increasing the number of reading
(b) is always greater than the internal estimate of uncertainty
(c) is usually obtained from the least count of the instrument
(d) is always symmetric about the mean value of the data.

(vii) The measurement accuracies of A and B are ±4% of A and ±3% of B, respectively. Then, the percentage accuracy in the parameter ($A \times B$) as percentage of ($A \times B$) is

(a) ±1% (b) ±5% (c) ±7% (d) ±12%.

(viii) The gradual departure of the instrument output caused by certain interfering input and component instabilities is termed as
(a) hysteresis (b) dead band (c) threshold (d) drift

(ix) The gain adjustment of a simple linear instrument gets increased by a small fixed amount. Then,
(a) both accuracy and precision decrease
(b) accuracy decreases while precision is unaffected
(c) precision decreases while accuracy remains unaffected
(d) accuracy, precision and resolution of the instrument increase proportionally.

(x) Maximum power is transmitted by an electrical transducer if the impedance of the external load
(a) is very low
(b) is very high
(c) matches with the internal impedance of the transducer
(d) increases from very low values to very high values

(xi) Repeatability of the instrument with respect to a given fixed input is
(a) accuracy (b) precision (c) resolution (d) sensitivity

(xii) Error caused by the act of measurement on the physical system being tested is
(a) hysteresis error (b) random error (c) systematic error (d) loading error

(xiii) Static sensitivity of the instrument is the
(a) least reading of the scale/range of scale
(b) least-reading of the scale/unit measurable quantity
(c) ratio of the change in output to the corresponding change in the input variable
(d) range of scale/least count of the scale

(xiv) The term backlash used in instrumentation means
(a) smallest increment in the measurand that can be detected by the instrument
(b) gradual departure of the measured value from its calibrated value
(c) maximum distance or angle through which any part of the mechanical system may be moved without causing any motion in the next part
(d) the ability of the instrument to give output reading close to each other, when the input is of fixed type

(xv) The radius of a sphere was estimated as (50 ± 0.5) mm. The estimated error in its mass is
(a) 3% (b) 1% (c) 0.1% (d) 10.0%

2.5 A dial gauge is used to measure the pressure in vessel. The pivot is not exactly centred and as a result the readings are subject to systematic error. It was found that the imperfection makes the readings too large in a linear fashion from state 1, i.e. 6.895 kN/m^2 for a dial readings of zero to state 2, i.e. 27.58 kN/m^2 for a dial reading of 150. What would be the value of pressure for a dial reading of 100?

2.6 The length of line is measured with a 50 m metallic tape and found to be 935.12 m at 45°C. The tape was standardised at 20°C and coefficient of thermal expansion is 0.000 0065 per °C. What is the correct length of the line to the nearest hundredth of a metre?

2.7 The surface area of a right circular cylinder is given by the formula $S = 2\pi rh$. Find the surface area and its error if r was measured as 5.00 ± 0.08 cm and h as 7.00 ± 0.03 cm.

2.8 Two resistors are placed in series and in parallel. If R_1 = (100 ± 0.12) Ω and R_2 = (50.00 ± 0.08) Ω, calculate the uncertainty in the combined resistance for both series and parallel arrangements.

2.9 If $z = 2x$ then it can be shown that $U(z) = 2U(x)$

Now if we put $z = x + x$ then,

$$U(z) = \sqrt{U^2(x) + U^2(x)} = \sqrt{2}\, U(x)$$

Which value of $U(z)$ is correct and why?

2.10 The resistance of certain length of wire R is given by

$$R = 4\rho l/\pi d^2$$

where ρ is the resistivity of the wire in Ω cm
l the length of the wire in cm
d the diameter of the wire in cm.

Determine the nominal resistance and the uncertainty in resistance of the wire with the following data:

$$l = 523.8 \pm 0.2 \text{ cm}$$
$$\rho = 45.6 \times 10^{-6} \pm 0.15 \times 10^{-6}\ \Omega \text{ cm}$$
$$d = 0.062 \pm 1.2 \times 10^{-3} \text{ cm.}$$

2.11 The impedance of the $R - L$ circuit operating on alternating current is given by:

$$Z = \sqrt{R^2 + \omega^2 L^2}$$

The resistance R is known to be 100 Ω with an uncertainty of 5%, L is known to be 2 H with an uncertainty of 10% and ω is known exactly $2\pi \times 50$. Determine the percentage uncertainty in the measurement of Z.

2.12 A rectangular steel rod of width b and depth d is supported at its ends and loaded at its centre by load W. If the length of the rod between the supports is l and y_c is the deflection at the centre, then $y_c = \dfrac{Wl^3}{4\,Ebd^3}$

where E is the modulus of elasticity. Measurement give

$$b = 4.942 \pm 0.042 \text{ cm}$$
$$d = 5.250 \pm 0.025 \text{ cm}$$
$$l = 1.000 \text{ m} \pm 0.5 \text{ cm}$$
$$y_c = 2.622 \pm 0.25\% \text{ of } y_c \text{ mm}$$
$$W = 15000 \text{ N (exact).}$$

(a) Determine the nominal value of the modulus of elasticity.
(b) Determine the percentage uncertainties in the various measured quantities.
(c) Compute the percentage uncertainty in the modulus of elasticity.

2.13 The decay constant of a ballistic galvanometer can be obtained from the equation

$$\theta_2 = \theta_1 e^{-kT}$$

where θ_1 and θ_2 are the deflections (on the same side) of the damped oscillation
T the period of vibration
and k the decay constant.

In an experiment T was measured as 6.8 ± 0.1 s, θ_1 and θ_2 were observed to be 28.1° and 18.3° with each measurement being uncertain to ± 0.2°. Determine the value of k and its uncertainty in the given measurements.

2.14 A McLeod gauge is used to measure vacuum. The vacuum pressure p in mm of Hg is given by

$$p = \frac{ay^2}{V}$$

where a is the area of cross-section of the capillary, y the length of the capillary occupied by the gas and V the volume of the bulb. The uncertainties associated with the parameters are

$V = (100 \pm 1)$ cc
d = diameter of capillary = (1 ± 0.01) mm, and
$y = 3$ cm $\pm$ 0.5 mm.

Determine the nominal vacuum pressure and uncertainty in the same.

2.15 A speed test of a car was conducted on a track of length 2000 m $\pm$ 10 cm. The time taken by the car to cover this track was (40.8 ± 0.2) s. Determine the speed of car and its uncertainty.

2.16 Using the conservation of mass principle, the mass flow rate in a pipe for a steady flow is given by $\dot{m} = \rho A V_{av}$.

Ten measurements of density of the oil flowing through the pipe gave the mean value as 782.5 kg/m^3 and the mean square deviation as 16.85 kg^2/m^6. Twelve measurements of diameter of the pipe gave the mean diameter as 100.2 mm with a mean square deviation of 0.45 mm^2. Further, 16 measurements of average velocity through a pipe gave the mean value as 3.6 m/s with a mean square deviation of 0.12 m^2/s^2.

Determine the nominal rate of mass flow of the oil through the pipe and its uncertainty.

2.17 A rectangular steel beam is used as a force measuring device by applying a force F at the free-end and correlating this with the deflection δ measured at the free-end by means of an LVDT. The dimensions of the beam are

length, $L = (50 \pm 0.5)$ mm
breadth, $b = (8.20 \pm 0.1)$ mm, and
thickness, $t = (4.1 \pm 0.1)$ mm

Determine the applied force F and its uncertainty if the deflection δ at the free-end way found to be (0.8 mm $\pm$ 2% of δ). Take the modulus of elasticity E of the beam as 2.07×10^{11} N/m^2 and use the following relationship for force and deflection at the free-end of the cantilever as

$$F = \frac{Ebt^3}{4L^3}\delta$$

2.18 A variable resistance potentiometric transducer having a resistance of 12 kΩ is connected to a dc voltage source of 60 V. The voltage output of the transducer is measured by means of a voltmeter of internal impedance of 120 kΩ. Determine impedance loading error at 25% position on the transducer and the actual voltage reading observed at this position.

2.19 The power amplifier shown in Fig. 2.7 has an internal resistance of 6 kΩ. It is connected in series with an electrodynamic vibrator with an impedance of 25 kΩ. Determine the percent loading error in the amplifier voltage output.

If the internal voltage generated by the amplifier is 100 V. Calculate the power output at the given loading conditions. What should be the maximum power output? Further, what power output would result if the external load (electrodynamic vibrator) has a resistance of 1000 Ω?

2.20 A transducer has a source impedance that varies from 5 to 2 kΩ during its operation and is used with different display instruments. Will the impedance loading be a serious problem, when transducer is used with:

(a) an oscilloscope having an input impedance of 1.0 MΩ, and
(b) an optical oscillograph having an input impedance of 100 Ω?

Answers

2.1 (ii) (3) (iii) (1) (iv) (9) (v) (10) (vi) (4)
(vii) (5) (viii) (12) (ix) (2) (x) (11) (xi) (8)
(xii) (7)

2.2 (i) true (ii) true (iii) false (iv) false (v) true
(vi) false (vii) false (viii) false (ix) false (x) false

2.3 (i) sensitivity (ii) noise (iii) backlash (iv) resolution
(v) normal distribution (vi) both accurate and precise
(vii) $q \times \sqrt{a^2 \frac{U_x^2}{x^2} + b^2 \frac{U_y^2}{y^2}}$ (viii) systematic (ix) matches (x) loading errors

2.4 (i) d (ii) c (iii) d (iv) d (v) c
(vi) c (vii) b (viii) d (ix) b (x) c
(xi) b (xii) d (xiii) c (xiv) c (xv) a

2.5 $p = 20.685$ kNm^2

2.6 $l = 935.272$ m

2.7 $S = 219.91 \pm 3.64$

2.8 $R_{\text{series}} = (150 \pm 0.14)\ \Omega$, $R_{\text{parallel}} = (33.33 \pm 00.04)\ \Omega$

2.9 $U(z) = 2\ U(x)$ is correct as x is the only independent variable of z.

2.10 $R = 7.911 \pm 0.307\ \Omega$

2.11 percentage uncertainty in $Z = 9.75\%$

2.12 (a) $E = 2 \times 10^{11}$ N/m^2
(b) $\frac{U_b}{b} = 0.85\%$, $\frac{U_d}{d} = 0.48\%$, $\frac{U_l}{l} = 0.5\ \%$
(c) $\frac{U_E}{E} = 2.26\%$

2.13 $k = 0.0631 \pm 0.0021$

2.14 $p = (7.068 \pm 0.329)\ 10^{-3}$ mm of Hg

2.15 $V = (49.02 \pm 0.24)$ m/s

2.16 $\dot{m} = (22.21 \pm 4.32\%)$ kg/s

2.17 $F = (187.17 \pm 8.24\%)$

2.18 2.44%, 14.634 V

2.19 19.4%, 0.26 W, 0.204 W

2.20 (a) No (b) Yes

Chapter

3

Dynamic Characteristics of Instruments

INTRODUCTION

When dynamic or time-varying quantities are to be measured, it is necessary to find the dynamic response characteristics of the instrument being used for measurement. The dynamic inputs to an instrument may be of the following types:

1. Periodic input—Varying cyclically with time or repeating itself after a constant interval, viz. *T* as in Fig. 3.1. The input may be of harmonic or non-harmonic type.
2. Transient input—Varying non-cyclically with time. As shown in Fig. 3.2, the signal is of a definite duration and becomes zero after a certain period of time.
3. Random input—Varying randomly with time, with no definite period and amplitude. This may be continuous, but not cyclic as in Fig. 3.3.

To mention examples of the above types of signals, vibration excitation due to unbalance of a rotating body is harmonic, pressure variation in an internal combustion engine is periodic, while forces due to an explosion are transient. Further, pressure fluctuations in fluid flow due to turbulence are of random type.

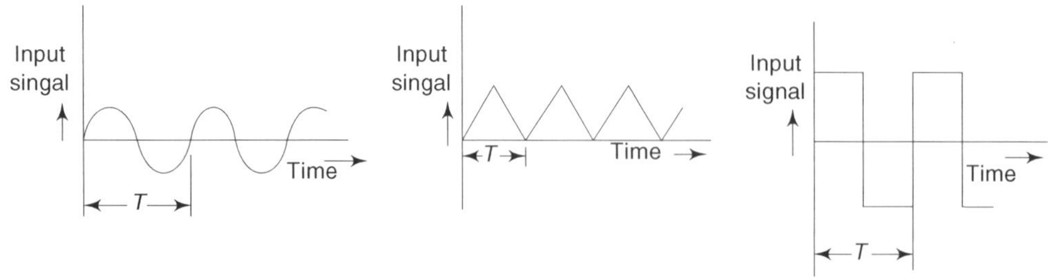

Fig. 3.1 *Periodic signals with time period T*

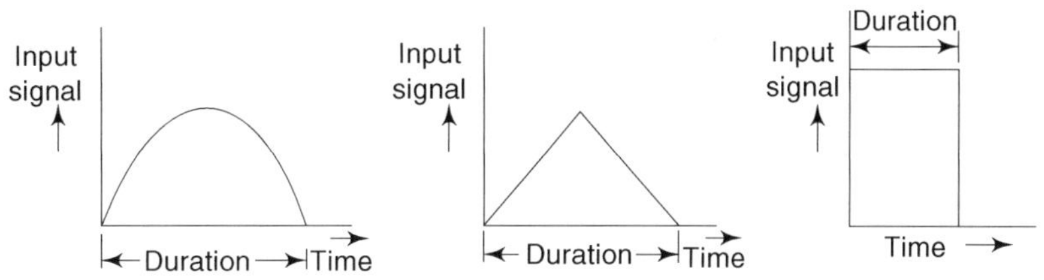

Fig. 3.2 *Transient signal*

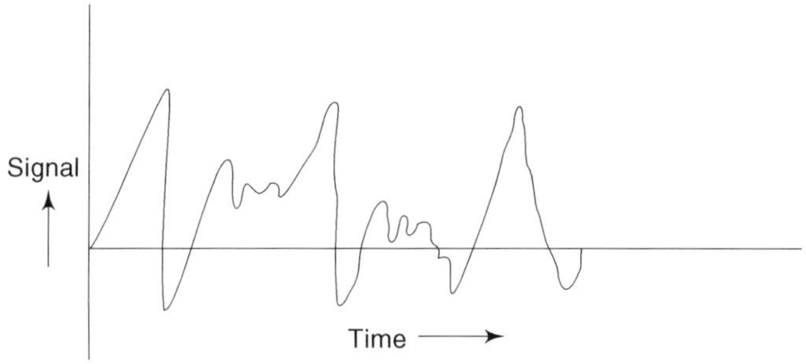

Fig. 3.3 *Random signal*

For studying the dynamic characteristics of an instrument or a combination of instruments, it is necessary to represent each instrument by its mathematical model, from which the governing relation between its input and output is obtained (Fig. 3.4). As will be shown later, such a relation between the output and input signals of a temperature measuring element like a thermometer or thermocouple, is

$$(1 + \tau D)x_o = Kx_i(t) \tag{3.1}$$

x_i and x_o being input and output signals, respectively, both being functions of time, while K and τ are system constants, D being the time derivative operator which is of first order and hence the instrument is called a first order one. Equation (3.1) may be represented by a block diagram, as in Fig. 3.5, with the term in the block, viz.

$$\frac{K}{1+\tau D} = \frac{x_o(t)}{x_i(t)}$$

The ratio of output signal to the input signal, viz. the term in the block, is known as the transfer function of the instrument. Similarly, for a second-order instrument, the block diagram is shown in Fig. 3.6, with a_0, a_1, a_2 and b_0 being constants of the instrument. An indicating instrument like an ammeter is seen to be of second-order type.

A zeroth order instrument has the transfer function as b_0/a_0 = 'K' where 'K' is a static sensitivity constant.

The dynamic characteristics of an instrument can be determined experimentally with a known dynamic input signal. For theoretical analysis, for any dynamic input, a solution of its governing equations, obtained from its mathematical model, is desirable. Thus, the following steps are essential for understanding the dynamic behaviour of an instrument.

1. To formulate its governing equations, relating dynamic input and output signals.
2. To obtain the dynamic output response, for the given input, by solution of the governing equations.
3. In case the output response is not satisfactory, it may be possible to improve the same by what is known as *compensation*.

Fig. 3.4 *Scheme for studying dynamic characteristics of an instrument*

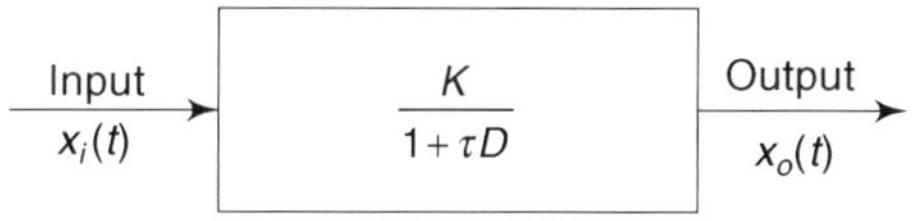

Fig. 3.5 *Block diagram of a first order instrument*

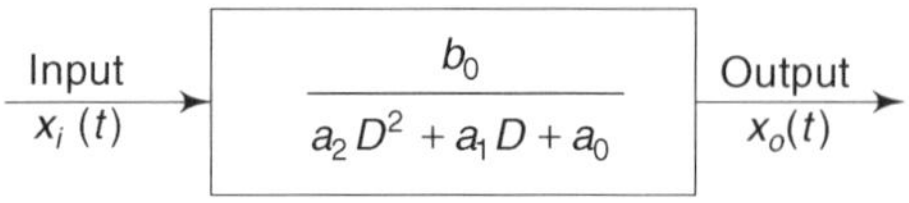

Fig. 3.6 *Block diagram of a second-order instrument*

3.1 FORMULATION OF SYSTEM EQUATIONS

The governing equations relating output and input signals of any instrument system, for dynamic input, will be obtained, for a few specific cases below.

Resistance Transducer Connected to Display Unit A resistance transducer, connected to a display instrument, which is taken as resistance and capacitance in parallel is shown in Fig. 3.7. To derive the equations relating its input and output voltages, we proceed as follows:

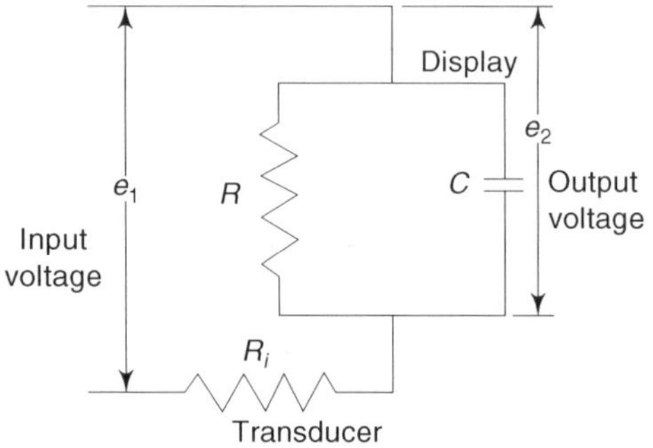

Fig. 3.7 *Resistance transducer connected to display unit*

Both input and output voltages e_1 and e_2, respectively, are functions of time t.

Now, $$\frac{e_2}{e_1} = \frac{Z}{R_i + Z} \tag{3.2}$$

where Z is the display impedance. This is given by

$$\frac{1}{Z} = \frac{1}{R} + \frac{1}{1/DC} \tag{3.3}$$

where $$D = \frac{\mathrm{d}}{\mathrm{d}t}$$

Thus, from Eqs. (3.2) and (3.3)

$$\frac{e_2}{e_1} = \frac{1}{1 + R_i\left(\frac{1}{R} + DC\right)} \tag{3.4}$$

In practice, R_i/R is small and may be neglected compared to unity.

Thus, Eq. (3.4) may be written as

$$(1 + \tau D)e_2 = e_1$$

where $$\tau = R_iC$$

This equation is a first order one, relating output and input. This can be written as

$$\tau\frac{\mathrm{d}e_2}{\mathrm{d}t} + e_2 = e_1(t) \tag{3.5}$$

Thermal Element It is required to find the equation relating medium temperature $\theta_0(t)$ and temperature $\theta(t)$ of a thermal measuring element, like a thermocouple or a thermometer bulb, as shown in Fig. 3.8.

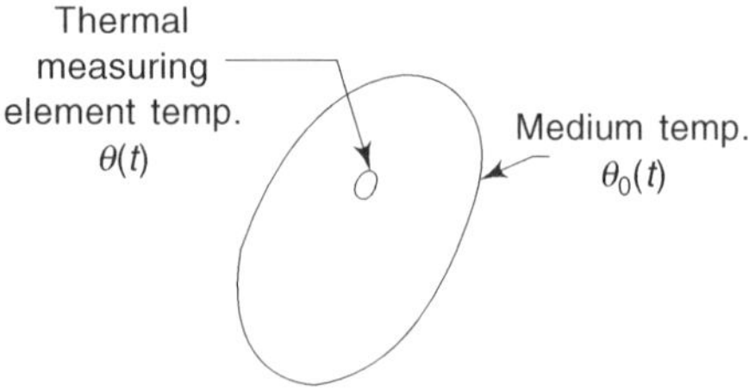

Fig. 3.8 *Thermal element*

Rate of heat flow into the thermal element

$$q = ha(\theta_0 - \theta) \tag{3.6}$$

where a is the exposed area of the thermal measuring element and h the heat transfer coefficient. q is also equal to rate of enthalpy gain by the element, viz.

$$q = mc\,\frac{d\theta}{dt} \tag{3.7}$$

where m is the mass of the thermal element and c is its specific heat. Thus, from Eqs. (3.6) and (3.7),

$$q = ha\,(\theta_0 - \theta) = mc\,\frac{d\theta}{dt} \tag{3.8}$$

or $\qquad \tau D\theta + \theta = \theta_0(t)$

where $\tau = mc/ha$ is known as the time constant of the system, which is of first order type.

U-Tube Manometer A U-tube manometer is shown in Fig. 3.9.

The input is pressure p to be measured, while the output is h, the level difference between the two tubes. To derive the governing equation, the inertia force of the liquid column is equated to the sum of the external forces due to pressure, gravity and friction between tube walls and fluid.

Taking A as the area of cross-section of the tube, L the length of the liquid column and p the mass density of the liquid.

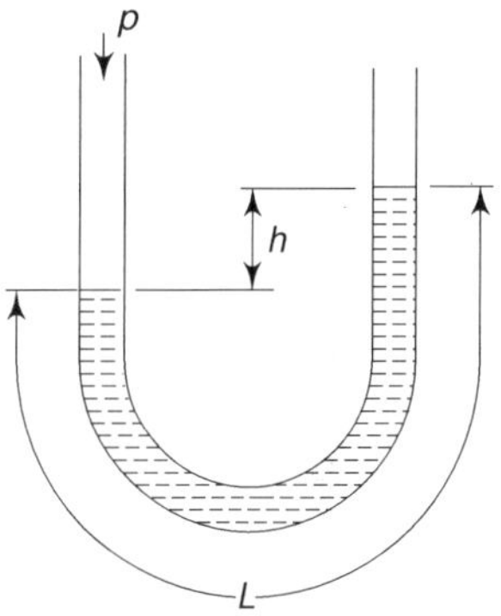

Fig. 3.9 *U-tube manometer*

$$\text{Inertia force of the liquid column} = AL\rho\frac{d^2 h}{dt^2} \tag{3.9}$$

$$\text{Pressure force} = pA \tag{3.10}$$

$$\text{Gravity force} = \rho g h A \tag{3.11}$$

$$\text{Friction force} = (\Delta p)\,A \tag{3.12}$$

where Δp is the pressure loss in the length L, obtained by assuming laminar flow. By the Hagen–Poiseuille equation

$$p = \frac{128\,\mu L Q}{\pi D^4} \tag{3.13}$$

where μ is the fluid viscosity, D the tube diameter, and Q, the volumetric flow rate = VA, V being the velocity = dh/dt and $A = \pi D^2/4$.

Substituting for Q = A dh/dt in Eq. (3.13), and using Eq. (3.12),

$$\text{Friction force} = 8\pi\mu L\,(dh/dt) \tag{3.14}$$

Using Eqs. (3.9), (3.10), (3.11) and (3.14),

$$AL\rho\,\frac{d^2 h}{dt^2} = pA - (\rho g A)h - 8\pi\mu L\,\frac{dh}{dt} \tag{3.15}$$

Equation (3.15) may be rewritten as:

$$L\rho\,\frac{d^2 h}{dt^2} + \frac{32\,\mu L}{D^2}\,\frac{dh}{dt} + (\rho g)h = p \tag{3.16}$$

This is of second order and the system may be classified as a second order instrument. After a long time of application of pressure $p\,(t)$, the various time derivatives become zero and we get the well known relation for static pressure, viz.

$$h = \frac{p}{\rho g}$$

Seismic Motion Transducer This type of transducer is attached to the object whose motion is to be measured. Thus, the input is the motion x of the base and output the motion z of the mass, relative to the frame. In practice, z is measured by an electromechanical transducer. Taking the mass as m, stiffness of spring as K and damping constant of the viscous damper as c, damping force being proportional to velocity of motion, the equation of motion is obtained from the inertial force of the mass and restoring forces of the spring and damper. Thus,

$$m \frac{d^2 y}{dt^2} + c \frac{d}{dt}(y - x) + K(y - x) = 0 \quad (3.17)$$

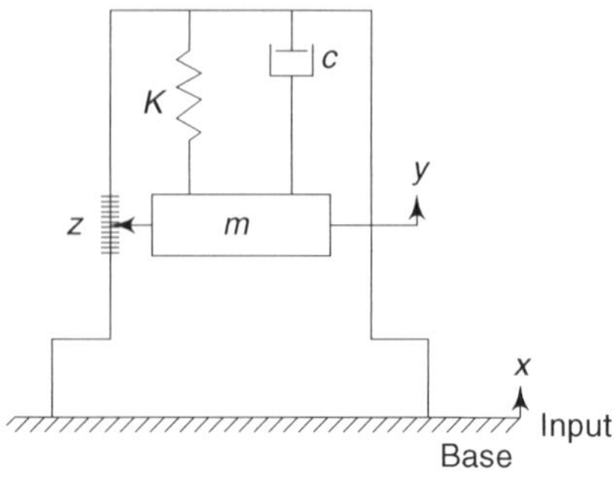

Fig. 3.10 *Seismic motion transducer*

where y is the absolute motion of the mass. Therefore, the relative motion

$$z = y - x \quad (3.18)$$

From Eqs. (3.17) and (3.18) and using D for d/dt, we get

$$(mD^2 + cD + K)z = -mD^2x \quad (3.19)$$

The above second order equation relates the input and output of the transducer.

3.2 ■ DYNAMIC RESPONSE

The output response of an instrument to various types of dynamic input signals can be obtained by solving its governing equation relating output and input, as given by its mathematical model. For various types of instruments, the following types of inputs are discussed:

1. Periodic input—harmonic type,
2. general periodic input—non-harmonic type,
3. transient type input, and
4. random input.

3.2.1 Periodic Input—Harmonic Signal

For the simple type of input varying harmonically with time, as shown in Fig. 3.11, the output response of the first, second and higher order systems, will now be obtained.

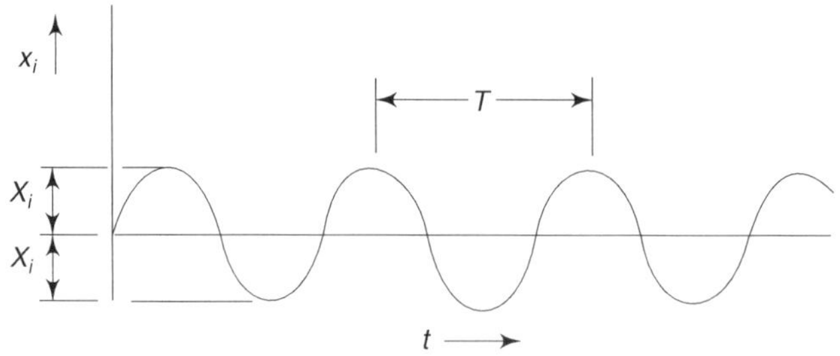

Fig. 3.11 *Periodic harmonic signal*

First-Order System The governing equation of a first-order system is

$$a_1 \frac{dx_o}{dt} + a_o x_o = b_o x_i(t)$$

or $(\tau D + 1)x_o = Kx_i(t)$ (3.20)

where $K = \dfrac{b_o}{a_o}$ is static sensitivity

$\tau = \dfrac{a_1}{a_o}$ is time constant

$D = \dfrac{d}{dt}$ or time derivative operator.

For the static case, $D = 0$ which is a zeroth order system for which

$x_0(t) = Kx_i(t)$, where K is the static sensitivity as defined here.

Taking $x_i = X_i \sin \omega t$ (3.21)

It is required to find the steady state response $x_o(t)$ of the output.

The input signal of Fig. 3.11 is of the form:

$$x_i(t) = X_i \sin \omega t \quad (3.22)$$

where X_i is the amplitude and ω the circular frequency, the latter being equal to $2\pi/T$, and T being the time period of harmonic input. Variation of output signal x_o against frequency ω is known as the *frequency response* of the output signal. This is a useful concept and describes the dynamic behaviour of an instrument. Ideally, an instrument should have a flat frequency response, as in Fig. 3.12(a). In such cases, the amplitude ratio of output and input signals does not vary with frequency ω and such an instrument is equally good for slow as well as fast varying signals. Figure 3.12(b) shows the typical response for an instrument that has a poor low frequency response while that of Fig. 3.12 (c) has a poor high frequency response.

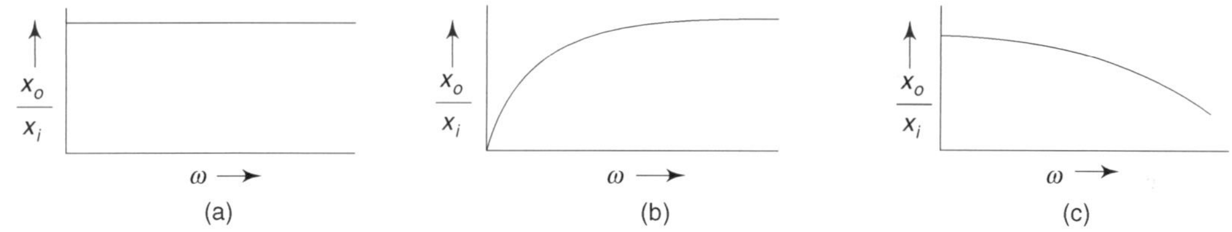

Fig. 3.12 *Typical frequency response curves*

Assuming steady state solution $x_o = A \sin \omega t + B \cos \omega t$ and substituting in Eq. (3.20), equating coefficients of sin ωt and cos ωt on both sides, we get

$$\tau\omega A + B = 0 \quad \text{and} \quad -\tau\omega B + A = KX_i$$

which gives

$$A = \frac{KX_i}{1+\tau^2\omega^2} \quad (3.23a)$$

$$B = \frac{-K\tau\omega X_i}{1+\tau^2\omega^2} \quad (3.23b)$$

Thus,
$$x_o = \frac{KX_i}{1+\tau^2\omega^2}[\sin \omega t - \tau\omega \cos \omega t] \quad (3.24)$$

Equation (3.24) can be written in the form

$$x_o = X_o \sin (\omega t + \varphi) \quad (3.25)$$

by substituting

$$\frac{KX_i}{1+\tau^2\omega^2} = X_o \cos\phi \tag{3.26}$$

and $$\frac{-\tau\omega KX_i}{1+\tau^2\omega^2} = X_o \sin\phi$$

From Eq. (3.26),

$$X_o = \frac{KX_i}{\sqrt{1+\tau^2\omega^2}} \tag{3.27}$$

and $$\tan\varphi = -(\tau\omega)$$

Equation (3.25) shows that the output is harmonic, with a frequency ω and has a phase difference of φ with the input signal. Equation (3.27) gives the expression for amplitude of output X_o and phase φ, which is negative indicating that the output lags behind the input.

An alternative method of obtaining the frequency response is to put $D = j\omega$ in Eq. (3.21), where j is complex operator = $\sqrt{-1}$. This is valid only for harmonic inputs. This gives

$$\frac{x_o}{x_i} = \frac{K}{1+j\omega\tau}$$, giving amplitude ratio:

$$\frac{X_o}{X_i} = \left|\frac{x_o}{x_i}\right| = \frac{K}{\sqrt{1+\omega^2\tau^2}}$$

and $$\varphi = \tan^{-1}(-\omega\tau)$$

which correspond with Eq. (3.27).

The frequency response curve of the instrument is plotted in Fig. 3.13. As ω increases, the amplitude ratio decreases. At a value of $\omega = 1/\tau$, $X_o/KX_i = 0.707$. This value of frequency is called the *break-point frequency.* If $K = 1$, output amplitude is 70.7% of the input one. Phase φ ranges from 0 to –90°, with the value being –45° at the break-point frequency.

It can be seen that the break-point frequency would be high or the frequency response would be flat for higher values of ω, the input frequency provided τ is small.

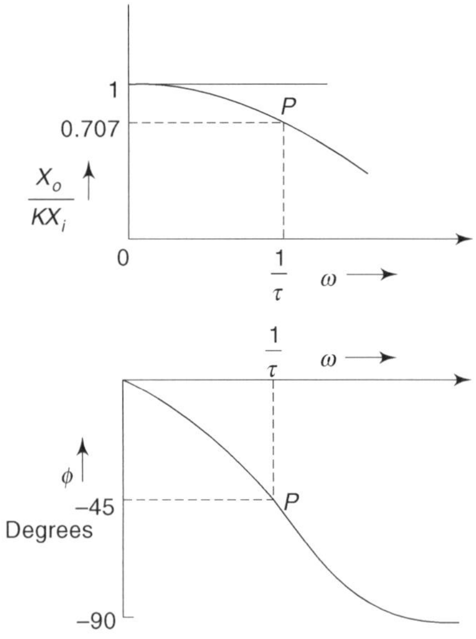

Fig. 3.13 *Frequency response of first order system*

Second Order System Taking a second order instrument system (Fig. 3.6), with the governing equation

$$a_2\frac{d^2x_o}{dt^2} + a_1\frac{dx_o}{dt} + a_o x_o = b_o x_i(t) \tag{3.28}$$

Input $x_i(t)$ is harmonic and = $X_i \sin\omega t$.

Equation (3.28) may be written in a dimensionless form given by

$$\left(\frac{D^2}{\omega_n^2} + \frac{2\xi}{\omega_n}D + 1\right)x_o = Kx_i(t) \tag{3.29}$$

where $$K = \frac{b_o}{a_o} = \text{static sensitivity}$$

$$\omega_n = \sqrt{\frac{a_o}{a_2}}, \text{ undamped natural frequency}$$

$$\xi = \frac{a_1}{2\sqrt{a_0 a_2}}, \text{ damping ratio and}$$

$$D = d/dt$$

The steady state solution $x_o(t)$ is obtained for Eq. (3.28) by putting $D = j\omega$ or by assuming $x_o = A \sin \omega t + B \cos \omega t$, as done for the first order system. It can be easily shown by a similar procedure that

$$x_o = X_o \sin (\omega t + \varphi) \tag{3.30}$$

where $$X_o = \frac{KX_i}{\sqrt{(1-r^2)^2 + (2\xi r)^2}}$$

$$\varphi = -\left(\tan^{-1} \frac{2\xi r}{1-r^2}\right)$$

$$r = \frac{\omega}{\omega_n}, \text{ frequency ratio}$$

Amplitudes ratio X_o/X_i and φ are plotted against ω for various values of damping ratio ξ in Fig. 3.14. It is seen that the frequency response is considerably affected by the value of damping ratio $\xi \cdot X_o/KX_i \approx 1$ only for low values of r. For $r < 1$, the curve is flat only for $\xi \approx 0.7$. The output is considerably reduced at higher values of frequency ratio r. Thus, in order to use second order instrument at high values of ω, it is important that its undamped natural frequency ω_n is high so that value of r is small. Further, the value of ξ should be high and around 0.7.

Problem 3.1 *A first-order instrument is to measure signals with frequency content up to 100 Hz with an amplitude inaccuracy of 5%. What is the maximum allowable time constant? What will be the phase shift at 50 Hz?*

Solution Using Eq. (3.27),

$$\frac{X_o}{KX_i} = \frac{1}{\sqrt{1+\tau^2\omega^2}}$$

$$\varphi = \tan^{-1}(-\tau\omega)$$

It is seen from Fig. 3.13 that with increase of ω, X_o/KX_i reduces.

Thus $$\frac{X_o}{KX_i} = 1 \text{ at } \omega = 0$$

$$= 0.95 \text{ at } \omega = 2\pi \times 100 \text{ rad/s} = 628.32 \text{ rad/s}$$

or $$0.95 = \frac{1}{\sqrt{1+\tau^2\omega^2}}$$

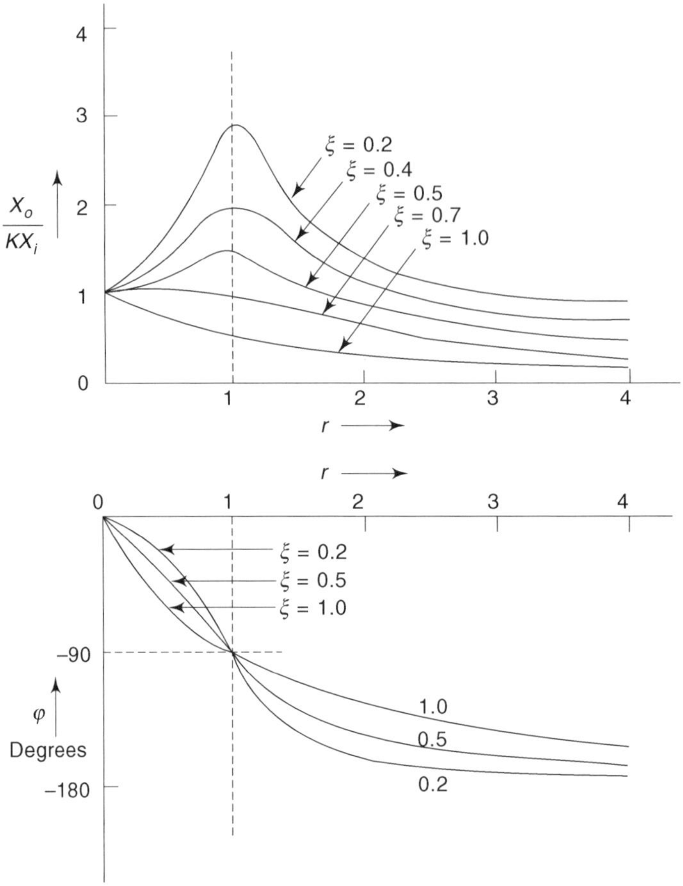

Fig. 3.14 *Frequency response of second-order system*

Putting $\omega = 628.32$ rad/s, we get

Time constant $\tau = 5.23 \times 10^{-4}$

Phase angle φ at 50 Hz $= -(\tan^{-1} 5.23 \times 10^{-4} \times 2\pi \times 50) = -9.33°$

Problem 3.2 *A second-order instrument is subjected to a sinusoidal input. Undamped natural frequency is 3 Hz and damping ratio is 0.5. Calculate the amplitude ratio and phase angle for an input frequency of 2 Hz.*

Solution The governing equation is

$$\left(\frac{D^2}{\omega_n^2} + \frac{2\xi}{\omega_n} D + 1\right) x_o = K x_i(t)$$

K is the static sensitivity. Its value is not given. It is assumed that x_o and x_i are of the same form and the static ratio is unity, viz. $K = 1$.

Thus, amplitude ratio (from Eq. (3.30)),

$$\frac{X_o}{X_i} = \frac{1}{\sqrt{(1-r^2)^2 + (2\xi r)^2}}$$

where $r = \dfrac{\omega}{\omega_n}$

In the problem, $\omega_n = 2\pi \times 3 = 18.85$ rad/s

$\omega = 2\pi \times 2 = 12.57$ rad/s

$r = \omega/\omega_n = 0.67$

$\xi = 0.5$

Using the above, $\dfrac{X_o}{X_i} = 1.153$

Using Eq. (3.30), 3 $\varphi = -50.6°$

Problem 3.3 *A temperature measuring system, with a time constant of 2s, is used to measure temperature of a heating medium, which changes sinusoidally between 350 and 300°C with a periodic time of 20s. Find the maximum and minimum values of temperature, as indicated by the measuring system and the time lag between the output and input signals.*

Solution Taking the system as a first order one, the governing equation is

$$(1 + \tau D)x_o = x_i(t)$$

The medium has a mean temperature of 325°C and amplitude $x_i = \dfrac{350 - 300}{2} = 25°C$ as shown in Fig. 3.15.

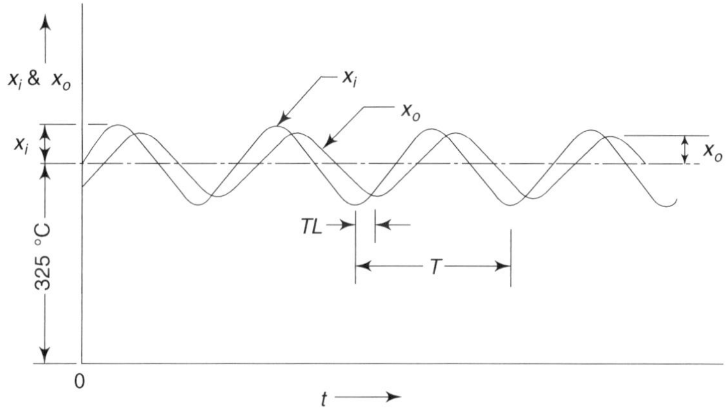

Fig. 3.15 *Figure for Problem 3.3*

Using $$\frac{X_o}{X_i} = \frac{1}{\sqrt{1+\tau^2\omega^2}}, \text{ with } \tau = 2.0 \text{ s}$$

$$\omega = \frac{2\pi}{T} = \frac{2\pi}{20} = 0.314 \text{ rad/s}$$

$$\frac{X_o}{X_i} = 0.847$$

For X_i = 25°C, X_o = 21.2°C. Thus, the indicated temperature will vary between 325 + 21.2 and 325 – 21.2°C, i.e. between 346.2 and 303.8°C.

Phase angle $\varphi = -\tan^{-1} \tau\omega = -32.1°$

or Time lag $TL = \frac{32.1}{360}(20) = 1.78$ s

High Order Systems When the order of the governing equation of a combination of instruments is high, it is convenient to plot the frequency response of the system by logarithmic plots, known as *Bode diagrams*. The advantage of this method is that the frequency response of a complex system can be obtained by adding the response due to various first and second order terms occurring in the transfer function of the system, e.g. in Fig. 3.16, frequency response of a component j is $M_j \angle\varphi_j$, where $j = 1, 2, 3$. M_j is the amplitude ratio of output to input for component j, and φ_j the phase difference. Thus, for the case of Fig. 3.16,

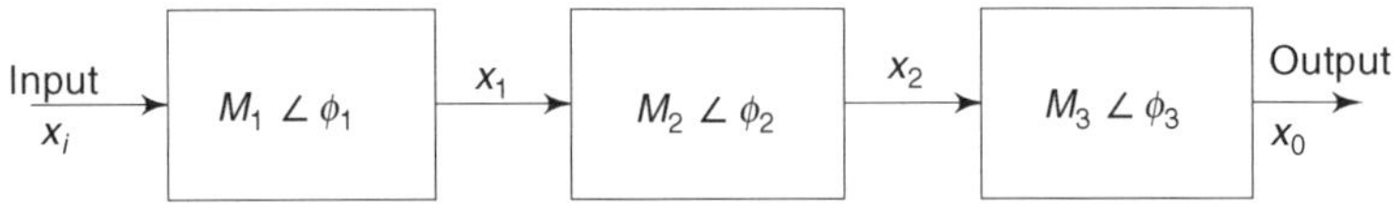

Fig. 3.16 *Frequency response of high-order systems*

$$\frac{x_1}{x_i}(\omega) = M_1 \angle\varphi_1 \tag{3.31}$$

$$\frac{x_2}{x_1}(\omega) = M_2 \angle\varphi_2 \tag{3.32}$$

$$\frac{x_o}{x_2}(\omega) = M_3 \angle\varphi_3 \tag{3.33}$$

From Eqs. (3.31), (3.32) and (3.33), we get

$$\frac{x_o}{x_i}(\omega) = M_1M_2M_3 \angle(\varphi_1 + \varphi_2 + \varphi_3) \tag{3.34}$$

$$= M\angle\varphi$$

where $M = M_1M_2M_3$

$$\angle\varphi = \angle(\varphi_1 + \varphi_2 + \varphi_3) \tag{3.35}$$

M and $\angle\varphi$ in Eq. (3.35) give the overall frequency response of the combination. Taking the logarithm of both sides in Eq. (3.35)

$$\log M = \log M_1 + \log M_2 + \log M_3 \tag{3.36}$$

Thus, it is seen from Eq. (3.36) that the frequency response of the total system may be obtained in logarithmic form by adding the logarithms of the amplitude ratios for each component. This applies to a system with any number of components.

Usually, decibel notation is used to express the amplitude ratios.

$$\text{Decibel value} = 20 \log M \tag{3.37}$$

The overall phase angle φ is obtained by adding the phase φ_i due to various components.

Further, it is convenient to use a log scale for the abscissa, viz. frequency ω in order to cover wide ranges of frequencies, which may be several decades. A decade implies that the ratio of the highest to the lowest frequency in a range is 10:1. Thus, Bode diagrams or plots are drawn for 20 log M in dB (decibels) against log ω and φ against log ω.

Bode diagrams for first- and second-order systems are shown in Figs 3.17 and 3.18 respectively. For the first-order system, with governing equation

$$(1 + \tau D)x_o = Kx_i(t)$$

and for $x_i(t) = X_i \sin \omega t,$

Amplitude ratio $M = \dfrac{X_o}{X_i} = \dfrac{K}{\sqrt{1+\tau^2\omega^2}}$

$$M \text{ in decibels} = 20 \log \left\{ \frac{K}{\sqrt{1+\tau^2\omega^2}} \right\}$$

$$= 20 \log K - 10 \log \{(1 + \tau^2\omega^2)\}$$

Taking $K = 1$ say,

$$M = -10 \log \{(1 + \tau^2\omega^2)\} \text{ dB}$$

For $\tau\omega \ll 1, \quad M = 0 \text{ dB}$

$\tau\omega = 1, \quad M \approx -3 \text{ dB}$

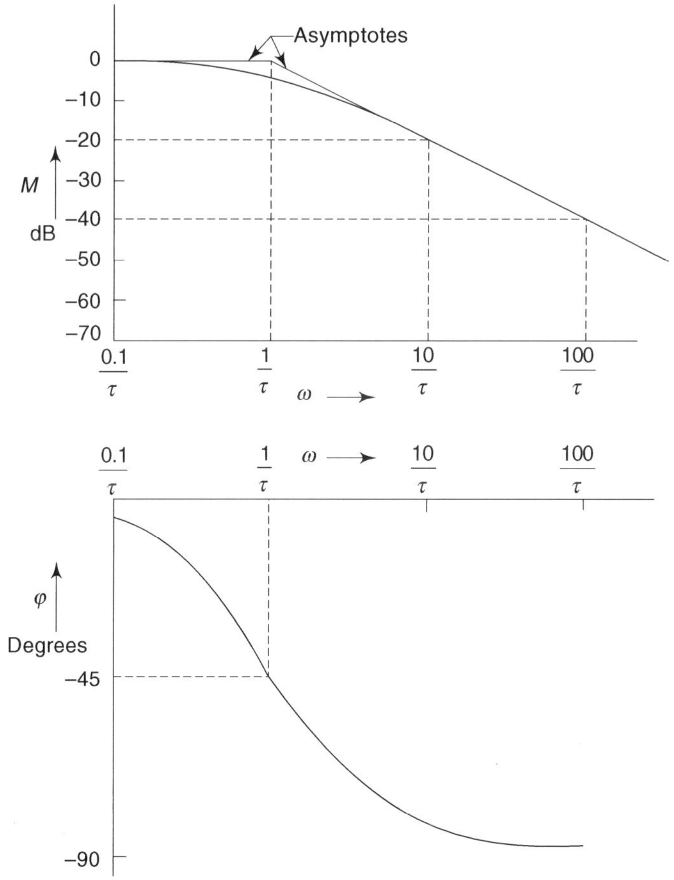

Fig. 3.17 *Bode diagram for first-order system*

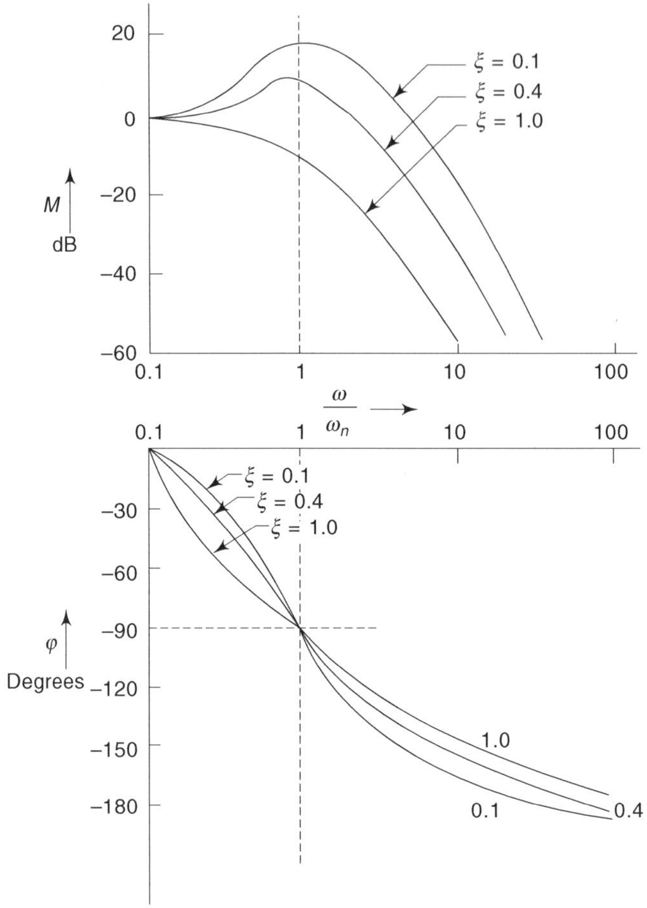

Fig. 3.18 *Bode diagram for second-order system*

$\tau\omega = 10, \qquad M \approx -20$ dB

$\tau\omega = 100, \qquad M \approx -40$ dB

Further, for every decade of frequency, there is a reduction of 20 dB. The plot is shown in Fig. 3.17. The plot of φ against ω is also shown. ω is on log scale in both the plots.

In a similar way, the Bode diagram for a second order system with governing equation

$$\left(\frac{D^2}{\omega_n^2} + \frac{2\xi}{\omega_n} D + 1\right) x_o = x_i(t)$$

$$= X_i \sin \omega t$$

can be obtained and is shown in Fig. 3.18. For the above,

$$M = 20 \log \frac{x_o}{x_i} = 20 \log \left\{ \frac{1}{\sqrt{(1 - r^2)^2 + (2\xi r)^2}} \right\}$$

$$= -10 \log \{(1 - r^2)^2 + (2\xi r)^2\}$$

r being frequency ratio $= \omega/\omega_n$.

For $r \gg 1$, a change of frequency ratio by a decade, changes M by 40 dB.

$$\text{Phase angle } \varphi = -\tan^{-1}\left(\frac{2\xi r}{1 - r^2}\right)$$

$\varphi = -90°$ for $r = 1$ and the range is from 0 to $-180°$.

For a system involving a combination of first and second order systems, the plots of Figs. 3.17 and 3.18 can be added, in order to get the overall frequency response of the combination. This is illustrated in Porblem 3.4.

Problem 3.4 *An instrument consists of a first-order sensing element and a second-order data presentation device. The time constant of the first-order element is 0.01s and static sensitivity is 4 mV/°C. The second order device has an undamped natural frequency of 100 rad/s and damping ratio of 0.5, with static sensitivity of 5 mm/mV. Draw the Bode diagram, giving the frequency response of the system.*

Solution The block diagram of the instrument is shown in Fig. 3.19.

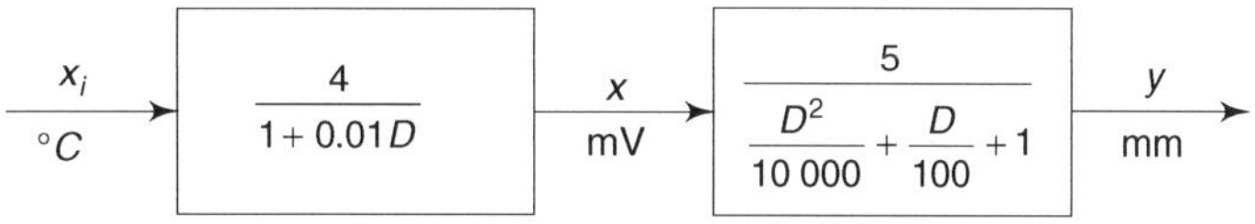

Fig. 3.19 *Figure for Problem 3.4*

Taking $\tau = 0.01$ s, $\omega_n = 100$ rad/s and $\xi = 0.5$

$$\frac{x}{x_i} = \frac{K_1}{1 + \tau D} = \frac{4}{1 + 0.01D} \text{ mV/°C}$$

$$\frac{y}{x} = \frac{K_2}{\frac{D^2}{\omega_n^2} + \frac{2\xi D}{\omega_n} + 1} = \frac{5}{\frac{D^2}{10\,000} + \frac{D}{100} + 1} \text{ mm/mV}$$

For the complete system, the transfer function is

$$\frac{y}{x_i} = \frac{20}{(1 + 0.01\,D)\left(\frac{D^2}{10\,000} + \frac{D}{100} + 1\right)} \text{ mm/°C}$$

The Bode diagram is drawn by putting $D = j\omega$ and adding the contributions due to the following:

$$M_1 = 20 \log [20] \text{ dB}$$

$$M_2 = 20 \log \left[\frac{1}{\sqrt{(1 + 0.01\,\omega)^2}}\right]$$

$$M_3 = 20 \log \left[\frac{1}{\sqrt{\left[1 - \frac{\omega^2}{10\,000}\right]^2 + \left(\frac{\omega}{100}\right)^2}}\right]$$

Thus, $M(\omega) = 20 \log y/x_i = M_1 + M_2 + M_3$

Similarly, $$\varphi = -\tan^{-1}(0.01/\omega) - \tan^{-1}\frac{\dfrac{\omega}{100}}{\left(1 - \dfrac{\omega}{10\,000}\right)^2}$$

In Fig. 3.20, M_1, M_2, M_3 and M have been plotted against ω, which has been varied from 10 to 10^4 rad/s. In a similar way, φ can be plotted against ω.

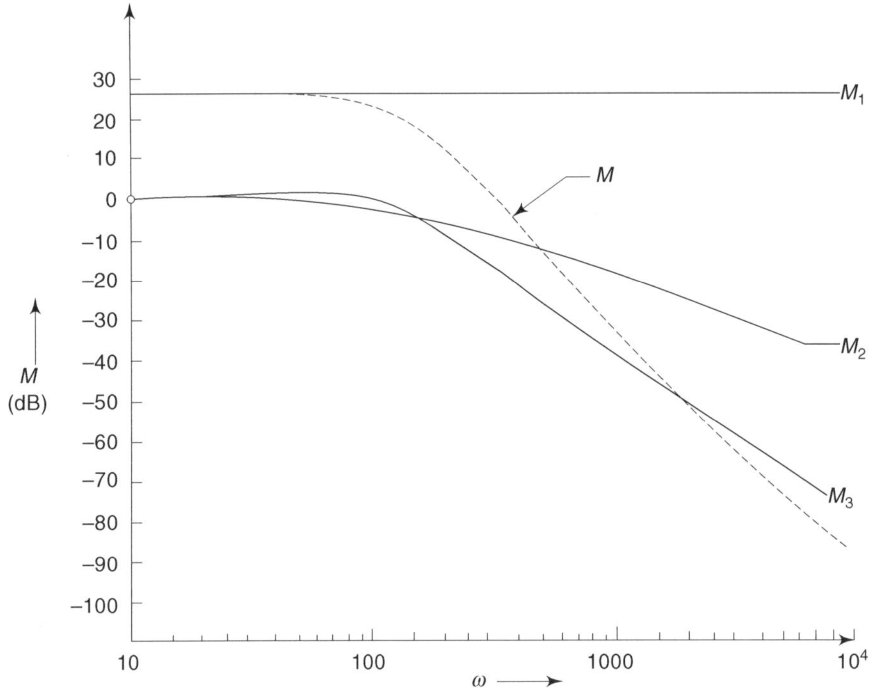

Fig. 3.20 *Figure for Problem 3.4*

3.2.2 Periodic Input—Non-harmonic Signal

If the input signal is a function of time such that it repeats itself at constant intervals and is not harmonic, e.g. a square wave, it may be represented by sinusoids of varying frequencies using Fourier series. The output can be determined in terms of each of the above sinusoidal components.

Fourier series for a periodic function $x_i(t)$ is

$$x_i(t) = \frac{1}{2}a_o + \sum_{n=1}^{\infty}\left(a_n \cos\frac{2\pi}{T}nt + b_n \sin\frac{2\pi}{T}nt\right) \tag{3.38}$$

where T is the time period.

In Eq. (3.38)

$$a_o = \frac{2}{T}\int_{-T/2}^{T/2} x_i\,(\mathrm{d}t)\,\mathrm{d}t$$

$$a_n = \frac{2}{T}\int_{-T/2}^{T/2} x_i\,(t)\cos\left(\frac{2\pi}{T}nt\right)\mathrm{d}t \tag{3.39}$$

$$b_n = \frac{2}{T}\int_{-T/2}^{T/2} x_i(t)\sin\left(\frac{2\pi}{T}nt\right)\mathrm{d}t$$

For the square wave signal shown in Fig. 3.21

$$x_i(t) = -C,\ \frac{T}{2} < t < 0$$

$$= C,\ 0 < t < \frac{T}{2} \tag{3.40}$$

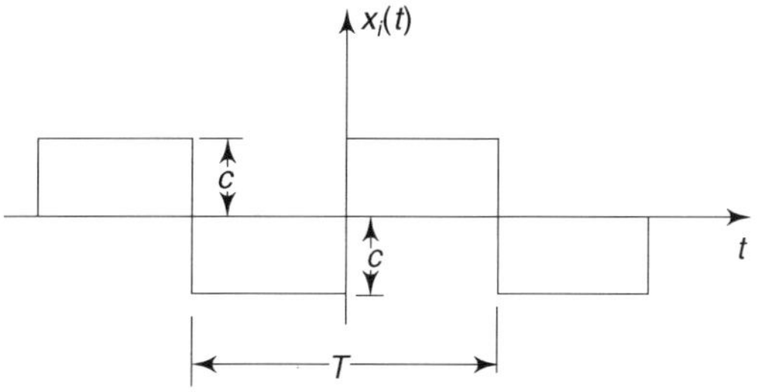

Fig. 3.21 *Periodic square wave input*

Using the relations given earlier, for the function given by Eq. (3.40)

$$a_0 = \frac{2}{T}\left[\int_{-T/2}^{0} -C\mathrm{d}t + \int_0^{T/2} C\mathrm{d}t\right] = 0$$

$$a_n = \frac{2}{T}\left[\int_{-T/2}^{0} -C\cos\frac{2\pi}{T}nt\,\mathrm{d}t + \int_0^{T/2} C\cos\frac{2\pi}{T}nt\,\mathrm{d}t\right] = 0$$

$$b_n = \frac{2}{T}\left[\int_{-T/2}^{0} -C\sin\frac{2\pi}{T}nt\,\mathrm{d}t + \int_0^{T/2} C\sin\frac{2\pi}{T}nt\,\mathrm{d}t\right]$$

$$= \frac{C}{n\pi}\left[2 - \cos n\pi - \cos(-n\pi)\right]$$

$$= \frac{4C}{n\pi},\quad \text{for}\quad n = 1, 3, 5, \ldots$$

$$= 0,\quad \text{for}\quad n = 2, 4, 6, \ldots$$

Thus
$$x_i(t) = \sum \frac{4C}{n\pi}\sin\frac{2\pi nt}{T},\quad \text{for}\quad n = 1, 3, 5, \ldots \tag{3.41}$$

In practice, the function may be represented by a finite number of terms and the output for each term, for the instrument, may be obtained. The total output would be the summation of all the terms so obtained.

In practice, using Eq. (3.39) to calculate the coefficients is quite tedious except for simple inputs. For such cases, instead, a finite number of the coefficients are computed such that the sum of the chosen terms in the Fourier series is exactly equal to $x_i(t)$ at a finite number of chosen points of time t. There will be some error at other values of t, depending on the number of points chosen and their locations. However, this approximate method is simple to use, as illustrated in Prob. 3.5.

Problem 3.5 *(a) Write a Fourier series approximation for the function of Fig. 3.22, such that the error is zero at t = 0, T/3, T/2, 2T/3 and T. Draw the approximate function thus obtained.*

(b) It the function obtained in (a) is used as an input signal to a second order instrument with ω_n = 10 rad/s, ξ = 0.3, K = 5 mV/mm, find the steady state output of the instrument.

Solution

(a) Using the first five terms in Eq. (3.38)

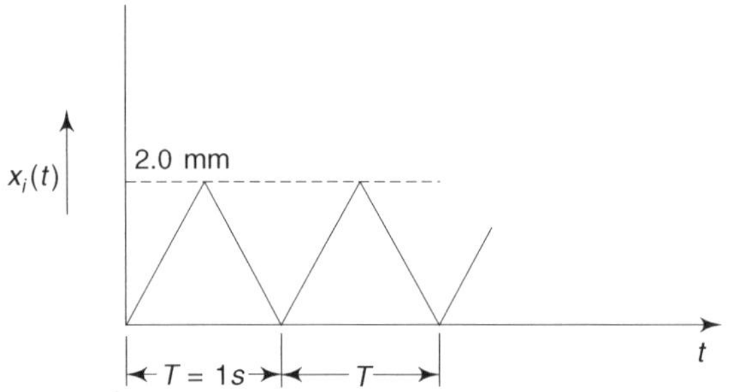

Fig. 3.22 *Figure for Problem 3.5*

$$x_i(t) = \frac{1}{2} a_0 + a_1 \cos \frac{2\pi t}{T} + a_2 \cos \frac{4\pi t}{T} + b_1 \sin \frac{2\pi t}{T} + b_2 \sin \frac{4\pi t}{T} \tag{3.42}$$

From Fig. 3.22, exact values of $x_i(t)$ are obtained as 0, 4/3, 2, 4/3, 0 at $t = 0$, $T/3$, $T/2$, $2T/3$ and T, respectively.

Using Eq. (3.42) for each of the above points, we get

$$0 = 0.5\ a_0 + a_1 + a_2 \tag{3.43}$$

$$\frac{4}{3} = 0.5\ a_0 - 0.5\ a_1 - 0.5\ a_2 + 0.87\ b_1 - 0.87\ b_2 \tag{3.44}$$

$$2 = 0.5\ a_0 - a_1 + a_2 \tag{3.45}$$

$$\frac{4}{3} = 0.5\ a_0 - 0.5\ a_1 - 0.5\ a_2 - 0.87\ b_1 + 0.87\ b_2 \tag{3.46}$$

$$0 = 0.5\ a_0 + a_1 + a_2 \tag{3.47}$$

Solving the above, we get

$$a_0 = \frac{16}{9},\ a_1 = -1,\ a_2 = \frac{1}{9},\ b_1 = b_2$$

It is not possible to get values of b_1 and b_2 since Eqs. (3.43) and (3.47) are identical, and we have four equations for five unknown. Thus, we shall take one of the values arbitrarily say $b_1 = b_2 = 1$.

Equation (3.42) becomes

$$x_i(t) = \frac{8}{9} - \cos \frac{2\pi t}{T} + \frac{1}{9} \cos \frac{4\pi t}{T} + \sin \frac{2\pi t}{T} + \sin \frac{4\pi t}{T} \tag{3.48}$$

Equation (3.48) is plotted in Fig. 3.23. It is seen that this only approximates the exact function $x_i(t)$. If b_1 and b_2 are taken equal to another arbitrary value say equal to 0.2, Eq. (3.42) becomes

$$x_i(t) = \frac{8}{9} - \cos \frac{2\pi t}{T} + \frac{1}{9} \cos \frac{4\pi t}{T} + 0.2 \sin \frac{2\pi t}{T} + 0.2 \sin \frac{4\pi t}{T} \tag{3.49}$$

Equation (3.49) is also plotted in Fig. 3.23. It is seen that this approximates $x_i(t)$ better than Eq. (3.48).

(b) Using the governing equation for the second-order system:

$$\left(\frac{D^2}{\omega_n^2} + \frac{2\xi D}{\omega_n} + 1 \right) x_o = K x_i(t)$$

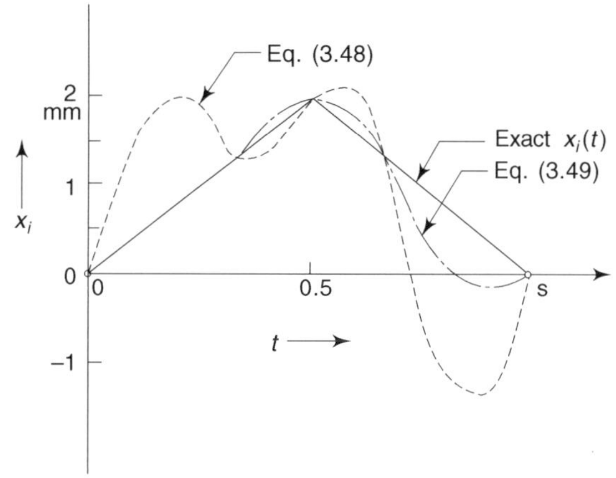

Fig. 3.23 *Figure for Problem 3.5*

Putting $\omega_n = 10$ rad/s, $K = 5$ mV/mm, $\xi = 0.3$, we get

$$(D^2 + 6D + 100)x_o = 500\, x_i(t) \tag{3.50}$$

Using Eq. (3.49) for $x_i(t)$ and putting

$T = 1$s, we get

$$(D^2 + 6D + 100)x_o = \frac{4000}{9} - 500 \cos 2\pi t + \frac{500}{9} \cos 4\pi t + 100 \sin 2\pi t + 100 \sin 4\pi t \tag{3.51}$$

The steady state solution for x_o will be obtained as the sum of solutions due to each term on RHS of Eq. (3.51). Due to the constant term, $x_o = 40/9$.

It can be easily proved that the steady state solution of the equation

$$(D^2 + 6D + 100)x_o = F_1 \cos \omega t,$$

is given by

$$x_o = A \cos \omega t + B \sin \omega t \tag{3.52}$$

where

$$A = \frac{(100 - \omega^2)\, F_1}{(100 - \omega^2)^2 + (6\omega)^2}$$

$$B = \frac{(6\omega)\, F_1}{(100 - \omega^2)^2 + (6\omega)^2} \tag{3.53}$$

Also, for the equation

$$(D^2 + 6D + 100)x_o = F_1 \sin \omega t,$$

solution is:

$$x_o = A \sin \omega t - B \cos \omega t \tag{3.54}$$

where A and B are as defined in Eq. (3.53).

Using the above expressions, the solution due to each of the sine and cosine terms on the RHS of Eq. (3.51) can be obtained. Total solution for x_o is given by

$$x_o = 40/9 - 6.69 \cos 6.28t - 2.51 \sin 6.28t - 1.84 \cos 12.56t - 0.68 \sin 19.56t \tag{3.55}$$

Equation (3.53) is plotted in Fig. 3.24. This gives the output response $x_o(t)$.

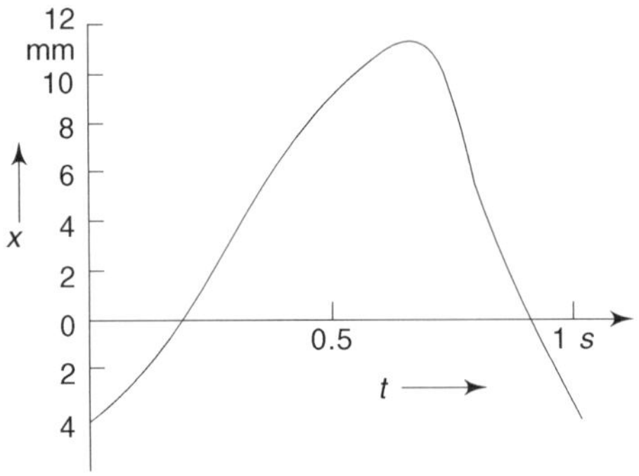

Fig. 3.24 *Figure for Problem 3.5*

3.2.3 Response to Transient Inputs

For transient input signals, the output response of an instrument, may be obtained, by either differential equation solution method or by Fourier transform method.

Differential Equation Solution Method This involves determination of the solution of the governing differential equation of the instrument, by classical or Laplace transforms method. Only the former method will be discussed here.

Step response of first-order system We take a first-order system, with the governing equation

$$\tau D x_o + x_o = K x_i(t),\ K \text{ being the static sensitivity} \quad (3.56)$$

We take input $x_i(t)$ to be of step type, as shown in Fig. 3.25 or $x_i(t) = x_s$. It is required to find $x_o(t)$ of the instrument system. Thus, Eq. (3.56) becomes

$$\tau D x_o + x_o = K x_i(t) = K x_s \quad (3.57)$$

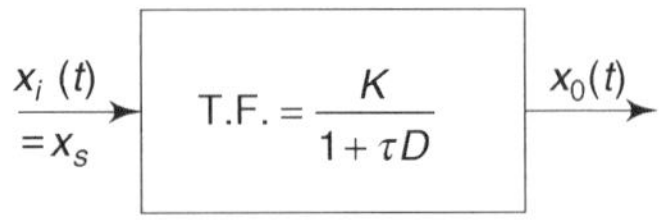

Fig. 3.25 *Step input on a first order system*

The solution of Eq. (3.57) is in two parts: complementary solution and particular solution.

The former is of the form Ae^{rt}. Substituting in Eq. (3.57) with RHS = 0 gives

$$r = -\frac{1}{\tau}$$

Particular solution for RHS = Kx_s is given by $x_o = Kx_s$. Thus, the solution is

$$x_o = Ae^{-t/\tau} + Kx_s$$

To find the constant A, we apply the initial conditions, viz. $x_o = 0$ at $t = 0$. This gives $A = -Kx_s$

$$x_o = [1 - e^{-t/\tau}]\,Kx_s \quad (3.58)$$

Thus x_o/K is plotted as in Fig. 3.26. It is seen that output x_o slowly increases with time t. At time $t = \tau$, the output value is 63.2% of the input-value. At $t \approx 5\tau$, the difference between output and input is almost zero.

Thus, the time constant τ denotes the time lag or sluggishness of the instrument. For dynamic measurements, its value should be as small as possible. Further, the difference between x_i and x_o is called "dynamic error" at any instant.

Step response of second order system Consider the second order system with governing equation, given earlier as Eq. (3.29), viz.

K being the static sensitivity, ω_n the undamped natural frequency and ξ the damping ratio. The solution for step input, viz. $x_i(t) = x_s$ is needed.

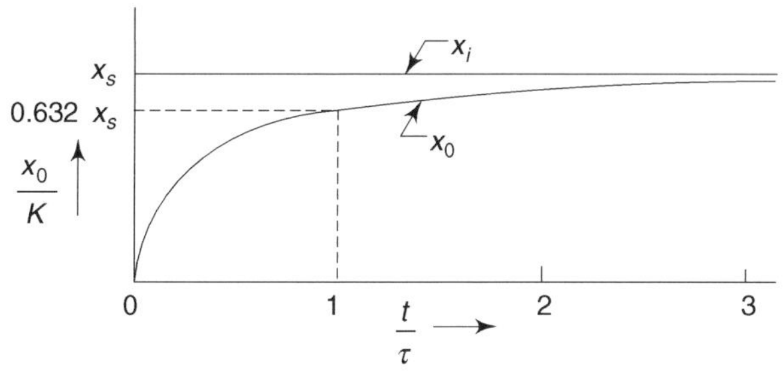

Fig. 3.26 *Transient output response of a first-order system*

$$\left(\frac{D^2}{\omega_n^2} + \frac{2\xi}{\omega_n} D + 1\right) x_o = K x_i(t) \tag{3.59}$$

Thus,
$$\left(\frac{D^2}{\omega_n^2} + \frac{2\xi}{\omega_n} D + 1\right) x_0 = K x_s \tag{3.60}$$

The solution as shown below, is the sum of two parts:

(1) Complementary solution, with the RHS of the above equation taken as zero.

(2) Particular solution which by inspection is equal to Kx_s.

For the complementary solution, take $x_0 = Ae^{rt}$, substituting Eq. (3.60) with RHS = 0, we get

$$(r^2 + 2\xi\omega_n r + \omega_n^2)\ Ae^{rt} = 0$$

Since Ae^{rt} cannot be zero, $r^2 + 2\xi\omega_n r + \omega_n^2 = 0$. The two roots of the above quadratic equation are:

$$r_{1,2} = -\xi\omega_n \pm \omega_n\sqrt{\xi^2 - 1}$$

Thus, complementary solution is

$$x_o(t) = A_1 e^{r_1 t} + A_2 e^{r_2 t}, \text{ where } A_1 \text{ and } A_2 \text{ are constants.}$$

There are three cases possible depending on the value of ξ.

Case 1: Over damped case, where $\xi > 1$. This gives roots r_1 and r_2 as real.

Thus, complementary solution is

$$x_o(t) = A_1 \exp\left[(-\xi\omega_n + \omega_n\sqrt{\xi^2 - 1})t\right] A_2 \exp\left[(-\xi\omega_n - \omega_n\sqrt{\xi^2 - 1})t\right]$$

$$= \exp(-\xi\omega_n t)\left[A_1 \exp(\omega_n\sqrt{\xi^2 - 1}t) + A_2 \exp(-\omega_n\sqrt{\xi^2 - 1}t)\right] \tag{3.61}$$

Case 2: Critical damping case or $\xi = 1$. This gives roots r_1 and r_2 as equal.

The solution cannot be defined by Eq. (3.61) since $r_1 = r_2 = -\xi\omega_n$ and two arbitrary constants are needed. It can be shown that for this case,

$$x_o(t) = A_1 \exp(-\xi\omega_n t) + A_2 t \exp(-\xi\omega_n t) \tag{3.62}$$

Equation (3.62) can be seen to satisfy Eq. (3.60), with RHS = 0.

Case 3: $\xi < 1$. This gives roots r_1 and r_2 as complex, which are

$$r_{1,2} = -\xi\omega_n \pm j\xi\omega_n\sqrt{1-\xi^2}$$

Complementary solution is

$$x_o(t) = \exp(-\xi\omega_n t)\left[A_1 \exp(j\omega_n\sqrt{1-\xi^2}\,t) + A_2 \exp(-j\omega_n\sqrt{1-\xi^2}\,t)\right]$$

$$= \exp(-\xi\omega_n t)\left[c_1 \cos \omega_n\sqrt{1-\xi^2}\,t + c_2 \sin \omega_n\sqrt{1-\xi^2}\,t\right] \quad (3.63a)$$

where $c_1 = A_1 + A_2$

$c_2 = j(A_1 - A_2)$

Case 3 is of practical interest since $\xi < 1$ for most practical systems. This is also called under-damped case.

The particular solution for Eq. (3.60) applicable to all cases it can be easily seen to be

$$x_o = Kx_s \quad (3.63b)$$

Adding the above value of particular solution to the complementary solution in each case to the particular solution, we get the total solution for x_0

Initial conditions $x_0 = 0$ and $\frac{dx_0}{dt} = 0$ at $t = 0$ have to be applied, to get the constants A_1 and A_2 or C_1 and C_2.

Case $\xi > 1$ Using the above procedure, the final equation obtained is

$$\frac{x_o(t)}{Kx_s} = 1 + \left[\frac{\xi + \sqrt{\xi^2-1}}{2\sqrt{\xi^2-1}} \exp\left[(-\xi + \sqrt{\xi^2-1})\,\omega_n t\right] - \frac{\xi - \sqrt{\xi^2-1}}{2\sqrt{\xi^2-1}} \exp\left[(-\xi - \sqrt{\xi^2-1})\,\omega_n t\right]\right] \quad (3.64)$$

Case $\xi = 1$ For this case, the final equation is obtained as

$$\frac{x_o(t)}{Kx_s} = 1 - (1 + \omega_n t) \exp(-\omega_n t) \quad (3.65)$$

Case $\xi < 1$

This case is of practical interest and the solution is derived in Appendix A-2, viz.

$$\frac{x_o(t)}{Kx_s} = 1 - \exp(-\xi\omega_n t)\left[\cos \omega_n \sqrt{1-\xi^2}\,t + \frac{\xi}{\sqrt{1-\xi^2}} \sin \omega_n \sqrt{1-\xi^2}\,t\right] \quad (3.66)$$

Equation (3.66) may also be written in the form

$$\frac{x_o(t)}{Kx_s} = \left[1 - \frac{1}{\sqrt{1-\xi^2}} \exp(-\xi\omega_n t) \sin(\sqrt{1-\xi^2}\,\omega_n t + \phi)\right] \quad (3.67)$$

where $\phi = \sin^{-1}\sqrt{1-\xi^2}$.

Equation (3.67) is plotted in Fig. 3.27. The output has frequency $\sqrt{1-\xi^2}\ \omega_n$, with gradual decay of its amplitude. An increase in value of damping ratio ξ reduces the oscillations but slows the response. ω_n is an indication of the speed of response since doubling its value say will reduce the time t to half its value for achieving a given output response. Further, the peak value of x_o over and above the value of x_i, called *peak overshoot,* should be limited in order to avoid damage to the instrument. Thus, a compromise has to be made while choosing value of ξ during design stage, in order to achieve a reasonably fast response and small peak overshoot.

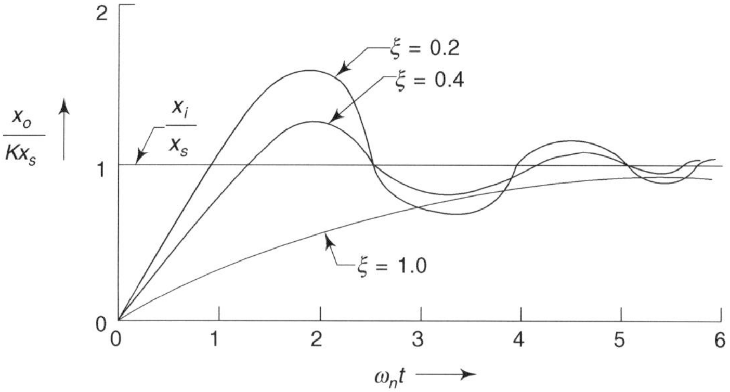

Fig. 3.27 *Transient output response of a second-order system*

Problem 3.6 *A thermocouple having first-order response characteristics is used to measure a single temperature pulse having a half wave sinusoidal shape, as shown in Fig. 3.28. Find the output response of the thermocouple for (a) T = 4τ, and (b) T = 10τ, τ being the time constant of the thermocouple.*

Solution If the temperature to be measured is $\theta_i(t)$ and thermocouple temperature is $\theta_o(t)$, the governing equation is

$$(\tau D + 1)\theta_o = \theta_i(t) \tag{3.68}$$

From Fig. 3.28, it is seen that

$$\theta_i(t) = E \sin \frac{\pi t}{T}$$

$$= E \sin \omega t$$

where $\omega = \dfrac{\pi}{T}$

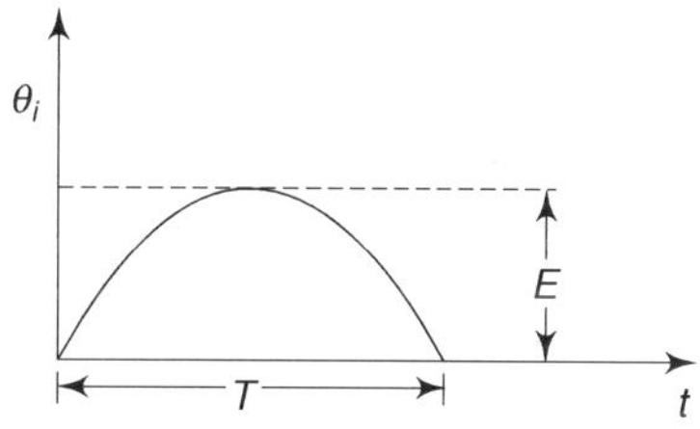

Fig. 3.28 *Figure for Problem 3.6*

Equation (3.68) then becomes

$$(\tau D + 1)\theta_o = E \sin \omega t \tag{3.69}$$

$$\theta_o = \text{complementary solution + particular solution.}$$

Complementary part of solution, being the solution of Eq. (3.69) with RHS = 0, is given by $Ae^{-t/t}$ as discussed earlier. The particular solution may be assumed to be:

$$\theta_o = B \sin \omega t + C \cos \omega t \tag{3.70}$$

Substituting Eq. (3.70) in (3.69) and equating coefficients of sin ωt and cos ωt on both sides of the equation gives the values of B and C. Thus,

$$\theta_o = Ae^{-t/\tau} + \frac{E}{1+\omega^2\tau^2}\,[(\sin\,\omega t - \tau\omega\cos\,\omega t)] \tag{3.71}$$

Applying the initial condition $\theta_o = 0$ at $t = 0$ gives

$$A = \frac{E\tau\omega}{1+\omega^2\tau^2}$$

or

$$\theta_o = \frac{E}{1+\omega^2\tau^2}\,[\tau\omega\,e^{-t/\tau} + \sin\,\omega t - \tau\omega\cos\,\omega t] \tag{3.72}$$

Case (a) For $T = 4\tau$ or $\omega = \dfrac{\pi}{4\tau}$

Equation (3.72) becomes

$$\theta_o = E\left[0.486\,e^{-t/\tau} + 0.618\sin\frac{\pi t}{4\tau} - 0.486\cos\frac{\pi t}{4\tau}\right] \tag{3.73}$$

Equation (3.73) is plotted and is shown in Fig. 3.29.

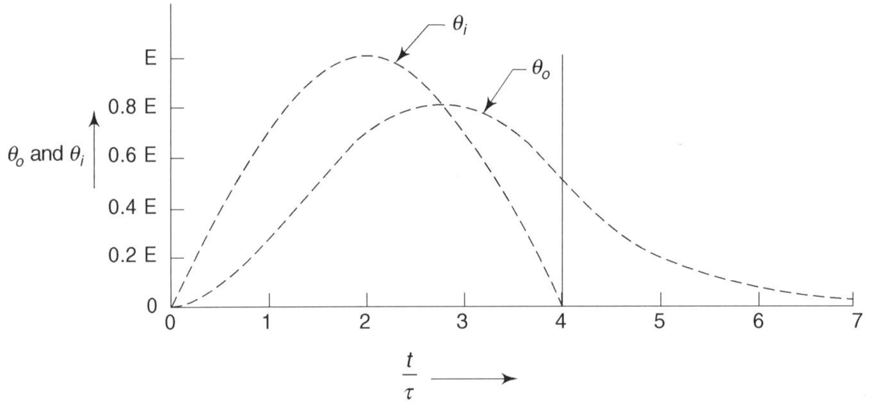

Fig. 3.29 *Figure for Problem 3.6*

Case (b) For $T = 10\tau$ or $\omega = \dfrac{\pi}{10\tau}$

Equation (3.72) becomes

$$\theta_o = E\left[0.286\,e^{-t/\tau} + 0.91\sin\frac{\pi t}{10\tau} - 0.286\cos\frac{\pi t}{10\tau}\right] \tag{3.74}$$

The plot of Eq. (3.74) is shown in Fig. 3.30.

It may be seen that both Eqs. (3.73) and (3.74) are valid for $t < T$.

For $t > T$, Eq. (3.69) becomes

$$(\tau D + 1)\,\theta_o = 0 \tag{3.75}$$

The solution of Eq. (3.75) is

$$\theta_o = Ae^{-t/\tau} \tag{3.76}$$

To find $\theta_o(t)$ applicable to $t > T$, we apply the initial condition to Eq. (3.76), viz. at $t = T$, $\theta_o = 0.495$ E. This is obtained from Eq. (3.73) for case (a). Using the initial condition, we get

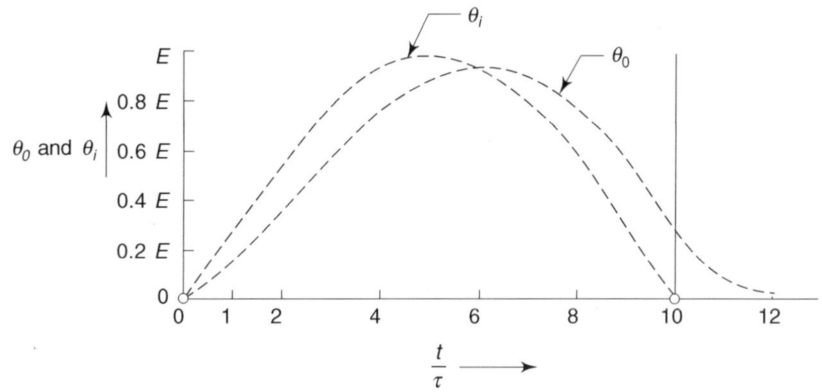

Fig. 3.30 *Figure for Problem 3.6*

$$A = 27.03$$

or $$\theta_o = 27.03e^{-t/\tau} \tag{3.77}$$

Similarly for case (b),

$$\theta_o = 6300e^{-t/\tau} \qquad \text{for } t > T \tag{3.78}$$

For $t > T$, Eqs. (3.77) and (3.78) have been used while plotting Figs. 3.29 and 3.30.

Comparing plots for $\theta_o(t)$ in Figs. 3.29 and 3.30, it may be seen that in the latter, viz. when $T = 10\tau$, the output approximates the input more closely than in the former case of $T = 4\tau$. Thus, in the case of transient inputs, the duration of input should be $\gg$ time constant of the first order instrument.

Problem 3.7 *A pressure transducer shown in Fig. 3.31 has a natural frequency of 30 rad/s, damping ratio of 0.1 and static sensitivity of 1.0 μV/Pa. A step pressure input of 8 × 10⁵ N/m² is applied. Determine the output of the transducer.*

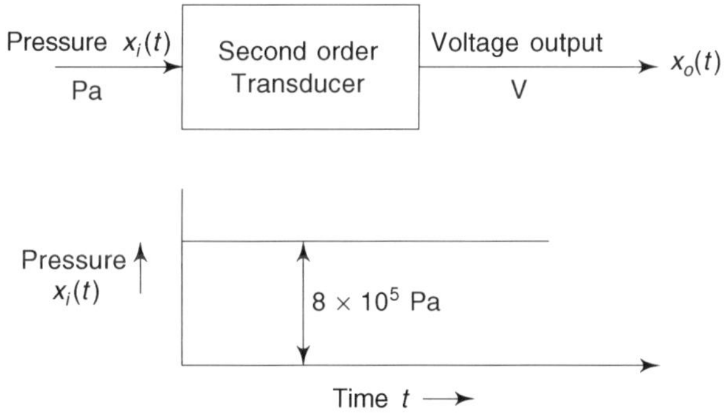

Fig. 3.31 *Figure for Problem 3.7*

Solution Using Eq. (3.59),

$$\left(\frac{D^2}{\omega_n^2} + \frac{2\xi}{\omega_n}D + 1\right) x_o = Kx_i(t)$$

Undamped natural frequency $\omega_n = 30$ rad/s
Damping ratio $\xi = 0.1$
Static sensitivity $K = 1\ \mu$V/Pa
Step pressure input $x_i(t) = 8 \times 10^5$ Pa

or $$\left[\frac{D^2}{(30)^2}+\frac{0.2}{30}D+1\right]x_o(t)=\frac{8\times10^5}{10^6}=0.8$$

x_o being in volts.

Solution of the above equation is obtained from Eq. (3.67) viz.

$$x_o(t) = K\left[1-\exp(-\xi\omega_n t)\left(\cos\omega_n\sqrt{1-\xi^2}\,t+\frac{\xi}{\sqrt{1-\xi^2}}\sin\omega_n\sqrt{1-\xi^2}\,t\right)\right]$$

$$= 0.8[1 - e^{-3t}(\cos 29.85t + 0.1005 \sin 29.85t)] \qquad (3.79)$$

Equation (3.79) is shown plotted in Fig. 3.32.

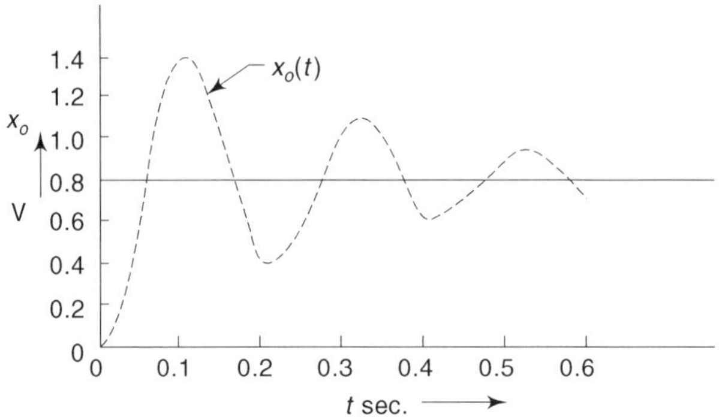

Fig. 3.32 *Figure for Problem 3.7*

Problem 3.8 *A thermocouple with a time constant of 0.3 s and a static sensitivity of 0.04 mV/°C is used to the measure temperature of a medium, which varies as shown in Fig. 3.33. If the initial temperatures of the thermocouple reference and measuring junctions was 25°C, find the output of the thermocouple at t = 0.6 and 30 seconds.*

Solution $\tau = 0.3$ secsonds

$K = 0.04$ mV/°C

The governing equation is

$$(\tau D + 1)V_o = K\theta_i(t) \qquad (3.80)$$

V_o being the output voltage.

For $t \le 0.6$ s, change in $\theta_i(t) = 100 - 25 = 75$ °C. Voltage output V_o would occur because of this change.

Solution of Eq. (3.80) for step input $\theta_i(t)$ is

$$V_o = K(75)[1 - e^{-t/\tau}]$$
$$= 3(1 - e^{-t/0.3}) \text{ mV}$$

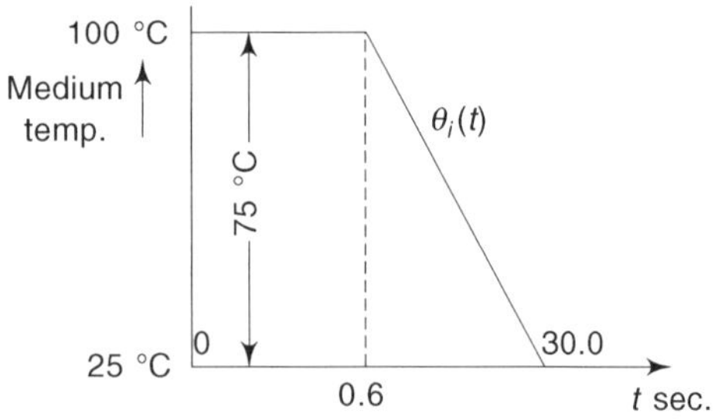

Fig. 3.33 *Figure for Problem 3.8*

At $t = 0.6$ s, $V_o = 2.6$ mV.

From Fig. 3.33, it can be shown that for $t > 0.6$ s, the change in $\theta_i(t) = -2.55t + 76.5$, which is obtained from the equation of the straight line. Thus for $t > 0.6$ s, the governing equation is

$$(\tau D + 1)\, V_o = (-2.55\, t + 76.5)\, 0.04 \qquad (3.81)$$

Complementary solution for Eq. (3.81) is:

$$V_o = Ae^{-t\tau}$$

Particular solution may be assumed to be:

$$V_o = Bt + C$$

Substituting in Eq. (3.81) and equating coefficients of t and constant terms on both sides, we get

$$B = -0.102 \quad \text{and} \quad C = 3.09$$

Thus,

$$V_o = Ae^{-t/0.3} - 0.102\,t + 3.09$$

In order to get A, we use the condition, viz. at $t = 0.6$ s, $V_o = 2.6$ mV.

This gives

$$A = 0.9$$

Thus from above V_o at $t = 30$ s is equal to 0.03 mV.

Fourier Transform Method In this method, the input signal, a function of time, is converted into a function of frequency, by means of Fourier transforms, as shown in Fig. 3.34. Using the input signal thus obtained as a function of frequency and the frequency response function of the instrument for harmonic inputs, the output can be obtained as a function of frequency. In case the output signal is desired to be obtained as a function of time, inverse Fourier transforms have to be employed.

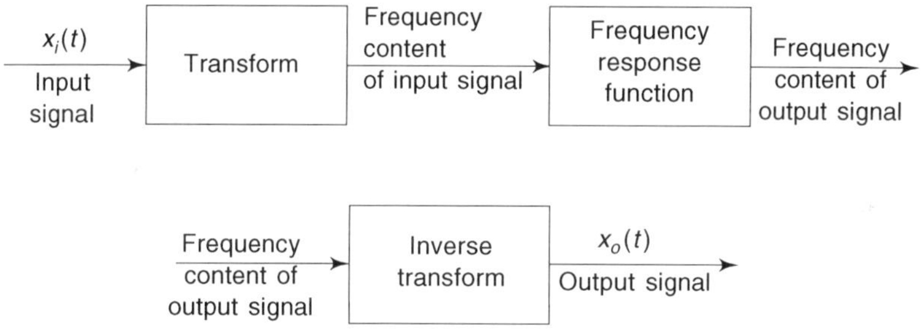

Fig. 3.34 *Application of Fourier and inverse transforms*

For transformation, the input signal must satisfy certain conditions, namely, it should be sectionally continuous and defined over a specified interval. Its integral must exist over the interval and it may be defined at each point of discontinuity by its mean value.

The exponential Fourier transform of a function $x_i(t)$ is defined as

$$\left.\begin{aligned} f(\omega) &= \int_0^{\infty} x_i(t)e^{-j\omega t}\,dt \\ &= L\int_0^{\infty} x_i(t)\cos\omega t\,dt - j\int_0^{\infty} x_i(t)\sin\omega t\,dt \end{aligned}\right\} \quad \text{for } 0 \le \omega \le \infty \tag{3.82}$$

$f(\omega)$ thus obtained is a continuous function of frequency.

The inverse transform is given by

$$x_i(t) = \frac{1}{2}\int_0^{\infty} e^{-j\omega t} f(\omega)\,d\omega \tag{3.83}$$

Consider the rectangular pulse shown in Fig. 3.35, viz.

$x_i(t) = E$ for $0 < t < T = 0$ for $t > T$

Using Eq. (3.82)

$$F(\omega) = \int_0^T Ee^{-j\omega t}\,dt$$

$$= \frac{E}{\omega}\sin\omega T + \frac{jE}{\omega}(\cos\omega T - 1)$$

$$= R \angle \varphi \tag{3.84}$$

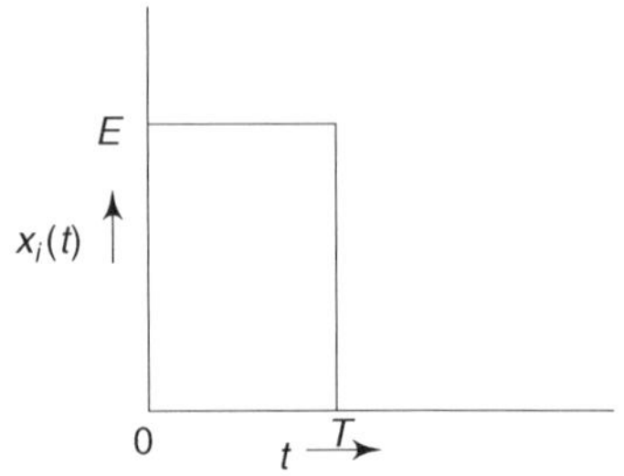

Fig. 3.35 *Rectangular pulse*

where $$R(\omega) = \sqrt{\left(\frac{\omega}{E}\sin\omega T\right)^2 + \left(\frac{E}{\omega}(\cos\omega T - 1)\right)^2}$$

$$= \frac{E}{\omega}(2 - 2\cos\omega T)^{1/2}$$

$$\varphi(\omega) = \tan^{-1}\left(\frac{\cos\omega T - 1}{\sin\omega T}\right)$$

$R(\omega)$ and $\varphi(\omega)$ are shown plotted in Fig. 3.36.

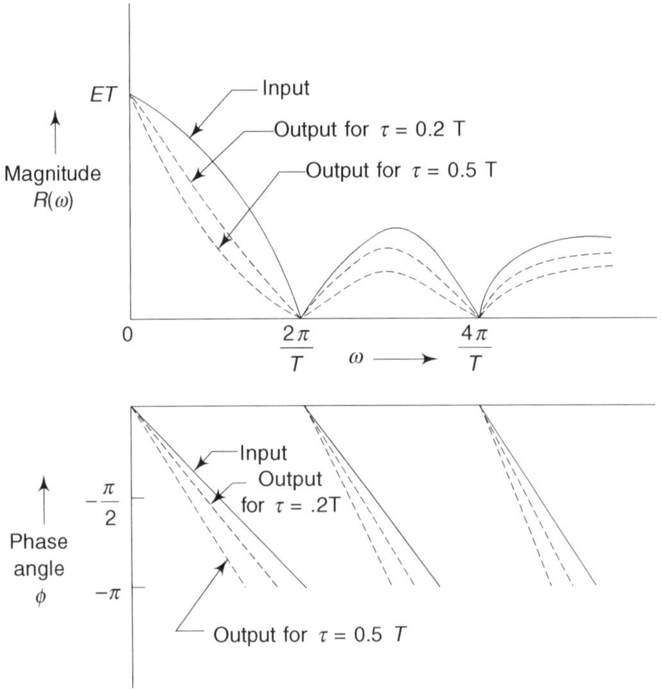

Fig. 3.36 *Frequency content of the signals*

If this rectangular pulse is applied to say a first-order instrument with governing equation

$$(\tau D + 1)x_o = Kx_i(t) \tag{3.85}$$

output x_o as a function of frequency, can be obtained as below.

For harmonic input of frequency ω as proved earlier:

$$x_o = \frac{Kx_i}{\sqrt{1+\tau^2\omega^2}} \angle \varphi_1$$

where $\varphi_1 = \tan^{-1}(\tau\omega)$

Using $f(\omega)$ as given by Eq. (3.84)

Output
$$x_o(\omega) = K\,\frac{E}{\omega}\,\frac{(2-2\cos\omega T)^{1/2}}{\sqrt{1+\tau^2\omega^2}} \angle (\varphi + \varphi_1)$$

$$= \bar{R} \angle \bar{\varphi} \tag{3.86}$$

where
$$\bar{R} = \frac{KE(2-2\cos\omega T)^{1/2}}{\omega\sqrt{1+\tau^2\omega^2}}$$

$$\angle \bar{\varphi} = \angle(\varphi + \varphi_1)$$

$\bar{R}$ and $\bar{\varphi}$ refer to the output signal's Fourier transform and are also shown plotted in Fig. 3.36, for various values of τ. Thus, the instruments' output characteristics can be studied in the frequency rather than the time domain and the manner in which the output and input spectra differ are analysed.

Problem 3.9 *A second-order pressure transducer has a natural frequency of 1000 rad/s and damping ratio of 0.3. Its static sensitivity is 0.02 V/bar. A pressure pulse, as shown in Fig. 3.37 is applied to the transducer. Using Fourier transform method, plot the frequency response of the input signal and the output voltage.*

Solution Using Eq. (3.84) and putting $E = 5$ bar $= 5 \times 10^5$ N/m^2 (1 bar $= 10^5$ N/m^2) and $T = 0.02$ s, the frequency components of the input signal are given by:

$$R(\omega) = \frac{E}{\omega}(2 - 2\cos\omega T)^{1/2}$$

$$= \frac{7.07}{\omega}\left(1-\cos\frac{\omega}{50}\right)^{1/2}$$

$$\varphi(\omega) = \tan^{-1}\left(\frac{\cos 0.02\,\omega - 1}{\sin 0.02\,\omega}\right)$$

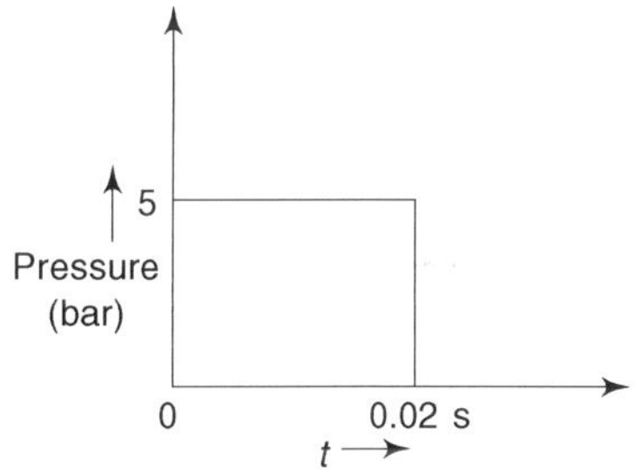

Fig. 3.37 *Figure for Problem 3.9*

For the second order system, using

$$\omega_n = 1000 \text{ rad/s}, \qquad \xi = 0.3$$

in the frequency response function, which is:

$$M(\omega) = \frac{K}{\sqrt{(1-r^2)^2+(2\xi r)^2}}, \qquad r = \frac{\omega}{\omega_n}$$

$$= \frac{0.02}{\sqrt{\left\{1-\left\{\left(\frac{\omega}{1000}\right)^2\right\}^2+\left(\frac{\omega}{1667}\right)^2\right\}}}$$

$$\varphi_1(\omega) = -\tan^{-1}\left(\frac{2\xi r}{1-r^2}\right) = -\tan^{-1}\left(\frac{\omega/1667}{1-\left(\frac{\omega}{1000}\right)^2}\right)$$

From the above, frequency components of output signals are given by $\bar{R}(\omega) \angle \bar{\varphi}(\omega)$,

where $\bar{R} = R(\omega)\, M(\omega)$

$$= \frac{0.1414\left(1-\cos\frac{\omega}{50}\right)^{1/2}}{\omega\left[\left(1-\frac{\omega^2}{10^6}\right)^2+\left(\frac{\omega}{1667}\right)^2\right]^{1/2}}$$

$$\angle\bar{\varphi} = \angle(\varphi + \varphi_1)$$

$$= \tan^{-1}\left(\frac{\cos 0.02\,\omega - 1}{\sin 0.02\,\omega}\right) - \tan^{-1}\left(\frac{\omega/1667}{1-\frac{\omega^2}{10^6}}\right)$$

R, φ, $\bar{R}$ and $\bar{\varphi}$ are shown plotted against ω in Fig. 3.38.

3.2.4 Response to Random Signal Input

A random signal (Fig. 3.39) does not have a definite time period or amplitude and has to be described statistically. Only stationary random signals will be discussed here. It is possible to describe such signals statistically over a certain time period. The statistical properties do not change with time. The statistical properties that are of relevance are: (i) mean or average value of the random signal, (ii) rms value, (iii) mean square spectral density, and (iv) auto-correlation function.

Mean or average value of the random signal

$$\bar{x}_i(t) = \lim_{T\to\infty}\frac{1}{2T}\int_{-T}^{T} x_i(t)\,dt \tag{3.87}$$

In practice, the signal is averaged over a time interval that is increased till a constant mean value is obtained and any further increase of time interval, does not appreciably change this value.

Mean square value of random singal

$$\bar{x}_i^2(t) = \lim_{T\to\infty}\frac{1}{2T}\int_{-T}^{T} x_i^2(t)\,dt \tag{3.88}$$

rms or root mean square value is the square root of the mean square value $\bar{x}_i^2(t)$.

In order to experimentally determine the amplitudes corresponding to any signal, the following steps are needed:

1. The signal has to be fed into a filter which allows a component of a single centre frequency ω_c, with a small bandwidth $\Delta\omega$ to pass through it (Fig. 3.40). Ideally $\Delta\omega$ should be as small as possible.

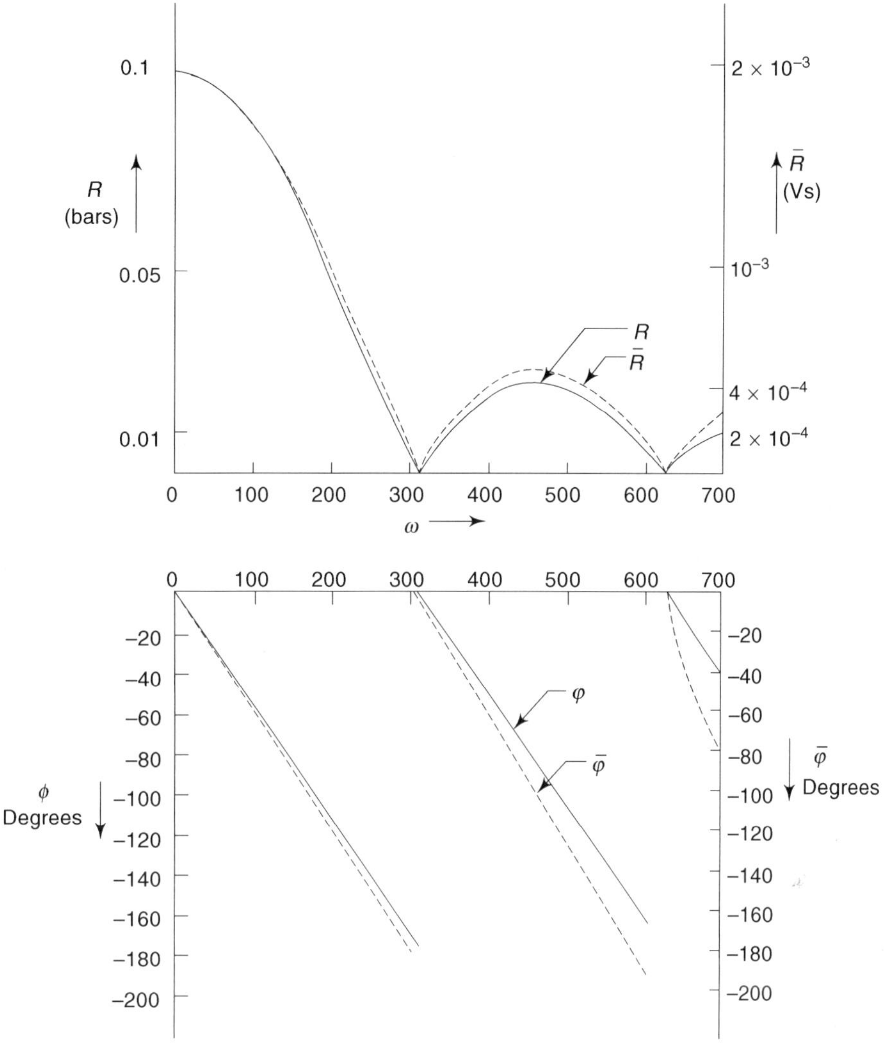

Fig. 3.38 *Figure for Problem 3.9*

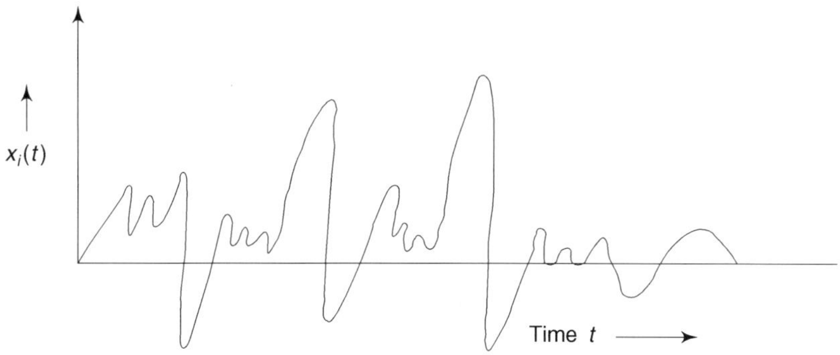

Fig. 3.39 *Random signal*

2. The filter output is then squared and averaged over a certain time interval. Thus, we get the mean square value $[\bar{x}_i^2]_{\omega_c}$ of the component of a certain frequency ω_c.
3. By varying the centre frequency, the mean square value corresponding to various frequency components is obtained.

From the above, the mean square spectral density $S(\omega)$ is defined as

$$S(\omega) = \frac{[\bar{x}^2]_{\omega_c}}{\Delta\omega} \tag{3.89}$$

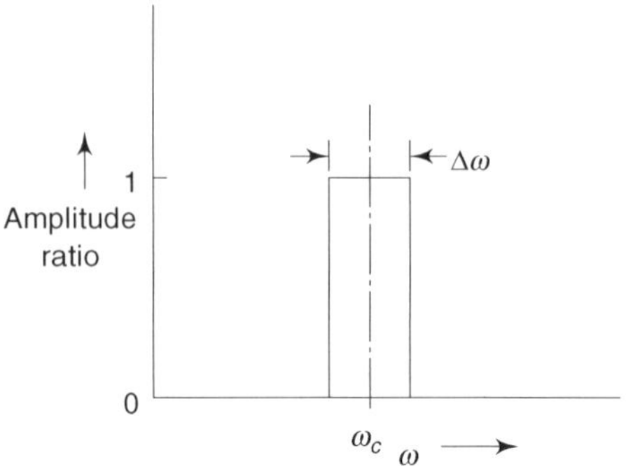

Fig. 3.40 *Filter characteristics*

$S(\omega)$ represents the density, i.e. the amount per unit frequency band width of the mean square value. Area $S\Delta\omega$ gives $[\bar{x}_i^2]_{\omega_c}$ viz. mean square value of the signal corresponding to frequency ω. Since the total mean square value $\bar{x}_i^2(t)$ is equal to sum of individual mean square values of various terms, it can be seen that the area under the curve in Fig. 3.41 gives the total mean square value $\bar{x}^2(t)$ of the signal.

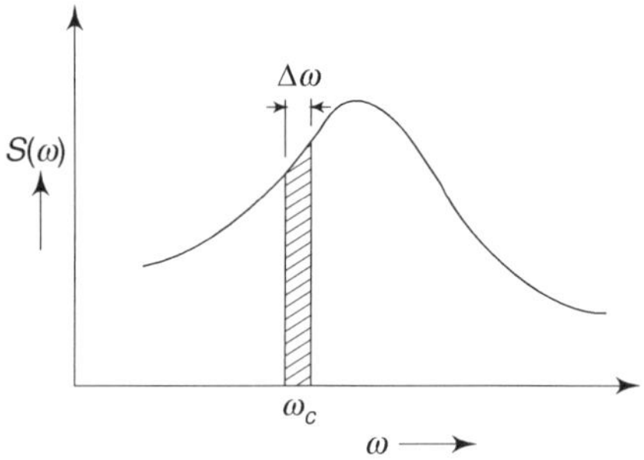

Fig. 3.41 *Mean spectral density*

Another useful concept for describing a random signal is *auto-correlation function,* which is related to $S(\omega)$. This is described in Appendix A.3.

We shall describe the random signal by its mean spectral density $S(\omega)$. Then, using the frequency response function of any instrument, it is possible to determine the output response of an instrument when an input signal of a certain $S(\omega)$ is applied to it as shown below:

Consider an instrument with frequency response function $M(\omega)$ and input signal

$$x_i(t) = X_i \sin \omega t$$

Output response

$$x_o(t) = M(\omega)\, X_i \sin \omega t$$

Mean square value of input signal

$$\bar{x}_i^2(t) = \frac{1}{T}\int_0^T X_i^2 \sin^2\omega t \, dt$$

T being the period of the signal.

Mean square value of output signal

$$\bar{x}_0^2(t) = \frac{1}{T}\int_0^T [M(\omega)]^2 X_i^2 \sin^2 \omega t \, dt = [M(\omega)]^2 \{\bar{x}^2(t)\}$$

This concept can also be used in situations where the frequency spectrum is continuous, rather than discrete. Thus, if the input random signal has mean spectral density $S_i(\omega)$, the mean spectral density of the output signal $S_o(\omega)$ is given by

$$S_o(\omega) = [M(\omega)]^2 S_i(\omega) \tag{3.90}$$

As mentioned earlier, from a plot of $S(\omega)$ against ω, the value of the mean square value of the signal can be obtained from the area of the plot.

Problem 3.10 *A second-order instrument with damping ratio 0.3 and an undamped natural frequency of 800 rad/s is subjected to a random input signal, with mean spectral density S_i as shown in Fig. 3.42. Plot the mean spectral density of the output signal and find the rms values of input and output signals.*

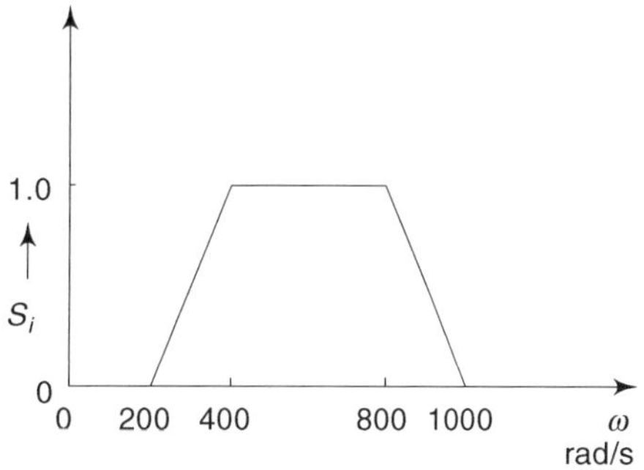

Fig. 3.42 *Mean spectral density of input signal*

Solution For a second-order system, the input-output relation is given by

$$\left(\frac{D^2}{\omega_n^2} + \frac{2\xi}{\omega_n}D + 1\right)x_o = x_i(t)$$

x_i being the input and x_o the output signals.

ω_n = Undamped natural frequency

and ξ = Damping ratio

Frequency response parameter

$$M = |x_o/x_i\,(j\omega)|$$

$$= \frac{1}{\sqrt{\left\{1-\left(\dfrac{\omega}{\omega_n}\right)^2\right\}^2 + \left(2\xi\dfrac{\omega}{\omega_n}\right)^2}}$$

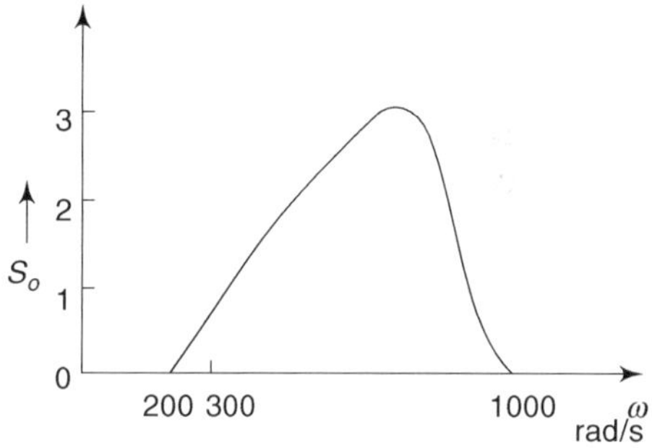

Fig. 3.43 *Mean spectral density of output signal*

Using Eq. (3.90), values of $S_o(\omega)$ have been obtained and plotted in Fig. 3.43, for various value of ω. From Figs. 3.42 and 3.43, areas under the curves are obtained. These are mean square values of input and output signals and are seen to be 600 and 1315 units. Thus, rms values of input and ouput signals are 24.5 and 36.2 units, respectively.

3.3 ■ COMPENSATION

In order to improve the dynamic characteristics of a measuring system, compensation is employed. This involves use of additional elements. As discussed earlier, the frequency and transient response of a first order element like a thermocouple, can be improved if its effective time constant can be reduced. Similarly, for a second order instrument, the dynamic performance can be improved by increasing the damping ratio ξ, if it is already small.

3.3.1 First-Order System Compensation

The governing equation of a first order system like a thermocouple is

$$(1 + \tau D)V_1 = K\theta_i(t) \tag{3.91}$$

τ is its time constant, K, the static sensitivity, $\theta_i(t)$ the input temperature to be measured, and V_1 the output voltage.

In order to reduce the effective value of the time constant, the voltage V_1 can be applied to a circuit (Fig. 3.44) whose output is V_2. The relation between V_2 and V_1 can be easily derived as under:

$$V_2 = i_3R = (i_1 + i_2)\, R \tag{3.92}$$

where i_1, i_2 and i_3 are currents as shown in Fig. 3.44

$$i_1 = \frac{V_1 - V_2}{R_o}$$

$$i_2 = CD(V_1 - V_2),\ D = \mathrm{d}/\mathrm{d}t$$

Substituting in Eq. (3.92)

$$V_2 = \left[\frac{V_1 - V_2}{R_o} + CD\,(V_1 - V_2)\right] R$$

This equation can be written as:

$$\frac{V_2}{V_1} = \alpha\left[\frac{1 + \tau_o D}{1 + \alpha\tau_o D}\right] \tag{3.93}$$

where $$\alpha = \frac{R}{R + R_o}$$

$$\tau_o = R_o C$$

If τ_o is so chosen, by choosing R_o and C, that

$$\tau_o = \tau \tag{3.94}$$

Equation (3.93) becomes

$$\frac{V_2}{V_1} = \frac{\alpha\,(1 + \tau D)}{(1 + \alpha\tau D)} \tag{3.95}$$

Substituting Eq. (3.91) in Eq. (3.95), we get

$$(1 + \alpha\tau D)\, V_2 = \alpha K \theta_i(t) \tag{3.96}$$

Equation (3.96) is the governing equation for the compensated system of Fig. 3.44. Comparing Eqs. (3.96) and (3.91), the latter being for an uncompensated system, it is seen that with compensation,

1. The effective time constant becomes $\alpha\tau$. Since $\alpha < 1$, the effective time constant is $< \tau$.
2. Static sensitivity becomes αK, which is less than K, the static sensitivity for the uncompensated system.

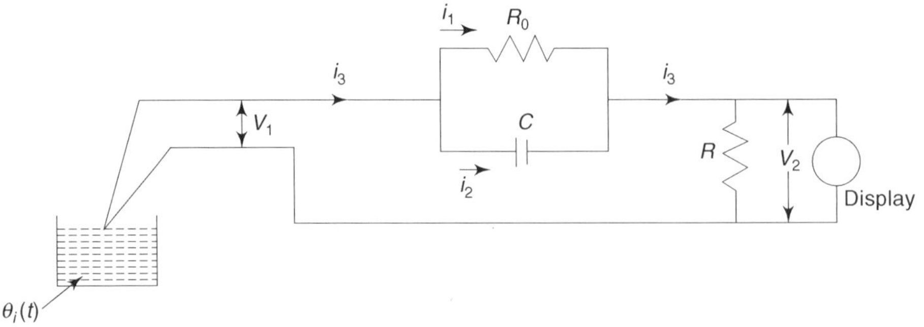

Fig. 3.44 *First-order system with compensation*

Thus, with compensation, the reduction in time constant is at the expense of sensitivity.

For a harmonic input $\theta_i = \bar{\theta}_i \sin \omega t$, the variation of output voltage amplitudes $\bar{V}_1$ and $\bar{V}_2$ for the thermocouple without and with compensation respectively with frequency ω would be, as shown in Fig. 3.45.

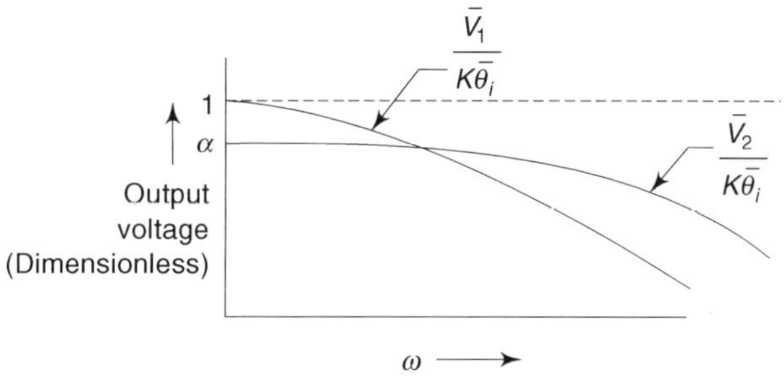

Fig. 3.45 *Frequency response of uncompensated and compensated systems*

3.3.2 Second-Order System Compensation

As discussed in Secs. 3.3.1 and 3.3.3, damping in a second order system affects the output response considerably. Usually, damping in practice is small and may be increased by additional means, like use of a compensating network. Figure 3.46 shows how a compensating network may be used along with a second order system.

In Fig. 3.46, V_1 is the output of the uncompensated second order system. This becomes the input to the compensating circuit, the output of which is V_2. The relation between V_1 and V_2 is derived as under:

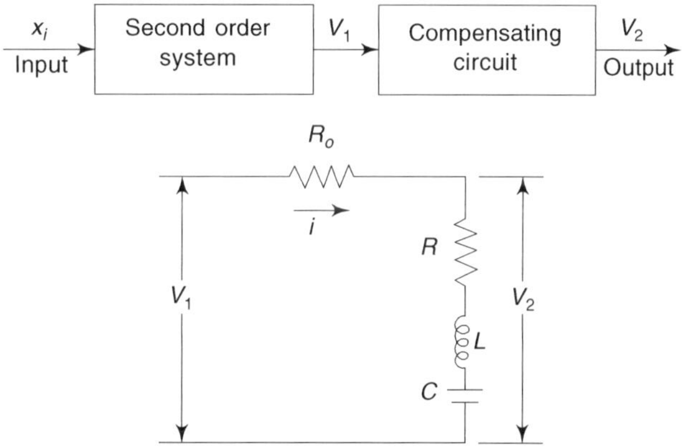

Fig. 3.46 *Second-order system compensation*

$$V_1 = \left(R_0 + R + LD + \frac{1}{CD} \right) i, \quad D \text{ being } \frac{d}{dt}. \tag{3.97}$$

i being the current through the circuit, R_0 and R resistances, L inductance and C the capacitance.

$$V_2 = V_1 - iR_0 \tag{3.98}$$

From Eqs. (3.97) and (3.98)

$$\frac{V_2}{V_1} = \frac{LCD^2 + RCD + 1}{LCD^2 + (R_0 + R)CD + 1}$$

$$= \frac{\dfrac{D^2}{\omega_1^2} + \dfrac{2\xi_1}{\omega_1}D + 1}{\dfrac{D^2}{\omega_1^2} + \dfrac{2\xi_2}{\omega_1}D + 1} \tag{3.99}$$

where $\omega_1 = \sqrt{\dfrac{1}{LC}}$

$$\xi_1 = \frac{R}{2\sqrt{L/C}}$$

$$\xi_2 = \frac{R + R_o}{2\sqrt{L/C}}$$

The relation between V_1 and x_i for a second order system is:

$$\left(\frac{D^2}{\omega_n^2} + \frac{2\xi}{\omega_n}D + 1\right)V_1 = Kx_i \tag{3.100}$$

K being the static sensitivity, ξ the damping ratio and ω_n the undamped natural frequency of the system.

If the compensating circuit is so designed that

$$\omega_1 = \omega_n$$

$$\xi_1 = \xi \tag{3.101}$$

then from Eqs. (3.99), (3.100) and (3.101), we get

$$\left(\frac{D^2}{\omega_n^2} + \frac{2\xi_2}{\omega_n}D + 1\right)V_2 = Kx_i \tag{3.102}$$

Comparing Eqs. (3.102) and (3.100), it is seen that the damping ratio for compensated system is ξ_2, while that for the uncompensated circuit is ξ. Since $\xi_2 > \xi_1$ as seen from the expressions for ξ_2 and ξ_1 from Eq. (3.99) and $\xi_1 = \xi$ (Eq. 3.101), we have $\xi_2 > \xi_1$. Thus, with compensation, the effective damping is increased, as desired.

Problem 3.11 *A thermocouple with a time constant of 0.3 s and a static sensitivity of 0.04 mV/°C is suddenly immersed in a bath of hot oil, which is at 100°C. The initial temperature of the thermocouple measuring and reference junctions was 25°C.*

(a) What is the output at t = 0.1, 0.3 and 1.0 s?

(b) Design a compensating circuit to be used along with the thermocouple so that the time constant is reduced to half its value. For this system, find the output at t, given in (a).

Solution

(a) The maximum difference between measuring and reference junctions

$$= 100 - 25 = 75°\text{C}.$$

For a step input of $\theta_i = 75°C$, the solution for voltage output, as obtained from Sec. 3.3.3 is

$$V_1 = K[1 - e^{-t/\tau}]\theta_i$$
$$= 0.04\ (75)\ [1 - e^{-t/0.3}] = 3[1 - e^{-t/0.3}]$$

From the above equation it is seen that at

$t = 0.1$ s, $V_1 = 0.85$ mV,
$t = 0.3$ s, $V_1 = 1.90$ mV, and
$t = 1.0$ s, $V_1 = 2.89$ mV,

(b) $\tau = 0.3$ (3.103)

With compensation,

effective τ or $\tau_{eff} = 0.15 = \alpha\tau$ (3.104)

Using the compensating circuit of Fig. 3.44, take

$$\tau = \tau_0 = R_0C = 0.3 \tag{3.105}$$

From Eqs. (3.103) to (3.105),

$$\alpha = 0.5 = \frac{R}{R + R_0} \tag{3.106}$$

$$\tau = 0.3 = R_0C \tag{3.107}$$

There are three unknowns of the cirucit: R_1, R_0 and C but only two equations, viz. Eqs. (3.106) and (3.107)

Assuming arbitrarily $R = 1\ M\Omega$

We get from the above $R_0 = 1\ M\Omega$

$C = 0.3\ \mu F$

With compensation $\tau_{eff} = 0.15$

$K_{eff} = \alpha K = 0.02$ mV/°C

Thus, for $\theta_i(t) = 75°C$, output voltage

$$V_2 = 0.02(75)\ [1 - e^{-t/0.15}] = 1.5[1 - e^{-t/0.15}]$$

From the above, we get at

$t = 0.1$ s, $V_2 = 0.729$ mV,
$t = 0.3$ s, $V_2 = 1.297$ mV, and
$t = 1.0$ s, $V_2 = 1.498$ mV.

Problem 3.12 *For a second-order instrument system, with $\omega_n = 1000$ rad/s, $\xi = 0.2$, $K = 10$ mV/mm, find the parameters of a compensating network so that its damping ratio is increased to 0.7.*

Solution Referring to Fig. 3.46 and Sec. 3.4.2, it is desired that $\xi_2 = 0.7$, given $\xi = 0.2$, $\omega_n = 1000$ rad/s. Using Eq. (3.101)

$$\omega_1 = \omega_n = 1000 \text{ rad/s}$$
$$\xi_1 = \xi = 0.2$$

Thus using Eq. (3.99)

$$\xi_1 = \frac{R}{2\sqrt{L/C}} = 0.2 \tag{3.108}$$

$$\xi_2 = \frac{R + R_0}{2\sqrt{L/C}} = 0.7 \tag{3.109}$$

$$\omega_1 = \sqrt{\frac{1}{LC}} = 1000 \text{ rad/s} \tag{3.110}$$

From Eqs. (3.108) to (3.110), four unknowns have to be obtained. Assuming one of them, say L arbitrarily = 10 H, we get from above $C = 10$ μF, $R = 4000$ Ω, $R_0 = 10$ kΩ.

Review Questions

3.1 Indicate true or false against each of the following:

(a) If the time constant of a temperature measuring system is increased, it would improve the frequency response of the system.

(b) In a second order instrument system, a small value of the damping ratio improves the dynamic characteristics.

(c) A first order system of time constant τ is subjected to a transient pulse of duration T. The output would closely correspond to the input if $T \gg \tau$.

(d) If the governing equation of an instrument and its frequency response are known, it is possible to find the response of the instrument to transient or random inputs.

(e) The mean square value of a random signal cannot be found from a plot of its mean square spectral density against ω.

(f) Compensation of an instrument system improves its dynamic characteristics.

(g) Compensation of a second order instrument is usually done to increase its damping ratio.

(h) For a good dynamic performance of a second order instrument, the ratio of the input signal frequency to the undamped natural frequency of the instrument should be very high.

(i) The concept of time constant is associated with a first order system subjected to a step input.

(j) A thin elastic type of beam for the measurement of force can be considered as a second order instrument.

(k) The amplitude ratio of a second order system is always less than unity.

(l) In a higher order instrument system, the overall amplitude ratio is the sum of amplitude ratios of all the subsystems.

(m) Response of a second order system is non-oscillatory when damping ratio is less than unity.

(n) In both first as well as second order systems, the amplitude ratio is close to unity at very low input frequencies.

3.2 Fill in the blanks in the following:

(i) If $a_2 \dfrac{d^2 y}{dt^2} + a_1 \dfrac{dy}{dt} + a_0 y = b_0 x_i(t)$ then,

static sensitivity constant k = ______________.
natural frequency of the system ω_n ______________ and
damping parameter of the system ξ ______________.

(ii) The time taken to reach 63.2% of the input value in a first order system subjected to step input is ____________.

(iii) The bead of a thermocouple has an approximate shape of a sphere (diameter 1.6 mm) and its properties may be taken as that of iron (density 7900 kg/m^3 and specific heat 0.452 kJ/kg/°C. If the bead is exposed to convection current with heat transfer coefficient of 95.2 W/m^2/°C, the time constant of thermocouple is ____________.

(iv) Consider a first order system $(1 + \tau D)\, x_0 = x_i$,

(a) if $x_i = \alpha t + \beta t^2$ then the complimentary function is ____________.

(b) if $x_i = \alpha e^{-t/\tau}$, then the particular solution is ____________.

(v) For a unit ramp input applied to a first order system, with time constant τ, the steady state error is ____________.

(vi) In a first order system, the amplitude ratio depends on ____________ and ____________.

3.3 Which of the following is the correct answer?

(i) The degree of damping most desirable for an analog-indicating instrument is

(a) zero (b) critical

(c) somewhat less than critical (d) slightly more than critical

(ii) Amplitude ratio of the first order instrument

(a) can be zero (b) is greater than one (c) lies between zero and one

(iii) A clinical thermometer is marked with 30 s as the time of insertion. The time constant of thermometer is

(a) 30 s (b) 18.96 s (c) 6 s

3.4 A transducer with a resistance of 100 Ω is connected to an oscilloscope, having a 1MΩ resistance, in parallel with a 0.07 μF capacitor. For that maximum sinusoidal frequency would the arrangement be usable if the permissible reduction in amplitude ratio is 3%?

3.5 A first order instrument has its time constant equal to 0.2 s.

(a) For sinusoidal input to the instrument, find the frequency at which the amplitude of the output signal is 0.8 times the amplitude of the input signal.

(b) For step input to the instrument, find the time at which the ratio of the output to input signals is 0.8.

3.6 For a first order instrument system, with a governing equation:

$$(1 + \tau D)\, V_0 = K\theta_i(t), \qquad D \text{ being } \frac{d}{dt},$$

with input $\theta_i(t)$ being of ramp type, viz

$$\theta_i(t) = ft,$$

where f is a constant, K the static sensitivity and τ the time constant. Derive the expression for the output $V_0(t)$. Also find the steady state error.

3.7 A measuring element with a time constant of 0.4 s and a static sensitivity of 0.05 mV/°C is used to measure the temperature of a medium, which changes from 20 to 60°C. Taking the output as zero at 20°C.

(a) Find the time taken for the output voltage to reach 80% of the steady state value, if the temperature change occurs suddenly.

(b) The output voltage at the end of 5 s, if the temperature change from 20 to 60°C, occurs at a constant rate in 5 s.

(c) Suggest a method of reducing the time constant to 0.1 s.

3.8 In a response test on a thermometer, which was suddenly put in a water bath kept at 100°C, the following data was obtained:

Time t s	0	1	3	6	8	11	15	18
Temp °C	30	50	65	80	90	95	98	99

(a) Find the time constant of the thermometer.

(b) Find the steady state error if the thermometer is used to measure temperature of a liquid cooling at a constant rate of 1°C every 6 s.

3.9 A velocity transducer generates 5 mV/cm/s. The output of the transducer is fed to an $R-C$ circuit with $R = 1$ MΩ and $C = 0.1$ μF. Find the amplitude and phase of voltage across 'R' relative to the input motion to the transducer, the amplitude of motion being 0.5 mm and frequency 10 Hz.

3.10 An electrodynamic transducer, whose output is proportional to velocity, has a sensitivity of 10 mV-s/cm. The output is fed to an RC circuit having resistance and capacitance in series and the voltage output across the capacitance is connected to an oscilloscope. If $R = 1$ MΩ, $C = 0.1$ μF,

(a) Find the amplitude of voltage output as displayed on the oscilloscope if the input displacement is harmonic, with an amplitude of 1 mm and frequency 5 HZ.

(b) Find the shape of the output voltage pulse as displayed on the oscilloscope, if the input transient signal is a shock pulse of half sine type, with peak value 0.1 mm and duration 0.5 s.

3.11 An instrument consists of a first order temperature sensing element and a second order data presentation device. The time constant of the first order element is 0.01 s and static sensitivity is 0.2 mV/°C. The second order device has an undamped natural frequency of 120 rad/s, damping ratio of 0.4 and static sensitivity 5 mm/mV. Draw the Bode diagram of the system. Also, find the output as seen on the data presentation device if the input signal is harmonic, with frequency 100 rad/s and maxium and minimum values of temperature as 100 and 40°C respectively.

3.12 A second order recording instrument is used to record a harmonic voltage signal. The recorder has a static sensitivity of 2 mm/mV, an undamped natural frequency of 100 Hz and viscous damping ratio of 0.7. Find the usable frequency range of the instrument over which the amplitude ratio of output to input signals does not change by more than 10%.

3.13 A pressure transducer has a natural frequency of 4 Hz, damping ratio of 0.2 and static sensitivity of 0.2 μV/Pa.

(a) For a step pressure input of 10^6 Pa, find the output at $t = 0.1$ and 1.0 s. Also, find the peak overshoot value.

(b) For a harmonic input of amplitude 5×10^5 Pa, at frequency 10 Hz, find the output amplitude and phase lag relative to the input.

3.14 A second order pressure transducer has an undamped natural frequency of 300 rad/s, damping ratio of 0.1 and a static sensitivity of 10^{-6} V/N/m^2. It is connected to a second order recording device which has an undamped natural frequency of 400 rad/s damping ratio of 0.5 and static sensitivity of 5 mm/mV. If the input is a harmonic pressure signal of amplitude 1000 N/m^2 and frequency 30 Hz, find the value of the output amplitude as recorded and the phase difference between the input and output signals.

3.15 A seismic motion transducer has a seismic mass of 50 g, spring of stiffness 2 N/cm and damper with damping constant 3.8 N-s/m. The relative motion of the seismic mass with respect to the frame of the transducer is converted to a voltage, by a first order system of time constant 0.01 s and static sensitivity 2 V/mm. Find the output voltage for an input motion of 0.51 mm at a frequency of 30 Hz.

3.16 Write the Fourier series expression for the periodic signal shown in Fig. 3.47. Take four terms in the series and draw the approximate functions thus obtained.

Using the above approximate function as an input to a first order instrument with time constant 0.5 s and static sensitivity K, find the steady state output of the instrument and plot the same for $t = 0$ to $t = 3$ s.

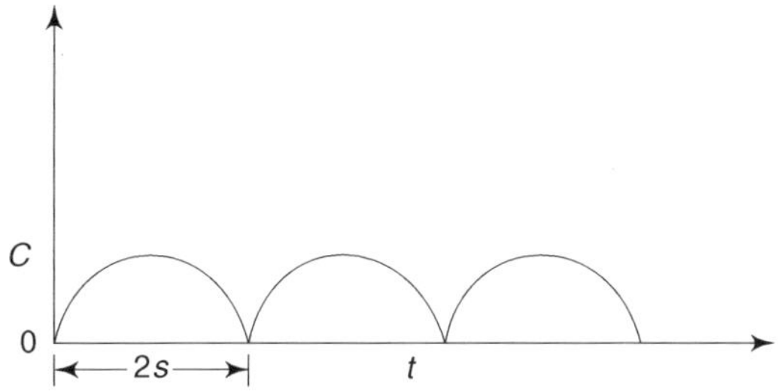

Fig. 3.47

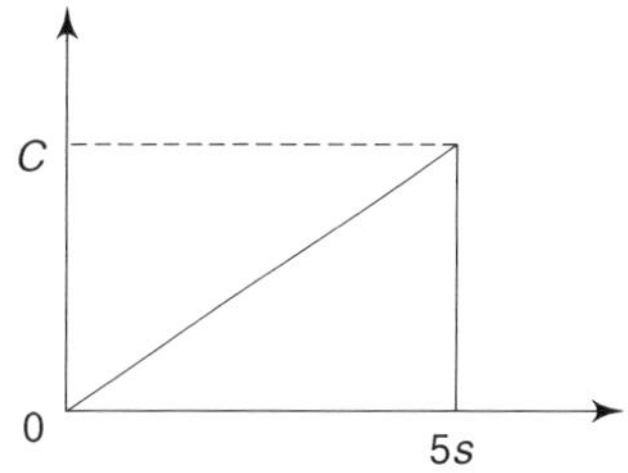

Fig. 3.48

3.17 Plot the frequency spectrum for the input signal shown in Fig. 3.48. If the signal is applied to a first order instrument with time constant 1 s and static sensitivity K, plot the frequency spectrum of the output signal.

3.18 A first order element of an instrument relates output and input voltages V_o and V_i respectively by

$$(1 + \tau D)\, V_o = V_i\,(t), \qquad D \text{ being } \frac{d}{dt}$$

$$\tau = 0.02 \text{ s.}$$

If the mean spectral density S_i of the random input signal is approximated as shown in Fig. 3.49, find the rms values of the output and input signals.

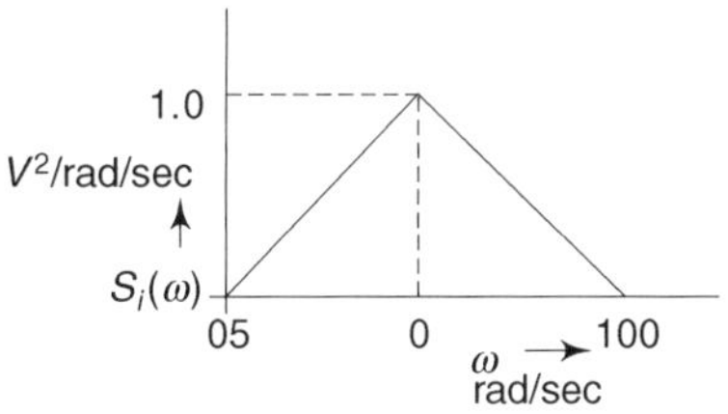

Fig. 3.49

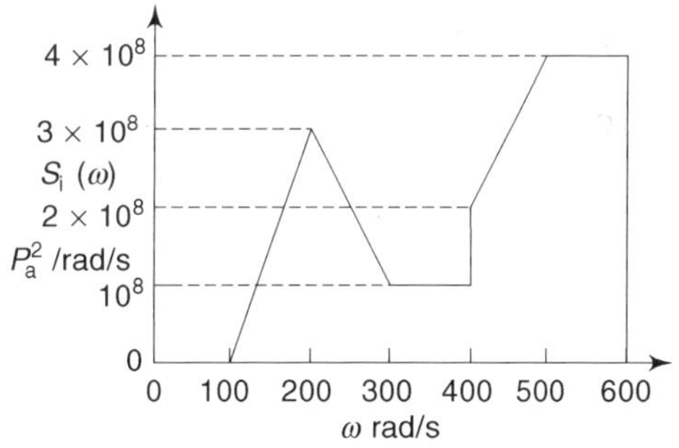

Fig. 3.50

3.19 A second order pressure transducer with static sensitivity 0.2 μV/Pa, undamped natural frequency 1000 Hz and damping ratio 0.5 is subjected to a random input signal with mean spectral density S_i as shown in Fig. 3.50. Plot the mean spectral density of the output signal and find the rms values of the input and output signals.

Answers

3.1	(a) F	(b) F	(c) T	(d) T	(e) F
	(f) T	(g) T	(h) F	(i) T	(j) F
	(k) F	(l) F	(m) F	(n) T	

3.2 (i) $\dfrac{b_0}{a_0}, \sqrt{\dfrac{a_0}{a_2}}, \dfrac{a_1}{2\sqrt{a_0\, a_2}}$ (ii) Time constant

(iii) 10 s (iv) (a) $Ae^{-t/\tau}$ (b) $\dfrac{\alpha t}{\tau} e^{-t/\tau}$

(v) $-\tau$ (vi) ω, τ

3.3 (i) (c) (ii) (c) (iii) (c)
3.4 5701 Hz
3.5 (a) 0.597 Hz (b) 0.32 s
3.6 $K\,[f\tau(e^{-t/\tau} - 1) + ft]$, $Kf\tau$
3.7 (a) 0.644 s (b) 1.84 mV
3.8 (a) 4.73 s (b) 0.79°C
3.9 15 mV, 9°
3.10 (a) 9.35 mV (b) $V = -0.45\, e^{-10t} + 0.45 \cos 6.28t + 0.283 \sin 6.28t$
3.11 Amplitude = 29 mm
3.12 0 – 70.7 Hz
3.13 (a) 2.78 V, 1.99 V (b) 0.0185 V, 10.78°
3.14 (a) 8.89 mm (b) –42.9°
3.15 Amplitude = 0.48 V
3.18 7.1 V, 5.2 V

Chapter

4

Transducer Elements

■ *INTRODUCTION* ■

As discussed in Ch. 1, transducer elements convert the input physical variable to usable form. In most cases, it is in the form of an electrical signal. The main advantages of getting an electrical signal as output are the following:

1. Inertia and friction effects are absent, unlike in transducers with mechanical outputs.
2. Amplification can be obtained with relative ease.
3. Indication or recording, especially at a distance, is greatly facilitated.

The transducers would be discussed according to the classification—analog or digital types—of transducers. In the former case, with the variation of input, there is a continuous variation of output while in the latter, the output is of digital or discrete type. The following types of analog transducers would be described:

1. Electromechanical types, comprising potentiometric resistance type, inductance, capacitive, piezoelectric, resistance strain gauge, ionisation and mechano-electronic types.
2. Opto-electrical transducers, comprising photo-emissive, photo-conductive and photo-voltaic types.

Digital transducers would be discussed under the following heads:

1. Frequency generating types
2. Digital encoder types

4.1 ■ ANALOG TRANSDUCERS

4.1.1 Electromechanical Types

In such transducers, an electrical output is produced due to an input of mechanical displacement or strain. The mechanical displacement or strain input in turn may be produced by a primary sensor due to the input physical variable which may be pressure, flow, etc.

Figure 4.1 shows the scheme for measurement using an electromechanical transducer. Some of the primary sensors used for measurement of force, pressure and temperature are shown in Fig. 4.2. In each case, the input physical variable results in a displacement of an elastic member. In Fig. 4.2 (a), the force applied results in a displacement of the elastic member. In Fig. 4.2 (b), an elastic diaphragm converts pressure input into deflection or displacement. Similarly, a bimetallic strip of brass and invar alloys gets deflected due to change of temperature, due to the difference in coefficient of expansion of the two materials. In each case, an electromechanical transducer may be employed to convert the displacement into an electrical signal, that can be related to the force, pressure or temperature input desired.

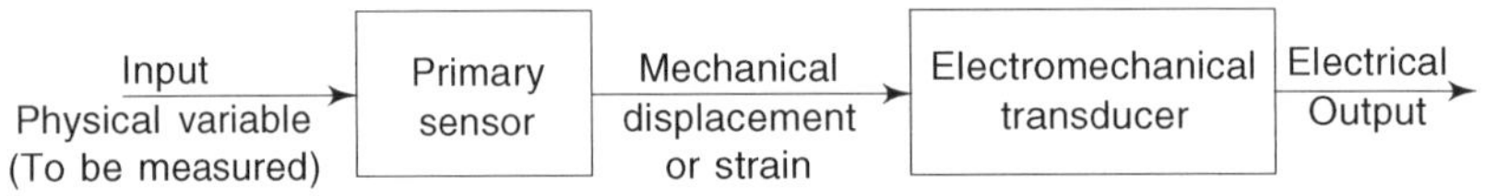

Fig. 4.1 *Scheme for measurement using electromechanical transducer*

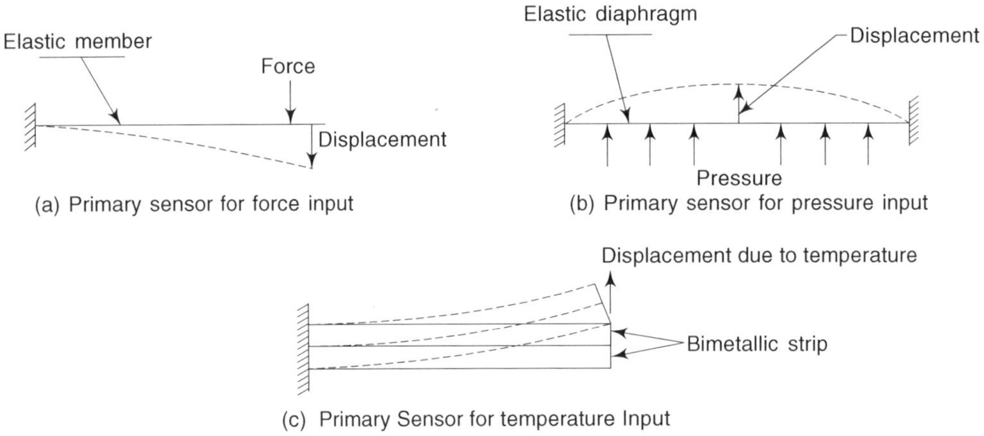

Fig. 4.2 *Conversion on input parameters to mechanical displacement*

Since displacement or motion is an input to an electromechanical transducer, this may be treated as a basic parameter, derivable from several types of physical inputs. The transducer discussed here are also known as *motion transducers*.

The following factors have to be kept in mind, while considering the selection of motion transducers, for a given application:

1. Magnitude of motion—whether the transducer is meant for measuring small, medium or large motions.
2. Type of input-output relation—whether the output is proportional to displacement motion, its rate of change with time, viz. velocity or rate of change of velocity, viz. acceleration.
3. Static and dynamic characteristics—whether the transducer can measure static or dynamic or both types of displacements.
4. Attachment or proximity type—whether the transducer has to be attached to the moving object or kept in close proximity to it.
5. Self-generating or external power source type—whether a power output is needed to energise the transducer or the same is generated due to the input motion itself, within the transducer.

6. Type of associated circuit—whether the circuit to be used along with the transducer for producing a measurable output is of a simple or complicated type.

Potentiometric Resistance-Type Transducer A wire-wound potentiometer may be used as a transducer for converting mechanical displacement to an electrical output. This may be of linear or angular type. As shown in Fig. 4.3, the motion of the object changes the effective resistance and hence the voltage output e_0 between points b and c. Thus, the output voltage appears as shown in Fig. 4.3 and is directly proportional to the dynamic displacement of the moving object.

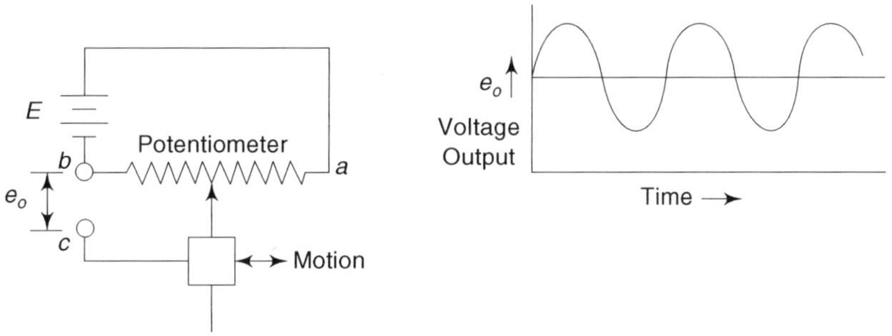

Fig. 4.3 *Potentiometric resistance transducer*

The value of the potential between points b and c, at the starting position, may be large, compared to the change in the potential due to motion. Due to this, a display on a display element like an oscilloscope, may not be possible and the signal may be taken off the screen. To bring the initial value of e_o to zero, a position resistor, as shown in Fig. 4.4 is used. Setting at B can be adjusted to give e_o as zero, depending on the initial position of the moving point A. Figure 4.4 also shows a protection resistor, which prevents the movable contacts of the potentiometer from burning off, in case the resistors are so placed that there is a short circuit across the battery.

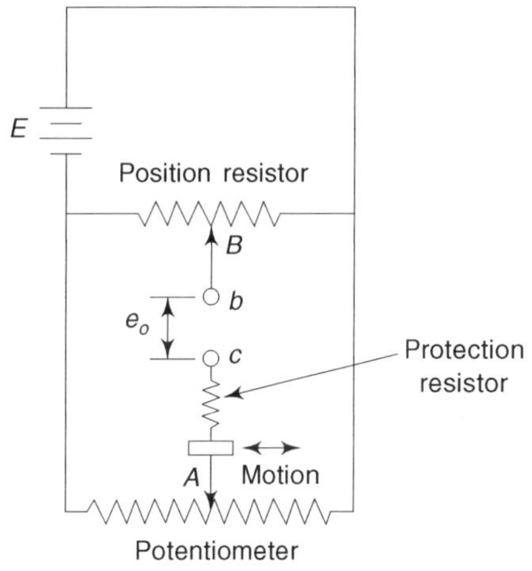

Fig. 4.4 *Modified potentiometric resistance transducer*

Inductive-Type Transducers In these types of transducers, the magnetic characteristics of an electric circuit change due to the motion of the object. These may be classified into two types:

1. Self-generating types, in which a voltage signal is generated in the transducer, because of relative motion of a conductor and a magnetic field. Electrodynamic, electromagnetic and eddy circuit types of transducers belong to this category.
2. Non-self-generating or external power source types of transducers, in which an external source is needed to energise a coil/coils, the inductance of which would change due to the motion of the object. The following types of transducers belong to this category: attachment type inductance transducer, air gap type, LVDT type and magneto-strictive type of transducer.

Self-generating types Figure 4.5 shows an electrodynamic type of transducer. A coil wound on a hollow cylinder of non-magnetic material moves in the annular space of a fixed magnet. The voltage generated in the coil is proportional to the rate of change of flux and hence the velocity of the moving object. The coil cylinder has to be attached to the moving object and thus this is a contact or attachment type transducer.

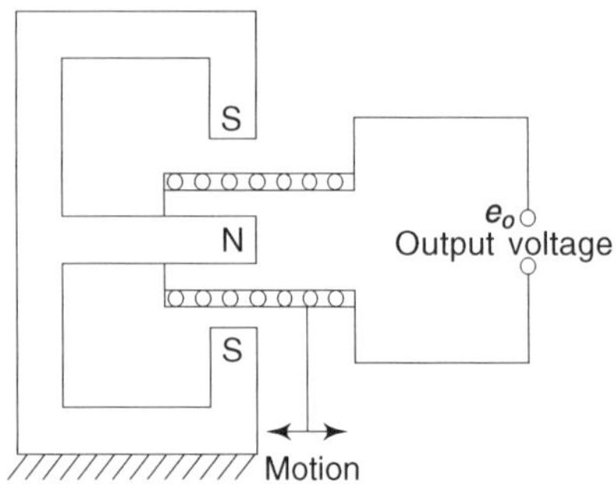

Fig. 4.5 *Electrodynamic transducer*

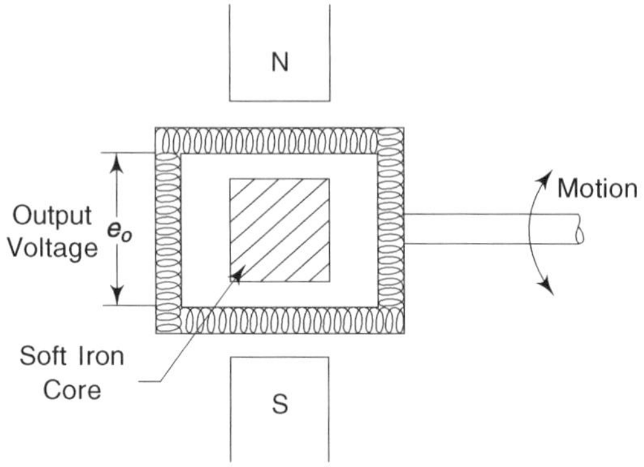

Fig. 4.6 *Electrodynamic transducer for rotary motion*

Figure 4.6 shows an electrodynamic transducer for measuring rotary motion. The coil moves in the annular space between a magnet and a soft iron core, generating a voltage in the coil.

Figure 4.7 shows an electromagnetic transducer, in which a voltage is induced in the coil when the magnetic flux about it is varied due to the motion of the object, which has to be for a ferromagnetic material. This is a proximity type velocity transducer and is linear only for small motions, as the flux intensity changes due to the change in air gap.

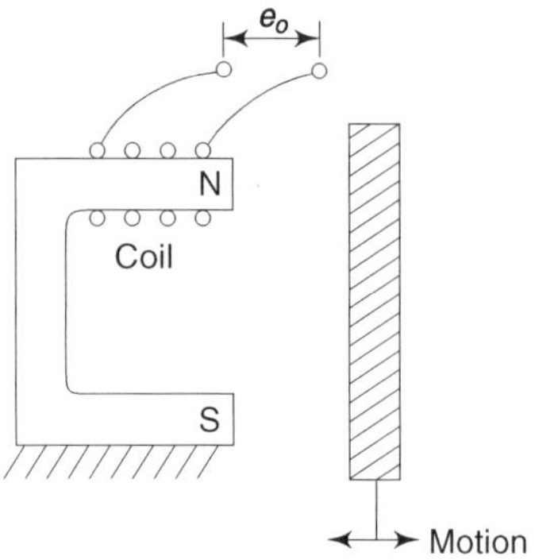

Fig. 4.7 *Electromagnetic transducer*

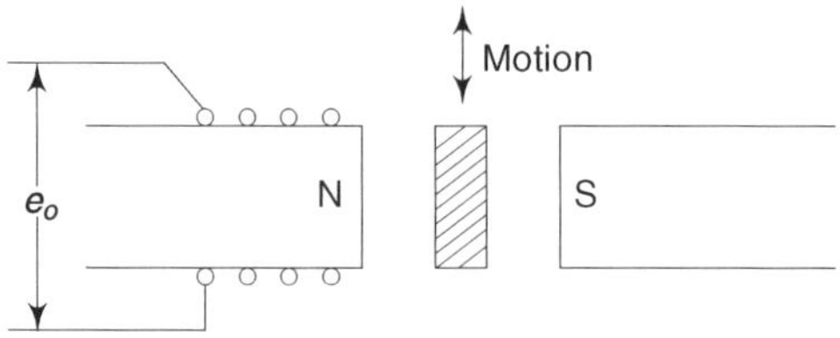

Fig. 4.8 *Eddy current transducer*

Figure 4.8 shows an eddy-current-type transducer. A non-ferrous plate moves in a direction perpendicular to the lines of flux of a magnet. Eddy currents are generated in the plate. These are proportional to the velocity of the plate. These eddy currents set up a magnetic field in a direction opposing the magnetic field that creates them. The output voltage e_o of the coil, shown in Fig. 4.8, is proportional to the rate of change of eddy current or the acceleration of the plate. Since the gap remains constant, the transducer has linear characteristics.

Non-self generating types An inductance transducer of attachment type is shown in Fig. 4.9(a). The core, made of high permeability steel, is attached to the moving object. The motion changes the length of the core inserted in the coil and thus the inductance of the coil gets changed due to the change of reluctance of the magnetic flux path. When the core moves up and down, the inductance of one half increases while that of the other half decreases. The two inductances L_1 and L_2 from the adjacent arms of a Wheatstone bridge, as in Fig. 4.9(b). The bridge output is modulated as shown in Fig. 4.9(c). The

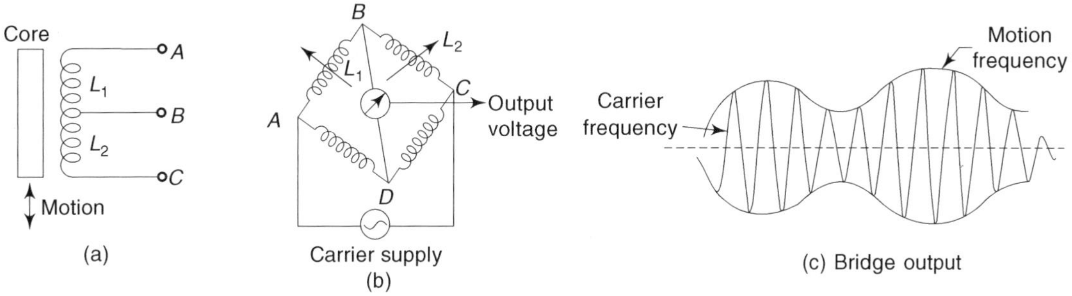

Fig. 4.9 *Variable inductance transducer*

output is supplied to a phase-sensitive demodulator which eliminates the carrier frequency and gives an output corresponding to the motion frequency, the value of the output voltage being proportional to the displacement. The carrier frequency is usually much higher compared to the motion frequency.

Figure 4.10 shows a variable inductance transducer for measuring angular displacement or torsional motion. One half of the core is made of a magnetic material while the other half of non-magnetic material. The inductances of the two halves of the coil depend upon the amount of magnetic material in their flux paths. The associated circuit for the same is identical to that in Fig. 4.9(b).

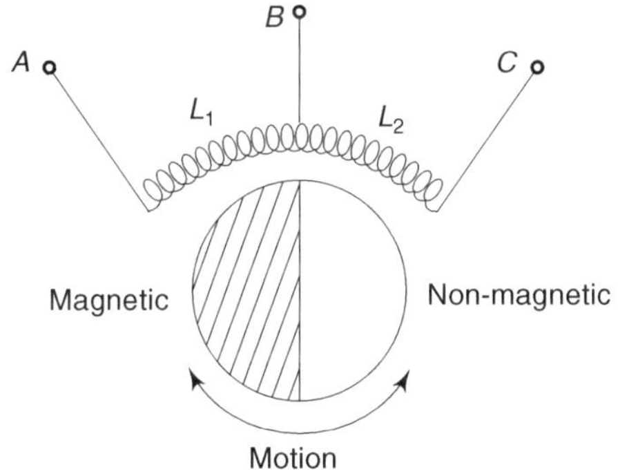

Fig. 4.10 *Variable inductance transducer for rotary motion*

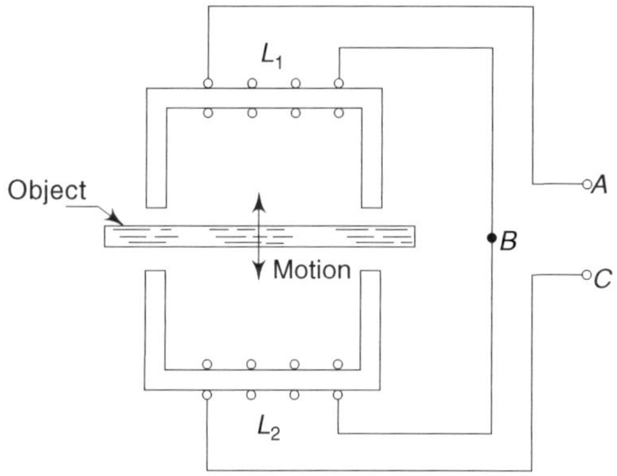

Fig. 4.11 *Proximity type inductance transducer*

In the variable inductance transducer, shown in Fig. 4.11, a small air gap in the magnetic flux path of an electromagnet is varied. The inductance of one coil increases while that of the other decreases. With the circuit being similar to that of other inductance transducers discussed earlier, the output is proportional to the displacement of the object.

In a linear variable differential transducer (LVDT) type of transducer, shown in Fig. 4.12, a soft iron core provides the magnetic coupling between a primary coil and two secondary coils, connected in series opposition. When the core is central and both secondaries are identical, the voltages across them are equal in magnitude. However, the output is zero as both the secondaries are in series opposition. As the core moves up or down, the induced voltage of one secondary coil increases while that of the other decreases. The output voltage, which is modulated, is the difference of the two, since secondaries are in opposition. The associated circuit is similar to that discussed earlier. The output is proportional to the displacement of the iron core. The device is very sensitive and is linear over a wide range of motion.

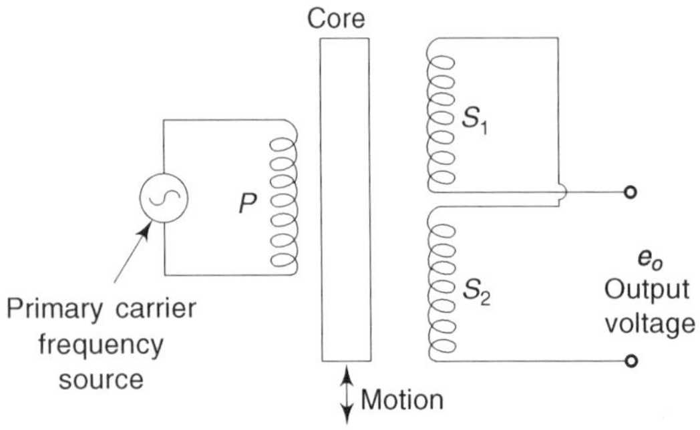

Fig. 4.12 *LVDT (Linear motion type)*

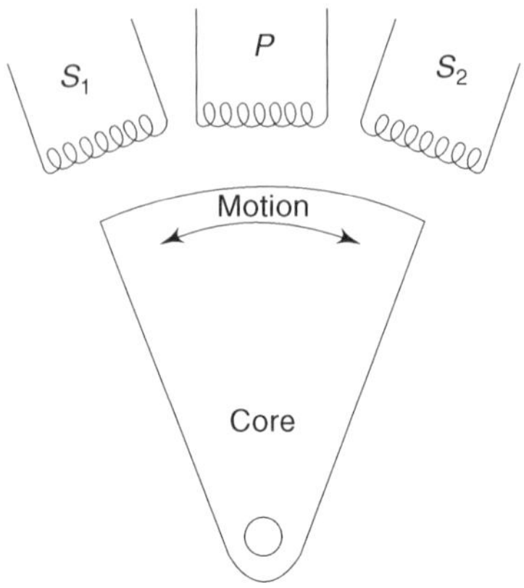

Fig. 4.13 *LVDT (Rotary motion type)*

Figure 4.13 shows a rotary type of transducer, based on the above principle.

Magnetostrictive type of transducer, shown in Fig. 4.14, is based on the principle that the magnetic permeability of a ferromagnetic material, like Ni, changes when the material is subjected to mechanical stress. The magnetic permeability of Ni increases when the material is subjected to compression and decreases due to tension. Thus, the inductance of the coil would change, due to compression or tension of the probe. The magnitude and frequency of the exciting current determines the coil inductance and a change in the same can be measured. Such transducers can be used for measurement of force, motion, etc. These have high mechanical impedance and thus resonant frequency is high, with a good dynamic response. However, these transducer need individual calibration due to the fact that these transducers depend on the change of a physical property of a material, which may differ.

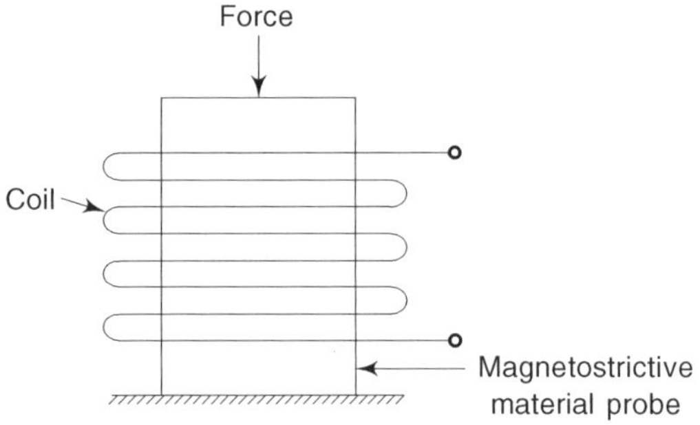

Fig. 4.14 *Magnetostrictive transducer*

Capacitive Type Transducer This is a displacement-sensitive transducer. Due to motion, there is a change in the capacitance between two plates. Suitable circuitry is used to generate a voltage, corresponding to the capacitance change.

The capacitance C between two plates is given by

$$C = \frac{1}{3.6\pi}\,\varepsilon\,\frac{A}{d} \tag{4.1}$$

where C is capacitance, pF

A is area of plates, cm^2

d is distance between plates, cm

ε is dielectric constant of the medium between the plates (= 1 for air).

Capacitance C between plates A and B may change due to the change of gap as shown in Fig. 4.15 or due to the change in area as shown in Fig. 4.16 as a result of motion of member A. Figure 4.17 shows to top view of an area-change type of capacitive transducer, which can be used for measuring rotational motion.

Figure 4.18 shows an associated circuit for capacitive transducers, using an ac carrier frequency oscillator, with the transducer forming one arm of a Wheatstone bridge. A change in the capacitance, causes modulation of the oscillator carrier frequency. A phase-sensitive demodulator is used to eliminate the carrier frequency signal.

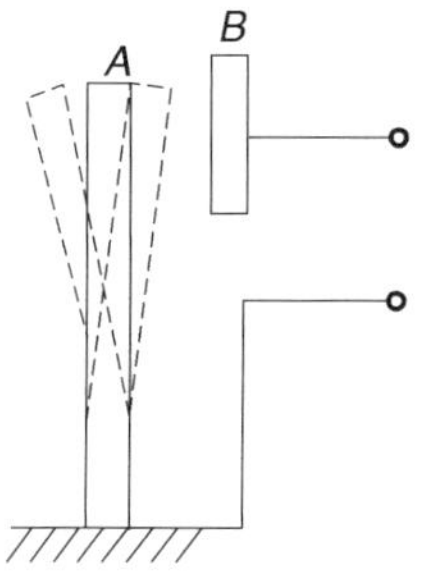

Fig. 4.15 *Gap-change type capacitive transducer*

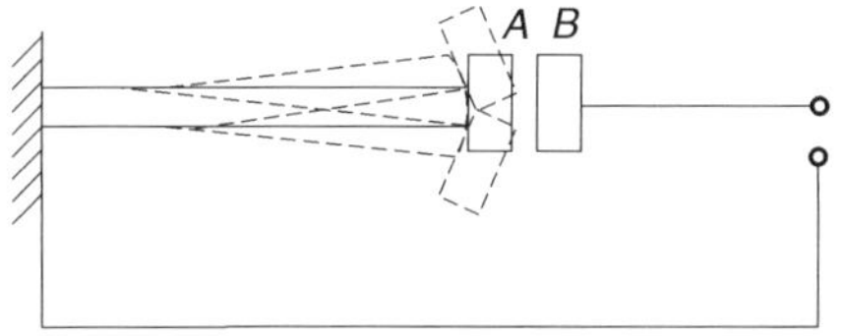

Fig. 4.16 *Area-change type capacitive transducer*

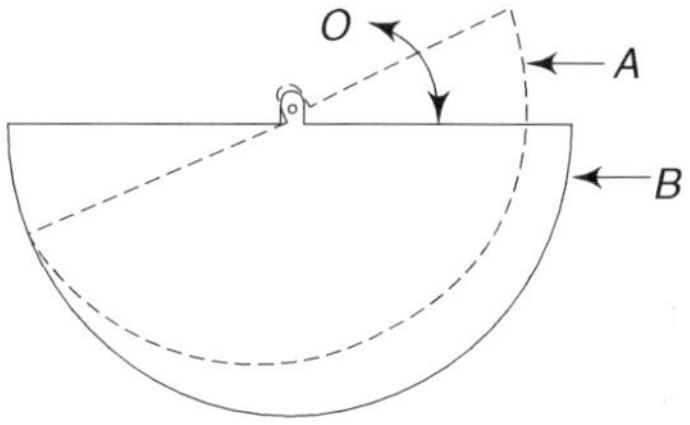

Fig. 4.17 *Area-change type capacitive transducer (for rotational motion)*

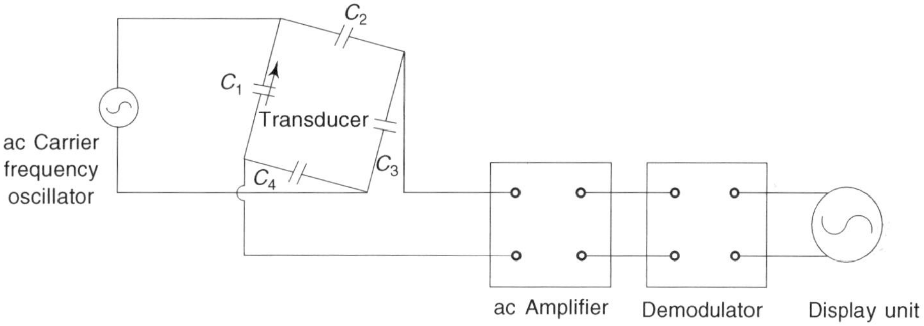

Fig. 4.18 *Associated circuit for capacitive transducer*

An alternative type of circuit of dc type, may be used along with a capacitive transducer, as shown in Fig. 4.19. This can only be used for dynamic measurements. Application of a dc potential V_o, results in a charge Q across the transducer with capacitance C_0, where

$$Q = C_0 V_0 \tag{4.2}$$

If changes in C_0 to C are made fast enough, due to motion, Q will not change appreciably. Thus

Voltage $$e_c = \frac{Q}{C} \tag{4.3}$$

Output $$e_o = V_0 - e_c \tag{4.4}$$

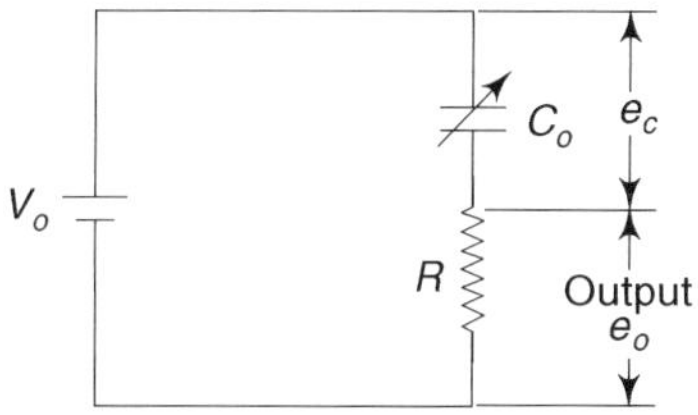

Fig. 4.19 *dc circuit for capacitive transducer*

Substituting from Eqs. (4.2) and (4.3) in Eq. (4.4), we get

$$\frac{e_o}{V_0} = \frac{C - C_0}{C} = \frac{\Delta C}{C} \tag{4.5}$$

ΔC being the change in capacitance.

The frequency response of the output voltage e_o can be shown to be as in Fig. 4.20. The output is very small at low values of frequency ω.

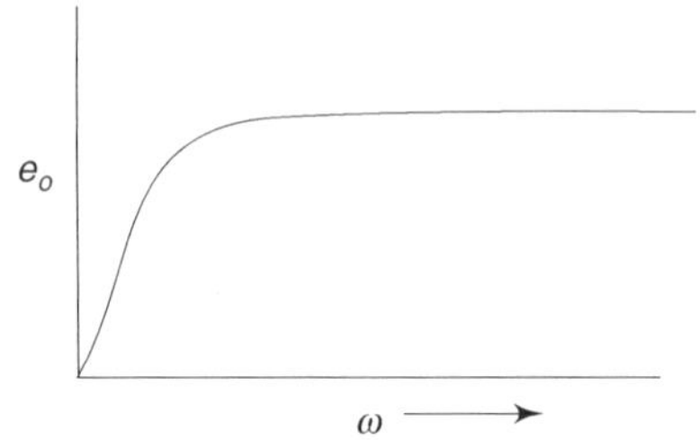

Fig. 4.20 *Output frequency response*

Problem 4.1 *A capacitive transducer consists of two plates of diameter 2 cm each, separated by an air gap of 0.25 mm. Find the displacement sensitivity.*

Solution Using Eq. (4.1), viz.

$$C = \frac{A\varepsilon}{3.6\pi d}$$

Sensitivity $$S = \frac{\partial C}{\partial d} = \frac{-A\varepsilon}{3.6\pi d^2}$$

Putting $$A = \frac{\pi}{4}(2)^2 = \pi \text{ cm}^2, \ \varepsilon = 1 \text{ for air}$$

$$d = 0.025 \text{ cm}$$

$$S = -444 \text{ pF/cm}$$

The negative sign indicates decrease of capacitance, with increase of air gap.

Piezo-Electric Transducer This operates on the principle that when a crystalline material like quartz or barium titanate is distorted, an electrical charge is produced.

Referring to Fig. 4.21, application of a force P causes deformation x_i, producing a charge Q, where

$$Q = K_1 x_i \tag{4.6}$$

K_1 is called the charge sensitivity constant.

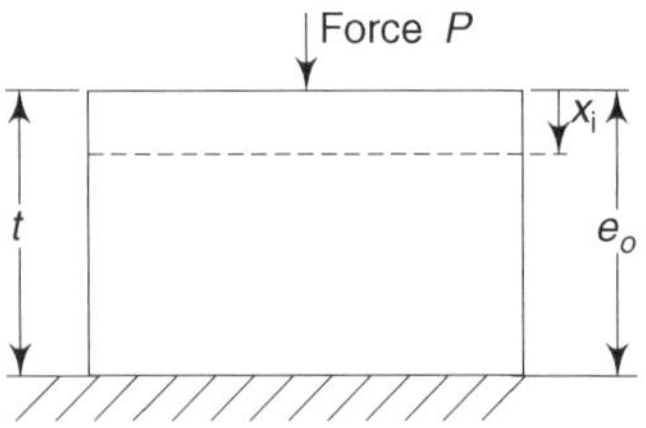

Fig. 4.21 *Piezo-electric crystal subjected to force P*

The crystal behaves as if it was a capacitor, carrying a charge across it. Voltage e_o across the crystal, is given by:

$$e_o = \frac{Q}{C} = \frac{K_1 x_i}{C} = K x_i \tag{4.7}$$

C being capacitance of the crystal, and K the voltage sensitivity constant equal to K_1/C.

As before, $$C = \frac{\varepsilon A}{3.6\pi t} \tag{4.8}$$

C being the capacitance of the crystal (pF), ε the dielectric constant of the crystal material, A its area (cm^2) and t its thickness (cm). If A is in square metre (m^2), t in metre (m) and C in farads (F), Eq. (4.8) becomes:

$$C = \frac{\varepsilon A}{1.13 \times 10^{11}\, t} \tag{4.9}$$

Relation between force P and deformation x_i is:

$$P = EA \frac{x_i}{t} \tag{4.10}$$

E being the Young's modulus of the crystal material.

Table 4.2 gives the properties of some typical piezo-electric materials.

Table 4.2 *Properties of some Piezo-electric Materials*

S.No.	Material	Charge sensitivity pC/N	Dieelectric constant ε	Young's modulus N/m²
1.	Quartz	2.0	4.5	9×10^{10}
2.	Tourmaline	1.9	6.6	16×10^{10}
3.	Barium titanate	150	1380	12×10^{10}
4.	Lead zirconate titanate	265	1500	7.9×10^{10}

Problem 4.2 *A quartz crystal has charge sensitivity of 2 pC/N. Its dielectric constant is 4.5 and Young's modulus is 9×10^{10} Pa. Find the voltage sensitivity constant.*

Solution Voltage sensitivity constant K is given by

$$K = \frac{K_1}{C} \tag{4.11}$$

Charge sensitivity constant K_1

$$= \frac{Q}{x_i} = \left(\frac{Q}{P}\right)\left(\frac{P}{x_i}\right) \tag{4.12}$$

$\frac{Q}{P}$ is given as 2×10^{-12} C/N.

Also, $\frac{P}{x_i}$ from Eq. (4.10) is EA/t.

Substituting in Eq. (4.12),

$$K_1 = 2 \times 10^{-12} \left(\frac{EA}{t}\right)$$

Using Eq. (4.9) for C and substituting the above value of K_1 in Eq. (4.11), we get:

$$K = 5.24 \times 10^9 \text{ V/m} = 5240 \text{ V/}\mu\text{m}$$

Dynamic characteristics of piezo-electric transducers In Fig 4.22 (a), a piezo-electric crystal is shown connected to an amplifier whose output voltage is e_o. It is desired to find a relation between the output voltage e_o and the deformation x_i of the crystal. An equivalent circuit for the arrangement is shown in Fig. 4.22(b). Here C is the combined capacitance due of the crystal, cables and the amplifier.

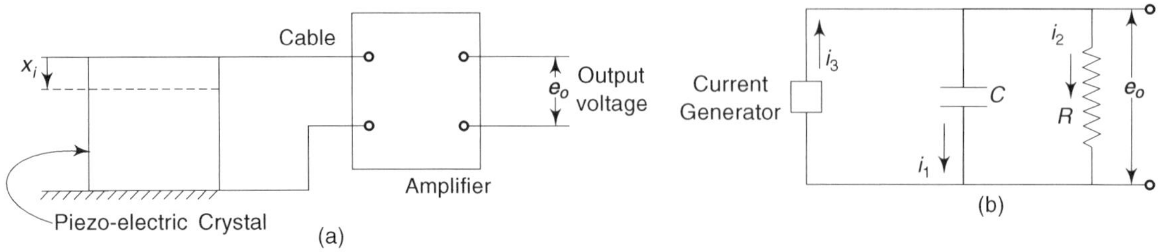

Fig. 4.22 *Piezo-electric crystal connected to an amplifier and equivalent circuit*

Thus $$C = C_{\text{crystal}} + C_{\text{cable}} + C_{\text{amplifier}} \tag{4.13}$$

Also, R is the effective resistance due to leak resistance of the crystal and the resistance of the amplifier, which can be taken to be in parallel. Usually $R_{\text{leak}} \gg R_{\text{amplifier}}$

and thus $$R \approx R_{\text{amplifier}} \tag{4.14}$$

Current generated by the crystal is

$$i_3 = \frac{dQ}{dt}, \; Q \text{ being the charge produced due to deformation } x_i.$$

Using Eq. (4.6)

$$i_3 = K_1 D x_i \tag{4.15}$$

where $$D = \frac{d}{dt}$$

Further, $$i_3 = i_1 + i_2$$

$$= CDe_o + \frac{e_o}{R} \tag{4.16}$$

From Eqs. (4.15) and (4.16), and substituting $K = K_1/C$, we get

$$(1 + \tau D)\, e_o = K\tau D x_i \tag{4.17}$$

where $\tau = RC$ = time constant of the arrangement

K = voltage sensitivity constant

This is a first order equation, though the equation is somewhat different from that derived in Ch. 3 for a first order system.

It may be seen from Eq. (4.17) that for constant x_i, viz. for static deformation of the crystal, voltage e would be zero. Thus, the piezo-electric transducer cannot be used for static conditions but may be used for dynamic inputs.

Frequency response of the arrangement can be obtained by substituting $D = j\omega$ in Eq. (4.17). This would give

$$\left|\frac{e_o}{x_i}\right| = \frac{K\tau\omega}{\sqrt{1+\tau^2\omega^2}} \tag{4.18}$$

This is plotted in Fig. 4.23. It may be seen that time constant τ should be large for a good frequency response over a wide frequency range.

Similarly, the transient response of the arrangement to a pulse input of rectangular type, can be obtained using

$$x_i = A, \quad \text{for} \quad 0 < t \leq T$$
$$= 0, \quad \text{for} \quad t > T$$

Substitution in Eq. (4.17) gives

$$(\tau D + 1)\, e_o = 0$$

The solution of this equation is

$$e_o = Be^{-t/\tau},$$

where B is a constant which depends on initial condition.

Using the initial condition,

at $t = 0$, $\qquad e_o = Kx_i = KA$

we get

$$B = KA$$

or $\qquad e_o = KAe^{-t/\tau} \qquad (4.19)$

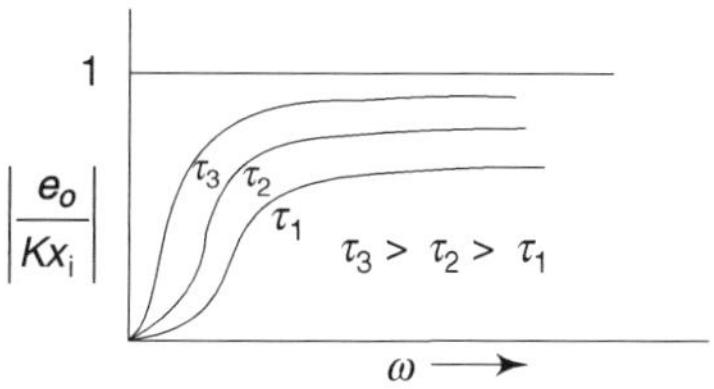

Fig. 4.23 *Frequency response of piezo-electric transducer*

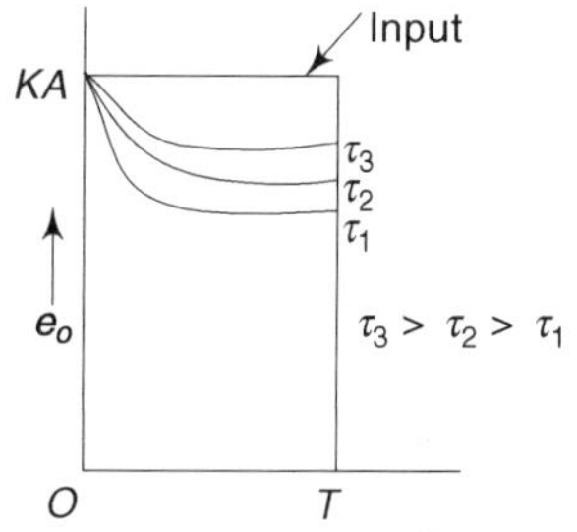

Fig. 4.24 *Transient response of piezo-electric transducer to pulse input*

Equation (4.19) is shown plotted in Fig. 4.24. It is seen that the shape of the output pulse e_o corresponds to that of the input pulse x_i provided the time constant τ is large. Thus, a high value of τ would improve both frequency and transient response of the transducer system. τ being equal to RC, the high value of τ may be achieved by having a high value of R or C or both. One method of obtaining this is to use an impedance converter device called cathode follower between the crystal and the amplifier, as shown in Fig. 4.25.

Fig. 4.25 *Use of cathode follower with a piezo-electric transducer*

A cathode follower unit, shown in Fig. 4.26, is essentially a triode, which presents a high impedance at its input to the crystal and a low impedance at its output. This avoids impedance loading problems, especially if the amplifier's input impedance is small compared to that of the crystal, which is usually the case.

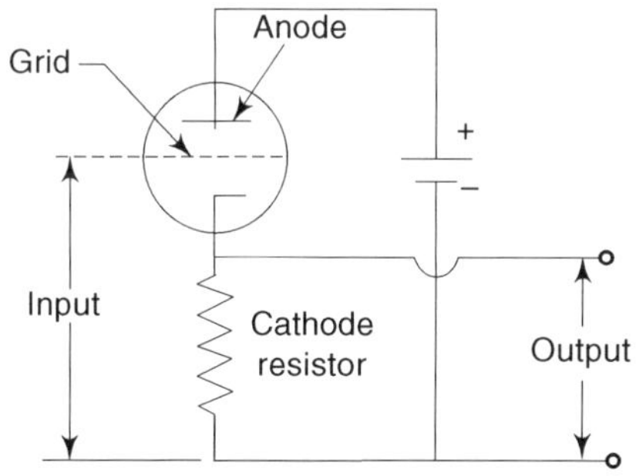

Fig. 4.26 *Cathode follower*

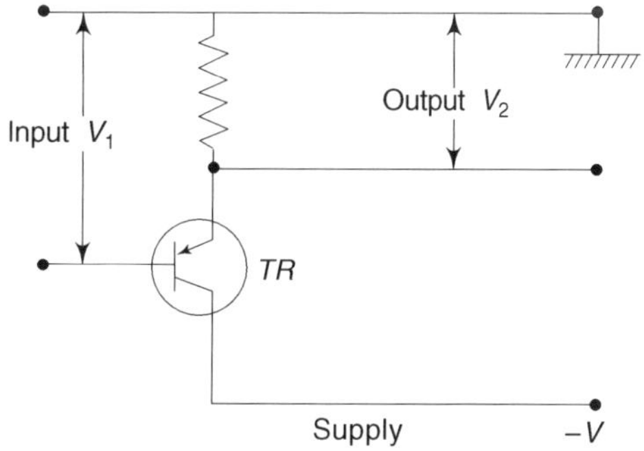

Fig. 4.27 *Emitter follower*

A similar unit using a transistor, in place of vacuum tube, is known as emitter follower and is shown in Fig. 4.27.

A charge amplifier using an operational amplifier may also be used for the above purpose, as discussed in Chapter 5.

Problem 4.3 *A piezo-electric transducer has the following characteristics:*

Capacitance of crystal $= 10^{-9}$ F

Capacitance of cable $= 3 \times 10^{-10}$ F

Charge constant of crystal $= 4 \times 10^{-6}$ C/cm

The oscilloscope used for read-out has a resistance of 1 MΩ in parallel with a capacitance of 10^{-10} *F. Find the amplitude of the output voltage, as displayed on the oscilloscope, if the crystal is subjected to a harmonic deformation of amplitude* 10^{-3} *mm and frequency 200 Hz.*

Solution

$$R = 10^6\ \Omega$$
$$C = 10^{-9} + 3 \times 10^{-10} + 10^{-10}\ \text{F}$$
$$= 1.4 \times 10^{-9}\ \text{F}$$
$$\tau = RC = 1.4 \times 10^{-3}$$
$$K_1 = 4 \times 10^{-6}\ \text{coulomb/cm}$$
$$K = \frac{K_1}{C} = 2857\ \text{V/cm}$$

K is the voltage sensitivity constant. Using Eq. (4.18), viz.

$$\left|\frac{e_o}{x_i}\right| = \frac{K\tau\omega}{\sqrt{1+(\tau\omega)^2}}$$

and substituting $\omega = 2\pi \times 200 = 1256.6$ rad/s and values of K and τ, we get for $x_i = 10^{-4}$ cm

$$e_o = 0.248\ \text{V}$$

Resistance Strain Gauges These types of transducers are based on the principle that if a conductor is stretched or compressed, its resistance will change, because of change in its length, area and resistivity. The resistance R of a conductor of cross-sectional area A, length L, made of a material of resistivity ρ is

$$R = \frac{\rho L}{A} \tag{4.20}$$

Gauge factor F of the conductor is defined as

$$F = \frac{\Delta R/R}{\Delta L/L} = \frac{\Delta R/R}{\varepsilon_a} \tag{4.21}$$

ΔR being change in resistance R due to axial strain ε_a, which is $\frac{\Delta L}{L}$.

With the application of mechanical strain, ρ, L and A may change as above. The corresponding expression for 'F' is derived as below:

$$\text{Change } \Delta R = \left(\frac{\partial R}{\partial L}\right)\Delta L + \left(\frac{\partial R}{\partial A}\right)\Delta A + \left(\frac{\partial R}{\partial \rho}\right)\Delta\rho$$

Substituting the expressions for derivatives from Eq. (4.20),

$$\Delta R = \left(\frac{\rho}{A}\right)\Delta L - \left(\frac{\rho L}{A^2}\right)\Delta A + \left(\frac{L}{A}\right)\Delta\rho$$

Dividing by expression for R from Eq. (4.20),

$$\frac{\Delta R}{R} = \frac{\Delta L}{L} - \frac{\Delta A}{A} + \frac{\Delta \rho}{\rho}$$

Area $A = CB^2$, where B is geometrical dimension of the strain gauge cross-section, and C is a constant whose value depends on the section, equal to $\pi/4$ for circular section of diameter B and 1 for square section of sides B each.

$$\Delta A = 2CB\Delta B$$

or

$$\frac{\Delta A}{A} = \frac{2CB\Delta B}{CB^2} = \frac{2\Delta B}{B}$$

Thus,

$$\frac{\Delta R}{R} = \frac{\Delta L}{L} - 2\frac{\Delta B}{B} + \frac{\Delta \rho}{\rho}$$

$$= \varepsilon_a - 2\varepsilon_t + \frac{\Delta \rho}{\rho},$$

ε_a being axial strain $= \dfrac{\Delta L}{L}$, ε_t being transverse strain $= -\nu\varepsilon_a$, where ν is Poisson's ratio.

Finally,

$$F = \frac{\dfrac{\Delta R}{R}}{\varepsilon_a} = 1 + 2\nu + \frac{\dfrac{\Delta \rho}{\rho}}{\varepsilon_a}.$$

For metallic strain gauges, the first two terms viz. $(1 + 2\nu)$ are higher than the third term while for semi-conductor strain gauges, the third term due to change in resistivity due to strain is much higher compared to $(1 + 2\nu)$. The change in resistivity due to strain is called Piezo-resistivity.

The value of F for Cu–Ni alloy gauge is 2 to 3 while that for semi-conductor is 100 to 200. In the latter case, the value of F is positive for silicon doped with small amounts of p type materials while it is negative for silicon doped with N type materials. The negative value implies decrease is resistivity with tensile strain.

In practice, the conductors used are in the form of thin wires or foils. Strain gauge transducers are of two types:

1. unbonded strain gauge, and
2. bonded strain gauge.

Unbonded strain gauges In an unbonded strain gauge, a resistance wire is stretched between two frames, one being the moving frame and the other, the fixed one as in Fig. 4.28. Typical dimensions of the wire are: 25 mm length and 25 μm diameter. The flexure plates act as springs between the two frames. The wires are under preload, which is greater than any compressive load expected. An input motion as shown in Fig. 4.28 would stretch wires 1 and 3 and reduce tensions in wires 2 and 4. Motion in the opposite direction does the reverse. The wires are connected in a Wheatstone bridge arrangement as shown in Fig 4.28(b). With this type of transducer one can measure very small motions, of the order 50 μm and very small forces. These transducers may be used to measure force, pressure, acceleration, etc.

Bonded resistance strain gauges Transducers, using bonded resistance gauges are widely used for measurement of several physical variables like strain, force, torque, pressure, vibrations, etc. These gauges may be of metallic or semiconductor materials, and are in the form of a wire gauge (about 25 μm diameter) or thin metal foil or small rods (in the case of semiconductor gauges), as shown in

Fig. 4.29. These gauges having paper or some other material backing, are cemented or bonded to the surface, whose strain is to be measured, as shown in Fig. 4.30. Once bonded, the gauges undergo the same strain as that in the member surface. These are very sensitive and when used with electronic equipments, strains as low as 10^{-7} may be measured.

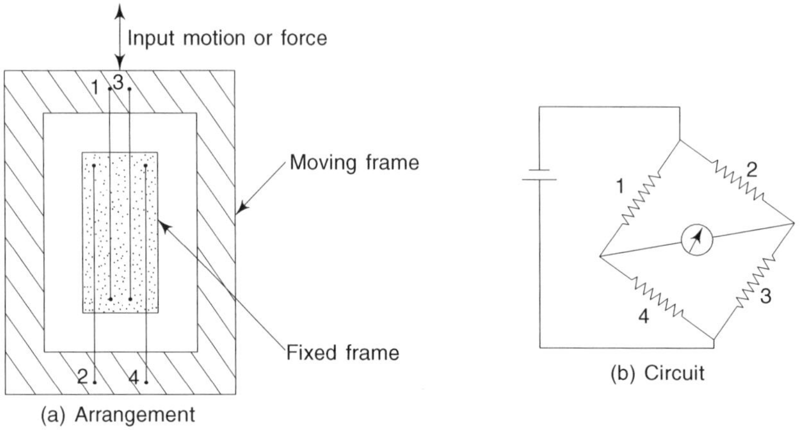

Fig. 4.28 *Unbonded strain gauge*

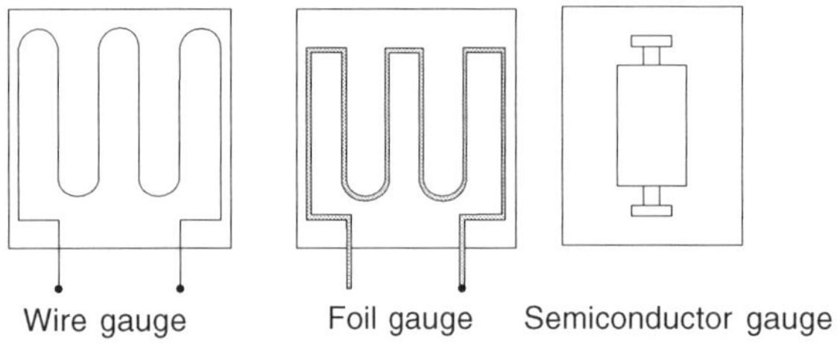

Fig. 4.29 *Types of resistance strain gauges*

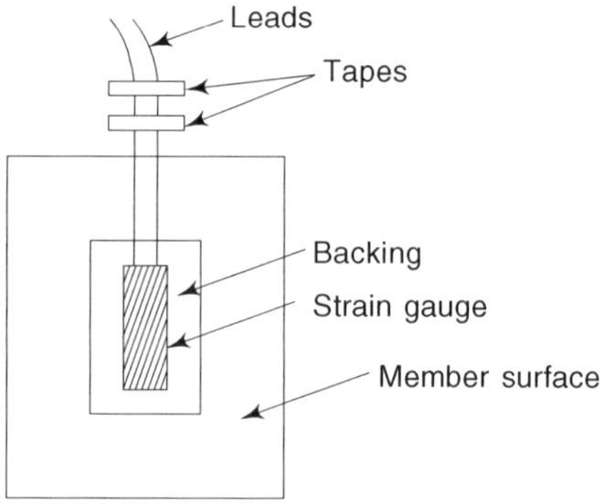

Fig. 4.30 *Strain gauge in bonded position*

Gauges made of copper–nickel alloys have a gauge factor of 2–3 while semiconductor gauges have gauge factors of 100–200.

Table 4.3 gives information about wire backing materials and adhesives:

Table 4.3

Gauge backing material	*Adhesive*	*Wire materials*	*Remarks*
Paper or silk	Nitrocellulose	Cu–Ni alloy	Useful up to 60°C
Bakelite	Epoxy	Cu–Ni alloy	Useful up to 200 °C
Glass weave	Ceramic cement	Ni–Cr alloy	Useful up to 400°C

Gauge current is usually limited to 10–30 mA, depending on the test duration, in order to prevent wire damage. Bakelite base gauges can withstand somewhat higher values of current.

Care has to be taken while bonding the gauges. The surface of the member has to be thoroughly cleaned. Later, the adhesive has to be applied and allowed to set, according to manufacturers' instructions. Then, the connecting leads are soldered to the gauge and securely fixed to the test member, as in Fig. 4.30. Finally, the gauge continuity and insulation resistance are checked.

In the subsequent figures, the strain gauges shown in Fig. 4.29 are represented by rectangles, with the longer side being along the length of the wire, foil or semiconductor.

Resistance strain gauge bridges The reistance strain gauge is normally made a part of a Wheatstone bridge so that the change in its resistance due to strain can either be measured or made to give an output which can be displayed or recorded. There are two types of bridge arrangements used for the purpose:

1. balanced bridge, and
2. unbalanced bridge.

In the balanced bridge arrangement (Fig. 4.31), strain gauge resistance R_1, shown as a rectangle in Fig. 4.31, forms one arm of the Wheatstone bridge while the remaining arms have resistance R_2, R_3 and R_4 as part of the bridge. The bridge is excited by a dc source, with voltage E.

R_G is the resistance of the galvanometer. The bridge is said to be balanced when there is no current through the galvanometer.

The condition for balance is well known and is

$$\frac{R_1}{R_4} = \frac{R_2}{R_3} \tag{4.22}$$

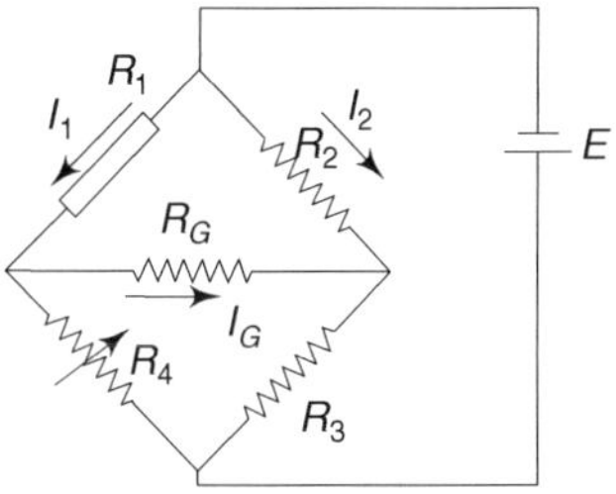

Fig. 4.31 *Balanced strain gauge*

If R_1 changes due to strain, the bridge, which may be initially balanced, becomes unbalanced. This may be balanced again by changing R_4 (or R_2), the change in which can be measured and used to indicate change in R_1, which is related to the strain. This technique can only be used for measuring static strains.

In the unbalanced bridge arrangement shown in Fig. 4.32, the current through the galvanometer or the voltage drop across it, is used to indicate the strain in the strain gauge. This can be used to measure dynamic as well as static strains. In order to derive the relation between the output unbalanced current and change in gauge resistance R_1, Eqs. (4.23)–(4.25) can be written, for the circuit in Fig. 4.32, using Kirchhoff's laws.

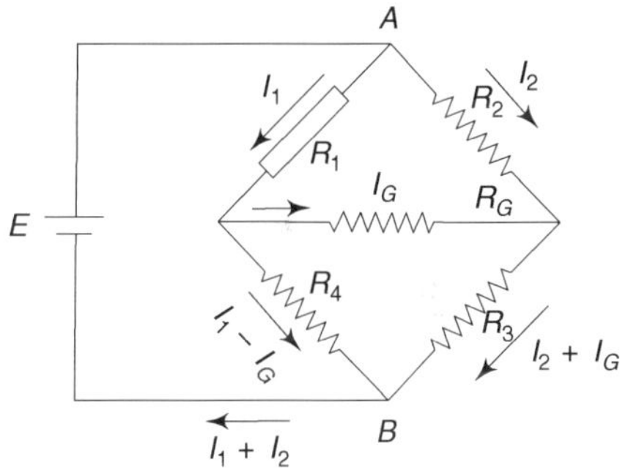

Fig. 4.32 *Unbalanced strain gauge bridge*

$$I_1R_1 + R_4(I_1 - I_G) = E \tag{4.23}$$

$$I_1R_1 + I_GR_G - I_2R_2 = 0 \tag{4.24}$$

$$I_GR_G + (I_2 + I_G)R_3 - (I_1 - I_G)\,R_4 = 0 \tag{4.25}$$

I_1, I_2 and I_G are currents as shown in Fig. 4.31. Solving Eqs. (4.23)–(4.25) for I_G, we get

$$I_G = \frac{E(R_2R_4 - R_1R_3)}{R_2(R_1 + R_4)(R_G + R_3 + R_4) + R_1R_3R_4 - R_2R_4^2 + R_GR_3(R_1 + R_4)} \tag{4.26}$$

It is seen from Eq. (4.26) that $I_G = 0$, when $R_2R_4 = R_1R_3$. This corresponds to Eq. (4.22).

From Eq. (4.26), the value of unbalanced current I_G can be calculated for any change in strain gauge resistance R_1. Taking a special case when $R_1 = R_2 = R_3 = R_4$ and if R_1 changes to $R_1 + \Delta R_1$, it can be seen from Eq. (4.26) that

$$I_G = \frac{-E\Delta R_1}{4R_1(R_1 + R_G)} \tag{4.27}$$

$$= \frac{-EF\varepsilon_1}{4(R_1 + R_G)} \tag{4.28}$$

where ε_1 is the strain which causes ΔR_1, and

$$\varepsilon_1 = \frac{\Delta R_1}{R_1 F}$$

F being the gauge factor of the strain gauge. Voltage output across $R_G = E_o = I_G R_G$.

$$= \frac{-E\varepsilon_1 R_G F}{4(R_1 + R_G)} \tag{4.29}$$

$$\text{Open circuit voltage output} = E_o' = \frac{-EF\varepsilon_1}{4} \tag{4.30}$$

This is obtained by letting $R_G \to \infty$ in Eq. (4.29).

Ratio E_o / E_o' is plotted against R_G/R_1 in Fig. 4.33. It is seen that R_G should be at least 100 times the strain gauge resistance R_1, so that maximum possible output E_o is obtained.

When more than one arm in the Wheatstone bridge contains strain gauges and their resistances change due to strains, the output is due to the combined effect of these changes, e.g. in Fig. 4.31, if R_2 changes to $R_2 + \Delta R_2$, then for the special case discussed, viz. for

$$R_1 = R_2 = R_3 = R_4$$

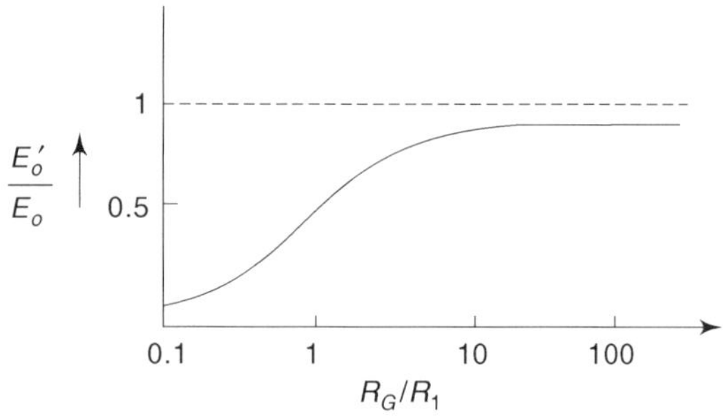

Fig. 4.33 *Plot of E_o'/E_o against R_G/R_1*

Equation (4.27) becomes

$$I_G = \frac{E\Delta R_2}{4R_2(R_1 + R_G)}$$

Thus, if R_1 changes to $R_1 + \Delta R_1$ and R_2 to $R_2 + \Delta R_2$, the combined effect results in

$$I_G = \frac{-E}{4(R_1 + R_G)}\left[\frac{\Delta R_1}{R_1} - \frac{\Delta R_2}{R_2}\right] \tag{4.31}$$

Similarly, if all the four arms have strain gauge whose resistances change due to strains, it can be shown that

$$I_G = \frac{E}{4(R_1 + R_G)}\left[-\frac{\Delta R_1}{R_1} + \frac{\Delta R_2}{R_2} - \frac{\Delta R_3}{R_3} + \frac{\Delta R_4}{R_4}\right] \tag{4.32}$$

Thus, it may be seen that the output current will be increased if the resistance change ΔR_1 is of opposite nature to ΔR_2 and ΔR_4 and of similar nature to ΔR_3. Thus, adjacent arms of the bridge should have strains of opposite nature, in order to have large output or sensitivity of the bridge. This can be taken care of while arranging the strain gauges on a member in such a way that if R_1 has tensile strain say, R_2 should have compressive strain, R_3 tensile strain and R_4 compressive. It is thus possible to define the signal enhancement factor of the bridge as the ratio of the maximum output due to changes in various strain gauges of the bridge to the maximum output obtainable with the use of only one strain gauge on the member. In other words, more than one active gauge (subjected to strain), suitably arranged, can lead to increased sensitivity or single enhancement.

Excitation of the bridges may be by dc or ac power source, circuits for which are given in Figs. 4.34 and 4.35, respectively. In each case, an amplifier is required—a direct coupled one in Fig. 4.34 and an ac amplifier in Fig. 4.35. In the latter case, the signal is amplitude modulated as shown in Fig. 4.35 (b) so that a demodulator is needed to filter the carrier frequency. The output being recorded is identical to the strain, being measured. Of the two types of bridges, ac bridges are free from drift and unwanted noise signals normally present in dc bridges. However, ac bridges are more expensive and their high frequency response is limited to only about one-fifth of the ac carrier frequency used.

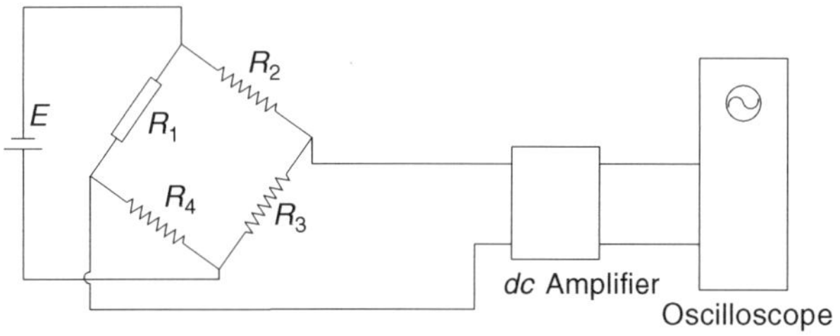

Fig. 4.34 *dc bridge*

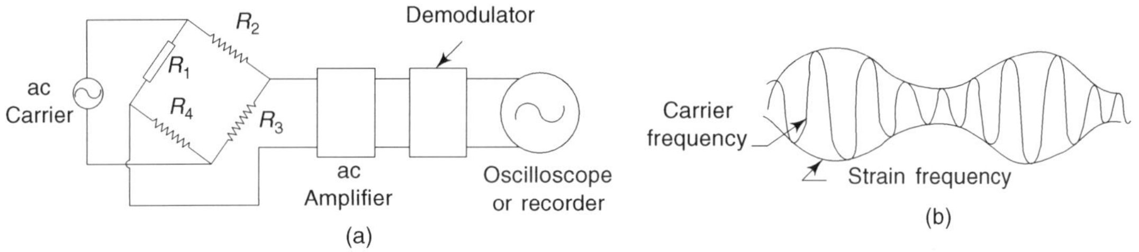

Fig. 4.35 *ac bridge*

Temperature compensation In addition to strain, temperature change would also change the resistance of a resistance strain gauge. Since it is inconvenient to calculate and apply temperature correction, temperature compensation is made in the experimental set-up itself. This is done by (a) using a dummy gauge or by (b) using more than one active gauge, with proper arrangement of the gauges.

In case of (a), as shown in Fig. 4.36, gauge resistance R_2 equal to R_1 is bonded on a piece of same material as the test specimen to which load is applied. However, the piece of material having R_2 is not strained but is subjected to the same temperature change as the test specimen.

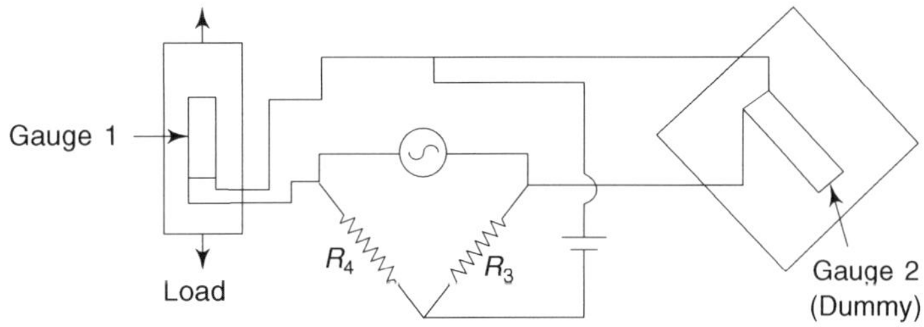

Fig. 4.36 *Temperature compensation with a dummy gauge*

R_1 is thus the active gauge, subjected to strain as well as temperature change while R_2 is the dummy gauge, subjected only to temperature change. Since R_1 and R_2 from adjacent arms of the bridge, the output due to temperature change is zero as both R_1 and R_2 change identically due to temperature.

In case of (b), gauge R_2 need not be a dummy gauge but may be suitably arranged on the test specimen so that it has strain of opposite nature to that in R_1. In this case, the system would be sensitive and temperature compensation would also be ensured. In fact, if all the four gauges of the bridge are active, as shown in Fig. 4.37, with adjacent arms having strains of signs shown, viz. opposite in nature, the signal enhancement would be maximum and temperature compensation would also be ensured. Of course, initial resistances of all the arms have to be equal.

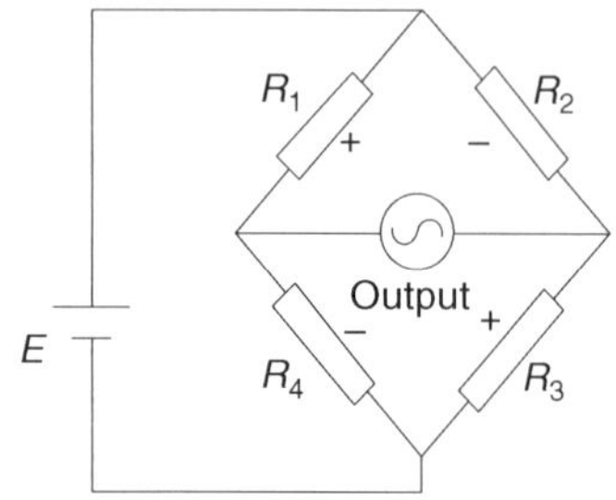

Fig. 4.37 *Strain gauge arrangement with four active gauges*

Temperature compensation is necessary when static strains are to be measured. Since a change of temperature causes only a drift in the output signals as such changes are normally slow, for measurement of dynamic strains, temperature compensation is not necessary.

Strain gauge arrangement The following two factors are kept in mind while deciding the arrangements of strain gauges on elastic members, for measuring various physical variables:

1. high sensitivity, and
2. temperature compensation.

The arrangement shown in Fig. 4.37 would be satisfactory from both considerations.

Taking the case of measuring direct force P applied to a member, the possible arrangements of gauges are shown in Figs. 4.37 (a)–(d). In the bridges, strain gauges are shown as rectangles. In Fig. 4.38 (a), one gauge R_1 is bonded on the elastic member to measure axial strain in the member. The system is not compensated for temperature and output is due to change in R_1 only.

In the arrangement of Fig. 4.38 (b), the two active gauges R_1 and R_2 are placed at right angles—Poisson's arrangement—so that R_1 changes due to axial tensile strain while R_2 changes due to transverse compressive strain in the member, the latter being ν times the former, where ν denotes Poisson's ratio. The signal enhancement factor is $(1 + \nu)$, since the two resistances R_1 and R_2 have strains of opposite nature. The system is compensated for temperature since temperature changes would affect R_1 and R_2 identically, thereby producing no output. The arrangement of Fig. 4.38 (c) in which both R_1 and R_3 are subjected to axial tensile strains of same amount, and in which R_1 and R_3 form opposite arms of the bridge, would give a signal enchancement factor of 2 and would not be compensated for temperature. Such an arrangement would need two dummy gauges, to ensure temperature compensation.

The arrangement of Fig. 4.38(d), in which four active gauges are used, with R_2 and R_4 arranged at right angles to R_1 and R_3, would give a signal enhancement factor of $2\ (1 + \nu)$ and would be compensated for temperature variation.

Taking another example force P can be measured using an elastic cantilever, as shown in Fig. 4.39. The strain gauges are bonded at the root of the cantilever, where bending strains are maximum. R_1 and R_3 have tensile strains while R_2 and R_4 have compressive ones. The bridge arrangement of Fig. 4.39(b) would give a signal enhancement factor of four and would ensure temperature compensation. An alternative arrangement is shown in Fig. 4.40, in which R_1 and R_2 are bonded according to Poisson's arrangement. This also applies to R_3 and R_4. With the bridge arrangement of Fig. 4.40(b), the adjacent arms have strains of opposite nature, though these are unequal in magnitude, unlike the arrangement of Fig. 4.39. The arrangement of Fig. 4.40, would give a signal enhancement factor of $2(1 + \nu)$, ν being Poisson's ratio and temperature compensation would be ensured.

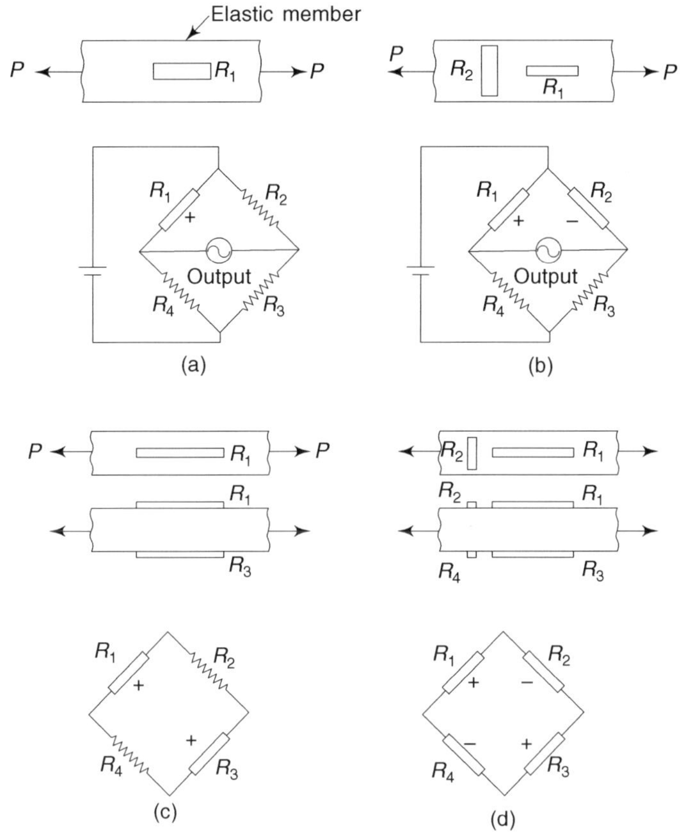

Fig. 4.38 *Possible arrangements for measurement of force P*

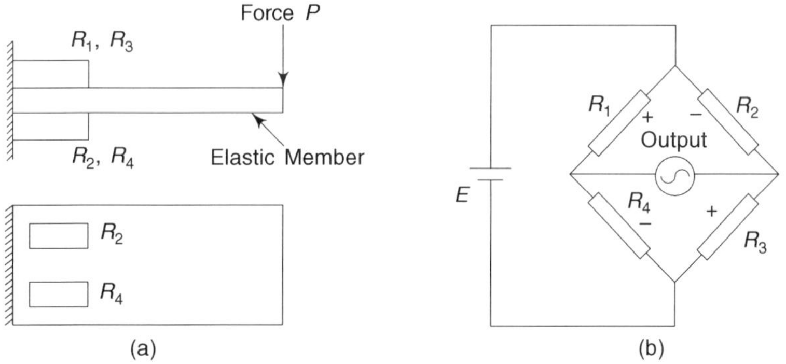

Fig. 4.39 *Strain gauge arrangement for measuring force P*

Problem 4.4 *A load cell (Fig. 4.41) is formed of a hollow steel cylinder loaded axially. The four strain gauges are so connected as to enhance the signal and compensate for temperature variation. The load cell has a cross-sectional area of 2 cm^2. Young's modulus of steel is 2.07×10^{11} N/m^2 and Poisson's ratio 0.3.*

Strain gauge resistance = 1000 Ω

Gauge factor = 2.1.

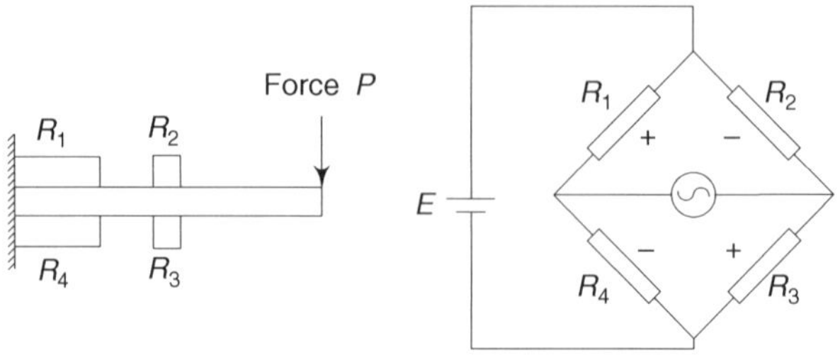

Fig. 4.40 *Alternative arrangement*

The current in each strain gauge is limited to 20 mA. Calculate (a) the bridge supply voltage and (b) the current in the detector arm if this consists of a microammeter of 500 Ω resistance, when the load cell is subjected to a force of 10^5 N.

Solution The arrangement of strain gauges is as shown in Fig. 4.41. R_1 and R_3 measure axial strains while R_2 and R_4 measure the circumferential strains, which are opposite in nature to the axial ones.

$$\text{Signal enhancement factor} = 2\,(1 + \nu)$$
$$= 2.6$$

Strain gauge resistance = 1000 Ω
Gauge factor = 2.1

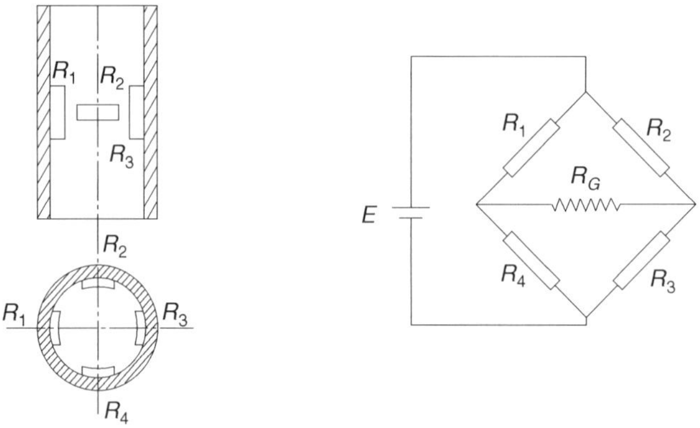

Fig. 4.41 *Figure for Problem 4.4*

For the initially balanced system, battery voltage

$$E = \frac{20}{1000}\,(R_1 + R_4)$$

Substituting $R_1 = R_4 = 1000\ \Omega$

$$E = 40\ \text{V}$$

or axial strain in the cylinder $= \dfrac{10^5 \times 10^4}{2.07 \times 10^{11} \times 2}$

$$= 2.415 \times 10^{-3}$$

If only R_1 were to change due to above strain, current in R_G, viz.

$$I_G = \frac{EF}{4(R_1 + R_G)}$$

$$= \frac{40 \times 2.1 \times 2.415 \times 10^{-3}}{4(1000 + 500)}$$

$$= 3.38 \times 10^{-5} \text{ A}$$

Since all the four resistances change, current

$$I_G = 2.6 \times 3.38 \times 10^{-5}$$
$$= 8.79 \times 10^{-5} \text{ A}$$

Strain Gauge Rosettes If more than one strain gauge is mounted in an area, with a view to finding principal strains, the arrangement is called strain gauge rosettes. In case the principal strain directions are known, for the case of plane stress, two-gauge rosette can be used. For the general case of plane stress, when such directions are not known, a three-gauge rosette is normally employed. Sometimes, a fourth direction gauge is used as a check. Two types of three-gauge rosettes, viz. Rectangular and Delta types and one type of four-gauge rosette, viz. T-Delta types are shown in Fig. 4.42.

Theory Since the direction of principal strains on the test object are generally not known, the gauges

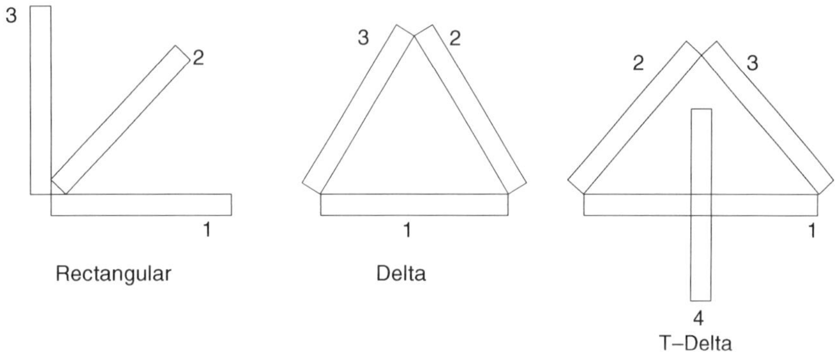

Fig. 4.42 *Types of rosettes*

are mounted in any direction in practice. From the measured strains, strains in any direction can be computed as shown below. The relation between strain in any direction and principal strains and that between principal strains and principal stresses are derivable from solid mechanics as in Appendix A-4. Thus a relation between measured strains and principal stresses can be obtained. At any location on a test object in plane strain, there may be deformations due to direct strains, ε_x, ε_y and shear strain γ_{xy}. The relation for direct strain, ε_θ measured by a strain gauge at angle θ to OX may be obtained as below, as in Figs. 4.43–4.45.

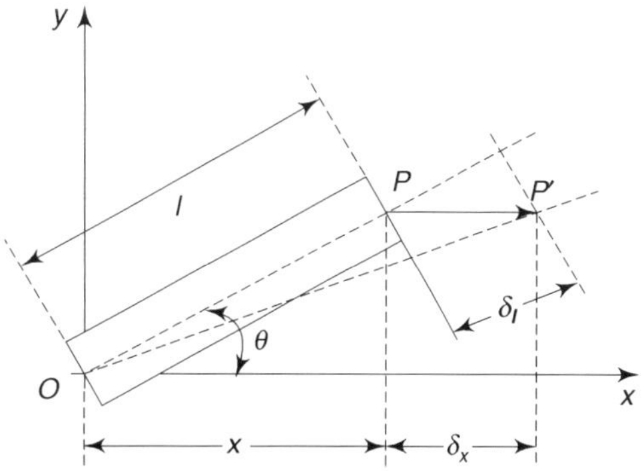

Fig. 4.43 *Relation between ε_θ and ε_x*

From Fig. 4.43, after deformation in X direction, OP increases to OP'. The gauge with length l along OP is

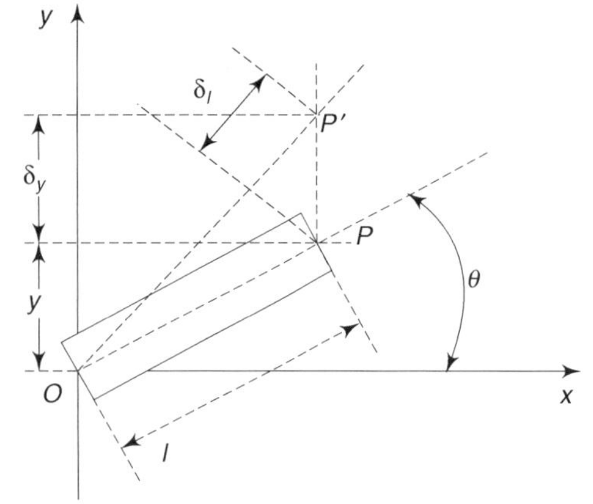

Fig. 4.44 *Relation between ε_θ and ε_y*

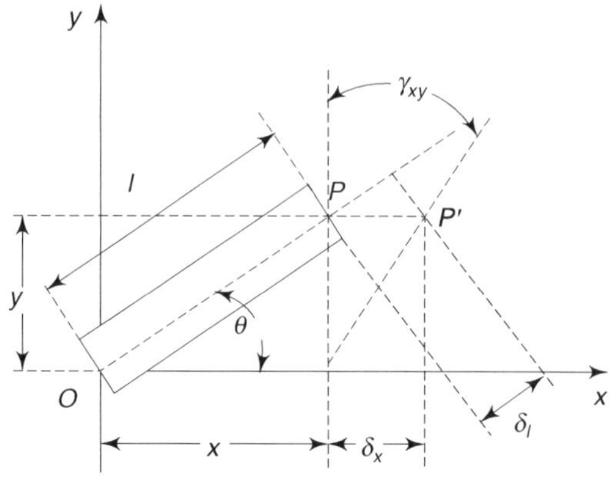

Fig. 4.45 *Relation between ε_θ and γ_{xy}*

along OP' after deformation.

Strain $\varepsilon_x = \delta_x/x$

Strain ε_θ measured by the gauge = $\delta l/l$

Since $\delta_l = \delta_x \cos\theta$

and $l = x/\cos\theta$

we get $$\varepsilon_\theta = \frac{\delta l}{l} = \varepsilon_x \cos^2\theta \quad (4.33)$$

Similarly from Fig. 4.44, $\varepsilon_\theta = \varepsilon_y \sin^2\theta$ (4.34)

In Fig. 4.45, if a shear strain, γ_{xy} is applied, the strain in the gauge would be

$$\varepsilon_\theta = \delta_l/l.$$

Assuming all deformations as small.

Since $\delta_l = \delta_x \cos\theta,\ l = \dfrac{y}{\sin\theta}$

and $\delta_x = y \tan\gamma_{xy} \approx y\,\gamma_{xy}$,

we get, $$\varepsilon_\theta = \frac{\delta l}{l} = \gamma_{xy}\sin\theta\cos\theta \quad (4.35)$$

If strains ε_x, ε_y and γ_{xy} act simultaneously,

$$\varepsilon_\theta = \varepsilon_x\cos^2\theta + \varepsilon_y\sin^2\theta + \gamma_{xy}\sin\theta\cos\theta \quad (4.36)$$

Equation (4.36) can be easily rewritten as:

$$\varepsilon_\theta = (\varepsilon_x + \varepsilon_y)/2 + [(\varepsilon_x - \varepsilon_y)/2]\,[\cos 2\theta] + (\gamma_{xy}/2)\sin 2\theta \quad (4.37)$$

Since ε_θ is measurable, in order to find three unknowns viz. ε_x, ε_y and γ_{xy}, we need three gauges 1, 2, and 3. Thus,

$$\varepsilon_1 = (\varepsilon_x + \varepsilon_y)/2 + (\varepsilon_x - \varepsilon_y)/2\cos 2\theta_1 + (\gamma_{xy}/2)\sin 2\theta_1 \quad (4.38)$$

$$\varepsilon_2 = (\varepsilon_x + \varepsilon_y)/2 + (\varepsilon_x - \varepsilon_y)/2\cos 2\theta_2 + (\gamma_{xy}/2)\sin 2\theta_2 \quad (4.39)$$

$$\varepsilon_3 = (\varepsilon_x + \varepsilon_y)/2 + (\varepsilon_x - \varepsilon_y)/2\cos 2\theta_3 + (\gamma_{xy}/2)\sin 2\theta_3 \quad (4.40)$$

Relations for rosettes

Rectangular rosettes Angles for the rectangular rosette, as given in Fig. 4.42 are: $\theta_1 = 0$, $\theta_2 = 45°$ and $\theta_3 = 90°$. Using Eq. (4.37) strains ε_1, ε_2 and ε_3 in gauges, 1, 2, and 3 respectively, are:

$$\varepsilon_1 = \varepsilon_x$$
$$\varepsilon_2 = (\varepsilon_x + \varepsilon_y)/2 + \gamma_{xy}/2$$
$$\varepsilon_3 = \varepsilon_y$$

Solving for ε_x, ε_y and γ_{xy}, we get,

$$\varepsilon_x = \varepsilon_1$$
$$\varepsilon_y = \varepsilon_3$$
$$\gamma_{xy} = 2\varepsilon_2 - (\varepsilon_x' + \varepsilon_3)$$

Substituting in Eqs. (A1), (A2) and (A3) of Appendix A-3 we can write expressions for angle θ_p, principal strains ε_{max} and ε_{min} respectively, in terms of ε_1, ε_2, and ε_3.

Using Eqs. (A4.4) and (A4.5) of Appendix A-4 we get expressions for principal stresses, which can be shown to be

$$\sigma_{max}, \sigma_{min} = E(\varepsilon_1 + \varepsilon_3)/[2(1-\mu)] \pm [E/2^{1/2}(1+\mu))]\,[\varepsilon_1 - \varepsilon_2)^2 + (\varepsilon_2 - \varepsilon_3)^2]^{1/2} \quad (4.41)$$

Similarly, maximum shear stress τ_{max} can be shown to be

$$\tau_{max} = E/2^{1/2}(1+\mu)\,[(\varepsilon_1 - \varepsilon_2)^2 + (\varepsilon_2 - \varepsilon_3)^2]^{1/2}$$

Delta rosette Referring to Fig. 4.42, $\theta_1 = 0$, $\theta_2 = 60°$ and $\theta_3 = 120°$. If strains in gauges 1, 2 and 3 are ε_1, ε_2 and ε_3 respectively, following the procedure for the previous case, we get expressions for ε_{max}, ε_{min} and θ_p as

$$\varepsilon_{max,\,min} = (\varepsilon_1 + \varepsilon_2 + \varepsilon_3)/3 \pm 2^{1/2}/3[(\varepsilon_1 - \varepsilon_2)^2 + (\varepsilon_3 - \varepsilon_1)^2]^{1/2} \quad (4.42)$$
$$\theta_p = 0.5\tan^{-1} 3^{1/2}(\varepsilon_3 - \varepsilon_2)/(2\varepsilon_1 - \varepsilon_2 - \varepsilon_3)$$

Angle θ_p will be in the first quadrant (i.e. 0–90°) when $\varepsilon_3 > \varepsilon_2$ and in the second quadrant (i.e. 90–180°) when $\varepsilon_2 > \varepsilon_3$.

Expression for σ_{max} and σ_{min} can be written using Eqs. A4 and A5 of Appendix A-3.

Maximum shear stress can be shown to be

$$\tau_{max} = 2^{1/2}E/3(1+\mu)\,[\varepsilon_1 - \varepsilon_2)^2 + (\varepsilon_2 - \varepsilon_3)^2 + (\varepsilon_3 - \varepsilon_1)^2]^{1/2}$$

Problem 4.5 *A rectangular rosette is bonded on to a steel plate and the three measured strains are*

$$\varepsilon_1 = 300\ \mu m/m,\ \varepsilon_2 = 200\ \mu m/m,\ \varepsilon_3 = -150\ \mu m/m.$$

Calculate the principal stresses and give their locations. For steel, $E = 2.3 \times 10^{11}$ N/m², $\mu = 0.3$.

Solution Using Eq. (4.41)

$$\sigma_{max,\,min} = E(\varepsilon_1 + \varepsilon_3)/[2(1-\mu)] \pm [E/(2^{1/2}(1+\mu)][(\varepsilon_1 - \varepsilon_2)^2 + (\varepsilon_2 - \varepsilon_3)^2]^{1/2},$$

we get,
$$\sigma_{max} = 7.03 \times 10^7\ \text{N/m}^2 \text{ and}$$
$$\sigma_{min} = -2.1 \times 10^7\ \text{N/m}^2$$

To find the angle θ of the maximum principal stress with the *X*-axis, we use:

$$\tan 2\theta = (2\varepsilon_2 - \varepsilon_1 - \varepsilon_3)/(\varepsilon_1 - \varepsilon_3)$$
$$= 0.55$$
$$\theta = 14.5° \text{ or } 104.5°$$

Since $\varepsilon_2 > (\varepsilon_1 + \varepsilon_3)/2$, $\theta = 14.5°$ for maximum principal stress. Further θ will be 104.5° for the minimum principal stress.

Balancing of Bridges The bridges need to be balanced initially since all the four arm resistances may not in practice be equal to the desired values. Any of the following methods may be used for this purpose.

1. Series balancing method, using an apex resistance, as shown in Fig. 4.46. Due to movement of apex resistance contact point, if R_2 increases by say r, R_3 decreases by the same amount and vice versa.
2. Parallel balancing method, in which motion of contact point P can balance the bridge, if desired (Fig. 4.47).

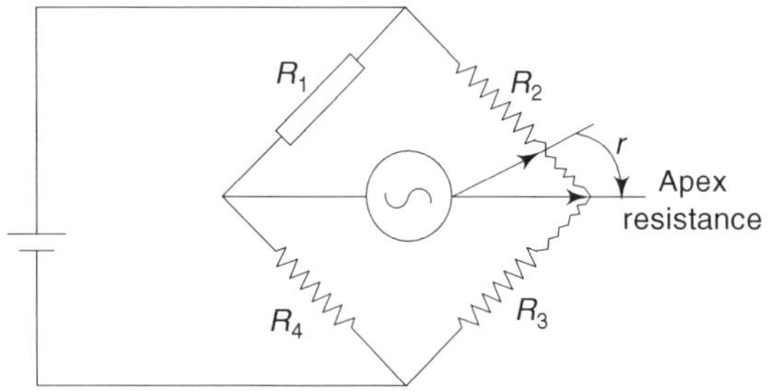

Fig. 4.46 *Use of apex resistance for balancing and calibration*

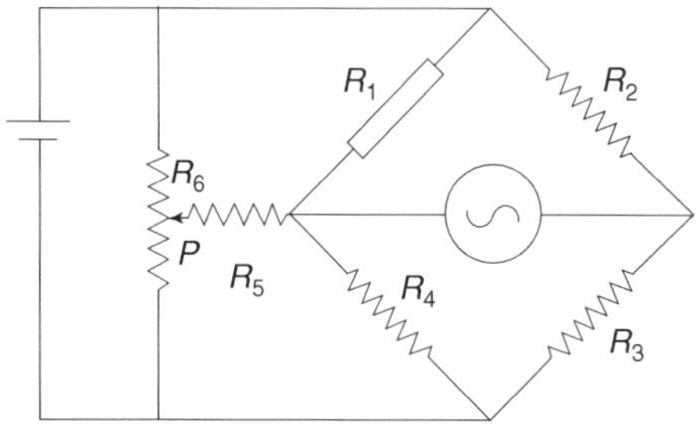

Fig. 4.47 *Parallel balancing arrangement*

Calibration For an unbalanced bridge arrangement, it becomes necessary to relate, in practice, the bridge output with the strain in the strain gauge. Either electrical or mechanical methods may be used, the former are usually built in the strain gauge bridges available commercially.

One of the electrical methods in which a change in arm resistance may be simulated is by using the apex resistor of Fig. 4.46. It is possible to calibrate the apex resistor in terms of strain ε in one of the gauges, say R_1. This is possible, shown as follows.

Referring to Fig. 4.46, if the bridge is initially balanced,

$$\frac{R_1}{R_4} = \frac{R_2}{R_3} = 1, \text{ say} \tag{4.43}$$

If the apex resistance is changed so that R_2 is increased to $R_2 + r$ and R_3 changes to $R_3 - r$, the bridge would be unbalanced. The value of $\Delta\, R_1/R_1$ corresponding to the above change would be

$$\frac{R_1 + \Delta R_1}{R_4} = \frac{R_2 + r}{R_3 - r} \tag{4.44}$$

Putting $x = r/R_3$, it is seen from Eqs. (4.43) and (4.44) that

$$\frac{\Delta R_1}{R_1} = \frac{2x}{1 - x} \tag{4.35}$$

Thus, the apex resistor can be calibrated in terms of $\Delta R_1/R_1$ or strain, if the gauge factor is known.

The second electrical method of calibration is by using a shunt resistor, as in Fig. 4.48. If a shunt resistor R_c is connected parallel to the strain gauge resistance R_1, the change ΔR_1 is

$$\Delta R_1 = \left(R_1 - \frac{R_1 R_c}{R_1 + R_c} \right)$$

or we would be simulating.

$$\frac{\Delta R_1}{R_1} = \frac{R_1}{R_1 + R_c} \approx \frac{R_1}{R_c}, \text{ if } R_c \gg R_1 \qquad (4.46)$$

which is usually the case, for simulating small values of $\Delta R_1/R_1$. Thus, the change in output as a result of connecting R_c parallel to any resistance, say R_1, can be measured and related to $\Delta R_1/R_1$ or the strain.

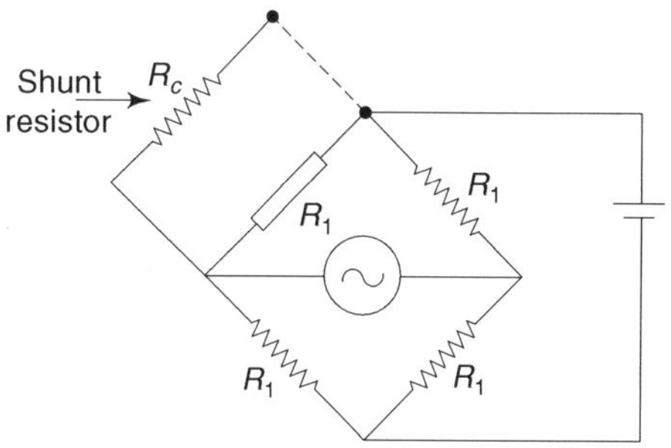

Fig. 4.48 *Shunt resistor method of calibration*

In the mechanical method of calibration, for static calibration, wherever possible, dead weights can be used directly and strain can be calculated and related to the output. For dynamic calibration, two methods are shown in Fig. 4.49 and Fig. 4.50, the former using a rotating eccentric and the latter a vibrator. In Fig. 4.49, a cantilever member is used as the elastic member. The actual strain can be calculated and the calibration factor relating output and strain can be computed. In Fig. 4.50, the frequency and amplitude of the vibrator can be changed. Due to the mass m, the amplitude of sinusoidal force applied to the elastic strip is $m\omega^2 x_o$, ω being the circular frequency of vibration and x_0 the amplitude of mass displacement, which needs to be measured by any of the motion transducers described earlier. Against this, the bridge output can be measured. In Fig. 4.50 both R_{1A} and R_{1B} are arranged in one arm of the Wheatstone bridge. This has been done to ensure the cancellation of any bending effect in the elastic strip. Bending would introduce similar changes of opposite nature in R_{1A} and R_{1B}, keeping the total arm resistance unchanged. Thus, the gauges would only respond to direct strains due to the dynamic force applied due to motion of m.

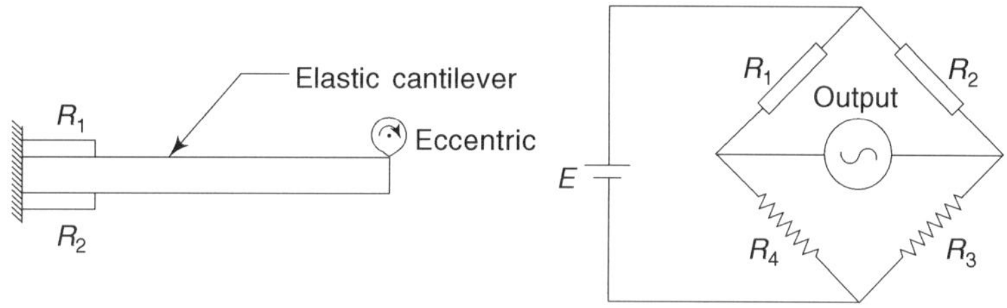

Fig. 4.49 *Dynamic calibration of strain gauges using an eccentric*

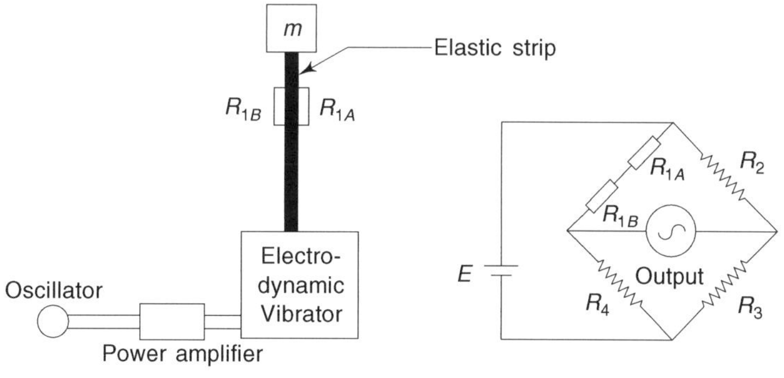

Fig. 4.50 *Dynamic calibration of strain gauges using a vibrator*

Problem 4.6 *A tensile force link is to be made by mounting two strain gauges back-to-back on a thin aluminium strip, as shown in Fig. 4.51. Four 120 Ω gauges are used, two on the strip and the other two as dummy gauges. Gauge factor = 2.2. Maximum force applied = 400 N. Maximum current through the gauges is limited to 25 mA. Young's modulus of strip material = 6.9 × 10^{10} N/m^2.*

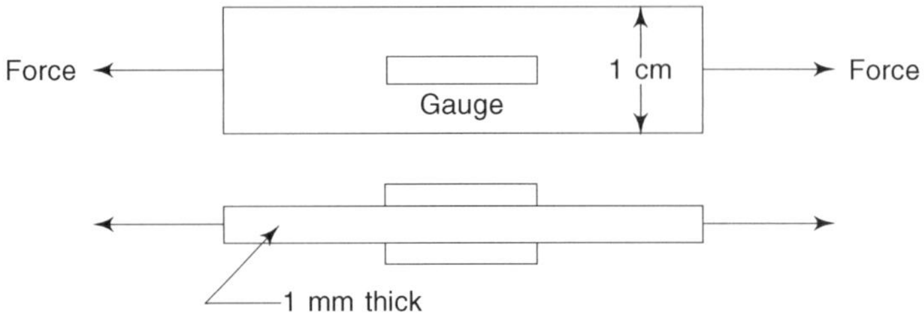

Fig. 4.51 *Figure for Problem 4.6*

(a) *If the output from the bridge is connected to an oscilloscope, which is set at 10 mV/cm of trace deflection, find the deflection of the trace at maximum load. The input impedance of the oscilloscope may be taken as infinity.*
(b) *When a 100 kW resistance is placed in parallel with one of the gauges, the trace shift is 1 cm. Find the trace deflection for the maximum load, corresponding to this setting of the oscilloscope.*

Solution Figure 4.52 shows the bridge arrangement. R_1 and R_3 are active gauges while R_2 and R_4 are dummy gauges. The signal enhancement factor is 2.

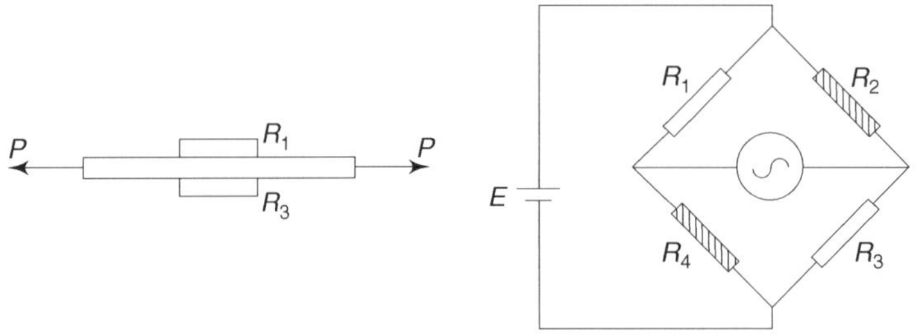

Fig. 4.52 *Bridge arrangement*

(a) Using Eq. (4.30), for an infinite impedance at output,

$$\text{Output voltage} = \frac{EF\varepsilon}{4}$$

for a change in resistance of one arm. Since the signal enhancement factor due to change of R_1 and R_3 is 2,

$$\text{Output voltage} = \frac{EF\varepsilon}{4} \times 2$$

$$= \frac{EF\varepsilon}{2} \tag{4.47}$$

$$\text{Battery voltage } E = \frac{25}{1000}(R_1 + R_4)$$

$$= \frac{25}{1000}(120 + 120) = 6 \text{ V}$$

$$F = 2.2$$

Area of strip $A = 10 \text{ mm}^2$

Strain ε in the strip $= \dfrac{P}{AY}$, Force $P = 400$ N, Young's modulus $Y = 6.9 \times 10^4$ N/mm^2

Y being Young's modulus.

Thus $$\varepsilon = \frac{400}{10(6.9 \times 10^4)} = 5.8 \times 10^{-4}$$

Substitution in Eq. (4.47) gives,

output voltage = 3.83 mV

or trace deflection $= \dfrac{3.83}{10} = 0.383$ cm

(b) Using Eq. (4.47),

$$\frac{\Delta R_1}{R_1} = \frac{R_1}{R_1 + R_c} = \frac{120}{120 + 10^5} = 1.199 \times 10^{-3}$$

This gives a trace shift of 1 cm.

$\dfrac{\Delta R_1}{R_1}$ due to strain ε of 5.8×10^{-4}, is equal to $2.2 \times 5.8 \times 10^{-4} = 1.276 \times 10^{-3}$

Since during calibration, the resistance of only one arm changes while during application of load, two arms change their resistances,

So, $$\text{Trace shift} = \frac{2 \times 1.276 \times 10^{-3}}{1.199 \times 10^{-3}} = 2.13 \text{ cm}$$

Problem 4.7 *Figure 4.53 shows a steel mass mounted on a force link. A piezo-electric accelerometer is mounted on top of the steel mass. The force link is of aluminium, with cross-sectional area 6 cm². Four 400 Ω gauges are bonded longitudinally at 90°. These form arms A and C while arms B and D have temperature compensating gauges. Gauge factor = 2, steel mass = 5 kg. Accelerometer is connected to one of the channels of an oscilloscope, with gain 50 mV/cm. The force link is connected to the second channel of the oscilloscope. A 10^6 Ω resistor, connected in parallel with arm A*

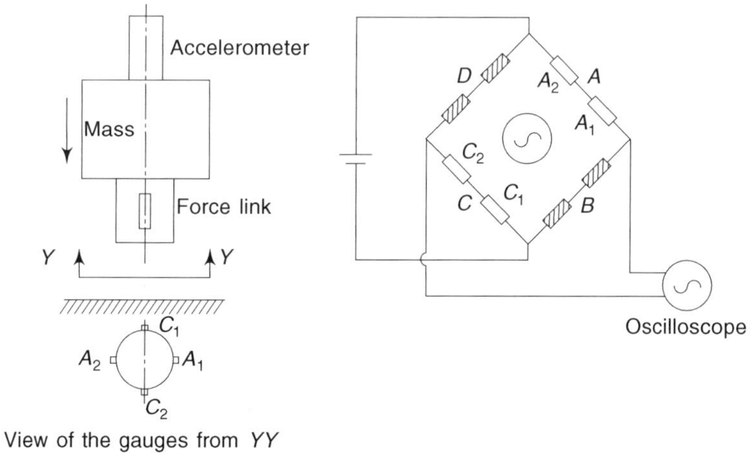

Fig. 4.53 *Figure for Problem 4.7*

produces a trace deflection of 5 cm. If during the experiment, the peak deflection of the force link trace is 3 cm and that of the accelerometer trace is 3.8 cm, find the accelerometer sensitivity in mV/g. Young's modulus of aluminium = 6.9 × 10^{10} N/m^2.

Solution Resistance of each arm

$$R_A = R_B = R_C = R_D = 800\ \Omega$$

In the force link channel,

$$\frac{\Delta R_A}{R_A} = \frac{R_A}{R_A + R_{\text{calib}}} = \frac{800}{800 + 10^6}$$

$$= 7.99 \times 10^{-4}$$

This gives a trace deflection of 5 cm.

During experimentation, the trace deflection is 3 cm. This is due to two active arms, viz. A and C. Thus, output due to change in each arm resistance due to strain = 3/2 = 1.5 cm.

$$F = 2.0$$

Corresponding strain in the force link

$$= \frac{1.5}{5} \times \frac{7.99 \times 10^{-4}}{2} = 1.1985 \times 10^{-4}$$

$$\text{Force applied} = 1.1985 \times 10^{-4} \times \frac{6}{10^4} \times 6.9 \times 10^{10}$$

$$= 4961.8 \text{ N}$$

or

$$\text{acceleration applied} = \frac{4961.8}{5} = 992.36 \text{ m/s}^2$$

$$= 101.2 \text{ g}$$

Output voltage from the accelerometer channel

$$= 50 \times 3.8 = 190 \text{ mV}$$

Thus, accelerometer sensitivity = 190/101.2 = 1.88 mV/g.

Ionisation Transducer This works on the principle of development of voltage across two electrodes placed in an ionised gas, the magnitude of which depends on the electrode spacing and state of balance, which can change due to the motion to be measured.

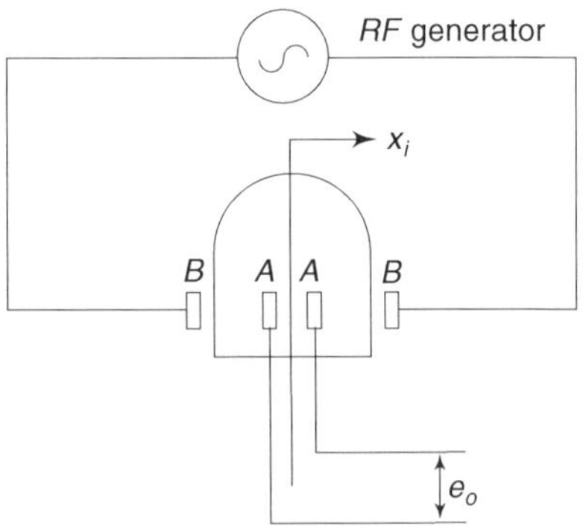

Fig. 4.54 *Moving tube type ionisation transducer*

The transducer consists of a glass tube (Fig. 4.54) containing gas under reduced pressure. A dc voltage is developed across the internal electrodes A, when the tube is subjected to an electric field due to external electrodes B, connected to a radio frequency (RF) voltage source. The gas in the tube gets ionised and the dc voltage produced depends on the electrode spacing, being zero at null position. As in Fig. 4.54, the motion x_i of the tube relative to the fixed external electrodes varies the output voltage. The balance between the electrodes may also be changed as in Fig. 4.55 by changing either capacitance C_1 or C_2 (C_1 in Fig. 4.55 shown), due to the motion x_i to be measured. This produces an output e_o.

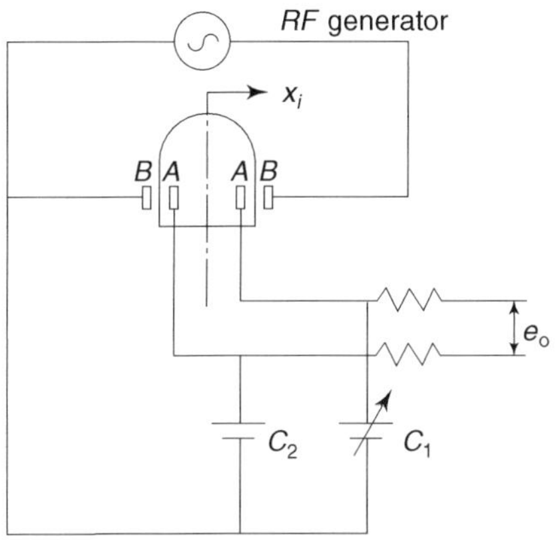

Fig. 4.55 *Ionisation transducer used with variable capacitance transducer*

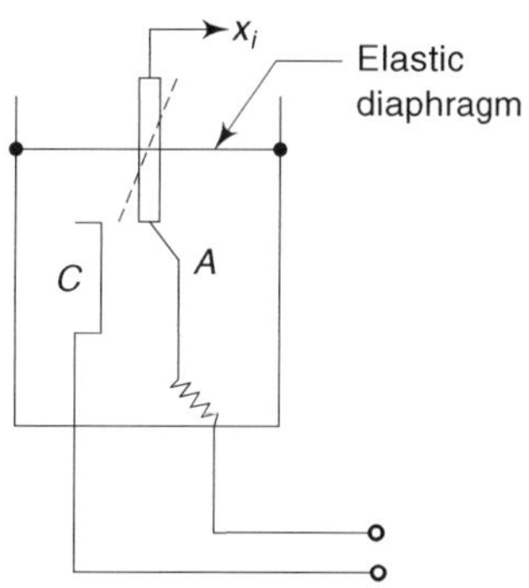

Fig. 4.56 *Mechano-electronic transducer*

Mechano-Electronic Transducer This type of transducer, which is of electronic displacement type, depends on the principle that the plate current depends on the spacing between an anode and a cathode in a diode or a triode. Figure 4.56 shows such a transducer, consisting of an evacuated tube, in which the cathode C is fixed and the position of the anode A can be changed by the input motion x_i, which causes deformation of an elastic diaphragm producing a change in plate current, which can be measured. This can be used for measuring displacement, pressure, force, etc.

Table 4.1 compares the above and other features of various motion transducers.

4.1.2 Opto-electrical Transducer

These transducers convert a light beam into an electrical signal. By proper interruption of the light signal due to motion input, the electrical signal produced can be related to the input. There are three types of photoelectric transducers—photo-emissive, photo-conductive and photo-voltaic transducers.

Photo-Emissive Transducer In this type of transducer shown in Fig. 4.57, light beam strikes a photo-emissive cathode, which releases electrons. These are attracted towards the anode producing a current I in the circuit. The cathode and anode are enclosed in an enclosure that is either evacuated or filled with an inert gas. Current I is seen to be proportional to the intensity of incident radiation, the sensitivity depending on the wavelength of the radiation as well. The cathode is made up of silver that is oxidised and covered with a layer of an alkali metal like cesium or alternatively of an alkali metal combined with antimony.

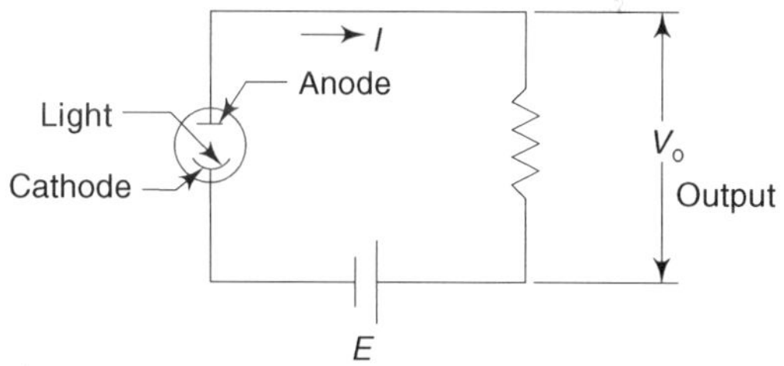

Fig. 4.57 *Photo-emissive transducer*

High gains are possible by using a photo-multiplier as in Fig. 4.58, wherein the number of electrons is increased. The emitted electrons from the cathode C are accelerated by voltage E_1 and focussed upon an electrode, the dynode D_1, where each incident electron causes the emission of secondary electrons. The electrons from D_1 get focussed on D_2. Finally, these are attracted by the anode A, producing current I. The number of stages may be more than those indicated in Fig. 4.58.

Photo-Conductive Transducer A photo-conductive material like PbS (lead sulphide) or InSb (indium antimonide) changes its resistance due to a change in the intensity of incident light. Thus, in a circuit of

Table 4.1 *Comparison of Features of some of the Motion Transducers*

Features \ Transducer	Potentiometric resistance type	Electro-dynamic Type	LVDT	Proximity inductance type	Capacitance type	Piezo-electric type	Strain Gauge type
1. Input or Measurand	Displacement	Velocity	Displacement	Displacement	Displacement	Acceleration	Displacement
2. Magnitude of input	Medium to large	Small to medium	Small to medium	Small	Small	Small	Small
3. Frequency response	dc to 5 Hz	2 – 200 Hz	dc to 2 kHz	dc to 10 kHz	dc to 10 kHz	1 Hz to 10 kHz	dc to 1 kHz
4. Attachment (AT) or Proximity (PR) type	AT	AT	AT	PR	PR	AT	AT
5. Self-generating (SG) External Power source (EP)	EP	SG	EP	EP	EP	SG	EP
6. Linearity	Linear	Linear	Linear	Non-Linear	Non-Linear	Linear	Linear metallic gauge; non-linear – semiconductor gauge.
7. Size	Medium	Medium	Medium	Small	Small	Small	Small
8. Associated circuit	Simple	Simple	Complicated (AC excitation, amplifier and demodulation)	Complicated, (AC excitation, amplifier and demodulation)	Complicated (AC excitation, amplifier and demodulation)	Somewhat complicated (Charge amplifier to prevent impedance loading)	Somewhat complicated (AC/DC excitation, amplifier)
9. Sensitivity	Medium	Medium	High	High	High	Medium	
10. Cost	Very Low	Low	Low	High	High	High	Medium
11. Miscellaneous	High friction, noise with signal, limited life due to wear	Used for low frequency vibration measurements	Widely used in metrological applications and level measurement	Used for measuring small motions	Used for gauging small thickness of paper, blades, surface flatness, etc.	Widely used in vibration measurements	Delicate and fragile, not usable in high temperature and corrosive environment used for force, torque and strain measurements

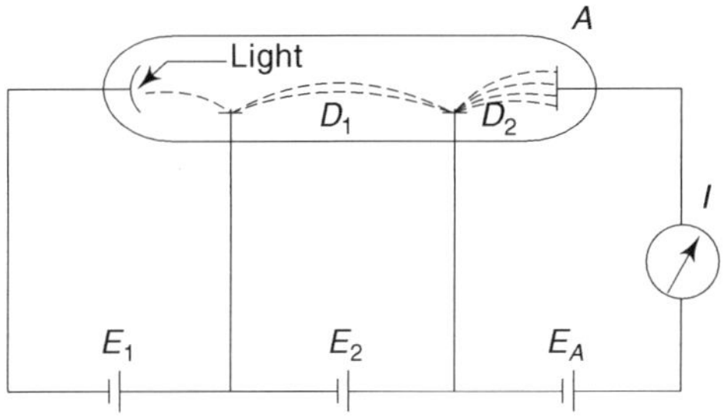

Fig. 4.58 *Photo multiplier*

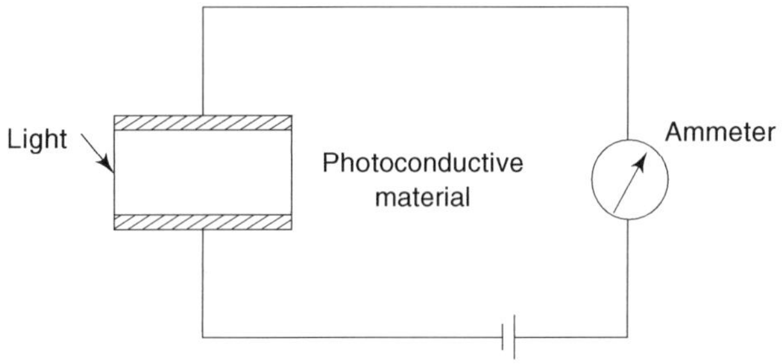

Fig. 4.59 *Photo-conductive material*

the type shown in Fig. 4.59, the current in the circuit would change due to a change in intensity of the incident light. The latter in turn can be changed by the motion, to be measured.

Photo-Voltaic Transducer This type of transducer has a sandwich construction, consisting of a metal base plate 3, a semiconductor material layer 2 (selenium) and a thin transparent metal layer 1, as shown in Fig. 4.60. A voltage output is generated due to incident light and can be suitably measured.

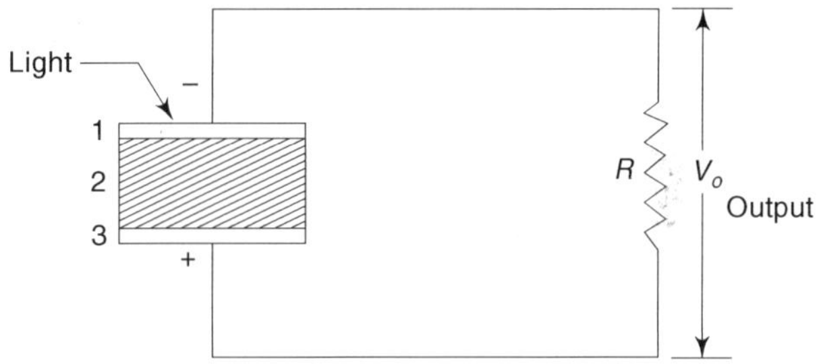

Fig. 4.60 *Photo-voltaic transducer*

4.2 ■ DIGITAL TRANSDUCERS

4.2.1 Introduction

The transducers described so far are analog ones, i.e. their output varies continuously according to the input. In digital transducers, however, the output is discrete and may give frequency type output or a digitally coded output, of binary or some other type:

The main advantages of digital transducers are

1. Use of digital computers, along with the transducers, for data manipulation, is made easier.
2. Digital signals—pulse count frequency or sequences of digitally codes outputs—are not dependent on signal amplitudes and are thus easy to transmit without distortion and external noise.
3. Increased accuracy in pulse count is possible.
4. There are ergonomic advantages in presenting digital data.

Digital transducers range from frequency domain or frequency generating types of transducers to digital encoders. Alternatively, an instrument may incorporate an analog transducer and an analog-to-digital (A-D) converter, giving a digital output. Figure 4.61 shows such an arrangement. Various types of A-D converters are described in Chapter 5.

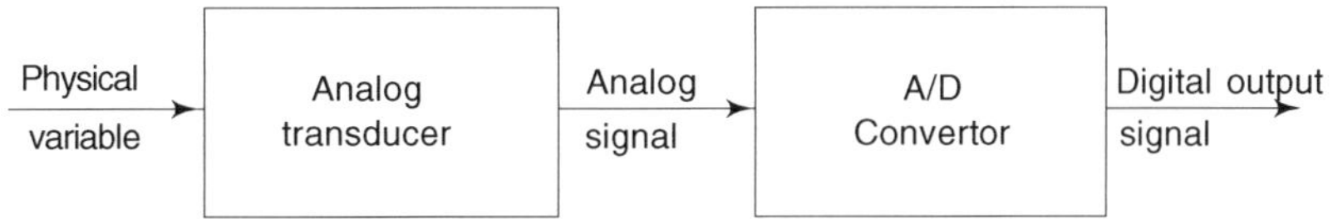

Fig. 4.61 *Digital instrument incorporating A-D converter*

4.2.2 Frequency Domain Transducers

In these transducers, the output is in the form of pulses or sinusoidal wave forms, the frequency of which is a measure of the magnitude of the physical variable. Frequency can be measured by a frequency or pulse counter. Three types of frequency domain transducers have been described below, viz. electromagnetic frequency domain transducer, opto-electrical frequency domain transducer and vibrating string transducer.

Electromagnetic Frequency Domain Transducer This type of transducer can be used for speed measurement, as shown in Fig. 4.62. The device consists of a permanent magnet or a solenoid. On the rotating shaft whose speed is to be measured, a gear of ferromagnetic material is attached. As each gear tooth passes in front of the magnet, the gap length changes. This changes the flux density and a voltage pulse is induced in the coil. Pulse frequency equals speed N times the number of teeth T. The form of the output signal is also shown in Fig. 4.62. Thus, pulse frequency is a measure of speed of rotation.

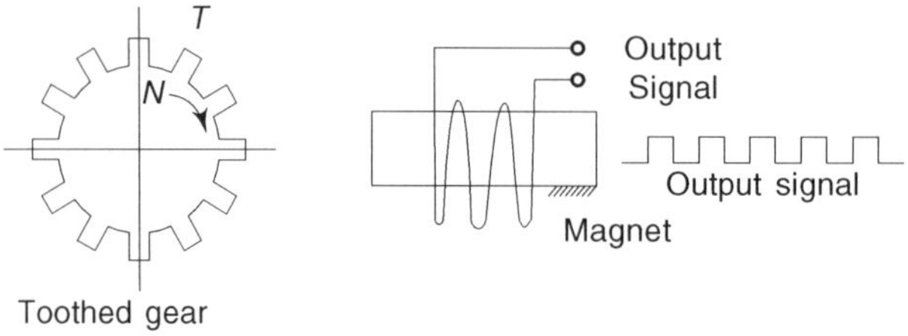

Fig. 4.62 *Electromagnetic frequency domain transducer*

Opto-Electrical Frequency Domain Transducer Figure 4.63 shows an opto-electrical frequency domain transducer for the measurement of speed of rotation of a shaft. The shaft has half dark and half white or shining portions. Every time the latter portion is in front of the light source, the reflected light, falling on the photo-electric transducer, gives an electrical pulse output. The frequency of the pulses is thus a measure of the speed of rotation.

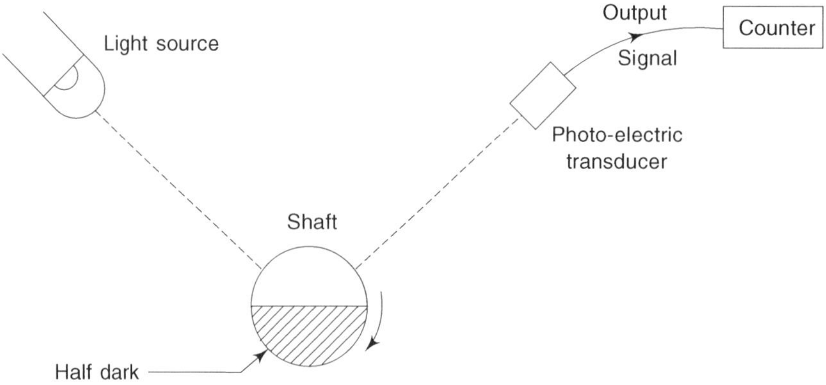

Fig. 4.63 *Opto-electrical frequency domain transducer for rotary motion*

For measurement of linear motion, an arrangement using the opto-electrical device is shown in Fig. 4.64. This uses a transparent scale with a grating. The moving object is attached to the transparent scale. Light from a source passes through the scale and a slit and then falls on a photo-electric transducer. The slit width is such that a motion equal to the pitch of the grating produces one complete cycle of light and darkness at the photo-electric cell. Thus, a pulse output is obtained. From the number of output pulses, the change in motion of the scale and the object attached to it can be determined.

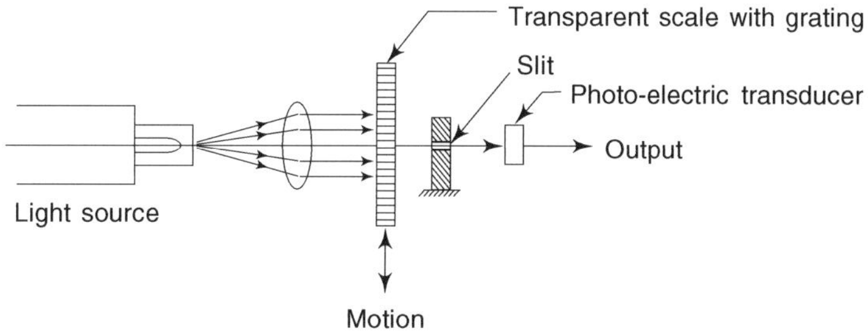

Fig. 4.64 *Opto-electrical frequency domain transducer for linear motion*

Vibrating String Transducer This is essentially used to measure the force applied to a metal string, which is kept vibrating, the frequency of which is dependent on the force applied. The natural frequency f of a string of length L and area of cross-section a is given by

$$f = \frac{1}{2L}\sqrt{\frac{P}{a\rho}} \tag{4.48}$$

where P is the force applied and ρ the mass density of the wire material.

The arrangement is shown in Fig. 4.65. One end of the string is fixed and the other can be moved relative to it, due to the force applied. An electromagnetic transducer picks up the vibrations; the output of which after amplification is fed to an electromagnetic vibration generator, which maintains the string vibration at its natural frequency f. Frequency f gets changed due to change in magnitude of force P. The frequency is measured by a frequency counter and is a measure of the force applied on the string. Initial string vibrations are obtained by an electro-magnetic device that plucks the wire on application of a pulse. The transducer can be used for force and displacement measurements.

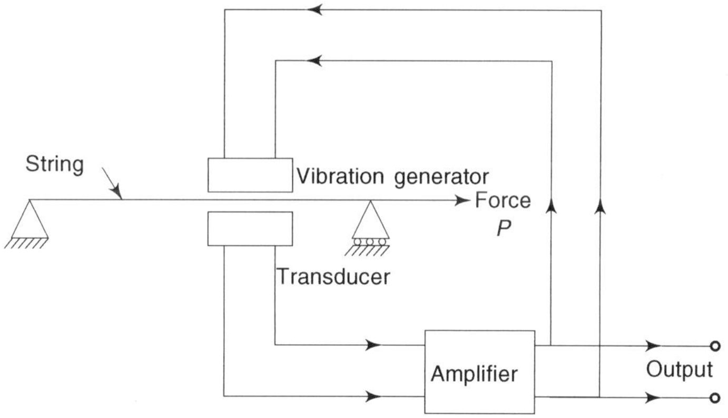

Fig. 4.65 *Vibrating string transducer*

Based on the above principle, vibrating diaphragm and cylinder types of transducers have been developed for measurement of pressure as well.

4.2.3 Binary Codes

The digital output is usually required to be in binary form, especially for use with a digital computer. Unlike the decimal system where there are ten possible states for a digit (0–9), the binary system has two states, viz. ON or OFF or 1 and 0, respectively. To form a number, these are arranged with least significant bit (LSB) on the right and most significant bit (MSB) on the left. In natural or straight binary code, the arrangement is as shown in Fig. 4.66.

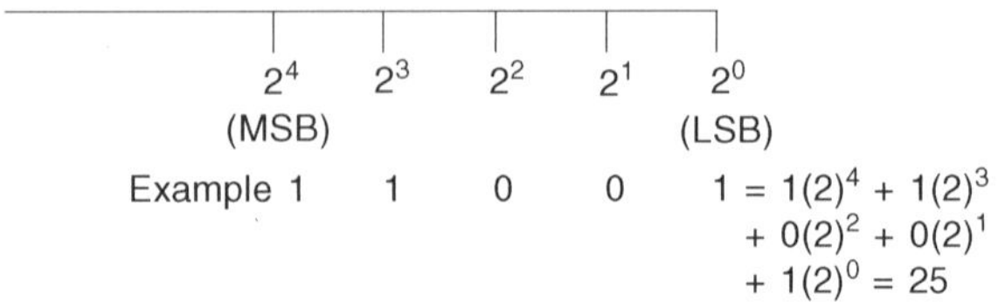

Fig. 4.66 *Binary code arrangement*

Thus, 11001 in binary system would correspond to the number 25 in the decimal system, as shown in Fig. 4.66.

The arrangement of Fig. 4.66 uses 5 bits and this can specify any number up to 31, in steps of 1 or the resolution is 1/31. More bits would be needed for smaller resolution. Each bit may be considered to be replaced by a switch, whose state ON would correspond to the state 1 and OFF to state 0. Thus, a number of bits or switches can specify any desired number.

Other types of codes as well are used for reasons of convenience. In a binary coded decimal code (BCD code), each decimal digit is specified by its binary equivalent. Since to represent numbers from 0 to 9, four bits would be needed, each decimal digit uses four bits, e.g. in BCD code,

$$387 = \underbrace{0011}_{3}\ \underbrace{1000}_{8}\ \underbrace{0111}_{7}$$

The BCD code is simple but not efficient, e.g. 12 bits can represent a number up to 999, while in the natural binary code, 12 bits can represent numbers up to 4095.

Another type of code, called the Gray code, is used to overcome an inherent disadvantage of the straight binary code. In a straight binary code, a change of number by 1 might result in a change of number of bits. Thus, a change from number 7 to 8 would need a change in all the four bits as 7 is represented by 0111 and 8 by 1000. Thus, an error of one bit in a large digital number, in practice, would result in a large error in decimal reconversion. In the case of a digital transducers, an increment in the measured variable would thus have to change all the bits and any error in a bit, would change the number considerably. This is overcome in the Gray code, in which with a change from one value to the next, only one bit is changed at a time. The conversion from natural binary to Gray code is done as follows, keeping the same number of bits:

(a) MSB is unchanged during conversion.
(b) Gray code bit for other bits in the natural binary is *same* if digit to the *left* in the natural binary is 0.
(c) The bit is changed if the bit to the left is 1, e.g. Fig. 4.67 shows the application of above rules for changing the number 10011 in natural binary code to 11010 in Gray code.

Similarly, in changing from Gray to natural code, the following rules are necessary.

(d) MSB is unchanged.
(e) For every other bit in Gray code, natural binary is *same* if the number of 1s to the left is *even.*
(f) The bit is *changed* if the number of 1s to the *left* is *odd*, e.g. 10011 in Gray code would be changed to 11101 in natural binary code, as shown in Fig. 4.68.

Table 4.4 gives the natural binary and Gray code equivalents for numbers up to 15. It can be seen that in giving an increment of unity in the decimal equivalent, only one bit is changed in the Gray equivalent.

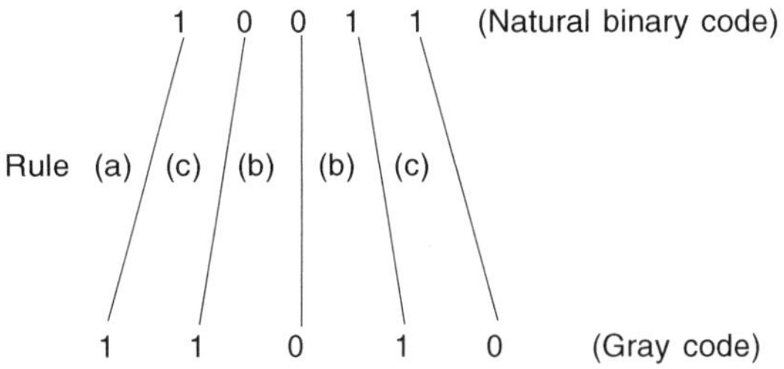

Fig. 4.67 *Natural binary to gray code conversion*

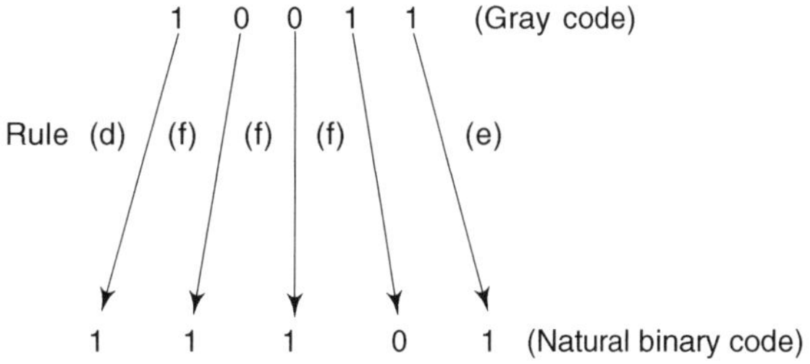

Fig. 4.68 *Gray code to natural binary conversion*

Table 4.4

Decimal Number	*Natural Binary Equivalent*	*Gray code equivalent*
15	1111	1000
14	1110	1001
13	1101	1011
12	1100	1010
11	1011	1110
10	1010	1111
9	1001	1101
8	1000	1100
7	0111	0100
6	0110	0101
5	0101	0111
4	0100	0110
3	0011	0010
2	0010	0011
1	0001	0001
0	0000	0000

Several other types of codes have also been devised for convenience but will not be discussed here.

4.2.4 Digital Encoders

These convert analog motion (rotary or linear) directly to a digital output form. Figure 4.69 shows a rotary encoder disc which can convert a rotary analog motion to a natural binary form of output. The encoder disc has four bits or tracks and is divided into conducting and insulating portions. Brushes *A*, *B*, *C* and *D* are stationary, brush *E* being on the common energising track, which is a conductor. When any of the brushes *A*, *B*, *C* or *D* is on the conducting portion, during rotary motion of the disc, the circuit is made and corresponding indicator *a*, *b*, *c* or *d* is ON (state 1), otherwise, it is OFF (state 0). The outermost track corresponds to the LSB track and the innermost to MSB track as shown in the figure.

Figure 4.69 also shows the natural binary numbers, with their decimal equivalents in brackets, corresponding to various possible brush positions, during the rotational motion of the disc. The disc shown has a resolution of 1/15. With increased number of tracks and brushes, the resolution can be reduced. The information can be supplied directly to the computer input switches, in binary digital form. In a similar way, an encoder can also be used for linear motion input.

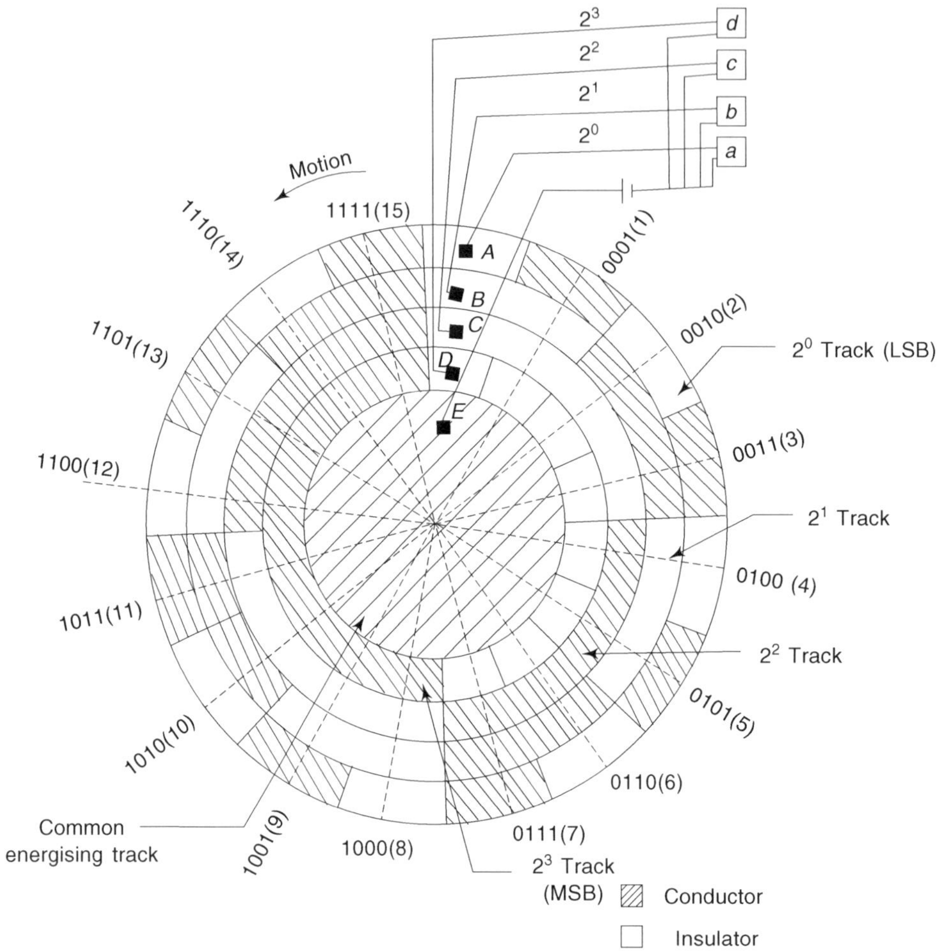

Fig. 4.69 *Basic encoder disc (Contact brush type)*

The encoder described above uses natural binary code. It can be seen that the motion from one position to the other may require a change in the number of brushes, from conducting to non-conducting surface or vice versa. If due to any misalignment during the change from one position to the other, all brushes do not change simultaneously as required, an error can occur in indication. This is overcome by using a different type of code, as discussed in Sec. 4.2.3.

The arrangement described above is of contact type as there is direct contact between the brushes and encoder disc. The disadvantages of such a system are: wear of brushes and disc and friction between the brush and disc. Some other types of arrangements are based on a different principle. These are of optical or magnetic types.

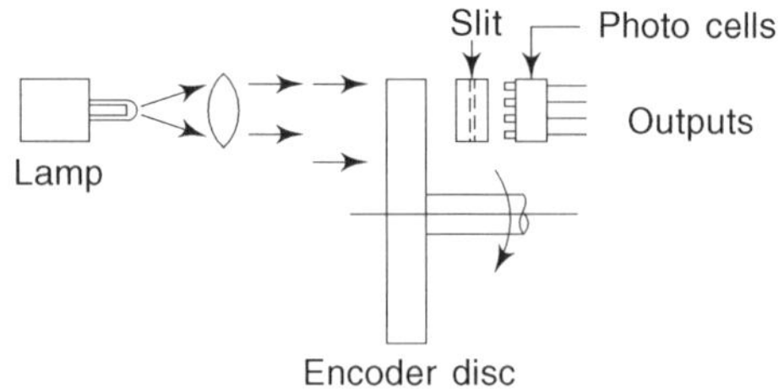

Fig. 4.70 *Optical encoder*

An optical type encoder is shown in Fig. 4.70. The disc has transparent and opaque areas, corresponding to the conducting and non-conducting ones respectively, in the contacting type encoder disc. A light source is used along with slits and photo cells. The photo cell, corresponding to a particular track, would produce an electrical output if the transparent portion is in front of the slit and light source, giving state ON (or 1) while no electrical output from a cell would correspond to OFF (or 0) state.

The magnetic type of encoder (Fig. 4.71), is also of non-contact type. This uses a number of small toroidal magnets with coils around them. The area corresponding to the conducting and non-conducting ones in the contact type encoder disc form the non-magnetic and magnetic areas on the disc. The presence or absence of such areas is detected by the coils which are in close proximity to each track on the disc. One of the coils in each toroidal magnet is energised with high frequency ac carrier signal. If the toroid is over a non-magnetic area of the encoder disc, a voltage due to transformer action is induced in the output coils (state 1) while a magnetic area would saturate the magnetic circuit and output is very small (state 0). The signal on the output windings are usually demodulated to remove the high frequency carrier signal.

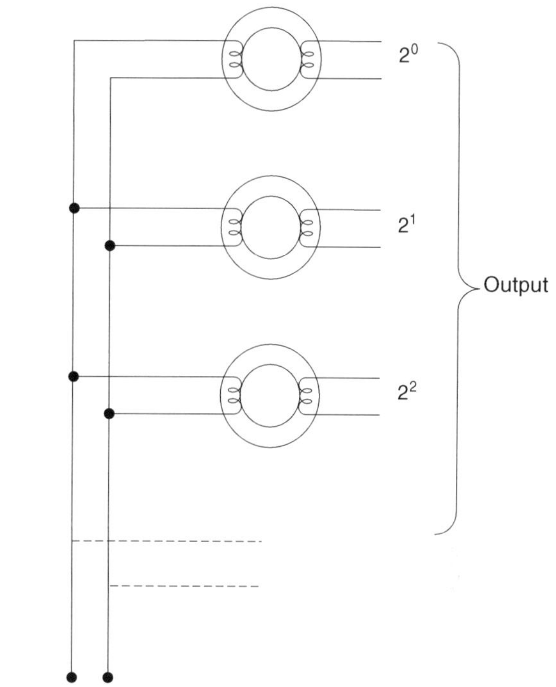

Fig. 4.71 *Magnetic encoder circuit*

Review Questions

4.1 State true or false against each of the following:
- (a) LVDT is a self-generating type of transducer.
- (b) In a Wheatstone bridge, if two adjacent arms have strains of opposite nature, the bridge output is increased.
- (c) For dynamic measurements using resistance strain gauges, temperature compensation is not necessary.
- (d) Eddy current type of transducer gives an output proportional to velocity.
- (e) A variable capacitance type transducer gives an output proportional to acceleration.
- (f) A piezo-electric transducer cannot be used to measure static variables.
- (g) A cathode follower is used to present high impedance to a piezo-electric transducer.
- (h) Digital output can be obtained from an electromagnetic transducer by suitable arrangement.

4.2 Choose the correct answer out of the following:
- (i) The relationship of the gauge factor and the material properties of a wire resistance strain gauge is:
  - (a) $F = 1 + v$

(b) $F = 1 + 2v$

(c) $F = 1 + \mu + \dfrac{\frac{d\rho}{\rho}}{\frac{dL}{L}}$

(d) $F = 1 + 2\mu + \dfrac{\left(\frac{d\rho}{\rho}\right)}{\left(\frac{dL}{L}\right)}$

where μ = Poisson's ratio, L = length and ρ = resistivity

(ii) The value of gauge factor for a semiconductor strain gauge used in practice can be approximately
(a) 0.48 (b) 2.05 (c) 3.5 (d) 150

(iii) The most usual value of resistance, suitable for a wire resistance strain gauge is
(a) 12 Ω (b) 50 Ω (c) 120 Ω (d) 2400 Ω

(iv) The calibration of strain gauge bridge circuit is carried out by
(a) heating the active gauge to a known temperature
(b) applying the known voltage across the dummy gauge
(c) applying a known mechanical strain on the active gauge
(d) shunting a known resistance across a dummy gauge

(v) Name the most sensitive type of sensing element for strain measurement
(a) potentiometric transducer (b) wire resistance strain gauges
(c) extensometer (d) semiconductor strain gange

(vi) The most common transducer for shock and vibration measurement is
(a) dial gauge (b) ring type of load cell
(c) LVDT (d) Piezoelectric pick-up

(vii) LVDT, used for displacement measurement is:
(a) an externally power operated transducer
(b) a self generating passive transducer
(c) a capacitive transducer
(d) a digital transducer

(viii) Wheatstrone bridge has got three resistances taken in one direction as 120.3 Ω, 119.2 Ω and 119.2 Ω. The value of the fourth resistance for null balance would be
(a) 120.3 Ω (b) 119.2 Ω (c) 120.0 Ω (d) 118.9 Ω

(ix) LVDT works on the principle of
(a) variable resistance (b) variable self-induction
(c) variable mutual induction (d) variable capacitance

(x) A solar cell is
(a) photo-voltaic transducer (b) photo-emissive transducer
(c) photo-conductive transducer (d) photo-resistive transducer

(xi) Which material out of the following has got the property of generating emf when subjected to mechanical strain
(a) strain gauge material (b) piezo-electric material
(c) steel conductor (d) thermosetting plastics

(xii) For which type of measurements, a piezo electric transducer is suitable
- (a) static pressures (b) dynamic pressures
- (c) high temperatures (d) low temperatures

(xiii) The abbreviation LVDT stands for
- (a) low voltage digital transducer (b) linear voltage differential transducer
- (c) least varying differential transformer (d) linear variable differential transformer

(xiv) Name the instrument for angular measurement
- (a) tachometer (b) strobometer (c) shaft encoder (d) LVDT

4.3 A capacitance transducer has two plates of area 5 cm^2 each, separated by an air gap of 1 mm thickness. Find the displacement sensitivity due to gap change.

4.4 A quartz crystal, with piezo-electric properties, has a diameter of 10 mm and thickness 2 mm. Its voltage sensitivity constant is 0.045 V m/N. Find the voltage output due to a force of 100 N. Young's modulus = 10^{11} N/m^2.

4.5 In the resistance strain gauge arrangement of Fig. 4.72 strain gauge resistance $R_1 = R_2 = R_3 = R_4 =$ 120 Ω E_b = 9 V. If the output voltage of the bridge is 20 mV, find the value of force applied to the strip. Thickness t of the strip = 1 mm. Width b of strip = 10 mm. Young's modulus E of strip material = 8 × 10^{10} N/m^2. Length L = 50 mm. Strain at root of cantilever = $6PL/Ebt^2$. Assume output resistance as infinity. Gauge factor = 2.4.

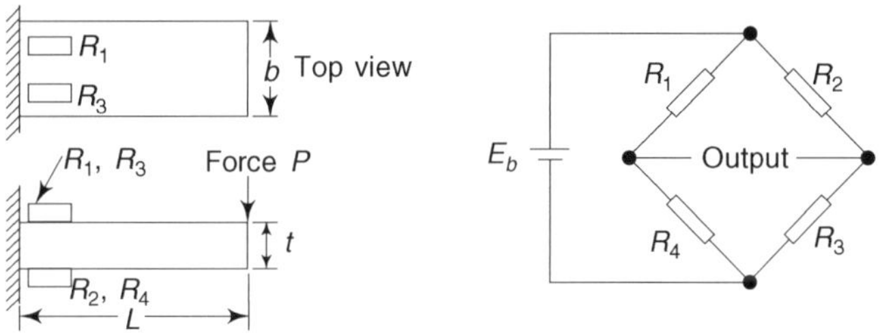

Fig. 4.72

4.6 A force measuring device, using resistance strain gauges, is shown in Fig. 4.73, R_1 is an active gauge and R_2 a dummy gauge for temperature compensation. $R_1 = R_2 = R_3 = R_4 =$ 120 Ω. Gauge factor = 2.5. Maximum gauge current is limited to 30 mA. Find the required battery voltage. If a calibrating resistance R_c = 1.2 MΩ is used and connected parallel to R_1, trace shift on oscilloscope = 5 cm. Find the corresponding trace shift due to the applied force P, which causes a strain of 10^{-5} at the root of the cantilever. If temperature compensation were not used, what will be the output voltage if temperature is increased by 10°C? Temperature coefficient of gauge material = 10^{-5} Ω/Ω°C. Take oscilloscope input impedance as infinity.

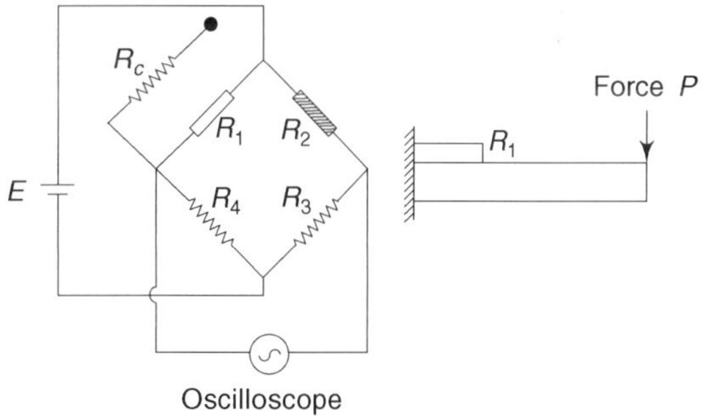

Fig. 4.73

4.7 In the arrangement shown in Fig. 4.74, A, B, C and D are resistance strain gauges bonded to a steel ring. These form the four arms of a Wheatstone bridge, so as to give temperature compensation and maximum possible sensitivity. When a force is applied, the compression of the rubber block

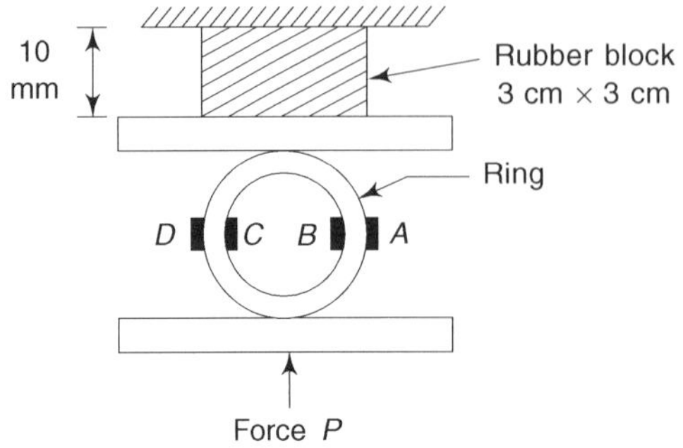

Fig. 4.74

is 1 mm and the shift on the display element due to output of the bridge is 4 cm. For calibration of the bridge circuit, a shunt resistance of 0.2 MΩ is connected parallel to A. This gives a shift of 1.2 cm on the display element.

Find the compression modulus of rubber used. Gauge factor = 2.2. Resistance of each gauge = 120 Ω. Young's modulus $E = 2 \times 10^5$ N/mm^2. r = 40 mm, b = 15 mm, t = 1.5 mm, where r is the mean radius of the ring, b its width and t its thickness. (For the ring, strain at the gauge location is given by 1.1 Pr/Ebt^2, P being the force applied on the ring.)

4.8 Figure 4.75 shows an arrangement for calibration of an accelerometer. The mass m is constrained to move vertically with harmonic motion. Resistance gauges R_1 and R_3 are bonded along the length of the strip while R_2 and R_4 are along the width of the strip. Strip width = 20 mm. Thickness of strip = 1mm. The gauges, each of 120 Ω, from the four arms of Wheatstone bridge. The output of the bridge is connected to one channel of an oscilloscope, while that from the accelerometer is connected to the second channel. Calibration is done by connecting a 1 MΩ resistance, parallel to R_1, giving a shift of 2 cm on the oscilloscope. During calibration, output of the acceleration corresponds to a shift of 5 cm while that of the strain gauge bridge gives a shift of 3.2 cm, in the respective oscilloscope channels. The gauge factor is 2. Oscilloscope sensitivity on accelerometer channel = 20 mV/cm.

Find the sensitivity of the accelerometer in mV/g. Young's modulus of strip material = 7×10^{10} N/m^2. Poisson's ratio = 0.3.

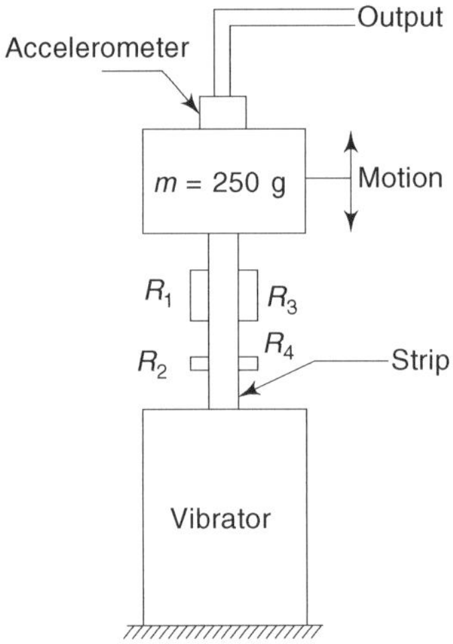

Fig. 4.75

4.9 A piezo-electric transducer has to following characteristics.

Capacitance of crystal = 5×10^{-9} F

Capacitance of cable = 5×10^{-10} F

A charge of 4×10^{-10} Coulomb is produced due to a force of 10 N. The output of the transducer is connected to a second order recording device which has at its input a resistance of 1 MΩ, connected in parallel with a capacitance of 5×10^{-10} F. The second order device has a damping ratio of 0.7, undamped natural frequency of 1000 Hz and static sensitivity of 0.2 cm/mV. Find the amplitude of the output as recorded if a sinusoidal force of amplitude 20 N at a frequency of 200 Hz is applied.

4.10 A piezo-electric crystal transducer is used to measure the force transmitted from a structure to its support as shown in Fig. 4.76. The crystal is in the form of a disc of diameter 2 cm and thickness 1 mm with its charge constant as 10^{-5} Coulomb/cm. Dielectric constant of the crystal = 5 and Young's modulus = 8×10^{10} N/m^2. Capacitance of the cable = 20 pF. The amplifier has at its input a resistance of 20 MΩ, in parallel with a capacitance of 50 pF. Amplifier gain = 50.

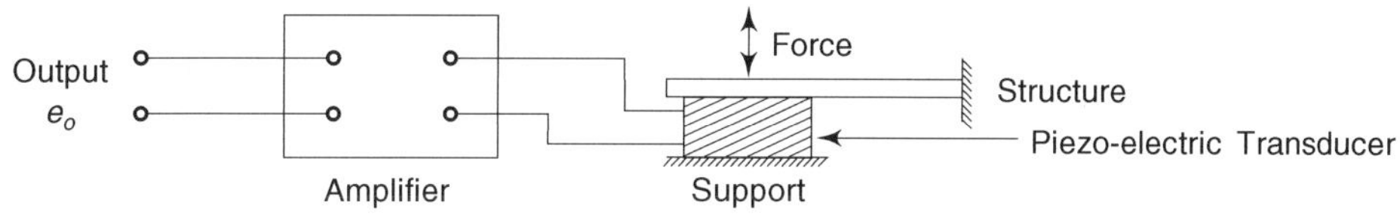

Fig. 4.76

Find the amplitude of the force transmitted if output signal e_o has an amplitude of 0.5 volt at a frequency of 100 Hz.

4.11 A delta type Rosette strain gauge arrangement is bonded on to an aluminium plate. The three measured strains are:

$$\varepsilon_1 = -100 \ \mu\text{m/m}$$
$$\varepsilon_2 = 150 \ \mu\text{m/m}$$
$$\varepsilon_3 = 250 \ \mu\text{m/m}$$

Calculate the principal stresses and their locations. For aluminium, $E = 8 \times 10^{10}$ N/m^2 and $\mu = 0.25$.

Answers

4.1	(a) F	(b) T	(c) T	(d) F	(e) F
	(f) T	(g) T	(h) T		
4.2	(i) (d)	(ii) (d)	(iii) (c)	(iv) (c)	(v) (d)
	(vi) (d)	(vii) (a)	(viii) (a)	(ix) (c)	(x) (a)
	(xi) (b)	(xii) (b)	(xiii) (d)	(xiv) (c)	

4.3 44.2 pF/cm   4.4 114.3 V   4.5 2.48 N   4.6 7.2 V, 1.25 cm, 0.18 mV
4.7 0.387 N/mm^2 4.8 4.75 mV/g   4.9 26.44 cm   4.10 350.6 N

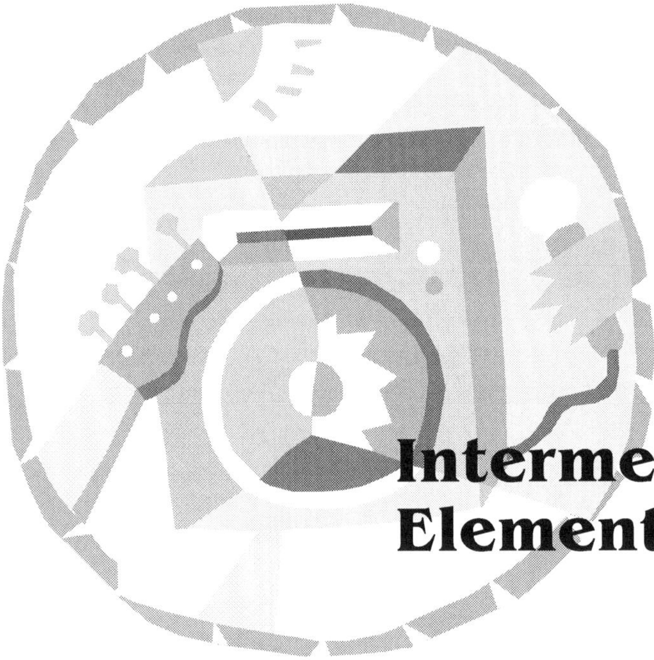

Chapter

5

Intermediate Elements

■ *INTRODUCTION* ■

The output signal of any transducer usually needs to be modified by elements known as intermediate elements, so that it can be displayed or recorded with convenience, as desired. These include:

1. Amplifiers for amplifying the transducer output, which may be small.
2. Compensating devices, in order to improve characteristics like frequency response, impedance loading, etc.
3. Differentiating or integrating elements, so that the output is proportional to the desired input which may, be, for example, displacement, velocity or acceleration, in any given situation.
4. Filters, for filtering out unwanted portions of the signal.
5. A–D/D–A converters, for converting analog type signals to digital form or vice versa.
6. Data transmission elements, which transmit the transducer output to a certain distance as desired, e.g. from a running machine in an inaccessible location, to a control room.

Some of the intermediate elements like amplifiers and compensating devices have already been discussed in Chapters 3 and 4.

5.1 ■ AMPLIFIERS

Amplifiers are intermediate elements that increase the magnitude of the signal from a transducer so that it can be conveniently displayed or recorded. These may be of mechanical, hydraulic, pneumatic, optical or eletrical/electronic types, depending on the type of transducer. In cases where a reduction is needed in the magnitude of the signal from the transducers, these are called attenuators.

5.1.1 Mechanical Amplifying Element

Mechanical amplifying elements are simple in operation, rugged type and inexpensive. They find a wide range of applications in the mechanical type of instruments. For example, Huggenburger extensometer employs mechanical amplification using a system of compound levers. A dial gauge with a least count of 1 μm employs a system of gears to amplify the displacement input. Further, a Bourden pressure gauge employs a combination of levers and gears to give the desired amplification to the transduced displacement signal. However, these types of amplifiers have the disadvantages of friction and stiction effects, backlash errors and inertial effects due to relatively higher mass. They are also affected by environmental temperature changes.

Figure 5.1 shows a lever mechanism that is a typical mechanical amplifier. As seen from the figure, the output signal X_o is given by:

$$X_o = \left(\frac{R}{r}\right) X_{\text{in}} \tag{5.1}$$

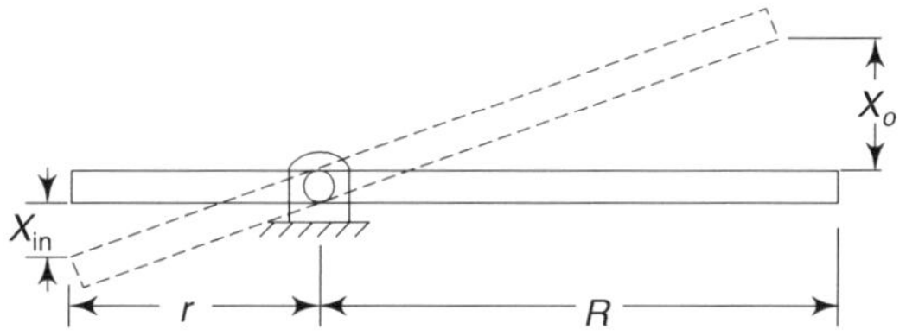

Fig. 5.1 *A typical mechanical amplifier (a lever type device)*

Similarly, for angular motion, using gears for mechanical amplification, the output signal θ_o is given by

$$\theta_o = \left(\frac{N}{n}\right) \theta_{\text{in}} \tag{5.2}$$

where N/n is the gear ratio.

5.1.2 Hydraulic Amplifying Element

Hydraulic amplifying elements find a wide range of applications in the form of hydraulic actuators in the control elements used in the automobile hydraulic brakes and hydraulic steering systems. The advantages is compactness for a specified force while the disadvantages are possible leakages and problems in dusty environments.

Figure 5.2 shows a typical hydraulic type element, in which the output signal X_o is given by

$$X_o = \left(\frac{D}{d}\right)^2 X_{\text{in}} \tag{5.3}$$

D and d being the respective diameters of the plunger and pipe, respectively (Fig. 5.2).

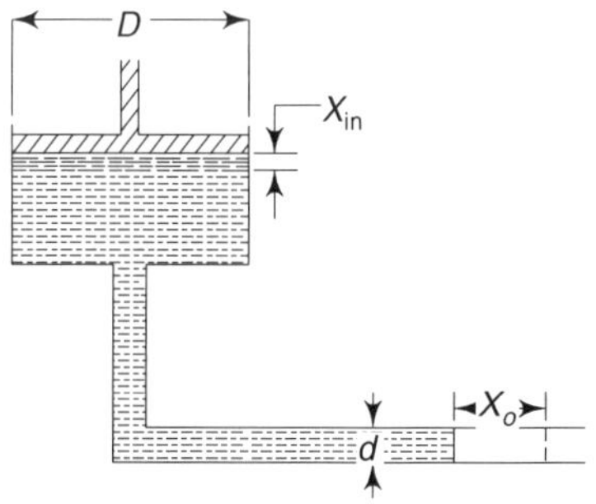

Fig. 5.2 *A typical hydraulic type of amplifier*

5.1.3 Pneumatic Amplifying Element

A pneumatic transducer, of flapper nozzle type, was described in Chapter 4, for converting mechanical displacement X_i to pressure p_2. This is shown in Fig. 5.3 as well.

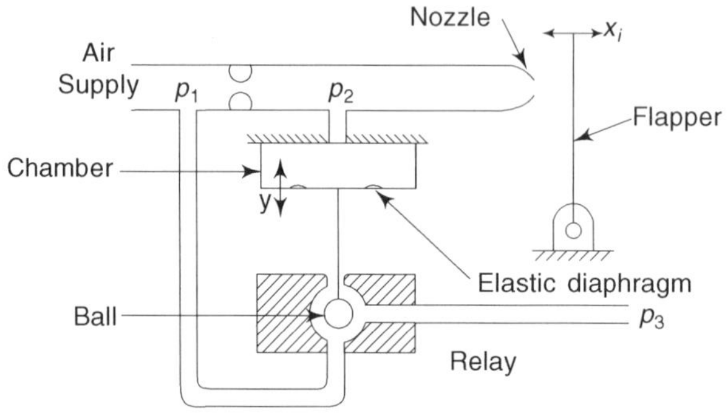

Fig. 5.3 *Pneumatic relay as amplifier*

In order to amplify pressure signal p_2, a ball type relay is shown which is operated by the motion of an elastic diaphragm which gets deflected due to p_2. If the ball is at the lowest position, pressure p_3 is atmospheric while at the topmost position, p_3 equals air supply pressure p_1. Thus, p_3 changes from zero gauge pressure to p_1 due to a small pressure change in p_2 and so the relay can be treated as a pneumatic amplifier. These are used in industrial environment where compressed air is easily available.

5.1.4 Optical Amplifying Element

One common application of the optical amplifying element is in the taut suspension type of the optical type of galvanometer which is a very sensitive type of instrument. The lamp and scale type of amplifier is relatively inexpensive and pnovides a large amount of amplification to the input signals. However, these amplifiers because of their inertia effects due to mirror mass connot be employed in the dynamic type of measurements.

In order to amplify the angular displacement θ_i, as shown in Fig. 5.4 an optical arrangement is used.

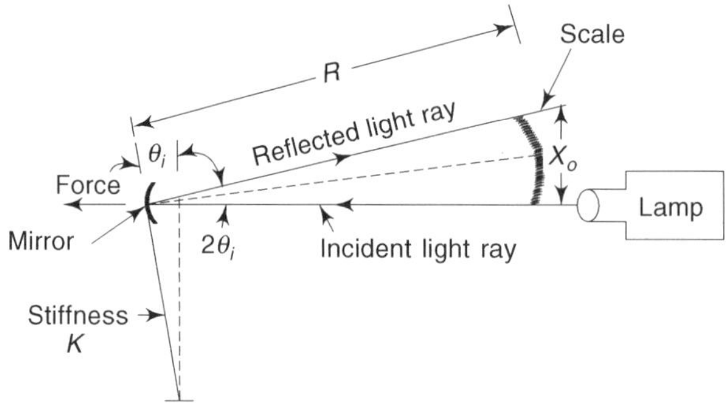

Fig. 5.4 *A typical lamp and scale arrangement*

A ray of light gets deflected by $2\theta_i$ from a mirror attached to the moving member and the distance X_o moved by the light spot on the scale is given by

$$X_o = 2R\theta_i \tag{5.4}$$

where R is the distance between the scale and the mirror. By proper choice of the distance R, X_o can be made as large as desired.

5.1.5 Electrical Amplifying Elements

Currently, most of the electrical amplifiers are either transistor based or employ suitable integrated circuits (ICs) or both. Vacuum tubes have now become obsolete and are employed in certain special applications only. Presently a wide variety of amplifiers are available to meet the specific requirements in the signal conditioning element of the instrument systems. The following are the characteristics of an *ideal amplifer* i.e. it should have:

(i) infinite input impedance, i.e. it should have no loading effect on the transducer.
(ii) zero output impedance
(iii) a very large gain (theoretically infinite) to improve resolution
(iv) zero output for zero input
(v) ability to filter spurious inputs and
(vi) excellent frequency response.

The above requirements are met by suitable operational amplifiers to a large extent and are discussed in Section 5.2.

In amplifiers, an external power source is invariably required, as shown in Fig. 5.5.

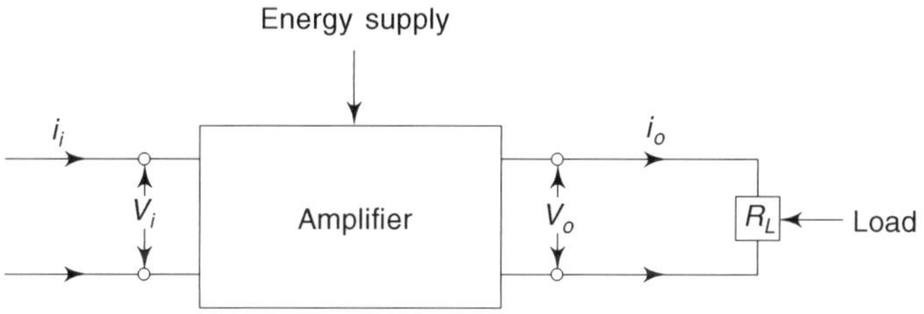

Fig. 5.5 *A typical electrical amplifying element*

The voltage input of transducer V_i is amplified to V_o. Thus, gain G of amplifier is

$$G = \frac{V_o}{V_i} \tag{5.5}$$

$$\text{Current amplification} = \frac{i_o}{i_i}$$

and

$$\text{power gain} = \frac{V_o}{V}\frac{i_o}{i_i}$$

It is usual to express the gain in decibels (dB), for convenience, with

$$\text{power gain (dB)} = 10 \log \frac{p_2}{p_1}$$

Since power is proportional to the square of voltages,

$$\text{Voltage gain (dB)} = 20 \log \frac{V_o}{V_i} \tag{5.6}$$

Amplifiers are of several types. The prominent ones used for instrumentation and measurement work are

1. ac and dc amplifiers,
2. carrier amplifiers, and
3. chopper amplifiers.

ac and dc Amplifiers Figure 5.6 shows a two-stage transistor amplifier. Coupling is through a capacitor in case of an ac amplifier. The circuit producing bias current is not shown in the figure.

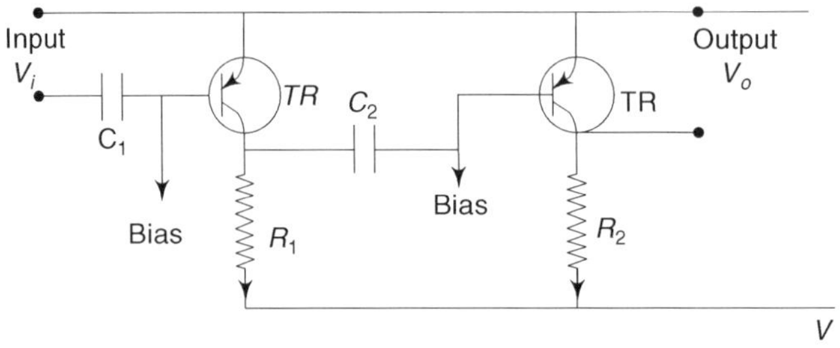

Fig. 5.6 *Two-stage transistor amplifier*

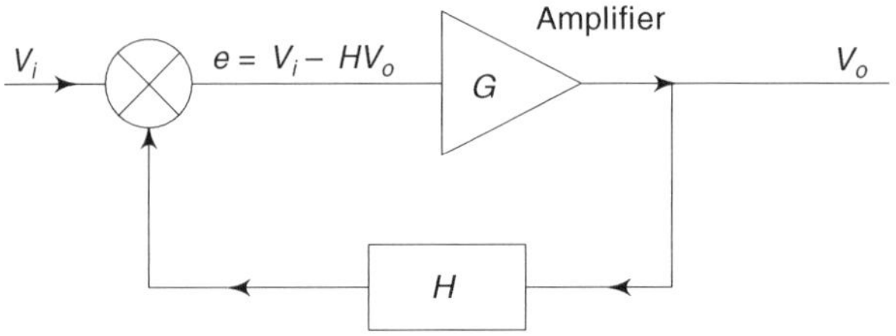

Fig. 5.7 *Feedback in electronic amplifier*

In order to improve the overall performance of electronic amplifiers, viz. improved frequency response, etc. negative feedback as shown in Fig. 5.7 is employed.

If the amplifier gain is G and a feedback device with gain H is employed such that input e to the amplifier is $(V_i - HV_o)$ then output

$$V_o = G(V_i - HV_o) \tag{5.7}$$

or

$$\frac{V_o}{V_i} = \frac{G}{1+GH} \tag{5.8}$$

If $GH >> 1$,

Overall gain $\dfrac{V_o}{V_i} = \dfrac{1}{H}$

Thus, the overall gain depends only on the factor H, which can be suitably selected so as not to have any dependence on frequency and so, frequency response with feedback would be flatter as in Fig. 5.8.

It may be seen from Eq. (5.8) that with the application of negative feedback, output voltage V_o is reduced. This may be explained as being the result of a corresponding reduction in input current for a given input voltage V_i or that there is an apparent increase in the input impedance of the amplifier. This is useful from the point of view of prevention of impedance loading effect.

Carrier Amplifiers The signal from the transducer may be superimposed on a high frequency signal called the carrier signal so that the carrier signal is modulated. Either its amplitude or frequency is changed, or the carrier is amplitude or frequency modulated. After amplification by the ac amplifier, the amplified transducer signal is recovered by demodulation. Such amplifiers have been described in Chapter 4.

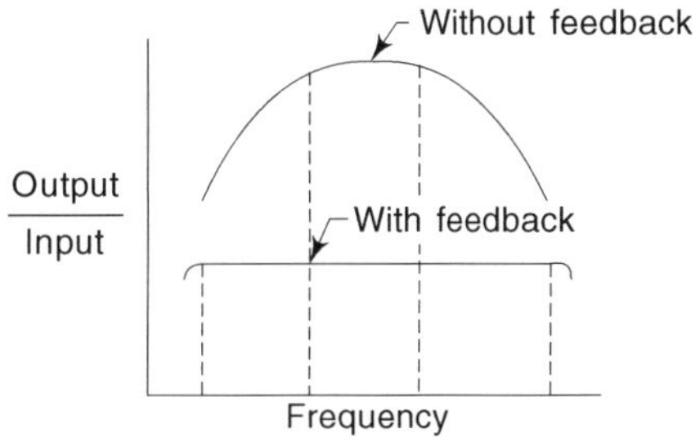

Fig. 5.8 *Effect of feedback on frequency response*

Chopper Amplifier In this type of amplifier, the input signal is, by a chopper, made to produce a square wave voltage, with amplitude proportional to the signal and frequency according to the frequency of the chopper, which is basically a mechanical switch or relay, driven magnetically. The chopped or modulated signal is amplified by an ac amplifier (Fig. 5.9) and the original signal recovered by demodulation and filtering. As shown in Fig. 5.8, for a dc voltage V_1, the chopped voltage V_i and V_o are of square type and after filtering, the original signal is recovered as V_2, with gain $G = V_2/V_1$.

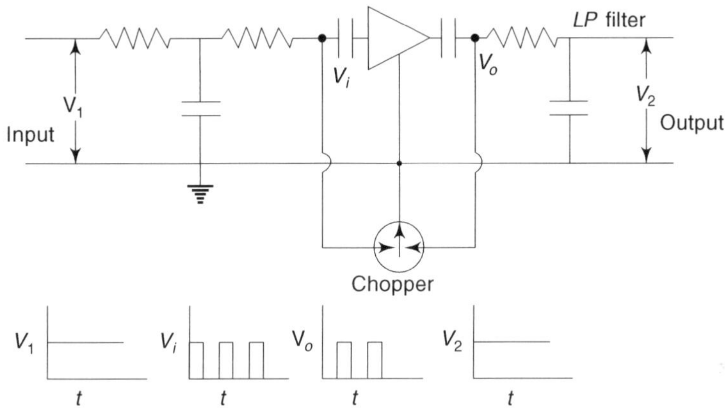

Fig. 5.9 *Chopper amplifier*

These amplifiers can be used for dc voltages and for ac signals up to a frequency of about one-tenth the chopper frequency.

5.2 ■ OPERATIONAL AMPLIFIERS

These are important signal conditioning elements of an instrument. An operational amplifier essentially, has the following characteristics:

- high gain dc difference amplifier type, having two inputs terminals and act on a difference in voltages at the terminals.
- available as an integrated circuit (IC)
- high input impedance
- low output impedance
- normally used along with outside resistance and capacitances

A typical operational amplifier IC has several elements e.g. an operational amplifier has 20 transistors, 12 resistors, 1 capacitor and the performance is gain up to 10^5, depending on frequency, power supply 18 V, maximum differential input 30 V, maximum signal ended input 15 V and input offset voltage 2 mV.

These amplifiers are used in several applications—amplification, summing and difference amplification, integrating, differentiating, charge amplifiers, sensor linearization, etc.

A typical representation of an operational amplifier is shown in Fig. 5.10.

The input voltages are e_1 and e_2 and output e_0, all measured relative to the ground. Input e_1 is called inverting input, written with a negative sign and e_2, the non-inverting one, written with a positive sign.

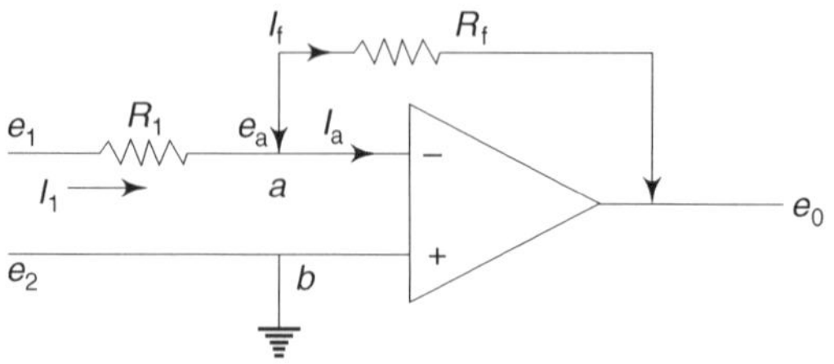

Fig. 5.10 *Representation of an operational amplifier*

5.2.1 Inverting Amplifier

This is shown in Fig. 5.11.

In the figure, resistances R_1 and R_f are used as shown. Since amplification is very high, the input voltage e_1, <<< e_0.

So $e_a \approx e_b \approx 0$

Summing the currents at a

$$I_1 = I_f + I_a, \qquad I_a \approx 0$$

or
$$\frac{e_1 - e_a}{R_1} = \frac{e_a - e_o}{R_f}, \qquad e_a \approx 0$$

or
$$\frac{e_o}{e_1} = \frac{-R_f}{R_1} \tag{5.9}$$

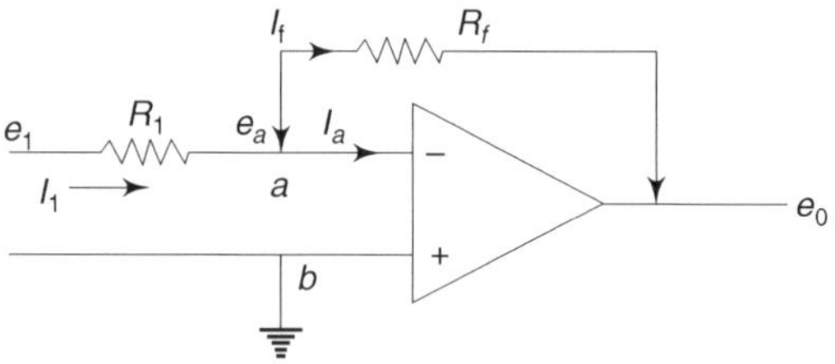

Fig. 5.11 *Inverting amplifer*

This is called inverting amplifier since the output is having 180° phase relative to input.

If $R_f = 5\ \text{M}\Omega \qquad R_1 = 0.5\ \text{M}\Omega$

Thus, voltage gain $G = \dfrac{e_o}{e_1} = -10$

5.2.2 Non-inverting Amplifier

In the case, given in Fig. 5.12.

$$e_a \approx e_1,$$

where,
$$e_a = \frac{R_1}{R_1 + R_f} e_o$$

$$\text{Voltage gain} = \frac{e_o}{e_1} = \frac{R_1 + R_f}{R_1} \tag{5.10}$$

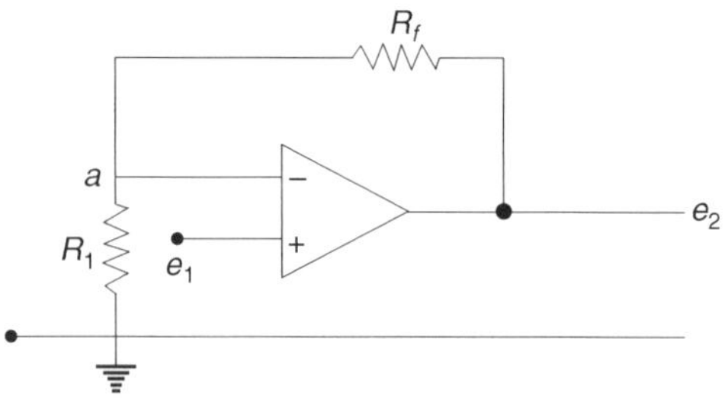

Fig. 5.12 *Non-inverting amplifier*

If $R_f = 0$, feed back loop is shorted. The arrangement is called voltage follower, where voltage gain = 1 and input impedance is infinity. This is used where impedance loading is to be avoided, since it presents high impedance at input and low impedance at output.

5.2.3 Summing Amplifier

This is shown in Fig. 5.13.

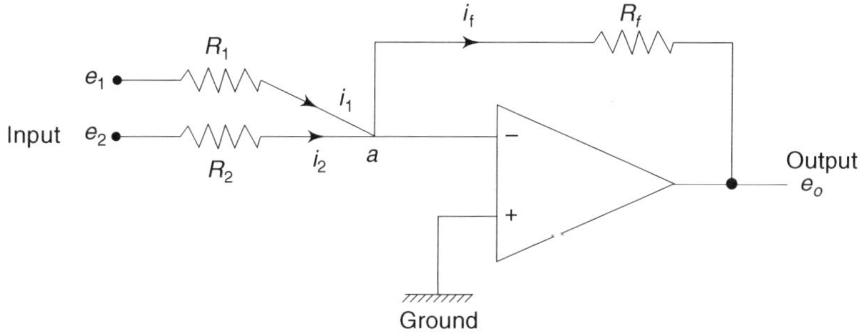

Fig. 5.13 *Summing amplifier*

In the summing amplifier shown in Fig. 5.13, the output is seen to be proportional to the sum of inputs, as shown below.

Summing currents at a,

$$i_2 + i_2 - i_f = 0 \tag{5.11}$$

since $\quad i_1 = \dfrac{e_1}{R_1},\ i_2 = \dfrac{e_2}{R_2}$ and

$i_f = \dfrac{e_o}{R_f}$, treating a to be at zero potential.

Substituting for i_1, i_2 and i_f in Eq. (5.11),

we get $\quad e_o = -\left(\dfrac{R_f}{R_1}e_1 + \dfrac{R_f}{R_2}e_2\right)$

If $R_1 = R_2$, than $e_o = \dfrac{-R_f}{R_1}(e_1 + e_2)$

Any number of input voltages may be added using the above arrangement.

5.2.4 Differential Amplifier

In the differential or difference amplifier shown in Fig. 5.14, the output is seen to be proportional to difference of the two inputs as shown below.

Since e_2 is across R_1 and R_2 in series,

$$e_b = \frac{R_2}{R_1 + R_2}e_2 = e_a \tag{5.12}$$

Current through R_2 is equal to that from e_1 through R_1, since the operational amplifier draws negligible current,

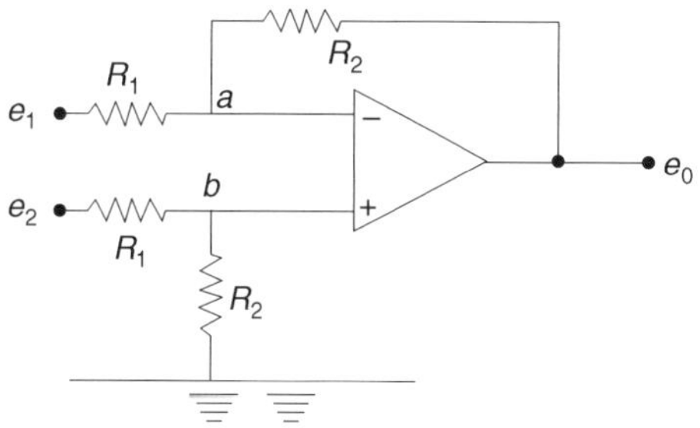

Fig. 5.14 *Differential amplifier*

or $$\frac{e_1 - e_a}{R_1} = \frac{e_a - e_o}{R_2}$$

or $$\frac{e_o}{R_2} = e_a\left(\frac{1}{R_1} + \frac{1}{R_2}\right) - \frac{e_1}{R_1}$$

Substituting for e_a from Eq. (5.12),

We get $$e_o = \frac{R_2}{R_1}(e_2 - e_1)$$

or output is a measure of difference between the two inputs.

5.2.5 Common Mode Rejection Ratio

If the two input channels are perfectly matched, operational amplifier has the same gain for both channels. In actual case, this is not so. If differential voltage between the two channels = $\Delta e = e_2 - e_1$,

$$\text{common mode voltage} = e_{\text{cm}} = \frac{e_1 + e_2}{2}.$$

The gain G_d for differential voltage would be different from gain G_{cm} for the common mode voltage,

or $$\text{Output } e_o = G_d\,\Delta e = G_{\text{cm}}\, e_{\text{cm}} \tag{5.13}$$

The gain G_{cm} should be much less than G_d, in order to minimize the effect of e_{cm} as output. The deviation of operational amplifier from the ideal situation is represented by common mode rejection ratio (CMRR), where

$$\text{CMRR} = \frac{G_d}{G_{\text{cm}}}$$

This is usually represented in decibels (dB),

If $$\frac{G_d}{G_{\text{cm}}} = 10{,}000$$

$$\text{CMRR} = 20 \log \frac{G_d}{G_{\text{cm}}} = 80 \text{ dB}$$

5.2.6 Charge Amplifier

Figure 5.15 shows a piezo-electric transducer connected to an operational amplifier and other elements. The purpose is to present a high impedance to the piezo-electric transducer to avoid impedance loading and to eliminate the influence of cable capacitance, etc.

As discussed earlier in Chapter 4, the piezo-electric crystal acts as a charge generator, due to deformation $x_i(t)$.

Charge $Q = K_1 x_i(t)$, where $x_i(t)$ is deformation of the crystal and K_1 is charge sensitivity constant,

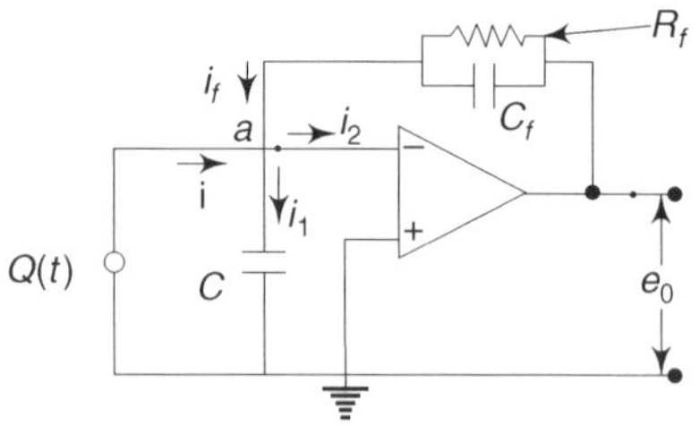

Fig. 5.15 *Piezo-electric transducer as charge generator, used with an operational amplifier*

Current $$i = \frac{dQ}{dt} = K_1 D x_i \tag{5.14}$$

C is Fig. 5.15 is the combined capacitance due to the crystal, cables etc.

In Fig. 5.15, $i_2 \approx 0$ due to high impedance of the operational amplifier and $i_1 \approx 0$ as point a is virtually grounded.

At a, $$i + i_f = 0 \tag{5.15}$$

$$i_f = \frac{e_o}{R_f} + C_f D\, e_0 \tag{5.16}$$

where R_f and C_f are feedback elements.

From Eqs. (5.14, 5.15, and 5.16), we get

Output $$e_o = \frac{-K\tau_f D x_i}{1 + D\tau_f} \tag{5.17}$$

where time constant, $\tau_f = R_f C_f$

and voltage sensitivity constant, $K = \dfrac{K_1}{C_f}$

As seen from Eq. (5.17), output e_o depends on C_f and R_f, which are constants and do not depend on cable lengths etc.

Equation (5.17) is similar to Eq. (4.17) except that in the latter equation, τ is used in place of τ_f. The value of τ_f can be chosen to be high for improved dynamic performance of the measuring system.

5.2.7 Linearizing Capacitance Transducer using Operational Amplifier

As described in Chapter 4, capacitance C of a capacitance transducer in pico-farads is given by:

$$C = \frac{1}{3.6\pi}\,\varepsilon\,\frac{A}{d}, \tag{5.18}$$

where A = plate area cm^2
D = Gap between plates (cm)
ε = dielectric constant

With change in gap d, C changes and with the transducer forming a part of a circuit with external supply voltage e, output can be obtained. It is seen from Eq. (5.18) that 'C' varies non-linearly with change of d. It will now be shown that this can be linearized by using an operational amplifier, as in Fig. 5.16.

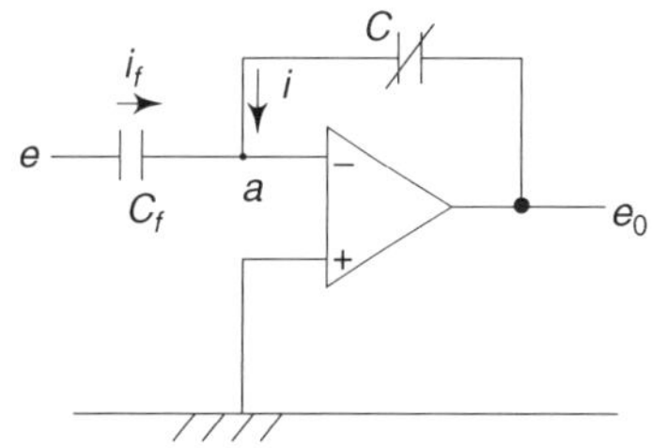

Fig. 5.16 *Capacitance Transducer used with an operational amplifier*

$$e = \frac{1}{C_f}\int i_f\, dt, \text{ ignoring the voltage at } a \tag{5.19}$$

or $$\frac{1}{C}\int i dt = e_o \tag{5.20}$$

$$i_f + i = 0 \tag{5.21}$$

Substituting Eqs (5.19) and (5.20) in (5.21), we get,

$$e_o = \frac{-C_f}{C} e$$

$$= \frac{-C_f de}{3.6\pi A\varepsilon}$$

$$= -kd \quad (5.22)$$

where $$k = \frac{-C_f e}{3.6\pi A\varepsilon}.$$

The parameters e, C_f, A, and ε being constants, output e_o varies linearly with gap d as seen from Eq. (5.22).

Problem 5.1 *Give the arrangement of an operational amplifier to produce an output voltage e_o, such that $e_o = 2 + 3e_i$, e_i being the input voltage.*

Solution Use a combination of a summing operational amplifier and an inverting amplifier. The summing amplifier may give a gain of $-(2 + 3e_i)$ and the inverting amplifiers gain may be -1. Figure 5.17 shows the combination, with suitable values of resistances chosen.

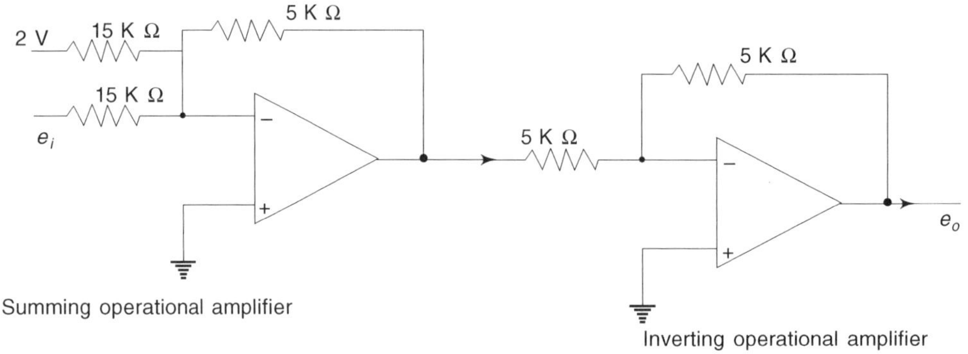

Fig. 5.17 *The arrangement to produce $e_o = 2 + 3\,e_i$*

5.3 ■ DIFFERENTIATING AND INTEGRATING ELEMENTS

There are certain applications where a transducer output needs to be differentiated or integrated, e.g. in an electrodynamic transducer, the output voltage is proportional to the velocity of the moving object. In case it is desired to measure the displacement and have an output proportional to it, it would be necessary to integrate the transducer output using an integrating element. Similarly, if the output is desired to be proportional to acceleration in the above case, a differentiating element is needed.

The circuit shown in Fig. 5.18 can be shown to give an output e_2 which is approximately a differential of the input e_1.

If i is the current,

$$e_1 = \left(R + \frac{1}{CD}\right) i \quad (5.23)$$

where R and C are the values of resistance and capacitance respectively, in the circuit and $D = d/dt$

Since $$e_2 = iR \tag{5.24}$$

we get from Eqs. (5.23) and (5.24)

$$(1 + RCD)e_2 = RCDe_1$$

If $$RCD \ll 1 \tag{5.25}$$

$$e_2 \approx RCD\, e_1 \tag{5.26}$$

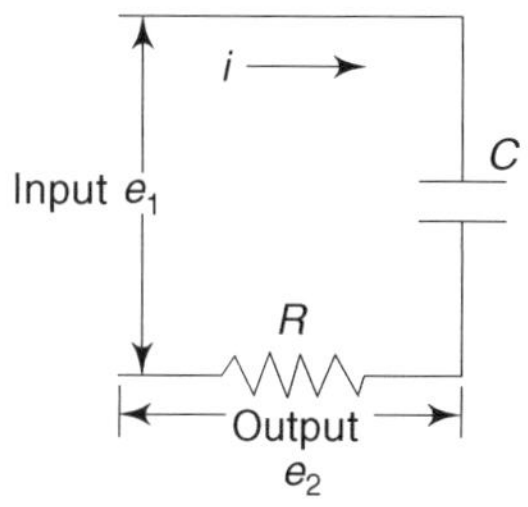

Fig. 5.18 *Differentiating circuit*

Equation (5.26) shows that voltage output e_2 is proportional to the derivative of e_1 provided Eq. (5.25) is satisfied. For a harmonic signal of circular frequency ω, Eq. (5.25) can be written in the following form, by putting $D = j\omega$,

$$j\omega RC \ll 1$$

or $$\omega \ll \left|\frac{1}{-jRC}\right| \ll \frac{1}{RC} \tag{5.27}$$

Similarly, it can be easily shown for the circuit of Fig. 5.19, that

$$e_2 \approx \frac{e_1}{RCD} \quad \text{or} \quad e_2 \text{ is proportional to integral of } e_1 \tag{5.28}$$

If $RCD \gg 1$

or $$\omega \gg \frac{1}{RC}, \tag{5.29}$$

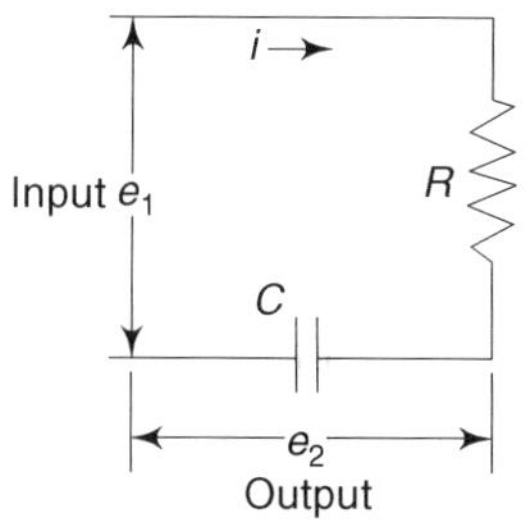

Fig. 5.19 *Integrating circuit*

for harmonic signals of circular frequency ω.

It may be seen from Eqs. (5.26) and (5.28) that (i) the magnitudes of the output voltages are reduced and (ii) a differentiating circuit would be effective at low frequencies, as determined by Eq. (5.27) while an integrating one at high frequencies, is given by Eq. (5.29).

Such systems are said to be passive. There is another class of differentiating and integrating elements of active type, in which an electronic amplifier is used. These systems are more accurate and their frequency range is not limited.

Figure 5.20 shows an active differentiator, in which a high gain operational amplifier with gain G is employed. The gain G is negative or the output is out of phase with the input.

If the gain G is very high, the potential at P can be considered to be very small and point P thus constitutes a virtual earth (zero potential). Thus, current $i_3 \approx 0$ and the amplifier can be taken to have an infinite input impedance. The current i_2 through R is given as

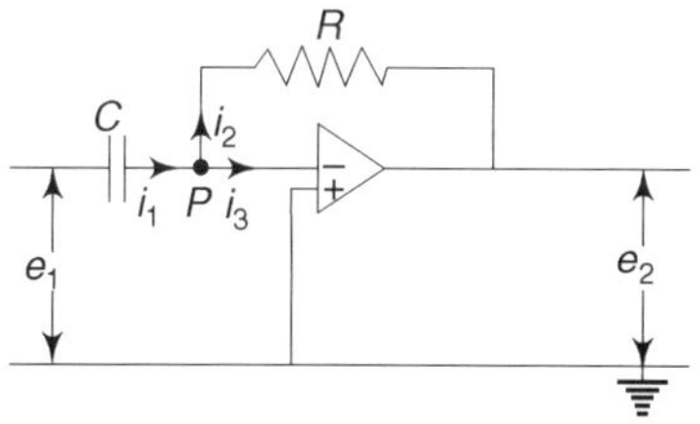

Fig. 5.20 *Differentiation using operational amplifier*

$$i_2 = -\frac{e_2}{R} \tag{5.30}$$

Current i_1 through C is

$$i_1 = C\frac{de_1}{dt} \tag{5.31}$$

Since $i_1 \approx i_2$, we find from Eqs. (5.30) and (5.31) that

$$e_2 = RC\frac{de_1}{dt} \tag{5.32}$$

Thus, e_2 is a differential of e_1,

It can similarly be proved that for the circuit of Fig. 5.21,

$$e_2 = \frac{1}{RC}\int_0^t e_1\,dt \tag{5.33}$$

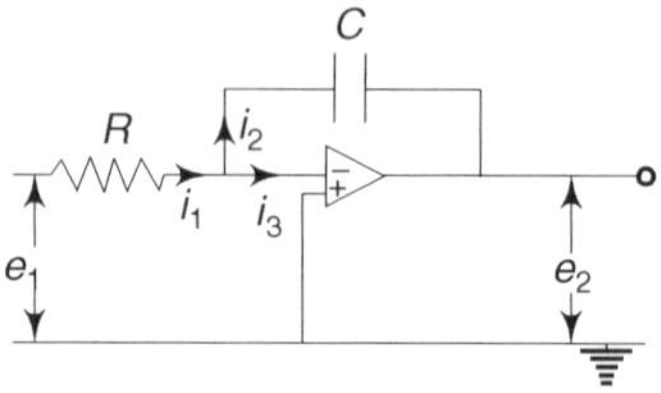

Fig. 5.21 *Integration*

5.4 ■ FILTERS

Filters are meant to remove unwanted signals from the desired transducer signal so as not to obscure the effect of the latter. Depending on the application, filters can be mechanical, electrical, pneumatic or hydraulic. Figure 5.22 shows a typical mechanical filter. The reference junction of a thermocouple is kept in ice, contained in a thermos flask, the insulation of which protects the system from ambient temperature changes. Similarly, in Fig. 5.23, a delicate instrument is shown mounted on vibration isolators, to isolate the instrument from disturbing vibration inputs.

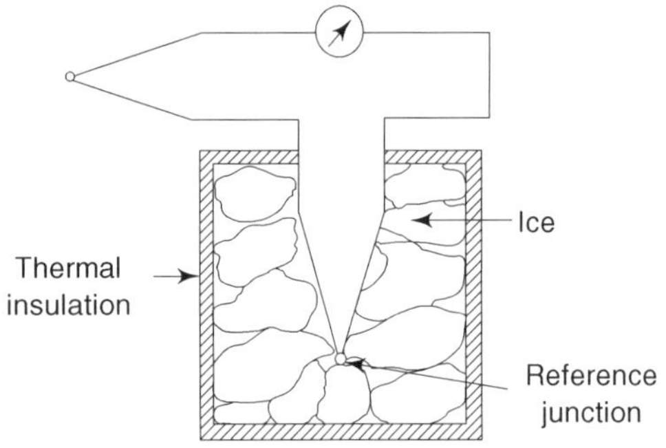

Fig. 5.22 *A typical mechanical filter (thermal insulation) provided in thermocouple device*

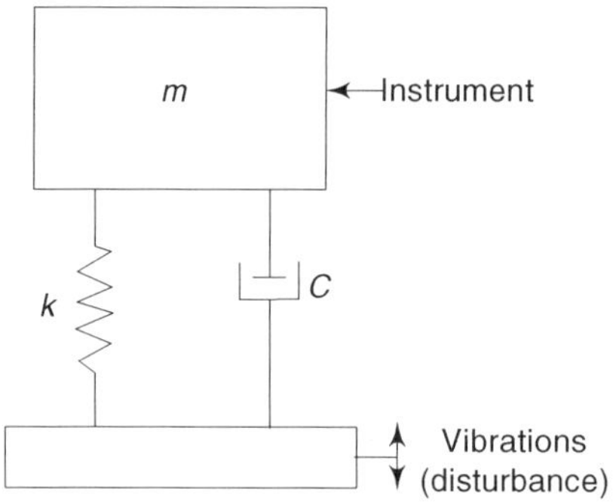

Fig. 5.23 *Isolators for filtering vibrations*

5.4.1 Classification of Filters

In general, filters may be classified as shown in Fig. 5.24. The amplitude ratio of output of the filter to the input is plotted against frequency. The low pass filter in Fig. 5.24(a) will pass signals of low frequencies till frequency f_1. The high pass filter of Fig. 5.24(b) will pass signals from frequency f_2 onwards while the band-pass filter of Fig. 5.24(c) will only pass signals which lie between frequencies f_3 and f_4. The notch type filter of Fig. 5.24(d) will block all signals that have frequencies between f_5 and f_6 while all others will appear in the output. Figure 5.24 shows the ideal filter characteristics, while in actual filters, the output to input amplitude ratio does not vary sharply.

Figure 5.25 shows a low pass electrical filter for which the ratio of output to input voltages, is given by

$$\frac{e_2}{e_1} = \frac{1}{1+\tau D} \tag{5.34}$$

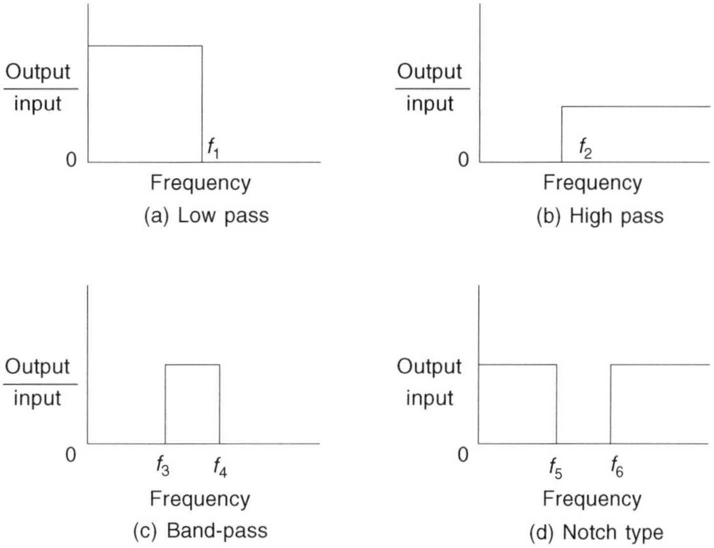

Fig. 5.24 *Various types of filters*

where $\tau = RC$. The frequency response of the above filter is similar to the plot in Fig. 5.24 (a), with the sharpness of cut-off depending on the value of τ. In Fig. 5.26, a low pass filter has been used with a resistance strain gauge bridge for filtering the unwanted noise signals of high frequency.

The low pass pneumatic filter shown in Fig. 5.27 has a governing relation similar to Eq. (5.34). If m is the mass flow rate through the restriction, with fluid resistance R, flowing to a space of pneumatic capacitance C, then according to the definitions of R and C,

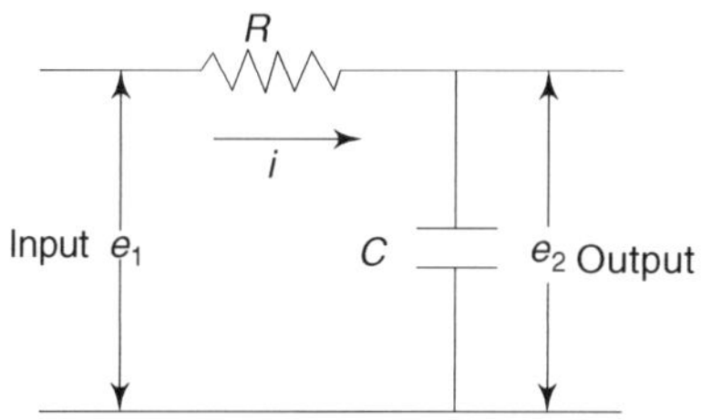

Fig. 5.25 *Low pass electrical filter*

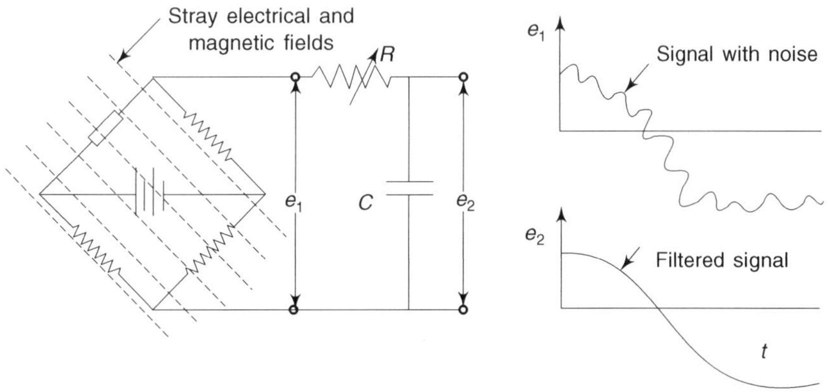

Fig. 5.26 *A low pass filter used with a resistance strain gage bridge*

$$\frac{p_1 - p_2}{R} = m = CDp_2$$

This gives

$$\frac{p_2}{p_1} = \frac{1}{1+\tau D} \tag{5.35}$$

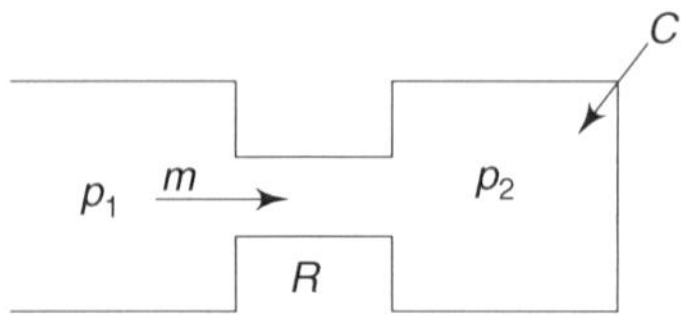

Fig. 5.27 *Low pass pneumatic filter C*

where $\tau = RC$. With such a filter, high frequency pulsations can be removed from the pressure signal p_1. Figure 5.28 shows a simple high pass filter, which can be shown to have the output-input relation

$$\frac{e_2}{e_1} = \frac{\tau D}{1+\tau D} \tag{5.36}$$

where $\tau = RC$. This has characteristics similar to those in Fig. 5.24(b).

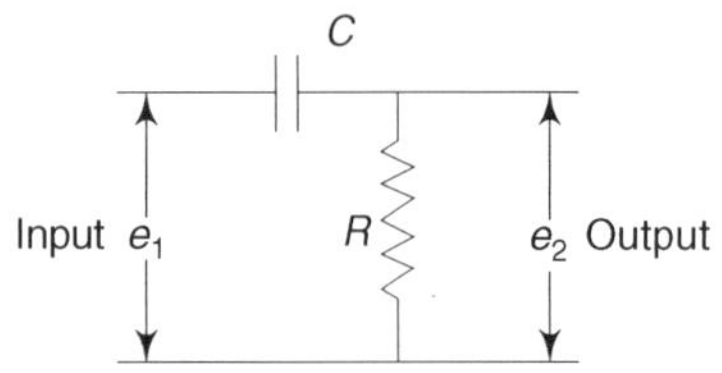

Fig. 5.28 *High pass electrical filter*

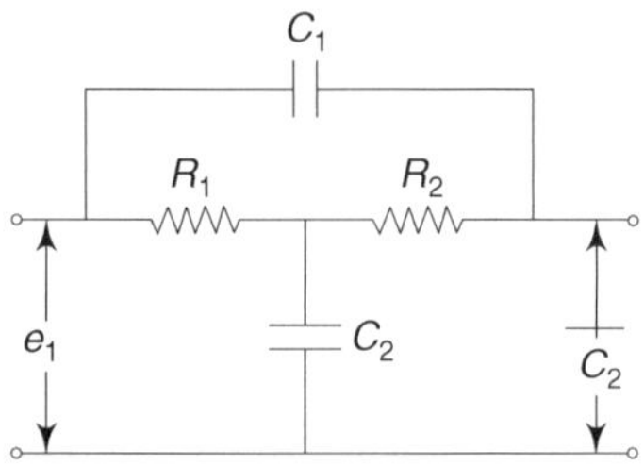

Fig. 5.29 *Notch type electrical filter*

The use of a low pass filter along with a high pass filter will result in a band-pass filter. Similarly, the filter shown in Fig. 5.29 can be shown to have characteristics of a notch type filter, corresponding to the characteristics of Fig. 5.24(d).

5.5 A–D AND D–A CONVERTERS

An analog output from any electromechanical transducer, may in certain cases, have to be converted to a digital form especially where a digital computer has to be used. This is done by using an A–D converter (analog to digital) between the transducer and the computer or a digital recording element. The reverse is also possible using D–A converters.

5.5.1 A–D Converters

The simplest type of an A–D converter is of potentiometric type and is shown in Fig. 5.30. This employs comparison elements that are in the form of semiconductor relays or magnetic cores. These elements change their state (ON/OFF), if the input signal is greater or less than a fraction of the reference voltage to which they are connected. The potentiometer resistances are weighted so that the reference potentials are according to binary values say 1, 2, 4, 8, etc. An automatic switching unit compares the analog input value with the reference voltages, starting with the highest and the comparison elements change their state accordingly, thus giving a digital output.

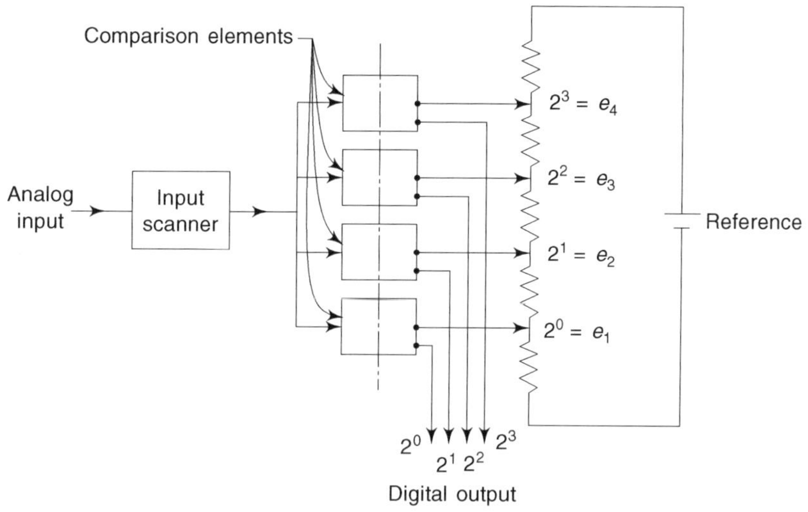

Fig. 5.30 *Potentiometric type A–D converter*

Based on a similar principle, a successive approximation type A–D converter, employing a D–A converter is commonly used, as shown in Fig. 5.31.

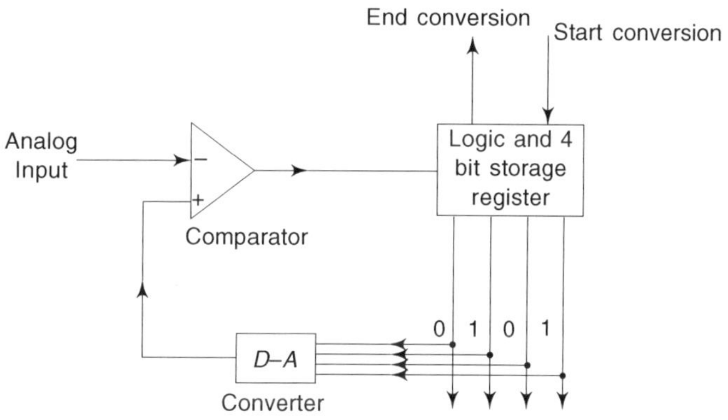

Fig. 5.31 *Successive approximation type A–D converter*

In the above converter, the analog input from a sensor is compared with the voltage generated by a D–A converter from a logic and multiple bit storage register successively in a certain sequence. When the two are equal, the conversion is ended and an acceptable digital output is available.

The sequence usually employed consists of selecting the MSB that is less than the analog value, then adding successively the less significant bits for which the total does not exceed the analog input, e.g. if 4 bits are used, we may start with 1000 and by the logic shown in Table 5.1, we get the acceptable values corresponding to various cases involving different values of analog signals.

In a similar way, the trails for other analog values in the range 0–15 can be written. If the number of bits were 8, the values in the range 0–255 can be obtained by trial. In this method, an n-bit word takes n steps for comparison and time taken to obtain the bit word is n/f, where f is the clock frequency.

Table 5.1

Cases	*Serial nos. of successive trials*	*D–A converter input*	*D–A converter Output*	*Comparator output*	*Resulting value*
A	1	1 0 0 0	8	Large	—
	2	0 1 0 0	4	Small	—
	3	0 1 1 0	6	Large	—
	4	0 1 0 1	5	Nil	5 Acceptable
B	1	1 0 0 0	8	Large	—
	2	0 1 0 0	4	Small	—
	3	0 1 1 0	6	Small	—
	4	0 1 1 1	7	Nil	7 Acceptable
C	1	1 0 0 0	8	Large	—
	2	0 1 0 0	4	Large	—
	3	0 0 1 0	2	Small	—
	4	0 0 1 1	3	Nil	3 Acceptable

Another type of A–D converter, is based on a different principle. This is called a counting type A–D converter and is shown in Fig. 5.32. Here, the analog input voltage e_i is compared against a ramp voltage e_r generated by a ramp generator. As shown in Fig. 5.33, the time taken to reach equal values is a measure of the analog input voltage e_i, since the slope of the ramp voltage is constant. In order to get a digital output proportional to the time taken for the two voltages to be equal (which in turn is proportional to e_i), an electronic gate is opened to permit the output from a frequency oscillator to be fed to a counter. The gate is open for the times t_1, t_2, t_3, etc. as shown in Fig. 5.33, viz. the times taken for e_i to be equal to e_r. Since the frequency of the oscillator is constant, the number of oscillations counted by

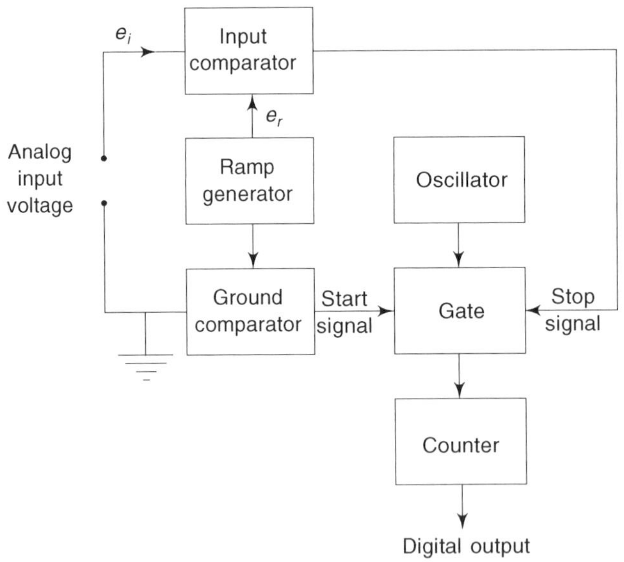

Fig. 5.32 *Counting type A–D converter*

the counter is thus a digital measure of the analog input voltage e_i. The counter can be arranged to give a binary output if required. The opening of the gate, according to the beginning of ramp and closing of the gate when $e_i = e_r$, are controlled as shown in Fig. 5.32.

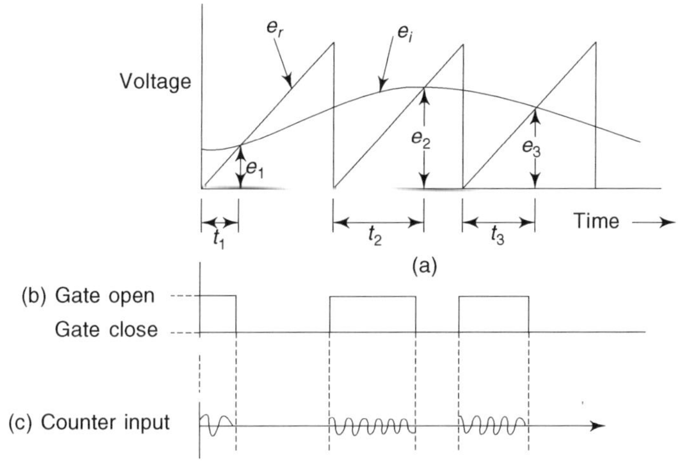

Fig. 5.33 *Voltage, gate positions and counter inputs*

Yet another type of A–D converter, known as the dual slope integrating type converter, is shown in Fig. 5.34(a).

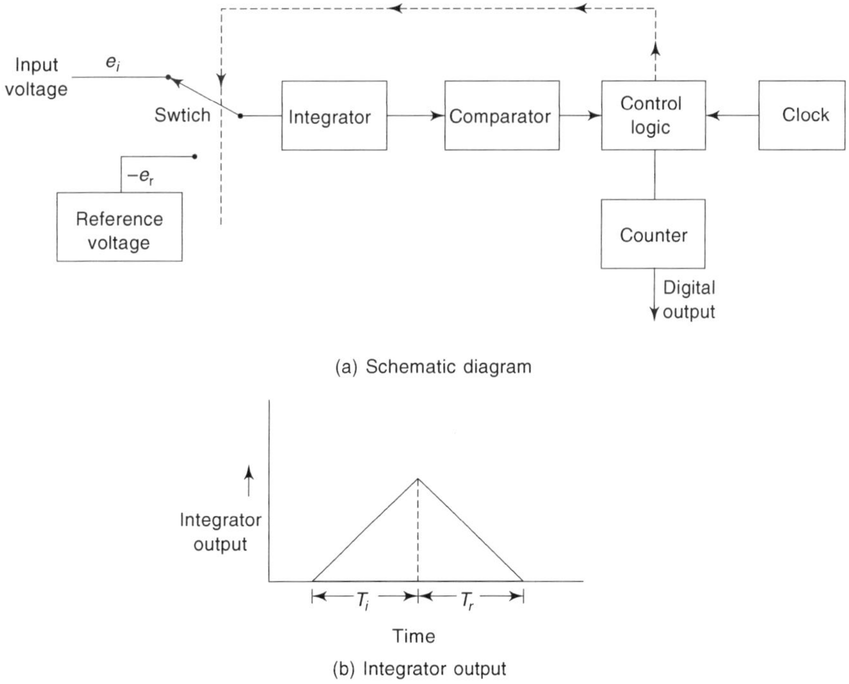

Fig. 5.34 *Dual slope integrating type A–D converter*

In this, the input voltage e_i is integrated for a fixed time T_i and later, an internal reference votage of opposite polarity is gated as the input. The switching is done by the control logic and the counter is reset to zero. At the end of time T_r, the integrator output reaches the threshold value of the comparator and the counter is stopped as shown in Fig. 5.34(b). The clock output is gated to the counter which counts up to the desired time T_i and T_r. It can be easily seen that input voltage

$$e_i = \frac{T_r}{T_i} e_r$$

This type of converter is quite accurate and stable as its performance is affected by the reference source whose accuracy and stability can be ensured.

5.5.2 D–A Converter

Whenever digital signals have to be converted to analog ones, the switching defining the digital signal value has to be changed to an equivalent voltage. Figure 5.35 shows one type of D–A converter, similar to the potentiometric A–D converter discussed earlier.

The switches are opened when a particular digital value is present, otherwise these are closed. Thus, the feedback resistance of the amplifier shown, at any instant, is the sum of the resistance values that are not shorted out. If the resistance in the feedback citcuit are binary weighted, with resistance values say 2^0, 2^1, 2^2, 2^3, etc. and if a digital input of value 7 is present, the switches for digits 2^0, 2^1, 2^2 would be open, giving

$$e_0 = -e_r \frac{(2^0 + 2^1 + 2^2)}{R_r}$$

$$= -7 \frac{e_r}{R_r}$$

The analog output voltage would vary in steps but is smoothed out by filtering.

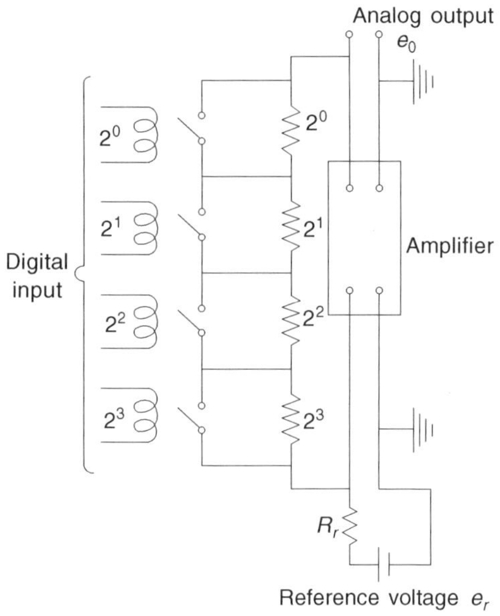

Fig. 5.35 *D–A Converter*

5.6 ▮ TERMINOLOGY AND CONVERSIONS

A–D and D–A conversions relate analog and digital values. In natural binary code, a number is represented by:

$$a_n 2^n + a_{n-1} 2^{n-1} + \ldots + a_1 2^1 + a_0 2^0.$$

The coefficients of a_n may have the value 1 or 0. The most significant bit (MSB) in the digital code is put on the LHS and the least significant bit (LSB) on the RHS in digital representation. MSB has a weight of $\frac{1}{2}$ of full scale or span of the converter, the second bit has a weight of $\frac{1}{4}$ of full scale or span and LSB has the same as $\frac{1}{2^n}$ of full scale or span of the converter, which may be a suitable voltage value.

$$\text{Resolution of the converter} = \frac{\text{Full scale or span}}{2^n}$$

n being the number of bits. It is also the weight of L.S.B.

The span is the difference between the maximum and minimum analog values while the offset is the minimum analog value. It is seen that analog value = digital number × step size + offset

Problem 5.2 *In a system, the analog range is –3 to + 5 V and there are 8 bits in the digital system. Specify the span, offset and step size. Also find the bit weights for each of the digital bits and the analog value corresponding to 10010001.*

Solution

$$\text{Span} = 5 - (-3) = 8 \text{ V}$$

$$\text{Offset} = -3 \text{ V}$$

$$\text{Step size} = \frac{8}{2^8} \text{ V} = \frac{8}{256} \text{ V}$$

$$= \frac{1}{32} \text{ V}$$

Digital bit	*Bit weight (V)*
7	$8 \times 2^{-1} = 4.0$
6	$8 \times 2^{-2} = 2.0$
5	$8 \times 2^{-3} = 1.0$
4	$8 \times 2^{-4} = 0.5$
3	$8 \times 2^{-5} = 0.25$
2	$8 \times 2^{-6} = 0.125$
1	$8 \times 2^{-7} = 0.0625$
0	$8 \times 2^{-8} = 0.03112$

$$\text{Analog value} = 1\,(8 \times 2^{-1}) + 1(8 \times 2^{-4}) + (8 \times 2^{-8})1 + (-3)$$

$$= 4 + 0.5 + 0.03112 - 3 = 1.53112 \text{ V}$$

5.7 ■ DATA TRANSMISSION ELEMENTS

When the measured variables have to be transmitted over long distances from the measuring points to a location for display or recording of data, data transmission elements are employed. These are also called telemetry units or systems and may be classified into two categories:

1. Land-line or cable type transmission elements.
2. Radio-frequency (RF) type data transmission elements.

In the former units, data is transmitted by wires or pipes while in the latter it is transmitted by radio waves. The former finds applications in data transmission in process plants, power generating stations, etc. and includes electrical, pneumatic and position type elements while the latter is used in aero-space systems.

5.7.1 Electrical-Type Data Transmission Elements

In these elements, the input measured variable, usually a motion signal, is made to change an electrical quantity, the effect of which is transmitted by wires to the receiving end, for record or display.

Figure 5.36 shows two such elements. In Fig. 5.36(a), the position of contact C on a variable resistance AB is adjusted by the input motion, changing the value of the current through the lines. The current at the receiving end is a measure of the input variable. In this type, resistance changes due to temperature changes introduce errors. In Fig. 5.36(b), instead of measuring the current at the receiving end a potentiometer is used, which is balanced so that no current flows, as indicated by G. Thus, the setting of the potentiometer gives an indication of the input signal. In this case, the effect of change of line resistance due to temperature, etc. is eliminated.

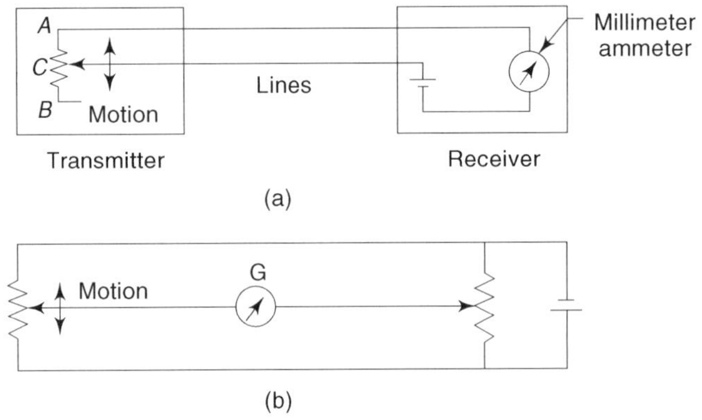

Fig. 5.36 *Data transmission by change of electric quantity*

5.7.2 Pneumatic-Type Transmission Elements

These are also a land-line type. A typical arrangement is shown in Fig. 5.37, and uses the flapper-nozzle arrangement. The signal to be transmitted is converted into the form of a motion signal x. With change in x, pressure p_2 changes as shown. The pressure p_2 gets transmitted to the receiving end and may, by use of an elastic element, be converted to motion for recording.

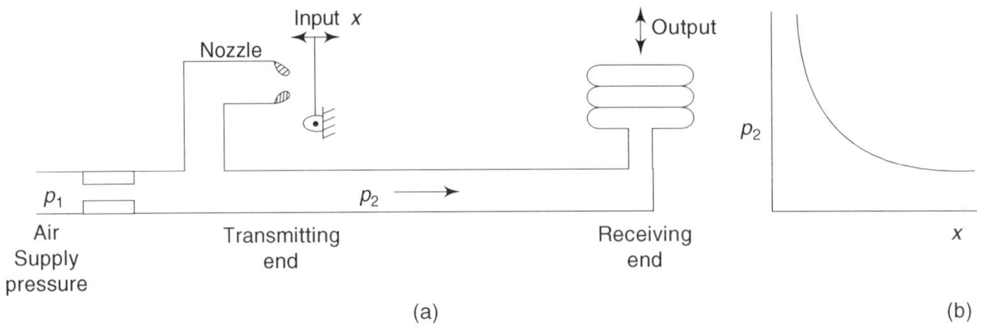

Fig. 5.37 *Pneumatic transmitter*

Figure 5.38 shows a flow transmitter of the pneumatic type, employing a system similar to that of Fig, 5.37. The flow signal results in a pressure difference across the restrictor (which may be an orifice or venturi or a nozzle). The pressure difference results in the deflection of an elastic diaphragm and the motion signal is transmitted, by being converted to the pressure signal p_2. These elements are not linear, as shown in Fig. 5.37 (b).

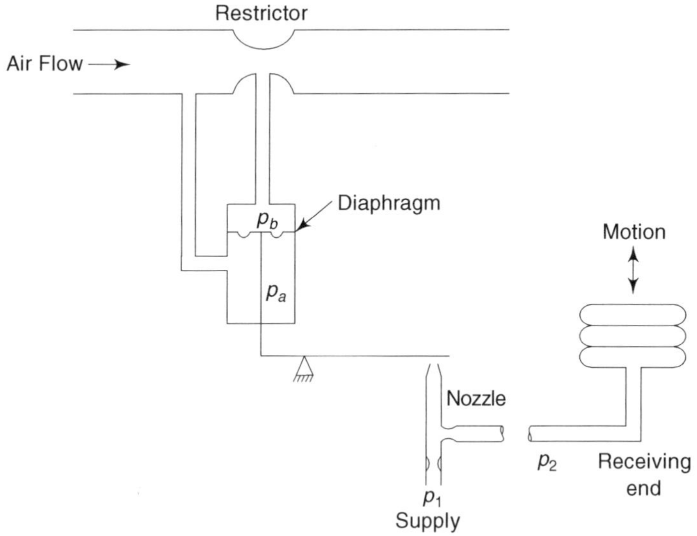

Fig. 5.38 *Pneumatic flow transmitter*

Another type of pneumatic transmission system called force-balance type, has better linearity characteristics. This is shown in Fig. 5.39.

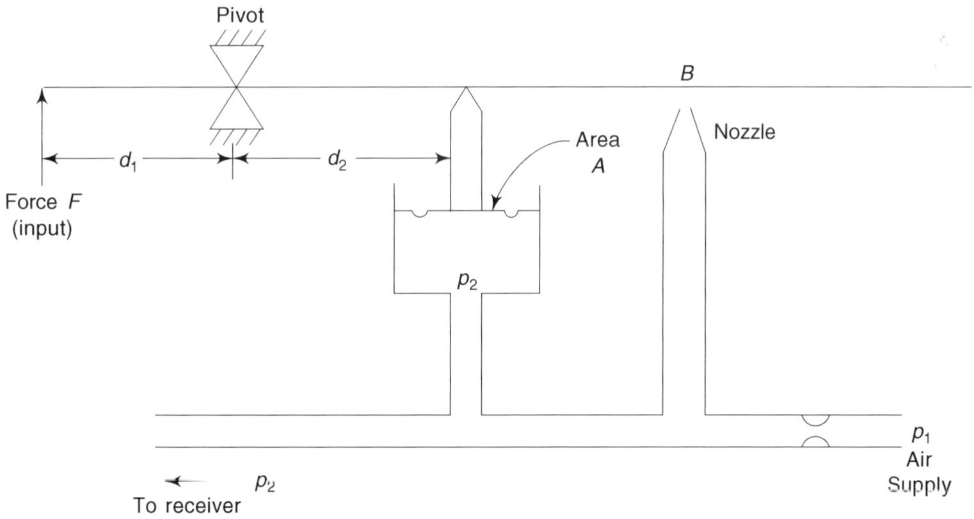

Fig. 5.39 *Force-balance type pneumatic transmitter*

The input signal is in the form of a force signal F applied at the end of a pivoted lever (a motion input signal may be applied through a spring, resulting in force F). Application of F rotates the level clockwise as shown, decreasing the gap between the level extension and the nozzle and thus increasing pressure p_2 till the lever balances and no further building up of pressure occurs. At balance,

$$p_2 A\, d_2 = F d_1 \tag{5.37}$$

A being the area of the diaphragm shown and d_1 and d_2, the distances from the pivot. Equation (5.37) shows that p_2 is linearly related to the input F.

In pneumatic type transmitters, there is a pressure drop in transmission piping, resulting in a reduction in the signal transmitted. In practice, these are used for transmitting the signal over a few hundred metres.

5.7.3 Position-Type Data Transmission Elements

In these types, the motion signal (like rotation of a pointer) is transmitted over long distances, by use of synchros. Two synchros–a transmitter and a receiver are employed (Fig. 5.40). A synchro consists of a stator with three coils at 120°, inside which is a rotor, which is free to rotate within the stator windings. The transmitting synchro is energised by an ac power source. If the two rotors are in identical positions, the voltages in stators have the same magnitude but opposite sense and no current flows in the stator wires. If the input rotor is turned, making the two rotor positions different, current flows in the stator wires producing a resulting torque which would align the two rotors making $\theta_0 = \theta_i$. Thus, the angular motion θ_i is transmitted to the receiving end.

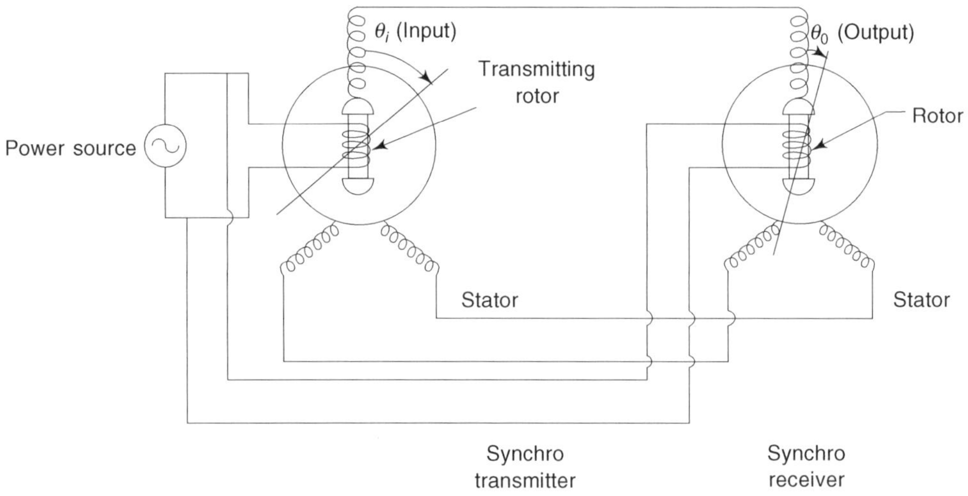

Fig. 5.40 *Synchros for data transmission*

5.7.4 Radio-Frequency Transmission System

Such data transmission systems use radio-frequency waves for data transmission and no wires or cables are needed between the transmitting and receiving ends. In large systems like aero-space systems, a number of input signals like temperature, pressure, vibrations, etc. may be transmitted by such units. Figure 5.41 shows such a radio-frequency telemetry system. The transducer outputs are first modified by signal conditioners, into suitable form and magnitude for each (sub-carrier oscillator SCO) whose output gets

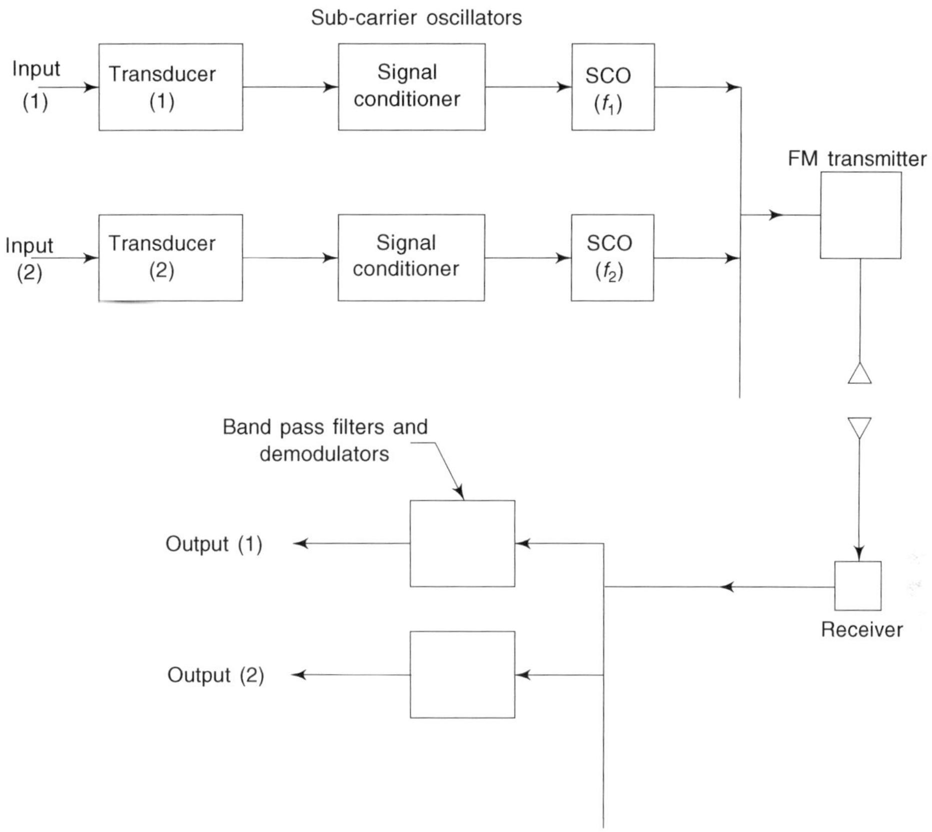

Fig. 5.41 *RF telemetry for data transmission*

modulated due to the input signals. The frequency corresponding to each SCO is f_1, f_2, etc. The outputs of sub-carrier oscillators are mixed and then applied to a radio-frequency oscillator for transmission. The signal is received by a suitable antenna at the receiving end and after suitable demodulation and filtering, the original signals are recovered.

Review Questions

5.1 Indicate true or false against each.

(i) A dc amplifier cannot amplify ac signals.

(ii) dc amplifiers suffer from drift.

(iii) Use of feedback in amplifiers improves the frequency response.

(iv) Carrier amplifiers cannot amplify dc signals.

(v) Compensating elements are used to improve dynamic characteristics.

(vi) An *RC* circuit can be used both as an integrating and differentiating element.

(vii) Condition for integrating a harmonic signal of frequency ω by an *RC* circuit is

$$\omega \ll \frac{1}{RC}$$

(viii) A low pass filter will filter out unwanted signals of low frequencies.

(ix) An A–D converter cannot be used for dynamic signals.

(x) Synchros can be used for data transmission.

5.2 Fill in the blanks:

(a) The most desirable characteristic of an amplifier is __________.

(b) Charge amplifier is used to amplify the import signal of __________.

(c) Operational amplifiers are used for __________.

5.3 In the R–C circuit shown in Fig. 5.42., (R = 0.1 MΩ, C = 0.1 μF), Find the frequency range over which this can be used as an integrating circuit, for harmonic input e_1.

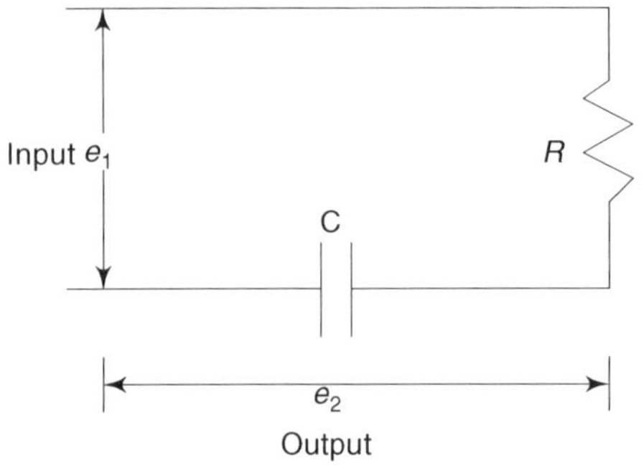

Fig. 5.42

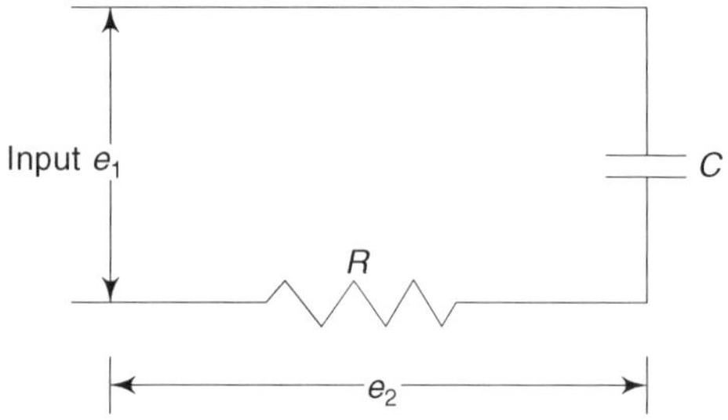

Fig. 5.43

5.4 For the circuit of Fig. 5.43, find for a harmonic input e_1, the frequency range over which it can be used as a differentiating circuit. (R = 1000 Ω, C = 1 μF).

5.5 Design a input high impedance amplifier to give a voltage gain of 50.

5.6 A 4 bit D–A converter has an analog output range of –2 to 4 V. Find the offset, span, step size and analog value corresponding to the input 1010.

Answers

5.1 (i) F (ii) T (iii) T (iv) F (v) T
(vi) T (vii) F (viii) F (ix) F (x) T

5.2 (a) high input impedance
(b) piezo-electric transducer
(c) amplification, integration, differentiation, summing

5.3 Frequency ≫ 15.9 Hz

5.4 Frequency ≪ 159 Hz

5.5 R_1 = 1 kΩ, R_f = 49 kΩ (op. Amp.)

5.6 –2 V, 6 V, 0.375 V, 1.75 V

Chapter

6

Indicating, Recording and Display Elements

INTRODUCTION

The final stage in a measurement system comprises an indicating and/or a recording element, which gives an indication of the input being measured. These elements may also be of analog or digital type, depending on whether the indication or recording is in a continuous or discrete manner. Conventional voltmeters and ammeters are the simplest examples of analog indicating instruments, working on the principle of rotation of a coil through which a current passes, the coil being in a magnetic field. Vacuum tube voltmeters use a similar principle but have amplifiers built in for enhanced sensitivity. Such instruments measure a wide range of ac or dc voltages. These instruments are well known and are not described here. Such instruments have circuits for measuring peak, average or rms values of voltage signals.

Digital voltmeters (DVMs) are commonly used as these are convenient for indication and are briefly described here. Cathode ray oscilloscopes (CROs) have also been widely used for indicating these signals.

Recording instruments may be galvanometric, potentiometric, servo types or magnetic tape recorder types. In addition to analog recorders, digital recorders including digital printers, punched cards or tape recording elements are also available.

In large-scale systems, data loggers incorporating digital computers are extensively used for data recording. The present day availability of memory devices has made the problem of data storage simpler than was hitherto possible.

6.1 DIGITAL VOLTMETERS (DVMs)

Digital voltmeters convert analog signals into digital presentations which may be as an indicator or may give an electrical digital output signal. DVMs measure dc voltage signals. However, other variables like ac voltages, resistances, current, etc. may also be measured with appropriate elements preceding the input of the DVM. The A–D converters discussed in Ch. 5, with appropriate indication, would be examples of DVMs.

In general, DVMs can be classified into two types—(a) non-integrating and (b) integrating types.

Figure 6.1 shows a potentiometric type of DVM, based on the non-integrating principle. Figure 6.1(a) shows a manual type, in which the input voltage e_i to be measured is compared with the voltage obtained from an internal reference, which is applied to a potentiometer. The null indicator indicates when the two voltages become equal. Thus, the position of the slider on the potentiometer would be an indication of the voltage e_i. The operation is made automatic in a servo type potentiometric DVM, as shown in Fig. 6.1(b), position x being an indicator of voltage e_i, and comparison of voltage e_i and voltage from the potentiometer being carried out automatically, by feedback action.

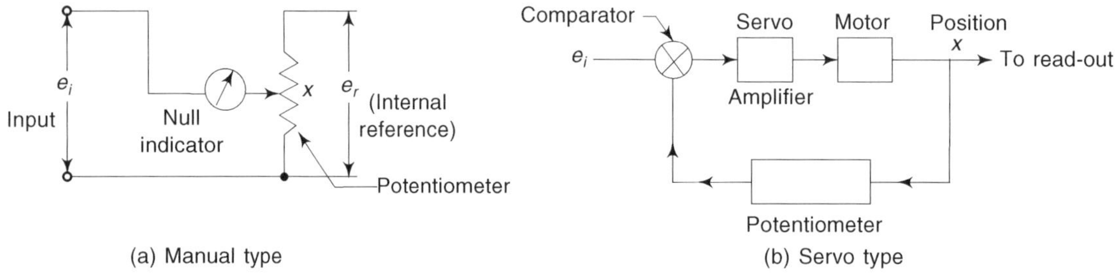

Fig. 6.1 *Potentiometric type DVM*

Figure 6.2 shows an outline of the integrating type of DVM. This is essentially a voltage-to-frequency converter, in which a dc signal is converted to a periodic signal of frequency proportional to the magnitude of the signal. Figure 6.2(a) shows the block diagram of the instrument. The input signal to be measured is e_i. The amplifier network, with the capacitances and resistances shown, is an integrating network. Thus, the voltage e_o would increase linearly with time if voltage e_i was constant. Voltage e_o is compared with a set value e_t as shown in Fig. 6.2(b). A trigger pulse is produced whenever the two voltages become equal. As seen from Fig. 6.2(b), the voltage e_o changes slowly with the level of e_i in a linear fashion (since the former is proportional to the integral of the latter). Thus, the frequency of the pulses would be less if the level of e_i is lower. These pulses are counted by a counter and give an indication of e_i. At the same time, the pulse trigger operates a pulse generator, which produces a pulse of short duration. This tends to restore the output voltage of the amplifier to zero, for starting the next cycle. Thus, pulses continue to be produced at a rate proportional to the magnitude of e_i.

6.2 ■ CATHODE RAY OSCILLOSCOPES (CROs)

As an indicating element, a CRO is widely used in practice. It is essentially a high input impedance voltage measuring device, capable of indicating voltage signals from the intermediate elements as a function of time.

Figure 6.3 shows the block diagram of a cathode ray oscilloscope. Electrons are released from the cathode and accelerated towards the screen by the positively charged anode. The position of the spot on the phosphorescent screen is controlled by voltages applied to the vertical and horizontal plates. The impingement of the electron beam on the screen results in emission of light and thus the signal becomes visible.

As seen in Fig. 6.3, the following are the essential components in a CRO:

1. display device, viz. the tube,

2. vertical amplifier,
3. horizontal amplifier,
4. time base,
5. trigger or synchronizing circuit, to start each sweep at a desired time, for display of signal, and
6. power supplies and internal circuits.

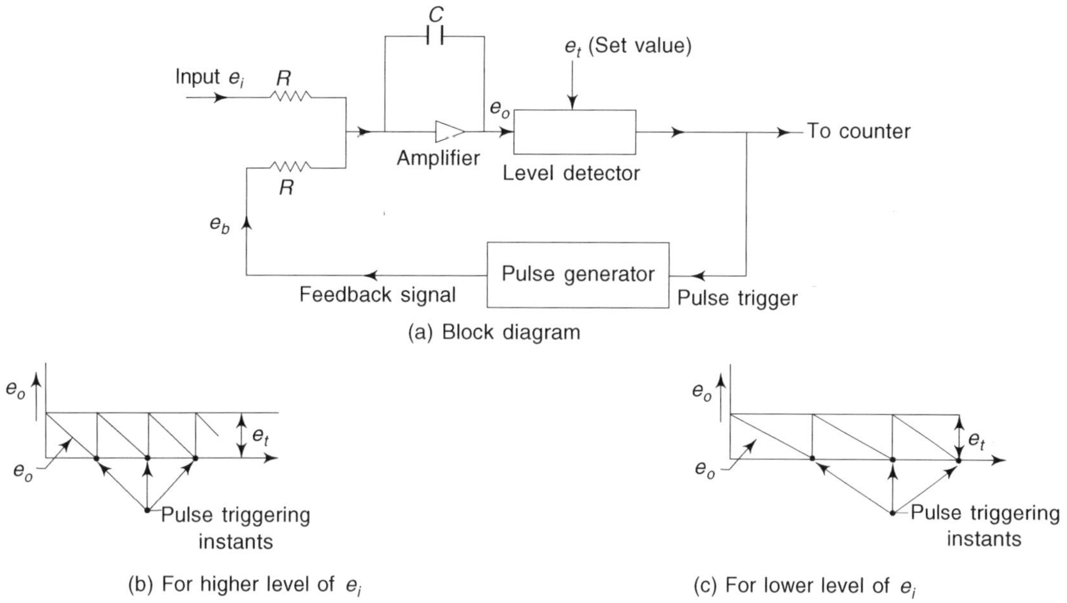

Fig. 6.2 *Voltage to frequency integrating DVM*

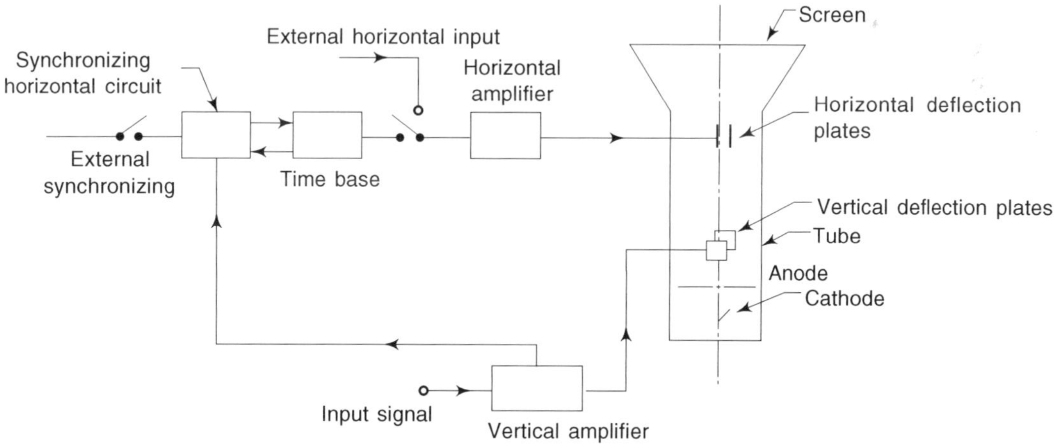

Fig. 6.3 *Block diagram of a CRO*

The time base signal is a ramp type sawtooth wave which provides a horizontal deflection to the electron beam. The frequency can be adjusted according to that of the vertical signal so that a stationary signal versus time plot is visible on the screen. The time base may be cut off and the external signal applied to the horizontal plates. An amplifier may be used to magnify the horizontal sweep. This permits magnification of a portion of the normal sweep for closer examination.

A delayed sweep may be used for the same purpose. A small portion of the normal display may be selected, to be shown at a selected value of sweep rate. This magnifies the portion selected. This is useful for phase measurement between the two signals, applied to the vertical and the horizontal plates as shown in Fig. 6.4.

It is shown in Chapter 20 that in case the two signals have the same frequency ω and have phase ϕ, the locus of the electron beam is an ellipse, and $\sin \phi = y_2/y_1$, where point o is at the centre of the ellipse, y_2 and y_1 are the distances shown in Fig. 6.4; y_2 is measured along OY and y_1 is the distance between the tangents at the extreme points of the ellipse, parallel to OX.

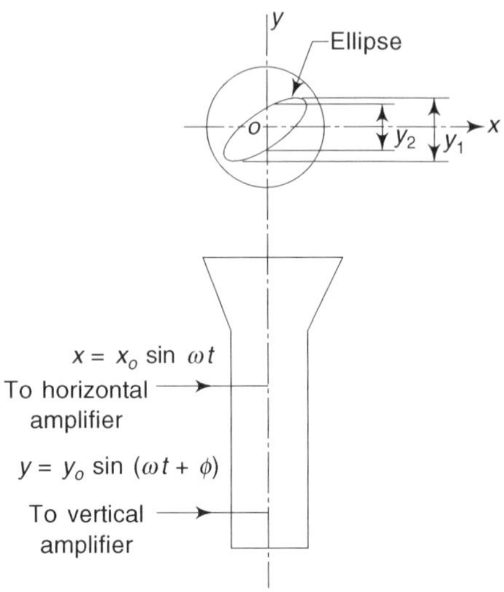

Fig. 6.4 *Use of CRO for phase measurement*

The patterns displayed on the screen by the two signals, applied to the vertical and the horizontal plates, are known as Lissajous diagrams. Some typical diagrams are shown in Fig. 6.5, for different ratios of vertical input frequency to the horizontal input frequency. These shapes can be used to compare two signals with different frequencies. The shapes would vary somewhat depending on the phase relation between the two input signals.

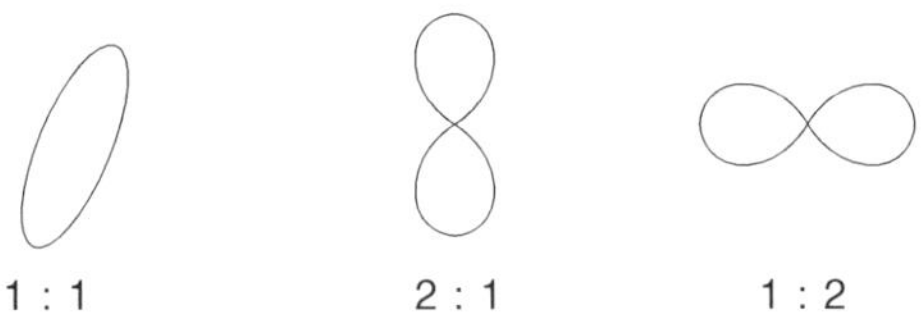

Fig. 6.5 *Typical Lissajous diagrams*

There are several ways, in which CROs can be classified. Some of the important specifications are:

1. Single or double beam oscilloscopes depending on the number of input signals that can be applied to the vertical plates, for simultaneous indication.
2. Frequency range, which may vary from dc (zero frequency) to several MHz, determined primarily by the frequency response of the amplifier used.
3. Sensitivity, viz. the voltage needed for unit deflection of the beam. The sensitivity can be adjusted and the range may vary from a few μV/cm to several V/cm.
4. Whether a signal can be stored (as in a storage oscilloscope) or not.
5. Timing resolution of the horizontal scan.
6. Shortest rise time possible for a signal.
7. Vertical sensitivity, accuracy and resolution.
8. Update rate to get ready for new triggering.
9. Whether it is analog or digital type of CRO.
10. Whether the signal can be stored or not.

In a conventional CRO, the persistence of the phosphor is for a few seconds. However, in a storage oscilloscope, the display can be stored for several hours. The storage oscilloscope can capture and store a transient signal that occurs only once.

A digital CRO can store analyse and display a signal. It uses digital acquisition system including an analog to digital converter (ADC), digital memory, digital to analog converter (DAC), and also a display system as shown in Fig. 6.6. In the memory unit, the stored signal can be analysed and manipulated as desired.

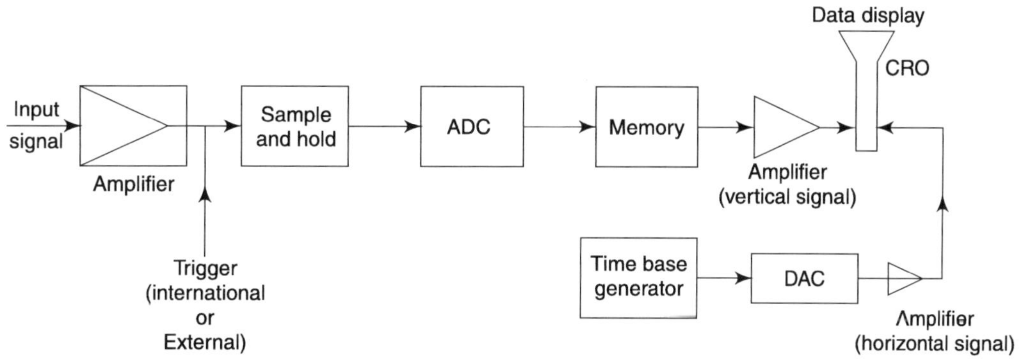

Fig. 6.6 *Digital Storage oscilloscope*

The frequency response of the digital storage oscilloscope is limited by the sampling frequency, the maximum signal frequency being about one-fourth of the sampling frequency. The hold in the sample and hold device is meant to keep the instantaneous signal waiting till the previous signal is being digitized in the ADC. The memory can store the signal as long as needed till it is deleted as in a personal computer.

6.3 ■ GALVANOMETRIC RECORDERS

These are based on the simple principle of rotation of a coil through which current due to the input signal to be recorded, flows while the coil is in a magnetic field, as shown in Fig. 6.7.

An ink pen attachment to the coil can be used to trace the signal on a paper wrapped around a rotating drum. The system acts like a second order instrument and the frequency response is limited to 200 Hz or so, due to the inertia effects of the pen and the coil. A pen recorder is shown in Fig. 6.7(a). In Fig. 6.7(b), the pen attachment is replaced by a light beam from a high-pressure mercury lamp source, with the light getting reflected from a small mirror attached to the coil. Due to rotation of the coil, the light beam gets deflected and a trace is made on the light sensitised paper. The high-frequency response is good till several kHz.

In some commercial instruments, there are plug-in types of galvanometers, with different natural frequencies depending on the type of coil suspensions.

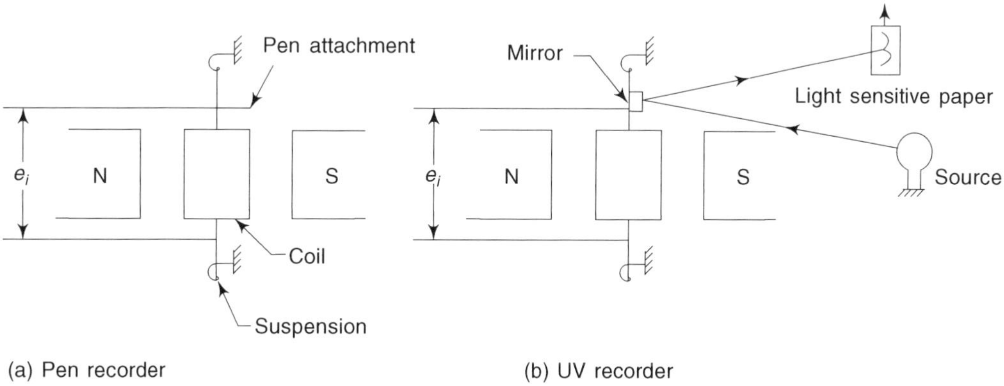

Fig. 6.7 *Galvanometric oscillograph*

6.4 ■ SERVO-TYPE POTENTIOMETRIC RECORDERS

These types of recorders, also known as self-balancing types of potentiometers, are commonly used in industrial situations, as they are quite rugged and not as delicate as the galvanometric recorders. Further, there is no limitation as far as the power required to move the pointer mechanism is concerned.

Figure 6.8(a) shows the block diagram and Fig. 6.8(b) shows the outline diagram of a servo type potentiometric recorder. The voltage to be recorded e_i is compared with a reference voltage e_r which is proportional to the output motion x_o. The difference or error voltage e_e is amplified and applied to a drive motor which, through a rack and pinion, moves the pointer through motion x_o. The system can be at rest only if e_e is zero. This occurs only when $e_t = e_r$. Since e_r is proportional to x_o, e_i is also proportional to x_o under steady conditions or the motion of the output pointer x_o is proportional to the voltage input e_i. If e_i varies, x_o tends to follow it.

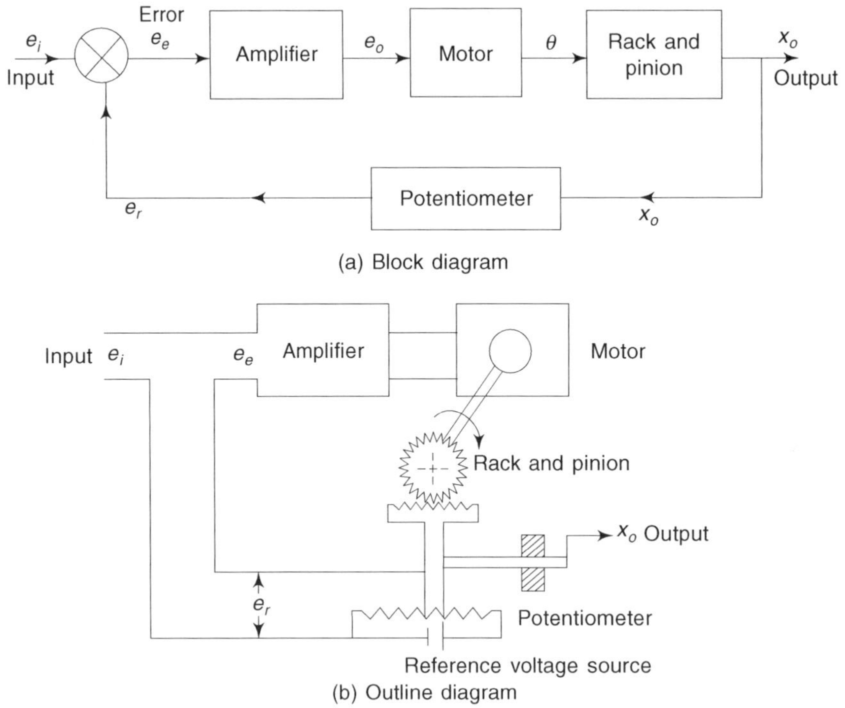

Fig. 6.8 *Servo type potentiometric recorder*

This type of system can also give a digital output by using a digital encoder. In some situations, the use of two independent servo systems allows one variable to be plotted against another, e.g. pressure–volume plot of an internal combustion engine. Such arrangements are known as *x–y* plotters.

The frequency response of servo type recorders is flat only at low frequencies. So, such types of recorders are normally employed for static measurements.

6.5 ■ MAGNETIC TAPE RECORDERS

Of late, a magnetic tape recorder has been used increasingly for recording data. The magnetic tape is made of a thin plastic material, coated with oxide particles, which become magnetised when the tape passes across a magnetising head which acts due to an input signal. The signal is recovered from the tape by a reproduce head. There are several types of magnetic recording systems, viz. direct recording, frequency modulated (FM), pulse duration modulation (PDM) and digital recording systems.

Figure 6.9(a) shows the block diagram of a direct recording system and Fig. 6.9(b) a typical magnetic head. The current in the record head winding is proportional to the input voltage to be recorded. This produces a magnetic flux at the gap and the tape is magnetised. If the tape speed is v and f is the frequency of the input signal, then the wavelength of magnetization variation is v/f. If this becomes equal to or less than the gap width, the average magnetisation tends to be zero and the variation in magnetisation cannot be detected. This may happen at high frequencies. Further, the output of the reproduce head is proportional to the rate of change of flux in the gap. The output is reduced considerably for low frequency signals and for dc at the input head the output at the reproduce head would be zero. Then, the frequency response is flat between say 50 Hz to about 50 kHz or so.

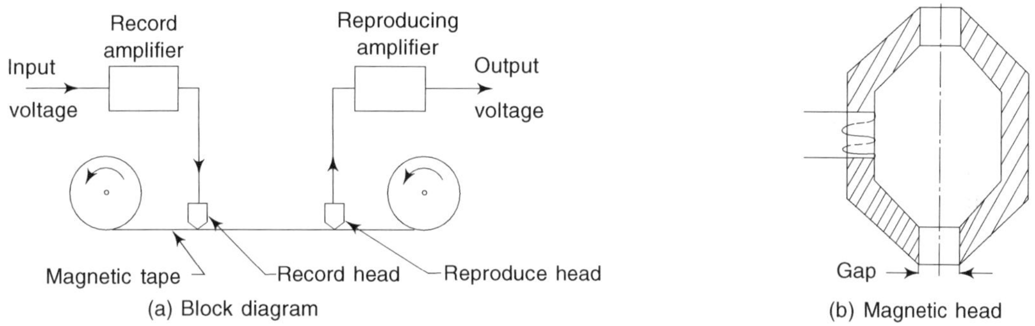

Fig. 6.9 *Direct recording system*

In the FM recording system shown in Fig. 6.10, the input signal to be recorded is made to modulate a carrier signal. The frequency modulated signal, as recorded, is shown in Fig. 6.10(c). The signal from the reproduce head is demodulated to get the output signal as shown in Fig. 6.10(b). FM recording system have the advantage that they can be used for dc voltages as well, viz. from zero frequency onwards. The higher limit is dictated by the carrier frequency used and can be till 10 kHz or so in available recorders. The disadvantages of the FM system are due to the complexity of the circuitry required. A number of channels may be used using a different carrier frequency for each channel.

Figure 6.11 show a PDM recording system. In this case, the duration of pulses from the keyer is made proportional to the amplitude of the signal to be recorded. The pulses are transformed to spikes as shown in Fig. 6.11(b) and (c), which define the beginning and end of the pulses. The positive and negative spikes are recorded on the tape. The original waveform is reconstructed on play back, using a dekeyer as shown in Fig. 6.11(d). A number of signals (more than in FM recording) can be easily recorded using a commutator before the keyer.

In digital recording systems, the tape is magnetised to saturation in either one of the two possible directions. Digital recording systems are essentially of the types return to zero (RZ) and non-return to zero (NRZ) types. The signal, after sampling, is a converted into binary code. In the RZ type of recording, one state of saturation (+) is for binary digit 1 while (–) is for 0, as shown in Fig. 6.12.

In the NRZ digital recording system (Fig. 6.13), each time the digit 1 is to be recorded, the state of magnetisation is reversed and remains constant for recording the digit 0. The NRZ technique is more common and efficient since it is possible to record twice the number of digits for the same number of pulses or reversals of magnetisation.

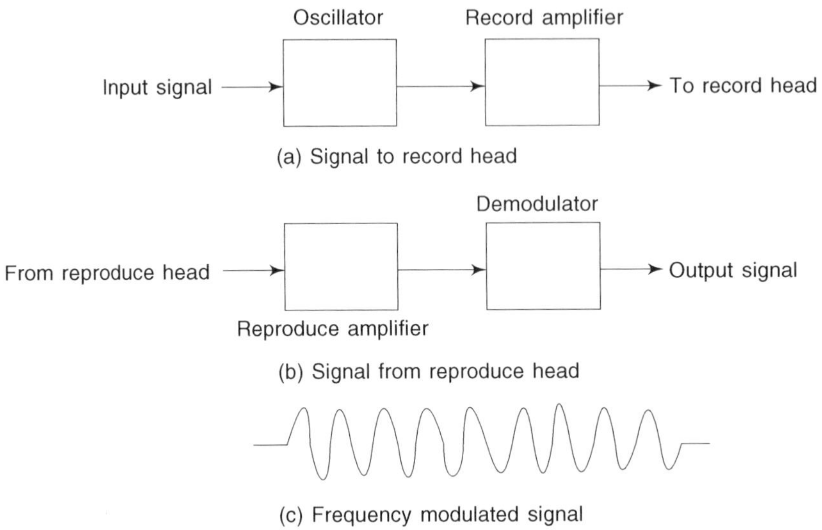

Fig. 6.10 *FM recording system*

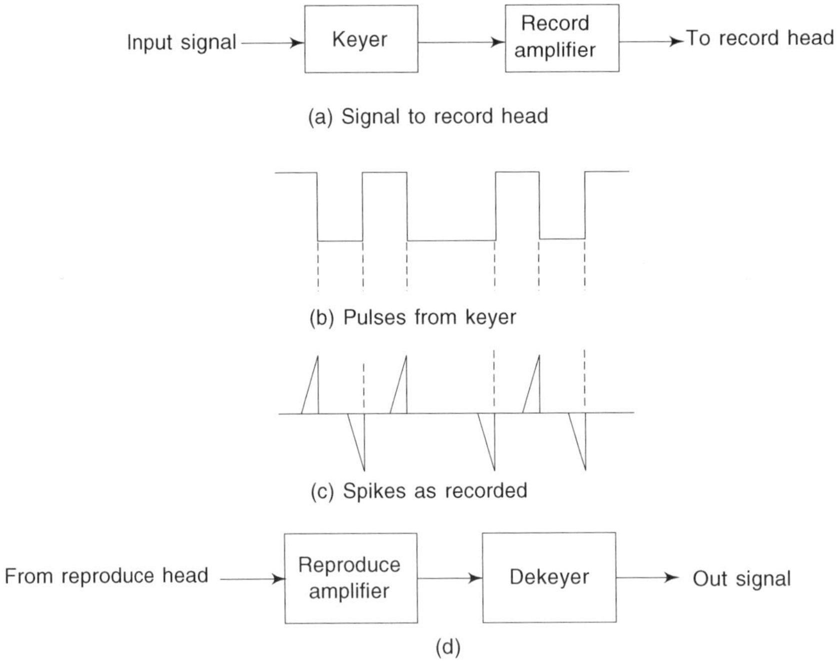

Fig. 6.11 *PDM recording system*

6.6 ■ DIGITAL RECORDER OF MEMORY TYPE

Another development in digital recording is to replace the magnetic tape with a large semiconductor memory, as shown in Fig. 6.14.

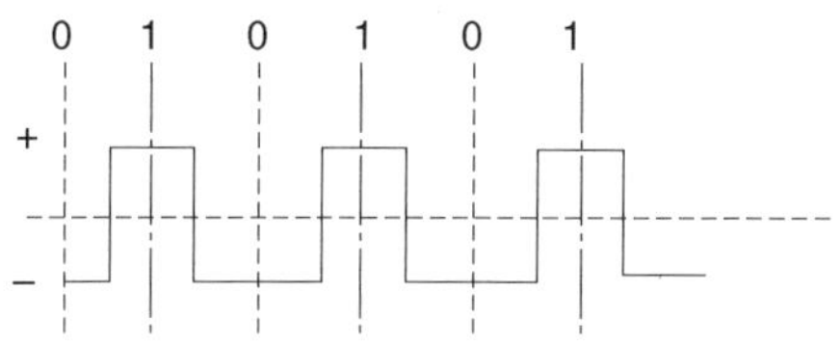

Fig. 6.12 *Tape magnetisation for RZ digital recording*

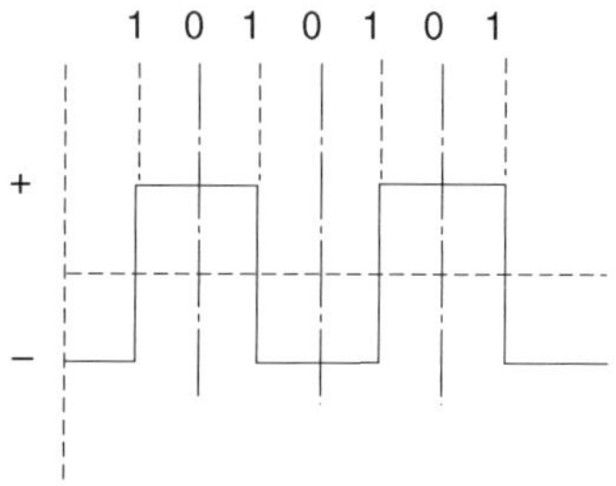

Fig. 6.13 *The magnetisation for NRZ digital recording*

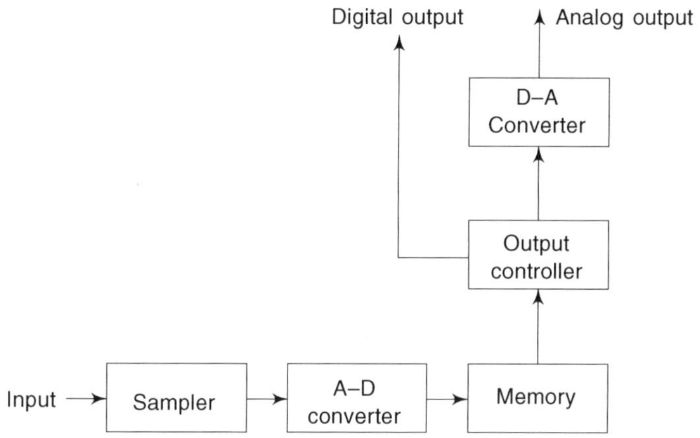

Fig. 6.14 *Digital waveform recorder with memory*

The analog input signal is sampled and converted to digital form by an A–D converter. The signal is stored in the memory and converted to analog or digital outputs for presentation as desired.

6.7 ■ DATA ACQUISITION SYSTEMS

For large-scale data recording, data acquisition systems or loggers are employed, e.g. in a power plant, the input signals, like temperatures, pressures, speeds, flow rates, etc. from a number of locations, may have to be recorded periodically or continuously. In such cases, such systems are employed.

The data acquisition systems used are usually of digital type using a digital computer and may have multiple channels for measurement of various physical variables, the number of channels may be upto 100 or even more.

Figure 6.15 shows a large-scale data acquisition system with the sensor being of analog types. After signal conditioning including amplification, a multiplexer is used, which is essentially a switching device, enabling each input to be sampled in turn. A sample and hold (*S* and *H*) device is used where an analog-to-digital converter (A–D converter) is employed and where the analog signal might change during conversion. The *S* and *H* device employs a capacitor, which is charged up to the analog signal value which is held at its value, till called by the A–D converter.

The computer controls the addressing and data input and processes the signals as desired, for display, printing and storage.

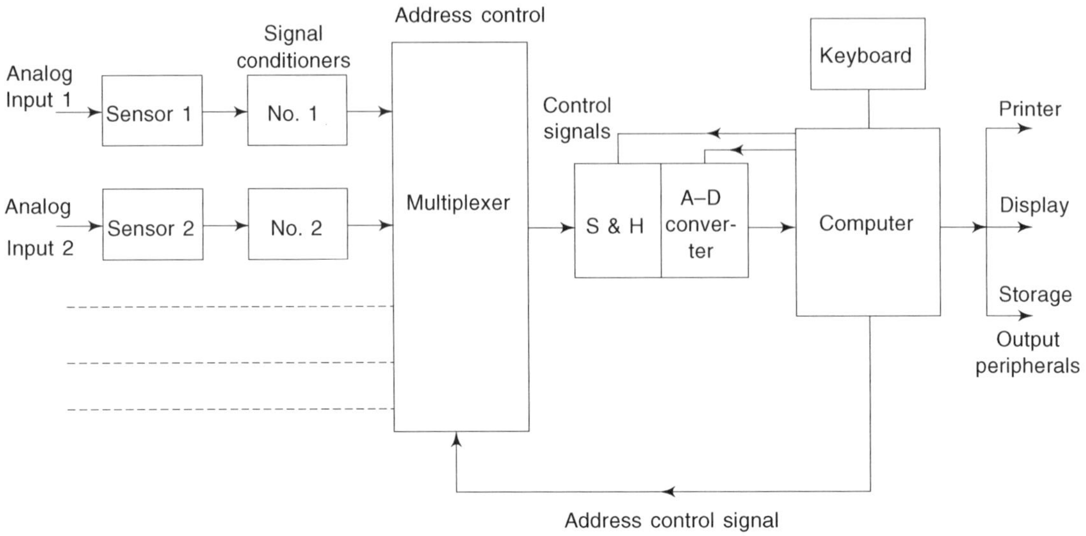

Fig. 6.15 *Data acquisition system*

The computer monitor unit is used for display, a laser or inkjet or dot matrix printer for permanent record as per the software used with computer and the measurement data may be stored in the hard disk and/or floppy disk for record or communication, where needed.

Details of the computer aided measurements using a data acquisition board at the back-end of a microcomputer, are given in Chapter 18, using a software for the control of the measurement and display operations.

The sampling rate of data for input to A–D converter has to be much higher than the signal frequency as discussed in Chapter 14.

6.8 ■ DATA DISPLAY AND STORAGE

The data may be in analog or digital form as discussed earlier and may be displayed or stored as such. The display device may be any of the following types:

1. Analog indicators, comprising motion of a needle on a metre scale.
2. Pen trace or light trace on chart paper recorders.
3. Screen display as in cathode ray oscilloscopes or on large TV screen display, called visual display unit (VDU).
4. Digital counter of mechanical type, consisting of counter wheel, etc.
5. Digital printer, giving data in printed form.
6. Punches, giving data on punched cards or tapes.
7. Electronic displays, using light emitting diodes (LEDs) or liquid crystal displays, (LCDs) etc. In LEDs, light is emitted due to the release of energy as a result of the recombination of unbound free electrons and holes in the region of the junction. The emission is in the visible region in case of materials like Gallium Phosphide. LEDs get illuminated ON or OFF, depending on the output being binary 1 or 0. In a microcomputer, the status of data, address and control buses may be displayed.

Using LEDs, a seven–segment diplay as in Fig. 6.16, can be made, which would display most of the desired characters. LCDs are made from organic molecules, which flow like liquids and have crystal-like characteristics, appearing dark or bright, depending on the application of a certain voltage range across the crystal. The seven–segment displays may also be made up of LCDs.

Fig. 6.16 *Seven–segment display*

8. The storage of data may be on cards, magnetic tapes, disks core memories, etc. Figure 6.17 shows a floppy disk storage system, which is of magnetic type.

   The digital data on the disk is recorded in concentric-circles, known as tracks. The disk is divided into sectors which are numbered and can hold a number of characters. The formatting of the disk is done to identify the tracks and the sectors. A reference hole is shown for numbering the start of the tracks.

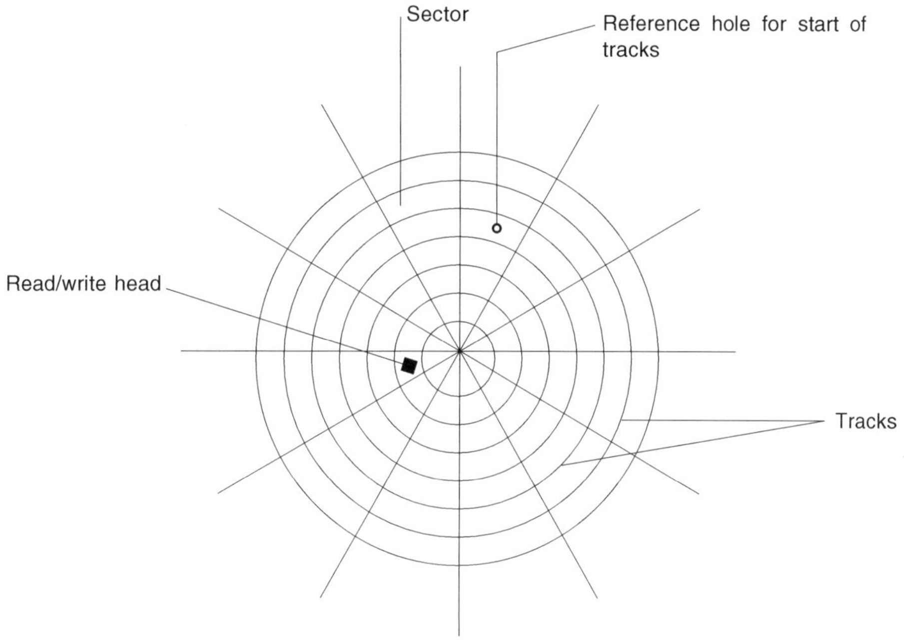

Fig. 6.17 *Floppy disk storage system*

A read/write head is used for each disk surface and heads and moved by an actuator. The disk is rotated and data is read or written. In some disks, the head is in contact with the disk surface which in others, there is a small gap.

The hard disks are sealed unit and have a large number of tracks and sectors and store much more data

9. The permanent record of data from a computer may be made on a dot matrix or inkjet or laser printer.

The dot matrix printer is of impact type where dots are formed by wires, controlled by solenoids pressed on ink ribbons onto the paper. The inkjet printer is of non-impact type, in which a stream of fine ink particles are produced. The particles can get deflected by two sets of electrodes is the horizontal and vertical planes. The image of the characters is thus formed.

The laser printer has high resolution and works according to the principle as shown in Fig. 6.18. The drum is coated with an organic chemical coating which is an insulator and gets charged as it passes the charging wire (1). The laser light is reflected from the white regions of the image or the characters to be produced, to the drum, making these portions conducting. The toner gets attracted to the charged regions of the drum. The paper is given a charge by the charging wire (2), which is higher than that on the drum, transferring the toner to the paper, creating the impressions of the character or images. Further, the impression get permanent by heating.

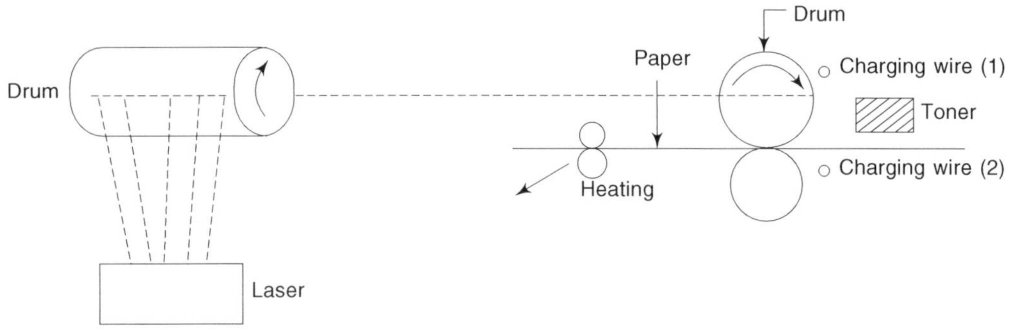

Fig. 6.18 *View of a laser printer*

Review Questions

6.1 Indicate true or false against each of the following:
   (a) Digital voltmeters measure dc voltage signals.
   (b) A digital voltmeter is essentially an A–D converter.
   (c) A CRO has usually a very low impedance at its input.
   (d) The time base signal generated in a CRO is of sinusoidal type.
   (e) It is not possible to measure dc voltages with a CRO.
   (f) A CRO can be used to measure the magnitude of two independent signals and phase between them.
   (g) A servo type potentiometric recorder has a frequency response better than that of a galvanometric recorder.
   (h) A signal recorded on a magnetic tape recorder can be displayed on a CRO.
   (i) A dc signal cannot be reproduced if recorded on a direct recording system.
   (j) Data loggers are usually of digital types.
   (k) A data logger has a hold device to hold the analog signal at its previous value till it is digitized.
   (l) The sampling rate should be less if the signal changes fast.
   (m) The data can be stored in the computer.
   (n) A floppy disk cannot store instrument data in digital form.
   (o) Ink jet printer has higher resolution compared to laser printer.

(p) A dot matrix printer is of non-impact type.
(q) A laser printer is of impact type.

Answers

6.1	(a) T	(b) T	(c) F	(d) F	(e) F	(f) T
	(g) F	(h) T	(i) T	(j) T	(k) F	(l) T
	(m) F	(n) F	(o) F	(p) F	(q) F	

Part 2

MEASUREMENTS, METHODS AND APPLICATIONS

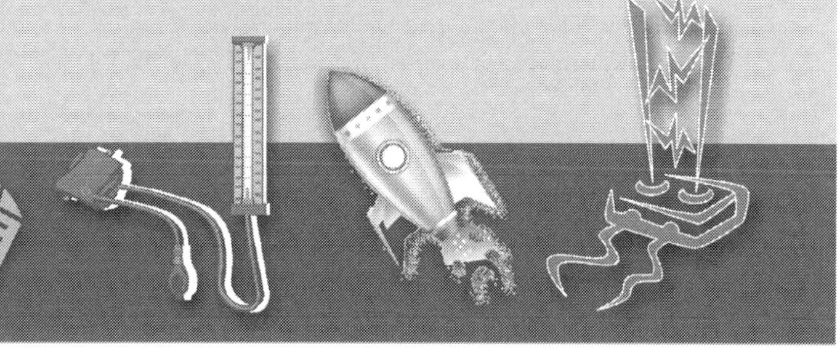

Chapter

7

Motion and Vibration Measurements

■ *INTRODUCTION* ■

Motion is an important basic parameter for measurement. This includes measurement of

- displacement or deformation of elastic members due to application of physical parameters like pressure, temperature, force, torque, etc., the motion may be static or dynamic, i.e., varying with time
- displacement, velocity or acceleartion due to vibratory motion of flexible structures or machines
- rigid-body motion of objects like aerospace vehicles or robot arms which may be rotary or linear type and is usually dynamic

The motion measuring devices are of two types:

- Relative motion devices, and
- Absolute motion devices.

In the case of the relative motion devices, the motion is measured with respect to a fixed reference. In such devices, one terminal of the instrument is attached to a point fixed in space and the other terminal is attached (mechanically or electrically) to the point whose motion is to be measured as shown in Fig. 7.1. The motion may be static or dynamic, the latter including vibrating motion as well.

In the case of absolute motion devices, also known as seismic devices or transducers, the only terminal is the base of a spring mass system. The base is attached to the point whose motion is to be measured, as shown in Fig. 7.2. In such cases, no fixed reference is available, e.g. to measure the vibration of a bridge or an aircraft wing in flight, one uses such types of transducers. From the motion of the mass with respect to the base, it is possible to find the absolute motion of the moving object. The motion in this case is dynamic or vibratory motion.

7.1 ■ RELATIVE MOTION OR VIBRATION MEASURING DEVICES

These devices are of several types, notably electromechanical types, optical or electro-optical types, etc.

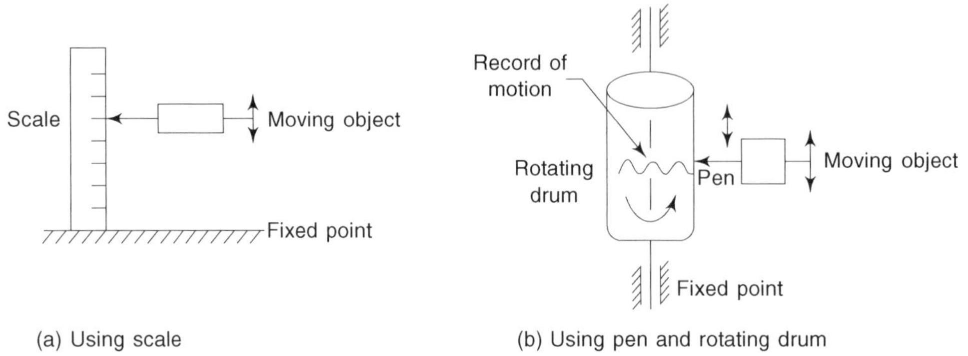

Fig. 7.1 *Relative motion or fixed reference type of motion transducer*

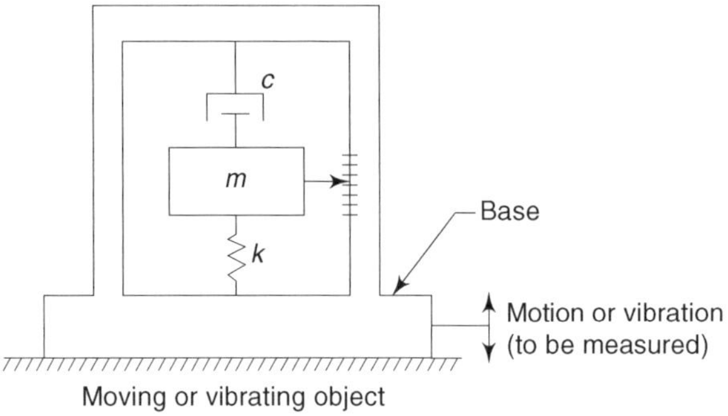

Fig. 7.2 *Seismic type of motion or vibration transducer*

7.1.1 Electromechanical Devices

These have been discussed in detail in Chapter 4, and are of the following types:

1. resistance type,
2. inductive type,
3. capacitance type,
4. piezo-electric type, and
5. resistance strain gauge type.

Electromechanical devices are very commonly used for motion measurement. They are versatile and may be designed for use in almost any application. Usually, these are simple in construction and can be made to give a large output which can be displayed or recorded with ease.

They can be used for translational or rotational motion, by suitably varying the construction of the transducer or device. Some of the inductive and capacitance types can be of proximity types, while the remaining types are usually of contact type. Except for the piezo-electric and electrodynamic types, which can be used only for dynamic motion measurement, others may be used for both static and dynamic motion measurements as may be seen from Table 4.1 in Chapter 4. Resistance type (potentiometric) transducer is usable for large motions. The same is true of electrodynamic and LVDT types of inductive transducers. The capacitance, piezo-electric and strain gage transducers are normally used for measuring small motions.

It can also be seen from Table 4.1 that dynamic motion or vibrations of an object like structure or machine can be measured by all the electro-mechanical transducers included.

The motion of the moving object relative to fixed point as in Fig. 7.1 may be measured using an electromechanical device or transducer. Similarly, the dynamic motion or vibration of an object attached to the base as in Fig. 7.2 for a seismic device or transducer can be obtained from the relative motion of the base with respect to mass m using an electro-mechanical transducer.

7.1.2 Optical Devices

These include microscopes, telescopes, interferometers, photo-electric devices and Moire fringe method based devices. Microscopes and telescopes are used for measuring small and large motions, respectively, provided there are proper graduations on these devices. For measuring very small motions, devices based on optical interference, called interferometers, are used. These will be discussed first, followed by devices that are strictly electro-optical.

Interferometers The principle of optical interference can be explained using Fig. 7.3. Two light waves of the same wave length, starting from different points and travelling the different paths, meet at O. In Fig. 7.3(a), the two waves are out of phase and cancel each other, producing a dark spot. In Fig. 7.3(b), the two waves are in phase at O and add, producing a bright spot. If the lengths of the two paths are l_1 and l_2, the number of wavelengths of light in each path are l_1/λ and l_2/λ, respectively, λ being the wavelength of light. If the difference between the number of wavelengths, viz. $(l_1 - l_2)/\lambda$ is a whole number (even multiple of 1/2) the two waves are in phase. In case, however, $(l_1 - l_2)/\lambda$ is an odd multiple of 1/2 like 1/2, 3/2, 5/2 . . . , the two waves are out of phase.

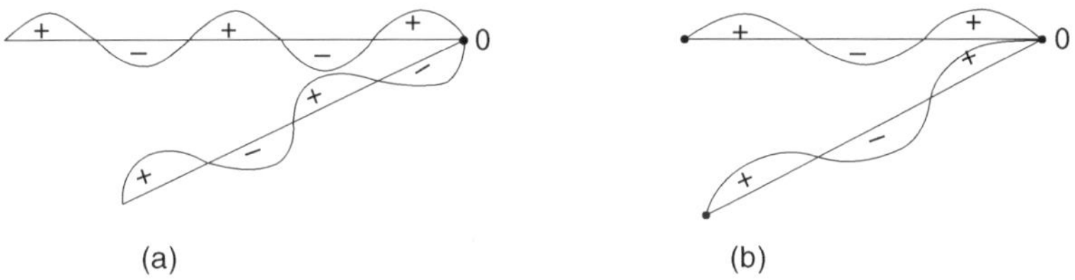

Fig. 7.3 *Interference of light waves*

Figure 7.4 shows the principle, based on the above, which is employed in an interferometer. The two rays starting from source S, get reflected at A and combine at C to travel to the eye at E. The distance between the top glass and bottom reflecting surface is h. Thus, the difference in path lengths of the two rays is $2h$. If $2h/\lambda = 1, 2, 3, \ldots$, the eye will see darkness. If h was to change due to motion of one of the surfaces, the eye will see alternately brightness and darkness, for every change in $2h/\lambda$ equal to 1/2. Since λ for yellow light is approximately 5460 Å (1Å = 10^{-8} cm), this principle can be used to measure very small motion.

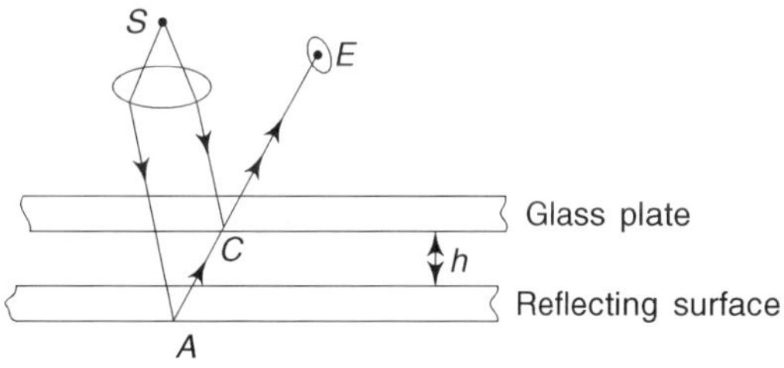

Fig. 7.4 *Principle of interferometers*

Based on the above principle, an interferometer known as the Michelson interferometer is widely used for motion measurement.

Figure 7.5 shows the constructional features of a Michelson interferometer, which can be used to measure the motion of N. The mirror at M is half reflecting and half transmitting.

The two waves after reflection from the N and C, meet at M and can the be seen from E. If, to start with, distance $MC = MN$, the two waves are in phase and eye E will see brightness. If due to motion of N, the difference between MC and MN is $\lambda/4$, the difference in paths travelled by the two waves would be $\lambda/2$, and thus the two waves will be out of phase, resulting in darkness at E. Another motion of $\lambda/4$ would make light appear at E. Thus, from a count of the number of illuminations or darknesses, the motion of N can be calculated.

For measurements requiring very high precision, a laser interferometer is used. λ of light changes somewhat due to change of air pressure, temperature and humidity. The change is much less in case of lasers.

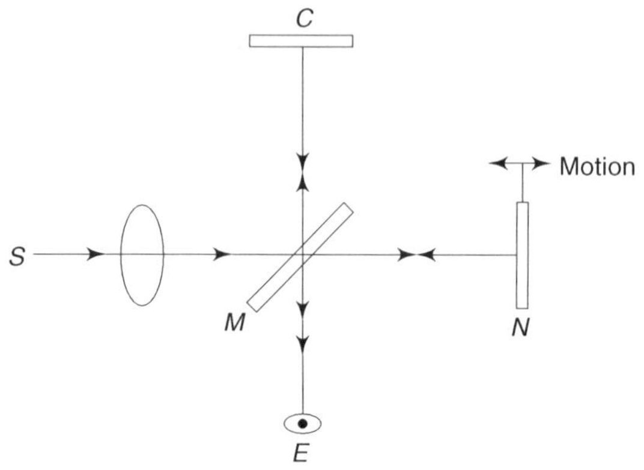

Fig. 7.5 *Michelson interferometer*

Photo-electric Devices These have already been discussed in Chapter 4. Any of the three types, viz. photo-emissive, photo-conductive or photo voltaic-types of cells may be used whereby the intensity of light falling on the cell may be changed by the motion to be measured.

Moire Fringe Method This is an optical method of amplifying displacement and uses two identical gratings. The gratings consist of a number of slits on an opaque screen, the slits being transparent. The pitch of the gratings is quite small, and may be between 1/200 and 1/2000 cm.

Figure 7.6(a) shows two gratings of the same pitch, mounted face-to-face, with the rulings inclined at $\angle\theta$ to each other. As shown in the figure, a set of dark bands, called Moire fringes, is obtained, with fringe spacing $\gg$ pitch of the grating. A movement of the grating in a direction perpendicular to the gratings would move the fringes in a direction perpendicular to the fringes by a larger amount. The movement of the fringes can be measured and from the same, movement of the grating can be calculated. Alternatively, the number of fringes passing through a given point can be counted, using a photo-electric transducer. This method of displacement measurement is used for measuring the movement of work in machine tools with an accuracy of $\pm$ 0.001 % over a large range. It can also be used for measuring angular displacements, in an identical way.

Moire fringes are formed in a direction which bisects the obtuse angle between the line of the two gratings, as shown in Fig. 7.6(b). If the angle between the gratings is θ, line OQ, which bisects the obtuse angle $(180 - \theta)$ is the direction of the fringe. It may be seen that $\angle POQ = \angle ROQ = \theta/2$, where OP and OQ are perpendicular to the directions of the two grating respectively.

The relation between the pitch of gratings and fringe spacing may be found from Fig. 7.7. OB and OD are perpendiculars to the two grating directions respectively and p is the pitch of the gratings measured perpendicular to the inclined grating direction AO. Fringe spacing, measured parallel to the horizontal, viz. direction of one of the gratings $= AC$, OC being fringe direction with $\angle BOC = \angle COD = \theta/2$, as mentioned earlier.

Thus, triangles BOC and COD are congruent, with the three angles equal and one side common.

Thus $$OB = OD = p$$

or $$\text{fringe spacing } AC = x_1 + x_2$$

$$= \frac{OB}{\tan\theta} + OB \tan\frac{\theta}{2}$$

$$= \frac{p}{\tan\theta} + p \tan\frac{\theta}{2}$$

$$= p\left(\frac{\cos\theta}{\sin\theta} + \frac{1-\cos\theta}{\sin\theta}\right)$$

$$= \frac{p}{\sin\theta}$$

Since θ is small, fringe spacing $\gg p$.

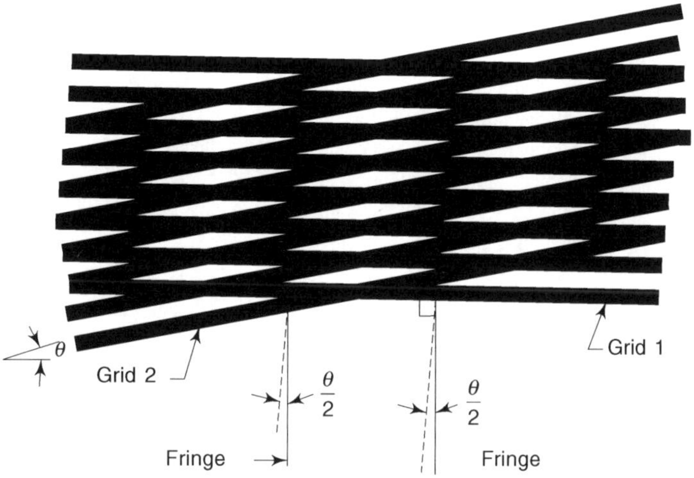

(a) Moire fringes

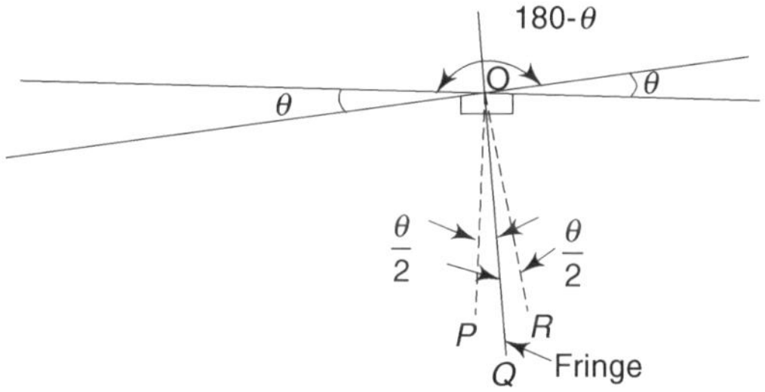

(b) Direction of fringes

Fig. 7.6 *Moire fringe effect*

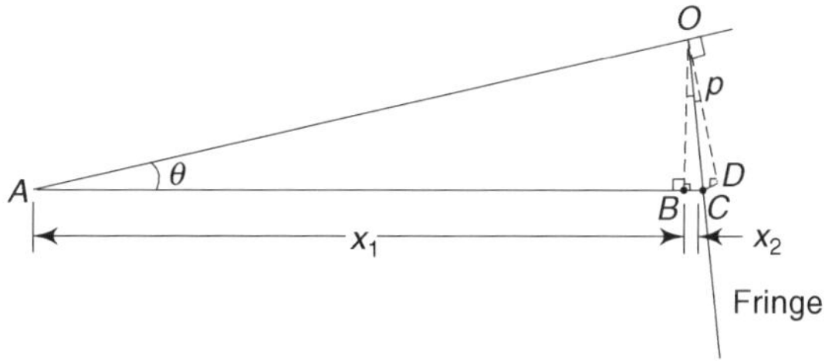

Fig. 7.7 *Calculating fringe spacing*

7.2 ■ ABSOLUTE MOTION OR VIBRATION DEVICES

7.2.1 Seismic Devices (Spring-Mass Type)

In these devices the base of the device or transducer is attached to the object whose motion or vibration is to be measured, as shown in Fig. 7.8. Inside the transducer, is a mass m supported on a spring of stiffness k and viscous damper, with damping coefficient c. The motion of the mass relative to the frame or base, gives an indication of the motion of the object and is the output of the instrument. The equation of motion of the system has already been derived in Ch. 3 and has been given again here.

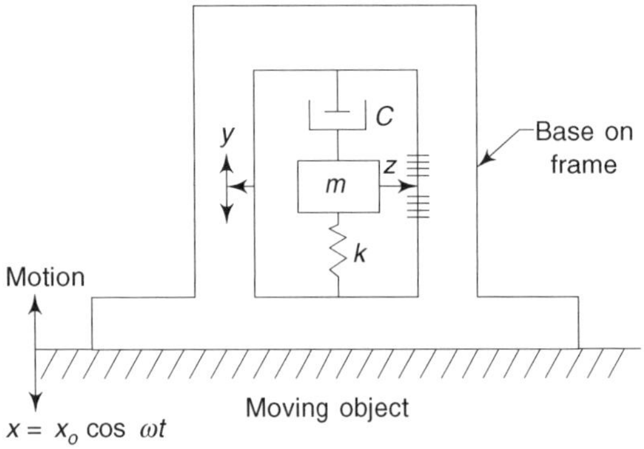

Fig. 7.8 *Seismic instrument*

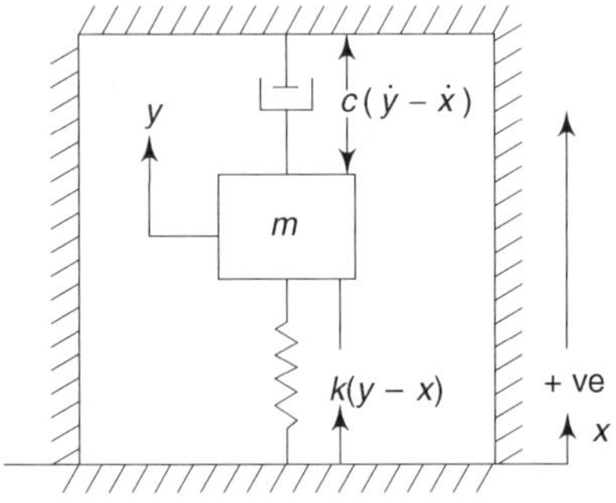

Fig. 7.9 *Seismic instrument showing forces acting on the mass*

Let $x = x_0 \cos \omega t$ be the harmonic motion of the object, which is to be measured. x_0 is the amplitude of dynamic motion or vibration, ω is the circular frequency of motion vibration ($\omega = 2\pi f$, f being the frequency in Hz and 'ω' in rad/sec). 't' stands for the time variable If y is the absolute motion of the mass at any instant, when the object motion is x, as shown in Fig. 7.9, the equation of motion of the mass m is:

$$m\ddot{y} = -c(\dot{y} - \dot{x}) - k(y - x) \tag{7.1}$$

where · denotes differentiation with respect to time t. Writing $y - x = z$, where z is the motion of the mass relative to the frame, Eq. (7.1) may be written as:

$$m\ddot{z} + c\dot{z} + kz = -m\ddot{x} = m\omega^2 x_0 \cos \omega t \tag{7.2}$$

(substituting $x = x_0 \cos \omega t$.)
Steady state solution of Eq. (7.2) is

$$z = z_0 \cos(\omega t - \varphi)$$

where amplitude

$$z_0 = \frac{m\omega^2 x_0}{\sqrt{(k - m\omega^2)^2 + (c\omega)^2}} \tag{7.3}$$

Expression for amplitude ratio z_0/x_0 may be written using Eq. (7.3)

$$\frac{z_0}{x_0} = \frac{r^2}{\sqrt{(1 - r^2)^2 + (2\xi r)^2}} \tag{7.4}$$

where $r = \omega/\omega_n$, called frequency ratio

$$\omega_n = \sqrt{\frac{k}{m}}, \text{ called undamped natural frequency}$$

$$\xi = \frac{c}{2\sqrt{km}}, \text{ called damping ratio}$$

ω = circular frequency of motion of the moving object

The plot of Eq. (7.4) is shown in Fig. 7.10. It may be seen that the ratio of the output to input motion amplitude is $z_0/x_0 \approx 1$ and the curves are fairly flat when the frequency ratio $r > \sqrt{2}$, and $\xi = 0.7$. Thus, the above device may be used for displacement measurement at frequencies ω which are greater than $\sqrt{2}\,\omega_n$, if the damping ratio of the instruments is around 0.7. If low frequency displacement are to be measured, ω_n should be small or a soft spring should be used in the instrument. Equation (7.3) may be written as

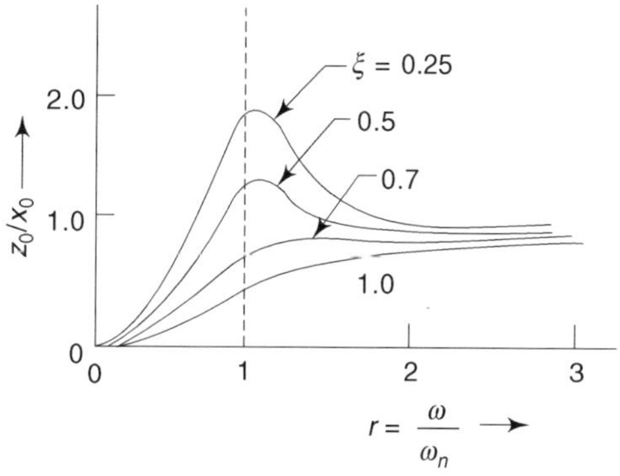

Fig. 7.10 *Displacement response of seismic transducer*

$$\frac{\omega_n^2 z_0}{\omega^2 x_0} = \frac{1}{\sqrt{(1-r^2)^2 + (2\xi r)^2}}$$

or

$$\frac{\omega_n^2 z_0}{A_0} = \frac{1}{\sqrt{(1-r^2)^2 + (2\xi r)^2}} \tag{7.5}$$

where $A_0 = \omega^2 x_0$, viz. acceleration amplitude of the object.

The plot of Eq. (7.5) is shown in Fig. 7.11. It may be seen that the relative motion z of the mass with respect to the instrument frame is a measure of the object acceleration if $r < 0.4$ and $\xi = 0.7$.

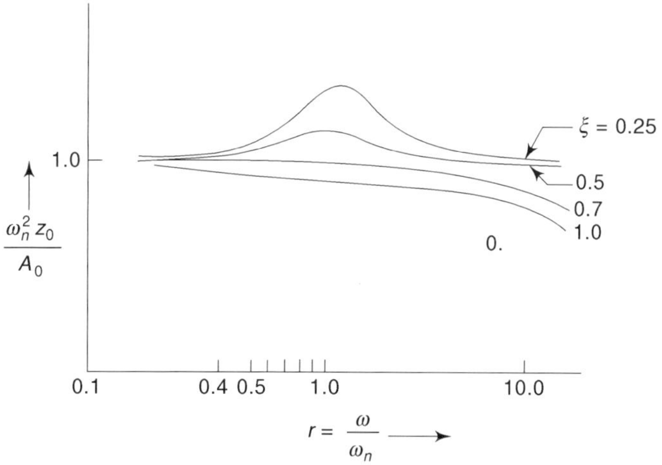

Fig. 7.11 *Acceleration response of seismic transducer*

Thus, the instrument may be used to measure acceleration at frequencies $\omega < 0.4\ \omega_n$. In case the value of frequency ω of the object is high, ω_n of the instrument should be high or a stiff spring should be used in the instrument.

Applications of measurement and analysis of dynamic motion or vibration, are given in Chapters 15 and 16.

Problem 7.1 *In a seismic instrument, mass m = 100 g, spring stiffness = 1 N/mm. Damping ratio = 0.4.*

(a) Find the amplitude of recorded motion if the motion of be measured is 3 sin 200 t (mm)

(b) Find the maximum frequency for which the instrument can be used as an accelero-meter if the error is not to exceed 10%.

Solution

$$m = 0.1 \text{ kg} \qquad k = 1000 \text{ N/m}$$
$$\xi = 0.4 \qquad \xi_0 = 3 \text{ mm}$$
$$\omega = 200 \text{ rad/s}$$

(a)

$$\omega_n = \sqrt{\frac{k}{m}} = \sqrt{\frac{1000}{0.1}} = 100 \text{ rad/s}$$

$$r = \frac{\omega}{\omega_n} = 2$$

Using Eq. (7.4), we get

$$z_0 = 3.53 \text{ mm.}$$

(b) Using Eq. (7.5), viz.

$$\frac{\omega_n^2 z_0}{A_0} = \frac{1}{\sqrt{(1-r^2)^2 + (2\xi r)^2}} = 1 \pm 0.1$$

It is seen that for RHS = 1.1, we get

$$r = 1.1048,\ 0.373$$

Also, for RHS = 0.9, we get $r = 1.23$. Thus maximum r, for which the instrument should be used as an accelerometer is $r = 0.373$ or $\omega = 37.3$ rad/s (as error will be higher for higher values of 'x'.

Problem 7.2 *An accelerometer has been developed by embedding a single strain gauge in plastic as shown in Fig. 7.12. Find the vertical sensitivity of the accelerometer in cm/cm-g. Also find the natural frequency of the accelerometer and its useful frequency range. Given: mass density of brass = 8.8 gm/cm³, mass density of plastic = 1.6 gm/cm³ and Young's modulus of plastic = 3 × 10⁵ N/cm².*

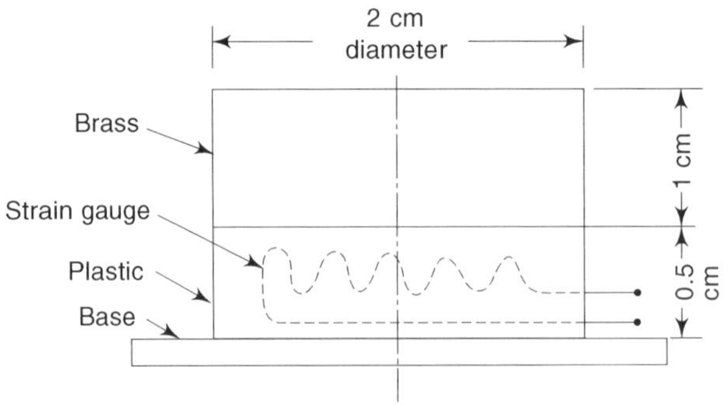

Fig. 7.12 *Figure for Problem 7.2*

Solution In this problem, plastic will be treated like a spring, with the brass piece as the accelerometer mass. It is well known from vibrations theory that 1/3rd of spring mass is to be added to the concentrated mass, for finding natural frequency. Thus,

$$m = \frac{8.8}{1000}\left(\frac{\pi}{4} \times 4 \times 1\right) + \frac{1}{3} \times \frac{1.6}{1000}\left(\frac{\pi}{4} \times 0.5 \times 4\right)$$

$$= 2.34 \times 10^{-2} \text{ kg}$$

Stiffness k of plastic piece = EA/l, E being Young's modulus, A the area and l the length.

$$k = \frac{3 \times 10^5 \times \pi \times (2)^2}{4 \times 0.5} = 18.85 \times 10^5 \text{ N/cm}$$

$$\omega_n = \sqrt{\frac{k}{m}} = \sqrt{\frac{18.35 \times 10^7}{2.34 \times 10^{-2}}} = 8.15 \times 10^4 \text{ rad/s}$$

Useful frequency range = 0 to 0.4 ω_n
= 0 to 5189 Hz

For the usable frequency range of the accelerometer,

$$\frac{\omega_n^2 z_0}{A_0} \approx 1, \ A_0 \text{ being acceleration amplitude}$$

Taking $A_0 = 1 \text{ g} = 9.81 \text{ m/s}^2$

Relative motion $z_0 = \dfrac{A_0}{\omega_n^2} = \dfrac{9.81}{(8.15 \times 10^4)^2}$

$$= 1.48 \times 10^{-9}\text{m} = 1.48 \times 10^{-7} \text{ cm}$$

Thus, Sensitivity $= \dfrac{1.48 \times 10^{-7}}{0.5} = 2.96 \, x \, 10^{-7}$ cm/cm-g

Constructional Features Basically a seismic transducer consists of a mass supported on a spring and with damper, with a relative motion transducer, as shown in Fig. 7.13, to measure motion of mass relative to the frame.

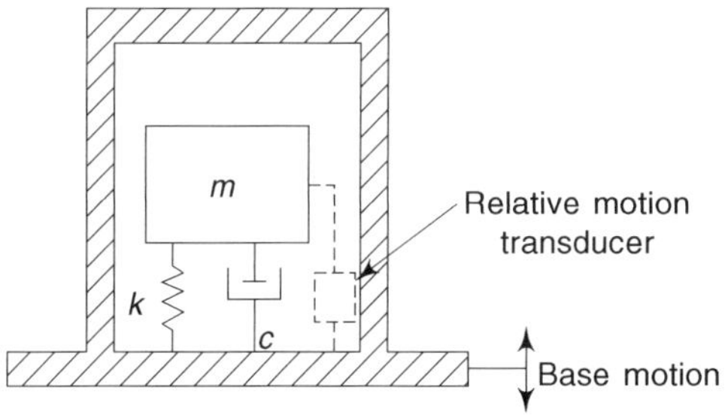

Fig. 7.13 *Use of a relative motion transducer in an absolute motion type transducer*

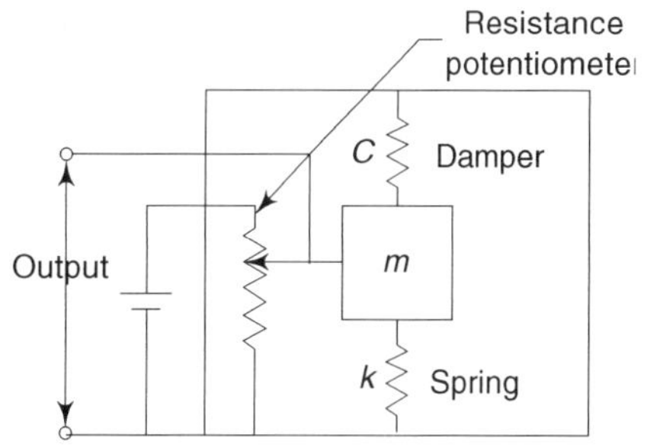

Fig. 7.14 *Seismic transducer using resistance transducer*

Figure 7.14 shows a seismic transducer using a resistance potentiometer as a relative motion transducer.

In Fig. 7.15, the relative motion induces strain in a cantilever spring which is measured by using resistance strain gauges. Since resistances R_1 and R_2 would have strains of opposite nature due to bending of the cantilever, the two are arranged in adjacent arms of the bridge in Fig. 7.15 to increase the output.

An alternative arrangement using resistance strain gauges is shown in Fig. 7.16. In this arrangement, the longitudinal strain induced in the strip is measured. Since the strips are quite stiff in the longitudinal direction compared to a cantilever of the same size, the arrangement of Fig. 7.16 would have a high ω_n and thus would act as an accelerometer (for frequencies < 0.4 ω_n).

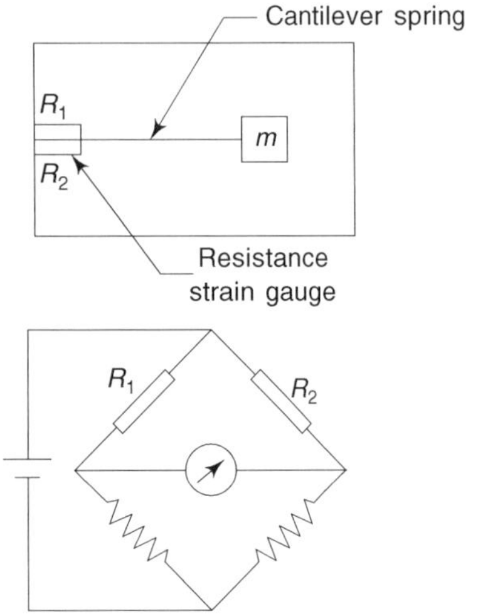

Fig. 7.15 *Seismic transducer using resistance strain gauges*

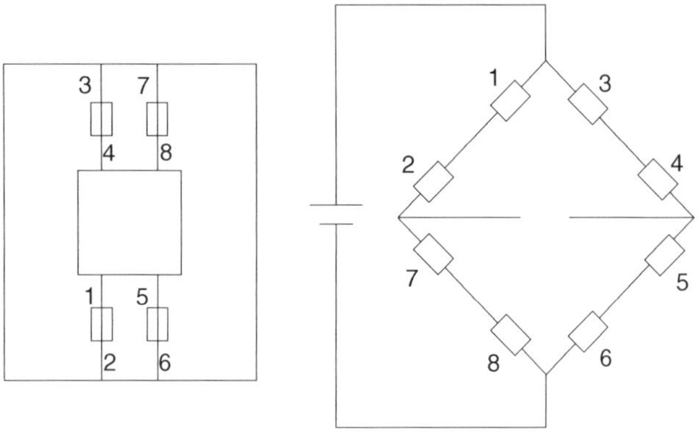

Fig. 7.16 *Accelerometer using resistance strain gauges*

Figure 7.17 shows the constructional features of another seismic motion transducer or device in which a magnet forms the seismic mass m. The relative motion of m with respect to the coil on the casing results in a voltage output from the coil, which is proportional to the rate of change of magnet motion relative to that of the coil or the base. Thus, for soft springs, the output is proportional to the velocity of the base, hence a velocity transducer.

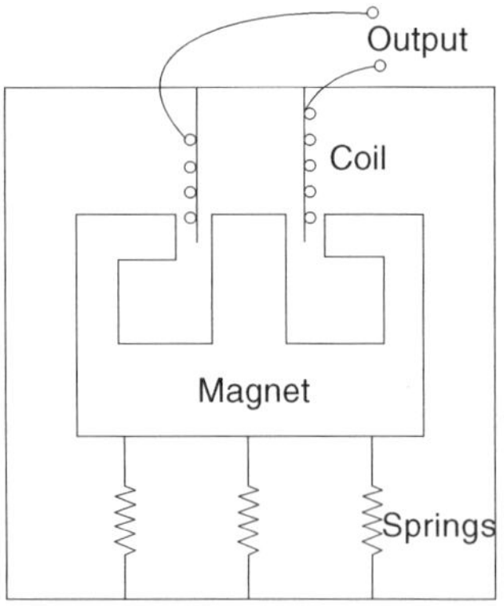

Fig. 7.17 *Velocity transducer*

Figure 7.18 shows a torsional angular motion sensor, based on the above principle. The magnet, which provides the seismic inertia, is free to revolve in angular motion while the core is attached to the shaft whose torsional motion is to be measured. A torsional spring is used between the core and the magnet. Due to torsional motion or vibrations of the shaft, there is a relative angular motion between the magnet and the cores which gives a voltage output in the coil attached to the core. The voltage output is proportional to torsional angular velocity.

Figure 7.19 shows a seismic transducer using a piezo-electric crystal. This is commonly used for shock and vibration measurements. The crystal is fairly stiff and held in compression by a spring. Because of the motion of the frame due to the moving object to which it is attached, an output voltage proportional to the acceleration of the moving object is obtained. Hence, the device is known as piezo-electric

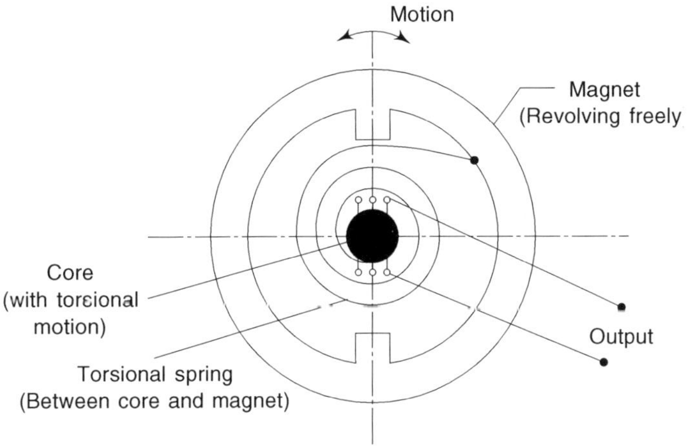

Fig. 7.18 *Torsional angular motion transducer*

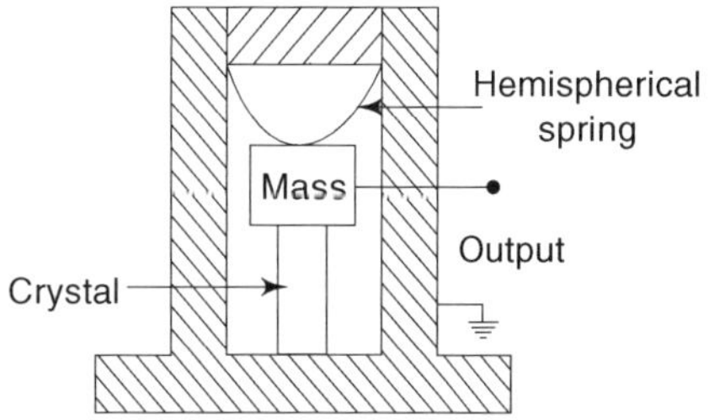

Fig. 7.19 *Piezo-electric accelerometer*

accelerometer. Such devices are very sensitive and light-weight and can be used over a wide frequency range. For improving the low frequency response, a cathode follower is used at the output, as discussed in Chapter 4.

Problem 7.3 *A piezo-electric accelerometer (Fig. 7.20) uses two discs, so connected that the charges produced are added up. Disc dia = 5 mm and thickness of each disc = 0.6 mm. Young's modulus of disc material = 1.08 × 10^5 N/m^2. Mass m = 1.1 g. Dielectric constant of crystal material = 10^{-10} F/cm. Piezo-electric constant = 2 × 10^{-10} C/N. Find the natural frequency of the instrument, its charge sensitivity (C/g) and voltage sensitivity (V/g).*

Solution Total stiffness of the crystal is found by taking the two discs to be in series. For each disc, stiffness

$$= \frac{EA}{t} = \frac{1.08 \times 10^5 \times \pi (5)^2}{4 \times 6.6}$$

$$= 3.53 \times 10^6 \text{ N/mm}$$

$$\text{Total stiffness } k = \frac{3.53 \times 10^6}{2}$$

$$= 1.765 \times 10^6 \text{ N/mm}$$

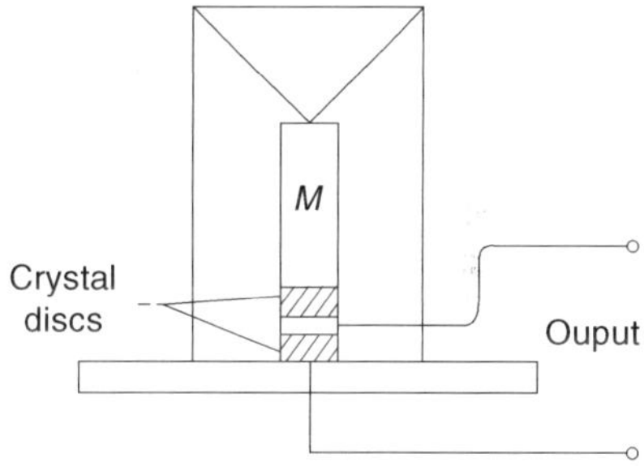

Fig. 7.20 *Figure for Problem 7.3*

$$\text{Natural frequency } \omega_n = \sqrt{\frac{k}{m}}$$

$$= \sqrt{\frac{1.765 \times 10^6 \times 1000 \times 1000}{1.1}}$$

$$= 1.27 \times 10^6 \text{ rad/s} = 0.202 \times 10^6 \text{ Hz}$$

$$\text{Capacitance } C = \frac{\varepsilon A}{3.6 \pi t}$$

ε being the dielectric constant (F/cm), A the crystal area (cm^2) and t the crystal thickness (cm). Substituting for ε, A and t, we get

$$C = 0.29 \times 10^{-10} \text{ F}$$

Total voltage V produced

$$= \frac{2Q}{C}, \; Q \text{ being charge due to each crystal.}$$

$$Q = \beta P$$

β being piezo-electric constant

$$= 2 \times 10^{-10} \text{ C/N}$$

and P being the force in N.

Thus $\qquad Q = \beta m A$

A being acceleration. For A = 1 g = 9.81 m/s^2

$$\text{Charge sensitivity} = 2Q$$

$$= 2 \times 2 \times 10^{-10} \times \frac{1.1}{1000} \times 9.81$$

$$= 4.32 \times 10^{-12} \text{ C/g}$$

$$\text{Voltage sensitivity} = \frac{4.32 \times 10^{-12}}{0.29 \times 10^{-10}} = 0.149 \text{ V/g.}$$

Problem 7.4 *The crystal described in Problem 4.1 is used in a piezo-electric accelerometer, along with a mass such that the natural frequency of the system is 10000 Hz and damping ratio = 0.2. Find the voltage output for an input acceleration amplitude of 500 g at a frequency of 2000 Hz.*

Solution We use

$$\frac{\omega_n^2 z_0}{A_0} = \frac{1}{\sqrt{(1-r^2)^2 + (2\xi r)^2}}$$

Substituting,

$$A_0 = 500 \times 9.81 = 4905 \text{ m/s}^2$$
$$\omega_n = 2\pi \times 10000 = 6.28 \times 10^4 \text{ rad/s}$$
$$\omega = 2\pi \times 2000 = 1.256 \times 10^4 \text{ rad/s}$$
$$\xi = 0.2$$
$$r = \frac{\omega}{\omega_n} = 0.2$$

we get,

$$z_0 = 1.291 \times 10^{-4} \text{ cm}$$

Sensitivity (from Problem 4.1) = 2857 V/cm. Thus,

Output voltage amplitude = $2857 \times 1.291 \times 10^{-4} = 0.369$ V

7.2.2 Force Balance Type Seismic Device

These are similar to seismic devices except that there is no mechanical spring used here and the restoring force is provided by a feedback arrangement, as shown in Fig. 7.21.

A feedback loop is used to produce a force due to the electrical output of the relative motion transducer, using an electromagnetic actuator and the force is applied to the mass m. A state of balance or equilibrium is reached, with the output current I or voltage E, across a resistor R being the output and thus a measure of input motion x.

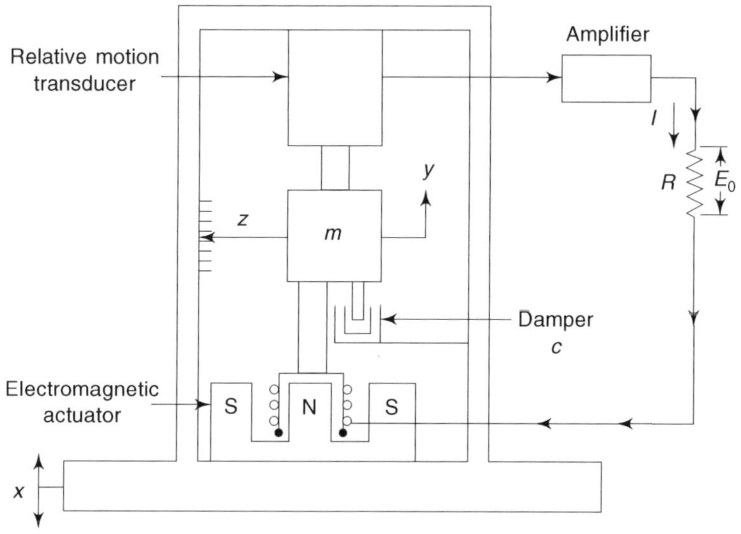

Fig. 7.21 *Force balance type motion transducer*

A relation between the input motion x and E_0, is now derived:

$$z = y - x \tag{7.6}$$

y being the absolute motion of mass.

$$k_1 I + m\ddot{y} + c\dot{z} = 0 \tag{7.7}$$

k_1 being force constant of actuator

$$\text{Current } I = k_2 z \tag{7.8}$$

k_2 being constant of the relative motion transducer.

Replacing time derivative by D, we get from Eqs. (7.6) and (7.7)

$$I = \frac{-mD^2(z + x) - cDz}{k_1} \tag{7.9}$$

$$\text{Output } E_0 = IR \tag{7.10}$$

From Eqs. (7.7)–(7.9) we get

$$(mD^2 + cD + k_1k_2)\, E_0 = -mk_2RD^2x \tag{7.11}$$

Equation (7.11) is of the same form as Eq. (7.2).

These types of motion measuring devices are used in inertial navigation systems. It is possible to get higher accuracy and increased stability as effects like hysteresis, non-linearity, temperature effects, etc. associated with mechanical springs are absent here. As mentioned earlier, there is no mechanical spring in such devices for providing the restoring force, though mechanical springs are introduced for the purpose of guiding mass motion only.

7.2.3 Gyroscopic Devices

These are commonly used in aero-space vehicle systems for sensing the absolute angular motion of the vehicle. There are two types of gyroscopic devices:

1. free gyros, and
2. restrained gyro.

In the latter, the gyros are restrained by dampers and/or springs while in the former, no such restraint exists.

The gyroscopic principle is well known. A gyroscopic torque is developed due to the precession of a spinning rotor, the three axes—spin, precession and torque—being mutually at right angles, with torque $T = I\omega\omega_p$, as shown in Fig. 7.22, with ω being the angular velocity of the rotor about the spin axis, ω_p, the angular velocity about the precession axis and I, the mass moment of inertia, of the disc about the spin axis,

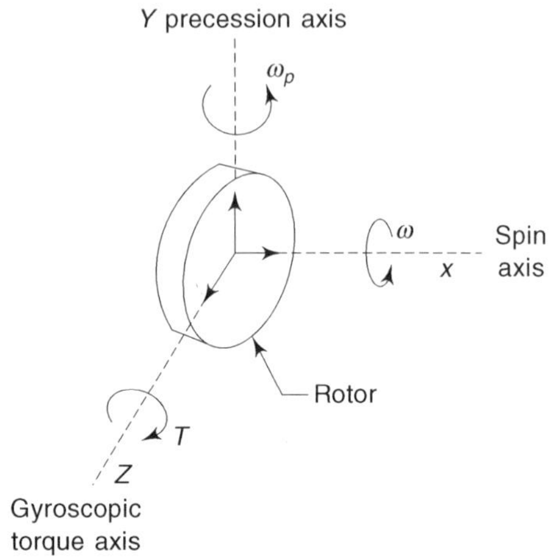

Fig. 7.22 *Gyro torque on a spinning rotor*

Free Gyros These are based on the principle that the rotor spin axis always points in the same direction unless it is acted on by a torque. Thus, a fixed reference is provided in the vehicle in which the gyro is used. In order to measure the three angular motions—pitch, roll and yaw—two gyros are necessary as shown in Fig. 7.23. The two gyros are called vertical and directional gyros, the latter is also called the horizontal gyro. The three angular motions relative to the vehicle are shown in Fig. 7.23(b).

Vertical gyros measure pitch and roll motions while the directional gyro measures yaw. Since the gyro and its gimbals remain fixed in direction in absolute space, any motion of the vehicle and various scales attached to the vehicle can be sensed by a relative motion transducer of the electromechanical type. Thus, a relative motion between the pitch scale (which rotates with the vehicle) and the pointer on the outer gimbal, gives an indication of the pitch motion of the vehicle. Similarly, the motion of the vehicle about the x-axis relative to the pointer in the inner gimbal gives an indication of the roll motion. A directional gyro would indicate motion about the y-axis as shown in Fig. 7.23(c) as the yaw scale would rotate with the vehicle about the y-axis while the pointer on the outer gimbal remains fixed in direction.

In a vertical gyro, the spin axis is aligned with the gravitational axis, with a reference like a pendulum position, while the spin axis in a directional gyro is aligned in N–S direction or slaved to a magnetic compass. In an actual system in aero-space applications, compensation is provided for drift due to friction, earth's rotation, etc. This is done by applying an external torque to realign the gyro wheel to the desired reference position by a servo arrangement.

Restrained Gyros In such gyros, the rotation of the gimbals due to precession of the spinning rotor is restrained by a damper and/or a spring and the rotation of the gimbals due to gyroscopic torque is a measure of the input motion (precession). There are two types of restrained gyros:

1. Rate gyro, and
2. Rate integrating gyro.

The rate gyro measures absolute angular velocity. The rate integrating gyro measures the absolute angular displacement and is used as a fixed reference in navigation and attitude control systems. The rate gyro is shown in Fig. 7.24. If ω_z = angular velocity of the gyro case about the z-axis, (which is to be measured), T = gyroscopic torque, tending to rotate the gyro through angle θ_z about x-axis = $H\omega_z$, H = angular momentum vector = $I\omega$, I beging moment of inertia about y-axis and ω angular velocity of the rotor. Ouput θ_z is converted to a voltage signal by an electromechanical relative motion transducer and is a measure of input motion ω_z.

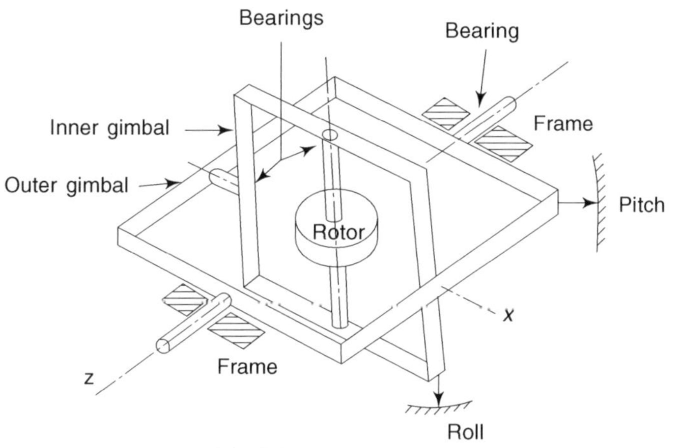

(a) Vertical gyroscope

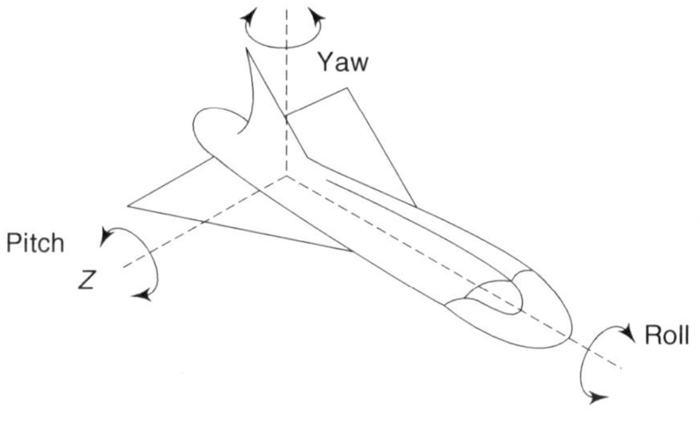

(b) Roll, yaw and pitch directions

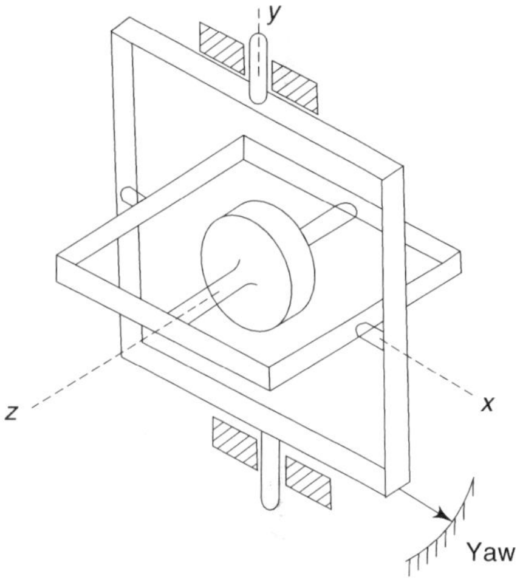

(c) Directional gyroscope

Fig. 7.23 *Free gyroscopes*

If k is the spring stiffness and c the damping coefficient of the damper between the case and gyro frames:

$$T = I\omega\, \omega_z = JD^2\theta_x + cD\theta_x + k\theta_x \tag{7.12}$$

J being the mass moment of inertia about the x-axis.

Thus

$$\frac{\theta_x}{\omega_z} = \frac{I_\omega}{JD^2 + cD + k} \tag{7.13}$$

For a constant input ω_z, the steady value of θ_x is:

$$\frac{\theta_x}{\omega_z} = \frac{I_\omega}{k} \tag{7.14}$$

Thus, output θ_x is proportional to input angular velocity ω_z.

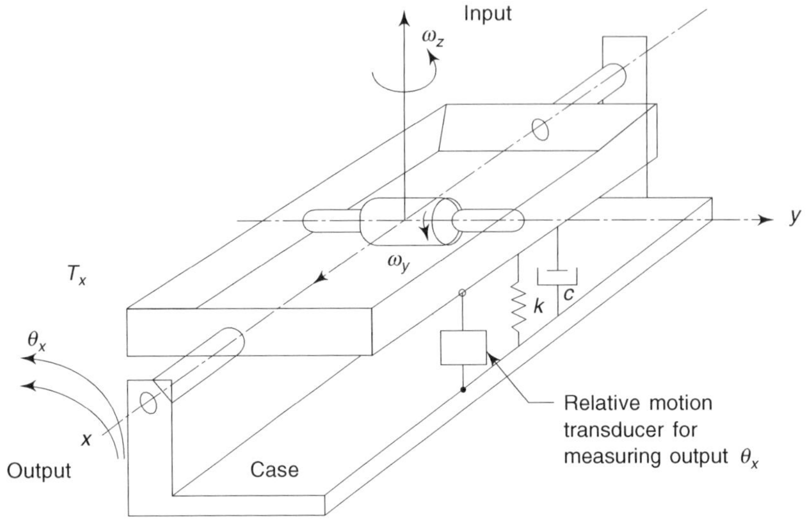

Fig. 7.24 *Rate gyro*

If the spring k is removed, the instrument becomes an integrating rate gyro. Equation (7.12) becomes

$$T = I\omega\ \omega_z = JD^2\theta_x + cD\theta_x \tag{7.15}$$

Thus

$$\frac{\theta_x}{\omega_z} = \frac{I\omega}{JD^2 + cD} \tag{7.16}$$

Since $\omega_z = d\theta_z/dt$, θ_z being the input angular displacement about the z-axis, we get

$$\frac{\theta_x}{\theta_z} = \frac{I\omega}{JD + c} \tag{7.17}$$

For a constant input θ_z, steady state value of θ_x is proportional to θ_z.

In practice, three such devices are needed for measuring the absolute angular motions about each of the three perpendicular axes.

7.3 ■ CALIBRATION OF MOTION OR VIBRATION MEASURING DEVICES

Static calibration of displacement measuring devices can be easily done by dial gauges, micrometers or slip gages. This can also be done by optical devices like a reading microscope or an interferometer. For dynamic calibration, the methods outlined here can be used.

7.3.1 Sinusoidal Motion or Vibration Method

For dynamic calibration of displacement, velocity or acceleration measuring devices, an electrodynamic shaker can be used (Fig. 7.25). The shaker is driven by a variable frequency oscillator and a power amplifier. The transducer to be calibrated is mounted on the shaker table and can be moved at circular frequency ω which can be changed by oscillator setting. The amplitude of harmonic motion can be changed by power amplifier setting. For motion

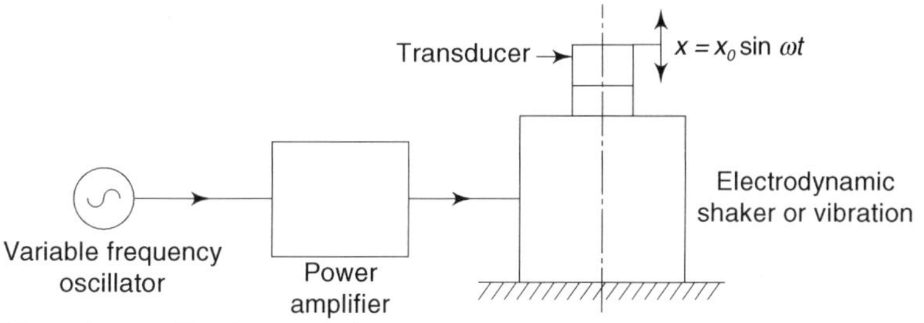

Fig. 7.25 *Sinusoidal motion or vibration method*

$$x = x_0 \sin \omega t$$

$$\text{Velocity } \dot{x} = \omega x_o \cos \omega t$$

and $$\text{Acceleration } \ddot{x} = -\omega^2 x_0 \sin \omega t$$

Thus, velocity amplitude of harmonic signal is ωx_0 and acceleration amplitude is $\omega^2 x_0$. Angular frequency ω is read from the oscillator setting while amplitude x_0 can be read by an optical device like a reading microscope.

Vibration transducers can be easily calibrated using this method.

7.3.2 Transient Motion Method

For dynamic calibration of accelerometers, the transient motion method can be used, especially when an accelerometer is required to be calibrated for high values of g. One of these methods has been explained in Problem 4.6. Other methods include the use of a drop test calibrator or a pendulum. These involve the application of an impact and measuring the output due to the impact.

If v is the velocity change of the transducer during calibration as a result of impact,

$$v = \int_0^T a\,dt \tag{7.18}$$

a being the acceleration and T the duration of the impact. For an accelerometer,

$$e = ka \tag{7.19}$$

k being the calibration constant of the transducer and c the voltage output.

From Eqs. (7.18) and (7.19), we get

$$k = \left[\int_0^T e\,dt\right] / v \tag{7.20}$$

In order to determine k, the numerator in Eq. (7.20) may be obtained from a record of e against time t while the denominator, viz. velocity change may be computed or obtained by measurement.

In the drop test calibrator (Fig. 7.26) the impact is produced due to the free fall of the accelerometer under gravity from a height h_1. If h_2 is the height of rebound,

$$v = (2gh_1)^{1/2} + (2gh_2)^{1/2} \tag{7.21}$$

In the case of a pendulum calibrator (Fig. 7.27) a pendulum mass is used to apply a sudden velocity change. The velocity change of the anvil mass due to the impact can be easily obtained by calculations or alternatively by the photo cell arrangement shown in Fig. 7.28. the grating attached to the anvil interrupts the light from the light source. From the output of photo cell plotted against time, the total velocity change v can be obtained.

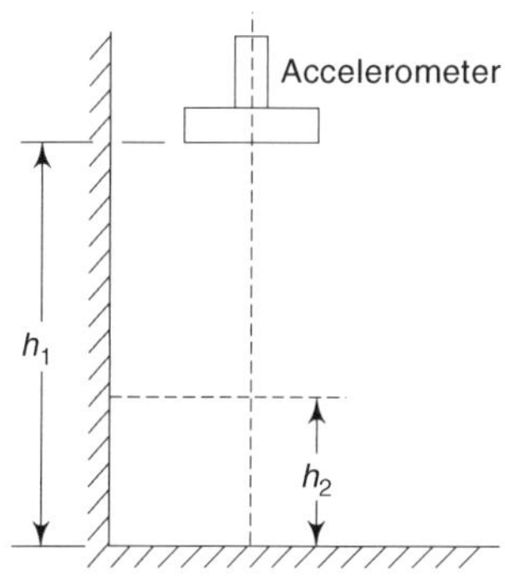

Fig. 7.26 *Drop test calibrator*

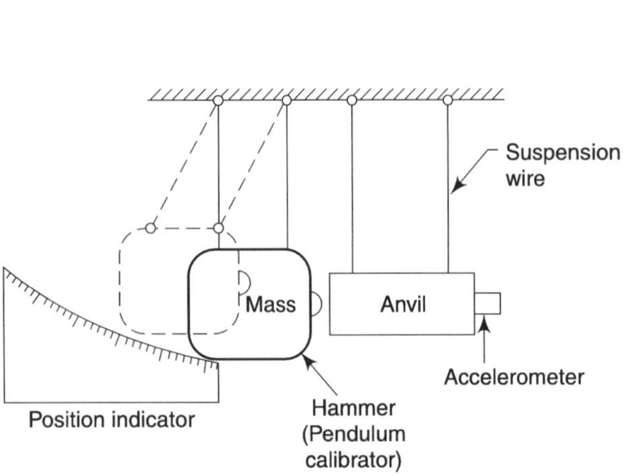

Fig. 7.27 *Pendulum calibrator*

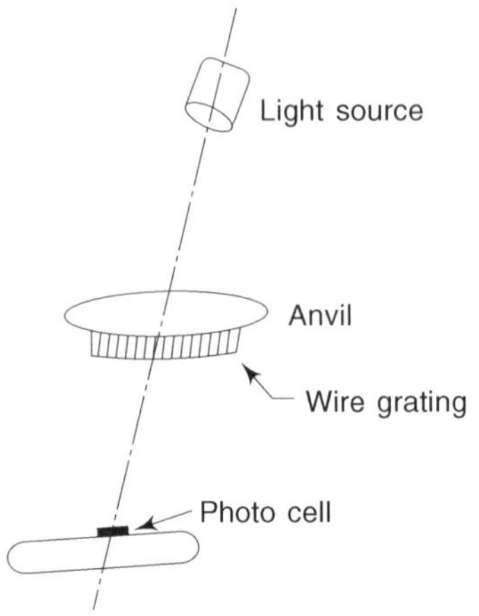

Fig. 7.28 *Electro-optical method for finding velocity change*

7.3.3 Comparison Method

This is a simple method in which a motion transducer can be calibrated against a standard calibrated transducer by comparing the output of the two transducers with a common input. The arrangement of Fig. 7.27 can be used with two transducers mounted on the same vibrator table so that they are subjected to the same motion during calibration.

Review Questions

7.1 Indicate True or False against each of the following:

(a) A seismic type displacement transducer is used for frequencies that are lower than the undamped natural frequency of the instrument.

(b) A piezo-electric accelerometer can only be used for static motion measurement.

(c) A seismic transducer incorporating a mass on a spring and damper can measure acceleration at low frequencies and displacement at high frequencies.

(d) It is necessary to reduce the damping ratio of a seismic instrument to as low a value as possible for good dynamic performance.

(e) An optical interferometer is useful for measuring extremely small motions.

(f) An optical interferometer cannot measure dynamic motions.

(g) A force balance type motion measuring device is more accurate than a spring mass damper type seismic device.

(h) A rate gyro is a relative motion measuring device.

(i) For measuring the vibrations of the engine body in a moving car, we have to use a relative motion transducer.

7.2 Find the open circuit voltage output of a strain gage type of seismic motion transducer shown in Fig. 7.29. Mass m weight 20 gm. Each of the steel strips on which resistance strain gages are mounted is 5 mm wide, 1 mm thick and 10 mm long. Each gage resistance = 120 Ω gage factor = 2, damping ratio = 0.7, Young's modulus of steel = 2.07×10^{11} N/m^2 and motion of frame (mm) = 0.6 sin 5000 t.

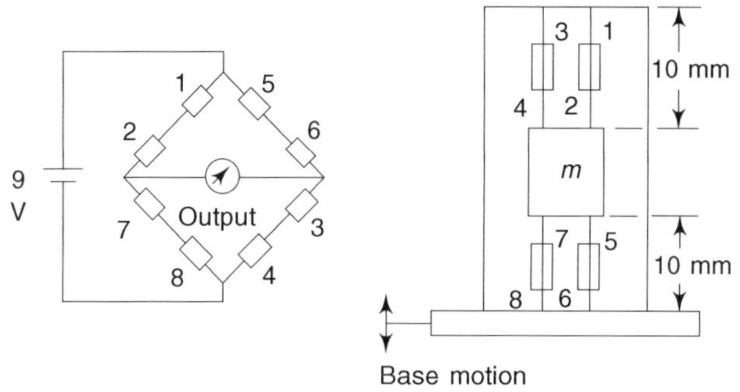

Fig. 7.29

7.3 A seismic instrument, used for motion measurement, of an object, has m of 100 gm, spring stiffness 1 N/mm and damping constant 8 N–s/m. If the recorded motion has an amplitude of 3.5 mm, find the amplitude of motion of the object if the frequency of motion is 30 Hz.

7.4 In a piezo-electric accelerometer, mass $m = 20$ gm, $k = 7 \times 10^7$ N/m and $c = 400$ N–s/m. If the voltage sensitivity constant of the crystal is 200 V/mm, find the voltage output for an input acceleration amplitude of 500 g at a frequency of 2000 Hz.

7.5 A seismic motion transducer has a mass of 50 gm. Its undamped natural frequency is 10 Hz and damping ratio is 0.6. The relative motion of the seismic mass with respect to the frame of the transducer is converted to voltage by a first order transducer which has a static voltage sensitivity of 2 V/mm and time constant of 0.01 s. Find the output voltage for an input motion of 0.5 mm at a frequency of 30 Hz.

7.6 A rate gyro has a spring of stiffness 100 N–cm/rad and mass moment of inertia of 1.2 kg–cm^2 about the output axis. The built-in relative motion transducer has a sensitivity of 100 mV/rad. Find the steady state voltage output for a rate of turn of 5 rad/s. The spin velocity of the rotor = 18000 rpm.

Answers

7.1 (a) F (b) F (c) T (d) F (e) T (f) F
(g) T (h) F (i) F (j) T (k) F

7.2 1.3 mV 7.3 4.2 mm 7.4 0.293 V

7.5 480 mV 7.6 113 mV

Chapter

8

Dimensional Metrology

INTRODUCTION

Dimensional measurements form an important part in the modern day engineering instrumentation techniques. The current trend in the various production techniques is accurate, precise and fast dimensional measurements of components. In other words, dimensional gauging of motor vehicles, washing machines, referigerator parts and a host of other devices is one of the foremost requirements in the day-to-day mass production techniques. This ensures that a high level of accuracy is maintained in the machine tools, jigs and fixtures in order to produce interchangeable components with in very close tolerances.

Quite often, the measurement of displacements as for example, caused due to motion, temperature, pressure, torque or vibratory motion may form an important parameter in a number of measurement situations. Such measurements are generally employed in the design, development and testing of a large number of industrial components. It is important to note that first and second derivatives of dimensional measurements and displacements with respect to time also form important design parameters. They represent the motion of objects in terms of velocity and accelerations. In this chapter we shall discuss the sensors and transducers of several types used to measure dimensional measurements. The selection of particular device for a particular application depends on several factors which have been listed below.

1. *Type of measurement* Whether linear or angular.
2. *Contact* Whether contacting or proximity type
3. *Output–input relationship* In terms of sensitivity, accuracy, precision, linearity, etc. and in addition other specifications relevant to the measurement situation
4. *Time dependance* Whether the measured quantity is static or dynamic
5. *Magnitude of displacement* Whether the magnitude of measured quantity is small or large. The small magnitude range may be of nano or micro scale and large values may be close to mega or tera range.

8.1 ■ LINEAR DIMENSIONAL GAUGING

Common linear dimensional measurements include measurement of lengths, widths and heights of components. In addition, quite often depths of holes and slots, etc. also need to be measured. In general, the dimensional linear gauging consists of comparing the unknown dimension of components by means of measuring tools which have been previously calibrated with the standard of known traceability. As discussed in Ch. 1, the international standard of length has been defined in terms of universally reproducible wavelength standard and has accuracy of 1 part in 10^8. In this, one metre length corresponds to 1,650,763.763 wavelengths of light emitted by Kr^{86} orange-red lamp. However, the National level dimensional metrology reference standards often adopted by various countries consist of very high quality slip gauges and length bars of hardened steel whose end faces are lapped flat and parallel to within ±10 nm. Therefore, these reference standards have accuracy specifications of the order of ±0.01 μm and are generally employed to check the calibrations of commonly used industrial instruments in dimensional metrology.

The dimensional gauging instruments can be classified as follows:

- Mechanical type
- Electro-mechanical type
- Pneumatic type
- Hydraulic type
- Optical type
- Special instruments like opto-electronic or fibre-optic type.

8.2 ■ MECHANICAL TYPE OF DIMENSIONAL GAUGING DEVICES

These devices have marked scales and can be conveniently obtained in the required accuracy specifications. It is easier and quicker to use them over the required range of dimensional measurements. However, the wear and tear due to long useage may introduce inaccuracies in measurements. Some commonly used mechanical types of dimensional gauging devices have been discussed.

8.2.1 Rulers and Tapes

Rulers and tapes are most commonly used tools in our day-to-day lives and shop floors. An engineer's steel rule is also termed *scale*, is a low cost and easy to use, length measuring device. It is made up of hardened steel and is generally available to measure dimensions up to 1000 mm i.e., 1 m. The accuracy of readings using the steel rule of 1 mm engravings is generally ± 0.5 mm, i.e., half the distance between millimeter markings using the judgement of interpolation by eye alone. However, improved type of steel rules marked with 0.5 mm engravings are also available. For such rulers the accuracy of measurements is of the order of ±0.25 mm.

For the measurement of larger dimensions up to 3000 mm or more, retractable type of the steel tapes are generally used. The end of the tape is usually provided with a small hook at 90° to the tape length for convenient placement with the wall for dimensional measurements of buildings/rooms, etc. The thickness of this hook is included in the tape and hence no correction or compensation for its thickness is necessary.

Advantages

1. They provide simplest, low cost, easy and quicker way of measuring a wide range of lengths.
2. They are useful shop floor instruments of measuring lengths where high levels of accuracies is not a requirement.

Disadvantages

1. They are not preferred where high levels of accuracies and precisions are required.
2. There may be wear of the leading edge of the scale which may lead to undersizing.
3. They are not suitable for measurements which are changing with time.

8.2.2 Vernier Caliper

The accuracy of measurement using a simple steel rule markings can be enhanced by providing a parallel sliding scale termed as vernier scale. The divisions of vernier scale are somewhat different than those of steel rule markings. The difference between the steel rule markings called the main scale and the sliding vernier scale is utilised in improving considerably the accuracy of measurements. In this instrument, shown in Fig. 8.1, the main scale is accurately graduated in millimeters and is joined with a fixed jaw. The second scale is of a moveable type and forms a part of moveable jaw. The moveable scale is a vernier scale. In the most commonly used vernier scale, a 9-mm length is equally divided into 10 parts. This way each graduation on the vernier scale is 0.9 mm in length. The least count or the resolution of the instrument is difference in length of one division of main scale division (i.e., 1 mm) and that of one division of vernier scale division (i.e., 0.9 mm). In other words, the accuracy of measurement is (1.0 – 0.9) = 0.1 mm.

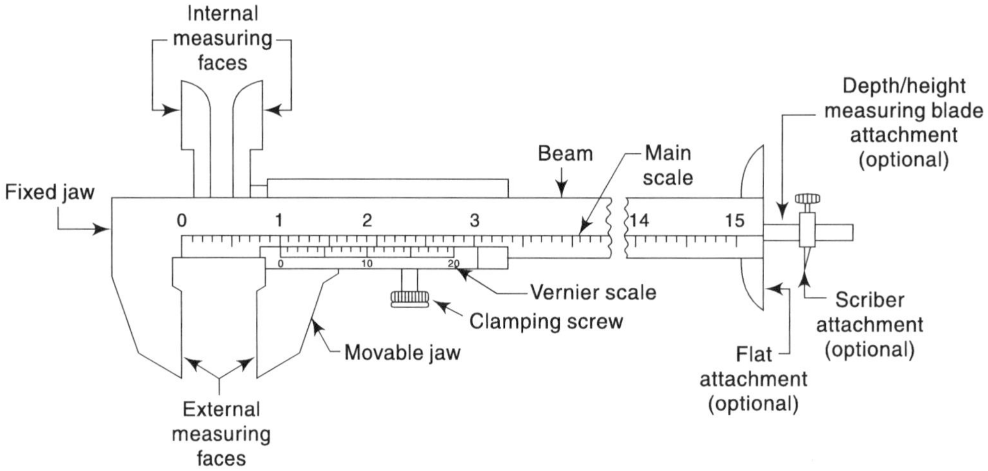

Fig. 8.1 *A typical shop floor vernier caliper*

The accuracy of a vernier caliper can be increased five times, i.e., up to a least count of 0.02 by graduating appropriately the main scale and vernier scales. In this case, the main scale divisions are marked in terms of 0.5 mm in place of the usual 1 mm divisions. Further, the distance of 24 graduations of the main scale (i.e., 12 mm distance) is divided equally in 25 parts in the sliding vernier scale. This way, one division vernier scale becomes 12/25 mm = 0.48 mm. Hence, the difference between one main scale division (i.e., 0.5 mm) and one vernier scale division (i.e., 0.48 mm) gives a least count of measurement of 0.02 mm.

Alternatively, a vernier scale of 49 mm length with 50 equally spaced vernier scale divisions a long with the main scale having 1 mm graduations also gives least count of 0.02 mm. (i.e., 1.0 mm – (49/50) mm = 0.02 mm).

Advantages

1. It is commonly used, low cost and rugged type of shop floor instrument, which is additionally quick and easy to use for accurate measurements of displacements.

2. Its least count is usually of the order of 0.1 mm. However, convenient vernier scales with least count of 0.02 mm are commonly available.
3. It is used in shop floors and in instrumentation applications for checking dimensions of products for quality control, checking the components of wear and tear, etc.

Disadvantages

1. The wear and warping of jaws introduces zero error, which needs to be accounted for in the measurements.
2. It is not suitable for measurements which change directions.

Vernier Height and Depth Gauge The conventional vernier caliper shown in Fig. 8.1 is provided with additional optional attachments for using the instrument for the measurements of the heights of objects or depths of slots, etc. The rear end of the vernier caliper can be conveniently fitted with a flat type of attachment with the beam of the fixed scale. In addition, a depth sensing rod type attachment is joined with the moving vernier scale. In case, the instrument is to be used for height measurements of the objects, then, a scriber attachment shown in Fig. 8.1 is fixed to the depth/height sensing rod of the instrument.

8.2.3 Micrometer Screw Gauge

Micrometer is a device in which an accurately machined screw threads are associated with circular scale divisions. These are used to indicate parts of primary scale divisions with high degree of accuracy. Figures 8.2(a) and (b) show respectively a typical micro-meter screw gauge for general type of measurements and modified type of the same gauge for the measurements of depths of holes, slots, etc.

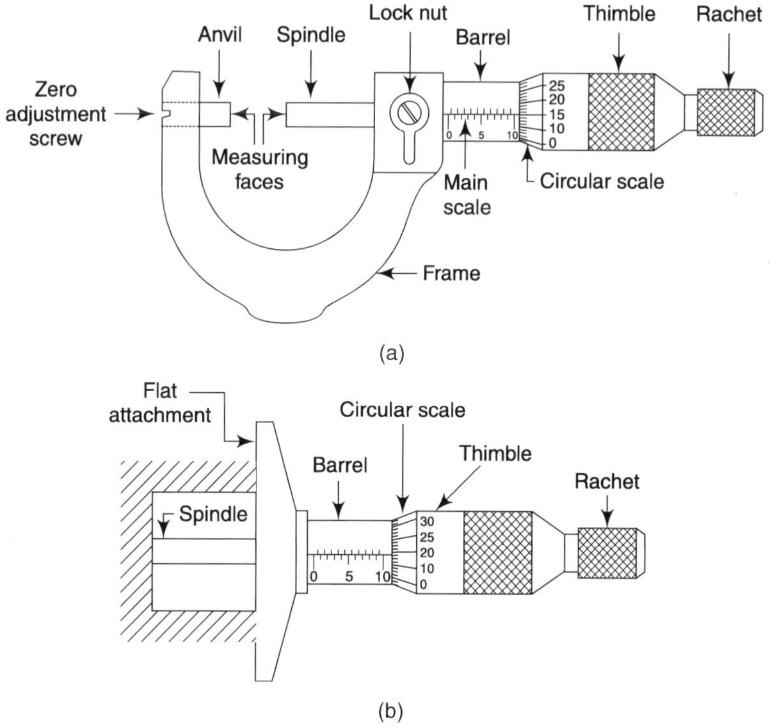

Fig. 8.2 *(a) A typical micrometer screw gauge and (b) A modified micrometer screw gauge for depth measurements*

In the micrometer screw gauge, a high quality machine screw having a uniform pitch of 0.5 mm is attached to the spindle. It is turned by the movement of a thimble with a rachet provided at the end of the spindle. The workpiece is held between the measuring faces of anvil and the spindle for dimensional gauging of the same. In order to obtain consistent measurements, a micrometer screw is tightened to the work piece by a constant pressure by means of a spring loaded rachet. For initial zero adjustment of the instrument, a screw is provided for moving the anvil in the micrometer frame.

The micrometer thimble has 50 circular scale divisions which are etched around its circumference. Since the pitch of the screw is 0.5 mm, therefore each circular scale division corresponds to 0.01 mm. In the measurements, using micrometer screw, the readings can be observed to an accuracy of half division of circular scale by interpolation by eye alone. Therefore, the least count of the instrument is of the order of 0.05 mm.

Presently, intelligent micrometers with digital read outs are available. They have self-calibration capability and have least count of 0.001 mm (i.e., 1 micron).

Advantages

1. It is low cost, rugged and easy to use instrument for accurate and precise dimensional gauging of components.
2. Its least count, i.e., its resolution is 0.01 mm, which is higher than that of vernier caliper. Further, micrometers of digital type with resolution of the order of 1 micron (i.e., 1 μm) are easily available.
3. It is commonly used in industrial amplications and is available in various sizes and ranges for checking the dimensions of products, for machine settings, depths and widths of holes and slots etc.

Disadvantages

1. The wear of screw with use may introduce error due to backlash.
2. If the faces of the end of spindle and anvil are not truly flat then the error due to incorrect positioning of the measured object gets introduced.
3. The instrument is not suitable for the measurement of time-varying displacements (i.e., dynamic displacements).

8.2.4 Micrometer Dial Gauge

Micrometer dial gauge is also commonly known as Dial Test Indicator (DTI) and has been shown in Fig. 8.3. It is a very commonly used shop floor instrument for measuring displacements for a wide range of applications.

The body of the instrument is kept fixed in relation to the object whose displacement needs to be measured. A spring-loaded displacement sensing rod, attached to a ball-end-contact point touches the object. The measured displacement is fed to the gear rack which in turn is connected to the gear train. The gear train acts as amplifier to the input displacement. The output is read by means of a pointer and scale. A movable scale on the dial can be rotated to obtain a suitable reference point.

To improve the accuracy of reading usually two dials are used. The larger dial having 100 equal divisions indicates the displacements in small increments of the order of 0.01 mm. Further, the smaller dial displays the displacements in larger units, i.e., in terms of millimetres. In other words, one complete revolution of pointer on a bigger dial (i.e., 100 divisions of 0.01 mm each) equal to 1 mm of displacement.

Presently intelligent type of dial test indicators are available. They are equipped with digital display in place of pointer on a circular scale. They have self-calibration capability and have least count of 0.001 mm (i.e., 1 μm).

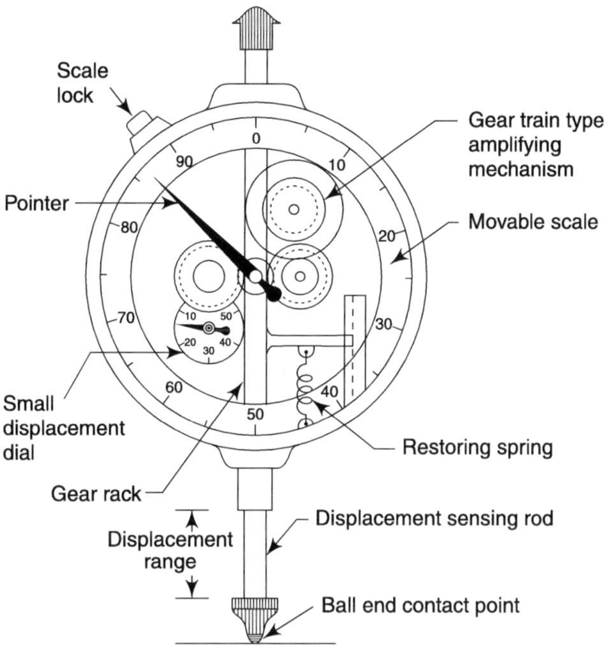

Fig. 8.3 *A typical dial gauge indicator*

Advantages

1. Even though it is a mechanical device like vernier calipers and micrometer screw gauge, but its main advantage is its quickness to indicate measured displacements.
2. It is easy to use and is accurate in displacement measurements.
3. Commonly available gauges have very good resolution of 0.01 mm. However, special digital gauges have extremely high resolution of 0.001 mm (i.e., 1 micron).
4. Its use is quite widespread in industry. Common examples are checking the dimensions of products for quality control, checking the components for wear, when used with prooving ring, it becomes a force transducer, etc.

Disadvantages

1. The instrument is suitable only for static measurements i.e., it cannot measure displacements which change directions with time.
2. The displacement to be measured should be accessibe to the plunger.
3. Spring force acting on the plunger even though small in magnitude has to be borne by the measured object. Therefore, it may not be suitable for delicate parts/components.

8.2.5 Gauge Blocks and Length Bars

Gauge block sets also known as *slip gauges* are extensively used as industries dimensional standard. They are routinely used for very accurate and precise dimensional gauging as well as for checking the accuracies of other shop-floor displacement devices. A gauge block (or a slip gauge) is a rectangular block made up of hardened steel. Its both sides are carefully machined and lapped to make them flat and parallel with in extremely fine tolerances. This way, the distance between the faces represents a very high standard of accuracy because of high level of surface finish and flatness.

The highly polished and extremely flat surfaces of gauge blocks have an infinitesimally thin layer of lubricant film. When two gauge blocks are slid together with a slight contact pressure, the contact surfaces adhere together with significant force. This process of combining the gauge blocks is termed *wringing*. During the wringing procedure, the adhesion between various blocks is quite substantial. The interblock gap between the various blocks is few molecules thick lubricant film, which is of the order of 1 nm (i.e., 0.001 μm). This value is very close to zero and hence there is no significant error caused in the displacement gauging process.

For measuring larger dimensions in the range of 100 mm to 1000 mm gauge block set is provided with additional set of length bars. These length bars are made up of high quality hardened steel and may be either cylindrical or rectangular shaped. Further, these bars are available for fixed lengths ranging from 100 mm to 1000 mm.

Gauge blocks and length bars are usually available with the following specifications:

- Reference grade
- Calibration grade
- Grade I
- Grade II

The reference and calibration grade have highly accurate and flat surfaces. The accuracy specification of the former is better than the latter. Both are used where the highest level of accuracy is required in the manufacturing practice. On the other hand, Grade I and Grade II are normal shop-floor standards. The former is used for high accuracy jobs. Further, the latter are used as general purpose devices for checking the jobs in which relatively higher tolerances may be acceptable.

Advantages

1. They form an important component in workshop practice for dimensional gauging of interchangeable type manufactured parts.
2. The least count of these gauges is quite low. It is of the order of ±0.002 μm for the calibration grade instruments.
3. They do not require any skilled operator and are quite convenient tools for dimensional gauging.
4. These can be used for periodic checking of the accuracies of other shop-floor dimensional gauging devices.

Disadvantages

1. These instruments cannot be used in which dimensional measurements are changing rapidly with time.
2. These devices are usually calibrated at a particular temperature. If the gauge is used at a different ambient temperature, then proper correction due to temperature change needs to be applied.

Thickness or Feeler Gauge A commonly used portable and pocket-type device based on gauge block principle is 'Thickness or feeler gauge' (see Fig. 8.4). It consists of a number of thin blades usually of

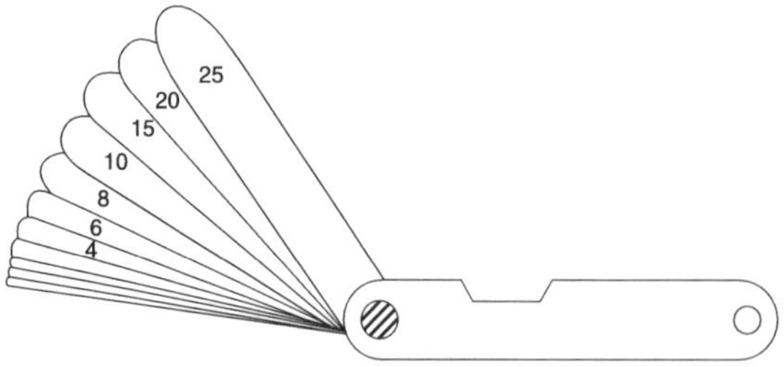

Fig. 8.4 *A portable type thickness or feeler gauge*

shop floor specifications of either grade I or grade II. The thickness of each blade is marked on each blade. The sum of thicknesses of selected blades is a measure of dimension to be measured. The main advantage of this device is that it is a rugged and low cost device. Further, it can be conveniently used for gauging narrow spaces or for checking small clearances.

Problem 8.1 *A gauge block and length bar set has the following available sizes:*

Size (mm)	*Increment (mm)*	*Number of pieces*
1.0005	—	1
1.001 to 1.009	0.001	10
1.01 to 14.9	0.01	49
0.5 to 9.5	0.5	19
10 to 90	10	9

Choose the appropriate gauge blocks and length bars to construct a length of 64.8975.

Solution

(i) We initially choose the block finished to four decimal accuracy from first row i.e., 1.0005.
(ii) Next, we choose the block from the second row to satisfy the third decimal place i.e., 1.007.
(iii) After this, we choose a block from the third row to satisfy, first and second decimals, i.e., 1.89.
(iv) Finally, the balance 61 mm can be composed of 1 mm from the 4th row and 60 mm from the fifth row.

This way the required length of 64.8975 mm can be obtained from the given set of gauge blocks and length bars.

8.2.6 Angular Dimensional Gauging using Sine Bar

A sine bar is a useful tool for accurate setting out of angles by converting angular measurements to linear ones. The sine bar when used in conjunction with slip gauges constitutes an excellent device for precise measurements of angles. The principle of operation of angle measurements depends on the accurate and precise measurements of height h and hypotenuse l of a right-angled triangle. The relationship generally employed is

$$\sin \theta = \frac{\text{height of a right angled triangle}}{\text{hypotenuse of a right angled triangle}}$$

$$= \frac{h}{l} \tag{8.1}$$

Figure 8.5 shows a typical test set-up for angle measurements of an inclined plane, using the sine bar. The sine bar shown in the figure is made up of high-carbon hardened steel. Two cylinders of equal diameters are attached at each end. The axes of these cylinders are mutually parallel to each other. In addition, they have equal distance from the upper surface of the sine bar. The distance between the axis of cylinders is kept exactly 100 mm or 200 mm or 300 mm depending on its size. Further, sine bars are available with specification of A grade or B grade. A grade sine bars have guaranteed accuracy of 0.01 mm/m whereas B grade sine bars have accuracies of the order of 0.02 mm/m.

The stack of slip gauges along with length bars on both sides are adjusted on both cylinders after placing the test component (i.e., the inclined plane) on the sine bar. The parallelism of the top surface with the base may be checked with dial test indicator or with the precision type of spirit level. The

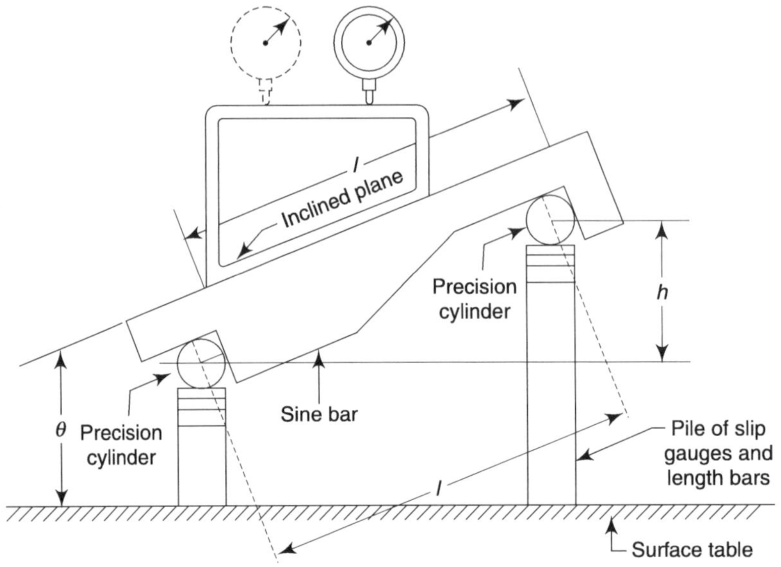

Fig. 8.5 *A typical test set-up for angle measurement using sine bar along with slip gauges and length bars*

difference of heights h is known from values of slip gauges and length bars stacked on both sides. The value divided by the standard length sine bar gives very accurate and precise value of sin θ from which the angle of inclination can be found accurately using the trigonometric tables.

Advantages

1. It is a very convenient method of accurate and precise measurements and checking unknown angles of manufactured parts/components with the use of calibration grade linear measurement devices, i.e., slip gauges.
2. The procedure is quite simple and does not involve skilled operation.

Disadvantages

1. Sine bars cannot be used for conveniently for measuring angles more than 60° because of slip gauge adjustment problems.
2. Misalignment of work piece with sine bar may sometimes introduce considerable errors.

8.3 ■ ELECTROMECHANICAL DIMENSIONAL GAUGING DEVICES

These have been discussed in detail in Ch. 4 of Transducer Elements (see section 4.2.1) and are of the following type.

- Potentiometric (resistance) type
- Inductive type
- Capacitive type
- Piezo-resistive (strain gauge) type
- Ionisation transducer type
- Optical-electrical transducers of
  - (i) Photo-emissive type
  - (ii) Photo-conductive type
  - (iii) Photo-voltaic type

Electro-mechanical dimensional gauging are of versatile type and are commonly used in a host of industrial applications. Some of these devices are also capable of measuring angular displacements. Usually, these are simple in construction and can be designed to give large outputs which can be displayed or recorded with ease. Some of the inductive and capacitive devices can be proximity type while the others are usually of contact type. It may be noted that piezo-electric type can only be used for dynamic measurements, while the order may be used for both static and dynamic measurements.

Generally variable inductance, capacitance and mechano-electric devices are suitable for small displacements of the order of a few millimetres. The differential transformer and ionisation transducers are used in the intermediate range of displacements, say from few millimeters to several centimetres. On the other hand, the potentiometric type of displacement transducers are comparatively less sensitive to small displacements. However, practically, there is no limit on the maximum value of measured displacement for which they may be used.

8.4 ■ PNEUMATIC DIMENSIONAL GAUGING TECHNIQUE

In pneumatic type of devices, the displacement signal is converted to a pressure signal. The device shown in Fig. 8.6 is a typical pneumatic displacement gauge and this is also commonly termed as flapper-nozzle device. The micro-motion x of the flapper controls the back pressure P_b. It has been observed

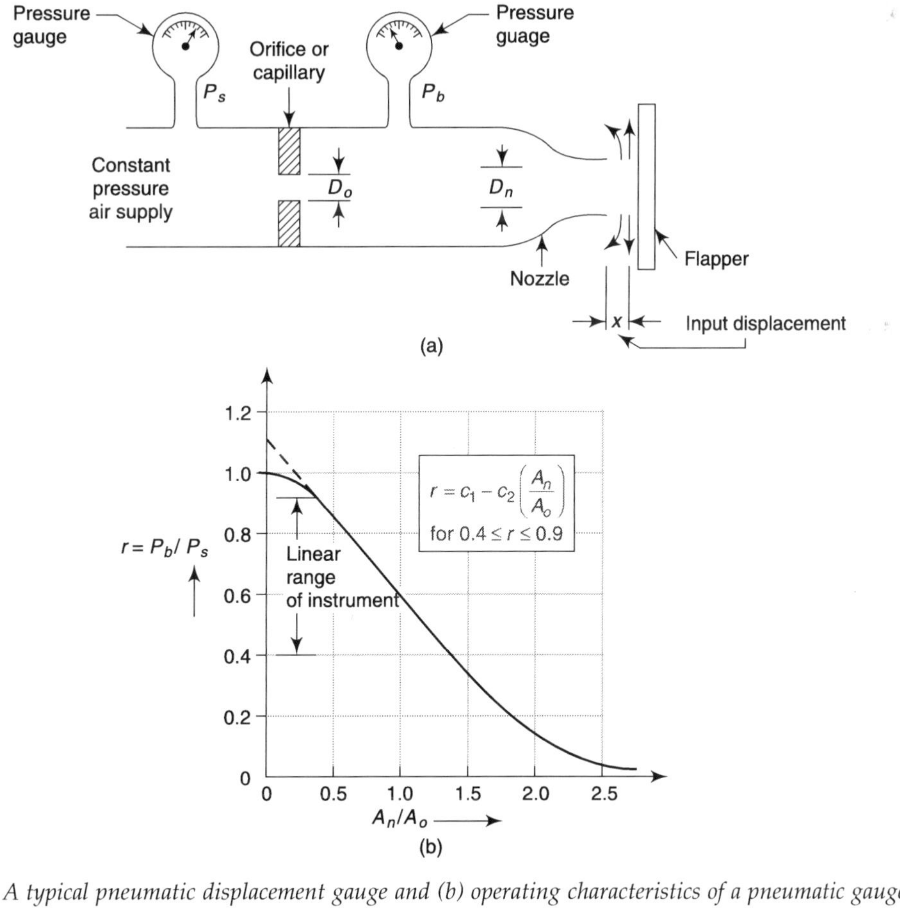

Fig. 8.6 *(a) A typical pneumatic displacement gauge and (b) operating characteristics of a pneumatic gauge*

experimentally that a linear relationship exists between P_b and x over a limited range of x. However, the sensitivity in this measured range is quite high. Pressure P_s and P_b are gauge pressures can be measured by the pressure measuring devices.

The approximate relationship between pressure P_s, P_b and measured displacement x due to the motion of the flapper is derived as follows:

The flow is assumed incompressible for small pressure changes and also the areas of orifice and nozzle are taken as small as compared to the area of cross-section of the pipes at supply to the orifice and nozzle.

Now mass flow rate of air through the orifice

$$\dot{m} = k_1 A_0 \sqrt{2(P_s - P_b)/\rho} \tag{8.2}$$

where k_1 is the constant of the orifice meter

ρ is the density of fluid i.e., air and

A_o is the area of the orifice $= \frac{\pi}{4} D_0^2$

The same mass-flow rate passes through the exit nozzle to the atmosphere having nozzle diameter as D_n. Taking the area of flow from nozzle to outside as that of a cylindrical periphery, the exit area of nozzle A_n becomes

$$A_n = \pi D_n x \tag{8.3}$$

Now, the mass-flow rate of air through the exit nozzle $\dot{m} = k_2 A_n \sqrt{2(P_b)/\rho}$ (8.4)

where k_2 is the constant of the nozzle and

P_b is the gauge pressure.

Taking k_1 and k_2 as nearly equal and equating the mass flow rates at steady state we get,

$$\frac{P_b}{P_s} = \frac{1}{1 + \frac{A_n^2}{A_0^2}} \tag{8.5}$$

This is a non-linear relation and when plotted experimentally, a typical operating characteristics similar to that shown in Fig. 8.6(b) is obtained. It usually has a nearly linear operating range between the pressure ratio $r = (P_b/P_s)$ and area ratio (A_n/A_o). for $0.4 \le r \le 0.9$. The slope of the curve is negative and approximate relationship of the following form can be fitted for the pneumatic displacement gauge.

$$r = C_1 - C_2\left(\frac{A_n}{A_0}\right) \tag{8.6}$$

It may be noted that the values of C_1 and C_2 can be determined experimentally and these values are nearly 1.1 and 0.5 respectively.

Substituting the values of A_n and A_o, Eq. (8.6) becomes

$$r = C_1 - C_2 \frac{(\pi D_n x)}{\left(\frac{\pi}{4} D_0^2\right)}$$

$$= C_1 - 4C_2\left(\frac{D_n}{D_0^2}\right)x$$

$$= C_1 - C_2' x \tag{8.7}$$

where $$C_2' = 4C_2\left(\frac{D_n}{D_0^2}\right) \text{ and}$$

r varies between 0.4 to 0.9

In actual practice, known input displacements are given to the pneumatic displacement gauge and the corresponding values of ratio of pressure $r = P_b/P_s$ are recorded for each displacement. Then, the experimental values are represented in the form of r vs x to obtain the calibration graph which is very similar to Fig. 8.6(b). From this graph, the values of slope and intercept, namely, C'_2 and C_1 for linear range are obtained. After calibration, the gauge gives a very accurate and precise values of r in terms of unknown displacements which can be evaluated using the relation (8.7).

Advantages

1. It is non-contact type of dimensional gauging device.
2. It is nearly linear instruments in its operating range i.e. $0.4 \le r \le 0.9$ for supply pressures ranging between 200 kPa to 500 kPa.
3. If the supply pressure can be kept constant, then the back pressure indicating pressure gauge can be calibrated in terms of displacements.
4. The resolution of gauge better than ± 5 μm can be easily obtained in its linear operating range.
5. It is widely used in industry for continuous measurements or control as well as in comparators for dimensional measurements.

Disadvantages

1. The linear relationship between P_b and x is valid for limited range of x.
2. The instrument has to be calibrated invariably using accurate and precise feeler gauges or any other high quality dimensional calibration device.
3. The calibrations of the back pressure gauge is non-linear when measurements close to zero displacements are measured.

Problem 8.2 *A pneumatic displacement gauge having orifice and nozzle diameters as 2 mm and 1 mm respectively has been calibrated using the calibration grade slip gauges. The experimentally obtained linear operating characteristics with usual notations was*

$$r = 1.15 - 0.52\,(A_n/A_o) \text{ for } 0.4 \le r \le 0.9.$$

(a) Determine the linear displacement measurement range of the instrument.

(b) If the gauge has a constant supply pressure of 40 kPa and back pressure P_b can be read with accuracy of ± 0.5 mm of Hg on the mercury manometer, calculate the resolution of dimensional gauging using this instrument.

Solution

(a) Substituting the values of A_n and A_o in the governing equation of the gauge we get,

$$r = 1.15 - 0.52\,\frac{\pi\,(0.001)\,x}{\frac{\pi}{4}(0.002)^2}$$

$$= 1.15 - 520\,x$$

(i) When $r = 0.9$ then $x = x_{\min} = 4.8 \times 10^{-4}$ m $= 0.48$ mm

(ii) When $r = 0.4$ then $x = x_{\max} = 1.44 \times 10^{-3}$ m $= 1.44$ mm

Hence the linear range of instrument lies for $0.48 \text{ mm} \le x \le 1.44$ mm

(b) The hydrostatic equation with usual notations gives:

$$p = \rho g H$$

$\therefore$ pressure corresponding to $\pm$ 0.5 mm of Hg = $(100 \times 13.6) \times 9.81 \times 0.5 \times 10^{-3} = \pm 66.7$ Pa

The non-dimensional uncertainty U_r in pressure ratio $r = \pm \dfrac{66.7}{40 \times 10^3}$

$$= 1.66 \times 10^{-3}$$

Using Eq. 2.7 of Ch. 2 in the governing equation of instrument mentioned above, we get

$$U_r = \pm\sqrt{\left(\frac{\partial r}{\partial x}\right)^2 U_x^2} = \pm\left|\frac{\partial r}{\partial x}\right| U_x$$

or $$1.66 \times 10^{-3} = \pm 520\ U_x$$

or $$U_x = \pm\frac{1.66 \times 10^3}{520} = \pm 3.2 \times 10^{-6} \text{ m}$$

$$= \pm 3.2\ \mu\text{m}$$

Hence the resolution of the pneumatic displacement gauge is $\pm$ 3.2 μm.

8.5 ■ HYDRAULIC DIMENSIONAL GAUGING TECHNIQUE

A simple, low cost, quick and convenient way of dimensional gauging uses the displacement of incompressible hydraulic fluid say water or light mineral oil by the movement of a diaphragm or bellows due to input of measured displacement. The displaced fluid is sent to a narrow tube attached to the fluid container, which magnifies the input displacement because of small area of cross section. In case higher magnification is required then the working fluid mercury may be used and further the connecting tube should be replaced by capillary of extremely small cross-sectional area. The use of mercury is necessitated as it is free from surface tension effects.

Figure 8.7 shows a typical set up of hydraulic displacement gauge used as comparator for checking the manufactured component with respect to standard component of desired dimensions. Herein, the ratio of (X_o/X_i) gives the magnification factor of the hydraulic displacement gauge.

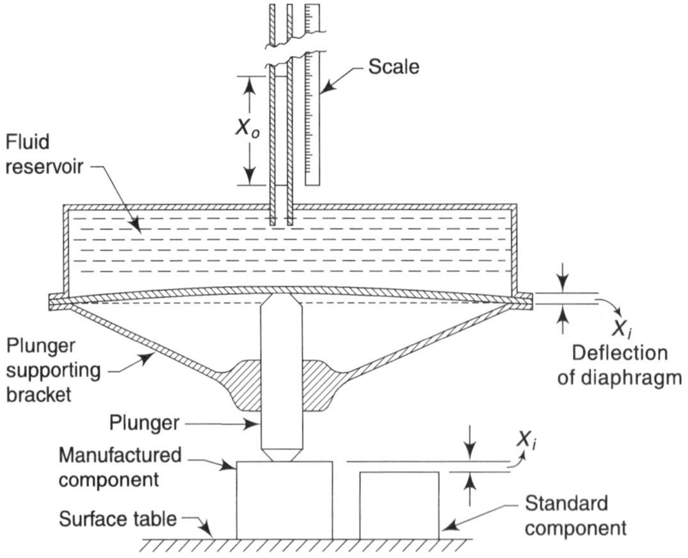

Fig. 8.7 *A typical set up of hydraulic displacement gauge used as comparator*

Advantages

1. It is an inexpensive device and does not require skilled operation.
2. It is a useful shop-floor dimensional gauging device. Its main advantage is, its quickness to indicate measured displacement.
3. Once calibrated, it gives accuracies of the order of micrometer screw gauge i.e., up to ±0.01 mm.

Disadvantages

1. It is a bulky device and cannot be used for machine settings.
2. The gauge can be used only for those measurements which are accessible to the plunger.
3. It has much less sensitivity as compared to pneumatic displacement gauge.

8.6 ■ OPTICAL DIMENSIONAL GAUGING

A wide variety of optical techniques are routinely used for the accurate and precise measurements of small and large displacements. These include microscopes, telescopes, interferometer and Moire technique. The main advantage of these devices is that they are non-contact type.

8.6.1 Interferometric Technique using Optical Flats

The principle of interferometry is used by employing monochromatic light (i.e., light of one colour or one wavelength) for measuring micro-level linear dimensions with a least count of ±1 μm or a lower value. Monochromatic light is conveniently obtained by passing the light through an optical filter. In practice, monochromatic light is used in conjunction with optical flat for comparing unknown dimensions by using the standard gauge block dimensions. Optical flat is made up of strain free glass in which at least one surface is lapped and polished to very high flatness tolerance, of the order of ± 0.1 μm or lower.

In general, light waves from a single source are passed through the optical flat and the reflective work piece is placed below that. The air gap between the optical flat and test piece causes reflection of parts of incident light at the both edges of the air gap. In other words, reflection of one part of the order of 30% to 40% of the incident light takes place at the top edge of the air gap, i.e., at the lower surface of the optical flat. The reflection of the remaining incident light takes place at the lower portion of the air gap i.e., from the reflecting surface of test piece. Both the reflected beams have different optical path lengths and interfere with each other. Due to this the inferometric fringes are formed which can be seen on the optical screen. In case, the beams happen to be out of phase by half a wavelength (i.e., 180° out of phase) with each other, then, the positive amplitude of one wave cancels with the negative amplitude of the second wave and vice versa. This way a dark spot or fringe is produced as shown in Fig. 8.8. On the other hand, if both the beams are in phase then they produce an enhanced brightness which represents in the production of a bright spot or a bright fringe.

It may be noted that if the distance of the air gap between the lower surface of the optical glass and polished reflecting test sample is $\frac{\lambda}{4}$, the incident and reflected beam in the air gap travels $\frac{\lambda}{4}$ both ways. This way the difference in path covered by two reflected beams becomes $\frac{\lambda}{2}$. In other words, the phase difference between the interference beams is $\frac{\lambda}{2}$ (i.e., 180° out of phase). As explained above, this would cause a dark fringe to be a observed. Now next, $\frac{\lambda}{4}$ increase in the air gap would produce a bright fringe. Again, next $\frac{\lambda}{4}$ i.e., $\frac{3\lambda}{4}$ from the initial distance would give the second dark fringe.

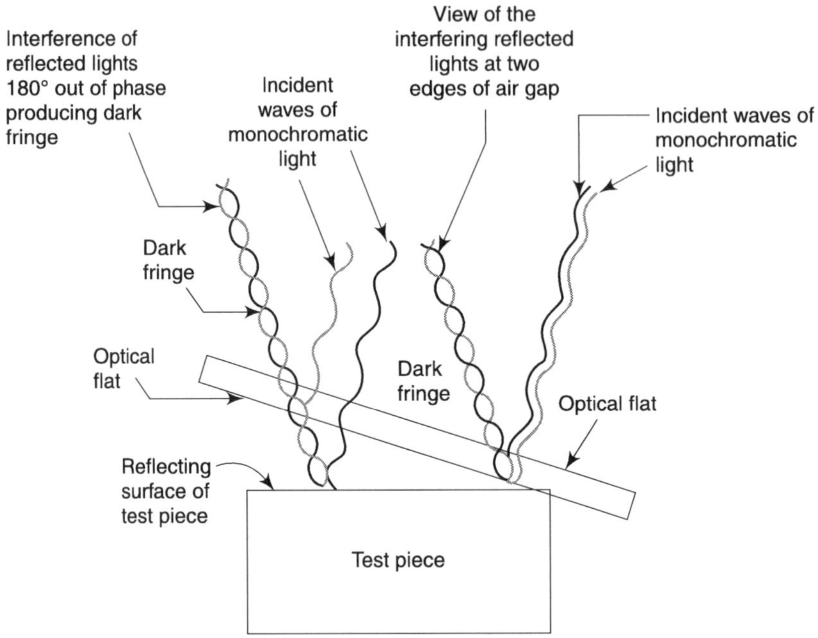

Fig. 8.8 *Dimensional gauging of air gap using light interference technique*

This way the distances of $\frac{\lambda}{4}, \frac{3\lambda}{4}, \frac{5\lambda}{4}$ .... waves correspond to first, second, third, ..., and so on dark fringes observed on the optical flat.

In general, we identify the start of the contact point and assign it as fringe of zero order. After that the fringes are counted sequentially and each fringe is assigned a fringe order which represents its number, when converted in sequence from a fringe of zero order. The distance of the air gap between the first two successive fringes dark fringes is $\frac{\lambda}{2}$. This is because, the difference of air gap distance of $\frac{3\lambda}{4}$ for second fringe and $\frac{\lambda}{4}$ of the first fringe gives the distance as $\frac{\lambda}{2}$. In fact the difference of elevation between any two successive fringes would be $\frac{\lambda}{2}$.

Thus, the distance of the air gap between two successive fringes is given by

$$[dD]_{\text{between two successive fringes}} = \frac{\lambda}{2} \tag{8.8}$$

Distance of air gap of interference fringe of *n*th order is

$$[dD]_{\text{fringe of } n\text{th order}} = n\frac{\lambda}{2} \tag{8.9}$$

The fringe pattern become visible on the top side of the optically flat glass. Some typical interferograms using the optical flat has been presented for different surface finishes of reflecting type of test specimens. Figure 8.9(a) shows the interferogram of circular fringes which would be observed if we place an optical flat on a convex surface. It may be noted that each adjacent fringe interval represents a change in elevation which is equal to half the wavelength of monochromatic light used. For a sodium vapour lamp source, the value of $\lambda/2$ would be [0.5 × 0.593 μm], i.e., 0.297 μm. A similar pattern is observed if the test surface is concave in nature. A simple check to test whether the test sample is convex or concave is done by gently pressing the optical flat at the centre. If the surface is convex then the fringe circles at the centre would tend to move apart. On the other hand for a concave shape the slight application of force on the test sample in middle would cause the circular fringes at the periphery to move apart. This is because, the central pressure on the optical glass would get applied on the periphery and the central point of test piece shall not be in contact with the optical glass because of its concave shape. Similarly, Fig. 8.9(b) shows the series of interference fringes patterns observed when the interferometer is used for determining accurately the height difference between a test component and standard length generated by a pile of slip gauges.

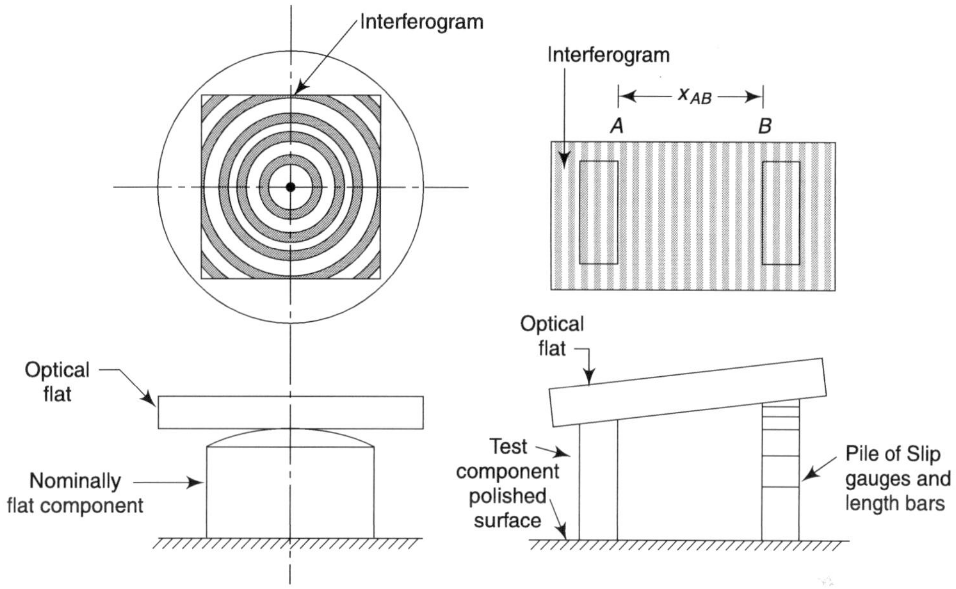

(a) (b)

Fig. 8.9 *(a) Interferogram of a nominally flat cylindrical component as seen on a circular flat. (b) Interferogram of a test set up of height difference between a component (top side polished) and slip gauge pile (polished slip gauge).*

It may be noted that for the measurements requiring precision, a laser interferometer is used. λ of the ordinary light changes somewhat due to change of air pressure, temperature, humidity, etc. These changes do not affect the λ of laser beam.

A Michelson interferometer described in chapter 7 for motion measurement is useful for calibration work in metrological applications.

Advantages

1. It is a very accurate and precise method of measuring extremely small displacements.
2. The resolution of measurement is of order of $\pm \lambda/4$ where λ is the wavelength light used. For example, a green coloured monochromatic light has $\lambda = 0.546$ μm and with this light resolution of interferometer = $\pm$ 0.14 μm.
3. It is commonly used in the calibration of slip gauges and other and other shop floor displacement gauging devices.

Disadvantages

1. Optically flat mirrored surfaces are expensive.
2. Optical alignments and adjustments are somewhat time consuming and require a skilled operation.

Problem 8.3 *A cadmium lamp employs an optical filter to generate monochromatic light of wavelength $\lambda = 0.509\ \mu m$. A set of available slip gauges wrung together have been used as a comparator of height of a test component which has been placed a distance of 10 cm from the pile of slip gauges. An optically flat glass plate was placed as shown in Fig. 8.10 and it formed a tilted surface. The number of fringes observed in the 10-cm separation distance were found to be 12.*

(a) Determine the height difference between slip gauges and test sample, and

(b) The angle of tilt of the optically flat glass plate at this test position.

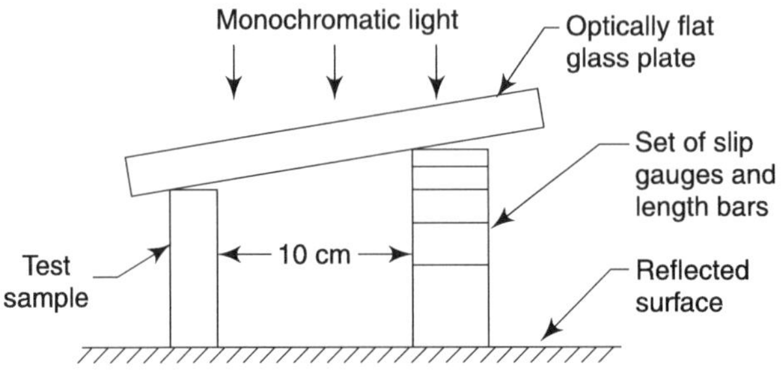

Fig. 8.10 *Test set up of a comparator for height measurements*

Solution

(a) The number of fringes n over a distance of 10 cm = 12

The height difference between slip gauges and test sample $= n\dfrac{\lambda}{2}$

$= (12)\ (0.5)\ (0.509)\ \mu m$

$= 3.054\ \mu m.$

(b) The tilt angle θ of glass plate $= \tan^{-1}\left[\dfrac{3.054 \times 10^{-6}}{10 \times 10^{-2}}\right]$

$= 3.054 \times 10^{-5}$ radians.

or $= 1.75 \times 10^{-3}$ degrees.

8.7 ■ SURFACE ROUGHNESS MEASUREMENT

The undulations or roughness heights in the form of protuberances and valleys are caused by inaccuracies of tool feed and tool chatter. In addition, another major causative factor is the vibration of a machine tool structure and moving parts at various frequencies. The combined effect of these factors result in more or less regular pattern which may repeat over a length, giving waviness as shown in a typical surface profile of component.

A very convenient device, namely, Taylor Hobson Talysurf, schematically shown in Fig. 8.11, is employed to visualize two-dimensional magnified surface profile either along a line across a surface or parallel to the axis of the cylinder. A diamond-tipped stylus having a nose radius of 2 microns is mounted at one end of the armature. The armature is hinged on the central limb of the E-shaped soft iron.

The outer limbs of this E-shaped head have two coils which form two arms of the inductance bridge in which other two arms are resistive type. A small air gap is provided between the outer limbs and the armature. The movement of the stylus follows the surface profile and it causes a rocking movement of the armature. The air variation of air gap in the outer limbs causes the variations of reluctances in the magnetic paths. This in turn varies the impedances of the two coils proportional to the movement of the stylus. These coils form adjacent arms of the inductance bridge in which an oscillator inputs the carrier frequency. The movement of a stylus controls or modulates the carrier frequency. This signal can then be amplified suitably in steps of 500 up to 30,000 times. The output signal on pen recorder of the instrument gives a true graph of the surface texture of the test component.

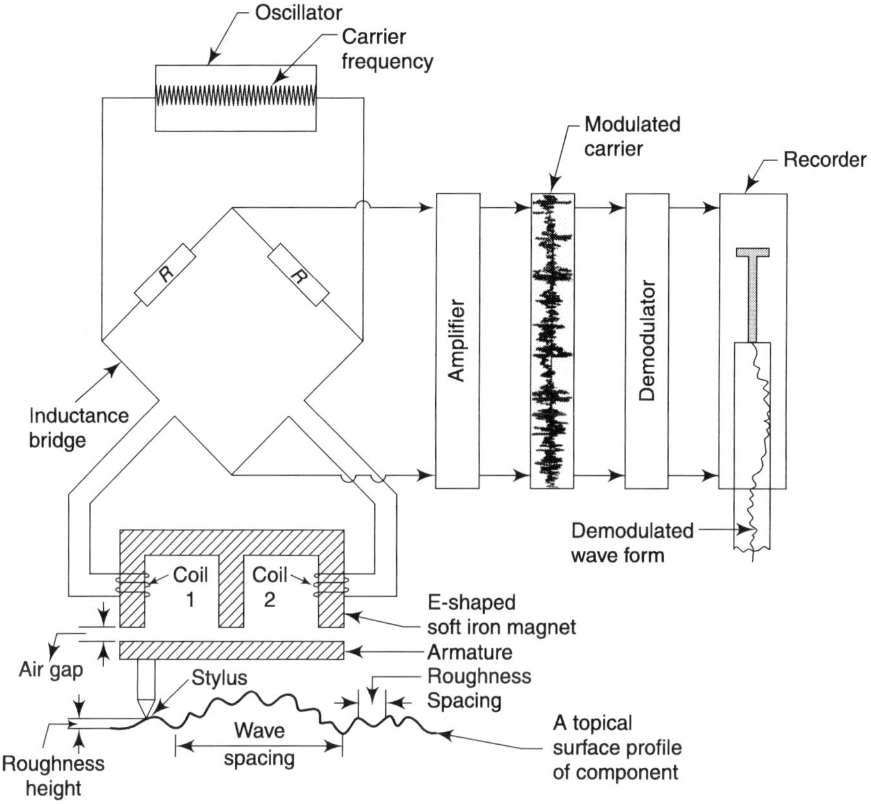

Fig. 8.11 *A schematic diagram of Taylor-Hobson Talysurf for surface roughness measurement*

Advantages

1. Quantitative estimates of maximum roughness, average roughness, waviness of surface texture can be conveniently evaluated from the talysurf trace.
2. The output of the instrument is electrical and, therefore, the output of instrument can be made micro-processor based for the analysis of surface texture parameters.

Disadvantages

1. The skidding of the stylus on the test surface would give erroneous results.
2. The technique does not give three-dimensional information of the surface profile.

8.8 ■ MEASUREMENT OF AREA USING POLAR PLANIMETER

Polar planimeter is a simple mechanical type of area integrating device. It consists of a circular wheel with attached revolution counter and two arms, namely, polar arm and tracer arm shown in Fig 8.12(a). This is conveniently used in evaluating the area of irregular planer figures. In this instrument, one end of the polar arm is fixed and is known as the fixed pivot point of this arm. The other end of the polar arm is hinged to the tracer arm on the adjustable hinge block fitted on the tracer arm. One end of the tracer arm has a tracer point which is moved along the outer periphery of measured area. The other end of the tracer arm is provided with a wheel which has a circular scale. The rotation of the wheel is provided with a suitable counter to count the number of revolutions traversed by the wheel. In addition, a vernier scale is also provided to read part of the small circular scale division of the wheel.

Let L be the length of the polar arm, R be the distance between the hinge point and tracer point in the tracer arm and the distance of the wheel from the hinge point is length a as shown in Fig. 8.12(a). When the planimeter is used for the measurement of small measured areas, then the pivot point of the polar arm O is kept outside the area again as shown in Fig. 8.12(b). On the other hand, if larger areas have to be measured then the pivot point of the polar arm is located inside the measured area.

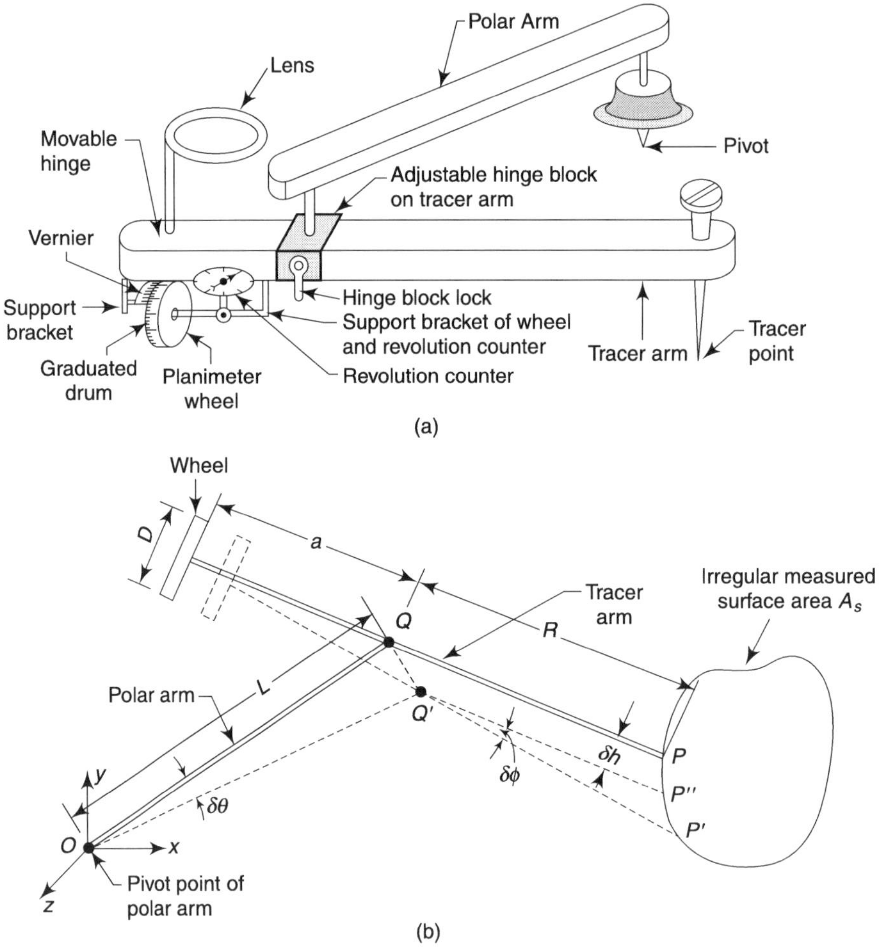

Fig. 8.12 *(a) A pictorial sketch of a polar planimeter (b) Plan view of polar planimeter with polar arm pivot point constrained outside the irregular surface area*

For the measurement of area, the starting point of tracer point is P and hinge point of tracer arm and polar arm end is Q. Further O is the pivot point of the pivot end of polar arm which is fixed to ground reference x,y,z [see Fig. 8.12(b)]. The tracer arm is moved and elemental distance δs from the initial point P to P'. This causes movement of polar arm from Q to Q' and the wheel moves to a dotted position as shown in the figure. A line $Q'P''$ is drawn parallel to QP which gives a distance δh which the wheel would move due to linear displacement of QP. The angle turned by the tracer arm and polar arm are $\delta\phi$ and $\delta\theta$ respectively as shown in the figure. The angular displacment $\delta\phi$ of tracer arm about point Q causes additional displacement in the wheel.

The elemental area traced by polar arm $OQ = \frac{1}{2} L^2 \delta\, \theta$ (8.8)

The elemental area traced by tracer arm $QP = \frac{1}{2} R^2 \delta\phi + R\, \delta\, h.$ (8.9)

Therefore the total area traversed by both arm is

$$\delta A_s = \left(\frac{1}{2} L^2 \delta\theta\right) + \left(\frac{1}{2} R^2 \delta\phi + R\delta h\right) \quad (8.10)$$

The elemental distance δW moved by the wheel is the sum of distances moved by the translation equal to δh of tracer arm OP and $(-a\delta\phi)$ due to rotation of tracer arm by the angle $\delta\phi$. In other words

$$\begin{aligned} \delta W &= \delta h - a\delta\, \phi \\ &= \pi\, D(\delta N) \end{aligned} \quad (8.11)$$

where

D is the diameter of wheel

a is the distance of wheel from hinge point Q.

δN is the elemental number of rotations of the wheel.

When the tracer point moves from P to P' then,

From Eq. 8.11 we get,

$$\delta h = \pi\, D\delta N + a\delta\phi \quad (8.12)$$

Substituting this in Eq. 8.10 we get,

$$\delta A_s = \frac{1}{2} L\delta\phi + \frac{1}{2} R^2 \delta\phi + R(\pi D\delta N + a\delta\phi) \quad (8.13)$$

Case I: When O lies outside the area When the anchor point O is constrained and kept outside the measured surface area A_s [see Fig 7.12(b)] then as the tracer point is moved along the outer periphery of the area, both the tracer arm and polar arms simply move up and down. In other words both the arms never complete a revolution, i.e.,

$$\int \delta\, \theta = \int \delta\phi = 0$$

Now integrating Eq. 813 we get,

$$\int \delta A_s = \frac{1}{2} \int L^2 \delta\theta + \frac{1}{2} \int R^2 \delta\phi + R\pi D \int \delta N + \int aR\delta\phi$$

$$\therefore \quad \begin{aligned} A_s &= \pi\, DRN \\ &= CN \end{aligned} \quad (8.14)$$

where C is calibration constant of planimeter = πDR

Case II: When O lies inside the area When, on the other hand, the anchor point O in constrained (i.e., fixed) in the measured surface area A_s (see Fig. 8.13) then as the tracer point is moved along the outer periphery of the area, both the tracer arm and polar arm complete one revolution. In other words, both arms traverse an angle of 2π radians i.e.,

$$\int \delta\theta = \int \delta\phi = 2\pi \text{ radians.}$$

Now integrating Eq. 8.13, we get,

$$\int \delta A_s = \frac{1}{2} \int_0^{2\pi} L^2 \delta\theta + \frac{1}{2} \int_0^{2\pi} R^2 \delta\phi + R\pi D \int \delta N + \int_0^{2\pi} aR\delta\phi$$

$$\therefore \quad \begin{aligned} A_s &= \pi(L^2 + R^2 + 2aR) + \pi RDN. \\ &= A_o + CN \end{aligned} \quad (8.15)$$

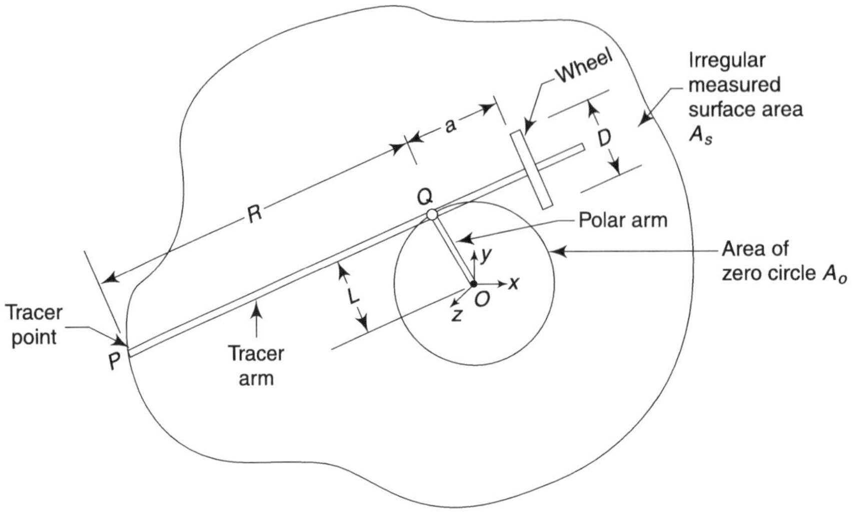

Fig. 8.13 *Plan view of polar planimeter with polar arm pivot print inside the irregular surface area*

where, the term $\pi(L^2 + R^2 + 2aR)$ is the area of zero circle, A_o and is constant for a given setting of planimeter
D is the diameter of the wheel
N is the number of revolutions of the wheel, and
C is the calibration on constant of planimeter $= \pi DR$.

Problem 8.4 *A planimeter with a given tracer arm setting was used to measure known circular areas of 5 cm and 25 cm diameters with polar arm pivot point outside and inside the areas respectively. The difference between final reading and initial reading in both cases was found to be 2.62 revolutions and 15.87 respectively.*

(a) Determine the calibrations constant C (i.e., πDR) and A_o (area of zero circle) of the given planimeter.

(b) Determine the magnitude of two unknown irregular areas when measured with the same planimeters. The measured areas gave the number of revolutions as 3.82 and 16.29 for pole/pivot point outside and inside the areas respectively.

Solution

(a) (i) When the pivot/pole point is outside the 5 cm diameter circular area then

$$A = (\pi DR)N = CN$$

$$C = \frac{\pi}{4}(5)^2/2.62$$

$$= 7.49 \text{ cm}^2\text{/revolution}$$

(ii) When the pivot/pole point is inside 25 cm diameter circular area then,

$$A = A_o + CN$$

or $$\frac{\pi}{4}(25)^2 = A_o + 7.49(15.87)$$

$\therefore$ area of zero circle $A_o = 372.0 \text{ cm}^2$

(b) (i) The unknown area say A_1 measured by the planimeter when the pivot point is outside the measured area is given by

$$A_1 = CN$$

$$= 7.49\ (3.82) = 28.61 \text{ cm}^2$$

(ii) The unknown area say A_2 measured by the planimeter when the pivot point is inside the area is given by

$$A_2 = A_o + CN$$
$$= 372.0 + 7.49(16.29) = 494.01 \text{ cm}^2$$

Review Questions

1. State whether the following statements are true or false. If false, rewrite the correct statement:
   (a) Rulers and tapes provide simple, low cost, easy and quick way of measuring wide range of lengths where high levels of accuracies is not a requirement.
   (b) The difference in length of one main scale division and one vernier scale division is the least count of the vernier caliper.
   (c) Both micrometer screw gauge and dial test indicator are quite suitable for measurement of dynamic displacements.
   (d) The slip gauges of highest quality are generally of grade I type.
   (e) The gauge blocks and length bars should preferably be used in dust free and air-conditioned environments which have been maintained at a constant temperature.
   (f) Sine bar is generally used for linear displacement gauging.
   (g) Pneumatic displacement device is capable of measuring macro displacements of several hundred metres.
   (h) In hydraulic dimensional gauging is based on the principle of incompressibility of hydraulic fluids.
   (i) Taylor–Hobson talysurf techniques gives qualitative estimate of surface roughnesses along the traversed path of stylus on a test component.
   (j) When the polar arm of the planimeter is outside the measured area then calibration constant of the instrument is obtained by the product of the length of the tracer arms (measured from hinge point) and circumference of planimeter wheel.
   (k) Two infering nanochromatic type of optical beams, which are 180° out-of-phase with each other, give rise to production of a bright fringe.
   (l) Circular fringes are observed if nanochromatic light is directed on a optical flat glass plate placed on either a convex or a concave surface.

8.2 Which of the following instruments is most accurate?
   (i) (a) vernier caliper (b) micrometer screw gauge
   (c) dial test indicator (d) slip gauges
   (ii) The least count of a metric vernier caliper having 50 divisions on a vernier scale, matching with 49 divisions of the main scale, (1 main scale division = 1 mm) is
   (a) 0.01 mm (b) 0.02 mm (c) 0.025 mm (d) 0.005 mm
   (iii) A feeler gauge is used to check the
   (a) pitch of a screw (b) surface roughness
   (c) thickness of clearnace (d) flatness of a surface
   (iv) A pneumatic gauge is used for
   (a) roundness measurements (b) surface roughness measurements
   (c) surface parallelism (d) dimensional gauging

(v) Taylor–Hobson talysurf technique is used for the measurement of
(a) clearance between surfaces (b) gear tooth thickness
(c) surface hardness (d) surface roughness

(vi) Optical flats used for dimensional gauging employs which of the following techniques?
(a) total internal reflection of light (b) interferometry
(c) Doppler shift of the frequency of light (d) photo-electric phenomenon

(vii) The commonly used dimensional gauging device LVDT works on the principle of variation of
(a) capacitance (b) resistance
(c) mutual inductance (d) self-inductance

(viii) Capacitive type of dimensional measurement gauges operate on the principle of
(a) variation of over-lapping areas of parallel plates
(b) variation of separation distance of parallel plates
(c) variation of the di-electric constant between the parallel plates
(d) all of the above

(ix) A set of high grade slip gauges and gauge blocks are kept in a factory and are not put to general use. This can be treated as
(a) international standard (b) primary standard
(c) secondary standard (d) working standard

(x) The minimum air gap distance between the polished test piece and optical flat to produce a dark fringe of first order is
(a) $\lambda/8$ (b) $\lambda/4$ (c) $\lambda/2$ (d) $3\lambda/4$
where λ is the wavelength of monochromatic light.

(xi) Which of the following is generally employed for checking the accuracy of mechanical type of shop floor dimensional gauging instruments like micrometer, vernier caliper, dial test indicator.
(a) clinometer (b) sine bar (c) auto-collimator (d) slip gauges

(xii) Which of the following components of the micrometer screw gauges is provided for maintaining uniform pressure on the test piece during dimensional measurement?
(a) ratchet (b) thimple (c) barrel (d) spindle

8.3 Fill in the blanks with appropriate word/words.

(i) One vernier scale division is the ____________ of the instrument.

(ii) Micrometers employ the principle of the movement of accurately machined screw moving in the accurately machined ____________.

(iii) Sine bars are not suitable for the measurements of angle greater than ____________ degrees.

(iv) Optical flats when used in dimensional gauging give ____________ magnifications.

(v) Pneumatic displacement gauge is linear in the range $0.9 \leq r \leq$ ____________.

(vi) Taylor–Hobson telesurf technique is used for the measurement of ____________.

(vii) The process of combining slip gauges and gauge block is known as ____________.

(viii) LVDT works on the principle of change of ____________ with change in displacement.

(ix) The constant of planimeter is obtained by the multiplication of length of tracer arm and ____________ of planimeter wheel, when the pole is outside the measured area.

(x) ____________ light is used in the interferometric dimensional gauging technique.

8.4 A 30 m steel tape is calibrated to measure true distance when temperature is 15°C. An architect used this tape in summer season at 40°C and measured distance between two points was found to be 19.837 m.

(a) Determine the true distance between the given points. Take the coefficient of linear expansion of the material of the steel tape to be 12×10^{-6} m/m/°C.

(b) What would be the observed distance between these points if ambient temperature was –5°C in winter season?

8.5 The diameter of a steel ball was measured using a vernier caliper. The main scale division was found to be 8.6 mm and the number of graduations of the vernier scale which coincided with the main scale was 17. Determine the diameter of the ball bearing if 12 mm of the main scale has 24 graduation and the corresponding vernier scale has 25 graduations.

8.6 A gauge block and length bar set has the following available sizes

Size (mm)	*Increment (mm)*	*Number of pieces*
1.0005	—	1
1.001 to 1.009	0.001	9
1.01 to 1.09	0.01	9
1.1 to 1.9	0.1	9
1 to 9	1	9
10 to 90	10	9

Choose the appropriate gauge blocks and length bars to construct a length of 35.7385.

8.7 A pneumatic displacement gauge has the measuring nozzle and control orifice diameters as 0.5 mm and 1.5 mm. Determine.

(a) the range of linear measurements

(b) the uncertainty in the measured value of displacement if the supply pressure is 200 kN/m^2 (2 bar) gauge pressure and the back pressure Bourdan pressure gauge can be read with resolution of ± 500 Pa gauge pressure

Assume the operating characteristics of the instrument with usual notation as follows:

$$r = 1.10 - 0.50\ (A_n/A_o) \text{ for } 0.4 \le r \le 0.9$$

8.8 A planimeter with a tracer arm setting of 12 cm has a wheel circumference of 5.0 cm. It was used to measure the area of an irregular plane area. When the pole was outside the area, the difference between the initial and final readings was 1.04 revolutions. Determine

(a) the magnitude of the irregular area.

(b) the area of the zero circle if the wheel records 2.52 revolutions when the pole is placed inside the measured area.

(c) the reading of the planimeter if it is used for measuring the area of the circle with 15 cm radius, when the pole is inside the measured area.

8.9 A planimeter tracer point was moved carefully over a standard square area of 5 cm × 5 cm with the pole outside the area. It gives the difference between the initial and final readings as 2.743 revolutions. It was then used in the same configuration to measure the irregular area of an indicator diagram of an IC engine with the pole again outside the measured area. If the initial and final readings of the planimeter were 1.238 and 2.096 revolutions, determine the area of the indicator diagram.

8.10 A hydraulic dimensional gauging set up was calibrated putting the pile of slip gauges of 1.1234-mm height under the plunger connected to the diaphragm of the gauge. The difference between the final and initial levels of the working fluid of the gauge was found to be 438.5 mm. Later it was used to determine the thickness of sheet metal used in the manufacture of the door frames of an automobile. If the observed reading, i.e., the difference between the initial and final readings was found to 292.5 mm, determine the thickness of steel sheet test sample.

8.11 Two gauge block sets of equal width w had a height difference of 0.005 mm. They were placed side by side. How many fringes would be observed by placing an optical flat glass plate on the gauge blocks using the helium light source with a wavelength of 0.589 μm.

8.12 Two accurately machined circular rollers A and B were tested for dimensional accuracy using the interferometric technique by viewing the optical fringes on the flat glass plate using the cadmium red light of wavelength 0.644 μm. If 12.5 fringes were observed in a distance x shown in the Fig. Prob. 8.12, determine the difference in the diameter of circular cylinders.

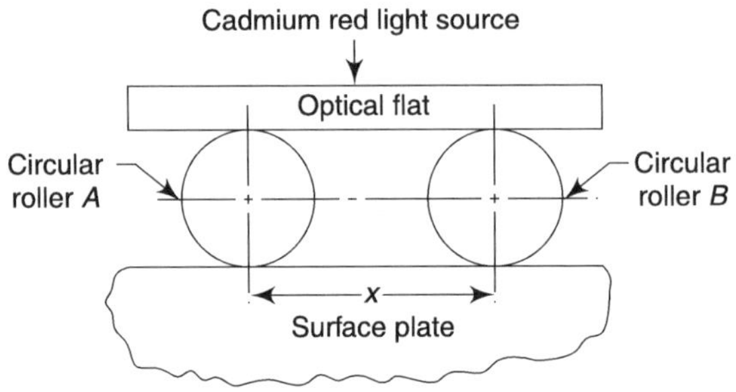

Fig. Prob. 8.12

Answers

8.1 (a) T
(b) T
(c) F, both micrometer screw gauge and dial test indicator are not suitable for the measurements of dynamic displacements
(d) F, the slip gauges of highest quality are generally of reference grade type.
(e) T
(f) F, sine bar is generally used for angular displacement gauging
(g) F, pneumatic displacement device is capable of measuring micro-displacements of the order of few millimeter.
(h) T
(i) F, Taylor–Hobson talysurf technique gives quantitative estimate of surface roughness along the traversed path of the stylus on a test component.
(j) T
(k) F, two interfering monochromatic type of optical beams, which are 180° out of phase with each other, give rise to production of a dark fringe.
(l) T

8.2 (i) (d) (ii) (b) (iii) (c) (iv) (d) (v) (d)
(vi) (b) (vii) (c) (viii) (d) (ix) (d) (x) (b)
(xi) (d) (xii) (a)

8.3 (i) least count (ii) nut (iii) 60 (iv) very large (v) 0.4
(vi) surface roughness (vii) wringing (viii) mutual inductance
(ix) circumference (x) monochromatic

8.4 (a) 19.843 m (b) 19.832 m

8.5 8.94 m

8.6 Many combinations are possible. A typical combinations from available 6 sizes is (3 × 10 + 1 + 1.7 + 1.03 + 1.008 + 1.0005) = 35.7385

8.7 (a) 0.450 mm $\leq x \leq$ 1.575 mm (b) Resolution of pneumatic gauge = ±5.6 μm
8.8 (a) 62.4 cm^2 (b) 88.8 cm^2 (c) 1.465 revolutions
8.9 7.82 cm^2
8.10 0.75 mm
8.11 8.5 fringes
8.12 8.05 μm

Chapter

9

Force Measurement

■ *INTRODUCTION* ■

In this chapter, various devices or transducers for measurement of force will be described. These devices are sometimes called *load cells*. Both static and dynamic force measurements will be considered, the latter including vibratory forces as well. The magnitude of the static or dynamic forces being considered may vary from a fraction of a Newton to several mega Newtons and thus special devices may be required, especially for extreme cases. For dynamic measurements, electromechanical transducers are often used. Calibration of such devices is an important feature which has to be carefully considered.

The following types of devices will be discussed here:

1. balance,
2. hydraulic load cells,
3. pneumatic load cells, and
4. elastic type devices.

Quite often, in applications like machine tool dynamometers, wind tunnel testing, etc. the direction of resultant force to be measured is not known. In such cases, it is required to find the magnitude of force components (this may also include torque components) along the three rectangular axes. Special techniques for such applications will also be discussed.

9.1 ■ BALANCE

A simple lever system, shown in Fig. 9.1, called a balance, has long been used as a force measuring device. To measure the unknown force F at a distance L from the pivot, a mass m at a distance l from the pivot is used. The system is in equilibrium when

$$FL = mgl \tag{9.1}$$

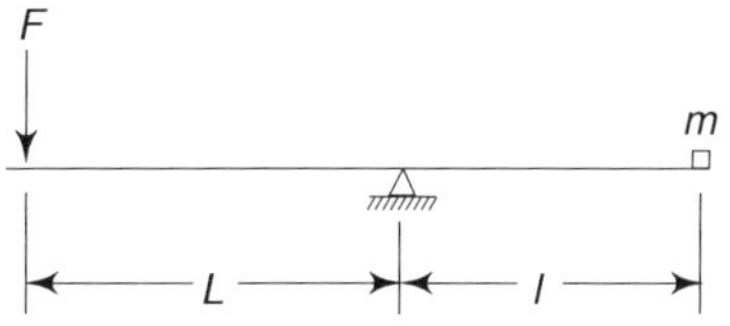

Fig. 9.1 *Balance principle*

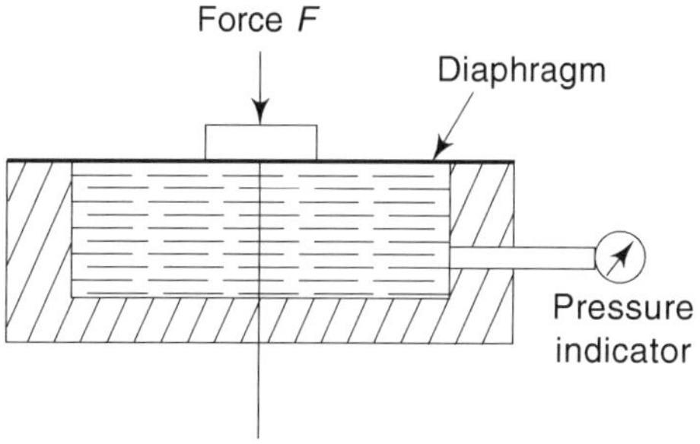

Fig. 9.2 *Hydraulic load cell*

With the knowledge of the other parameters, viz. lengths L and l, mass m and gravitational constant g, the force F can be calculated. Several versions of such balances are available for various force ranges and degrees of accuracy.

9.2 ■ HYDRAULIC LOAD CELL

In this type of device (Fig. 9.2), hydraulic pressure is used to indicate the force F, applied to a diaphragm or some other type of force transmitting element. When a force F is applied, pressure is developed in the fluid which is normally an oil. This can be measured by a pressure indicating device like a Bourdon gauge. Such a device can be used up to very large forces, of the order of millions of Newtons.

9.3 ■ PNEUMATIC LOAD CELL

In this type of load cell, shown in Fig. 9.3, air is supplied under pressure to a chamber having a diaphragm at one end and a nozzle at the other. Application of force to the diaphragm deforms it and changes the gap between the extension of the diaphragm and the nozzle, thus changing the pressure in the chamber. If the force F increases, the gap reduces and this increases pressure P_2 in the chamber. This increase in pressure produces a force tending to return the diaphragm to its original position. For any force F, the system attains equilibrium and pressure P_2 gives an indication of the force F. This type of load cell is used up to 20 kN.

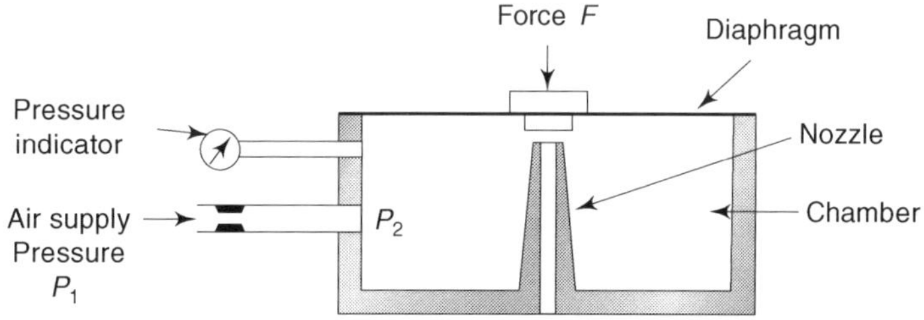

Fig. 9.3 *Pneumatic load cell*

9.4 ■ ELASTIC FORCE DEVICES

These are important devices for the measurement of both static and dynamic forces. In such devices, the force applied to the elastic member results in a displacement or strain in the elastic member, which is sensed by mechanical or electromechanical means. The elastic members may be in the form of rings, diaphragms, strips, cylinders, etc. Relations between strain and stiffness for some of the members are given here.

1. Axially Loaded Elastic Member For the member shown in Fig. 9.4, strain ε in axial direction and stiffness k in the same direction are

$$\varepsilon = \frac{P}{AE} \tag{9.2}$$

$$k = \frac{EA}{L} \tag{9.3}$$

where P is the force, E the Young's modulus, A the area of cross-section and L the length of member.

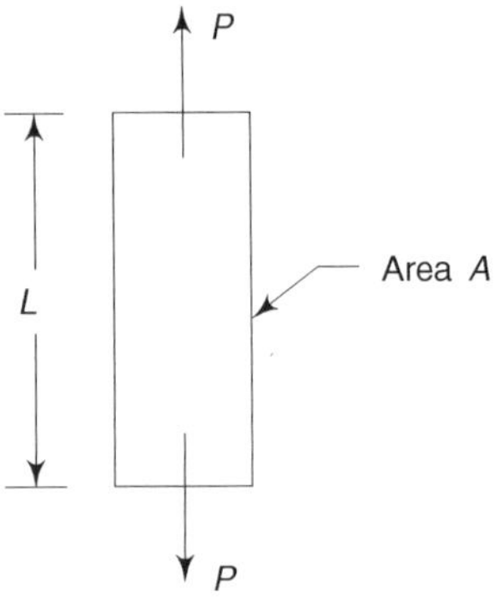

Fig. 9.4 *Axially loaded elastic member*

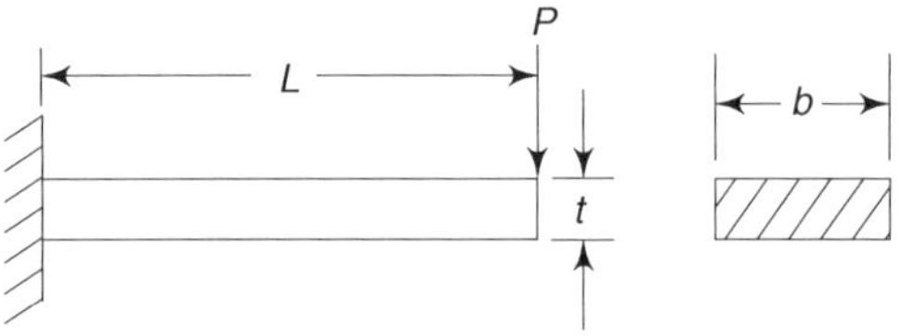

Fig. 9.5 *Cantilever type elastic member*

2. Cantilever-type Elastic Member Strain ε at the root of the cantilever is given by

$$\varepsilon = \frac{6\,PL}{Ebt^2} \tag{9.4}$$

$$\text{and stiffness } k = \frac{\text{Force } P}{\text{Deflection at free end}}$$

$$= \frac{Ebt^3}{4\,L^3} \tag{9.5}$$

where b and t are the width and thickness of the cantilever, respectively.

3. Ring-type Elastic Member Due to the forces V and H, strain ε at $\theta = 90°$ is given by

$$\varepsilon_{90°} = 1.09\,\frac{Vr}{Ebt^2} \tag{9.6}$$

Further, at $\theta = 39.6°$,

$$\varepsilon_{39.6°} = 2.31\,\frac{Hr}{Ebt^2} \tag{9.7}$$

where r = mean radius of the ring.

It is seen from Eqs. (9.6) and (9.7) that at $\theta = 90°$, the strain is due to force V only and at $\theta = 39.6°$, the strain is due to force H only. Thus, the two components can be separately measured.

Expressions for deflections in the directions of V and H are:

$$\delta_V = 9.42\,\frac{Vr^3}{Ebt^3} \tag{9.8}$$

$$\delta_H = 1.79 \frac{Hr^3}{Ebt^3} \tag{9.9}$$

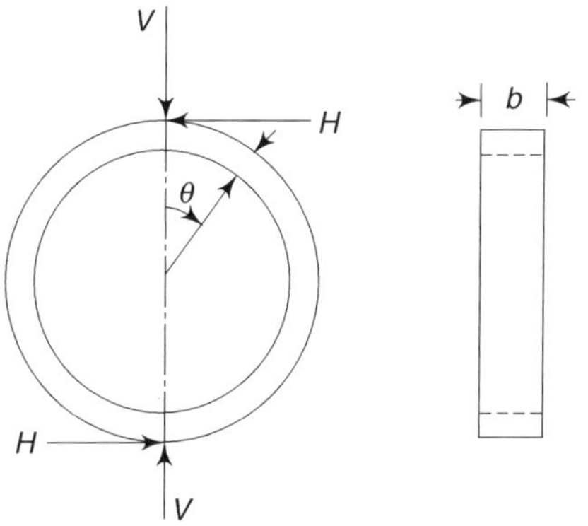

Fig. 9.6 *Ring-type elastic member*

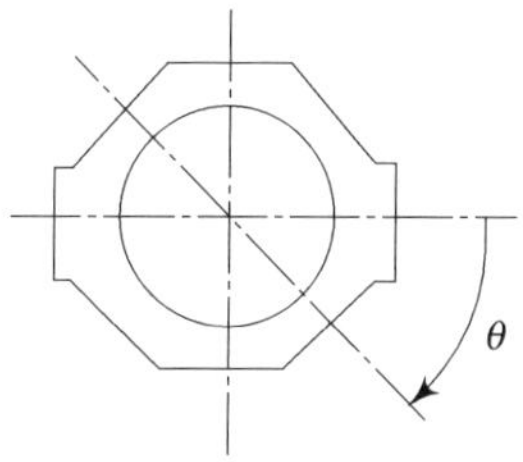

Fig. 9.7 *Octagonal type ring*

In order to prevent rolling of the ring due to force H, an octagonal type ring as shown in Fig. 9.7 is used. In such a case, the strain at $\theta = 90°$ is only due to force V while that at $\theta \approx 50°$ is only due to force H.

9.4.1 Mechanical Method

In order to measure the deflection of the elastic element due to the applied force, a mechanical method involving the use of a dial gauge may be used as shown in Fig. 9.8. The elastic ring, called the *proving* ring, has on its inside an accurate dial gauge that can be calibrated in terms of the force. Different proving rings are available for various force ranges.

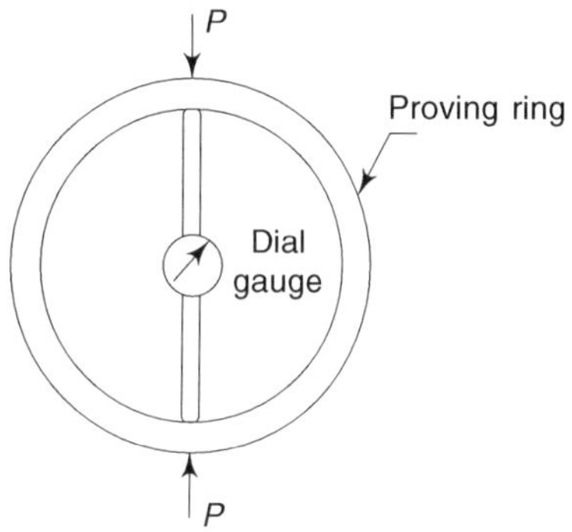

Fig. 9.8 *Proving ring*

9.4.2 Electromechanical Methods

For measuring displacement or strain, these methods use electromechanical transducers discussed in Chapter 4.

A piezo-electric transducer employing a piezo-electric crystal is a useful device for dynamic force measurement. The details of the same have been discussed in Chapter 4. Such transducers are very sensitive and can be used for measurement of forces over a wide range of values.

Figure 9.9 shows an LVDT type of force transducer for force measurement. The deflection of the elastic diaphragm due to force P is measured by a linear variable differential transducer, the principle of which has already been discussed in Chapter 4. This type of device can be used for static as well as dynamic force measurements.

Figures 9.10–9.12 illustrate the use of resistance strain gauges for force measurement. Figure 9.10 shows a cantilever type load cell, with four strain gauges bonded at the root such that strains in R_1 and R_3 are opposite in nature to those in R_2 and R_4. In Fig. 9.11, the resistance gauges are bonded on the inside and outside of an elastic ring which is a ring type load cell. In Fig. 9.12, a hollow cylinder is loaded in axial direction, inducing both longitudinal and circumferential strains. R_2 and R_4 measure longitudinal strains while R_1 and R_3 measure circumferential strains. The gauge arrangement has been shown with adjacent arms in the Wheatstone bridge having strains of opposite nature.

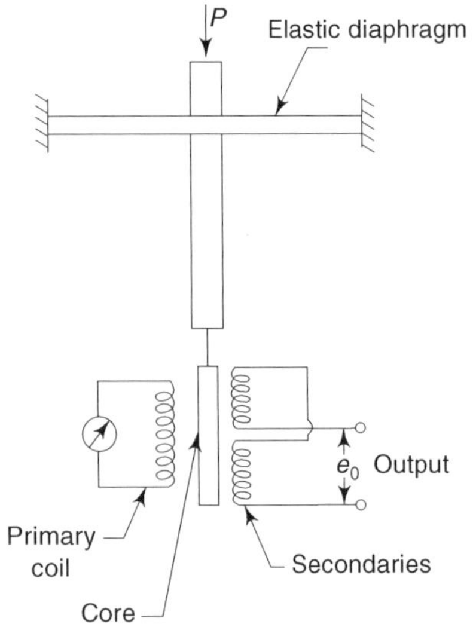

Fig. 9.9 *LVDT type force transducer*

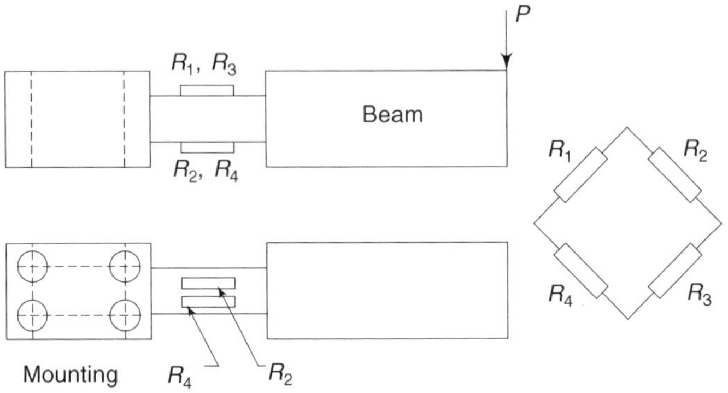

Fig. 9.10 *Cantilever type load cell*

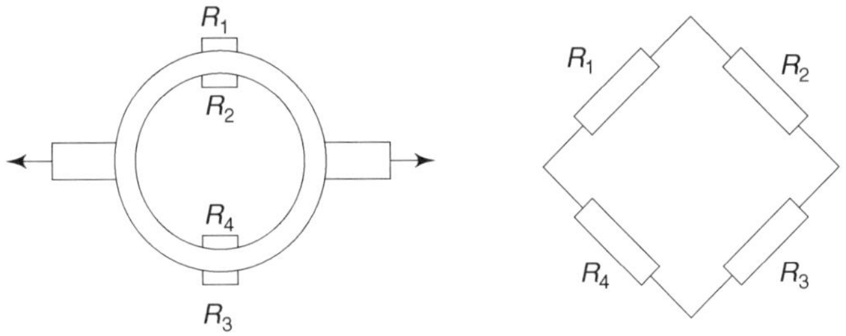

Fig. 9.11 *Force measurement-ring type cell*

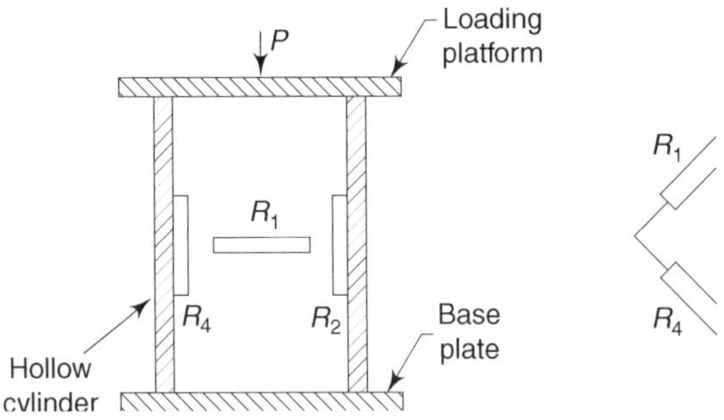

Fig. 9.12 *Force measurement-cylindrical type cell*

Problem 9.1 *A piezo-electric crystal transducer is used for measuring the force transmitted from a structure to its support as shown in Fig. 9.13. The crystal is in the form of a disc of diameter 2 cm, thickness 1 mm, with a charge constant of 10^{-5} C/cm. The dielectric constant of crystal is 5 and its Young's modulus 8×10^{10} N/m^2. Capacitance of the cable is 20 pF. The amplifier has at its input a resistance of 20 MΩ in parallel with a capacitance of 50 pF, and has a gain of 50. Find the amplitude of force transmitted if the output signal e_0 has an amplitude of 0.5 V at a frequency of 100 Hz.*

Solution Capacitance of crystal $C\,(\text{pF}) = \dfrac{A}{3.6\pi t}$

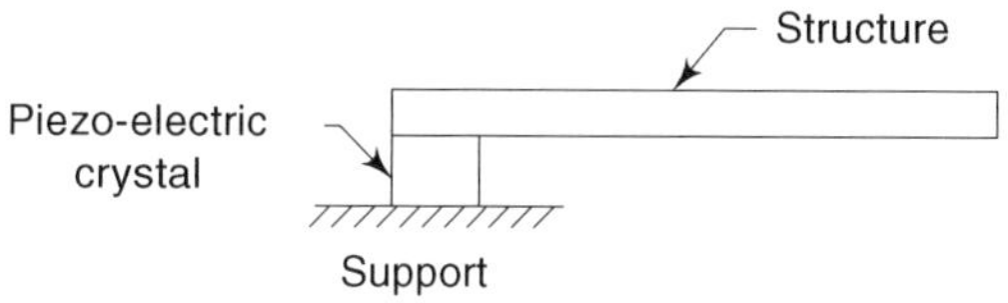

Fig. 9.13 *Figure for Problem 9.1*

where area A and thickness t are in cm^2 and cm, respectively.

$$C_{\text{crystal}} = \frac{5 \times \pi (2)^2}{4 \times 3.6\pi \times 0.1} = 13.9 \text{ pF}$$

$$\text{Total } C = 13.9 + 20.0 + 50.0 = 83.9 \text{ pF}$$

$$\tau = RC = \frac{20 \times 10^6 \times 83.9}{10^{12}} = 1.678 \times 10^{-3} \text{ s}$$

$$\text{Output voltage amplitude } e_0 = \frac{GK\tau\omega x_0}{\sqrt{1 + \tau^2 \omega^2}}$$

$$G = \text{gain of amplifier} = 50$$

$$\text{Voltage sensitivity } K = \frac{10^{-5}}{83.9 \times 10^{-12}} \text{ V/cm} = 1.192 \times 10^5 \text{ V/cm}$$

$$\omega = 2\pi \times 100 = 628.3 \text{ rad/s}$$

$$\tau\omega = 1.0538$$

Using the above equation, we get $x_0 = 1.156 \times 10^{-7}$ cm

$$\text{Stiffness of crystal } k = \frac{EA}{L}$$

$$= \frac{8 \times 10^6 \times \pi (2)^2}{4 \times 0.1} = 2.51 \times 10^8 \text{ N/cm}$$

Thus, transmitted force amplitude = kx_0 = 29.05 N

Problem 9.2 *Figure 9.14 shows a force measuring device using two thin steel rings, on which eight resistance strain gauges are bonded, so as to give temperature compensation and high sensitivity. Resistance of each strain gauge is 120 Ω, gauge factor 2.0 and battery voltage for the bridge 9 V. The resistance of the output device is 1000 Ω. For each ring, radius is 30 mm, width 15 mm and thickness 2 mm. Young's modulus E = 2.1 × 10⁵ N/mm².*

Sketch the bridge arrangement and find the value of the force applied corresponding to a bridge output of 1.6 mV.

Solution The equation to be used is

$$\text{Output voltage} = \frac{EF\varepsilon N}{4\left(1 + \dfrac{R_1}{R_G}\right)}$$

with $E = 9$ V

$F = 2.0$

N = single enhancement factor = 4

$R_1 = 120\Omega$, $R_G = 1000\Omega$

ε = strain in each gauge

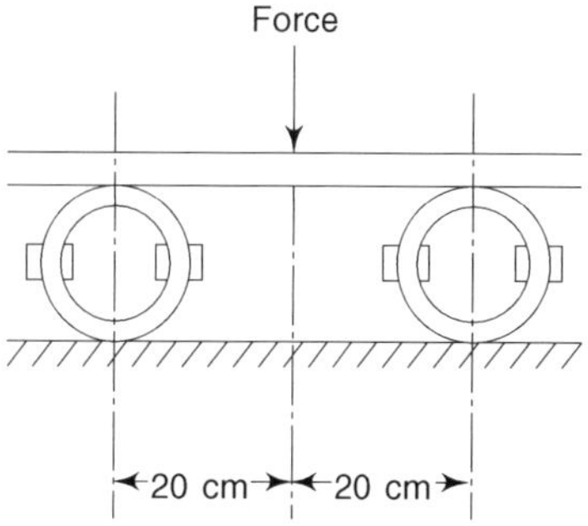

Fig. 9.14 *Figure for Problem 9.2*

We thus get for an output voltage equal to 1.6 mV,

$$\varepsilon = 9.96 \times 10^{-5}$$

Using Eq. (9.6), viz.

$$\varepsilon = \frac{1.09\,Pr}{Ebt^2}, \text{ we get}$$

For each ring force $P = 38.0$ N

or Total force = 76.0 N

9.5 ■ SEPARATION OF FORCE COMPONENTS

As mentioned in Sec. 9.1, in several applications it is required to find the components of the force along specified axes. In such cases elastic force transducers, especially strain gauge types, are widely used. An example is given in Fig. 9.15, in which it is desired to measure forces H and V in the horizontal and vertical directions, respectively. These forces are made to act at the end of a cantilever and strain gauges are bonded as shown. Gauges R_{1H}, R_{2H}, R_{3H} and R_{4H} are meant to measure force H, with a signal enhancement force of 2 (1 + ν), ν being Poisson's ratio. The bridge arrangement is such that no output results from this bridge due to application of V. Due to application of H, as in Fig. 9.15, R_{2H} and R_{4H} have compressive strains while R_{1H} and R_{3H} have tensile strains. Application of force V results in tensile strain in R_{2H} and a compressive strain of equal amount of R_{4H}, compressive strain in R_{1H} and a tensile one in R_{3H}, thus giving no output from the bridge. Similarly, it can be seen that the second bridge with gauges R_{1V}, R_{2V}, R_{3V} and R_{4V} gives an output with signal enhancement factor of 4 due to application of force V but gives no output due to force H (all gauges get compressed by the same amount due to H). Thus, the two bridges shown respond to forces H and V separately and it is thus possible to measure both the force components independently.

Problem 9.3 *Figure 9.16 shows a machine tool dynamometer, for measuring horizontal and vertical cutting forces.*

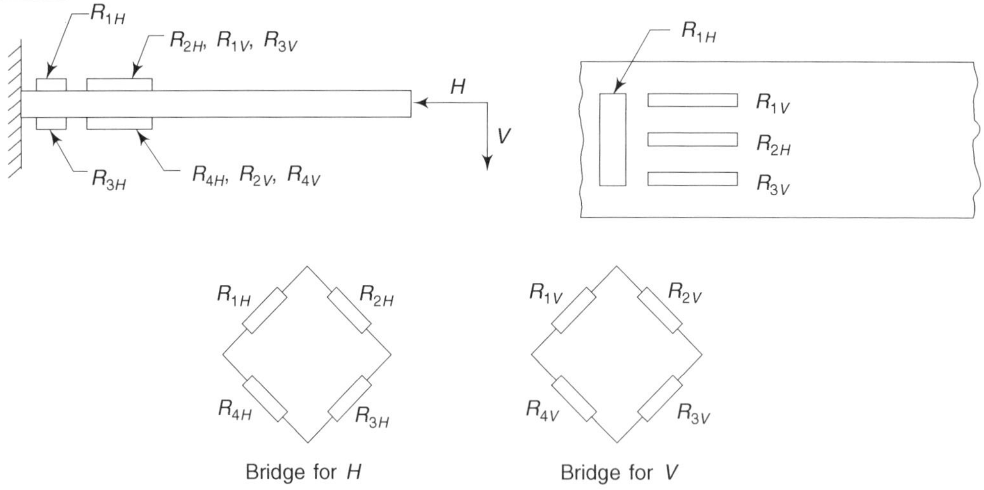

Fig. 9.15 *Separation of forces*

Strain gauges are fixed on the rings, both inside and outside to measure the circumferential strains. Four rings A, B, C and D are used for the purpose. Gauges 1 – 8 are to be used for measuring horizontal forces and gauges 9 – 16, for vertical forces.

Show how the strain gauges may be used in two different bridges for measuring the horizontal and vertical forces independently. This should ensure temperature compensation and bridge sensitivity.

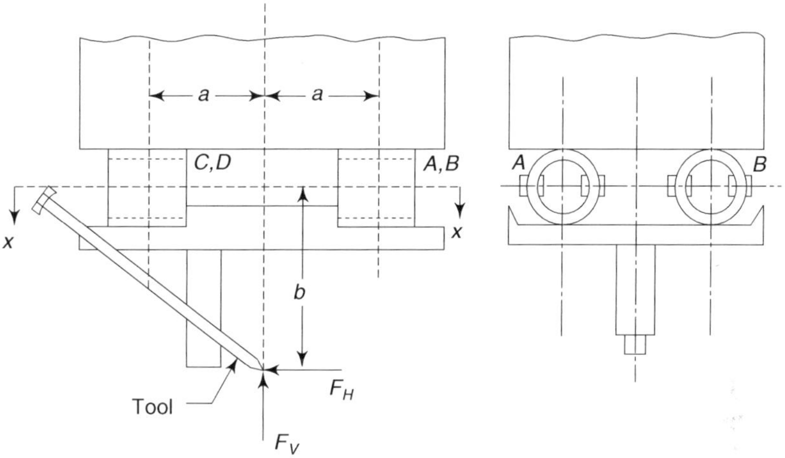

Fig. 9.16 *Figure for Problem 9.3*

Solution Due to force F_V, all the four rings, *A, B, C* and *D* are in compression, as a result of which the strain gauges on the outside of the rings will have tensile strain (+) while those on the inside of rings will have compressive strains (–). Due to force F_H, a clockwise moment is applied about the axes of the rings, as a result of which rings *C* and *D* are in compression while *A* and *B* are in tension. Thus, the strains in various gauges corresponding to the arrangement used in Fig. 9.17 are as given in Table 9.1.

With the gauges forming the two bridges as shown in Fig. 9.17, it is seen that the bridge with gauges 9–12 gives output due to F_V only and no output due to F_H. Due to F_V, gauges in all the adjacent arms have strains of opposite nature and thus the output adds up while due to F_H, opposite arms have strains of opposite nature and thus no output is obtained from the bridge.

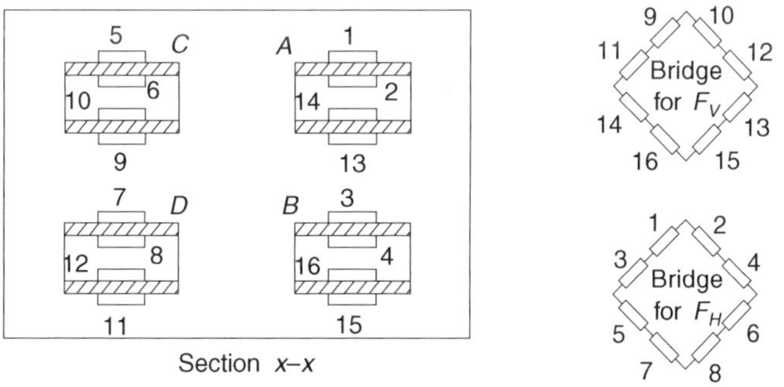

Fig. 9.17 *Solution for Problem 9.3*

Table 9.1

Strain Gauge No. and Ring	*Type of strain due to F_V*	*Type of strain due to F_H*
1 (A)	+	–
2 (A)	–	+
3 (B)	+	–
4 (B)	–	+
5 (C)	+	+
6 (C)	–	–
7 (D)	+	+
8 (D)	–	–
9 (C)	+	+
10 (C)	–	–
11 (D)	+	+
12 (D)	–	–
13 (A)	+	–
14 (A)	–	+
15 (B)	+	–
16 (B)	–	+

Similarly, the bridge with gauges 1–8 gives output due to F_H only and no output due to F_V for identical reasons.

9.6 CALIBRATION

Static calibration of force transducers is often carried out using dead weights. The procedure is simple and is applicable over a wide range of force values. For very large forces, a known force may be applied by hydraulic pressure. The pressure reading multiplied by the area on which the pressure acts gives the magnitude of the force applied.

A possible system for dynamic calibration is shown in Fig. 9.18. An electrodynamic shaker provides the dynamic force of harmonic type, whose frequency can be varied by the oscillator and force can be

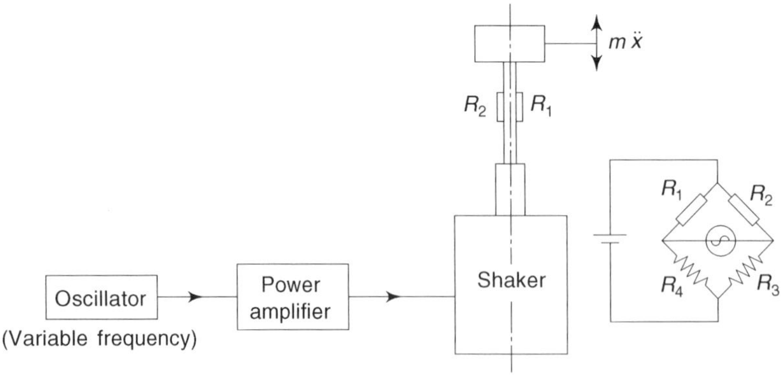

Fig. 9.18 *Dynamic force calibration*

varied by the power amplifier. If the device to be calibrated is a strip with resistance gauges as shown, a mass m can be attached on top of the strip.

$$\text{The dynamic force applied} = m\ddot{x}$$
$$= -m\omega^2 x_0 \sin \omega t \qquad \text{for } x = x_0 \sin \omega t$$

$\ddot{x}$ being the acceleration of the mass, ω the circular frequency and x_0 the amplitude of mass displacement. Thus, the amplitude of harmonic force applied is $m\omega^2 x_0$.

x_0 can be measured by other means, like a reading microscope or another calibrated displacement transducer.

Review Questions

9.1 Indicate true or false against the following:
   (a) Very small forces of the order of 1 N can be easily measured with hydraulic load cells.
   (b) Dynamic response of a resistance strain gauge type load cell is very good over a wide range of frequencies.
   (c) A piezo-electric type load cell is preferred over other devices for measuring static forces.
   (d) A pneumatic type load cell would be non-linear over a wide range of input force.
   (e) Static force calibration can be carried out with dead weights.
   (f) A force measuring system using an elastic element can be treated as a second order system.

9.2 Using a cylindrical cantilever beam and three different bridges, give an arrangement for independent measurement of forces along three perpendicular axes.

9.3 Design a strain gauge load cell of cantilever type capable of measuring a force of 5 N to an accuracy of 5%,
Gauge factor = 2, battery voltage = 6 V,
Material of cantilever is mild steel, and
Read-out resolution = 10^{-5} V.

9.4 Using a thin circular ring, draw a sketch showing the arrangement of strain gauges so that vertical and horizontal forces can be measured independent of each other.

9.5 A dynamic force measuring device can be represented by a second order system, with an effective mass of 200 gm, damping ratio of 0.7 and an undamped natural frequency of 10 Hz.
   (a) For a unit step force input, find the peak overshoot and the time taken for the output to be within 5% of its steady state value.
   (b) If the force to be measured were to increase linearly with time, find the percentage error after 5 s.

9.6 A load cell deflects 0.05 mm due to a machine tool slide whose mass is 200 kg. Find the highest frequency of force that may be measured with a 5% accuracy.

9.7 It is required to measure the dynamic and static vertical forces transmitted to the ground, due to a running centrifugal pump. The pump is mounted at all the four mounting points connecting the pump with the ground. Choose suitable types of transducer and instruments to be used and sketch the arrangement. Also, suggest a suitable calibration method for the transducers.

9.8 It is required to accurately measure a static force upto 10,000 N applied by a hydraulic jack in vertical direction to a roller bearing under test. Choose a suitable measuring system and suggest a calibration arrangement also.

Answers

9.1 (a) F (b) T (c) F (d) T (e) T (f) T

9.5 (a) 4.6%, 0.076 s (b) 0.446%

9.6 41.66 Hz

Chapter

10

Torque and Power Measurements

INTRODUCTION

Instruments for measuring torque and power in rotating systems have been covered in this chapter. The power transmitted can be calculated from the torque, using the equation $P = \omega T$, where P is the power (W), T the torque (N-m) and ω the angular speed (rad/s). Devices used for power measurement are also known as dynamometers and may be classified into three types, depending on the nature of machine arrangement, for which torque or power is to be measured. The three types are the following:

1. *Transmission types dynamometer*, in which the power being transmitted through the device is measured. The device is neither a power generator nor a power absorber and is used on the shaft transmitting power, between the prime mover and the load (Fig. 10.1).
2. *Driving-type dynamometer*, in which drive is obtained from the dynamometer itself or the dynamometer is the power generator (Fig. 10.1) like an electric motor.
3. *Absorption type dynamometer*, in which the mechanical energy is absorbed after it is measured (Fig. 10.1). The power generator may be an engine or a motor.

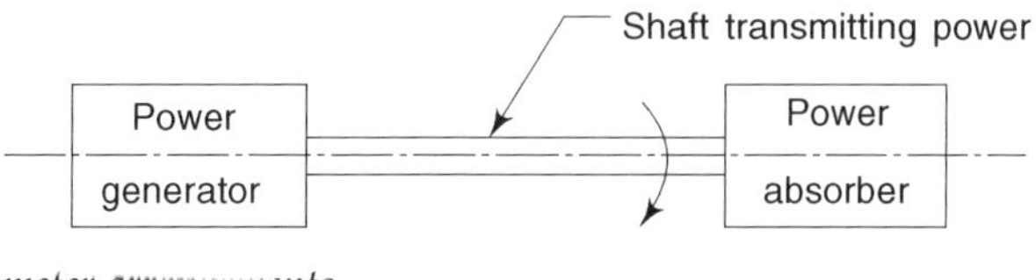

Fig. 10.1 *Various dynamometer arrangements*

10.1 ■ TRANSMISSION DYNAMOMETERS

These are of the following types:

1. Torsion dynamometers, in which either the torque is directly measured using resistance strain gauges or the angle of twist in the shaft due to the torque is measured, the former being called a torque meter and the latter a torsion meter.
2. Belt dynamometer, and
3. Gear dynamometer.

10.1.1 Torsion Dynamometers

In a solid shaft of diameter d, subjected to torque T, the following relations exist:

$$T = f_s \left(\frac{\pi}{16} d^3 \right) \tag{10.1}$$

$$\gamma = \frac{f_s}{G} \tag{10.2}$$

$$\varepsilon_{45^\circ} = \frac{\gamma}{2} \tag{10.3}$$

$$T = \frac{G\theta\pi d^4}{32L} \tag{10.4}$$

where f_s is the shear stress induced in the shaft, G the shear modulus, γ the shear strain, ε_{45° the longitudinal strain in the shaft at 45° to the axis and θ the twist in the shaft (rad) over a shaft length L. If a hollow shaft of outer radius r_o and inner arc r_i is used, the relations for ε_{45° and θ are:

$$\varepsilon_{45^\circ} = \frac{Tr_o}{\pi G(r_o^4 - r_i^4)} \tag{10.5}$$

$$\theta = \frac{2TL}{\pi G(r_o^4 - r_i^4)} \tag{10.6}$$

Thus, either ε_{45° may the measured by resistance strain gauges or θ may be measured by optical or electromechanical means, in order to measure T. In order to increase the sensitivity, the shaft section has to be reduced at the point of measurement of strain in the shaft.

Torque Meter During torsion of cylinder, the principal strains (tensile or compressive) exist at 45° to the axis. These can be measured by bonded resistance gauges, as shown in Fig. 10.2. The output is increased by using four gauges so that the adjacent arms have strains of opposite nature. For taking signals in and out of the rotating shaft, slip rings and brushes are used, as shown in Fig. 10.3. The device is useful for both static and dynamic torque measurement. Torques over a range of 100 to 100 000 N-cm can be measured with this device.

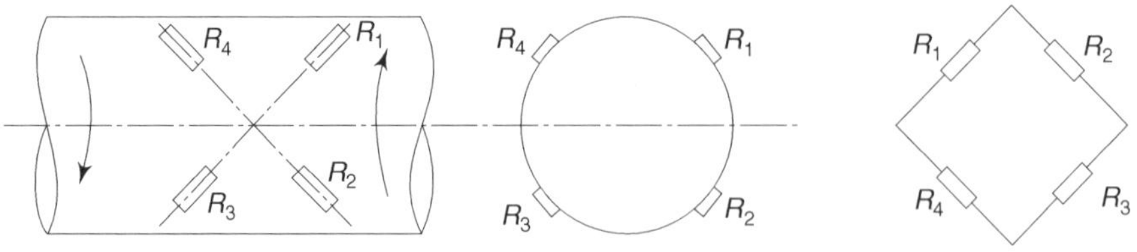

Fig. 10.2 *Strain-gauge torque transducer*

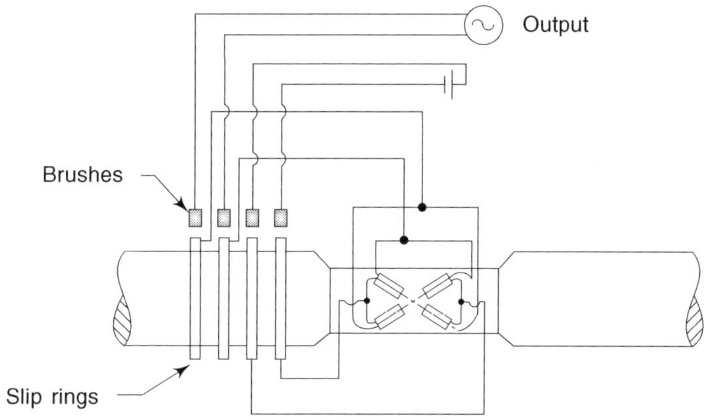

Fig. 10.3 *Slip ring arrangement in torque transducer*

It is easier to measure bending strains rather than strains due to torque at 45° and so an arrangement using beams (Fig. 10.4) may be employed, in which the transmitted torque results in bending the beams.

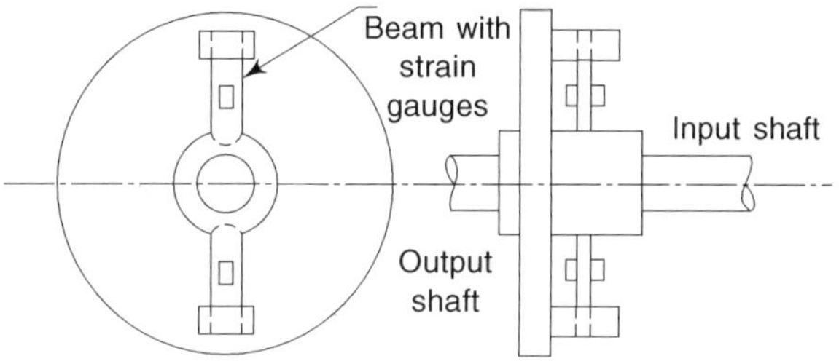

Fig. 10.4 *Strain gauge transducer using beams in bending*

A slip-ring arrangement, results in noise due to changes in contact resistance. A non-contacting type of arrangement, such as the one in Fig. 10.5, is thus preferred. The bridge supply and output signals are transmitted between the rotating and stationary members through transformers. An ac bridge arrangement is employed, which gets modulated due to torque. A demodulator is employed to eliminate the carrier frequency of the oscillator.

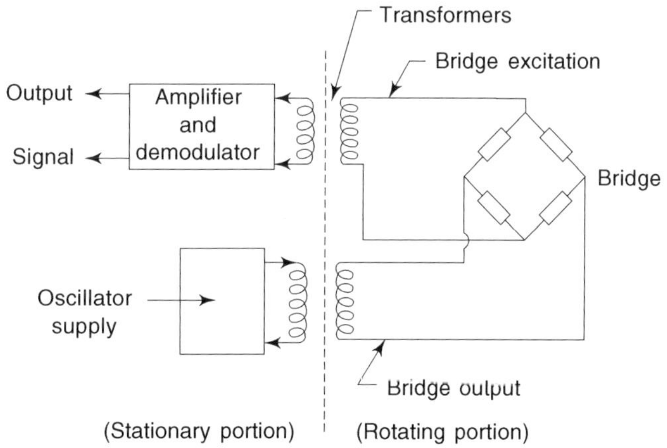

Fig. 10.5 *Torque transducer-non-contacting type*

Problem 10.1 *For measuring the torque transmitted by a shaft, four resistance strain gauges of 120 Ω each are used at 45° to the shaft axis. These form the four arms of a dc Wheatstone bridge, with output resistance of 1000 Ω. Find the output voltage of the bridge if the bridge supply voltage is 9 V. Power transmitted is 5 hp, speed of shaft 900 rpm, diameter 2 cm, modulus of rigidity of shaft material 8×10^{10} N/m² and gauge factor 2.5.*

Solution

$$\text{Power} = 5 \times 746 \text{ W} = \frac{2\pi \times 900}{60} T$$

$$\text{Torque } T = 39.6 \text{ N–m}$$

Using Eq. (10.1), viz.

$$T = f_s\left(\frac{\pi}{16} d^3\right), \text{ with } d = 2 \text{ cm},$$

we get

$$f_s = 2521 \text{ N/cm}^2.$$

Using Eqs. (10.2) and (10.3)

$$\varepsilon_{45°} = 1.575 \times 10^{-4}$$

Output voltage

$$E_o = \frac{EF\varepsilon_{45°} \times N}{4\left(1 + \frac{R_1}{R_G}\right)}$$

where N number of active gauges = 4, gauge factor $F = 2.5$, E = 9 V, $R_1 = 120\ \Omega$ and $R_G = 1000\ \Omega$. This gives $E_o = 3.16$ mV.

Torsion Meter Angular twist θ in a shaft due to torque is given by Eq. (10.4). After measurement of θ, the torque T can be easily calculated from the previously given relations. Either an optical or an electrical arrangement can be used to measure θ.

An optical arrangement is shown in Fig. 10.6. Due to the transmission of torque, the two discs A and B mounted at distance L on the shaft move relative to each other through an angle θ. This is recorded by the observer. Disc C is only to help in viewing the relative angular displacement of A and B.

Figure 10.7 shows an arrangement using toothed wheels and proximity sensors of electro-mechanical

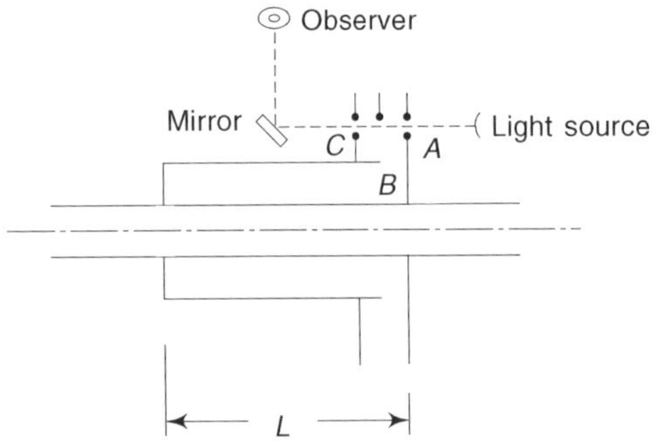

Fig. 10.6 *Optical torsion measuring device*

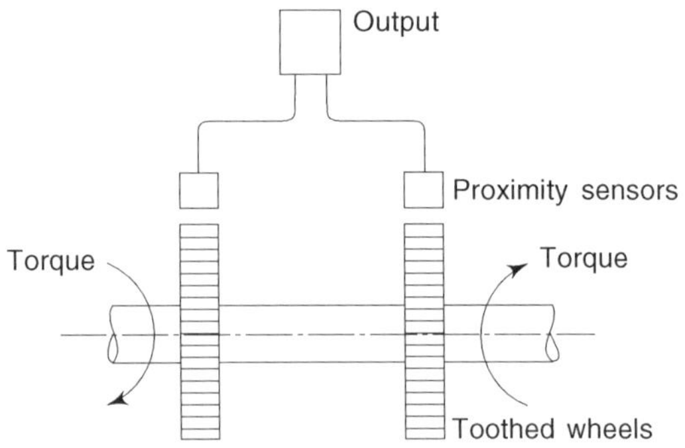

Fig. 10.7 *Use of proximity sensors for torque measurement*

types. Two identical toothed wheels are fixed on the shaft at a certain distance. The two proximity sensors produce output voltages with phase difference proportional to torque. Alternatively, an arrangement using photo-cells and a light source may be employed.

10.1.2 Belt Dynamometer

A transmission type belt dynamometer is shown in Fig. 10.8. 1 is the driving pulley, 2 and 3 are guide pulleys mounted on a lever pivoted at P and 4 is the driven pulley. If the effective belt tensions during running conditions are T_1 and T_2, as shown in the figure, and m the mass at the lever required for balance,

$$mg\,z = (T_1 - T_2)\,2x \tag{10.7}$$

$$\text{Output torque } T = (T_1 - T_2)\,R_4 \tag{10.8}$$

In this type of dynamometer, there is some loss of power due to bending and slipping of the belt.

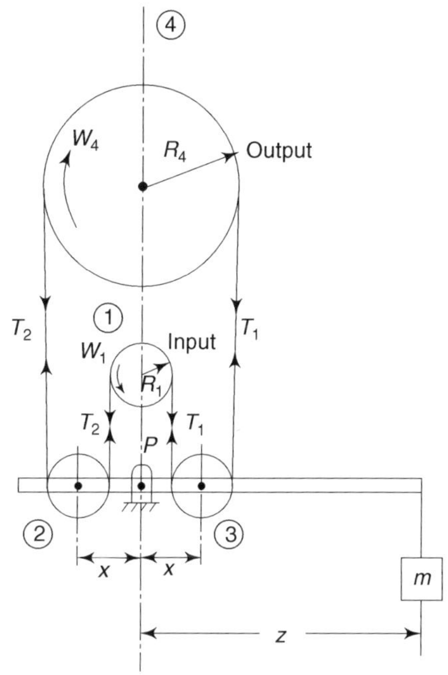

Fig. 10.8 *Belt dynamometer*

10.1.3 Gear Dynamometer

This type of arrangements shown in Fig. 10.9, is used for small power measurement. The input and output shafts are coaxial and rotate in the same direction. The gear box casing is mounted on trunnions on the shafts. In order to prevent the casing from rotating, an external torque is applied by a mass m at distance R.

If T_i and T_o are torques on the input and output shafts, respectively, and the corresponding speeds are N_i and N_o, with η, as the overall mechanical efficiency, then

$$T_o N_o = \eta T_i N_i \tag{10.9}$$

$$T_o - T_i = mg\,R \tag{10.10}$$

From Eqs. (10.9) and (10.10), we get

$$T_i = \frac{mgR}{\eta \dfrac{N_i}{N_o} - 1} \tag{10.11}$$

Thus, the input torque T_i can be obtained using Eq. (10.11).

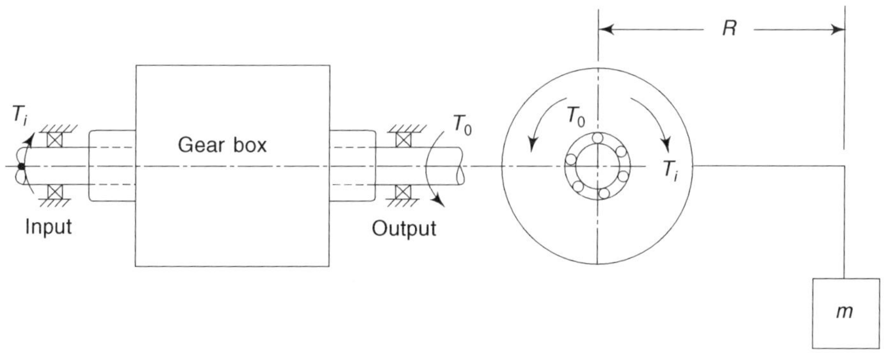

Fig. 10.9 *Gear dynamometer*

Problem 10.2 *In a gear box type transmission dynamometer, the input and output shafts are co-axial and rotate in the same direction (Fig. 10.9) at speeds of 1600 and 400 rpm, respectively. An external torque is applied to the casing to prevent it from rotating using a mass of 120 kg at a distance of 30 cm from the axis. The overall mechanical efficiency is 90%. Find the power at the input shaft.*

Solution Using Eq. (10.11),

$$T_i = \frac{mgR}{\eta \dfrac{N_i}{N_o} - 1}$$

$$= \frac{120 \times 9.81 \times 0.3}{0.9 \times 4 - 1} = 135.83 \text{ N-m}$$

$$\text{Input power} = \frac{2\pi \times 1600}{60} T_i \text{ Watt} = 22.76 \text{ kW}$$

10.2 ■ DRIVING TYPE DYNAMOMETERS

When a motor drives a driven member like a pump or a compressor, the motor can be used as both a driver and a dynamometer, as in Fig. 10.10.

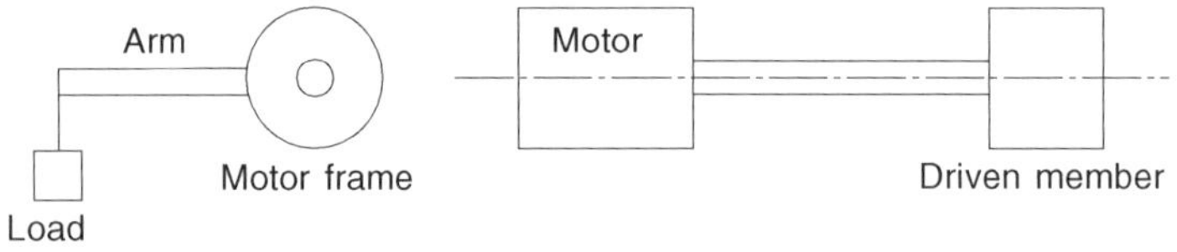

Fig. 10.10 *Driving dynamometer*

Often, dc motors are used for the purpose as speed regulation is easier. The frame of the motor is mounted freely on trunnion bearings. The torque reaction of the arm is balanced by external load. The load times the lever arm, gives the magnitude of the driving torque supplied by the motor for driving the driven member. If e and i are the voltage and current inputs to the dc motor, power supplied to the driven member is $ei\eta$, η being the efficiency of the drive.

10.3 ■ ABSORPTION DYNAMOMETERS

These types of dynamometers absorb the energy after measuring it. These are of electrical generator type, eddy current type, mechanical and hydraulic types.

10.3.1 Electrical Generator

In this type of absorption dynamometer, shown in Fig. 10.11, a driver member like an engine or a motor under test, drives an electric generator, the stator of which is freely mounted on trunnions. An arm attached to the stator frame balances the same. The armature supplies to variable resistances that can be varied to change the load on the driver. These may be used for high powers, up to 5000 hp or even higher.

10.3.2 Eddy-Current-Type Dynamometer

This consists of a toothed rotor attached to the shaft, running inside the bore of a stator, the latter carrying field windings. The windings, when excited by dc current, set up a magnetic field that is concentrated

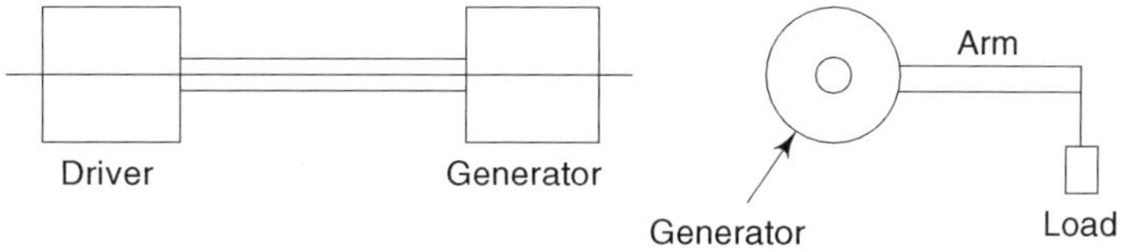

Fig. 10.11 *Electrical generator as dynamometer*

near the rotor teeth. When the rotor rotates, eddy currents are induced in the stator. The eddy current field reacts with the main field, setting up a force on the rotor which tends to resist the driving torque. The stator is mounted on trunnions and the stator has a torque measuring arm. The power produced by the driver is dissipated as heat, which is removed from the dynamometer by a flow of cooling water.

10.3.3 Mechanical Absorption Dynamometer

In such dynamometers, the friction of rope, band or block brake absorbs the energy. Heat is dissipated by cooling the brake with water.

Figure 10.12 shows a rope brake, one end of which is connected to a mass while the other end is connected to a spring balance. Due to rotation, frictional force is generated, inducing tensions at the two ends of the rope. If s is the force in the spring balance,

$$\text{torque } T = (mg - s)r$$

r being the pulley radius.

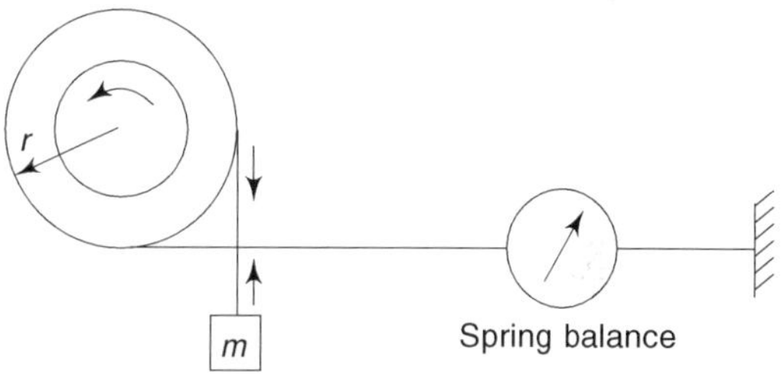

Fig. 10.12 *Mechanical absorption dynamometer*

Band or block brakes also operate on the above principle. In the former, a band brake with friction lining material, passes over a pulley while in the Prony type of block brake (Fig. 10.13), the curved blocks are clamped on the pulley rim. Load is adjusted by varying tensions in the adjoining bolts. The lever floats between the stops, as shown. If spring balance forces are S_1 and S_2, torque

$$T = S_1x_1 + S_2x_2$$

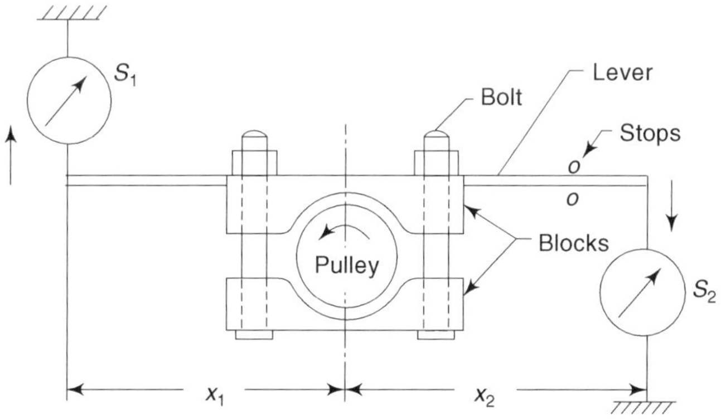

Fig. 10.13 *Prony brake*

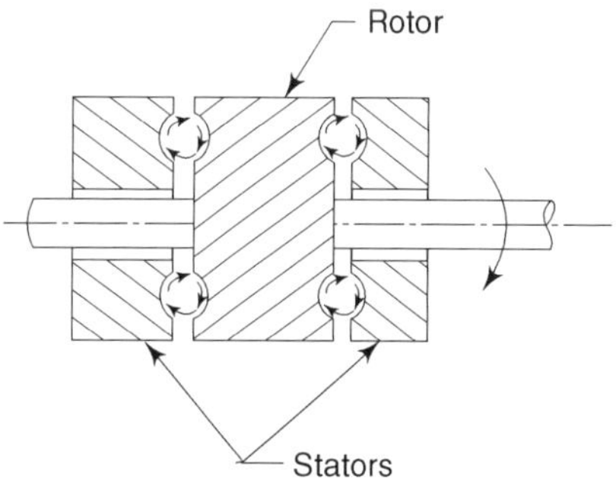

Fig. 10.14 *Hydraulic absorption dynamometer*

In a hydraulic absorption dynamometer, shown in Fig. 10.14, power is absorbed by fluid friction due to the braking action of the vortex produced by flow of water along a helix. This is brought about by the relative motion of the rotor with respect to stator or the casing of the dynamometer. The rotor and stator have cup-shaped pockets such that the path of the water is helix. The tendency of the stator to rotate is opposed by an arm on the stator with a balancing mass. The stator is freely pivoted on the bearings. The load is controlled by control of sluice gates in the spaces between stator and rotor pockets. This control can be effected from outside and changes the braking effect between the rotor and the stator. These dynamometers are commonly used in engine testing and high-speed-high-power applications. There is no need of a separate cooling arrangement as the control fluid, usually water, acts as the coolant.

In order to determine the available range of absorption dynamometers, the following characteristics have to be considered:

1. Friction and windage torque characteristics, with speed, as shown in Fig. 10.15. The useful range for use is above this line in Fig. 10.15.
2. Maximum torque, as limited by the mechanical design of the dynamometer structure, as shown in Fig. 10.15. The useful range is below this line in the Fig. 10.15.
3. Cooling limit characteristics of the dynamometer, the value of available torque is always lower than that indicated by the corresponding torque–speed line, as in Fig. 10.15.

The shaded area is the available range of the dynamometer. In addition, the maximum allowable speed would also limit the available range.

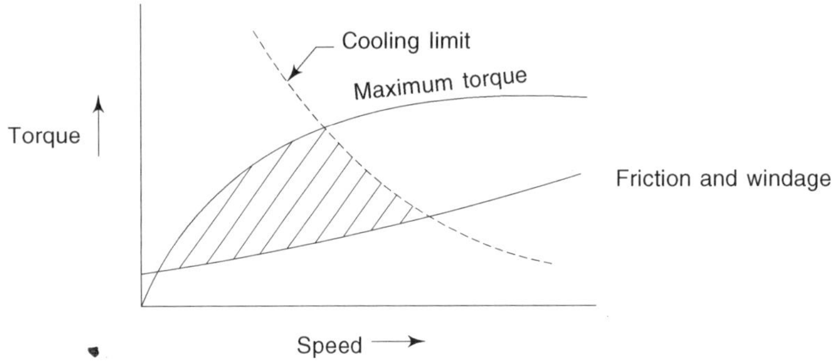

Fig. 10.15 *Finding available range of absorption dynamometer*

10.4 ■ CALIBRATION

Static calibration of a torque measuring device can be easily carried out by dead weights applied at a suitable lever arm. If the force required to be applied is large, this may be applied hydraulically or by other means like a spring and measured by a suitable load cell.

Review Questions

10.1 Indicate true or false against each of the following:

(a) A dc electric motor may be used as a driving dynamometer as well as an absorption one.
(b) Gear dynamometer may be used for very high powers.
(c) Frictional losses in a belt dynamometer are very low.
(d) Slip rings result in electrical noise due to variation of contact resistance.
(e) The angular twist in most cases in a torsion dynamometer does not exceed 1–2°.
(f) Dynamic characteristics of a strain gauge torque transducer are poor at low frequencies.

(g) Cooling water is not required in the case of a hydraulic absorption dynamometer.
(h) For measuring the torque of a rotating shaft, use of slip rings is absolutely necessary.
(i) An eddy current dynamometer is of absorption type.
(j) A transmission dynamometer is a passive device.
(k) An electric generator is essentially an absorption type dynamometer.
(l) Angular twist in a torsion dynamometer cannot be measured by an electromechanical transducer.
(m) For finding power transmitted by a shaft, it is necessary to measure both speed and torque.

10.2 A mild steel shaft is used to connect a motor drive to a constant load torque. To measure this torque, a resistance strain gauge with resistance of 120 Ω and gauge factor of 2, is mounted on a shaft with its active axis at 45° to the shaft axis. Shear modulus of mild steel is 8×10^{10} N/m^2. Shaft diameter is 3 cm and change in strain gauge resistance due to load is 0.2 Ω. Find the load torque.

10.3 A hollow steel cylinder is used as a torque sensing element with $r_i = 2.5$ cm, $r_o = 3.3$ cm and $L = 15$ cm. Calculate the angular deflection for an applied torque of 30 N–m. Also find the strain indicated by a resistance gauge bonded at 45° to the axis.

10.4 A torque-sensing element with torsional stiffness 100 N–m/rad is used between an electric motor and a hydraulic pump. Mass moment of inertia of motor as well as pump is 2×10^{-2} kg m^2 each. If the motor has an oscillatory torque component at 50 Hz, determine if this measuring system will be satisfactory or not. If not, suggest means to make it satisfactorly.

10.5 Suggest a suitable measurement device for obtaining torque against rotation angle (measured from a reference on the crankshaft) for a reciprocating engine driving a generator. Explain with a sketch the various measuring instruments and calibration method to be adopted.

Answers

10.1	(a) T	(b) F	(c) F	(d) T	(e) T
	(f) F	(g) T	(h) F	(i) T	(j) T
	(k) T	(l) F	(m) T		

10.2 706.8 Nm.

10.3 4.35×10^{-5} rad., 4.785×10^{-6}.

10.4 Not satisfactory.
Increase torsional stiffness.

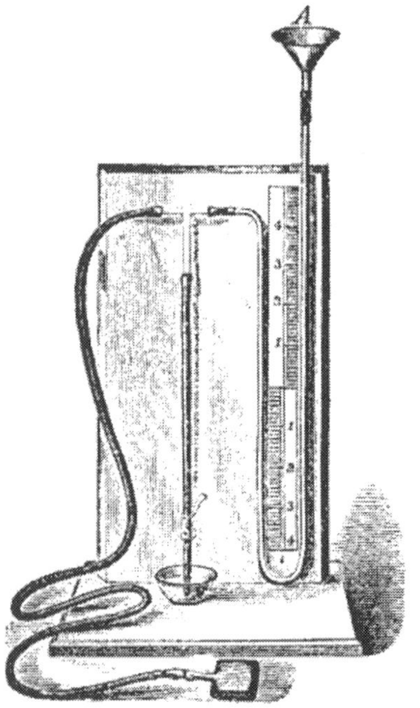

Chapter

11

Pressure Measurement

INTRODUCTION

Pressure means force per unit area, exerted by a fluid on the surface of the container. Absolute pressure means the fluid pressure above the reference value of a perfect vacuum or the absolute zero pressure. Gauge pressure represents the value of pressure above the reference value of atmospheric pressure. It is the difference between the absolute and local atmospheric pressures. The atmospheric pressure, at sea level, is nearly 14.7 lb/in^2. (psi) or 1.013×10^5 N/m^2 (Pa) or 760 mm of Hg. Figure 11.1 shows the various terms used to express pressure. Vacuum represents the amount by which the atmospheric pressure exceeds the absolute pressure. Thus, from Fig. 11.1, it can be seen that the gauge pressure, corresponding to absolute pressure $P_1 = P_1 - p_a$ and vacuum corresponding to absolute pressure $P_2 = P_a - P_2$.

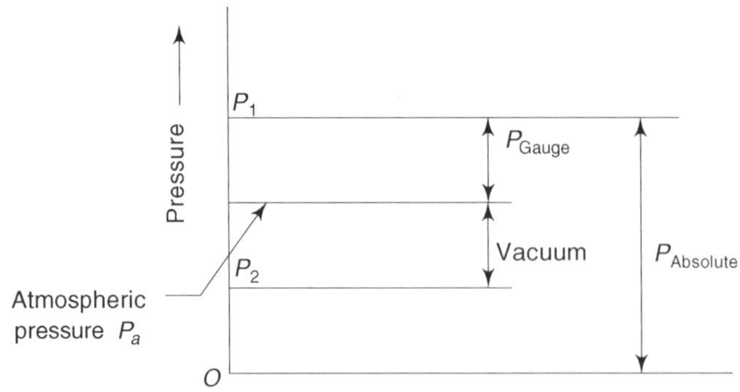

Fig. 11.1 *Various pressure terms used*

The techniques for pressure measurements are quite varied, depending on whether the pressure is moderate, very high or very low and also whether it is static or dynamic. Pressures higher than 1000 atm are usually regarded as very high while those of the order of 1 mm of Hg or below are regarded as very low.

For dynamic pressure measurements, like those in reciprocating compressors or engines, it is usual to convert it into displacement by an elastic element and then use one of the electromechanical transducers for an electrical output. The accuracy of measurements, in the case of dynamic pressures, is influenced by the characteristics of the elastic elements, the electromechanical transducer as well as those of the fluid.

11.1 ■ MODERATE PRESSURE MEASUREMENT

Essentially, there are two types of devices, which can be included in this category:

1. manometers, and
2. others using elastic elements.

Manometers are meant for measuring static pressures, while devices using elastic elements may be used for both static and dynamic measurements.

11.1.1 Manometers

A manometer is the simplest device for measuring static pressure. A simple U-tube manometer (Fig. 11.2) uses water, mercury or any other suitable fluid. The difference in levels h between the two limbs is an indication of the pressure difference $(p_1 - p_2)$ between the two limbs. If one of the pressures, say that applied to limb 2, is atmospheric, the difference gives the gauge pressure applied to limb 1.

$$h = (p_1 - p_2)\,\rho g \tag{11.1}$$

ρ being the mass density of the liquid used in the manometer.

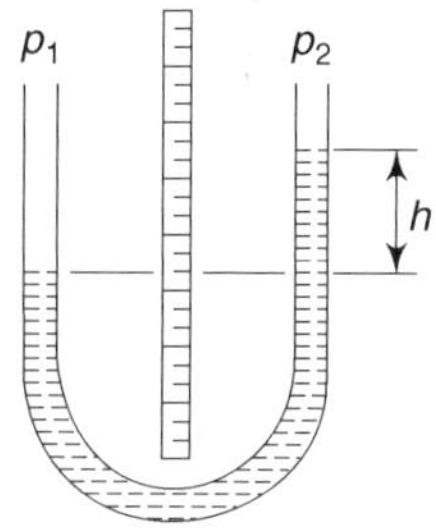

Fig. 11.2 *U-tube manometer*

The desirable characteristics of a manometer fluid are:

1. it should be non-corrosive and not have any chemical reaction with the fluid whose pressure is being measured,
2. it should have low viscosity and thus ensures quick adjustment with pressure change,
3. it should have negligible surface tension and capillary effects.

Several types of modified manometers are available which have the advantages of ease in use and high sensitivity.

One such device that is convenient to use is the cistern or well type manometer, shown in Fig. 11.3. In this type, the well area is large compared to that of the tube. Thus, only a single leg reading may be noted and the change in level in the well may be ignored. If p_1 and p_2 are absolute pressures applied as shown, force equilibrium gives:

$$p_1A - p_2A = Ah\rho g$$

ρ being mass density of the liquid.

$$\frac{p_1 - p_2}{\rho g} = h$$

If p_2 is atmospheric, h is a measure of the gauge pressure applied at the well.

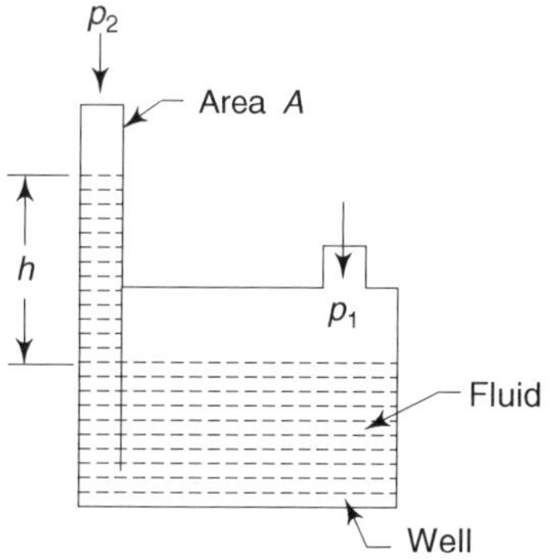

Fig. 11.3 *Cistern or well type manometer*

An inclined type manometer (Fig. 11.4) is another device, which is sensitive and convenient to use. In such a manometer, the length l along the inclined tube is read as a measure of the pressure difference $(p_1 - p_2)$. The relation between $(p_1 - p_2)$ and l is derived as follows:

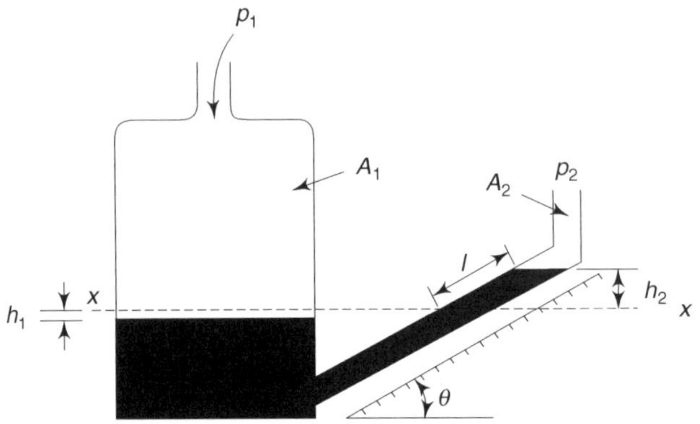

Fig. 11.4 *Inclined tube manometer*

When pressures in the two limbs are the same, the levels of the liquid are at equilibrium position xx. On application of pressures p_1 and p_2, difference in levels between the two limbs is

$$h_1 + h_2 = \frac{p_1 - p_2}{\rho g} \tag{11.2}$$

If A_1 and A_2 are the respective areas of the two limbs,

$$A_1 h_1 = A_2 l \tag{11.3}$$

$$h_2 = l \sin \theta \tag{11.4}$$

From Eqs. (11.2)–(11.4), we get

$$p_1 - p_2 = \rho g l \left(\frac{A_2}{A_1} + \sin \theta \right) \tag{11.5}$$

If $A_1 \gg A_2$ or A_2/A_1 is negligible,

$$p_1 - p_2 = \rho g l \sin \theta = \rho g h_2 \tag{11.6}$$

If $\theta = 30°$, $l = 2h_2$ and thus it would be more accurate to read l rather than h_2 as the output. Since $A_1 \gg A_2$, the reading on one side only, viz. l is required.

For increased accuracy in reading the output of the manometer, i.e., liquid displacements, can be measured with micrometer heads (Fig. 11.5). The contact between the micrometer movable points and the liquid may be sensed electrically or visually.

In order to minimise capillary and meniscus errors, the meniscus is returned to a reference position marked on the inclined transparent portion of the manometer tube (Fig. 11.6). This is done by a motorised screw (Fig. 11.7), whose rotation is converted to a digital read-out. This type of manometer is quite sensitive, accurate and easy to use.

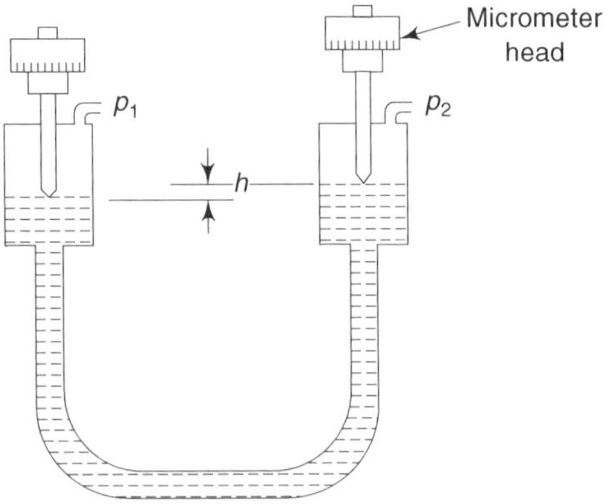

Fig. 11.5 *Micrometer type manometer*

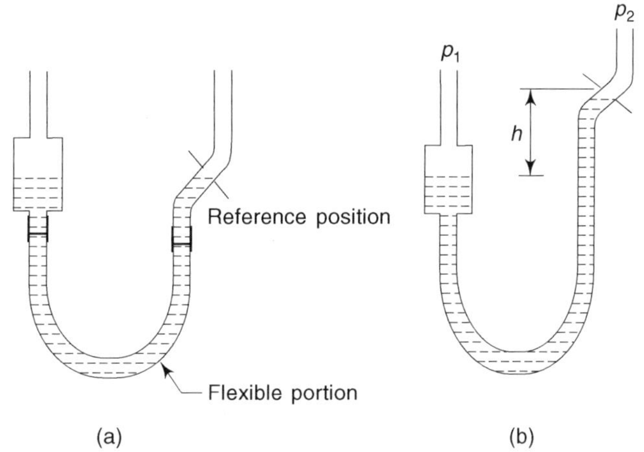

Fig. 11.6 *Movable tube type micromanometer*

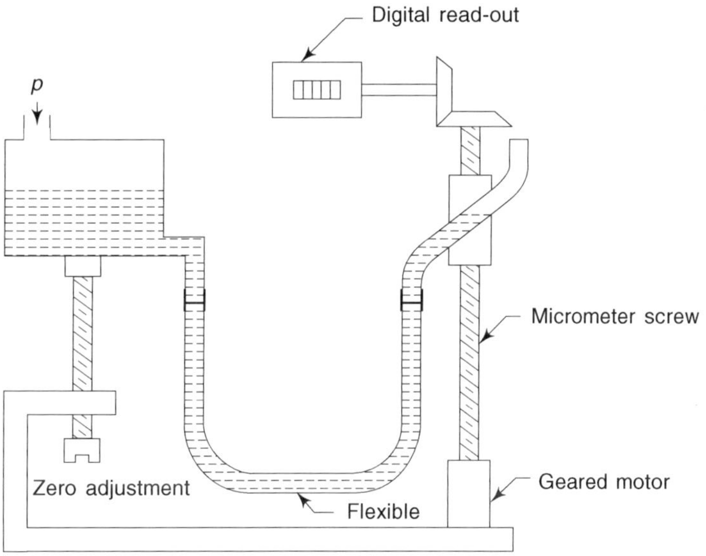

Fig. 11.7 *Micromanometer with motor drive and digital read-out*

Under dynamic conditions, manometers can be modelled as second order systems as discussed in Chapter 3. These are, however, used for static measurements as the dynamic response is not satisfactory due to, the inertia of the liquid.

Problem 11.1 *A mercury manometer of the type shown in Fig. 11.8 is to have a float in the left-hand chamber. An electromechanical transducer is used to measure the motion of the fluid. The float motion is 5 mm for a gauge pressure of 50 kN/m^2. If the diameter of the float chamber is 40 mm, find the required diameter for the right-hand chamber. For mercury, density ρ = 13600 kg/m^3. Assume that the other end of the manometer is open to the atmosphere.*

Solution If due to pressure differential Δp, the displacements from equilibrium position in the two chambers are y_1 and y_2

$$y_1 A_1 = y_2 A_2 \quad (11.7)$$

where A_1 and A_2 are the respective chamber areas. Also,

$$\Delta p = \rho g(y_1 + y_2) \quad (11.8)$$

Substituting

$$A_1 = \frac{\pi}{4}(0.04)^2 \text{ m}^2$$

$$y_1 = 0.005 \text{ m}$$

$$\Delta p = 50\ 000 \text{ Pa}$$

and $\rho = 13600$ kg/m^3, we get from Eqs. (11.7) and (11.8)

$$A_2 = 1.7 \times 10^{-5} \text{ m}^2$$

which gives a diameter 4.65 mm for the right-hand chamber.

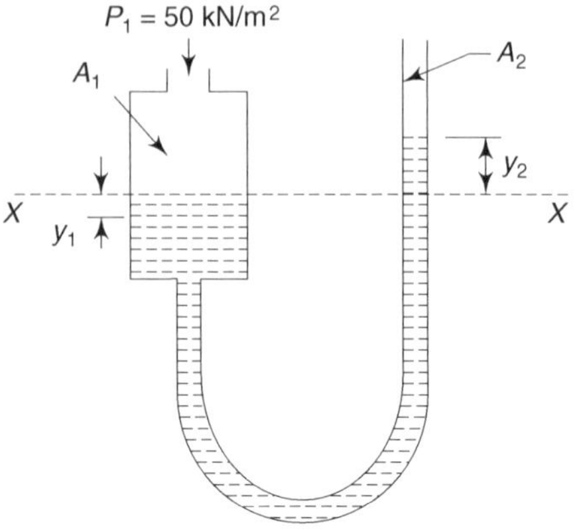

Fig. 11.8 *Figure for Problem 11.1*

11.1.2 Elastic Transducers

Elastic elements, when subjected to pressure, get deformed. The deformation, when measured, gives an indication of the pressure. These elements may be in the form of diaphragms, capsules, bellows, Bourdon or helical tubes (Fig. 11.9). The deformation may be measured by mechanical or electrical means.

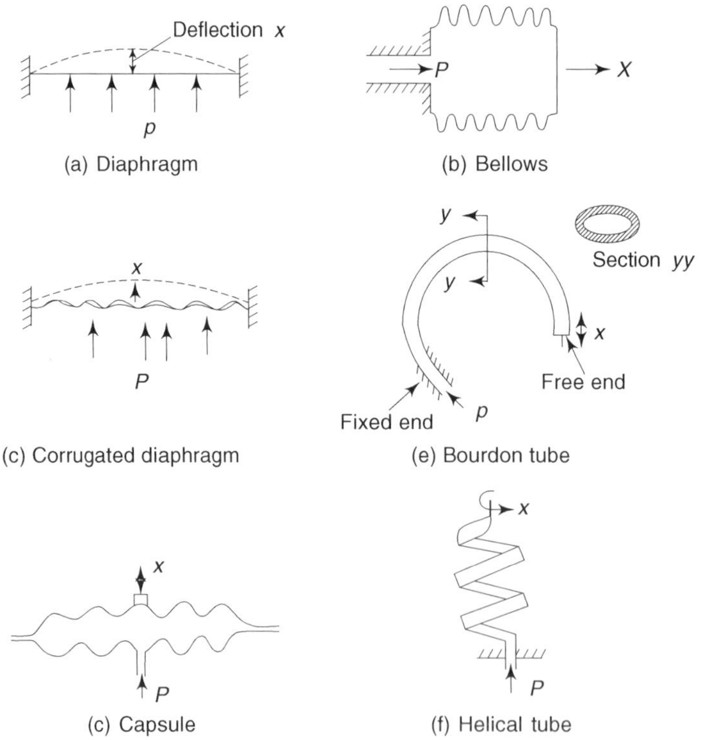

Fig. 11.9 *Elastic elements*

These devices are convenient to use and can cover a wide range of pressures, depending on the design of the elastic elements.

Types of Measuring Devices A Bourdon gauge is commonly used for measuring pressure (Fig. 11.10). This incorporates an elastic Bourdon tube. The elliptical cross-section of the tube, due to pressure, tends to become round shaped. The tube uncoils since the inner and outer arc lengths remain almost equal to their original lengths. The motion of the end of the tube is amplified and indicated by a pointer moving on a calibrated scale.

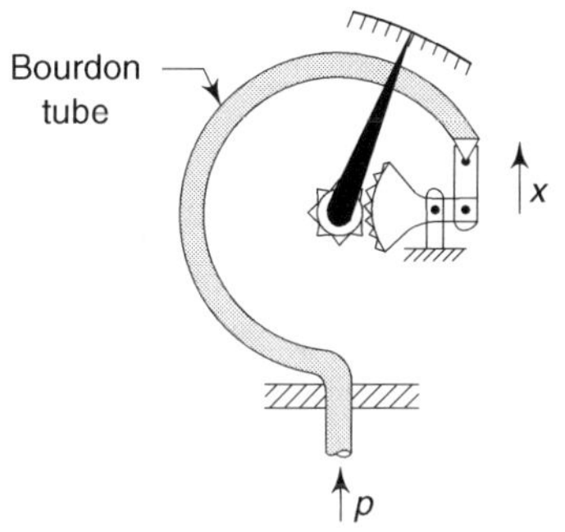

Fig. 11.10 *Bourdon pressure gauge*

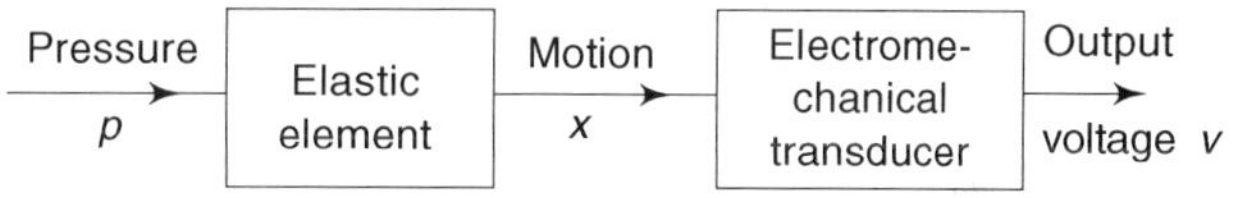

Fig. 11.11 *Pressure measurement using an elastic element and an electromechanical transducer*

Usually, an electromechanical transducer is used along with the elastic element, especially when dynamic pressures are to be measured (Fig. 11.11). The output voltage can be indicated or recorded by a suitable instrument like an oscilloscope or a recorder. Figure 11.12 shows a linear variable differential transducer (LVDT) type of pressure transducer, the principle of which has already been discussed in Ch. 4. The motion of the bellows is communicated to the core, whose motion gives an output voltage proportional to it. Since the bellows motion is proportional to pressure p, the output voltage is also proportional to p.

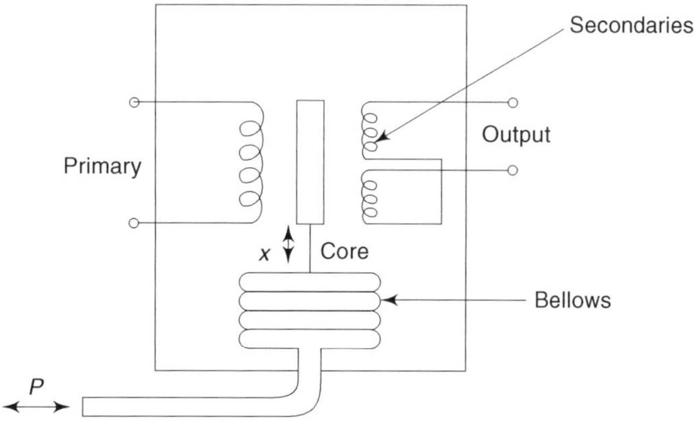

Fig. 11.12 *LVDT-type pressure transducer*

Figure 11.13 shows a variable capacitance type pressure transducer. Due to a pressure p, the elastic diaphragm deflects, changing the capacitance between it and the fixed electrode. The output of the bridge is proportional to the pressure p. Similarly, a piezo-electric crystal can be used (Fig. 11.14) for measuring

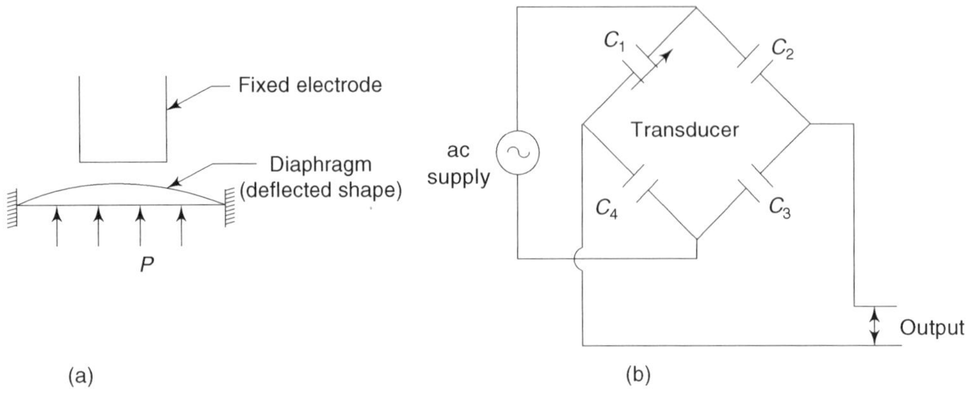

Fig. 11.13 *Capacitance-type pressure transducer*

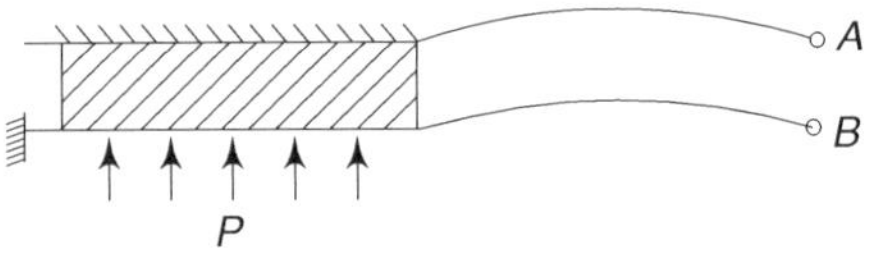

Fig. 11.14 *Piezo-electric type pressure transducer*

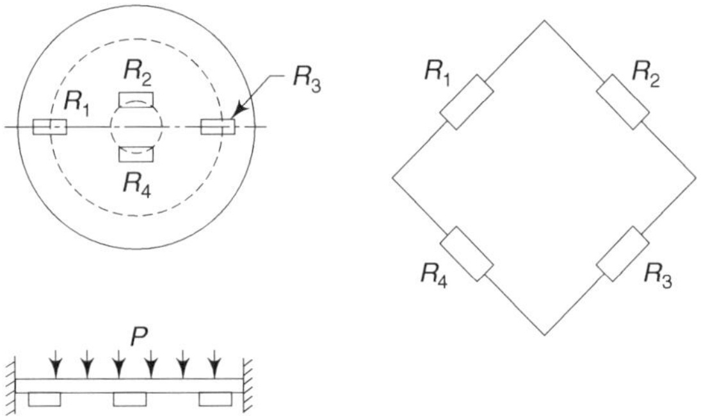

Fig. 11.15 *Diaphragm type strain gauge pressure transducer*

the dynamic pressure. A voltage is developed between A and B, according to the pressure p applied. As discussed earlier, static variables cannot be measured with this type of transducer.

Three types of pressure transducers, using resistance gauges are shown in Figs. 11.15, 11.17 and 11.18. In Fig. 11.15, resistance gauges R_1 and R_3 are bonded, so as to measure radial strain near the outer radius of the diaphragm while R_2 and R_4 are bonded near the centre and measure tangential strains, Expressions for radial stress σ_r and tangential stress σ_t at a radius r are

$$\sigma_r = \frac{3pR^2 v}{8t^2}\left(\frac{1}{v}+1\right)-\left(\frac{3}{v}+1\right)\left(\frac{r}{R}\right)^2$$

$$\sigma_t = \frac{3pR^2 v}{8t^2}\left(\frac{1}{v}+1\right)-\left(\frac{1}{v}+3\right)\left(\frac{r}{R}\right)^2 \qquad (11.9)$$

t being the diaphragm thickness, v the Poisson's ratio of diaphragm material and p the pressure on the diaphragm. A plot of radial and tangential stresses (Fig. 11.16) shows that stresses in R_1 and R_3 would be opposite in nature to those in R_2 and R_4. Figure 11.17 shows a strain gauge pressure transducer in which the resistance gauges are bonded on a member which are strained longitudinally. This type of transducer can be used for high pressures. Figure 11.18 shows a strain gauge pressure transducer, meant for measuring the pressure of a fluid in a pipe, without disturbing the fluid. Strain gauge R_1 measures the hoop strain in the pipe due to the pressure of the fluid.

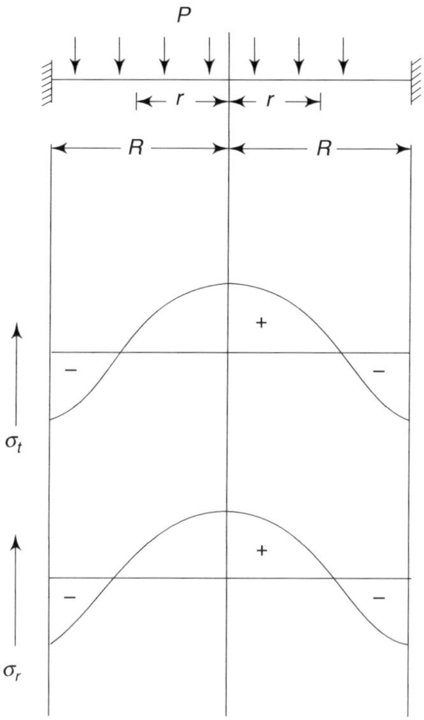

Fig. 11.16 *Plot of radial and tangential stresses*

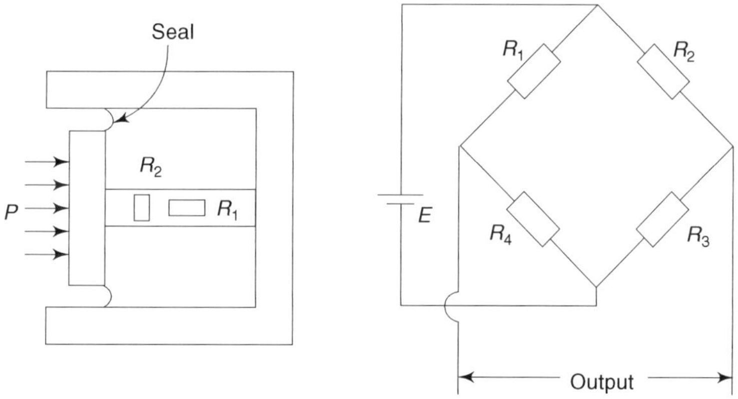

Fig. 11.17 *Piston type strain gauge pressure transducer*

Except for the pizeo-electric type, other transducers discussed here can be used for both static and dynamic measurements. The advantage of the piezo-electric type transducer lies in its increased sensitivity and thus higher output for a given pressure change.

Elastic Element Characteristics Elastic elements used for pressure transducers have to be carefully designed for

1. deflection due to pressure, and
2. dynamic considerations.

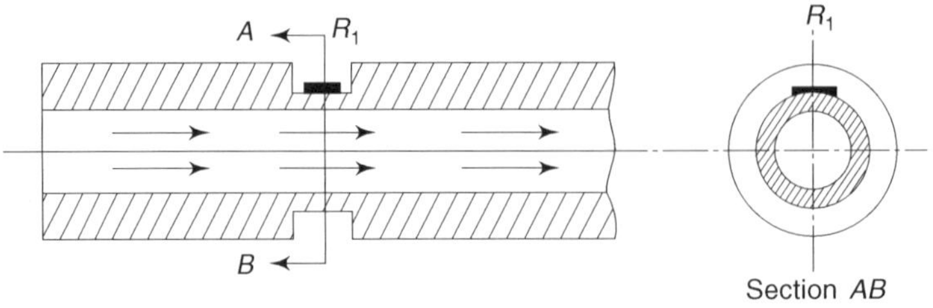

Fig. 11.18 *Strain gauge pressure transducer for measuring fluid pressure in a pipe*

It is known that an elastic diaphragm would remain linear for small deflections only and for this purpose, maximum deflection y should be $< t/3$, t being the diaphragm thickness.

From strength and elasticity of materials, it is seen that deflection y at radius r of a circular diaphragm clamped at its outer periphery (Fig. 11.19) is given by

$$y = \frac{3}{16} p \frac{(1-\nu^2)}{Et^3} (R^2 - r^2)^2 \qquad (11.10)$$

where p is the pressure on the diaphragm of radius R and thickness t, E being Young's modulus of the diaphragm material and ν its Poisson's ratio.

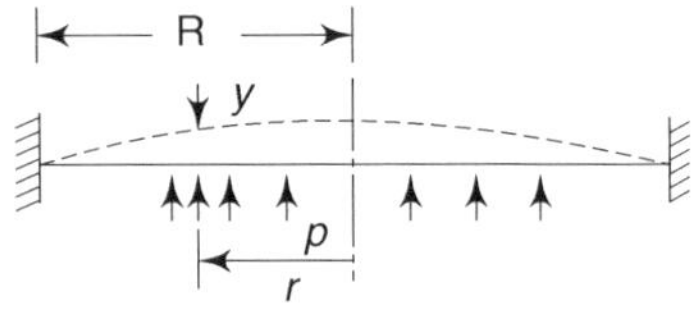

Fig. 11.19 *Configuration of a deflected diaphragm*

Thus

$$y_{\max} \text{ (at } r = 0) = \frac{3}{16} \frac{p}{Et^3} R^4 (1 - \nu^2) \qquad (11.11)$$

For dynamic considerations, it is important to check that the fundamental frequency of vibrations of the elastic element is higher than the exciting frequency due to fluctuating pressure. For a circular diaphragm fixed at its periphery, the fundamental bending natural frequency is given by

$$\omega_n = \frac{10.21}{R^2} \sqrt{\frac{Et^2}{12(1-\nu^2)\rho}} \text{ rad/s} \qquad (11.12)$$

ρ being the mass density of the diaphragm material. Remaining symbols have been defined earlier.

It may be further seen that the dynamic characteristics of a pressure measuring system are dependent not only on the characteristics of the elastic transducing element but also upon the characteristics of the pressure transmitting fluid, and the connecting tubing. As seen from Fig. 11.20, the effective mass of the moving system depends on the mass of the fluid that moves with deflected elastic diaphragm. Similarly, the damping action depends on the fluid friction. A consideration of the above factors for predicting the dynamic response is thus desirable.

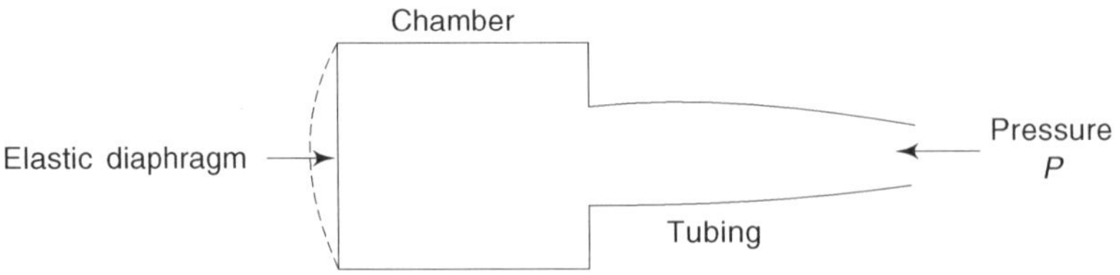

Fig. 11.20 *Elastic diaphragm with chamber and tubing*

Problem 11.2 *Figure 11.21 shows a pressure transducer, using a clamped diaphragm. Strain gauges 2 and 4 are meant to measure the tangential strain while gauges 1 and 3 measure the radial strain. Find the open circuit sensitivity in mV/Pa.*

Resistance of each gauge = 120 Ω, gauge factor = 2, radius R = 7 cm, r_o = 1 cm and r_i = 6 cm. Thickness t of the diaphragm = 1 mm.

Young's modulus	*E = 2.07 × 10^5 N/mm²*
Poisson's ratio	*ν = 0.25*

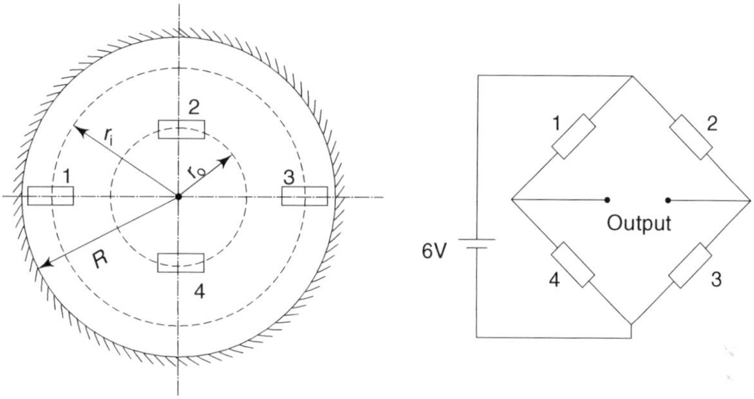

Fig. 11.21 *Figure for Problem 11.2*

Solution Radial strain e_r at any radius r is

$$\varepsilon_r = \frac{\sigma_r - \nu\sigma_t}{E} \tag{11.13}$$

σ_r and σ_t being the radial and tangential stresses, respectively, at radius r. Similarly, tangential strain e_t is

$$\varepsilon_t = \frac{\sigma_t - \nu\sigma_r}{E} \tag{11.14}$$

Expressions for σ_r and σ_t are given in Eq. (11.9) and are:

$$\sigma_r = \frac{3pR^2\nu}{8t^2}\left[\left(\frac{1}{\nu}+1\right)-\left(\frac{3}{\nu}+1\right)\left(\frac{r}{R}\right)^2\right]$$

$$\sigma_t = \frac{3pR^2\nu}{8t^2}\left[\left(\frac{1}{\nu}+1\right)-\left(\frac{1}{\nu}+3\right)\left(\frac{r}{R}\right)^2\right]$$

For r = 60 mm, using Eqs. (11.9) and (11.13) and taking $E = 2.07 \times 10^{11}$ N/m²

$$\sigma_r = -2090.6p$$

$$\sigma_t = -65.6p$$

$$\varepsilon_r = -1.002 \times 10^{-8}\, p = \varepsilon_1 = \varepsilon_3 \tag{11.15}$$

ε_1 and ε_3 are strains in resistance gauges 1 and 3, respectively. For r = 10 mm, using Eqs. (11.9) and (11.15)

$$\sigma_r = 2175p$$

$$\sigma_t = 2231.25p$$

$$\varepsilon_t = 8.15 \times 10^{-9}\, p = \varepsilon_2 = \varepsilon_4 \tag{11.16}$$

where ε_2 and ε_4 are strains in resistance gauges 2 and 4, respectively. Open circuit output voltage expression, as deduced earlier

$$e_o = \frac{E_b F}{4}\,[\varepsilon_2 + \varepsilon_4 - \varepsilon_1 - \varepsilon_3] \tag{11.17}$$

E_b being the battery voltage, given as 6 V and F the gauge factor. Substituting from Eqs. (11.15) and (11.16) in Eq. (11.17) and taking p = 1 Pa

$$\text{Output voltage} = 1.086 \times 10^{-7}\text{ V}$$

or $\quad$ Sensitivity of diaphragm type pressure gauge = 1.086×10^{-7} V/Pa

Problem 11.3 *A variable capacitance pressure gauge (Fig. 11.22) has the following specifications: diameter of clamped diaphragm = 20 mm, diameter of fixed electrode = 15 mm, thickness of diaphragm = 1 mm, Young's modulus E of diaphragm material = 2.07 × 10⁵ N/mm², Poisson's ratio v = 0.3, and initial air gap = 1 mm. The variable capacitance C due to change of air gap, forms a part of circuit of Fig. 11.22(b). Find the sensitivity (V/Pa) of the instrument, given V = 12 V and R = 10⁵Ω.*

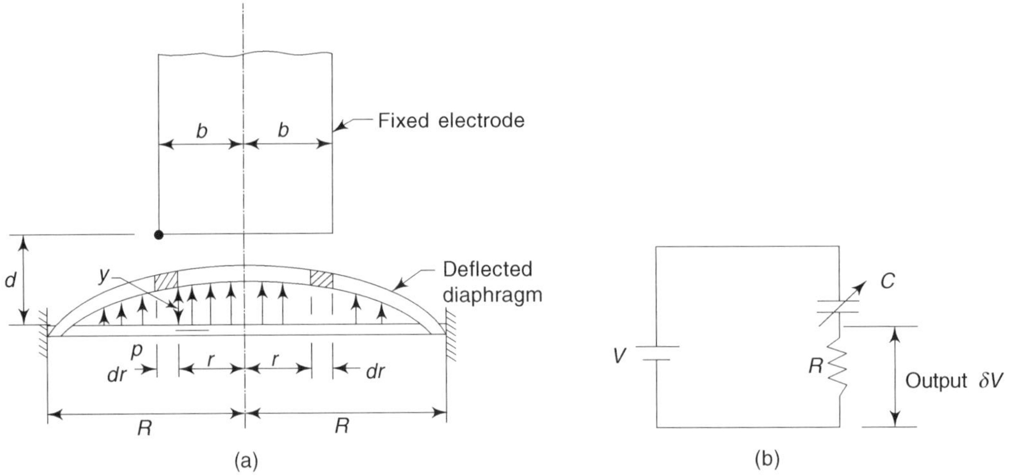

Fig. 11.22 *Figure for Problem 11.3*

Solution As seen from Fig. 11.22(a), the change in the gap between the diaphragm and the fixed electrode, due to pressure p is not uniform but is a function of the radius r.

With initial air gap d, initial capacitance C_0 between the diaphragm and the fixed electrode is obtained using Eq. (4.1) as

$$C_0 = b^2/(3.6d) \tag{11.18}$$

A being the area of the electrode (cm²) = πb^2 and d the air gap (cm).

After application of pressure p, as seen from Fig. 11.22(a), gap at radius $r = d - y$, y being the deflection of diaphragm, given by Eq. (11.10), viz.

$$y = B(R^2 - r^2)^2 \tag{11.19}$$

$$B = \frac{3}{16}\,\frac{p(1-v^2)}{Et^3}$$

The capacitance C between the diaphragm and the fixed electrode is

$$C = \int_0^b \frac{2\pi r dr}{3.6\pi\,(d-y)}$$

$$\approx \frac{1}{1.8d}\int_0^b \left(1+\frac{y}{d}\right) r dr \qquad \text{(for } y/d \text{ being small).} \tag{11.20}$$

Substituting from Eq. (11.19) in Eq. (11.20) and integrating

$$C = \frac{b^2}{3.6d} + \frac{0.0174\,(1-\nu^2)\,p}{Ed^2 t^3}\,[b^6 + 3R^2b^2\,(R^2 - b^2)] \tag{11.21}$$

From Eqs. (11.18) and (11.21)

$$\frac{\delta C}{C_0} = \frac{C - C_0}{C_0} = \frac{0.0625(1-\nu^2)\,p}{Edt^3}\,[b^4 + 3R^2(R^2 - b^2)] \tag{11.22}$$

where δC is the change is capacitance due to pressure p. For the circuit of Fig. 11.22(b), as proved in Chapter 4,

$$\frac{\delta V}{V} \approx \frac{\delta C}{C_0} \tag{11.23}$$

Using Eqs. (11.22) and (11.23) and substituting the values of the various quantities,

$$\text{Sensitivity of capacitive type pressure gauge} = \frac{\delta V}{p} = 5.36 \times 10^{-12}\ \text{V/Pa}$$

11.1.3 Dynamic Effect of Connecting Tubing

Usually, a connecting tube is used between the pressure measuring instrument and the location where the fluid pressure is to be measured (Fig. 11.23). Such a tubing would cause a lag under dynamic conditions. The lag may be determined as below. The inertia effect due to fluid motion is ignored.

Referring to Fig. 11.23, in which a manometer is the measuring instrument, it is seen that the mass rate of fluid flow in the tubing would be affected due to

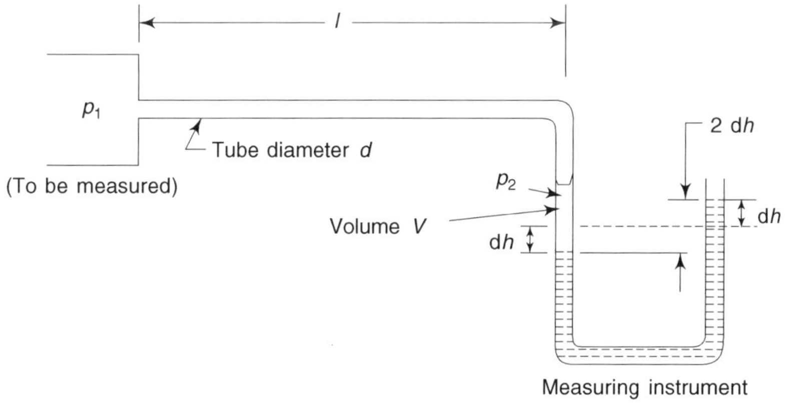

Fig. 11.23 *Effect of connecting tubing*

1. change of fluid density due to pressure change, and
2. change of volume of fluid as a result of change of liquid level of the manometer.

Effect 1 would be absent in the case of liquids while both effects 1 and 2 would be involved in case of gases.

If p_1 and p_2 are absolute pressures at the beginning and end of the tubing respectively, using Eq. (3.13) for pressure loss in the tubing, viz.

$$p_1 - p_2 = \frac{128\,\mu l Q}{\pi d^4},$$

μ being viscosity of the fluid and Q the volume flow rate. Mass flow rate of the fluid through the tubing

$$\rho Q = \frac{\rho \pi d^4 (p_1 - p_2)}{128\,\mu l} \tag{11.24}$$

ρ being the mass density of the fluid mass. Flow rate is also equal to d/dt (ρV), V being the volume of space

$$= \frac{\rho \mathrm{d}V}{\mathrm{d}t} + \frac{V \mathrm{d}\rho}{\mathrm{d}t} \tag{11.25}$$

The first term on RHS of Eq. (11.25) is due to volume change of fluid as a result of the motion of the liquid in the manometer while the second term is due to change of mass density due to pressure change. Now,

$$\mathrm{d}V = A\mathrm{d}h$$

A being area of cross-section of the manometer tube and dh the liquid motion in the manometer arm. Also,

$$2\mathrm{d}h = \frac{\mathrm{d}p_2}{\rho_0 g}$$

ρ_0 being mass density of liquid in manometer. Thus,

$$\mathrm{d}V = \frac{A}{2\rho_0 g}\,\mathrm{d}p_2$$

or
$$\frac{\mathrm{d}V}{\mathrm{d}t} = \frac{A}{2\rho_0 g}\frac{\mathrm{d}p_2}{\mathrm{d}t} \tag{11.26}$$

Further, due to pressure change dp_2,

$$\mathrm{d}\rho = \left(\frac{\rho}{p_2}\right)\mathrm{d}p_2$$

or
$$\frac{\mathrm{d}\rho}{\mathrm{d}t} = \frac{\rho}{p_2}\frac{\mathrm{d}p_2}{\mathrm{d}t} \tag{11.27}$$

Equating Eqs. (11.24) and (11.25) and substituting from Eqs. (11.26) and (11.27), we get

$$\tau \frac{\mathrm{d}p_2}{\mathrm{d}t} + p_2 = p_1 \tag{11.28}$$

where
$$\tau = \frac{128\,\mu l}{\pi d^4}\left[\frac{A}{2\rho_0 g} + \frac{V}{p_2}\right]$$

τ is the time constant of the first order relation representing the lag of the system. In order to reduce τ; l, A and V should be small and d should be large.

11.2 ■ HIGH PRESSURE MEASUREMENT

For pressures above 1000 atm, special techniques have to be used. One such technique is based on the electrical resistance change of a manganin (alloy of Cu, Ni, Mn) or gold chrome wire, with hydrostatic pressure, due to bulk compression effect. Figure 11.24 shows an outline diagram of this type of device. Usually, the coil is enclosed in a flexible bellows (not shown in the figure) filled with kerosene, for transmitting the pressure to be measured to the coil. The change in the resistance of the wire between A and B is measured by usual methods like Wheatstone bridge, etc.

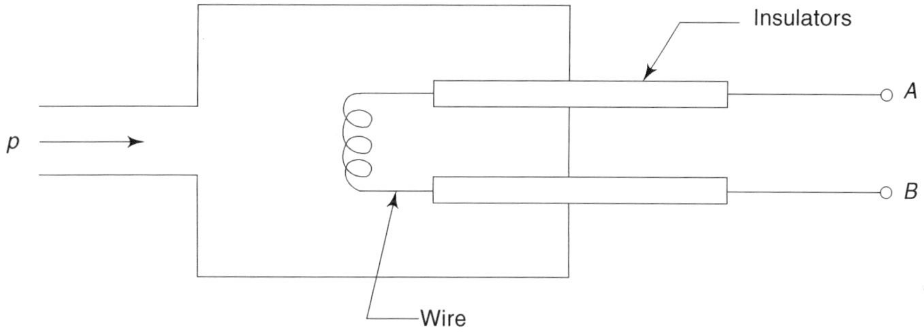

Fig. 11.24 *High-wire pressure transducer*

For manganin, the sensitivity is 2.5×10^{-11} Ω/Ω-Pa while for gold–chrome, the same is 9.85×10^{-12} Ω/Ω-Pa. Even though gold–chrome is less sensitive, it is preferred to manganin, since the former is less temperature sensitive than the latter.

An expression for sensitivity for the arrangement of Fig. 11.24 may be deduced as under: For a wire of diameter D and length L, the resistance

$$R = \frac{4\rho L}{\pi D^2}$$

ρ being resistivity constant. From the above, we can write

$$\frac{dR}{R} = \frac{dL}{L} - \frac{2\,dD}{D} + \frac{d\rho}{\rho} \tag{11.29}$$

Relations between strains ε_x, ε_y and ε_z in the three perpendicular directions, in terms of stresses σ_x, σ_y and σ_z are

$$\varepsilon_x = \frac{1}{E}\,[\sigma_x - \nu(\sigma_y + \sigma_z)]$$

$$\varepsilon_y = \frac{1}{E}\,[\sigma_y - \nu(\sigma_x + \sigma_z)]$$

$$\varepsilon_z = \frac{1}{E}\,[\sigma_z - \nu(\sigma_x + \sigma_y)]$$

ν being Poisson's ratio and E Young's modulus.

The wire in the arrangement of Fig. 11.24 is subjected to a bi-axial state of stress as the ends will not be subjected to pressure.

Thus, taking $\sigma_x = \sigma_y = -p$ and $\sigma_z = 0$ we get from Eq. (11.30)

$$\varepsilon_x = \varepsilon_y = \frac{dD}{D} = -\frac{p}{E}(1 - \nu)$$

$$\varepsilon_z = \frac{dL}{L} = \frac{2\nu p}{E}$$

Substituting in Eq. (11.29)

$$\frac{dR}{R} = \frac{2p}{E} + \frac{d\rho}{\rho}$$

Thus,

$$\text{sensitivity} = \frac{dR/R}{p} = \frac{2}{E} + \frac{d\rho/\rho}{p} \tag{11.31}$$

11.3 ■ LOW PRESSURE (VACUUM) MEASUREMENT

Units of vacuum measurements are Torr and micron. One Torr is a pressure equivalent of 1 mm of Hg while 1 micron is 10^{-3} Torr.

Manometer and elastic element gauges can be used to about 0.1 Torr. Below these ranges, other types of vacuum gauges are needed.

McLeod Gauge This is used between 0.01 and 1000 microns and is a modified mercury manometer. It operates on the principle of compressing a known volume of low-pressure gas to a higher pressure and measuring the resulting volume change.

The unknown pressure source is connected to the gauge as shown in Fig. 11.25 and the mercury level adjusted so that the pressure source fills the bulb *B* and capillary *C*. Mercury is then forced out of reservoir *A* up in the bulb and reference column R. When the level reaches the cut-off point *F*, a known volume of gas is trapped in the bulk and capillary.

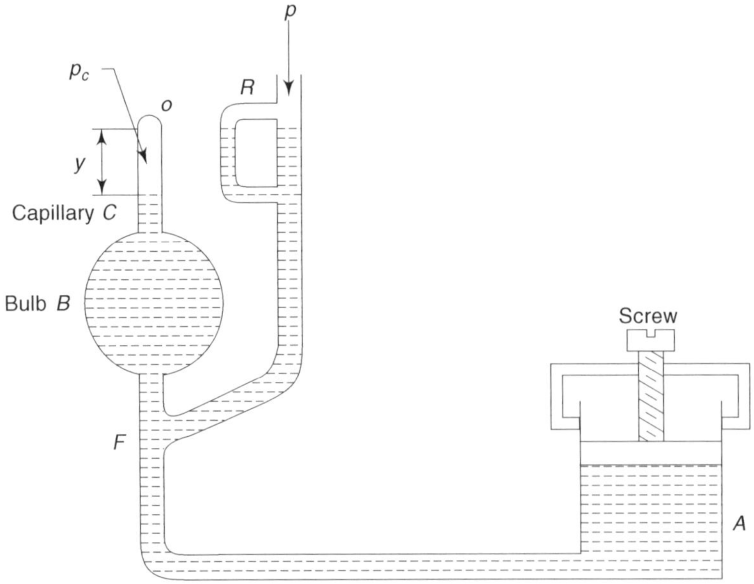

Fig. 11.25 *McLeod gauge*

The mercury level is raised till it reaches the zero reference point O in R. Under these conditions, the volume remaining in the capillary is read directly from the scale and the difference in heights y is a measure of the pressure p. Pressure p is then obtained using Boyle's law,

If p = unknown pressure

A = area of cross-section of capillary

V_c = Volume of gas in capillary

$= Ay$ (11.32)

P_c = pressure of gas in the capillary c after compression

V_F = volume of capillary and bulb till F

then,

$$p = P_c \frac{V_c}{V_F} \tag{11.33}$$

where $$y = P_c - p \tag{11.34}$$

the pressure being expressed in terms of heights of mercury.

From Eqs. (11.32)–(11.34), we get

$$p = \frac{Ay^2}{V_F - Ay} \tag{11.35}$$

The capillary can be directly calibrated in terms of pressure p.

Thermal Conductivity Gauge or Pirani Gauge It consists of a platinum filament enclosed in a chamber, as shown in Fig. 11.26. The wire forms an arm of a Wheatstone bridge. The temperature of the wire, for a given magnitude of current, depends on the rate of heat dissipation, which in turn depends on the conductivity of the surrounding medium and hence its pressure. Thus, with change in pressure of the medium, the temperature and hence the resistance of the wire changes, which can be measured by using the Wheatstone bridge.

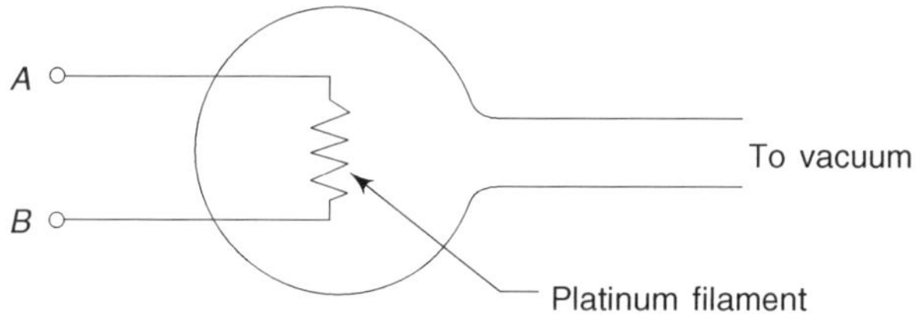

Fig. 11.26 *Thermal conductivity gauge*

Ionisation Gauge This is used for measurement of very low pressures, of the order of 1 micron and below.

The gauge consists of a triode vacuum tube (Fig. 11.27). The heated cathode emits electrons which are accelerated by the positively charged grid. As the electrons move towards the grid, they ionise the gas molecules through collisions. The plate is maintained at a negative potential so that positive ions collect there, producing a plate current i_1. The electrons and negative ions are collected by the grid, producing grid current i_2. It is found that the pressure of the gas is given by

$$p = \frac{1}{K} \frac{i_1}{i_2} \tag{11.36}$$

where K is termed as the sensitivity of the gauge.

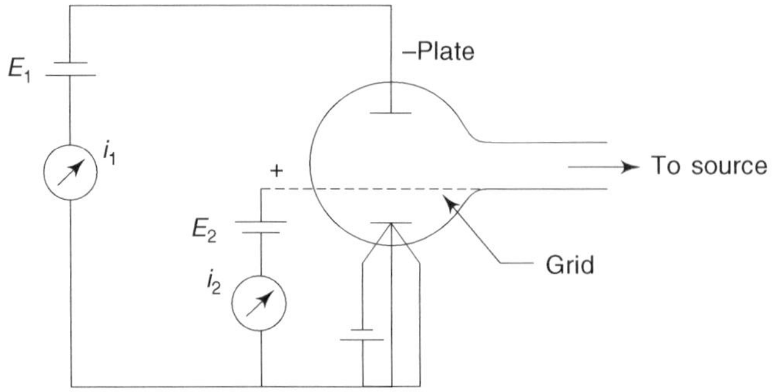

Fig. 11.27 *Ionisation gauge*

Knudsen Gauge In this type of gauge (Fig. 11.28) the gas chamber contains fixed plates F_1 and F_2 heated to and maintained at temperature T. Near these plates is a restrained movable vane V, such that the gap between the vane and fixed plates is less than the mean free path of the gas whose pressure is to be measured. The vane is at gas temperature T_0. The gas molecules rebound from plates F_1 and F_2 with greater momentum than from V, thus imparting a net force to V which may be measured by measuring the angular displacement of mirror M. It is seen that

$$p = \frac{KF}{\sqrt{\dfrac{T}{T_0} - 1}} \tag{11.37}$$

where p is gas pressure, F the force and K is a constant.

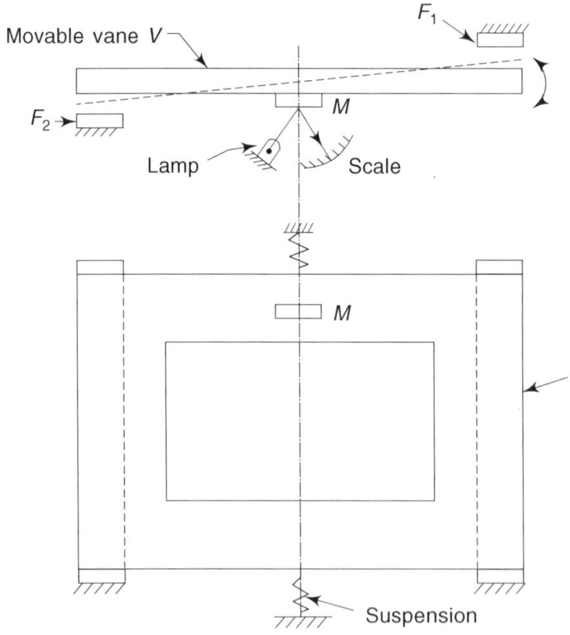

Fig. 11.28 *Knudsen gauge*

The Knudsen gauge gives an absolute measurement of pressure p, which is independent of gas composition. It can cover a range of pressure from 10^{-8} to 10^{-2} Torr.

Problem 11.4 *A McLeod gauge has volume of bulb, capillary and tube down to its opening equal to 90 cm³ and a capillary diameter of 1 mm. Calculate the pressure indicated by a reading of 3 cm on the capillary tube.*

Solution Using Eq. (11.35) and

Substituting $V_F = 90{,}000$ mm³, $A = \frac{\pi}{4}(1)^2 = 0.785$ mm² and $y = 30$ mm, we get

$$p = 7.85 \times 10^{-3} \text{ mm of Hg.}$$

Problem 11.4 *A vacuum gauge is to use a differential transformer combination, which has a resolution of 2.5 microns. The diaphragm is to be constructed of steel ($\rho = 7.9 \times 10^{-6}$ kg/mm³, $E = 2.07 \times 10^5$ N/mm², $\nu = 0.3$), with a diameter of 15 cm. Calculate the diaphragm thickness so that the maximum deflection does not exceed one-third of its thickness. Also find the lowest vacuum, which may be sensed and the frequency limit of the instrument.*

Solution Using Eq. (11.11), viz.

$$y_{max} = \frac{3}{16}\frac{p}{Et^3} R^4 (1 - \nu^2), \text{ } R \text{ being the radius of diaphragm} \tag{11.38}$$

Also $$y_{max} = \frac{t}{3}, \text{ where '}t\text{' is the diaphragm thickness} \tag{11.39}$$

p corresponding to y_{max} can have a maximum value equal to the atmospheric pressure, since in the vacuum gauge the end of diaphragm would be subjected to atmospheric pressure and the other end to the vacuum to be measured.

Taking $p = 1.013 \times 10^5$ N/m² and using Eqs. (11.38) and (11.39), we get, $t = 1.68$ mm. Thus, $y_{max} = 0.56$ mm.

Lowest vacuum, that can be sensed, would correspond to the resolution of the instrument, viz. $y = 2.5 \times 10^{-3}$ mm. Thus,

$$\text{Lowest vacuum} = \frac{1.013 \times 10^5}{0.56} \times 2.5 \times 10^{-3}$$
$$= 452 \text{ N/m}^2 = 3.40 \text{ mm of Hg}$$

The frequency limit of the instrument is given by the fundamental frequency of bending vibrations of the diaphragm, which is given by Eq. (11.12).

Using Eq. (11.12), and putting $R = 0.075$ m, $E = 2.07 \times 10^{11}$ N/m², $\rho = 7.9 \times 10^3$ kg/m³

and $t = 1.68 \times 10^{-3}$ m, we get

$$\omega_n = 3644.3 \text{ rad/s} = 580 \text{ Hz}$$

11.4 CALIBRATION AND TESTING

For static calibration of pressure measuring devices, a dead weight tester is commonly used (Fig. 11.29). The pressure in the tester is built up till the weights are seen to float, when the fluid gauge pressure equals the dead weight divided by the piston area.

Calibration of pressure measuring instruments can also be checked against monometers. These can be used for moderate pressures. For low pressures, McLeod gauge can be used. Special methods have to be used for very low or very high pressures.

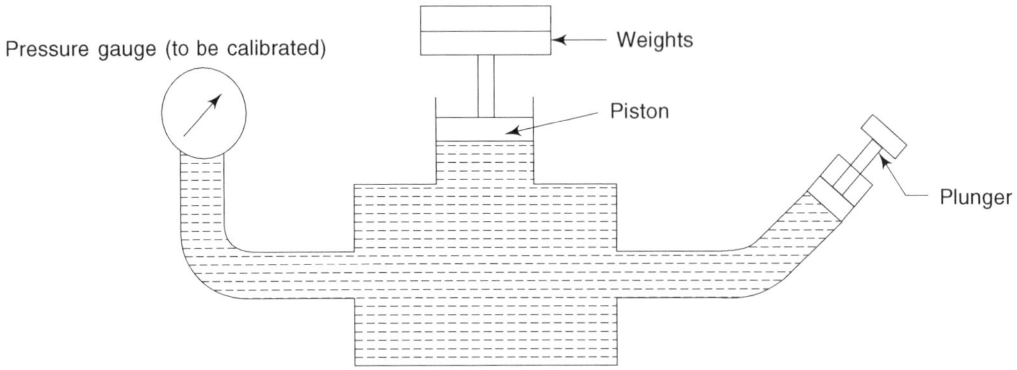

Fig. 11.29 *Dead weight tester*

Dynamic calibration of pressure measuring instruments is quite involved. In order to determine the usefulness of these instruments under dynamic conditions, special testers based on the frequency response or transient types of tests, have to be used. Figure 11.30 shows a harmonic steady state test apparatus. The shaker, which may be of electrodynamic or mechanical type, is used to create sinusoidal variations in pressure at desired frequencies. The reference transducer should have a flat frequency response over the test frequencies and the performance of the test transducer can be compared against the same.

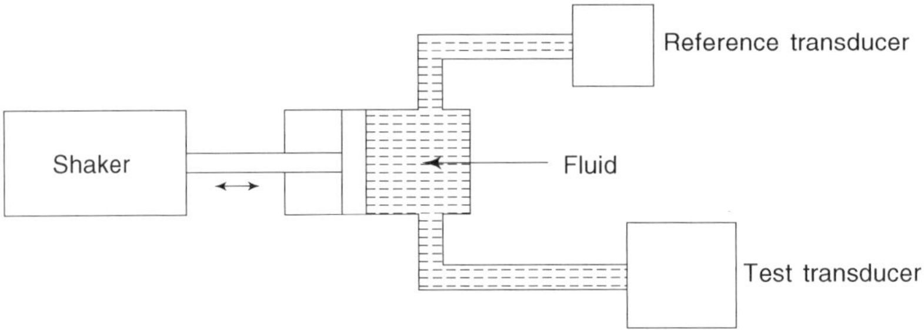

Fig. 11.30 *Frequency response tester*

Transient tests can be carried out using a shock tube or by using steel balls, dropped on to the elastic member of the pressure transducer. In a shock tube (Fig. 11.31) the bursting of a thin diaphragm, subjected to gas pressure, creates a nearly step pressure. The diaphragm is burst by either pressure differential or an external probe.

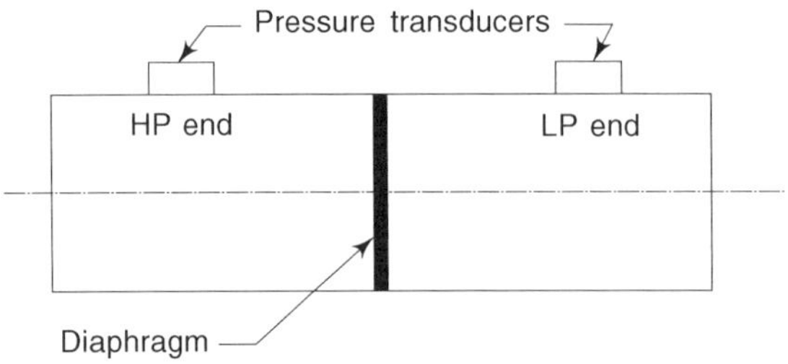

Fig. 11.31 *Shock tube*

11.5 SUMMARY

Table 11.1 gives the approximate range for various types of pressure measuring devices discussed in this chapter.

Table 11.1

Type of measuring device	*Pressure range*
Manometer	10 to 10^6 Pa
Bourdon gauge	10^3 to 5×10^8 Pa
Elastic diaphragm with LVDT or capacitance or resistance strain gauge transducer	100 to 10^8 Pa
Piezo-electric transducer	10^4 to 10^8 Pa
Hydrostatic compression gauge	100 to 10^5 atm
McLeod gauge	10^{-5} to 1 Torr
Pirani gauge	10^{-4} to 1 Torr
Ionisation gauge	10^{-12} to 10^{-3} Torr
Knudsen gauge	10^{-8} to 10^{-2} Torr

Review Questions

11.1 Indicate true or false against each of the following:

(a) Fluid pressure is independent of momentum exchange between the molecules of the fluid and a containing wall.

(b) Pressure readings can be negative if measurements are taken in 'gauge pressure' scale.

(c) In a U-tube manometer, the location of the pressure source does not affect the pressure measurements.

(d) In a well type manometer, the ratio of cross-sectional areas of the well and the tube is not the factor that determines the scale of the manometer.

(e) Accuracy of a manometer is affected by the shape or size of the tubes.

(f) In Bourdon pressure gauge, incorrect readings may be encountered due to hysteresis.

(g) For dynamic pressure measurement, the natural frequency of an elastic diaphragm should be lower than the frequency of the pressure signal.

(h) A piezo-electric pressure transducer can be used for measuring both static and dynamic pressures.

(i) A resistance strain gauge type pressure transducer cannot be used for dynamic pressure measurement.

(j) Calibration of the McLeod gauge can be directly related to the physical dimensions of the gauge, independent of nature of the gas being used.

(k) Vacuum levels lower than 1 micron can be measured with an ionisation gauge.

(l) Pressures higher than 10 atm are usually regarded as very high.

(m) Pressures of the order of 1 mmHg or below are usually regarded as very low.

(n) An inclined tube manometer is more accurate if the inclination of the tube with the vertical is high.

(o) Any metallic coil may be used as a sensor for measuring high pressure.

(p) In a bulk compression effect type of the high-pressure measurement device, the sensitivity increases with the increase in Young's modulus of the wire.

(q) Knudsen vacuum gauge is used for measuring pressure of the order of 1 Torr.

11.2 A mercury manometer has one arm in the shape of a well and the other as a tube inclined at 30° to the horizontal. The well is 4 cm in diameter and the tube 5 mm in diameter. Find the percentage error if no area correction factor is used.

11.3 For a McLeod gauge, with a capillary of 1 mm diameter and effective bulb volume of 80 cm^3, find the reading as indicated by mercury column due to a pressure of 10 Pa.

11.4 A variable capacitance transducer is shown in Fig. 11.32. The housing is made of steel. Find the pressure sensitivity (μF/F per Pa), fundamental natural frequency of vibrations of the diaphragm and maximum allowable pressure for which the transducer may be used.

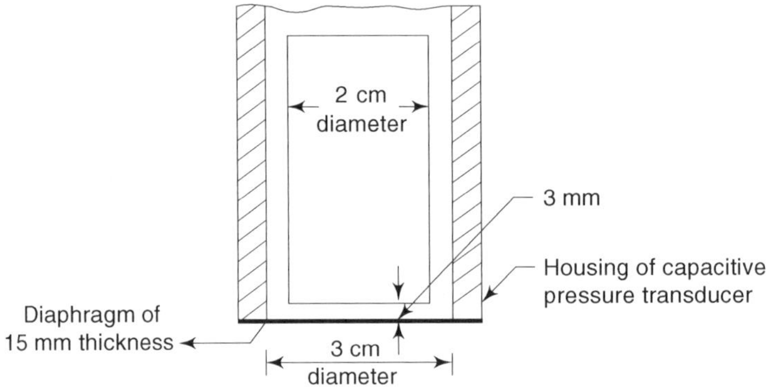

Fig. 11.32

11.5 A very high-pressure gauge, using a manganin element is to measure a maximum pressure of 10^8 Pa. The wire diameter is 25 μm, length is 3 cm. Pressure sensitivity of wire material is 2.5 × 10^{-11} Ω/Ω-Pa, resistively 45 × 10^{-6} Ω-cm. The wire forms one arm of a Wheatstone bridge, with resistances of all arms being equal. If the supply voltage is 12 V, find the output voltage due to maximum pressure.

11.6 A pressure tapping is taken from a measuring point by means of a tubing 50 cm long, 5 mm bore to a U-tube manometer using water as the working liquid. The tubes are of 10 mm bore. To start with, the water level is 30 cm below the top of the tube. Taking viscosity of air as 1.78 × 10^{-5} kg/m-s, find the time constant of the system if a small change occurs in the pressure which is around atmospheric.

11.7 Design a resistance strain gauge type pressure transducer, with the following specifications:
Maximum pressure, to be measured = 2 bar
Diameter of steel diaphragm = 8 cm
Size of resistance gauges used = 3 mm × 6 mm
Resistance of each gauge = 120 Ω
Gauge factor = 2
Find the thickness of the diaphragm, sensitivity of the transducer and other parameters of the transducer.

11.8 It is required to measure and plot the pressure in the cylinder of a single cylinder reciprocating air compressor, at several positions in a cycle. Suggest a suitable measuring system, and mention the type of sensor, signal conditioning and recording elements. Draw a sketch and suggest a calibration system.

Answers

11.1	(a) F	(b) T	(c) F	(d) F	(e) F
	(f) T	(g) F	(h) F	(i) F	(j) T
	(k) T	(l) F	(m) T	(n) T	(o) F
	(p) F	(q) F			

11.2 3.03%.

11.3 8.74 cm.

11.4 271 μF/F – Pa
1.04×10^5 Hz
4.04×10^7 Pa

11.5 7.5 mV

11.6 4.65×10^{-3} s

11.7 1.06 mm
1.134×10^{-10}/Pa

Chapter

12

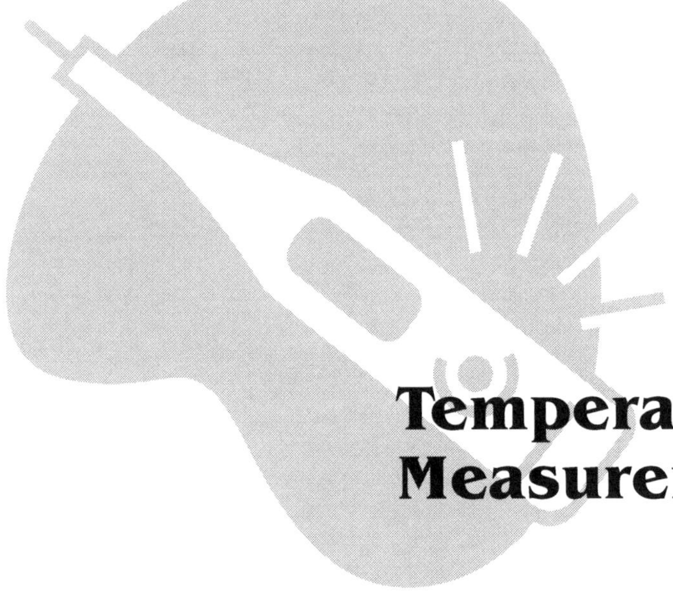

Temperature Measurement

INTRODUCTION

Temperature is a very widely measured and frequently controlled variable used in numerous industrial applications. In general, chemical reactions in the industrial processes and products are temperature dependent and the desired quality of a product is possible only if the temperature is accurately measured and maintained. Further, it forms an important governing parameter in the thermodynamic and heat transfer operations like steam raising, gas turbines in power generations and also in numerous propulsion systems. In addition, in the heat treatment of steel and aluminium alloys, temperature measurement and control plays a crucial role in incorporating the desired material properties in the finished heat-treated products. The other areas where measurement and control of temperature is essential are: plastic manufacture, nuclear reactor components, milk and dairy products, plant furnace and molten metals, heating and air-conditioning systems, space shuttle components, blades of gas turbines, etc.

A number of definitions of temperature have been proposed. In a layman's language one could define this as *the degree of hotness or coldness of a body or an environment measured on a definite scale.* Another simplified definition of temperature is based on its equivalence to a driving force or potential that caused the flow of energy as heat. Thus, we can define temperature as a *condition of a body by virtue of which heat is transferred to or from other bodies.* Further, in the kinetic theory of gases and in statistical thermodynamics, it is shown that temperature is related to the average kinetic energy of molecules or atoms of which the material is made of. Finally, the definition of temperature in themodynamic sense is based on the ideal Carnot cycle. According to this, it is defined as a quantity whose difference is proportional to the work obtained from a Carnot engine operating between a hot source and a cold receiver.

It may be noted that there is a marked difference between the quantities temperature and heat. Temperature may be defined as 'degree' of heat whereas heat is taken to mean as 'quantity' of heat. For example, a bucket of warm water would melt more ice than a small spoon of boiling water. The warm water in the bucket obviously contains greater quantity of heat than that in the spoon containing boiling water. But its temperature is lower than the boiling water, a fact that is readily apparent if a finger is dipped in both the vessels.

Temperature is a fundamental quantity, much the same way as mass, length and time. The law that is used in temperature measurement is known as the *Zeroth law of thermodynamics*. This states that if two bodies are in thermal equilibrium with a third body, then they are all in thermal equilibrium with each other. In other words, all the three bodies would have the same temperature. Thus, if one can set up a reproducible means of establishing a range of temperatures, unknown temperatures of other bodies may be compared with the standard by subjecting any type of *thermometer* successively to the standard and unknown temperature and allowing thermal equilibrium to reach in each case. In other words, the thermometer is calibrated against a standard and is subsequently used to read unknown temperatures.

12.1 ▌ TEMPERATURE SCALES

Two temperature scales in common use are the Fahrenheit and Celsius scales. These scales are based on a specification of the number of increments between freezing point and boiling point of water at the standard atmospheric temperature. The Celsius scale has 100 units between these points, while the Fahrenheit scale has 180 units. The Celsius scale is currently more in use because of the adoption of metric units. However, the absolute temperature scale based on the thermodynamic ideal Carnot cycle has been correlated with the Celsius and Fahrenheit scales as follows:

$$\text{K (Absolute temperature, Kelvin scale)} = {}^\circ\text{C} + 273.15 \tag{12.1}$$

where °C is temperature on Celsius scale.

$$\text{R (Absolute temperature, Rankine scale)} = {}^\circ\text{F} + 459.69 \tag{12.2}$$

where °F is the temperature on the Fahrenheit scale.

The zero points on both the scales represent the same physical state and the ratio of two values is the same, regardless of the absolute scale used, i.e.

$$\left[\frac{T_2}{T_1}\right]_{\text{Rankine}} = \left[\frac{T_2}{T_1}\right]_{\text{Kelvin}} \tag{12.3}$$

The boiling and freezing points of water at a pressure of one atmosphere (101.3 kN/m^2) are taken as 100° and 0° on the Celsius scale and 212° and 32° on the Fahrenheit scale. The following relationships between Fahrenheit and Celsius and Rankine and Kelvin scales can be easily derived:

$${}^\circ\text{F} = 32 + \frac{9}{5}\ {}^\circ\text{C} \tag{12.4}$$

$$\text{R} = \frac{9}{5}\text{K} \tag{12.5}$$

With SI units, the kelvin temperature scale (which is also termed as absolute temperature scale or 'thermodynamic' temperature scale) is used in which the unit of temperature is the kelvin (K). It may be noted that the degree symbol (°) is not used in this scale.

12.2 ▌ INTERNATIONAL PRACTICAL TEMPERATURE SCALE (IPTS)

To enable the accurate calibration of a wide range of temperatures in terms of the Kelvin scale, the International Practical Temperature Scale (IPTS-68) has been devised. This lists 11 primary 'fixed' points which can be reproduced accurately. Some typical values are

Table 12.1 *Typical Values of Primary 'Fixed' Points*

Primary 'fixed' points	*Temperature (K)*	*Temperature (°C)*
1. Triple point of equilibrium hydrogen (equilibrium between solid, liquid, and vapour phases of equilibrium hydrogen)	13.18	–259.34
2. Boiling point of equilibrium hydrogen	20.28	–252.87
3. Triple point of oxygen	54.361	–218.789
4. Boiling point of oxygen	90.188	–182.962
5. Triple point of water (equilibrium between solid, liquid and vapour phases of water)	273.16	0.01
6. Boiling point of water	373.15	100
7. Freezing point of zinc	692.73	419.58
8. Freezing point of silver	1235.58	961.93
9. Freezing point of gold	1337.58	1064.43

In order to establish a scale completely, a means of interpolation between the fixed points is required. Resistance thermometers are used for this purpose for temperatures below 630°C and thermocouples between 630 and 1064°C and above this temperature range, radiation methods are used to extrapolate the scale.

Apart from the primary standard points, there are 31 secondary points on the International Practical Temperature Scale which forms the convenient working standard for the workshop calibration of the temperature measuring devices. Some typical values of these points are given in Table 12.2.

Table 12.2 *Typical Values of Secondary Points*

Secondary points	*Temperature (K)*	*Temperature (°C)*
1. Sublimation point of carbon dioxide	194.674	–74.476
2. Freezing point of mercury	234.288	–38.862
3. Equilibrium between ice and water (ice point)	273.15	0
4. Melting point of bismuth	544.592	271.442
5. Melting point of lead	600.652	327.502
6. Boiling point of pure sulphur	717.824	444.674
7. Melting point of antimony	903.87	630.74
8. Melting point of aluminium	933.52	660.37
9. Melting point of copper	1357.6	1084.5
10. Melting point of platinum	2045	1772
11. Melting point of tungsten	3660	3387

12.3 ■ MEASUREMENT OF TEMPERATURE

Temperature cannot be measured directly but must be measured by observing the effect that temperature variation causes on the measuring device. Temperature measurement methods can be broadly classified as follows:

1. non-electrical methods,
2. electrical methods, and
3. radiation methods.

12.4 ■ NON-ELECTRICAL METHODS

The non-electrical methods of temperature measurement can be based on any one of the following principles:

1. change in the physical state,
2. change in the chemical properties, and
3. change in the physical properties.

As mentioned before, the temperatures at which a number of pure substances change their physical states are used for the calibration of temperature scales. But these devices give one particular value of a unique temperature corresponding to the change in the state of the substance. Hence, these are not suitable for measuring a range of temperatures. Further, for using the change in chemical properties, we have to look for a reversible chemical process in order to obtain a repeatable/reproducible scale with respect to change in temperature. However, such processes are very few in nature and thus this principle of temperature measurement is hardly used. In fact, most non-electrical methods of temperature measurement are based on the change in a physical property, namely the simple thermal expansion due to the change in temperature. Since these devices have no electrical connections, therefore they are commonly used in areas where there is a risk of explosion, e.g., to provide display of temperature in petrol-storage tanks, etc. Further, although the magnitude of expansion in such thermometers is small, yet they produce enough power for direct operation of the mechanical indicating devices.

12.4.1 Bimetallic Thermometer

This type of thermometer also employs the principle of solid expansion and consists of a 'bimetal' strip usually in the form of a cantilever beam [Fig.12.1(a)]. This comprises strips of two metals, having different coefficients of thermal expansion, welded or riveted together so that relative motion between them is prevented. An increase in temperature causes the deflection of the free end of the strip as shown in Fig. 12.1(b), assuming that metal *A* has the higher coefficient of expansion. The deflection with the temperature is nearly linear, depending mainly on the coefficient of linear thermal expansion. Invar is commonly employed as the low expansion metal. This is an iron-nickel alloy containing 36% nickel. Its coefficient of thermal expansion is around 1/20th of the ordinary metals. Brass is used as high expansion material for the measurement of low temperatures, whereas nickel alloys are used when higher temperatures have to be measured. A plain bimetallic strip is somewhat insensitive, but the sensitivity is improved by using a longer strip in a helical form as shown in Fig. 12.2. One end of the helix is anchored to the casing and the other end which is free is conveniently connected to the pointer which sweeps over a circular dial graduated in degrees of temperature. In response to the temperature change, the bimetal expands and the helical bimetal rotates at its free end, thus turning the stem and the pointer to a new position on the dial. The length of the stem may be up to about 0.6 m, allowing the bimetal element (enclosed in a protective sheath) to be submerged in a hot substance without the indicator itself being subjected to excessive temperatures.

Bimetallic thermometers are usually employed in the range of –30 to 550°C. Inaccuracies of the order of ± 0.5 to ± 1.0% of full-scale deflection are expected in bimetallic thermometers of high accuracies. In addition to temperature indication, bimetal elements find a wide range of applications in the combined sensing and control elements in temperature control systems, mainly of the on–off type (thermostats). Movement of the strip has sufficient force to actuate the control switches employed in domestic ovens, electric irons, car winker lamps and the refrigerators. In addition, these devices are also used as compensating elements for the ambient temperatures in the pressure thermometers, aneroid barometers and as balance wheel compensators in some watches.

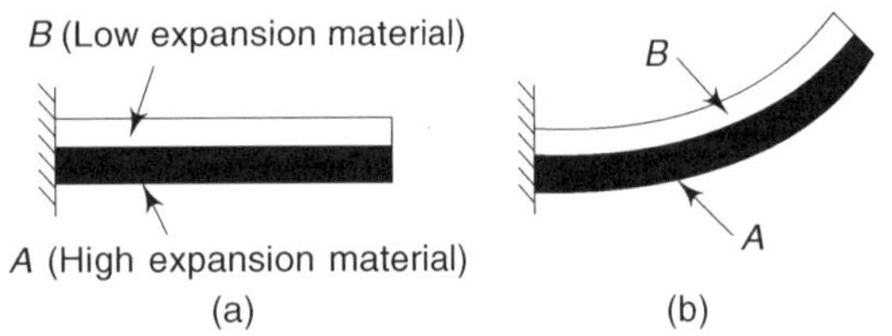

Fig. 12.1 *Bimetallic thermometer*

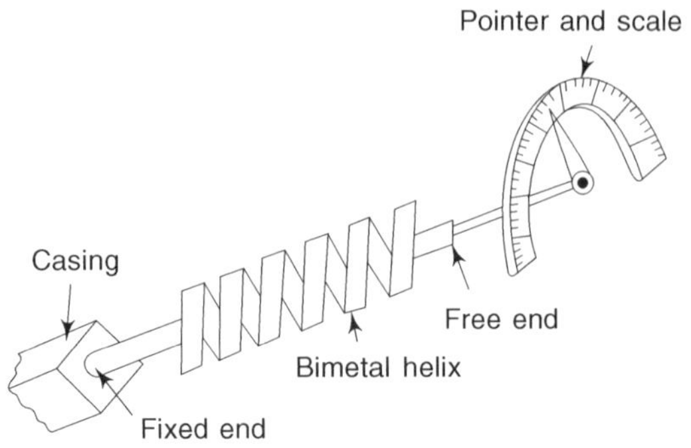

Fig. 12.2 *Bimetal helix thermometer*

The bimetallic strip has the *advantage* of being self-generating type instrument with low cost practically no maintenance expenses and stable operation over extended period of time. However, its main *disadvantage* is its inability to measure rapidly changing temperatures due to its relatively higher thermal inertia.

12.4.2 Liquid-in-Glass Thermometer

The *liquid-in-glass thermometer* is one of the most common temperature measuring devices. Both liquid and glass expand on heating and their differential expansion is used to indicate the temperature. The lower temperature limit is –37.8°C for mercury, down to –130°C for pentane. The higher temperature range is 340°C (boiling point of mercury is 357 °C) but this range may be extended to 560°C by filling the space above mercury with CO_2 or N_2 at high pressure, thereby increasing its boiling point and range. The precision of the thermometer depends on the care used in calibration. A typical instrument is checked and marked from two to five reference temperatures. Intermediate points are marked by interpolation. The calibration of the thermometer should be occasionally checked against the ice point to take into account the aging effects. Precision thermometers are sometimes marked for partial or total immersion and also for horizontal or vertical orientation. The accuracy of these thermometers does not exceed 0.1°C. However, when increased accuracy is required, a Beckmann range thermometer can be used. It contains a big bulb attached to a very fine capillary. The range of the thermometer is limited to 5 – 6°C with an accuracy of 0.005°C.

Liquid-in-glass thermometers have *notable qualities* like low cost, simplicity in use, portability and convenient visual indication without the use of any external power. However, their use is limited to certain laboratory applications. It is not preferred in industrial applications because of its fragility and its lack of adaptability to remote indication. Further, it introduces time lag in the measurement of dynamic signals because of relatively high heat capacity of the bulb.

12.4.3 Pressure Thermometers

Pressure thermometer is based on the principle of fluid expansion due to an increase in the pressure in a given volume of the temperature measuring system. It is one of the most economical, versatile and widely used devices in industrial temperature measurements. It has a relatively large metal bulb (often stainless steel) instead of glass. This results in a robust, easy-to-read thermometer that may be read remotely by connecting the bulb to a Bourdon gauge or any other pressure measuring device by means of a capillary tube as illustrated in Fig. 12.3.

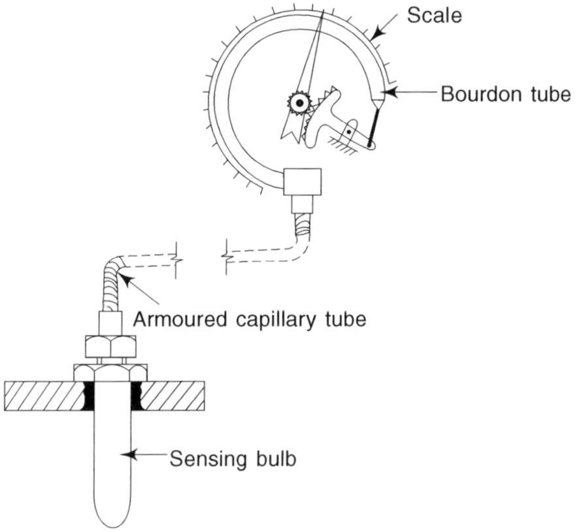

Fig. 12.3 *A schematic diagram of pressure thermometer*

The entire assembly of the bulb, capillary and gauge is calibrated directly on the basis of pressure change corresponding to the temperature change. The bulb of the thermometer may be filled with either a liquid (usually mercury) or gas or a liquid-vapour mixture and depending upon the type of fluid, the thermometer is termed as mercury-in-steel thermometer or constant volume gas thermometer or vapour pressure thermometer respectively.

Fluid expansion thermometers are low in cost, self-operated type, rugged in construction, with no maintenance expenses, stable in operation and accurate to $\pm 1\,^\circ$C. Further, the response of these instruments can be increased by using a small bulb connected to an electrical type of pressure sensor connected through a short length of capillary tube.

Mercury-in-steel Thermometer The mercury-in-steel thermometer has a near-linear scale. Sufficient power is available to operate a recording pen if required. As the total expansion of mercury is dependent not only on the bulb temperature but also on the temperatures of the capillary tube and Bourdon tube, the system is subject to ambient temperature errors. Further, the magnitude of these errors depends on the ratio of the volume of mercury in the capillary at the measured temperature to the volume of the bulb. A reduction of the error is obtained by making the sensing-bulb volume considerably greater than that of the capillary and Bourdon tubes. This inevitably increases the thermal capacity and hence the thermal lag. The bulb size is therefore a matter of compromise. The volumes of the capillary and Bourdon tube are made as small as the transmission distance and the required size of display could allow. Pressure thermometers are often connected with long capillary (connecting) tubes for remote measurements, of the order of 100 m.

Ambient temperature error may be reduced by suitable compensation techniques. One arrangement uses a second Bourdon tube and capillary, so arranged that the effect of the ambient temperature change on this compensating system cancels the corresponding effects on the measuring system.

Another potential cause of error is the change in pressure head which is introduced by any change in the relative levels of the bulb and the display. If the bulb is raised by a height h from the calibration elevation, then the Bourdon gauge experiences an increase in pressure equal to ρgh. This increase in level introduces an error in the indicator. This error is constant for any specified relative position of the bulb and display and may the removed by means of the zero adjustment of the indicating mechanism.

The temperature range over which mercury-in-steel thermometer may be used is –25 to 550°C when the mercury is filled under pressure in the steel bulb.

Constant Volume Thermometer The constant volume thermometer uses an inert gas (usually nitrogen) in place of mercury and the principle of its working is the increase is pressure of the gas with increase in temperature at constant volume. However, the volume of the system, i.e., that of bulb, capillary and the Bourdon tube, does not remain constant and increases slightly due to the increase in pressure and in addition the volume of the bulb also increases due to increase in temperature.

Gas filled systems operate over a range –130 to 540°C with linear ranges as large as 500°C. However, its disadvantage over the liquid filled system is that the pressure developed for a given temperature change is smaller and further ambient temperature compensation is more difficult. The accuracy of these instruments is of the order of ±1% at lower ranges, i.e. up to 300°C and ±2% above this range.

Vapour Pressure Thermometer The system in the vapour pressure thermometer is filled partly with liquid and partly with vapour of the same liquid so that there is a liquid-vapour interface in the bulb. The liquid-vapour system does not have any error as long as a free liquid surface exists in the sensing bulb. This is because such a system follows one of the Dalton's laws of partial pressure which states that if both liquid and vapour are present, there is only one saturation pressure corresponding to a given temperature. The general usefulness of the vapour pressure thermometers is restricted due to the limited number of liquids providing suitable saturation vapour pressure ranges. These include mostly hydrocarbon type of fluids like ethane, ethyl alcohol, ethyl chloride, methyl chloride, chlorobenzene, toluene, pentane, ether, acetone, etc. The scale range is usually of the order of 100°C and accuracy is up to ±1% of the differential range. Further, the temperature is roughly a logarithmic function of the temperature ($\log p = a - b/T$) and therefore the scale of the vapour pressure thermometers is noticeably non-linear.

Problem 12.1 *A mercury-in-steel thermometer employs a Bourdon pressure gauge which has a range of 0– 6.0 MPa for the pointer rotation from 0 to 270°C. In the temperature calibration process, the pointer movement was set to 0° rotation at 0°C and the instrument indicated 250° rotation corresponding to 200°C. Determine:*

(a) the sensitivity of the instrument in rad/°C,

(b) the error due to ambient temperature rise of 16 °C if the thermometer bulb has 8 times that of combined volume of capillary and the Bourdon tube, and

(c) the error in the observed temperature values if the bulb is raised by 60 cm from calibration elevation.

Solution

(a) Sensitivity of the mercury-in-steel thermometer

$$= \frac{\text{total angular displacement in radians}}{\text{temperature range}}$$

$$= \frac{250 \times \pi / 180}{200} = 2.18 \times 10^{-2} \text{ rad/°C}$$

(b) The volume change in the capillary plus the Bourdon tube caused by ambient temperature change introduces an error in the readings. This error is calculated as equivalent to the temperature rise in the bulb of the thermometer that causes the same volume change as before. Since the volume change in each case depends on the initial volume, temperature change and coefficient of volumetric expansion, we can write,

$$\gamma(8v)\,(\Delta T)_{\text{error}} = \gamma(v)\,(\Delta T)_{\text{ambient temp}}$$

where γ is the coefficient of volumetric expansion and v the combined volume of the capillary and Bourdon tube.

$$\therefore \qquad (\Delta T)_{error} = \frac{16}{8} = 2°C$$

Thus the error due to ambient temperature rise is 2°C or alternatively the correction to be applied to the observed readings would be –2°C.

(c) The pressure of 60 cm column of Hg = $\rho_{Hg}\,gh$

$$= (13.6 \times 10^3)\,(9.81) \times \frac{60}{100} = 80.05 \text{ kPa}$$

Angular movement in radius caused by 80.05 kPa pressure in the pressure gauge

$$= \frac{270 \times \pi / 180 \times 80.05 \times 10^3}{6 \times 10^6} = 0.06287 \text{ rad}$$

$\therefore$ Error due to elevation effect in the pressure thermometer

$$= \frac{\text{angular movement due to elevation}}{\text{sensitivity of thermometer}}$$

$$= \frac{0.06287}{2.18 \times 10^{-2}} = 2.88°C$$

12.5 ■ ELECTRICAL METHODS

Electrical methods are in general preferred for the measurement of temperature as they furnish a signal which can be easily detected, amplified or used for control purposes. There are two main electrical methods used for measuring temperature. They are:

1. Thermo-resistive type i.e., variable resistance transducers and
2. Thermo-electric type i.e., emf generating transducers.

12.5.1 Electrical Resistance Thermometers

In resistance thermometers, the change in resistance of various materials, which varies in a reproducible manner with temperature, forms the basis of this important sensing technique. The materials in actual use fall in two classes namely, conductors (metals) and semiconductors. In general, the resistance of the highly conducting materials (metals) increases with increase in temperature and the coils of such materials are called *metallic resistance thermometers*. Whereas the resistance of semiconductor materials generally (not always) decreases with increase in temperature. Thermo-sensitive resistors having such negative temperature characteristics are commonly known as NTC thermistors. Figure 12.4 illustrates the typical variation of specific resistance of the metals (platinum for example) and the NTC thermistor.

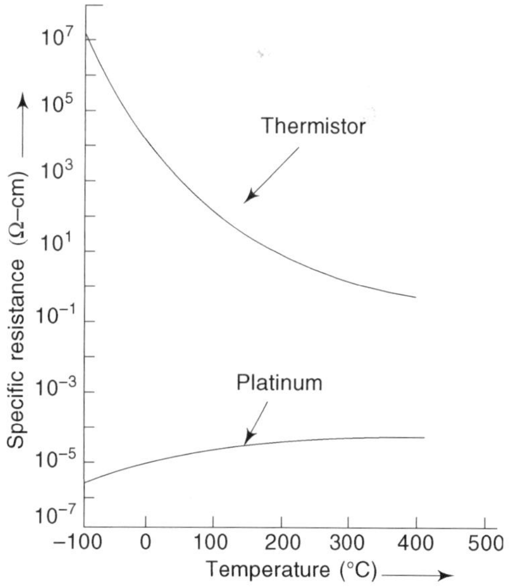

Fig. 12.4 *Resistance–temperature characteristics of platinum and a typical NTC thermistor*

Metallic Resistance Thermometers or Resistance—Temperature Detectors (RTDs) Metals such as platinum copper, tungsten and nickel exhibit small increases in resistance as the temperature rises because they have a positive temperature coefficient of resistance. Platinum is a very widely used sensor and its operating range is from 4K to 1064°C. Because it provides extremely reproducible output, it is used in establishing International Practical Temperature Scale from 13.81 K to 961.93°C. However for the measurement of lower temperatures up to 600°C, RTD sensor is made of nickel. Further, for ranges of temperature below 300°C, the sensing element is fabricated using pure copper wire. Metallic resistance thermometers are very suitable for both laboratory and industrial applications because of their high degree of accuracy as well as long-term stability. In addition, they have a wide operating range and have linear characteristics throughout the operating range. However, the limitations of the RTDs are low sensitivity, relatively higher cost as compared to other temperature sensors and their proneness to errors caused due to contact resistance, shock and accelerations.

Metallic resistance thermometers are constructed in many forms, but the temperature sensitive element is usually in the form of a coil of fine wire supported in a stress-free manner. A typical construction is shown in Fig. 12.5, where the wire of metal is wound on the grooved hollow insulating ceramic former and covered with protective cement. The ends of the coils are welded to stiff copper leads that are taken out to be connected in one of the arms of the Wheatstone bridge circuit. In some cases, this arrangement can be used directly in the medium whose temperature is being measured, thus giving a fast speed of response. However, in most applications, a protective metal sheath is used to provide rigidity and mechanical strength. Alternatively, RTD sensor's may be fabricated by depositing the thin films of platinum, nickel or copper on a ceramic substrate. These thin film sensors have the advantage of extremely low mass and consequently more rapid thermal response.

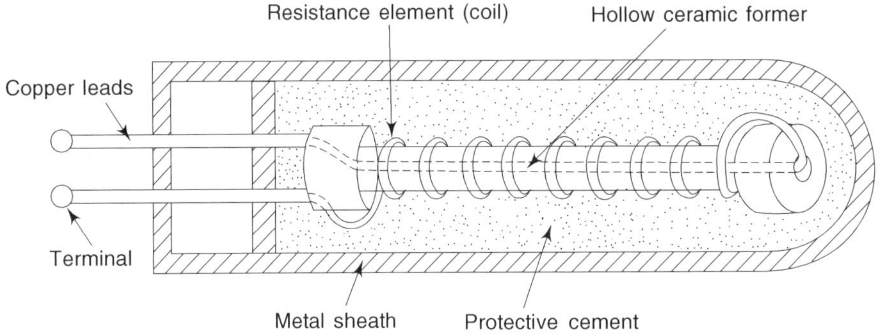

Fig. 12.5 *Construction of a platinum resistance thermometer (PRT)*

Platinum, in spite of its low sensitivity and high cost as compared to nickel and copper, is the most widely used material for metallic resistance element. This is because of the following:

1. The temperature–resistance characteristics of pure platinum are well defined and stable over a wide range of temperatures.
2. It has high resistance to chemical attack and contamination ensuring long-term stability.
3. It forms the most easily reproducible type of temperature transducer with a high degree of accuracy. The accuracy attainable with PRT is ±0.01°C up to 500°C and ±0.1°C up to 1200°C.

In general, the resistance relationship of most metals over a wide range of temperatures is given by the quadratic relationship:

$$R = R_0 [1 + aT + bT^2] \tag{12.6}$$

where R = resistance at absolute temperature T

R_0 = resistance at 0°C

a and b = experimentally determined constants.

However, over a limited temperature range around 0°C (273 K), the following linear relationship can be applied:

$$R_t = R_0 (1 + \alpha t) \tag{12.7}$$

where α = the temperature coefficient of resistance of material in (Ω/Ω)/°C or °C^{-1}.

R_0 = resistance at 0°C

t = temperature relative to 0°C

Some typical values of α are

copper = 0.0043°C^{-1}

nickel = 0.0068°C^{-1}

platinum = 0.0039°C^{-1}

If a change in temperature from t_1 to t_2 is considered, Eq. (12.7) becomes:

$$R_2 = R_1 + R_0 \alpha (t_2 - t_1)$$

Rearranging gives

$$t_2 = t_1 + \frac{R_2 - R_1}{\alpha R_0} \tag{12.8}$$

The variation of resistance of the sensing element is normally measured using some form of electrical bridge circuit which may employ either the deflection mode of operation or the null (manually or automatically balanced) mode. However, particular attention must be given to the manner in which the thermometer is connected into the bridge. Leads of same length appropriate to the situation are normally required and any resistance change therein due to any cause, including temperature, may be credited to the thermometer element. It is desirable, therefore, that the lead resistance be kept as low as possible relative to the element resistance. In addition, some modifications may be employed for providing the lead compensation. For more precise results, either the Siemen's three wire lead arrangement or Callender's four wire lead arrangement may be employed (Fig. 12.6). Further, it is essential that the thermo-electric emf's do not affect the system. These can be eliminated by utilising ac excitation or by manually varying the polarity of the dc supply.

Problem 12.2 *A platinum resistance thermometer has a resistance of 140.5 and 100.0 Ω at 100 and 0°C, respectively. If its resistance becomes 305.3 Ω when it is in contact with a hot gas, determine the temperature of the gas. The temperature coefficient of platinum is 0.0039°C^{-1}.*

Solution The temperature-resistance relationship of platinum is given by:

$$R_2 = R_1 + R_0 \alpha (t_2 - t_1)$$

$$\therefore \quad t_2 = 100 + \frac{305.3 - 140.5}{0.0039 \times 100}$$

$$= 100 + 422.56 = 522.56°\text{C}$$

Semiconductor Resistance Sensors (Thermistors) Thermistor (shortened form of the words: thermal resistor) is a thermally sensitive variable resistor made of ceramic-like semiconducting materials. They are available in a greater variety of shapes and sizes having cold resistance ranging from a few ohms to mega ohms. The size can range from extremely small bead, thin disc, thin chip or wafer to a large sized

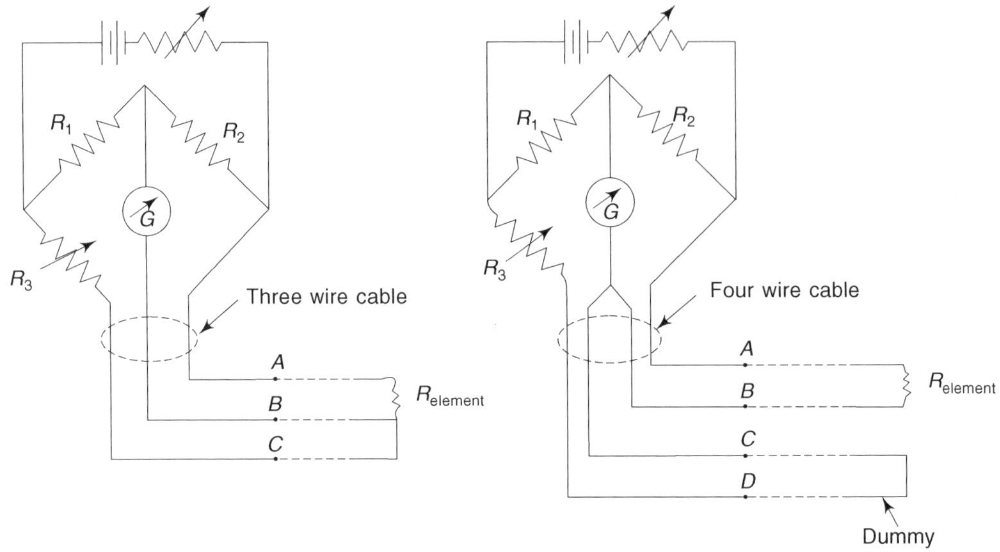

(a) Siemen's three wire lead arrangement (b) Callender's four wire lead arrangement

Fig. 12.6 *Cable compensation arrangements for platinum resistance thermometer*

rod as illustrated in Fig. 12.7. Unlike metals, thermistors respond negatively to temperature and their coefficient of resistance is of the order of 10 times higher than that of platinum or copper.

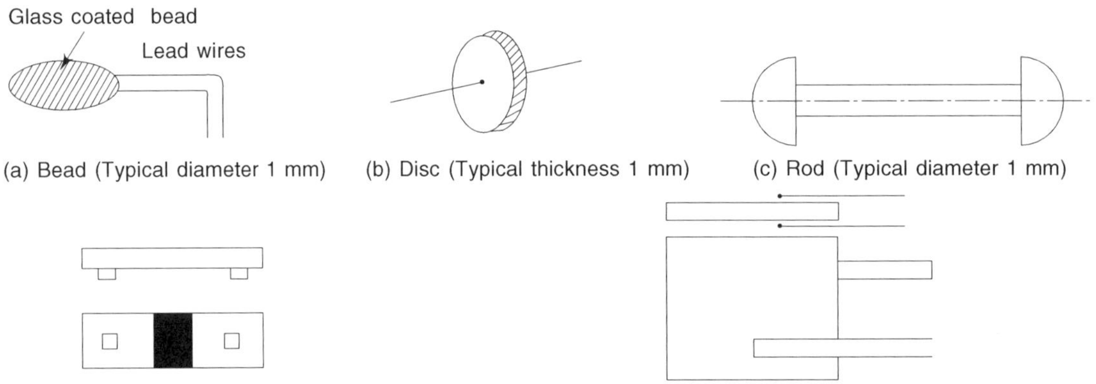

(a) Bead (Typical diameter 1 mm) (b) Disc (Typical thickness 1 mm) (c) Rod (Typical diameter 1 mm)

(d) I.C. Chip (Typical dimension 1.25 mm × 100 mm × 0.3 mm thickness) (e) Wafer (Typical dimension 1.25 mm to 15 mm × 0.3 mm thickness)

Fig. 12.7 *Range of thermistor forms*

Thermistors are fabricated from the semiconducting materials which include the oxides of copper, manganese, nickel, cobalt, lithium and titanium. These oxides are blended in a suitable proportion and compressed into desired shapes from powders and heat treated to recrystallise them, resulting in a dense ceramic body with the required resistance–temperature characteristics.

Thermistors have the following advantages for temperature measurements:

1. a large temperature coefficient which makes the thermistor an extremely sensitive device, thus enabling accuracy of measurement up ± 0.01°C with proper calibration,

2. ability to withstand electrical and mechanical stresses,
3. fairly good operating range which lies between –100 and 300°C,
4. fairly low cost and easy adaptability to the available resistance bridge circuits, and
5. the high sensitivity and the availability in extremely small sizes (of the size of a pin head) enable a fast speed of thermal response.

Thus, these devices are extremely useful for dynamic temperature measurements.

However, the *disadvantages* are a highly non-linear resistance–temperature characteristics and problems of self-heating effects which necessitate the use of much lower current levels than those with metallic sensors.

The temperature-resistance characteristics of a thermistor is of exponential type and is given by:

$$R = R_0 \exp\left[\beta\left(\frac{1}{T} - \frac{1}{T_0}\right)\right] \tag{12.9}$$

where R_0 is the resistance at the reference temperature T_0 (kelvin)
R is the resistance at the measured temperature T (kelvin)
β is the experimentally determined constant for the given thermistor material.

The values of β usually lie between 3000 and 4400 K depending on the formulation or grade.

Using Eq. (12.9) we can obtain the temperature coefficient of resistance as:

$$\frac{dR/dT}{R} = -\frac{\beta}{T^2} \tag{12.10}$$

Assuming $\beta = 4000$ K and $T = 298$ K

we get, $$\frac{dR/dT}{R} = -0.045 \text{ K}^{-1}$$

The value of $\frac{dR/dT}{R}$ for platinum is 0.0039 K^{-1}, indicating that the thermistor is at least 10 times more sensitive than the platinum resistance element.

Problem 12.3 *For a certain thermistor, $\beta = 3140$ K and the resistance at 27°C is known to be 1050 Ω. The thermistor is used for temperature measurement and the resistance measured is as 2330 Ω. Find the measured temperature.*

Solution The governing equation of the temperature-resistance characteristics of the thermistor is given by

$$R = R_0 \exp\left[\beta\left(\frac{1}{T} - \frac{1}{T_0}\right)\right]$$

The given data is

$$R_0 = 1050\ \Omega$$
$$T_0 = 273 + 27 = 300 \text{ K}$$
$$\beta = 3140 \text{ K}$$
$$R = 2330\ \Omega$$

Taking the logarithm of both sides of equation and rearranging we get,

$$\frac{1}{T} = \frac{\ln R - \ln R_0}{\beta} + \frac{1}{T_0}$$

$$= \frac{7.754 - 6.957}{3140} + \frac{1}{300} = 3.587 \times 10^{-3}$$

$$T = 278.77 \text{ K}$$
$$= 5.77°\text{C}$$

12.5.2 Thermo-electric Sensors

The most common electrical method of temperature measurement uses the thermo-electric sensor, also known as the *thermocouple* (TC). The thermocouple is a temperature transducer that develops an emf which is a function of the temperature between hot junction and cold junction. The construction of a thermocouple is quite simple. It consists of two wires of different metals twisted and brazed or welded together with each wire covered with insulation which may be either.

1. mineral (magnesium oxide) insulation for normal duty, or
2. ceramic insulation for heavy duty.

The basic principle of temperature measurement using a thermo-electric sensor was discovered by Seebeck in 1821 and is illustrated in Fig. 12.8. When two conductors of dissimilar metals, say A and B, are joined together to form a loop (thermocouple) and two unequal temperatures T_1 and T_2 are interposed at two junctions J_1 and J_2, respectively, then an infinite resistance voltmeter detects the electromotive force E, or if a low resistance ammeter is connected, a current flow I is measured. Experimentally, it has been found that the magnitude of E depends upon the materials as well as the temperature T_1 and T_2. Now, the overall relation between emf E and the temperatures T_1 and T_2 forms the basis for themo-electric measurements and is called the *Seebeck effect.* Thus, in practical applications, a suitable device is incorporated to indicate the emf E or the flow of current I. For convenience of measurements and standardisation, one of the two junctions is usually maintained at some known temperature. The measured emf E then indicates the temperature difference relative to the reference temperature, such as ice point which is very commonly used in practice.

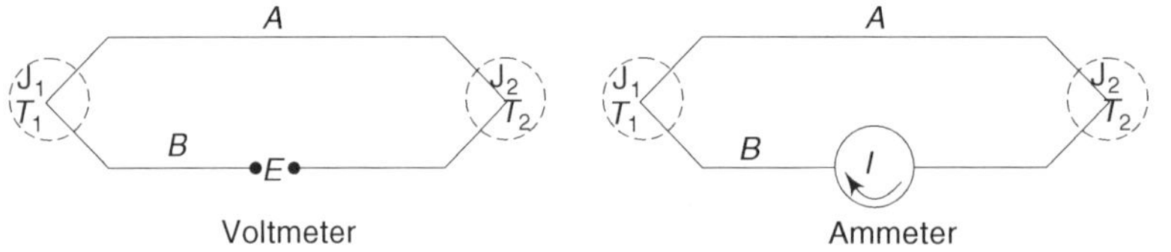

Fig. 12.8 *Basic thermo-electric circuit*

It may be noted that temperatures T_1 and T_2 of junctions J_1 and J_2 respectively are slightly altered if the thermo-electric current is allowed to flow in the circuit. Heat is generated at the cold junction and is absorbed from the hot junction thereby heating the cold junction slightly and cooling the hot junction slightly. This phenomenon is termed *Peltier effect.* If the thermocouple voltage is measured by means of potentiometer, no current flows and *Peltier* heating and cooling are not present. Further, these heating and cooling effects are proportional to the current and are fortunately quite negligible in a thermocouple circuit which is practically a millivolt range circuit. In addition, the junction emf may be slightly altered if a temperature gradient exists along either or both the materials. This is known as *Thomson effect.*

Again, the Thomson effect may be neglected in practical thermo-electric circuits and potentiometric voltage measurements are not susceptible to this error as there is no current flow in the circuit.

The actual application of thermocouples to the measurements requires consideration of the laws of thermo-electricity.

Law of Intermediate Temperatures This states that the emf generated in a thermocouple with junctions at temperatures T_1 and T_3 is equal to the sum of the emf's generated by similar themocouples, one acting between temperatures T_1 and T_2 and the other between T_2 and T_3 when T_2 lies between T_1 and T_2 (Fig. 12.9).

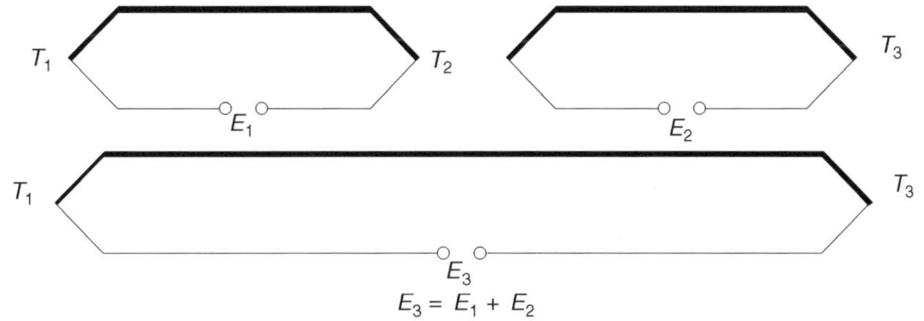

Fig. 12.9 *Law of intermediate temperatures*

This law is useful in practice because it helps in giving a suitable correction in case a reference junction temperature (which is usually an ice bath at 0°C) other than 0°C is employed. For example, if a themocouple is calibrated for a reference junction temperature of 0°C and used with junction temperature of say 20°C, then the correction required for the observation would be the emf produced by the thermocouple between 0 and 20°C.

Law of Intermediate Metals The basic thermocouple loop consists of two dissimilar metals *A* and *B* [Fig. 12.10 (a)]. If a third wire is introduced, then three junctions are formed as shown in Fig. 12.10(b). The emf generated remains unaltered if the two new junctions *B–C* and *C–A* are at the same temperature.

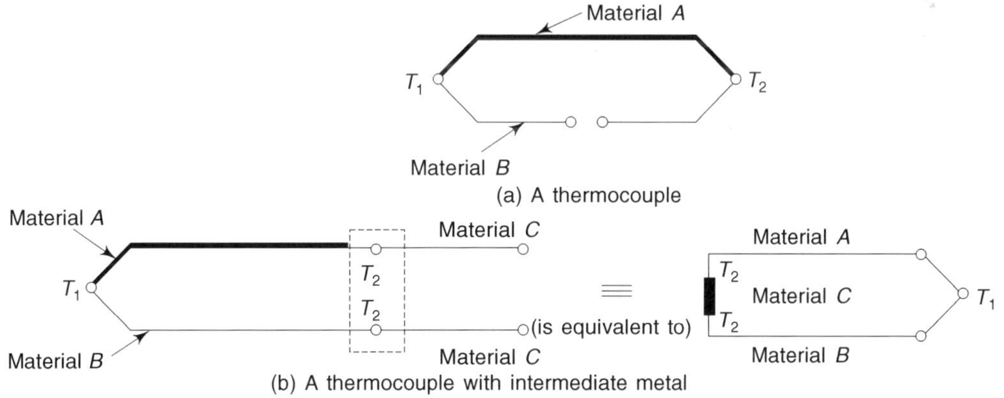

Fig. 12.10 *Law of intermediate metals*

It may be noted that extension wires are needed when the measuring instrument is to be placed at a considerable distance from the reference junction. Maximum accuracy is obtained when the leads are of the same material as the themocouple element [Fig. 12.11(a)]. However, this approach is not economical while using expensive themocouple materials. Further, a small inaccuracy is still possible if the binding post of the instrument is made of say copper and the two binding posts are at different temperatures. Therefore, it is preferable to employ the system shown in Fig. 12.11(b) to keep the copper-iron and copper-constantan junctions in the thermos flask at 0°C and provide binding posts of copper. This ensures maximum accuracy in the thermocouple operation.

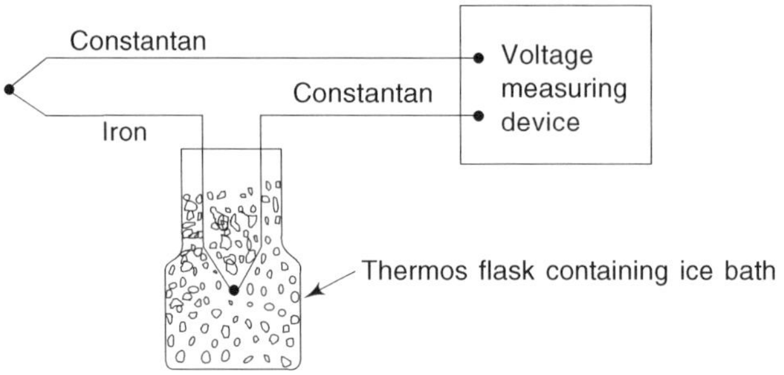

(a) A thermocouple without extension leads

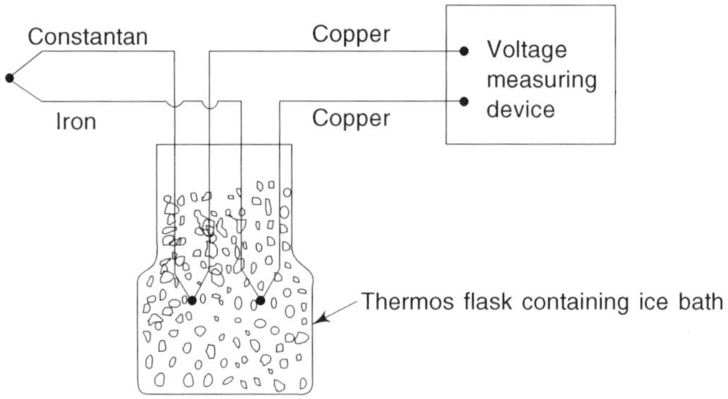

(b) Conventional method of establishing reference function temperature with copper extension leads

Fig. 12.11 *Schematics of Thermocouple circuits with and without extension leads in a typical iron–constantan thermocouple circuit*

Thermocouple Materials The choice of materials for thermocouples in governed by the following factors:

1. ability to withstand the temperature at which they are used,
2. immunity from contamination/oxidation, etc. which ensures maintenance of the precise thermo-electric properties with continuous use, and
3. linearity characteristics.

It may be noted that the relationship between thermo-electric emf and the difference between hot and cold junction temperatures is approximately of the parabolic form:

$$E = aT + bT^2 \tag{12.11}$$

Thermocouple can be broadly classified in two categories:

1. base-metal thermocouples, and
2. rare-metal thermocouple.

Base-metal thermocouples use the combination of pure metals and alloys of iron, copper and nickel and are used for temperature up to 1450 K. These are most commonly used in practice as they are more sensitive, cheaper and have nearly linear characteristics. Their chief limitation is the lower operating range because of their low melting point and vulnerability to oxidation. On the other hand, rare-metal thermocouples use a combination of pure metals and alloys of platinum for temperatures up to 1600°C and tungsten, rhodium and molybdenum for temperatures up to 3000°C.

Typical thermocouples with their temperature ranges and other salient operating characteristics, are given in Table 12.3.

Table 12.3 *Characteristics of Some Thermocouples*

S.No.	*Type*	*Thermocouples material**	*Approximate sensitivity in (μV/°C)*	*Useful temperature range (°C)*	*Approximate accuracy (%)*
1.	T	Copper–Constantan	20 – 60	–180 to +400	± 0.75
2.	J	Iron–Constantan	45 – 55	–180 to +850	± 0.75
3.	K	Chromel–Alumel	40 – 55	–200 to +1300	± 0.75
4.	E	Cheromel–Constantan	55 – 80	–180 to +850	± 0.5
5.	S	Platinum–Platinum/10% Rhodium	5 – 12	0 to +1400	± 0.25
6.	R	Platinum–Platinum/13% Rhodium	5 – 12	0 to +1600	± 0.25
7.	B	Platinum/30% Rhodium–Platinum/6% Rhodium	5 – 12	+100 to +1800	± 0.25
8.	W5	Tungston/5% Rhenium–Tungston/20% Rhenium	5 – 12	0 to +3000	± 0.15

*Constantan = copper/nickel; chromel = nickel/chromium; alumel = nickel/aluminium

For special purposes where high sensitivity is needed, thermocouples may be attached in series. The output is then the numerical sum of the voltages expected from each of the single couples. This is commonly known as *thermopile*. When connected in parallel, a group of thermocouples will give a reading that is the numerical average of the individual ones provided the resistance of each individual thermocouple being the same.

The following are the advantages of the TC sensors:

1. Thermocouple bead can be made of small size and consequently with low thermal capacity. In other words dynamic response of sensor is fairly good.
2. They cost considerably less as compared to other thermal sensors and further, they require no maintenance.
3. They are quite rugged type, i.e. they can withstands rough handling.
4. They cover wide range of temperature, i.e., from –200 to 3000°C.
5. Output signal is electrical and they can be used for indicating recording or microprocessor-based control systems.
6. Output signal, i.e., emf is independent of length or diameter of the wire.
7. They have good accuracy of the order of ±0.2 to ±0.75% of f.s.d.
8. They have excellent stability for a long period of time.
9. They can be conveniently mounted in a variety of temperature measurement situations.

The TC sensors, however, have the following limitations:

1. Inhomogeneity of composition of the thermocouple material and cold working of wires affect the sensitivity of the thermocouple.
2. The require insulation covering while using them in conducting fluids.
3. The output signal, i.e., emf requires amplification in most applications.

Problem 12.4 *A copper–constantan thermocouple was found to have linear calibration between 0 and 400°C with emf at maximum temperature (reference junction temperature 0°C) equal to 20.68 mV.*

(a) Determine the correction which must be made to the indicated emf if the cold junction temperature is 25°C.

(b) If the indicated emf is 8.92 mV in the thermocouple circuit determine the temperature of the hot junction.

Solution

(a) Sensitivity of the thermocouple= $\frac{20.68}{400-0} = 0.0517$ m V/°C

Since the thermocouple is calibrated at the reference junction of 0°C and is being used at 25°C, then the correction which must be made should be the thermo emf say E_{corr} between 0 and 25°C, that is:

$$E_{corr} = 0.0517 \times 25 = 1.293 \text{ mV}$$

(b) Indicated emf between the hot junction and reference junction at 25°C = 8.92 mV
Difference of temperature between hot and cold junctions

$$= \frac{8.92}{0.0517} = 172.53°\text{C}$$

Since the reference junction temperature is equal to 25°C
∴ hot junction temperature = 172.53 + 25 = 197.53°C

12.5.3 Solid State Temperature Sensors

Common I.C. devices like silicon diodes and transistors exhibit a stable and reproducible response to temperature. When a PN junction is forward biased by a constant current source, its governing equation between current and voltage is as follows:

$$V_{BE} = \frac{kT}{q} \ln\left(\frac{I_c}{I_{es}}\right) \tag{12.12}$$

where V_{BE} = base emitter voltage
I_c = collector current
I_{es} = emitter saturation current
K = Boltzmann constant (1.38×10^{-23} J/K)
q = electron charge (1.6×10^{-19} C)
T = absolute temperature (K).

Generally, the term within the parenthesis in Eq. (12.12) is constant and the emitter base voltage i.e., the output of the transducer becomes directly proportional to T which is the measured input. The *main advantage* of the solid state temperature sensors is their inherent linear operating characteristics with excellent accuracy of the order of ±1°C. In addition they have high levels of output signal which is capable of direct indication without any signal conditioning. The sensitivity of the silicon transistor within its useable range of –55 to 150°C is of the order of –2 mV/°C. Further, since the output is electrical, they have the capability of μp based control applications.

The *disadvantages* of these sensors are their limited temperature measuring range and their thermal mass which limits their response characteristics.

12.5.4 Quartz Thermometer

A piezo-electric crystal provides a highly accurate and sensitive method of temperature measurement based on the change in its resonant frequency which is directly proportional to the temperature change. Herein, the crystal is cut in the form of shear type *LC* cut, in which the change in resonant frequency is highly linear as well as repeatable. The associated electronic circuitry of this thermometer consists of frequency counters and digital read-out of the measured frequency.

The fundamental frequency f_0 depends on the thickness of the crystal and can be adjusted so as to give a sensitivity of the order of 1000 Hz for a temperature change of 1°C. In other words, the detection of change in frequency of oscillation of 1 Hz gives a resolution of 0.001°C. Further, temperature in the range of –40 to 230°C can be measured precisely and accurately by this method.

The advantages of the quartz thermometer are:

1. Highly linear output as the linearity error is ±0.5% of F.S.
2. Long-term stability and reliability.
3. High resolution of the order of 0.001°C.
4. Excellent repeatability in the measuring range of –40 to 230°C.

The limitations of the quartz thermometer are:

1. Limited measuring range i.e., –40 to 230°C.
2. Piezo-electric crystals have strong cross-sensitivity for pressure changes if they occur simultaneously in the temperature measuring systems.

12.6 ■ RADIATION METHODS (PYROMETRY)

All the temperature measuring deviced discussed so far i.e., pressure thermometer, thermistor or thermocouple, etc. require the thermometer to be brought into physical contact with the body whose temperature is to be measured. This means that the thermometer must be capable of withstanding this temperature. In the case of very hot bodies, the thermometer may melt at the high temperature. Secondly, for bodies that are moving, a non-contacting device for measuring the temperature is most convenient. Thirdly, if the distribution of temperature over the surface of an object is required, a non-contacting device can readily 'scan' the surface.

For temperatures above 650°C, the heat radiations emitted from the body are of sufficient intensity to be used for measuring the temperature. Instruments that employ radiation principles fall into three general classes: (a) total radiation pyrometer, (b) selective (or partial) radiation pyrometers, and (c) infrared (IR) pyrometer. The first is sensitive to all the radiation that enters the instrument and the second only to radiation of a particular wavelength. Further, the IR pyrometers employ the infrared portion of the spectrum by using a thermal detector to measure the temperature on the surface of the body.

12.6.1 Total Radiation Pyrometer

The total radiation pyormeter receives a controlled sample of the total radiation of a hot body (say a furnace) and focusses it on to a temperature sensitive transducer. The term 'total radiation' includes both visible (light) and invisible (infrared) radiations. It may be noted that the wave lengths of light in the visible range is from 0.3 to 0.72 μm, whereas the infrared radiations are associated with relatively large wavelengths of 0.72 to 1000 μm. They require special optical materials for focussing. Ordinary glass is unsatisfactory, as it absorbs infrared radiations. In fact, the practical radiation pyrometers are sensitive to a limited wavelength band of radiant energy, (i.e., from 0.32 to 40 μm) although theory indicates that they should be sensitive to the entire spectrum of energy radiated by the object.

Figure 12.12 shows a schematic diagram of the Fery's total radiation pyrometer. It consists of blackened tube T open at one end to receive the radiations from the object whose temperature is to be measured. The other end of the tube has a sighting aperture in which an adjustable eyepiece is usually fitted. The thermal radiations impinge on the concave mirror whose position can be adjusted suitably by a rack-and-pinion arrangement so as to get proper focussing of the thermal radiations on the detector disc *S*. The detector disc is usually of blackened platinum sheet/foil and is connected to a thermocouple/thermopile junctions or to a resistance thermometer bridge circuit. Leads from the detector are led out of the casing to a meter for measuring the thermoelectric emf or the variation the electric resistance of the platinum foil.

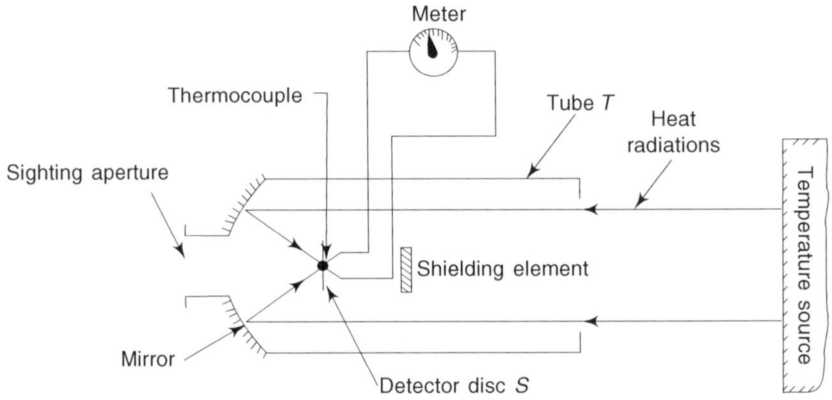

Fig. 12.12 *Schematic of Fery's total radiation pyrometer*

The theory underlying the operation of total radiation pyrometers is that the rate of radiation from a body A (the source) to a body B (the pyrometer), i.e. $E_{A/B}$ is given by the Stafan-Boltzmann law as follows:

$$E_{A/B} = C\varepsilon\sigma\left[T_A^4 - T_B^4\right] \tag{12.13}$$

were $E_{A/B}$ is the energy received by the pyrometer in W/m^{-2}

C is a geometrical factor to adjust the relative shapes of the two bodies

ε is the emissivity of the detector disc which varies from 0.05 to 1.0 for the theoretical black body

σ is the Stefan-Baltzmann constant and its value is 56.7×10^{-12} kW/(m^2 · K^4)

T_A and T_B are the steady state absolute temperature of the source and pyrometer detector disc.

Such pyrometers are usually calibrated against known temperatures in the range of 700 –2000°C where thermocouples and resistance thermometers cannot be employed. However, the errors arise from two sources in actual use. Any filtering material such as smoke, dust, gases, windows, etc. which were not present in the calibration will reduce the energy received hence cause an unknown error. Secondly, an error may be caused due to a surface having emissivity other than used in the calibration. Since the surface emissivities are not known very accurately and a change occurs with time due to oxidation, therefore the error due unknown emissivity is usually not known. To reduce such uncertainties, pyrometers calibrated from time to time in actual use.

In view of the troubles due to filtering and emissivity, the total radiation pyrometer is not a very accurate temperature indicator. However, it can be used to good advantage in fixed locations where the emissivity and optical paths are well known and constant. A typical use is a large furnace in metal industries. The signal is electrical and therefore can be used for control applications.

12.6.2 Selective Radiation Pyrometer

The principle of this instrument is based on Planck's law which states that the energy level in the radiations from a hot body are distributed in the different wavelengths. As the temperature increases, the emissive power shifts to shorter wavelengths. The planck's distribution equation is:

$$W = \frac{c_1 \lambda^{-5}}{e^{c_2 \backslash \lambda T} - 1} \tag{12.14}$$

where $c_1 = 3.740 \times 10^{-12}$ (W–cm^2)
$c_2 = 1.4385$ (cm –°C)
λ = wavelength (cm)
T = absolute temperature in (K)
W = energy level associated with wavelength at temperature T (W/cm^3)

The classical form of this optical pyrometer is the disappearing filament optical pyrometer (or the monochromatic brightness radiation pyrometer). It is most accurate of all radiation pyrometers; however, its use is limited to temperatures greater than about 700°C since it requires visual brightness match by a human operator. This instrument is used to realise the International Practical Temperature Scale above 1064°C.

It is obvious from Planck's distribution equation that for a given wavelength, the radiant intensity (brightness) varies with the temperature. In the disappearing filament instrument shown in the Fig. 12.13, an image of the target is superimposed on the heated filament. The tungsten lamp, which is very stable, is previously calibrated so that when the current through the filament is known, the brightness temperature of the filament is also known. A red filter that passes only a narrow band of wavelengths around 0.65 μm is placed between the observer eye and the tungsten lamp and the target image. The observer controls the lamp current until the filament disappears in the superimposed target image [Fig. 12.14(c)]. The temperature calibration is made in terms of the lamp heating current. Because of the manual null balancing principle, the optical pyrometer is not usable for continuous recording or automatic control applications. However, it is more accurate and less subject to large errors than the total radiation pyrometer. The accuracy of such pyrometers is usually ± 5°C in the range of 850–1200°C. Further, when used in the extended range of 1100 – 1950°C, its accuracy is better than ±10°C.

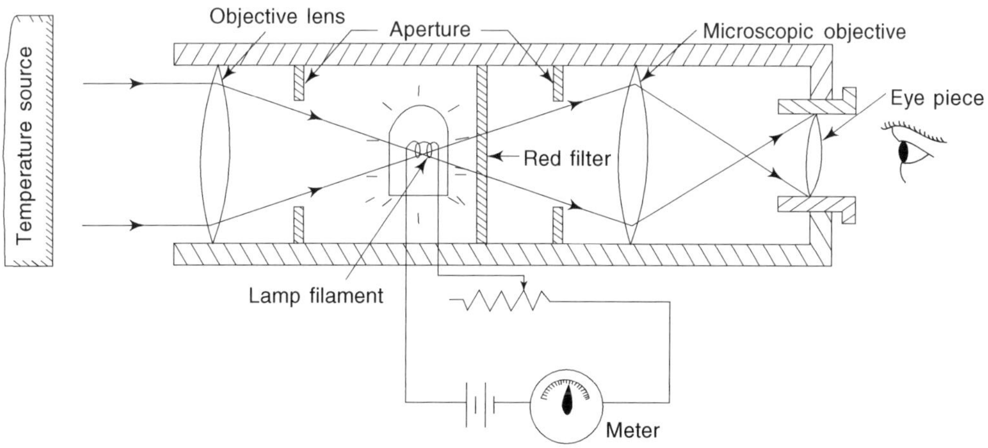

Fig. 12.13 *Schematic of the disappearing filament type of optical pyrometer*

(a) Filament too dark (b) Filament too bright (c) Equal brightness

Problem 12.4 *The power radiated from a hot piece of metal was measured by the radiation pyrometer and the temperature was determined as 820°C assuming a surface emissivity of 0.75. Later it was found that the accurate value of emissivity was 0.69. Find the error in the temperature determination.*

Solution From the Stefan–Boltzmann law, we get

$$T = \left[\frac{E_A}{\varepsilon\sigma}\right]^{1/4}$$

$$\therefore \quad \text{incorrect temperature } T_1 = \left[\frac{E_A}{\varepsilon_1\sigma}\right]^{1/4}$$

and

$$\text{actual temperature } T_{act} = \left[\frac{E_A}{\varepsilon_2\sigma}\right]^{1/4}$$

$$\therefore \quad \frac{T_{act}}{T_1} = \left[\frac{\varepsilon_1}{\varepsilon_2}\right]^{1/4}$$

$$T_{act} = T_1 \times \left[\frac{\varepsilon_1}{\varepsilon_2}\right]^{1/4}$$

$$= (820 + 273) \times \left[\frac{0.75}{0.69}\right]^{1/4} = 1116.0 \text{ K}$$

$\therefore$ Actual temperature = 843°C

Hence, the error in temperature determination = 843 – 820 = 23°C

Review Questions

12.1 Indicate which of the following statements is correct in the following:

(i) The basis for measuring the thermodynamic property, the temperature, is given by
- (a) zeroth law of thermodynamics
- (b) first law of thermodynamics
- (c) second law of thermodynamics
- (d) third law of thermodynamics.

(ii) A thermostatic cut-out works on the principle of
- (a) thermal expansion of fluids
- (b) variation of resistance with temperature
- (c) expansion due to air pressure
- (d) thermal expansion of metals

(iii) According to the Stefan–Boltzmann law, the amount of radiant energy per unit area is proportional to
(a) absolute temperature (b) square of absolute temperature
(c) fourth power of absolute temperature (d) cube of absolute temperature

(iv) Bimetallic strips made of two different materials bend during a rise in temperature on account of
(a) differences in coefficient of linear expansion
(b) differences in the elastic properties
(c) differences in the thermal conductivities
(d) none of the above

(v) The principle of working of the constant volume thermometer is based on
(a) Boyle's law (b) Charle's law
(c) Gay–Lussac's law (d) equation of state

(vi) The NTC thermistor is a thermally sensitive variable resistance semiconductor and its resistance as compared to metallic conductors
(a) increases linearly with temperature
(b) decreases linearly with temperature
(c) decreases exponentially with temperature
(d) increases exponentially with the temperature

(vii) The units of Stefan–Boltzmann constant as
(a) $W/cm^2–K$ (b) $W/cm^2–K^4$ (c) $W/cm^2–K^2$ (d) $W^2/cm–K^4$

(viii) The Stefan–Boltzmann law is applicable for heat transfer by
(a) conduction and radiation combined
(b) convection and radiation combined
(c) conduction, convection and radiation combined
(d) radiation alone
(e) convection alone

(ix) Optical pyrometer is used to measure
(a) light intensity (b) low temperatures
(c) high temperatures (d) light intensity and high temperatures

(x) At what temperature do the Fahrenheit and celcius scales coincide
(a) 0 (b) 20 (c) 40 (d) –40

(xi) Boiling point of water which is used as one of the fixed point in the International Practical Temperature Scale in K is given by
(a) 100 (b) 212 (c) 273.15 (d) 373.15

(xii) The approximate range up to which an ordinary mercury-in-glass thermometer can be used is
(a) 0 to 100°C (b) –20 to 340°C (c) –50 to 560°C (d) –100 to 500°C

(xiii) The primary transducer element in a pressure thermometer is
(a) Bourdon tube
(b) capillary tube
(c) bulb
(d) bulb together with capillary tube and Bourdon tube

(xiv) A thermocouple arrangement is to be used to measure a high temperature of 1400°C. Point out the pair of thermocouple that would be most suitable for this application.
(a) chromel–constantan (Type E) (b) iron–constantan (Type J)

(c) chromel–alumel (Type K) (d) copper–constantan (Type T)
(e) Platinum–13% rhodium/platinum (Type R)

(xv) The instrument which measures the temperature of the source without direct contact is:
(a) bimetallic cut-out (b) vapour pressure thermometer
(c) pyrometer (d) thin-film thermometer

(xvi) The sensing element of the industrial pressure thermometer is usually made of:
(a) brass (b) platinum (c) steel (d) constantan

(xvii) Thermistor are
(a) IC clips whose voltage output is directly proportional to temperature
(b) semiconductors which generally have negative coefficient of resistance
(c) non-contact type of temperature sensors
(d) thin-film metallic sensors.

(xviii) The disappearing filament type of optical pyrometer works on the principle of
(a) comparison of monochromatic component brightness of light from a radiating body with respect to a heated filament
(b) focussing the total radiations on a thermal sensor in the visible as well as in the infrared range
(c) measuring photon flux density on the surface of the hot body
(d) scanning the infrared radiations from the surface of the hot body

(xix) Which of the following thermocouples has the lowest measuring range
(a) iron–constantan (b) chromel–alumel
(c) copper–constantan (d) chromel–constantan

(xx) Which property of quartz crystal changes with the change in temperature
(a) capacitance (b) inductance
(c) resistance (d) resonant frequency

(xxi) Which of the following temperature sensors has excellent linear characteristics?
(a) RTD (b) Thermocouple
(c) Radiation pyrometer (d) Silicon-based IC chip

12.2 Indicate whether the following statements are true or false. If false, rewrite the correct statement.
(i) Temperature is related to the average kinetic energy of the molecules or atoms.
(ii) Bimetallic thermonmeter is unsuitable if the temperature changes are rapid.
(iii) Vapour pressure thermometers have uniformly divided scales.
(iv) Response and sensitivity of a thermocouple is improved by using heavy gauge wire (i.e. bigger diameter wire).
(v) One of the important points of optical pyrometers is that temperature readings may be made very quickly.
(vi) Conductors and semiconductor materials produce similar types of temperature effect with respect to resistance change.
(vii) To obtain accurate results using the liquid-in-glass thermometer, it should be calibrated under the conditions in which it has to be used.
(viii) The parallel arrangement of a group of thermocouples is called a thermopile.
(ix) Cable compensation arrangement of Siemen's type is a must while employing thermistor in temperature measurement.
(x) The platinum resistance thermometer is used as a standard for establishing Internal Temperature scale from –182.93 to 961.93°C.

12.3 Fill in the blanks in the following:

(i) The sensing element of the RTD for measuring temperature below 600°C is usually made of ____________.

(ii) Thermistors are ____________ which generally have negative coefficient of resistance.

(iii) The most suitable thermometer to measure the temperature of the wire of an incandescent bulb is ____________.

(iv) Type T thermocouple refers to a thermocouple made of ____________.

(v) The base type thermocouple that can measure a temperature of 1000°C is ____________.

(vi) The most suitable devices for measuring the time-varying temperatures are: (a) ____________, (b) ____________, and (c) ____________.

(vii) The temperature measuring instrument with a least count of the order of 0.001°C is ____________.

(viii) The volume of the bulb of a pressure thermometer is 10 times the combined volume of the capillary and Bourden tubes. If the ambient temperature falls by 20°C from the calibration temperature then, the error in temperature would be ____________.

(ix) The most convenient temperature sensor employed in the intensive care units of the hospitals is ____________.

(x) The primary fixed point, triple point of water gives a temperature of ____________.

12.4 The radius of curvature of the bimetallic thermometer is given by

$$r = t\frac{3(1+m^2)+(1+mn)(m^2+1/mn)}{6(\alpha_2-\alpha_1)(T-T_0)(1+m^2)}$$

where t is the combined thickness of the bonded strip $(t_1 + t_2)$
m is the ratio of thickness of low to high expansion materials, t_2/t_1
n is the ratio of modulii of elasticity of low to high expansion materials, E_2/E_1
α_1 is lower coefficient of thermal expansion
α_2 is higher coefficient of thermal expansion
T is operating temperature
T_0 is the initial bonding temperature

A bimetal strip was made of strips of nickel-chrome alloy and Invar bonded together at 20°C. Each material had a thickness of 2 mm and the composite element was fixed at one end with the other end kept free. The length of the cantilever was 50 mm. Determine the radius of curvature of the strip subjected to 100°C. Take the following data of the material properties:

for Invar $\alpha_1 = 1.7 \times 10^{-5}$°C^{-1}, $E_1 = 1.5 \times 10^5$ N/mm^2

for nickel-chrome alloy

$\alpha_2 = 12.5 \times 10^{-5}$°C^{-1}, $E_2 = 2.2 \times 10^5$ N/mm^2

12.5 A chromel-alumel thermocouple is assumed to have nearly linear operating range up to 1100°C with emf (reference temperature 0°C) 45.14 mV at this temperature. The thermocouple is exposed to a temperature of 840°C. The potentiometer is used as the cold junction and its temperature is estimated to be 25°C. Calculate the emf indicated on the potentiometer.

12.6 Radiant energy measurement was made from a hot outer surface of a furnace. The amount of energy emitted was 20.5 ± 0.5 kW/m^2 and the surface emissivity was estimated as 0.85 ± 0.1. Determine the surface temperature of the furnace and estimate the uncertainty.

12.7 For a certain NTC thermistor chip, the values of resistance for ice point steam point are 10,000 and 250 Ω, respectively. Determine the value of the constant β for the thermistor material.

12.8 It is required to design a resistance thermometer using a nickel wire of 0.02 mm diameter. The thermometer resistance at 0°C is to be 100 Ω. How long the wire should be? Take for nickel value of resistivity r as 8.7×10^{-6} Ω-cm at 0°C and the temperature coefficient of resistance $0.0068°C^{-1}$. Determine also value of resistance at steam point.

12.9 A thermistor has a resistance of 12 kΩ ± 7% at 25°C and 1.05 kΩ ± 5% at 100°C. Determine
(a) constant β and percentage uncertainty in β, and
(b) resistance and percentage uncertainty in resistance at 50°C.

12.10 A radiation pyrometer indicated the temperature of a furnace as 975°C, assuming a surface emissivity of 0.85. Subsequently, it was found that accurate value of emissivity was 0.78. Determine the error in the temperature measurement of the furnace.

12.11 An RTD has a resistance of 600 Ω and carries a current of 5 mA. The surface area of its bulb is 6 cm^2 and it is immersed in stagnant air where in the convective heat transfer coefficient h = 10 J/m^2–°C–s. Determine
(a) the self-heating error in air, and
(b) the self-heating error in water (when h = 1000 J/m^2–°C–s)

12.12 A mercury-in-steel bulb thermometer employs Bourden pressure gauge that has a range of 0 – 10 MPa, for a pointer movement of 300°. The dial of the pressure gauge is calibrated to read 0° at 0°C and 300° at 400°C. Determine
(a) the sensitivity of the device in rad/°C,
(b) if the bulb volume is eight times as large as the volume of capillary and Bourden tube, then what is the error if the ambient temperature rises by 24°C above that during calibration?
(c) if the bulb is raised by half a metre higher (relative to the Bourdon gauge), what is the apparent change in the temperature indicated by the thermometer?

12.13 A thermistor has R_0 = 2500 Ω at T_0 = 298 K. If β = 4150 K, determine the resistance of the thermistor at the following temperatures –100, –50, 0, 50, 100, 150, 200, 250 and 300°C.

12.14 Determine the error caused due to self-heating of a platinum RTD sensor of R_T = 100 Ω placed in one of the arms of a balanced Wheatstone bridge with excitation voltage of 6V. The self-heating factor of the sensor is F_{sh} = 0.05°C/mW.

Answers

12.1

(i) (a)	(ii) (d)	(iii) (c)	(iv) (a)	(v) (a)
(vi) (c)	(vii) (b)	(viii) (d)	(ix) (c)	(x) (d)
(xi) (d)	(xii) (b)	(xiii) (d)	(xiv) (e)	(xv) (c)
(xvi) (c)	(xvii) (b)	(xviii) (a)	(xix) (c)	(xx) (d)
(xxi) (d)				

12.2 (i) True
(ii) True
(iii) False, vapour pressure thermometer have non-linear scales of the type $\log p = (a - b/T)$
(iv) False, the sensitivity does not change, however, the response depends on the size of the thermocouple bead. It does not depend on the connecting wires.
(v) False, manual null balancing may take some time.
(vi) False, conductor have positive temperature coefficient, whereas thermistors generally have negative temperature coefficient
(vii) True
(viii) False, series arrangement is called thermopile

(ix) False, cable compensation is not employed in thermistor circuit
(x) True

12.3 (i) nickel
(ii) semiconductor materials
(iii) optical pyrometer
(iv) copper–constantan
(v) chromel–alumel
(vi) (a) thermistor (b) thermcouple(c) solid state thermometer
(vii) quartz thermometer
(viii) 2°C
(ix) thermistor
(x) 273.16 K

12.4 r = 3.92 m
12.5 emf = 33.44 mV
12.6 T = 807.6 ± 24.3
12.7 β = 3756 K
12.8 l = 36.11 cm, $R_{100°C}$ = 168 Ω
12.9 β = 3610.5
12.10 ΔT = 27°C
12.11 2.5°C, 0.025 °C
12.12 0.0131 rad/°C, 3 °C, 2.67 °C
12.13 5.86×10^9, 2.69×10^5, 8.94×10^3, 851.5, 152.02, 40.81, 14.5, 6.24, 3.13 Ω.
12.14 4.5°C

Chapter

13

Flow Measurement

■ *INTRODUCTION* ■

Flow measurements are essential in many applications such as transportation of solids as slurries, compressed natural gas in pipelines, water and gas supply systems to domestic consumers, irrigation systems and a number of industrial process control systems. The types of flows encountered in the measurements may be any one or combination of the following types:

- clean or dirty/opaque,
- wet or dry,
- hazardous/corrosive or safe,
- single-phase, two-phase or multiphase,
- laminar or transitional or turbulent,
- pressure may vary from vacuums to high pressures of many atmospheres,
- temperature may vary from cryogenic levels to hundreds of centigrades,
- flow rate may be of miniscule type, i.e., few drops per minute or massive type involving thousands of litres per minute.

The selection of a particular flow-measuring equipment depends primarily on the nature of the metered fluid and the demands of the associated plant. For example, an aircraft fuel meter requires to be compact and must not be affected by the changes in orientation, but has to handle only a clean and non-corrosive fluid. On the other hand, many industrial flow meters have to work with fluids which may be corrosive in nature or may contain foreign matters, but the equipment may be relatively large and of fixed type. Additionally, the other factors that govern the choice of a particular flow metering device are the various performance parameters like range, accuracy, repeatability, linearity, dynamic response, type of output like analog/digital, etc. Further, another requirement may be to indicate or record the rate of flow, total flow or may be both these quantities.

Flow measuring devices generally fall into one of the two categories, namely, primary devices or quantity meters and secondary devices known as rate meters. The distinction between the two is based on the character of the sensing element that interacts with the fluid flow. The output of the sensing element is then suitably modified so as to indicate or record the measured values. Quantity measurements, by mass or volume, are usually accomplished by counting successive isolated portions, whereas rate measurements are inferred from effects of flow rates on pressure, force, heat transfer, flow area, etc. It is quite often possible to obtain the rate of flow from a quantity meter by a suitable choice of modifying elements. Further, it is important to note that the quantity meters are generally used for the calibration of rate meters.

13.1 ■ PRIMARY OR QUANTITY METERS

Quantity or total flow measurement signifies the amount of fluid in terms of mass or volume that flows past a given point in a definite period of time. In other words, in this technique, the time required to collect a particular amount of fluid is determined accurately and then the average flow rate can be evaluated. Improved accuracy may be obtained by using longer and precise time measurement along with precise volume/weight measurements. This technique of flow measurement is not only simple and economical but also is extremely accurate. Because of this, it is routinely *used in the calibration of other flow measuring devices*. The flow meter calibration procedures using the quantity measurements fall into the following two categories.

13.1.1 Volumetric Method

In this technique, the fluid flowing in the flow meter which is being calibrated, is diverted into a tank of known volume. When the tank is completely filled, then this known volume is compared with the integrated, volumetric quantity registered by the flow meter under test.

13.1.2 Gravimetric Method

In this technique also, the fluid flowing in the flow meter, which is being calibrated, is diverted into a vessel which can be weighed either continuously or in the vessel after a pre-determined time. The weight of the liquid collected is compared with the gravimetric quantity registered by the flow meter under test.

13.2 ■ POSITIVE-DISPLACEMENT METERS

The term positive-displacement meter is applied to a flow measuring device so designed that the metered fluid is repeatedly filled and emptied from a space of known volume. The principle of this measurement is that the liquid flows through a meter and moves the measuring element that seals the measuring chamber into a series of measuring compartments each holding a definite volume. Each element is successively filled from the flow at the inlet and empited at the outlet of the meter. In other words, we could say that positive-displacement meters chop the flow into 'pieces' of known size and then count the number of 'pieces'.

Positive-displacement meters are widely used in low flow rate metering applications where high accuracy and repeatability under steady flow conditions are required. Further, they are easy to install and maintain and have moderate cost. These types of meters are generally used by the water and oil undertakings for accounting purposes. However, since there are moving parts in these devices, the wear of the components may alter the accuracy. Therefore, these instruments need calibration/adjustment over an interval of time. Another limitation of such meters is their suitability to clean fluids only. Further, these devices are generally flow totalizers and do not give instantaneous rate of flow.

In most positive-displacement meters, the transduction of the flow takes place in the form of rotary motion. In short, we can say that these meters are hydraulic or pneumatic motors whose cycles of motion are recorded by some form of a counter. Energy is extracted from the flow to drive these meters resulting in a pressure loss in the fluid system whose flow is being metered. But the energy required is extremely small, just enough to overcome the friction of the system. Some of the commonly used devices of this type are as follows.

13.2.1 Nutating Disc Meter

The nutating disc meter shown schematically in Fig. 13.1 consists of an eccentrically mounted disc which nutates or wobbles in the metering chamber which has spherical sides. The liquid enters the left side of the meter, alternately above and below the disc, forcing it to rock (nutate or wobble) in a circular path while rotating about its own axis. Thus the liquid from the inlet causes the disc to nutate before it goes to the outlet. Further, the viscosity of the liquid ensures both sealing and lubrication. A small spindle attached to the sphere traces a circular path and is used to drive the mechanical or electronic counter which can be calibrated in terms of liquid discharge.

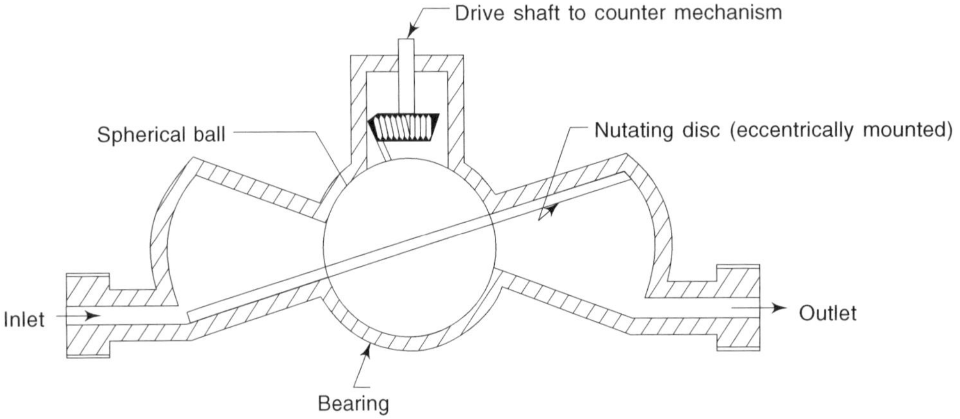

Fig. 13.1 *Nutating disc meter*

This type of flow meter has simple and rugged construction, low pressure drop, low cost and a good accuracy of the order of ± 1%. It can be used to meter wide variety of liquids. Moreover, it finds a wide application as domestic water meter.

13.2.2 Sliding-Vane-Type Meter

This type of meter has an accurately machined body having a rotor with four evenly spaced slots which form the guides for the vanes. The metered liquid entering the inlet revolves the rotor and the vanes around a cam causing the vanes to move radially. The vane nearest to the inlet port begins to move outwards and becomes fully extended at point A shown in Fig. 13.2. The vane ahead at point B is already fully extended and thus a measuring chamber of known volume is formed between the two vanes and the meter body. A continuous series of chambers at the rate of four per revolution are formed which deliver the flow at the outlet.

This type of meter has low pressure drop and can give an accuracy of the order of ± 0.2% of the measured values. An important application of sliding vane type meter is its use as positive displacement pump which can cause the flow rate and measure it simultaneously. Such a pump is routinely used for dispensing petrol/diesel or compressed natural gas in the fuel filling stations.

13.2.3 Lobed-impeller Meter

This type of meter consists of two rotors mounted on separate parallel shafts. The rotors are lobed and revolve in opposite directions in a close fitting chamber. The rotor lobes have cycloidal or involute shape which ensures accurate mating. The incoming fluid is trapped between the two rotors and is conveyed to the outlet as a result of the rotation of the rotors. For every rotation of each rotor, the swept volume

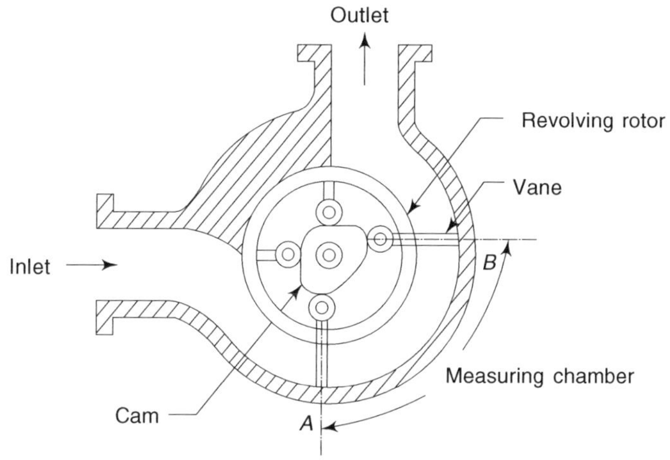

Fig. 13.2 *Sliding vane type meter*

corresponding to twice the area A shown in Fig. 13.3 is passed through the meter. Thus the number of revolutions of the rotor gives an indication of the volumetric flow. Since the speed of rotation is proportional to the volume flow rate, the rate of flow display can also be obtained by monitoring the speed of rotation of one of the rotor shafts.

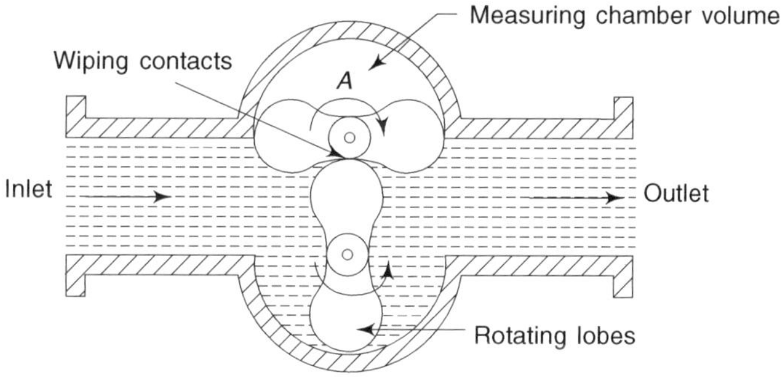

Fig. 13.3 *Lobed impeller flow meter*

The rotations of the lobbed impellers can be monitored by magnetic or photo-electric pick-up. The frequency of the output is proportional to the flow rate and the electronic integrating circuit gives the totalized flow. Alternatively, the rotations of the lobbed impeller are counted by connecting them to mechanical registers. Meters of this type are available for pressures up to 60 atm for the flow rates ranging from 10 to 10,000 m^3/h. Within these flow rates, the meters have an accuracy of $\pm$ 1.0% over a range of 10–100% of the rated capacity. Further, these meters have low pressure drop of the order of 50 mm of Hg at the rated capacity.

The *limitation* of the instrument is that it is relatively expensive because, for obtaining accurate results, the clearances between the rotors and the casing as well as between the rotors have to made very fine. Further, these close tolerance levels restrict the meter to clean fluid applications like metering of gas flows and refined petroleum products.

13.3 SECONDARY OR RATE METERS

The secondary or rate meters are also termed as *inferential* type of flow measuring devices. This is because of the fact that they do not measure the flow directly but instead measure another physical quantity which is related to the flow. These devices fall into two categories, namely, the flow rate meters and the velocity meters.

The transduction principle of some typical flow rate meters is as follows:

(i) *Variable head meters* These are also termed as obstruction type of meters in which the obstruction to the flow consists of an engineered constriction in the metered fluid which causes a reduction in the flow pressure.

(ii) *Variable area meters* Herein, the change in area causes change in the drag force of a body placed in the flowing fluid.

(iii) *Variable head and variable area meters* In these devices, a specified shaped restriction is placed in the path of the flow which causes a rise in the upstream liquid level, which is a function of the rate of flow.

(iv) *Constant head device* In this device a constant head is applied to cause a laminar flow in the capillary tube. In this device, the applied head is loss in fluid friction but it causes a flow rate which can be metered.

Quite a few secondary flow devices or rate meters are velocity meters which are also commonly termed as anemometers. Some of the devices measure the local velocity of the flow, while others provide the average velocity. In the former type, the flow rate can be obtained by integrating the measured data of velocities in different elemental areas, whereas in the later case, the average velocity is multiplied by the flow area to obtain the flow rate. Some typical secondary flow devices which are based on the flow velocity measurement principle are as follows:

(i) *Variable head meter* In this device the difference between the pitot head i.e., the stagnation pressure head and the static head gives the dynamic head which gives the indication of local velocity at the location of the sensor.

(ii) *Target flow meter* Herein, the flowing fluid impinging on the target is brought to rest. The pressure increases by the velocity head term, i.e., $V^2/(2g)$. Further, the force measured on the target is proportional to the velocity head of the flowing fluid.

(iii) *Turbine/propeller type meter* In this type of device, the fluid flow causes rotation of the curved vanes and its rotational speed is proportional to the average flow velocity passing through the vanes. Quite often these vanes are fitted inside a casing of a given area of cross-section. In that case, the rotational speed can be calibrated for the flow rate as, in such a case, the flow rate is obtained by the average velocity multiplied by a constant area of cross-section of the flow meter.

(iv) *Ultrasonic flow meter* In this type of flow meter, the transit time of the ultrasonic pulse between the transmitter and the receiver is obtained which is a function of the average velocity of flow.

(v) *Electromagnetic flow meter* In this type of device, a magnetic field is applied on the moving conducting fluid. It results in the production of emf which is proportional to the average velocity of flow and it is in the direction perpendicular to the plane of velocity and the magnetic field according to the Faraday's law of induction.

(vi) *Hot wire/hot film anemometer* In this device, the resistance of a thin heated wire changes due to the cooling effect of the flowing fluid stream. The change in resistance is the transduction principle for the measurement of local fluid flow velocity as well as the turbulence level.

(vii) *Laser Doppler anemometer* In this device, a laser beam gets scattered due to the presence of particle in a flowing fluid. The Doppler frequency shift of the scattered laser beam is directly proportional to the local fluid flow velocity as well as the turbulence level.

13.3.1 Variable Head Meters

These meters essentially introduce an engineered constriction in the flow passage. The, devices in general can be as termed *obstruction type* of flow meters. The term 'obstruction meter' applies to the devices that act as obstacles placed in the path of the flowing fluid, causing localised changes in the velocity. Concurrently with the velocity change, there is a corresponding pressure change in the flow. This variation in pressure change is correlated with the rate of flow of the fluid. It may be noted that these devices cause a loading error in the metered value because obstruction introduces extra resistance in the flow system consequently, the flow rate reduces somewhat. The main forms of restriction used in the flow are venturi tube, orifice plate and a nozzle, which are shown in Fig. 13.4.

The variation of pressure in these differential pressure devices is also indicated in Fig. 13.4. The position of minimum pressure is located slightly downstream from the restriction at a point where the stream is the narrowest and is called the *vena-contracta.* Beyond this point, the pressure again rises but does not return to the upstream value and thus there is a permanent pressure loss. The magnitude of this loss depends on the type of restriction and on the dimensions of device. The ratio of the diameter at the constriction to the diameter D of the pipe is called the diameter ratio. If this ratio is too small, the opening is narrow and the pressure loss becomes considerable and also the efficiency of the measurement is low. If the ratio is rather large, then the reduction in pressure is too small for accurate measurements. In practice, ratios in the range 0.2–0.6 are usually employed. The general expression for the rate of flow in these devices can be derived as follows:

Say, the pressure, velocity and area of fluid stream at point 1, upstream of obstruction are p_1, V_1 and A_1 and at point 2 just downstream of the obstruction are p_2, V_2 and A_2. Further, we assume the flow to be incompressible, i.e., its density does not vary in the flow field.

Applying the continuity equation in the flow we get

$$\text{Rate of discharge } Q = A_1V_1 = A_2V_2 \tag{13.1}$$

Applying Bernoulli's equation (assuming the flow to be ideal) we get,

$$p_1 + \frac{\rho V_1^2}{2} = p_2 + \frac{\rho V_2^2}{2} \tag{13.2}$$

The differential pressure head Δh is given by

$$\frac{p_1 - p_2}{\rho g} = \Delta h \tag{13.3}$$

Eliminating V_1 and V_2 from Eqs. (13.1) and (13.2) and substituting the value of Δh from Eq. (13.3) we get the ideal rate of discharge as

$$Q_{\text{ideal}} = \frac{A_1 A_2}{\sqrt{A_1^2 - A_2^2}} \cdot \sqrt{2g}\,\sqrt{\Delta h} \tag{13.4}$$

In actual practice, the actual rate of fluid flow is always less than Q_{ideal} as given by Eq. (13.4), because of the losses in the fluid flow due to friction and eddying motions. To account for this discrepancy, we define the term coefficient of discharge C_d as

$$C_d = \frac{Q_{\text{actual}}}{Q_{\text{ideal}}} \tag{13.5}$$

Thus, we can write the actual rate of fluid flow as

$$Q_{\text{actual}} = C_d \frac{A_1 A_2}{\sqrt{A_1^2 - A_2^2}} \sqrt{2g}\,\sqrt{\Delta h} \tag{13.6}$$

Equation (13.6) can be rewritten in the simplified form as

$$Q_{\text{actual}} = C_d K\,(\Delta h)^{1/2} \tag{13.7}$$

where K is the constant of flow obstruction device and

$$K = \frac{A_1 A_2}{\sqrt{A_1^2 - A_2^2}} \sqrt{2g}$$

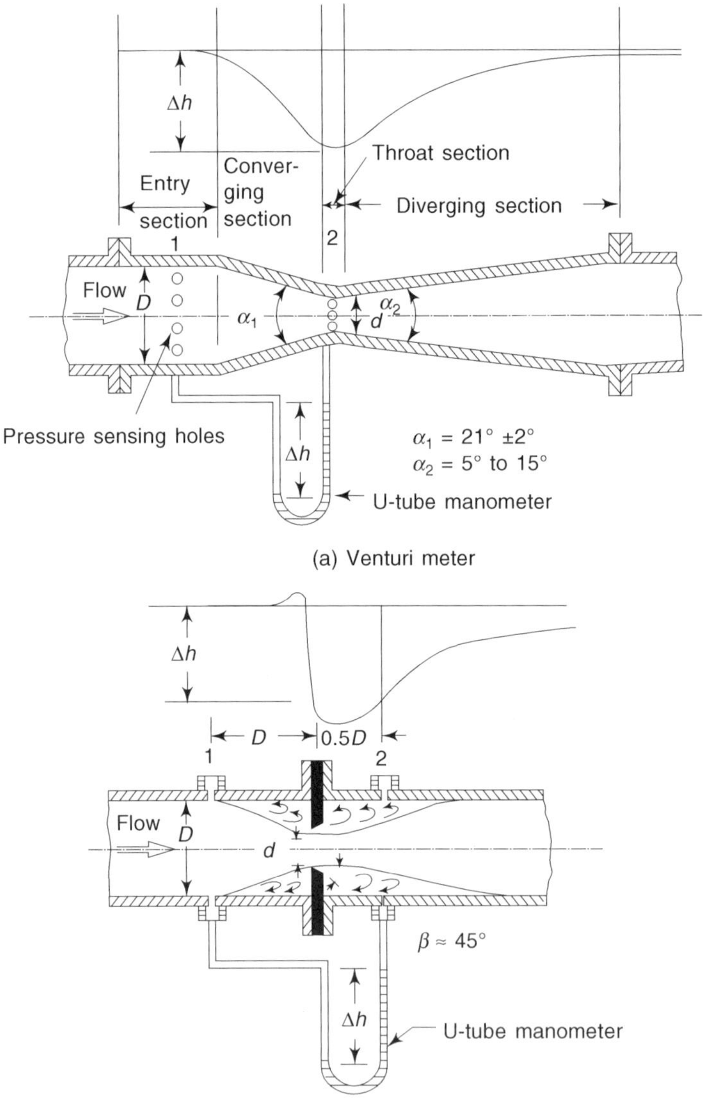

(a) Venturi meter

(b) Orifice meter

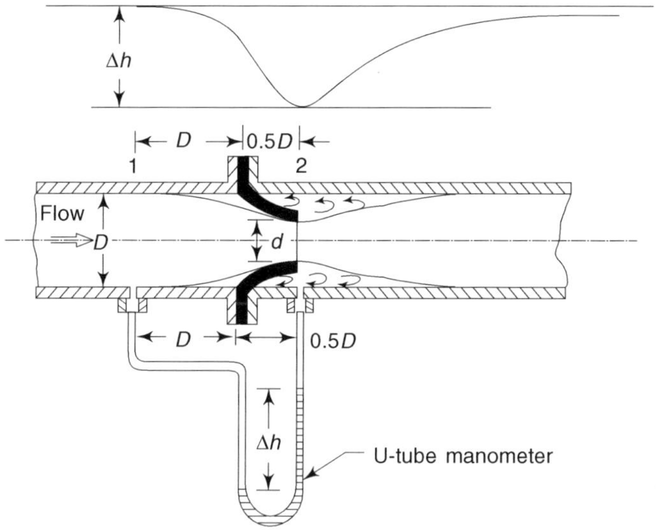

(c) Nozzle meter

Fig. 13.4 *Different types of variable head meters*

where C_d is the coefficient of discharge which depends on the type of flow, obstruction type configuration and also on the Reynolds number of the flow which is given by the expression $\rho VD/\mu$.

The venturimeter offers the best accuracy, least head loss as compared to both nozzle meter as well as the orifice meter. Because of the smooth surface, it is not much affected by the wear and abrasion from dirty fluids. Further, due to low value of losses, the coefficient of discharge is high and approaches unity under favourable conditions. However, it is expensive and occupies substantial space.

The nozzle flow meter offers all the advantages of the venturimeter to a lesser extent but it occupies considerably less space. The nozzle is however difficult to install and is further limited to moderate pipe sizes.

On the other hand, an orifice meter consists of thin orifice plate [Fig. 13.4(b) which may be clamped between pipe flanges. Since its geometry is simple, it is low in cost, easy to install or replace and takes almost no space. However, it suffers from a head loss which is of the order of 30–40%. Also, it is susceptible to inaccuracies resulting from erosion, corrosion, clogging, etc. due to flow of dirty fluids.

The variable head devices are widely used in practice, because they have no moving parts and require practically no maintenance. Further, they can be used without calibration if made to standard dimensions. However the *major disadvantage is the square-root relationship between the pressure loss and the rate of fluid flow*. Further, it is not practical to measure the flow below 20% of the rated meter capacity because of the inaccuracies involved in a very low-pressure differential measurements.

Problem 13.1 *A nozzle is fitted in a horizontal pipe diameter 15 cm, carrying a gas of density 1.15 kg/m³, for the purpose of flow measurement. The differential pressure head indicated by a U-tube manometer containing oil of specific gravity 0.8 is 10 cm. If the coefficient of discharge and diameter of nozzle are 0.8 and 5 cm, respectively, determine the flow of gas through the nozzle flow meter.*

Solution Pressure differential Δh indicated by the manometer = 10 cm of oil of sp. gr. 0.8

$$= \frac{10 \times 800}{1.15} \text{ cm of gas} = 6956.52 \text{ cm of gas}$$

Using Eq. (13.6) we get,

$$Q_{\text{actual}} = C_d \frac{A_1 A_2}{\sqrt{A_1^2 - A_2^2}} \cdot \sqrt{2g}\,\sqrt{\Delta h}$$

$$= 0.8 \frac{(176.71)(19.63)}{\sqrt{176.71^2 - 19.63^2}} \sqrt{2 \times 981 \times 6956.52}$$

$$= 58378.48 \text{ cm}^3/\text{s} = 58.378 \text{ l/s}$$

13.3.2 Variable Area Meters

In variable head meters, the constricted area of opening A_2 in Eq. (13.6) is fixed and the change in the volume rate of flow produces a corresponding pressure differential Δh. In the variable area meter, the area of the restriction can be altered to maintain a steady pressure difference.

A commonly used variable area flow meter is the *rotameter* shown in Fig. 13.5. In this device, the flow enters the bottom of a vertically placed tapered tube and causes the bob or 'float' (which has higher density than the fluid) to move upwards. The float will rise to a point in the tube where the drag force (upward direction) and buoyant force (upward direction) is balanced by the weight of float (downward direction). The position of the float in the tube is taken as an indication of the flow rate. Since the elevation of the float is dependent on the annular area between it and the tapered glass tube, it is also called the variable area orifice meter.

The force balance equation of the float is

$$F_{\text{drag}} + F_{\text{buoyancy}} = F_{\text{weight}}$$

$$A_f(p_d - p_u) + \rho_{ff}\, g\, V_f = \rho_f\, g\, V_f \tag{13.8}$$

$$(p_d - p_u) = \frac{V_f}{A_f}\, g(\rho_f - \rho_{ff}) \tag{13.9}$$

where ρ_f and ρ_{ff} are the densities of the float and flowing fluid, respectively
V_f is the volume of the float
p_d and p_u are the pressures at the downward and upward faces of the float, respectively.

Now a kind of constriction is formed between the downward surface and upward surface of the float. Using Eq. (13.6) we get the volume rate of flow:

$$Q_{\text{actual}} = C_d \frac{A_t(A_t - A_f)}{\sqrt{A_t^2 - (A_t - A_f)^2}} \sqrt{2g} \sqrt{\frac{(p_d - p_u)}{\rho_{ff}\, g}} \tag{13.10}$$

where A_t is the area of the tube at the float level $(A_t - A_f)$ is the minimum annular area between the tube and the float and C_d is the coefficient of the discharge.

Substituting the value of $(p_d - p_u)$ from Eq. (13.9) we get,

$$Q_{\text{actual}} = \frac{C_d(A_t - A_f)}{\sqrt{1 - (A_t - A_f)^2 / A_t^2}} \sqrt{2g} \sqrt{\frac{V_f}{A_f} \cdot \frac{(\rho_f - \rho_{ff})}{\rho_{ff}}} \tag{13.11}$$

If the variation of C_d with the float position is slight and if $(A_t - A_f)/A_t \ll 1$ then,

$$Q_{\text{actual}} = K(A_t - A_f) \tag{13.12}$$

where K is the constant of the rotameter.

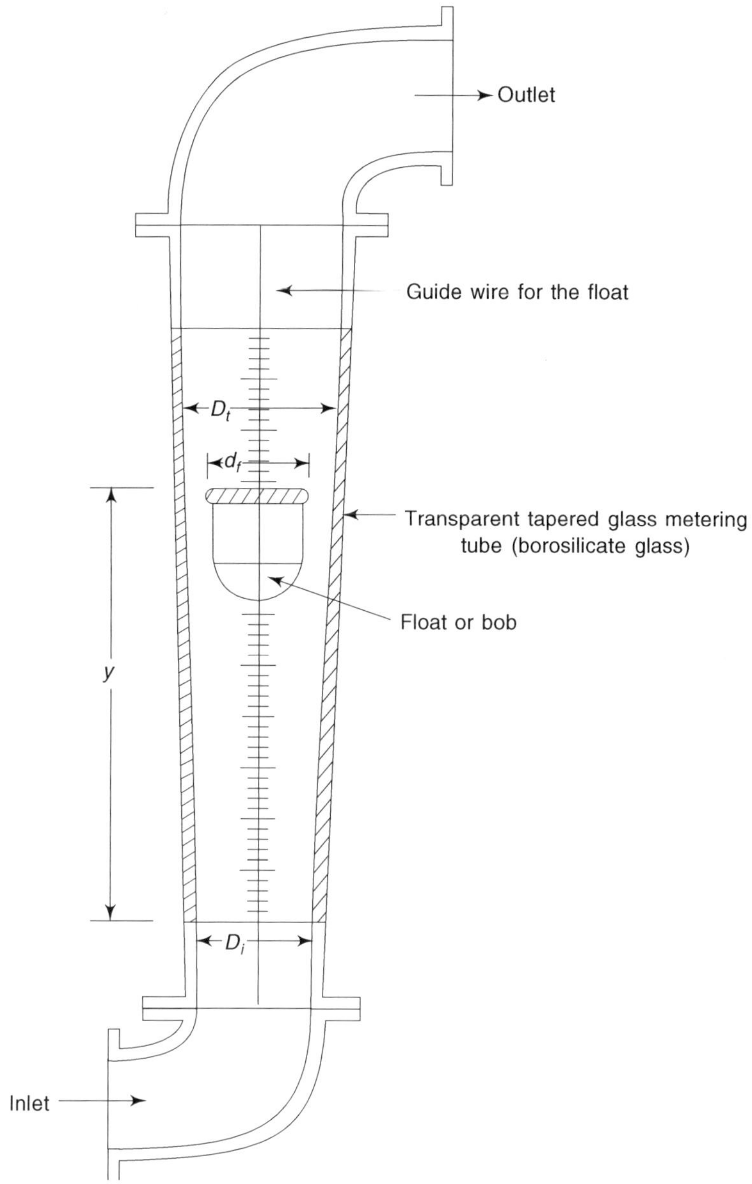

Fig. 13.5 *A rotameter*

If the angle of taper is θ (which is very small), then,

$$A_t = \frac{\pi}{4}(D_i + y \tan \theta)^2$$

$$= \frac{\pi}{4} D_i^2 + \frac{\pi}{2}\, y\, D_i \tan \theta \qquad (13.13)$$

where y is the float position with respect to inlet, and

D_i is the diameter at the inlet.

Substituting the value of A_t in Eq. (13.12) we get,

$$Q_{\text{actual}} = K \frac{\pi}{4} D_i\, y \tan\theta + K\left(\frac{\pi}{4} D_i^2 - A_f\right)$$
$$= K_1 y + K_2 \qquad (13.14)$$

where K_1 and K_2 are constants depending on the shape of the rotameter and float.

Thus, Eq. (13.14) shows that the rotameter gives a direct reading of the float on a linear scale.

The rotameter floats are made of different materials to obtain the desired density difference $(\rho_f - \rho_{ff})$ in Eq. (13.11) for metering a particular liquid or gas. For metering gases or air, a small sphere can be used in a narrow bore tube which requires no guiding in the tube. But for liquids, the floats are kept central by guide wires or internal ribs in the tube. The rotameter tube is often made of high strength borosilicate glass to allow direct observation of the float position. However, where greater strength is required metal tubes may be used and the float position is detected magnetically. Further, the float position can also be monitored electrically by using a suitable displacement transducer.

The *main advantage* of rotameters is that they give direct visual indication on a linear scale. Capacities of these instruments range from 0.1 ml/min to several hundred litres per minute. The accuracy is often within ±1% of the maximum flow rate, but very inexpensive units are available with accuracies of ±5%. Further, they can handle a wide variety of fluids including corrosive ones. However, the *limitations* of these units are that they must be mounted vertically and are subject to oscillations in pulsating flows.

Problem 13.2 *A rotameter is calibrated for metering a liquid of density 1000 kg/m^3 and has a scale ranging from 1 to 100 l/min. It is intended to use this meter for metering the flow of gas of density 1.25 kg/m^3 with a flow range between 20 and 2000 l/min. Determine the density of the new float, if the original one has a density of 2000 kg/m^3. The shape and volume of both floats is assumed to be the same.*

Solution Let the subscripts 1 and 2 refer to the liquid flow and gas flow, respectively, through the rotameter. Using Eq. (13.11) in simplified form we get the actual discharge as:

$$Q_1 = K\sqrt{(\rho_{f1} - \rho_{ff1})/\rho_{ff1}} \quad \text{for liquid flow}$$

and

$$Q_2 = K\sqrt{(\rho_{f2} - \rho_{ff2})/\rho_{ff2}} \quad \text{for gas flow}$$

where K is the constant of the rotameter.

The scale ratio between gas flow and liquid flow is

$$\frac{20}{1} = \frac{2000}{100} = 20$$

Therefore, $Q_2 = 20\, Q_1$ or $Q_2/Q_1 = 20$

Substituting the values of Q_1 and Q_2 from the rotameter discharge equation we get,

$$\frac{Q_2}{Q_1} = 20 = \left[\frac{(\rho_{f2} - \rho_{ff2})\,\rho_{ff1}}{(\rho_{f1} - \rho_{ff1})\,\rho_{ff2}}\right]^{1/2}$$

Substituting the values of ρ_{f1}, ρ_{ff1} and ρ_{ff2} and squaring we get,

$$400 = \frac{(\rho_{f2} - 1.25)\times 1000}{(2000 - 1000)\times 1.25}$$

Simplifying the above expression we get,

$$\rho_{f2} = \text{density of float for gas flows} = 501.25 \text{ kg/m}^3$$

13.3.3 Variable Head and Variable Area Flow Meters (Weirs)

Weirs are variable head, variable area flow meters used for measuring large volumes of liquids in open channels. These devices operate on the principle that if a restriction of a specified shape and form is placed in the path of the flow, a rise in the upstream liquid level occurs which is a function of the rate of flow through the restricted section.

Weirs have a variety of forms and are classified according to the shape of the notch or opening. The most commonly used weirs are the rectangular, the triangular or V-notch and the trapezoidal or cipolleti weir (Fig. 13.6). The rectangular weirs are quite suitable for measuring large flows, whereas the V-notch is used for smaller flows below 50 l/s.

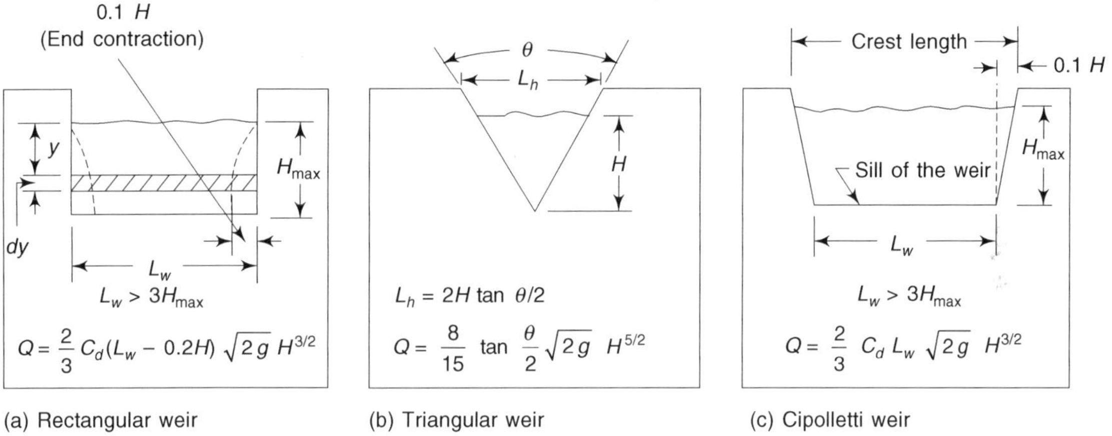

Fig. 13.6 *Measurement of flows in open channels*

The theoretical formula for the rectangular weir without end correction (or for suppressed weir with no end corrections) can be arrived at by considering a layer of fluid thickness *dy* at a depth of *y* from the top surface of the water level.

Velocity of layer of fluid $= \sqrt{2gy}$

Theoretical elemental discharge of the thin layer $= \sqrt{2gy}\; L_w dy$

Actual elemental discharge of the thin layer

$$= C_d \sqrt{2gy}\; L_w dy \tag{13.15}$$

where C_d is the coefficient of discharge with a value of between 0.57 and 0.64.

Integrating Eq. (13.15) for limits of *y* from 0 to *H*, we get the actual discharge as follows:

$$Q_{\text{actual}} = C_d L_w \sqrt{2gy} \int_0^H \sqrt{y}\; dy$$

$$= \frac{2}{3} C_d L_w \sqrt{2g}\; (H)^{3/2} \tag{13.16}$$

In actual practice, the effective crest length of the weir is less than the actual crest length L_w due to the formation of end contractions of 0.1 *H* on each side. Therefore, Eq. (13.16) is accordingly modified to take into account the effect of end contractions so that:

$$Q_{\text{actual}} = \frac{2}{3} C_d [L_w - 0.2\, H] \sqrt{2g}\; H^{3/2} \tag{13.17}$$

Alternatively, the effect of end contractions is taken care of by making the weir trapezoidal in shape so that the discharge Eq. (13.17) simplifies as:

$$Q_{\text{actual}} = \frac{2}{3} C_d L_w \sqrt{2g}\ H^{3/2} \tag{13.18}$$

The discharge equation for the triangular weir can also be derived in a similar manner by integrating the elemental discharge of a thin layer over the complete range of liquid level, i.e., from 0 to H. The discharge equation obtained would be

$$Q_{\text{actual}} = \frac{8}{15} C_d \tan\frac{\theta}{2} \sqrt{2g}\ H^{5/2} \tag{13.19}$$

The weirs are simple in construction, easy to instal and quite accurate devices that are used extensively for measuring large volume flows of liquids in open channels. They can be used for measuring dirty/muddy flows such as river flows, sewerage discharge, etc.

The *main disadvantage* of such devices is their non-linear characteristics, i.e., 3/2 power law for rectangular/Cippolletti weirs and 5/2 power law for triangular weir.

Problem 13.3 *A right-angled V-notch is employed to measure the discharge. If the head H above the sill is measured as 0.25 ± 0.01 m, estimate the discharge if C_d = 0.60.*

Solution The discharge through V-notch is given by Eq. (13.19) as

$$Q_{\text{actual}} = \frac{8}{15} C_d \tan\frac{\theta}{2} \sqrt{2g}\ H^{5/2}$$

$$= \left(\frac{8}{15}\right) (0.60) \tan 45° \sqrt{2 \times 9.81}\ (0.25)5/2$$

$$= 0.04429 \text{ m}^3/\text{s} = 44.29 \text{ l/s}$$

The discharge through the V-notch can written as

$$Q = CH^{5/2}$$

where C is constant.

Writing this equation in the differential form we get,

$$dQ = CH^{3/2}\, dH$$

Dividing the expression for dQ by Q we get,

$$\frac{dQ}{Q} = \frac{5}{2}\frac{dH}{H} = \frac{5}{2} \times \frac{0.01}{0.25} = 0.1$$

$\therefore$ $dQ = 0.004\,429 \text{ m}^3/\text{s} = 4.43 \text{ l/s}$

Hence, the discharge through the V-notch = (44.29 ± 4.43) l/s.

13.3.4 Linear Resistance Element Flow Meter

For very small flow rates or for highly viscous flows the linear resistance element flow meter (also known as capillary flow meter) is specially suitable. It is a constant head loss type of flow meter and its working principle is based on the well-known Hagen–Poiseuille equation for laminar flow (i.e., Re No. < 2200) in a tube, which is

$$Q = \frac{\pi D^4}{128\,\mu L} (p_1 - p_2) \tag{13.20}$$

where Q is the flow rate

D the inside diameter of the tube

L the length of the tube
μ the coefficient of viscosity
$(p_1 - p_2)$ the pressure drop along the length of the tube.

The total metered fluid is guided by means of flow straighteners to the metering element in the shape of a bundle of capillary tubes in honeycomb configuration (Fig. 13.7). Because of the high viscosity, small flow rate and small size diameter tube, the Reynolds number (Re No. = $\rho VD/\mu$) is quite small and is in the laminar range for which Eq. (13.20) is applicable.

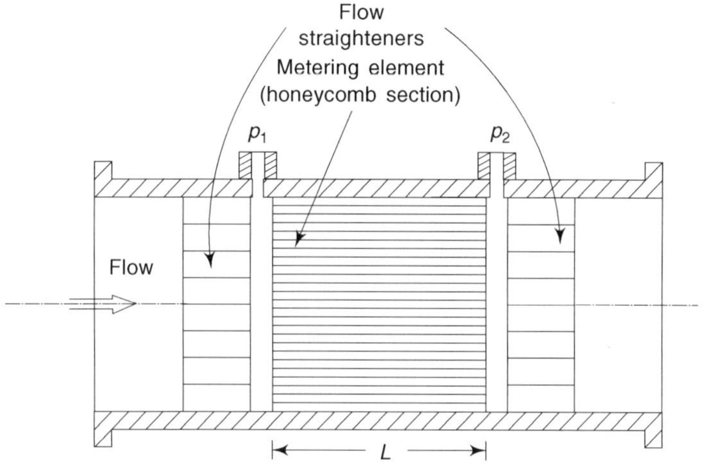

Fig. 13.7 *Linear resistance element flow meter*

The *principal advantage* of this type of device is that the flow rate is directly proportional to the pressure drop and that is why it is termed as linear resistance element flow meter. Further, it gives quite accurate average measurements of pulsating flows because of the good damping ability of the device. In addition, it can measure reversed flow with no difficulty. The *main disadvantage* of the device is that the metering element is subject to plugging if the metered fluid is not clean. Further, these units are relatively expensive and are associated with high-pressure losses of the metered flows.

Problem 13.4 *A single tube capillary flow meter has a capillary tube of 2 mm bore and 30 cm length. Determine the amount of discharge in 100 s if the flow of oil of sp. gr. 0.9 and coefficient of viscosity μ = 0.006 N-s/m^2 gives a pressure differential of 20 cm of oil in a capillary length of 30 cm.*

Solution Using Eq. (13.20) we get,

$$Q = \frac{\pi D^4}{128\,\mu L}(p_1 - p_2)$$

$$= \frac{\pi \times (0.002)^4}{(128)(0.006)}\,\frac{(900 \times 9.81 \times 0.2)}{0.30} = 3.852 \times 10^{-7}\ \text{m}^3/\text{s}$$

$$= 0.3852\ \text{cm}^3/\text{s}$$

Amount of discharge through the capillary in 100 s = 38.52 cm^3.

13.3.5 Pitot-Static Tube

In many experimental studies of the fluid flow phenomenon, it may be necessary to determine the local velocities in the stream or the velocity distribution profile in a cross-section of the flow field. From the knowledge of velocity profiles and the geometry of the passage, the flow rate can be determined by integrating the velocity data over the area of the flow passage.

A commonly used variable head velocity measuring device is the pitot-static tube, usually called simply pitot tube. It is perhaps better known as the air-speed indicator used in aircrafts.

Figure 13.8 shows a schematic diagram of the pitot-static tube. It consists of two concentrically arranged tubes bent at right angles. The inner tube, also known as stagnation tube, is open-ended and faces the incoming stream of fluid. The fluid impinging this open end is brought to rest and its kinetic energy is converted to the pressure head commonly known as velocity head. Thus, the pressure sensed by the stagnation tube (called the stagnation pressure) is greater than that in the free stream by velocity head. In other words, the stagnation pressure consists of the velocity head and the static pressure head of the free stream. Now, the outer or static tube of the pitot tube is closed at the nose of the tube. Further, it is shaped in the form of a streamlined body (ellipsoidal shape) at the nose of the tube so as to avoid flow separations. The accuracy in the free stream static pressure depends on the position of the sensing holes with respect to the nose of the tube and the main supporting stem. Streamlines next to the nose of the tube must be longer than in the undisturbed flow, indicating an increase in velocity. Therefore, acceleration effects tend to lower the static pressure at the nose. On the other hand, the right-angled stem stagnates the flow and tends to raise the static pressure in its vicinity. Thus, in a properly compensated static tube, the location of the static holes is such that the acceleration effects due to the nose and the stagnation effects due to the stem just balance each other in the plane of the static pressure holes. In practice, the compensated pitot-tubes have the static-pressure holes at least eight diameters downstream from the front of the probe. It may be noted that even though the static holes are placed downstream of the nose of the pitot tube, they in fact indicate the corrected/compensated value of the static pressure at the nose of the probe.

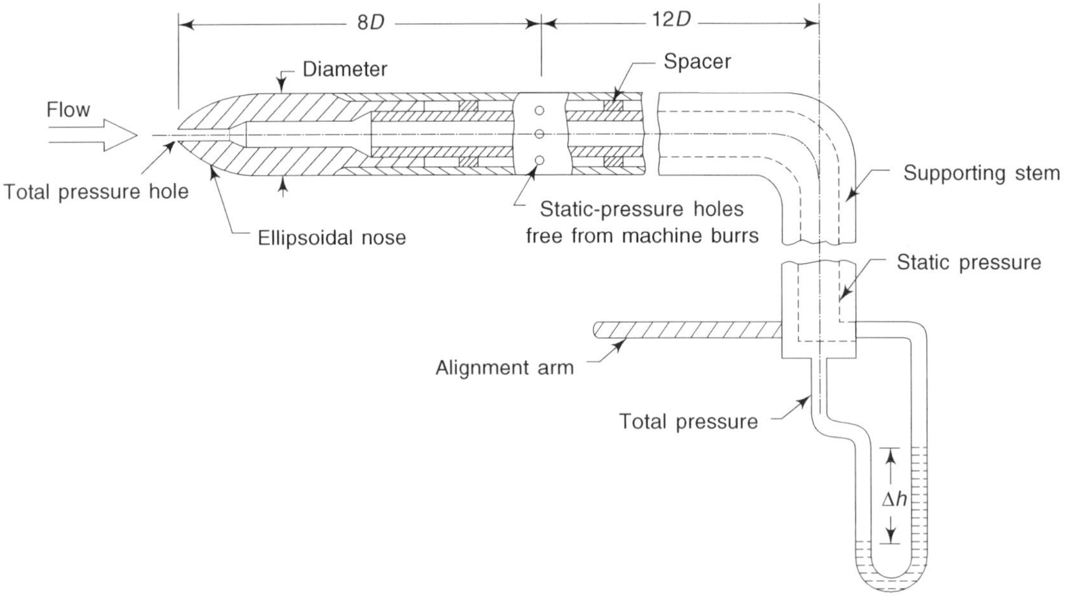

Fig. 13.8 *A pitot-static tube*

Assuming a steady, one-dimensional flow of an incompressible, frictionless fluid, we can derive the expression for free stream velocity by applying Bernoulli's equation between a point in the free stream and another point at the tip of the stagnation tube. Thus

$$\frac{p_{\text{stat}}}{\rho} + \frac{V^2}{2} = \frac{p_{\text{stag}}}{\rho}$$

which gives

$$V = \sqrt{\frac{2(p_{\text{stag}} - p_{\text{stat}})}{\rho}} \tag{13.21}$$

where V is the flow velocity
ρ is the density of the fluid
p_{stag} is the stagnation or total pressure of the free stream given by the stagnation tube
p_{stat} is the free stream static pressure given by static tube.

The pitot tube has the following advantages:

1. It is a simple and low-cost device,
2. It produces no appreciable pressure loss in the flow system,
3. It can be easily inserted through a small hole into the pipe or duct, and
4. It is very useful for checking the mean velocities of the flows in venturi, nozzle, orifice plate or any other complex flow field.

The limitations of this device are follows:

1. It is not suitable for measuring low velocities, i.e., below 5 m/s, because of difficulties in the accurate measurement of pressure differential.
2. It is sensitive to misalignment of the probe with respect to free stream velocity. Usually an angle of yaw or misalignment up to 5° has little effect on the velocity values but beyond 20° the error in the velocity determination is of the order of 2%.
3. It is not suitable for the measurement of highly fluctuating velocities, i.e., highly turbulent flows.
4. The use of pitot-tube is limited to exploratory studies. It is not commonly used in industrial applications as numerous pitot tube traverses are required for velocity distribution data which is quite tedious and time-consuming.

Problem 13.5 *A submarine moves horizontally in the sea and has its axis much below the surface of sea water. A pitot-tube properly placed just in front of the submarine is connected to a differential pressure gauge. The pressure differential between the pitot pressure and static pressure was found to be 20 kN/m². Find the speed of the submarine if the density of sea water is 1026 kg/m³.*

Solution The pressure differential between the pitot pressure and static pressure

$$= 20 \text{ kN/m}^2 = \frac{20 \times 10^3}{1026 \times 9.81}$$

$$= 1.987 \text{ m of sea water.}$$

This head is due to the velocity of fluid with respect to the submarine, i.e., due to the velocity of the submarine.

$$\therefore \quad \frac{V^2}{2g} = 1.987$$

or

$$V = \sqrt{1.987 \times 2 \times 9.81} = 6.24 \text{ m/s}$$

13.3.6 Target Flow Meter

The insertion of a suitably shaped body (obstruction) into the flow stream can serve as a flow meter. The drag force on the body becomes the measure of the flow rate after suitable calibration. The drag force F_d acting on the body immersed in a flowing fluid is given by

$$F_d = \frac{1}{2} C_d \rho g V^2 A \tag{13.22}$$

where C_d is the coefficient of drag
A is area of cross-section (in m^2)
ρ is the fluid density (in kg/m^3)
V is the fluid velocity (in m/s).

For a sufficiently high Reynolds number, the drag coefficient is reasonably constant. Therefore, for a given density, F_d is proportional to V^2 and thus square of volume flow rate. The drag force of the body can be measured by attaching the drag body to a suitable force measuring device. One such device commonly used in this type of meter is the cantilever beam arrangement with bonded strain gauges (Fig. 13.9). The *main advantage* of this type of flow meter is its very good dynamic response. However, when it is used for dynamic flow studies, the strain gauge output should preferably be linearised by incorporating a suitable square-root element. The overall accuracy of the instrument is ±0.5% and repeatability within ±0.1%.

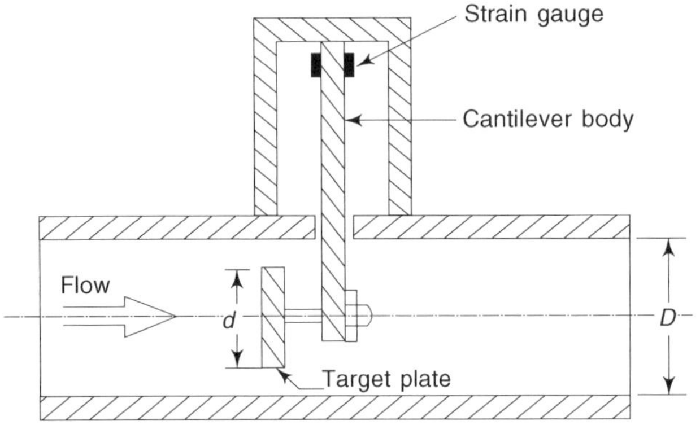

Fig. 13.9 *Target flow meter*

A small zero shift occurs if the meter is calibrated in the horizontal position and is used in the vertical position. This error can be taken care of by suitable zero adjustment. Further, with a symmetrically shaped drag body, the meter can work for flow in either direction.

It may be noted that drag force flow meter is specially suited to flows in which differential pressure measurements before and after the obstruction are not conveniently obtainable. Common examples of such flows are the highly viscous flows of hot asphalt, tar, oils or slurries at high pressures of the order of 100 bars.

13.3.7 Turbine Flow Meter

The turbine type of flow measuring devices incorporate some form of multi-vane rotor driven by the metered fluid. A very familiar device of this type is the cup-type anemometer employed by weather stations. Another somewhat similar type of device is used by civil engineers for measuring the flows in rivers, streams, channels, etc. in which the rotor is in the form of a propeller and the device is termed as propeller-type current meter. If the frictional torque is small, then the number of turns of the rotor per unit time gives a measure of the volumetric rate of flow over a wide range of flow rates.

Figure 13.10 illustrates a typical turbine flow meter with an electrical output suitable for measuring the flow in tubes and pipes. In this device, the rotor motion, proportional to the flow rate, is sensed by means of a proximity type of pick-up of reluctance type. A permanent magnet is encased in the rotor body and each time the rotating magnet passes the pole of the pick-up coil, the change in permeability of the magnetic circuit produces a voltage pulse at the output terminal. These voltage pulses are counted by means of an electronic digital counter to give the total flow. Alternatively, the frequency is converted into voltage and is fed to an analog/digital voltmeter to give the rate of flow. In fact, in a turbine flow meter it is possible to simultaneously get the measurement of total flow as well as the rate of flow with simple and easily available electronic components/devices.

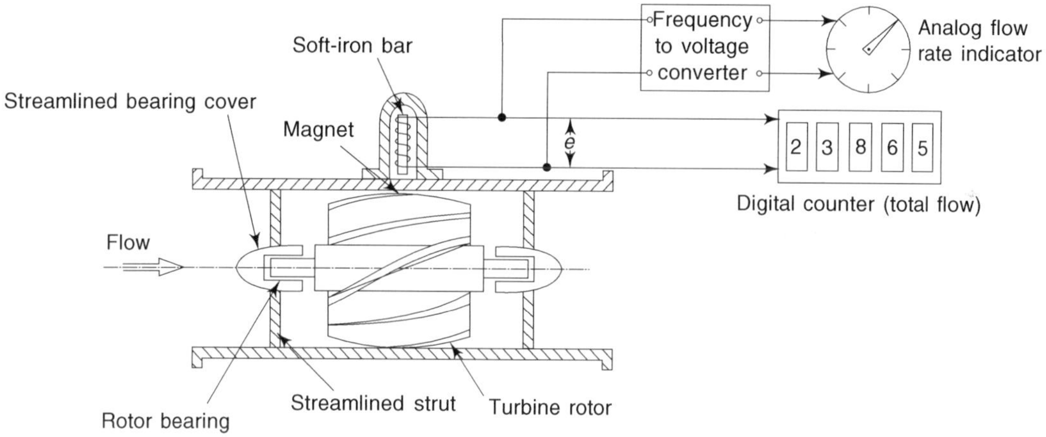

Fig. 13.10 *Turbine flow meter*

The *main advantage* of the turbine flow meter is the linear relationship between the volume flow rate and the angular velocity of the rotor. The relationship can be written as

$$Q = k\,n \tag{13.23}$$

where Q is the volume flow rate
n is the rotor angular velocity (in rad/s)
k is constant for any given meter and is independent of the fluid properties at high flow rates.

Turbine flow meters are commercially available with full-scale flow rates ranging from 0.5 to 150 000 1/min for liquids and 5 to 100 000 1/min for air. The accuracy of the instrument is ±1.0% over the design range. Further these meters can follow the flow transients quite accurately since their fluid/mechanical time constant is of the order of 2–10 ms. However, the bearing maintenance is a problem and accuracy drops off greatly at low flow rates.

13.3.8 Vortex-shedding Flow Meter

This type of flow meter operates on the principle that if a bluff body (non-streamlined body) or an obstruction is placed in a fluid stream, vortices are shed alternately on each side of the bluff body. In other words the wake behind the bluff body, which is termed as Karman Vortex street, consists of separated flow containing the eddies/vortices alternately shed from each side of the bluff body. Now, the vortex-shedding frequency of the bluff body is a measure of the average flow velocity of the fluid flow.

The fluid parameter which governs the operation of the vortex-shedding flow meter is the non-dimensional number, Strouhal number S which is defined as

$$S = \frac{f_s D}{V} \tag{13.24}$$

where f_s = vortex-shedding frequency
D = diameter or characteristic length of the bluff body
and V = average velocity of the flow

The type of the bluff body i.e., the shedder which is commonly employed in the vortex-shedding flow meter is a symmetrical shaped body in the form of a symmetrical triangular wedge and extends the diametral length of the pipe as shown in Fig. 13.11. This type of body is capable of generating strong and consistent vortex-shedding for a wide range of Reynolds numbers. The height of the bluff body 'h' is governed by the pipe size. In general, the ratio of the height of the bluff body to the pipe diameter should not be less than 0.2. In practice, it may be taken as one-third the diameter of the pipe. Further, the length L of the shedder is taken approximately as 1.3 h with a sharp edge just before the triangular portion, from where the vortices start shedding alternately from the top and bottom sides.

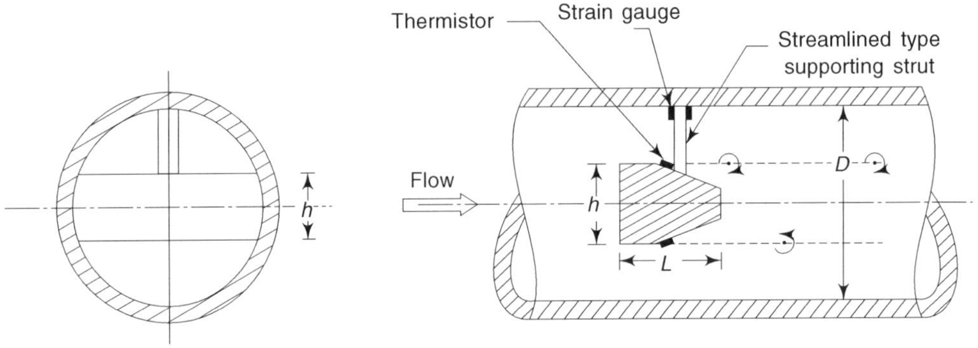

Fig. 13.11 *Schematic diagram of vortex-shedding flow meter*

The value of Strouhal number S for the triangular wedge-shaped body remains nearly at a constant value of 0.88 ± 0.01 over a very wide range of Reynolds numbers ranging from 10^4 to 5×10^6. Using Eq. (13.24), the governing equation of the flow meter becomes:

$$V = \frac{D f_s}{0.88} \quad \text{(i.e., when } S = 0.88)$$
$$= (1.136D) f_s$$
$$= K f_s \tag{13.25}$$

where K = constant or calibration factor of the flow meter = 1.136 D

It shows that as flow velocity increases with the increase in flow rate, the shedding frequency of the vortices also increases linearly, at the same rate. Further, the constant K of the flow meter does not depend on the state of the metered fluid i.e., it is same for air, water or two-phase flow (like wet-steam in a boiler). In addition, K is also independent of the fluid properties like flow density, pressure, temperature and viscosity.

The flow rate Q in the vortex flow meter can be evaluated as follows:

$$Q = \frac{\pi}{4} D^2 V_u = \left(\frac{\pi}{4} D^2 - hD\right) V_d \tag{13.26}$$

where V_u is the velocity of flow upstream of the shedder and V_d is the velocity of flow downstream of shedder.

Substituting the value V which is equal to V_d from Eq. (13.25) we get,

$$Q = (1.136\, D f_s) \left(\frac{\pi}{4} D^2 - hD\right) \tag{13.27}$$

The measurement of the vortex-shedding frequency is accomplished monitoring the alternating strain signal from the strain gauge or piezo-electric pick-up mounted on the cantilever type of supporting strut of the bluff body. The alternating strain signal is generated by the alternating type of vortices when they strike the supporting struts. Alternatively, a thermistor may be bonded either on the top face or the bottom face of the shedder to sense velocity fluctuations caused by the cooling effect of the vortices. The output of the flow meter in both cases is electrical and can be suitably processed to either indicate instantaneous flow or totalised flow.

The following are the advantages of the vortex-shedding flow meter:

1. It has no moving parts and causes very low pressure loss in the metered fluid.
2. The calibration constant remains constant in the wide range of Reynold's numbers i.e., 10^4 to 5×10^6. Hence, the instrument has linear characteristics with respect to the shedding frequency.
3. The calibration constant is same for all fluids which include hazardous or corrosive liquids/gases, slurries or cryogenic liquids or two-phase flows.
4. The calibration constant is not affected by the changes in the fluid properties like density, temperature, pressure, viscosity, etc.
5. The instrument is very accurate and precise. Accuracy and precision errors are of the orders of $\pm 0.5\%$ and $\pm 0.1\%$, respectively.
6. The flow meter is of portable type and can be fabricated as an integral piece consisting of shedder body fitted in a pipe along with an eddy frequency sensor. It can be then inserted in the flow field for which the average velocity or instantaneous flow or the totalised flow rate can be monitored.

The *main limitation* of the meter is its inability to give accurate results for Reynold's number $< 10^3$ or $> 10^7$. For these values, the value of strouhal number of the pulsating flow shows marked variation from the value of 0.88.

13.4 ■ SPECIAL METHODS

There are certain flow measuring devices that cannot be strictly categorised as fluid mechanical types of primary flow measuring devices or the indirect type of rate flow meters. Therefore, these are discussed as special type of devices where principle of operation is generally non-fluid-mechanical in nature. For example, ultrasonic time delay technique, electromagnetic effects of fluid flows, heat transfer from heated wires, Doppler frequency shift of scattered light, etc. are some of the principles used in the measurement of fluid flow parameters.

13.4.1 Ultrasonic Flow Meters

The operating principle of this type of instrument is based on the apparent change in the velocity of propagation of sound pressure pulses in a fluid with a change in velocity of the fluid flow. In practice, we employ short bursts of sinusoidal pressure pulse whose frequency is above the audio-frequency range (i.e., above 20 000 Hz). A typical frequency may be of the order of 10 MHz (10^7 Hz).

Figure 13.12(a) illustrates a typical single receiver-transmitter system of acoustic (ultrasonic) pulses. With zero flow velocity, the transit time t_0 of the pulse from the transmitter to the receiver is given by

$$t_0 = \frac{l}{V_s} \tag{13.28}$$

where l is the distance between transmitter and receiver
V_s the velocity of sound in the fluid

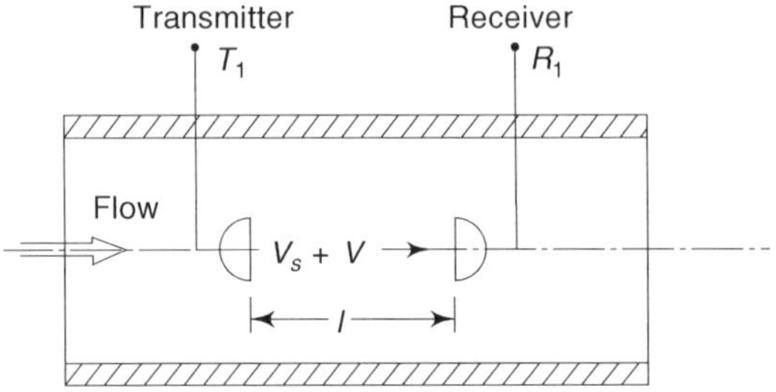

(a) Travel time difference method (single transmitter–receiver system)

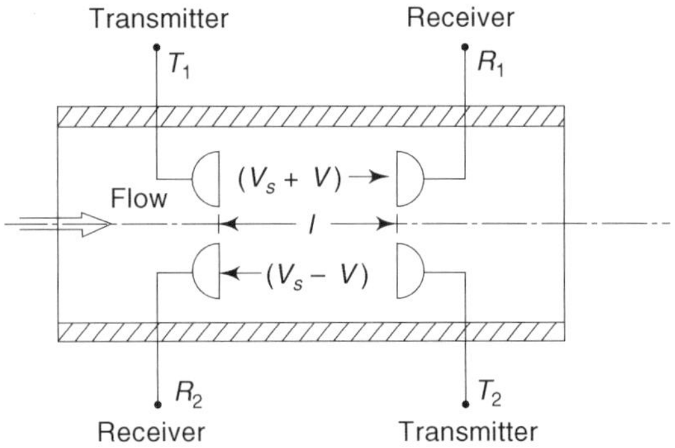

(b) Travel time difference method (twin transmitter–receiver system)

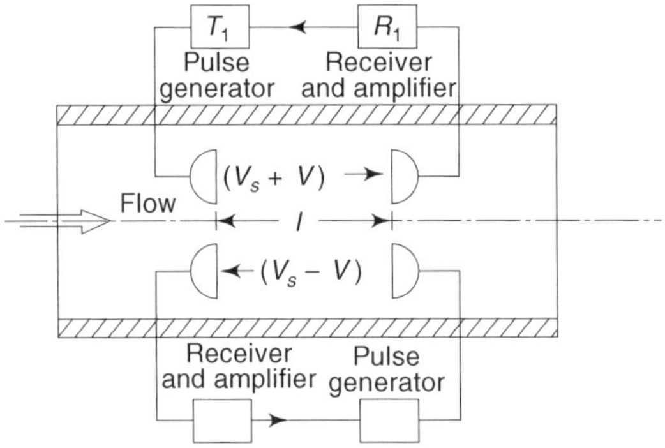

(c) Oscillating loop method

Fig. 13.12 *Ultrasonic flow meters*

Now, in a fluid moving with a velocity V, the transit time t becomes

$$t = \frac{l}{V_s + V}$$

$$= \frac{l(V_s - V)}{V_s^2 - V^2} \tag{13.29}$$

For flow velocities of liquids encountered in practice, the error involved would be negligible if we assume $V \ll V_s$. Therefore, Eq. (13.29) becomes

$$t \simeq \frac{l}{V_s}\left(1 - \frac{V}{V_s}\right)$$

$$= t_0\left(1 - \frac{V}{V_s}\right) \tag{13.30}$$

If we define $\Delta t = (t_0 - t)$ as the travel time difference, then Δt becomes

$$\Delta t = t_0 \frac{V}{V_s}$$

$$= \frac{lV}{V_s^2} \tag{13.31}$$

Since the measurement of t_0 is generally not provided for in the present arrangement, therefore, it is preferable to have an additional set of transmitter-receiver system along with the present system [Fig. 13.12(b)] to determine the transit time against the direction of flow. If t_1 is the transit time along the flow and t_2 the transit time against the flow, then we get the travel time difference Δt as

$$\Delta t = t_2 - t_1$$

$$= \frac{l}{V_s - V} - \frac{l}{V_s + V} = \frac{2lV}{V_s^2 - V^2}$$

$$\simeq \frac{2lV}{V_s^2} \qquad (\because \quad V \ll V_s) \tag{13.32}$$

Thus the output signal proportional to Δt is linear in V for constant V_s. However, the calibration constant is strongly dependent on V_s which depends on the temperature and pressure of the flow field.

Another approach in this techniques is the *oscillating loop* system, also called the frequency difference method. In this system, the effect of sonic velocity is eliminated by arranging each pair of transducers in an oscillating loop [Fig. 13.12(c)]. A pulse is emitted by the transmitting transducer T_1 (pulse generator) and is received by the receiving transducer R_1 after time t_1. This pulse is amplified and instantaneously fed back to the transmitting transducer for retransmission. This generates a train of pulses in each path whose time period equals the acoustical travel time. Now the repetition frequencies along and against the fluid flow are

$$f_1 = \frac{1}{t_1} = \frac{V_s + V}{l} \tag{13.33}$$

and

$$f_2 = \frac{1}{t_2} = \frac{V_s - V}{l} \tag{13.34}$$

Now the frequency difference or beat frequency becomes

$$\Delta f = f_1 - f_2$$

$$= \frac{2V}{l} \tag{13.35}$$

Thus the frequency difference is directly proportional to V which affords a useful linear relationship. In addition, the expression is independent of the value of sonic velocity V_s. However, this method has the drawback that the frequency difference is usually very small and a long counting interval is needed in actual practice resulting in relatively large response time for reasonably good resolution.

The following are the advantages of an ultrasonic flow meter.

1. offers negligible resistance to the metered fluid system,
2. has a reasonably good accuracy, of the order of ±2% of the full-scale value,
3. has a linear relationship between the velocity and the output,
4. is suitable for both liquids and gases, and
5. since the output is electrical, the read-out can easily be either analog or digital.

Because of the above-mentioned advantages, the device finds special applications, namely measurement of ocean currents, vessel speeds, water flows in large conduits, flows of various bio-medical and industrial fluids, etc.

The main *disadvantages* of the instrument are its relatively high cost due to close tolerances and high accuracies involved in the design and manufacture of mechanical, acoustical and electronic portions of the system. This has somewhat limited its wide use in industrial applications presently.

13.4.2 Electromagnetic Flow Meter

Rate of flow may be determined, for electrically conducting fluids, by measuring the emf induced across the fluid stream when it passes through the magnetic field. This technique cannot be used for electrically non-conducting fluids like gases. However, it gives quite satisfactory results for fluids with as low conductivity as water.

The principle of operation of the unit is directly analogous to Faraday's law of electromagnetic induction for solid conductors. The law states that whenever a conductor cuts lines of magnetic field, an induced emf is generated and the magnitude of this emf is proportional to the rate at which these lines are cut and the emf is perpendicular to the plane of conductor and the magnetic field. The direction of the induced emf is given by the Fleming's right hand rule [Fig. 13.13(a)].

The construction of the electromagnetic flow meter [Fig. 13.13(b)] consists of the following:

1. A permanent magnet or an electromagnet which may be either ac or dc around a non conducting pipe.
2. Two electrodes placed at right angles to the magnetic field for picking up the induced emf.
3. Fluid flow in the pipe which is at right angles to the plane of magnetic flux lines and the induced emf direction which is along the line joining the electrodes.

Now for the conducting fluid flows, the induced voltage E_0 generated according to Faraday's law is

$$E_0 = Blv \times 10^{-8} \text{ volts} \tag{13.36}$$

where B = magnetic flux density (in V-s/cm^2)

l = length of the conduction (in cm)

v = velocity of the conductor (in cm/s)

The effective length of the conductor corresponds to the inner diameter of the pipe and the velocity of the conductor is proportional to the mean flow velocity. The volume flow rate for the circular pipe is given by $Q = (\pi/4)d^2v$. Thus Eq. (13.36) can be modified as

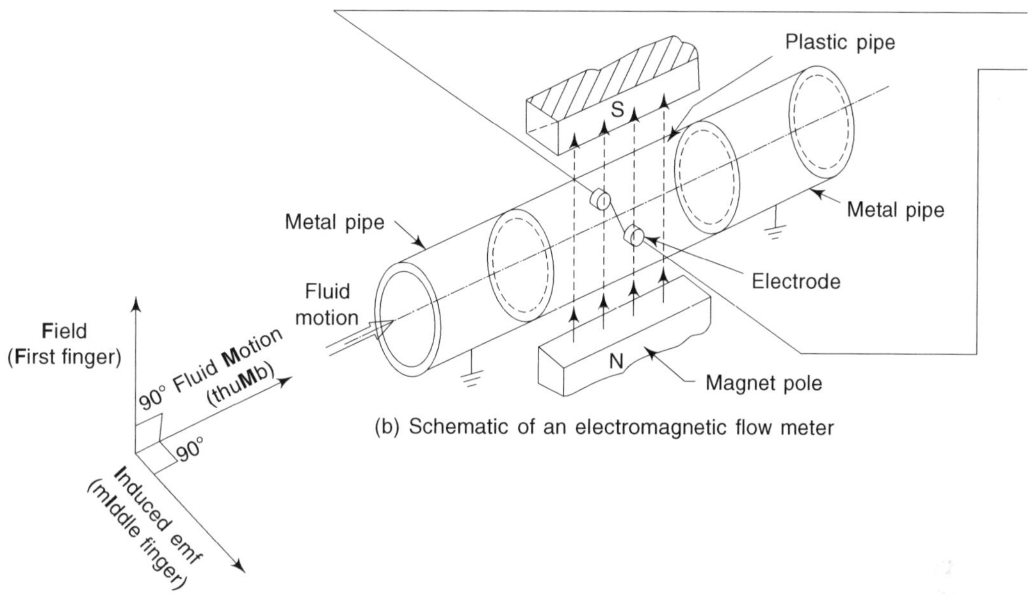

(b) Schematic of an electromagnetic flow meter

(a) Fleming's right hand rule

Fig. 13.13 *Application of Fleming's right hand rule in electromagnetic flow measurements*

$$E_0 = \frac{4B}{\pi d} Q \times 10^{-8} \text{ volts} \tag{13.37}$$

This shows that the volume flow rate Q is directly proportional to the induced emf E_0 as long as the flux density remains constant.

The emf is measured by means of two electrodes built into the non-magnetic length of pipe. Further, these electrodes are made to be completely flush with the inner surface so that they do not obstruct the flow and at the same time are in direct contact with the flowing liquid. Unfortunately, the emf generated by the instrument is very small, i.e., typically of the order of 1 mV for a fluid velocity of 1 m/s, and the resistance of the fluid is often very high. Therefore, the output of the instrument is generally amplified suitably.

In the practical equipment, alternating magnetic field is usually preferred. The main reason is that it prevents polarisation of the electrodes (i.e., collection of gas bubbles on the electrodes due to electrolytic action which form the insulating pockets in the vicinity of the electrodes).

The *main advantage* of the electromagnetic flow meter is that it causes no obstruction in the flow line of the metered fluid. This makes it particularly suitable for fluids containing solid matter. Further, the device is quite accurate and has a wider linear range with good transient response. That is why this instrument is widely used for metering corrosive acids, cement slurries, sewage, paper pulp, detergents, greasy and sticky fluids, etc. However, these meters are usually expensive and their use is limited to fluids having conductivity at least of the order of 1×10^{-6} ℧/cm.

13.4.3 Hot Wire Anemometer

Hot wire anemometer is a commonly used device for measuring the mean and fluctuating velocities in fluid flows. The flow sensing element is a short length of 5 μm diameter platinum–tungsten wire welded between two prongs of the probe and heated electrically as a part of Wheatstone bridge [Fig. 13.14(a)].

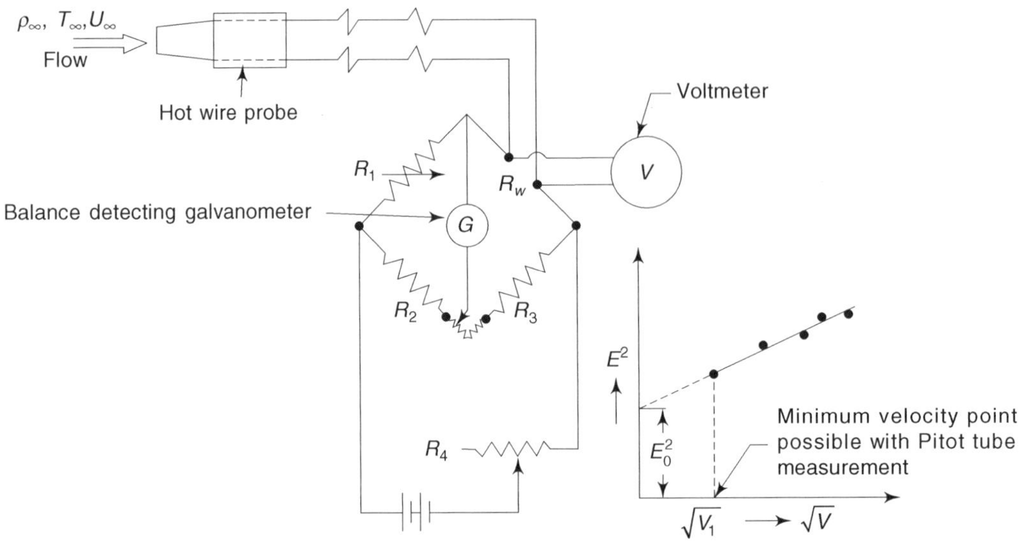

(a) Hot wire bridge circuit

(b) A typical calibration curve of hot wire anemometer

Fig. 13.14 *Schematic diagram of hot wire operation*

When the probe is introduced in the fluid stream, it tends to get cooled by the instantaneous velocity and consequently there is a decrease in its resistance. The rate of cooling of the wire depends on the following:

1. shape, size and physical properties of the hot wire,
2. difference of temperature between the heated hot wire and the fluid stream,
3. physical properties of flowing fluid, and
4. velocity of the fluid stream.

Generally, the first three conditions are effectively constant in the hot wire operation and the instrument response is then a direct measure of the flow velocity.

The hot wire anemometer may be used to measure the velocity in two ways. The first mode known as constant current method uses an electric circuit adjusted to feed constant current to the hot wire. A knowledge of this current and the resistance of the wire defines the power being fed to the wire, which is a function of the flow velocity. In this constant current operation, a large resistance is put in series with the hot wire and a thermal compensating circuit is also applied to the output ac voltage. The constant current system though simple in construction has a limited range of velocities. However, for electronic stability and amplification reasons, it can be used at higher frequencies and relatively smaller signals.

The second mode of hot wire operation is termed as the constant resistance or constant temperature method. In this the current in the hot wire sensor is continuously adjusted by means of a suitable servo system to maintain the wire resistance and hence the wire temperature at a constant value throughout the range of hot wire operation. The current (or the voltage across the wire) is then a measure of the heat transfer rates and consequently of the flow velocities. The constant temperature design has the inherent advantage that the wire is protected against burn out. However, the associated electronic circuitry, specially the dc feedback amplifier has an upper limit of the frequency response beyond which the system becomes unstable and the wire temperature oscillates.

The basic governing equation of the hot wire operation is based on the well-known King's law for the convective heat transfer from the heated wire, which is

$$\frac{hD}{k} = 0.30 + 0.5\left(\frac{\rho VD}{\mu}\right)^{1/2} \tag{13.38}$$

for $$\frac{\rho VD}{\mu} > 10^2$$

where h is convective film coefficient of heat transfer
k is the thermal conductivity of the hot wire
ρ is the density of the fluid.
V is the velocity of the fluid stream
μ is the coefficient of viscosity of the fluid
D is the diameter of the hot wire.

Since the flow properties and material parameters like diameter and k are fixed for a particular wire being used, therefore, we can simplify Eq. (13.38) as follows:

$$h = c_1 + c_2\sqrt{V} \tag{13.39}$$

For equilibrium condition in a hot wire, the electrical energy input is equal to the convective heat transfer in the flow. Therefore, writing the energy balance equation we get,

$$I^2R_w = K_chA\,[T_w - T_f] \tag{14.40}$$

where I is the current flowing in the hot wire
R_w the resistance of the wire
K_c the conversion factor from electrical to thermal power
A the heat transfer area
T_w the hot wire temperature
T_f the temperature of flowing fluid

Substituting the value of h from Eq. (13.39) in (13.40) and simplifying we get,

$$I^2R_w = K_cA\,(T_w - T_f)\,(C_1 + C_2\,\sqrt{V}\,) \tag{13.41}$$

The change in resistance from temperature of fluid T_f to hot wire temperature T_w is directly proportional to the temperature difference for the platinum–tungsten material. Therefore,

$$R_w - R_f = C_3\,(T_w - T_f) \tag{13.42}$$

Equation (13.42) now becomes

$$\frac{I^2 R_w C_3}{K_c A(R_w - R_f)} = C_1 + C_2\sqrt{V} \tag{13.43}$$

Introducing new constants A_1 and B_1 in place of C_1 and C_2 to account for constants C_3, K_c and A in Eq. (13.43) we get,

$$\frac{I^2 R_w}{(R_w - R_f)} = A_1 + B_1\sqrt{V} \tag{13.44}$$

For constant resistance operation, the factor $R_w/(R_w - R_f)$ is constant and the value of hot wire current I can be replaced by E which is directly proportional to I. Introducing new constants A_2 and B_2 to account for the constant $R_w/(R_2 - R_f)$ and proportionality constant between I and E we modify Eq. (13.44) to give the probe calibration equation as:

$$E^2 = A_2 + B_2\,\sqrt{V} \tag{13.45}$$

Equation (13.45) is applicable to constant current operation as well. This is because the factor $R_w/(R_w - R_f)$ remains nearly constant for a limited range of low velocities for which this mode of operation is applicable.

The hot wire probe calibrated against the pitot static tube by putting both probes side-by-side in a wind tunnel or any other controlled air stream. The calibration plot between E^2 and $\sqrt{V}$ is obtained. The curve is extrapolated in the lower range to give the value of E_0 for zero velocity [see Fig. 13.14(b)] and the calibration Eq. (13.45) is generally written in the form:

$$E^2 = E_0^2 + B\sqrt{V} \tag{13.46}$$

For turbulence level measurements, the ratio of the rms turbulence intensity to the local mean velocity can be obtained by differentiating Eq. (13.46), giving

$$2E\mathrm{d}E = \frac{1}{2}B(V)^{-1/2}\,\mathrm{d}V$$

or

$$\frac{B\mathrm{d}V}{\sqrt{V}} = 4E\mathrm{d}E \tag{13.47}$$

Substituting the value of B from Eq. (13.46) in Eq. (13.47) we get,

$$\frac{E^2 - E_0^2}{\sqrt{V}}\,\frac{\mathrm{d}V}{\sqrt{V}} = 4E\,\mathrm{d}E$$

or

$$\frac{\mathrm{d}V}{V} = \frac{4E\mathrm{d}E}{E^2 - E_0^2}$$

or

$$\text{turbulence level} = \frac{4Ee}{E^2 - E_0^2} \tag{13.48}$$

where e is the rms value of the fluctuations in the hot wire signal.

The hot wires used in practice are of platinum, nickel, tungsten or platinum-rhodium (90/10). The wire needs to be strong enough to resist the impinging dirt/dust particles, yet fine enough to give adequate resistance and extremely small thermal capacity in order to follow the fluctuations in velocities faithfully and with infinitesimal time lag. The usual sizes employed are of 1–2 mm length and 2–5 μm in diameter [Fig. 13.15(a)].

It may be noted that accumulated dirt on the wire can lead to serious heat transfer errors in only few minutes of operation at high velocities unless the flow is exceptionally clean. Further, the hot wire system is suitable for only gas flow measurements. For liquid flows, hot film probes [Fig. 13.15(c)] are employed but the associated electronic circuitry is the same as for the hot wire probe. The hot film is usually 5μm thick and is coated with a thin layer of epoxy to avoid short circuiting in liquid flows. Films are usually of gold, platinum or nickel which are made by suitable coating technique like vacuum deposition/sputtering etc. Pyrex or similar glass is a universal material for the support (substrate) for such probes.

The following are the advantages of the hot-wire anemometer:

1. The output of the instrument is electrical and therefore the read-out can be analog or digital.
2. The time constant of the instrument is of the order of 10^{-4} to 10^{-6}s. Therefore, the instrument exhibits excellent dynamic characteristics.
3. The instrument shows excellent accuracy of the order ±0.1% in the measurement of mean velocities and ±2.0% in the measurement of turbulence levels.
4. The instrument has extended measuring range from very low velocities to supersonic velocities.
5. Probe size is very small and it does not cause much disturbance or pressure loss to the flow.

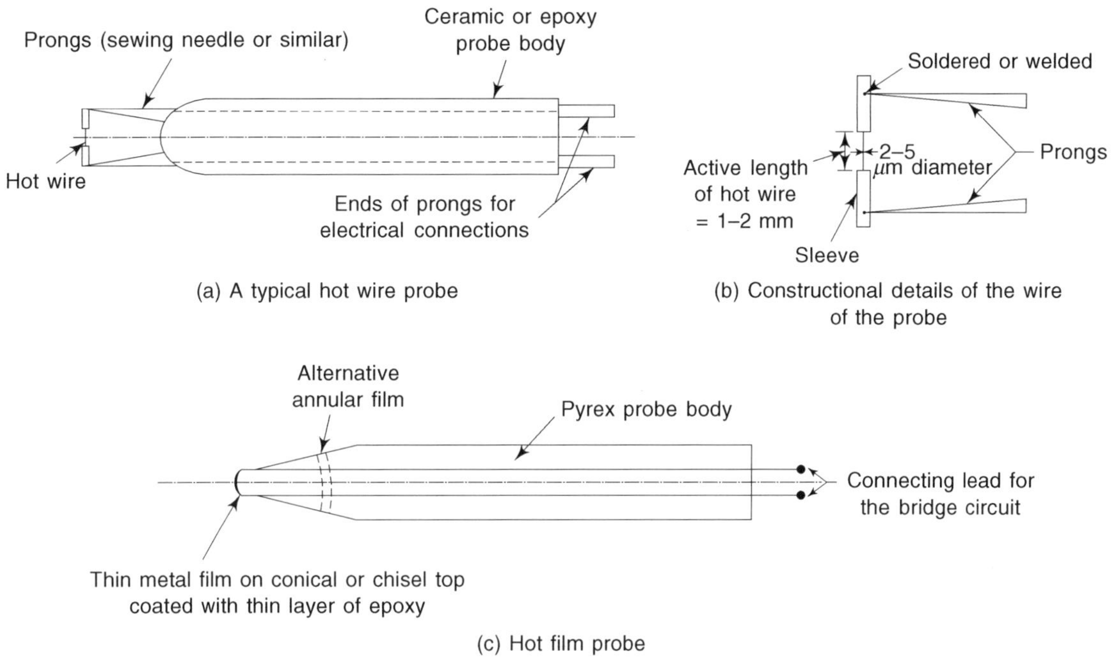

Fig. 13.15 *Typical forms of hot-wire/film anemometer probes*

6. It is suitable for measurement in both gases and liquids. In liquids, the probe has to be modified as the hot film probe.

In spite of several useful features, the hot wire anemometer has the following limitations:

1. The hot wire does not sense whether the flow is in the forward or reverse direction. Its output depends on the cooling of the wire which is same in both the cases.
2. Accumulation of dirt/dust on the hot wire changes its sensitivity. Therefore, it needs frequent calibrations.
3. The wire is very fragile and dust particle striking the hot wire can break the same. Therefore, hot wire use and operation has to be done with exceptionally clean fluids.
4. The operating characteristics are highly non-linear i.e., the square of the output voltage is proportional to the square root of the velocity.
5. It has expensive circuitry and also needs skilled operations.

13.4.4 Laser Doppler Anemometer

The Laser Doppler Anemometer (LDA) is known as the optical type of velocity meter. This instrument measures the instantaneous velocities of gases or liquids flowing in a transparent (usually glass) channel.

The operating principle of this device is based on the Doppler shift in frequency of the light scattered by an object moving relative to the radiating source. The technique basically consists of focussing laser beams at the point in the fluid where the velocity is to be measured. At this focal point, the laser light scattered from the fluid or fluid particles entrained in the fluid is sensed by a photo-detector. Signal processing of the photo-detector output yields the magnitude of Doppler frequency shift which is directly proportional to the instantaneous velocity of the flow.

The use of a LASER (Light Amplification by the Simulated Emission of Radiation) beam in place of the monochromatic light beam is necessitated because of the following:

1. The laser provides much higher quality of monochromatic (single frequency) light source.
2. Laser beam is coherent, i.e., it stays in phase with itself over long distances.
3. Its frequency is very stable and is precisely known to about 1 part in 10^7. This enables to accurately detect the relative Doppler frequency shift, i.e., $\Delta f/f$ which is of the order of 10^{-7} for flow velocities of the order 45 km/h.
4. Unlike a monochromatic beam, the wavelength of the laser beam is affected minimally by changes in the ambient pressure, temperature or humidity.

The following are some of the materials known as active media suitable for production of laser beams:

1. ruby (aluminium oxide crystal doped with a small amount of chromium),
2. Nd-glass (glass doped with neodymium),
3. Nd-YAG (a type of garnet stone dipped with neodymium),
4. carbon dioxide gas,
5. neon gas,
6. ionised argon gas,
7. helium–neon gas, and
8. semiconductor crystal gallium arsenide.

There are several optical arrangements for laser Doppler anemometer. The one most commonly employed is the dual beam or the fringe mode shown in Fig. 13.16(a). In this, the laser source employed is usually the helium-neon laser of 5–15 mW power when the measured flow is at a distance 10–20 cm. This laser operates at a wavelength of 632.8 nm ($\simeq 5 \times 10^{14}$ Hz). Alternatively, an intense beam of argon laser with an output of 1 W or greater is used when measurement are to be made at greater distances (i.e., remote anemometry). This laser beam is split into two equal parts by means of a beam splitter in

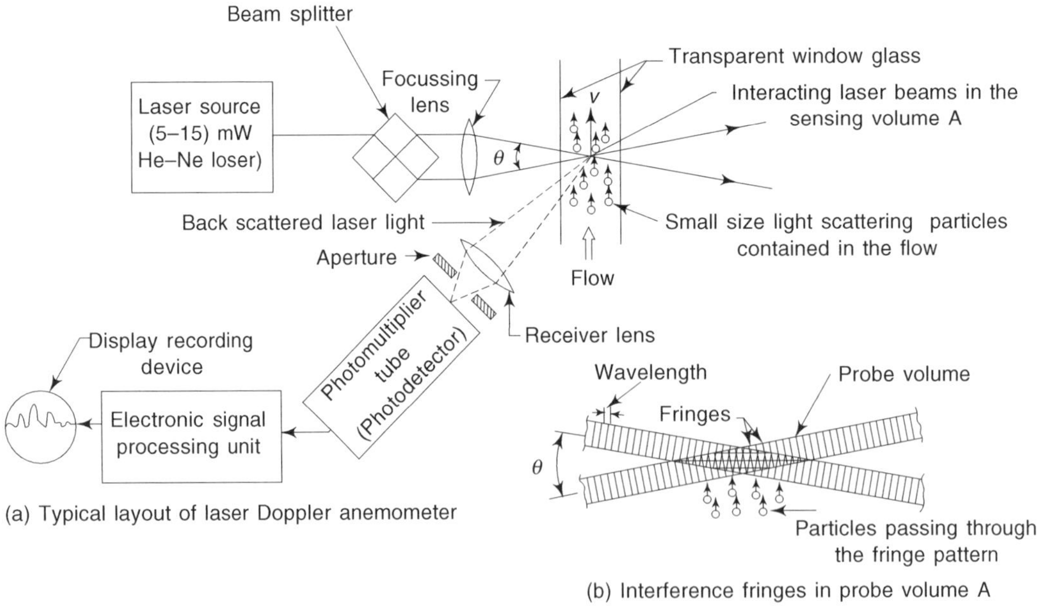

Fig. 13.16 *Details of laser Doppler anemometer in dual beam or fringe mode*

the form of either a rotating optical grating or an optical prism or a half-silvered mirror placed at 45° to the beam. The focussing lens is put in front of the beam splitter to focus the two beams at a point where the velocity of the fluid flow is to be measured. At the focal point, the two split laser beams cross each other and form an interference fringe pattern that consists of alternate regions of high and low intensity [Fig. 13.16(b)]. If the tiny tracer particles (usually microscopic dust or dirt particles present in tap water or air flows) pass through the region of high intensity, they would scatter light and cause a Doppler shift in the frequency of the scattered light. Thus, the light received by the photo-detector will show a varying electrical signal, the frequency of which is proportional to the rate at which the particles cross the interference fringes.

If the tracer particles (assumed to have a velocity equal to that of the fluid flow) pass across the fringes with a velocity v in a direction perpendicular to the fringes, the signal would experience a Doppler shift in frequency given by:

$$\Delta f = \frac{2v}{\lambda} \sin\left(\frac{\theta}{2}\right) \tag{13.49}$$

where λ is the wavelength of the laser beam in the fluid (which may be different from the vacuum wavelength by a factor equal to the index of refraction)

In the LDA system shown in Fig. 13.16(a), the photo-detector receives the back-scattered light. This enables the system to be single-ended, i.e., the laser and collector are near the same position. This is a most convenient configuration when remote measurements on the flow are to be carried out. On the other hand, when the flow measurement distances are small, the photo-detector may be located on the opposite side of the flowing fluid. This forward-scattered light collecting arrangement increases the signal and reduces the requirement of the laser power.

The laser Doppler anemometer has a number of significant advantages over the usual pitot tube and the hot wire velocity metering techniques. They are

1. There is no transfer function involvement, i.e., the output voltage of the instrument is directly proportional to the instantaneous velocities rather than its inference from pressure difference (pitot tubes) or heat transfer phenomenon (hot wire).
2. Non-contact type of measurements, i.e., no physical object is inserted in the flow field and thus the flow is undisturbed by the measurement.
3. Very high frequency response of order of megahertz is possible. Therefore, its accuracy in the determination of fluctuating velocities is far superior to that of hot wire anemometer because of practically no time lag in the measurements.
4. Very high accuracies of the order of ± 0.2%.
5. Measurements are possible in a miniature size of volume of the order 0.2 mm side cube.
6. Suitable for measurements in both gas and liquid flows.
7. The instrument has established proven superiority over other methods of measurement in the following areas:
   (i) investigation of boundary layers and shock wave interaction phenomenon for both laminar and turbulent flows,
   (ii) determination of three-dimensional wing tip vortices near the tips of the aircraft wings,
   (iii) measurements of flow between the blades of turbines,
   (iv) combustion and flame phenomenon in gas turbines and jet propulsion systems,
   (v) in vivo measurement of blood flows and
   (vi) remote sensing of wind velocities.

However, this instrument has the following disadvantages:

1. It involves the need for a transparent channel.

2. The measurement technique is not suitable for clean flows. For such flows, the tracer particles have to be seeded in the flow for scattering the incident light.
3. The instrument is quite expensive and requires a high degree of experience and skill in operation.

Problem 13.6 *A laser Doppler anemometer employs a He–Ne laser (λ = 632.8 nm) to measure the velocity of flow at a point in a dusty gas. A 160 mm focussing lens having θ = 12° is used to operate the LDA in the dual beam mode. Estimate the velocity of flow if the average Doppler shift in frequency was found to be 1.62. MHz.*

Solution Using Eq. (13.49)

$$V = \frac{\lambda}{2 \sin \frac{\theta}{2}} \Delta f$$

$$= \frac{(632.8 \times 10^{-9}) \times 1.62 \times 10^{6}}{2 \sin 6°} = 4.904 \text{ m/s}$$

Review Questions

13.1 Indicate correct statements in the following:

(i) A meter suitable for flow totalisation is
- (a) turbine flow meter
- (b) venturimeter
- (c) rotameter
- (d) orifice plate

(ii) An electromagnetic flow meter must be
- (a) mounted vertically
- (b) mounted horizontally
- (c) rotated at constant speed to develop proper emf
- (d) can be mounted in any position

(iii) A flow meter that is independent of fluid density is
- (a) rotameter
- (b) electromagnetic flow meter
- (c) venturimeter
- (d) orifice meter

(iv) Rotameter is a
- (a) drag force flow meter
- (b) variable area flow meter
- (c) variable head flow meter
- (d) rotating propeller type flow meter

(v) Which of the following is a positive displacement device?
- (a) Rotameter
- (b) Sliding vane flow meter
- (c) Turbine flow meter
- (d) Volumetric tank method

(vi) Which of the following are used for clean fluids only?
- (a) Ultrasonic flow meter
- (b) Turbine flow meter
- (c) Laser Doppler anemometer
- (d) Hot wire anemometer

(vii) Which of the following devices has a non-linear instrument output versus discharge characteristics?
- (a) Capillary tube flow meter
- (b) Turbine flow meter
- (c) Ultrasonic flow meter
- (d) Rotameter
- (e) None of these

(viii) Which of the following flow meters is capable of giving the rate of flow as well as the totalised flow?
(a) Nutating disc flow meter (b) Electromagnetic flow meter
(c) Orifice meter (d) Lobbed impeller flow meter

(ix) The tube of the rotameter
(a) tapers upwards (b) is of uniform cross-section
(c) tapers downwards (d) tapers in the horizontal direction

(x) An orifice meter ($C_d = 0.61$) measuring the flow rate of air in a pipe is substituted by a venturimeter ($C_d = 0.98$) having the throat diameter same as that of the orifice. For the same flow rate, the ratio of the pressure drops for the venturi to orifice is:
(a) 0.61/0.98 (b) $(0.61)^2/(0.98)^2$ (c) 0.98/0.61 (d) $(0.98)^2/0.61$

(xi) A rotameter can be used
(a) only in a vertical orientation (b) only in a horizontal orientation
(c) in any orientation (d) only for air

(xii) The head loss of an orifice meter is
(a) less than that of the venturimeter
(b) less than that of the nozzle flow meter
(c) greater than that of the venturimeter
(d) none of the above

(xiii) Electromagnetic flow meters are used to measure the
(a) flow of non-conducting fluids
(b) flow of non-conducting fluids in a metallic pipe
(c) flow of conducting fluids in a plastic pipe
(d) wind velocity

(xiv) The discharge over a V-notch is proportional to
(a) $H^{3/2}$ (b) $H^{5/2}$ (c) $H^{1/2}$ (d) $H^{-3/2}$

(xv) A nozzle flow meter indicates a pressure drop of 20 cm of water when the flow rate of air is 100 1/min. What would be the flow rate when the pressure drop is 40 cm of water?
(a) 200 1/min (b) 165 1/min (c) 141 1/min (d) 362 1/min

(xvi) Hot wire anemometer is a device used to measure
(a) pressure in gases (b) liquid discharge
(c) gas velocities (d) temperature

(xvii) In which flow meter the pressure drop remains nearly constant but the area changes?
(a) Pitot tube (b) Venturimeter (c) Rotameter (d) Orifice meter

(xviii) The transducer preferred to measure highly fluctuating velocities is
(a) hot wire anemometer (b) electromagnetic flow meter
(c) rotameter (d) turbine flow meter

(xix) In an electromagnetic flow meter, the induced voltage is proportional to
(a) flow rate (b) square root of flow rate
(c) square of flow rate (d) logarithm of flow rate

(xx) Which flow meter is used for measuring the flow rate in an open channel?
(a) Orifice meter (b) Ultrasonic flow meter
(c) Weir (d) Rotameter

13.2 Indicate whether the following statements are true or false:

(i) In the measurement of flow discharge with the help of a rotameter, it is essential that the meter be installed vertically.

(ii) Pitot tube is quite suitable for measuring very low velocities.
(iii) Hot-wire anemometer is an obstruction type of flow meter.
(iv) Volumetric flow equation $Q = C_d K \sqrt{H}$ for the variable head flow meter is independent of the fluid being used.
(v) Velocity of the liquid flowing is measured by means of pitot-static tube at a specific point. Velocity of fluid in this case can be taken to be exactly the same as the measured by the pitot tube.
(vi) Volumetric flow measurements done by means of an electromagnetic flow meter are independent of fluid properties such as density, pressure, temperature, viscosity and compressibility.
(vii) The method of measurement is not changed even though the conductivity of the fluid varies drastically, when the method being employed is the method of electromagnetic flow meter.
(viii) In a hot-wire anemometer, when the flow past the wire varies with time, the sensing element response would lag behind the actual fluctuation.
(ix) The laser Doppler anemometer gives good results when the flow is exceptionally clean.
(x) The domestic water meter uses the positive displacement method for flow measurements.
(xi) For the same flow rates for fluids of different densities, the rotameter will give different readings.
(xii) If the area of the orifice plate is halved for the same flow rate, the pressure difference at the orifice with respect to the main pipeline is doubled.
(xiii) Positive displacement meters give good results even when industrial fluids like slurries, etc are metered.
(xiv) Hot wire anemometer output signal is directly proportional to the velocity of the flow field.
(xv) The obstruction type of flow meters have a linear relationship between the pressure difference and the rate of fluid flow.
(xvi) Primary/quantity meters are employed for the calibration of flow meters.
(xvii) The average value of discharge coefficient for a venturimeter and an orifice meter are 0.97 and 0.62, respectively.
(xviii) Rotameter has a turbine type impeller for measuring the rate of flow and is suitable for measuring unsteady flows.
(xix) The square of the output voltage in the hot wire anemometer is proportional to the square root of gas velocity.
(xx) In a rotameter, the density of a float is always less than that of a flowing fluid.

13.3 Fill in the blanks with appropriate word/words
(i) Rotameter is a variable ____________ flow meter and it should always be installed in a ____________ pipeline.
(ii) Electromagnetic flow meter cannot be used for ____________.
(iii) The non-contact device for measuring the flow velocities is ____________.
(iv) The pitot-static tube measures the ____________ velocity whereas the turbine flow meter measures the ____________ velocity.
(v) The gasoline dispensing machine pumps as well as meters the fuel using ____________ type.
(vi) The flow meter which cause no obstruction in the flow of the metered fluid is ____________.
(vii) Hot wire anemometer is used for measuring velocities of gas flows in a variable ____________ transducer.

(viii) The flow meter which is linear and also gives direct visual indication of the flow is ________.

(ix) The obstruction type of flow meter with maximum pressure recovery is ________.

(x) ________ is a quantity meter whereas ________ is a rate meter.

13.4 A gas of density 0.52 kg/m^3 flows through a pipe of 8.0 cm diameter. The flow is measured by a venturi tube of 4.0 cm diameter throat and a U-tube manometer containing mercury. What is the flow for a manometer reading of 10 cm? Take coefficient of discharge $C_d = 0.95$.

13.5 A liquid of density 900 kg/m^3 flows in a horizontal pipe of radius 5 cm. In a section of the tube of restricted radius 3 cm, the liquid pressure is 15 kN/m^2 less than the main pipe. Calculate the average velocity of liquid in the main pipe. Take $C_d = 0.9$.

13.6 A miniature pitot tube is used to measure the velocity of blood flow and a differential pressure gauge records a pressure of 1 Torr (1 Torr = 1 mm of mercury pressure). If the density of the blood is taken to be 1020 kg/m^3, calculate the blood velocity.

13.7 A rotameter has been calibrated in l/min for water. It is to be used for metering brine solution of sp. gr. 1.15. For this purpose, the density of the float has been changed from 2000 to 2250 kg/m^3 without altering the shape and volume of the float. What correction factor should be introduced in the original scale in order to use the rotameter for the brine solution?

13.8 A gas (density = 0.8 kg/m^3) flows through a 20 cm diameter pipe at the rate of 1000 m^3/h. The flow is measured by a pitot tube located centrally in the pipe and connected to an inclined manometer (inclination 12 in 1). It contains an oil of sp. gr. 0.85 as the manometric liquid. If the average velocity of the gas is 0.8 of the maximum velocity, determine the manometric reading for this flow.

13.9 The inlet and the throat diameters of a horizontal venturimeter used for measuring the rate of flow of water are 30 cm and 10 cm, respectively. At a certain flow rate, the pressure intensity at the inlet and the throat was found to be 13.734 N/cm^2 and 37 cm of Hg (vacuum), respectively. Assuming that 4% of head loss occurs between the inlet and the throat, determine (i) the coefficient of discharge C_d and (ii) the rate of flow of water through the venturimeter.

13.10 A right-angled V-notch with $C_d = 0.62$ was used to measure the rate of discharge of water in a channel. Estimate the rate of discharge if the head measured above the sill was given as (25 cm ± 1.5 mm).

13.11 An electromagnetic flow meter having a flow tube of 150 mm diameter gives an output voltage of 60 mV for a magnetic flux density of 5000 V-s/cm^2. Determine the rate of discharge of liquid through the flow meter.

13.12 A symmetrical shaped triangular wedge of height 60 mm is mounted across the full diametral space of the pipe of 200 mm. Given is the Strouhal number of 0.88. If the vortex-shedding frequency is 40 Hz, determine the rate of discharge through the flow meter.

Answers

13.1	(i) (a)	(ii) (d)	(iii) (b)	(iv) (b)	(v) (b)	(vi) (d)
	(vii) (e)	(viii) (d)	(ix) (c)	(x) (b)	(xi) (a)	(xii) (c)
	(xiii) (c)	(xiv) (b)	(xv) (c)	(xvi) (c)	(xvii) (c)	(xviii) (a)
	(xix) (a)	(xx) (c)				
13.2	(i) true	(ii) false	(iii) false	(iv) false	(v) false	(vi) true
	(vii) true	(viii) true	(ix) false	(x) true	(xi) true	(xii) false
	(xiii) false	(xiv) false	(xv) false	(xvi) true	(xvii) true	(xviii) false
	(xix) true	(xx) false				

13.3 (i) area, vertical (ii) non-conducting fluids (iii) laser doppler anemometer (iv) local, average (v) positive displacement device of vane (vi) electromagnetic flow meter (vii) resistance (viii) rotameter (ix) venturimeter (x) Nutating disc flow meter, venturimeter.

13.4 $Q = 279.30$ l/s 13.5 $V_{av} = 2.00$ m/s 13.6 $V = 0.51$ m/s

13.7 Correction factor = 1.0225 13.8 Manometric reading = 7.03 cm

13.9 $C_d = 0.98$, $Q = 150$ l/s 13.10 $Q_{-} = (45.77 \pm 0.69)$ l/s 13.11 $Q = 14.14$ ls

13.12 $Q = 176.51/s$

Chapter

14

Acoustics Measurement

■ *INTRODUCTION* ■

Sound waves or acoustic pluses are vibrations that consist of a succession of rapid variations in air pressure, usually of a small magnitude. These are transmitted through the fluid (liquid or gaseous medium) and produce the sensation of hearing in the hearing meachanism. Sound forms an important part of man's environment. In some ways, it is quite useful to us. It makes speech communication possible, warns a person of danger, enables a person to enjoy music, etc. However, much of the sound one encounters unfortunately is unwanted sound termed *noise*. Some of the detrimental effects of noise are that excessive noise causes irritation, distraction or annoyance i.e., it hurts people psychologically and physiologically and tends to reduce man's efficiency. Further, it interferes with the communication between people and tends to mask audible warning signals thereby reducing the safety in workshops. In addition, prolonged exposure to noise can cause noise-induced hearing loss.

Acoustical measurements find wide application in our day-to-day life. For example, in noise control studies, i.e. the development of less noisy machinery and equipment, these days form an important area of the environmental pollution or noise pollution control. In addition, the other applications include the design and testing of sound recording and reproducing equipment, study of water-borne sound (sonar) for underwater direction and range finding, study of aerodynamic noise and its interaction with structures, etc.

14.1 ■ CHARACTERISTICS OF SOUND

Sound is characterised by the following parameters:

1. its intensity in W/m^2 or its pressure in μbar,
2. its frequency in cycles per second or Hz,

3. the nature of noise which may be either continuous or impulsive,
4. its loudness or loudness level, and
5. its *noise rating* which is also called *annoyance* rating based on the sound pressure level and frequency.

Characteristics (1), (2) and (3) are measured objectively by the instruments and do not depend on the subjective assessment of the observer; whereas (4) and (5) depend on the subjective assessment of the quality of sound by the human observer which may be based on measured results, set of standard tables, graphs, etc.

14.2 SOUND PRESSURE, POWER AND INTENSITY LEVELS

14.2.1 Sound Pressure Level

The audible range of human hearing mechanism usually measured at 1 kHz, extends from 0.0002 μbar at the *threshold of hearing* to 1 mbar at the *threshold of pain.* This represents an increase of 5×10^6. Because of this very large range, it is more convenient to express the magnitude of sound pressure in logarithmic form, in terms of decibels.

Thus, sound pressure level is often abbreviated as SPL or L_p and is defined in decibels as

$$\text{SPL} = 10 \log_{10} (p^2)/(p_{\text{ref}}^2) \qquad (14.1)$$

$$= 20 \log_{10} (p)/(p_{\text{ref}}) \text{ dB}$$

where p is the root mean square (rms) sound pressure and p_{ref} the rms reference pressure ($p_{\text{ref}} = 0.0002$ μbar $= 0.00002$ N/m^2 $= 0.0002$ dyn/cm^2). A reference pressure of 0.0002 μbar has been chosen as the smallest possible sound level at 1 kHz that a young adult can just hear (in quiet surroundings). In other words, the *threshold of hearing* is taken as the reference pressure of SPL measurements. Hence, Eq. (14.1) can be written as

$$\text{Sound pressure level} \quad (\text{SPL}) = 20 \log \frac{p}{0.00002} \text{ dB} \qquad (14.2)$$

where p is N/m^2.

The advantage of using logarithmic scale in defining the SPL is that it permits a large range of pressures to be described with convenience. Further, it also represents the non-linear behaviour of the ear more convincingly.

SPL is measured using a sound level meter. Figure 14.1 gives pictorially the typical values of sound pressure levels of some common sounds.

Problem 14.1 *If the pressure of sound measured at 1 m from an unmuffled motor cycle was found to be 12.0 N/m^2, determine the corresponding sound pressure level in decibels, p_{ref} is 0.0002 μbar.*

Solution Using Eq. (14.2) we get,

$$\text{SPL} = 20 \log \frac{12.0}{0.00002} = 115.56 \text{ dB}$$

14.2.2 Sound Power Level

Sound power is the total sound energy radiated by a sound source per unit time. The power in the sound emitted from a source can have an extremely large range. The power can be as low as 10^{-9} W at a faint

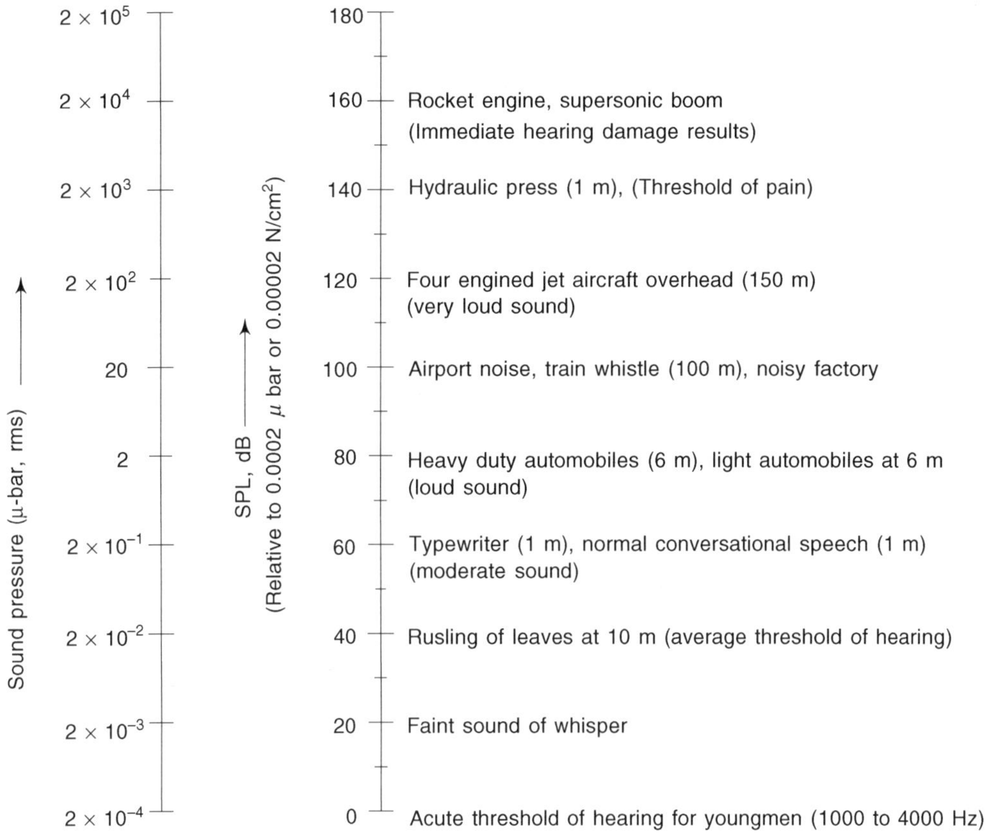

Fig. 14.1 *Typical values of sound pressure levels of common sounds with reference sound pressure of 0.002 μbar, the threshold of hearing at 1 kHz*

whisper or 10^{-5} W at normal conversational speech. Further, it could be as large as 50 kW for a jet liner or 10 MW for a saturn rocket. Because of this wide range of values of power in the various power sources, the sound power level is conveniently defined in the units of decibels. Sound power level is often abbreviated as PWL or L_w and is defined as:

$$\text{PWL} = 10 \log_{10} \frac{W}{W_{\text{ref}}} \text{ dB} \tag{14.3}$$

where W is acoustic power of the source and W_{ref} is the reference acoustic power ($W_{\text{ref}} = 10^{-12}$ W). Since the power ratio 10^{-12} means –120 dB, the above equation may be written as:

$$\text{PWL} = (10 \log_{10} W + 120) \text{ dB} \tag{14.4}$$

where W is in watts.

Figure 14.2 gives the typical values of sound power levels of some common sounds.

It is interesting to note that small sound power levels can have a significant effect on the ear. It is estimated that the acoustic power emitted by the roars of the football crowd during a match is just enough to heat a cup of coffee! Further, the power of a full symphony orchestra is only one acoustic watt or so!

Sound pressure level (SPL) of a machine, that is radiating noise, depends on the machine as well as on the surroundings. The same machine may show different SPLs when placed in different surroundings. It is for this reason that sound power level (PWL) has been defined. PWL cannot be directly defined but is found by the measurement of sound pressure levels or sound intensity levels.

When there are more than one acoustic power sources, acoustic power W of all sources can be added and the combined PWL can be found using Eq. (14.3).

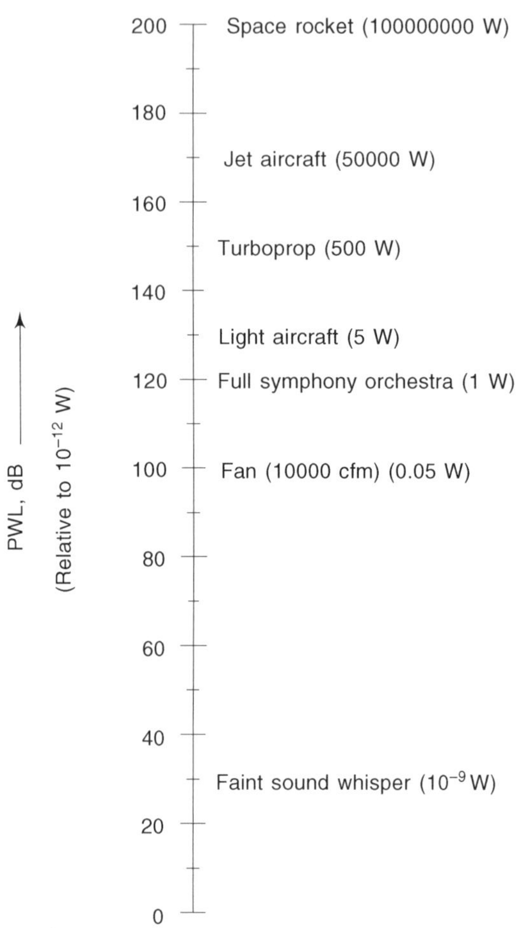

Fig. 14.2 *Typical values of sound power levels of common sounds with reference power of 10^{-12} W*

14.2.3 Sound Intensity Level

Sound intensity level I at a location in a sound field in a specified direction, is the average sound power passing through a unit area for which the level is being specified. Its unit is W/m^2.

Sound intensity level L_I is defined:

$$L_I = 10 \log_{10} \frac{I}{I_{\text{ref}}} \text{ dB} \qquad (14.5)$$

The value of I_{ref} is taken as 10^{-12} W/m^2 in Eq. (14.5).

I is the time averaged product of sound pressure p and particle velocity u or

$$I = p \cdot u \qquad (14.6)$$

In a free field, $$I = \frac{p_{\text{rms}}^2}{\rho c} \qquad (14.7)$$

where p_{rms}= is the root mean square sound pressure, ρ the density of air, c the speed of sound in the medium and ρc is called characteristics acoustic impedance of the medium.

14.2.4 Addition of Sound Pressure Levels

If there are two or more sources radiating noise, the total sound pressure level may be determined from the individual sound pressure levels of the sources. It may be done by addition based on energy or power levels which are dependent on mean square pressures. Thus, the SPL of each based may be converted from number of decibels to $\left(\frac{p^2}{p_{\text{ref}}^2}\right)$ values, by dividing the decibels by 10 and calculating antilogs. After adding the relative $(p^2/\,p_{\text{ref}}^2)$ values of various sources, the conversion to dB levels may be made for the combined sound pressure levels, using Eq. (14.1).

Let us consider two such sound pressure levels of X dB and Y dB.

$$\therefore \qquad \text{Resultant SPL} = 10 \log \left[a \log\left(\frac{X}{10}\right) + a \log\left(\frac{Y}{10}\right)\right]$$

Similarly, if there are several sources, with SPL's $X, Y, Z, \ldots$, etc. then

$$\text{Total SPL} = 10 \log \left[a \log\left(\frac{X}{10}\right) + a \log\left(\frac{Y}{10}\right) + a \log\left(\frac{Z}{10}\right) + \ldots \right. \quad (14.8)$$

Now if in Eq. (14.7) $X = Y$, i.e. there are two sound sources of equal SPL, then,

$$\text{Total SPL} = 10 \log \left\{ 2\, a \log\left(\frac{X}{10}\right) \right\}$$

$$= 10 \left\{ 0.301 + \left(\frac{X}{10}\right) \right\}$$

$$\simeq X + 3 \text{ dB} \quad (14.9)$$

i.e. if we have sounds of SPL of 100 dB each, then the resultant sound pressure level would be (100 + 3) = 103 dB.

Problem 14.2 *Two machines are working in noisy environments. The background noise when the machines are inoperative is 65 dB. If the two machines having individual sound pressure levels of 84 and 88 dB are switched on simultaneously, determine the combined sound pressure level of the machines along with the background noise.*

Solution Using Eq. (14.8) we get

Combined SPL of three sound sources

$$= 10 \log \left\{ \sum_{i=1}^{3} a \log\left(\frac{X_i}{10}\right) \right\}$$

$$= 10 \log \left\{ a \log\left(\frac{65}{10}\right) + a \log\left(\frac{84}{10}\right) + a \log\left(\frac{88}{10}\right) \right.$$

$$= 10 \log (3.162 \times 10^6 + 2.512 \times 10^8 + 6.3 \times 10^8)$$

$$= 10 \log (8.853 \times 10^8) = 89.47 \text{ dB}$$

Problem 14.3 *The following sound pressure levels were measured for a machine operating in a noisy environment.*

SPL of machine + background noise = 90 dB

SPL of the background noise = 80 dB

Determine the SPL of the machine alone.

Solution Using Eq. (14.8) we get,

$$\text{Resultant SPL} = 10 \log \left\{ \log\left(\frac{X}{10}\right) + a \log\left(\frac{Y}{10}\right) \right\}$$

Substituting the numerical values of SPL's in the problem we get,

$$90 = 10 \log \left\{ \log\left(\frac{X}{10}\right) + a \log\left(\frac{80}{10}\right) \right\}$$

where X dB is the SPL of the machine.

$$\therefore \quad a \log\left(\frac{X}{10}\right) = a \log\left(\frac{90}{10}\right) - a \log\left(\frac{80}{10}\right)$$

Taking log of both sides we get,

$$X = 10 \log \{a \log 9 - a \log 8\} = 89.54 \text{ dB}$$

i.e. the sound pressure level of the machine is 89.54 dB.

14.2.5 Variation of Intensity of Sound with Distance

It has been observed that as the distance of measurement from the sound source increases, the intensity of sound decreases. The simplest source of sound is the point source, where the sound is radiated equally in all direction from the apparent centre. For a point source, the relationship between the distance d and intensity I is given by

$$I = \frac{W}{4\pi d^2} \tag{14.10}$$

This applies to spherical wave fonts

or $$I \propto 1/d^2 \tag{14.11}$$

Equation (14.12) is commonly termed as inverse square law.

In practice, the industrial sound/noise sources have finite dimensions, i.e. they may be rectangular, cylindrical or other composite shapes. But they are rarely point sources. However, as a general rule of thumb, if the distance from the noise source is greater than five times the maximum dimension of the source, then the inverse square law is applicable.

Using Eq. (14.7), it can also be seen that $$p \propto \frac{1}{d} \tag{14.12}$$

Problem 14.4 *The sound pressure level measured at 10 m from an automobile horn is 110 dB. Determine the sound pressure level at distances of (a) 20 m and (b) 80 m. Assume that the inverse square law holds good between intensity and distance.*

Solution Using Eq. (14.11), we get,

$$I \propto 1/d^2$$

Therefore, if d is doubled than I becomes one-fourth.

(a) At 10 m, $\text{SPL} = 10 \log \dfrac{I_1}{I_0} = 110 \text{ dB}$

At 20 m, $\text{SPL} = 10 \log \dfrac{I_1/4}{I_0}$

$$= 10 \log \left(\frac{I_1}{I_0}\right) + 10 \log \frac{1}{4}$$

$$= 110 - 6 = 104 \text{ dB}$$

This shows that doubling the distance reduces the sound pressure level by 6 dB.

(b) At 80 m, $\text{SPL} = 10 \log \dfrac{I_1}{I_0} + 10 \log \dfrac{1}{64}$

$$= 110 - 18.1 = 91.9 \text{ dB}$$

14.2.6 Sound Measurement Conditions

There are three types of environments for sound power measurements. They are briefly described here.

Anechoic (Echo-free) rooms These rooms are near-prefect sound absorbers and the sound pressure level in such rooms follow the inverse square law, i.e., there will be a decrease of 6 dB SPL for each doubling of the length of the noise source. Usually, the obstruction-free outdoor locations approximate to anechoic conditions. These are also termed free-field conditions. Under such conditions, the relationship between sound pressure level and sound power level from a simple source (which radiates uniformly in all directions) is given by:

$$\text{PWL} = (\text{SPL})_d + 20 \log d + 10.9 \text{ dB} \tag{14.13}$$

where $(\text{SPL})_d$ is the sound pressure level at distance d (in metres) from the source.

Reverberant rooms Such rooms have hard walls that are near-perfect reflectors of sound. This results in perfect diffusion of the sound field and thus the SPL is everywhere nearly the same and sound waves travel in all directions with equal probability.

Ordinary rooms These rooms provide semi-anechoic conditions, i.e. conditions somewhere between anechoic and reverberant. The SPL decreases as distance is increased but the decrease is not 6 dB for doubling the distance. Normally, the noise testing of machinery and equipment is carried out in semi-anechoic conditions. Normally a large room (minimum volume 300 m^3 but preferably 1000 m^3) with a high ceiling and protected from external noises approximates to the semi-anechoic conditions.

14.3 ■ LOUDNESS

The physical response of the human ear to sound pressure level is different for different frequencies. The ear is sensitive in the range of frequencies between 500 and 5000 Hz. It is less sensitive at higher frequencies and much less sensitive at low frequencies. Therefore the SPL necessary at each frequency to produce the same loudness response in a listener is different. A higher SPL is required at low frequencies than that at higher frequencies to produce the same loudness sensation. Through the listening tests, the variation of sensitivity of the ear has been evaluated and equal loudness curves or equal sensitivity curves have been drawn. The unit of loudness is phon. It is defined as the value assigned to each curve which is the SPL at the point where the curve touches the 1000 Hz ordinate. All points lying on the same contour have the same loudness expressed in phons. In other words, a sound pressure level at 40 dB is 40 phons when heard at 1000 Hz. Figure 14.3 represents a typical representation of equal loudness conttours indicating sound pressure level at different frequencies. For example, if the magnitude of sound pressure level at 1000 Hz is 40 dB (loudness level equal to 40 phons), then for an equal loudness level at 100 Hz, the average person will require an SPL of 52 dB.

14.4 ■ TYPICAL SOUND—MEASURING SYSTEMS

14.4.1 Sound Level Meter

Sound pressure levels are measured by sound level meters, which convert acoustic pressure into a directly proportional voltage indicated on a meter. The meter is positioned at the desired location with no obstruction from the sound source and the reading is taken. The block diagram of the instrument is shown in Fig. 14.4. This includes a microphone, an electronic amplifier with frequency weighting network and a meter or recorder calibrated in decibels.

A rectifier circuit is incorporated to produce a signal proportional to root mean square value. The input signal generated by the sound pressure alternates from positive to negative levels. This signal is squared, which has the effect of producing all positive signals varying in amplitude. By taking the square root of this signal, it is possible to obtain the output of the meter in the form of averaged root mean square value of the signal.

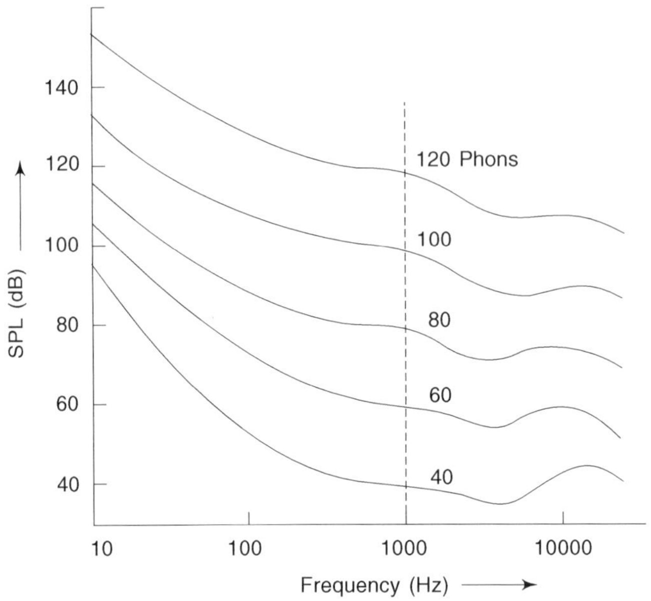

Fig. 14.3 *A typical representation of equal loudness contours indicating SPL at different frequencies*

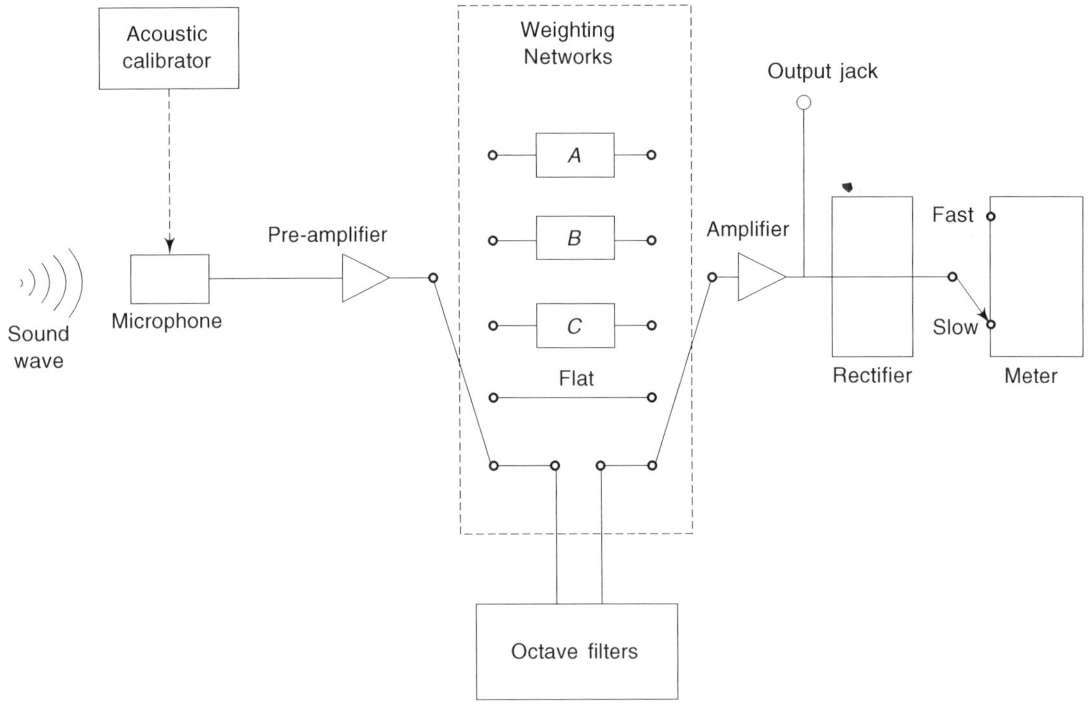

Fig. 14.4 *A block diagram of a typical Sound Level Meter (SLM)*

The frequency response of the unweighted system is generally flat or linear over the entire available range from 20 Hz to 20 kHz. Further, some commercially available instruments have the upper frequency measuring range of approximately 30 kHz. The results of the measurements using the linear response are quoted in dB or dB linear.

A frequency weighting network is used in the circuit to provide a response similar to that of the human ear. With the network, there is a different amount of amplification for each frequency so that the overall measurement made gives greater emphasis to some frequencies than to others. It is seen from the equal loudness contours that the ear is more responsive to frequencies between 500 and 5000 Hz than to those above or below these frequencies. The weighting networks, therefore allow greater amplification in this range, than at higher or lower frequencies. Three standard weighting networks *A*, *B* and *C* are used to approximate the equal loudness curves. The response of the *A* scale approximates to 40 phon equal loudness curve and is used for SPL below 53 dB. Results of measurements made with this scale are expressed as dB *A* or dB *A*. The *B* scale is used in the range 55–95 dB and the response approximates to the 70 phon equal loudness curve. The *C* scale, intended for use above 85 dB has approximately a flat response, except at the two ends. Figure 14.5 shows the linear and the various weighted responses of the sound level meters.

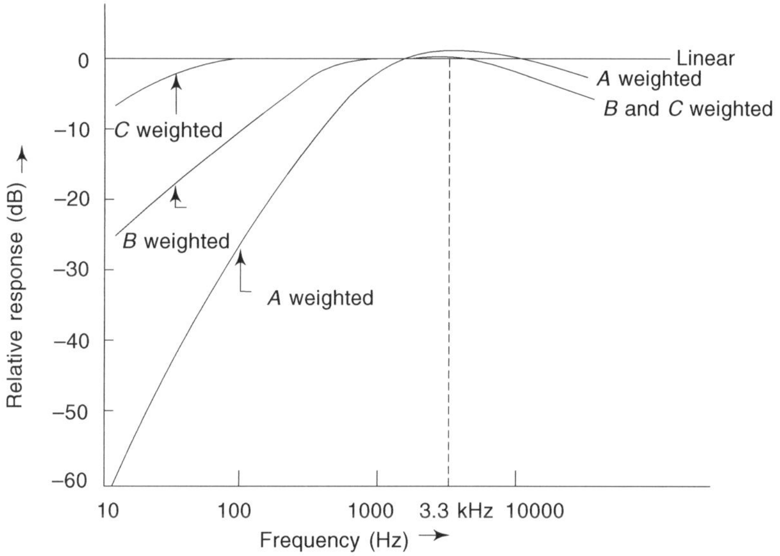

Fig. 14.5 *Linear and A, B and C weighted response of sound-level meters*

14.4.2 Frequency Analysis of Noise Signal

The frequency content or sound spectrum of noise signal is determined by frequency analysis. It is useful to the noise control people/analysts for determining the probable sources of noise and it also helps them to suggest means to control/confine them. Further, noise is a symptom of malfunctioning in the machinery. Frequency analysis can quite often describe various frequency spectrum-related malfunction. In addition, spectral characteristics are important in describing the transmission of sound, say through a wall or its absorption by some materials.

Frequency spectrum analysis of a sound signal consist of distribution of sound pressure level at various frequencies versus the frequencies ranging from low to high frequencies. This is achieved with the help of filters, each filter allowing only that part of the spectrum which lies inside the bandwidth of the filter. Frequency analysis can be performed by using either 1/1 octave band or ½ octave band or 1/3 octave band. A band of 1/1 octave represents a range of frequencies in which the upper frequency and the lower frequency is in the ratio of 2 : 1. In other words, if the upper frequency is f_u and the lower frequency is f_e then,

$$f_u = 2 f_1 \quad (14.14)$$

The procedure of measuring pressure levels using the 1/1 octave band analysis consists of setting the meter at the geometric mean frequency and take the reading. The geometric mean is the average of the logarithm of the upper and lower frequencies and is termed as the centre frequency f_c

Now, $$\log f_c \text{ (geometric mean)} = \frac{1}{2}(\log f_1 + \log f_u)$$

$$= \log \sqrt{f_1 \, f_u}$$

or $$f_c = \sqrt{f_1 \, f_u} \quad (14.15)$$

The centre frequencies of 1/1 octave band have been standardised internationally ranging from 31.5 to 16000 H_z and are given in Table 14.1. Many portable sound level meters (SLM_s) have such built-in filters and it is possible to measure sound pressure level for each band by changing the centre frequency.

Table 14.1 Standard Octave Band Frequencies in the Audio Frequency Range

Low band edge	*Centre frequency*	*High band edge*
22	31.5	44
44	63	88
88	125	176
176	250	353
353	500	706
706	1,000	1,414
1,414	2,000	2,828
2,828	4,000	5656
5,656	8,000	11,312
11,312	16,000	22,624

To obtain more accurate characterisation of the sound spectrum, the octave band may be subdivided into smaller bands. The band may be divided into two producing 1/2 octave band analysis or the band may be divided into three producing 1/3 octave band analysis. In these measurements, again the centre frequency which is the geometric mean of these smaller bands is used for reading the corresponding values of sound pressure levels.

Problem 14.5 *The lower and upper frequencies an octave band are 11,312 and 22,614 Hz. Determine:*
(i) the centre frequency of the 1/1 octave band,
(ii) the intermediate frequency of the 1/2 octave band, and
(iii) the intermediate frequencies of the 1/3 octave band.

Solution

(i) The centre frequency for 1/1 octave band

$$f_c = \sqrt{11{,}312 \times 22{,}624} = 15{,}998 \text{ Hz}$$

(ii) In the 1/2 octave analysis, the frequencies are:
f_l, f_{u1} and f_u
If x is the multiplying factor then,

$$f_u = xf_{u1} = x^2 f_l = 2 f_l$$

or $x^2 = 2$

$\Rightarrow$ $x = \sqrt{2}$

$\therefore$ $f_{u1} = 15998$ Hz.

(iii) In the 1/3 octave analysis, the frequencies are:
f_l, f_{u1}, f_{u2} and f_u.
If x is the multiplying factor in the intermediate frequencies, then:

$$f_{u1} = xf_l; f_{u2} = xf_{u1} \quad \text{and} \quad f_u = x f_{u2}$$

This gives $f_u = x^3 f_l$

or $x^3 = 2 \quad (\because f_u = 2f_l)$

or $x = (2)^{1/3} = 1.26$

Hence, the values of intermediate frequencies in 1/3 octave band are:

$$f_{u1} = (1.26) \times 11{,}312 = 14{,}253 \text{ Hz, and}$$

$$f_{u2} = (1.26)2 \times 11{,}312 = 17{,}959 \text{ Hz}$$

For narrow-band frequency analysis of acoustic signals, a Fast Fourier Transform (FFT) analyser, described in Chapter 15 may be used.

14.4.3 Sound Intensity Measurements

Sound intensity $I = p \cdot u$, as given in Eq. (14.6), where p is the sound pressure and u is the particle velocity. Through sound pressure measurement with microphones, have been commonly used, sound intensity measurements have been applied recently, with considerable benefits, in noise-source location and sound power determination.

The technique involves the use of a two-microphone probe, as shown in Fig. 14.6.

Commonly, the two microphones used are of 12 mm in diameter with a spacer 12 mm in length between the two microphones. Velocity u is estimated from the pressure gradient between the two microphones and the average pressure of the microphones p is used to find the intensity values, which is averaged over time. Alternatively, FFT analysis is used. In the method, sound intensity in the frequency domain is taken from the imaginary part of the cross-spectrum between two signals from closely spaced microphones, using a dual channel FFT analysis. It is seen that intensity I_r in r direction is

$$I_r = \frac{I_{mag.}(G_{12})}{w\rho\Delta r} \qquad (14.16)$$

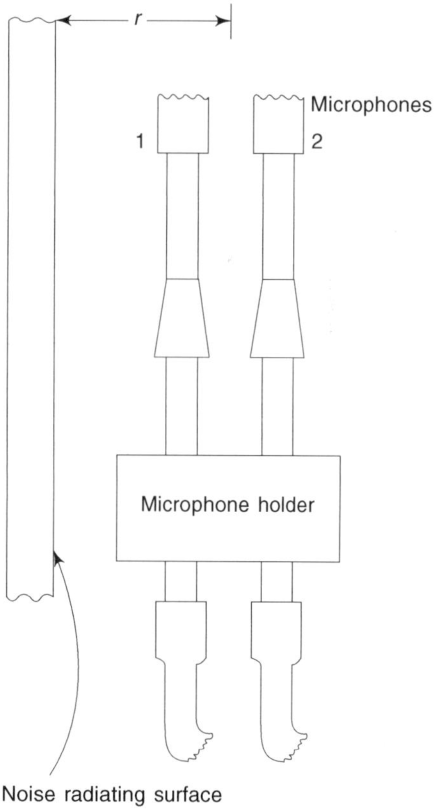

Fig. 14.6 *Sound intensity probe*

where 'ω is the angular frequency, I_{mag} (G_{12}) the imaginary part of cross-spectrum between the two microphone signals, ρ the medium density and Δr in the spacing between the microphones. A computer with software is normally used for calculations and plotting. The measurements are carried out with distance r from the noise radiating surface being 40 – 80 mm. For measurements, the noise radiating surfaces are divided into suitably spaced grades and intensity measurements are done at each point of the grid by direct use of Eq. (14.6), from which intensity maps can be obtained indicating the regions of high sound radiation. In another method called sweeping or scanning method, the probe is slowly swept (usually, 0.2 m/s) over the surface while the analyser is averaging the measurements.

Problem 14.6 *In a certain sound level meter, the output voltage was directly proportional to the sound pressure generated by a sound source. The instrument was calibrated with a sound pressure excitation of 0.1N/m² and the meter scale was set at –60 dB referenced to 1 V.*

For a given noise source, the output voltage developed by the instrument is 6 mV. Determine the SPL of the noise source.

Solution The output voltage of the sound level meter (SLM) for an excitation of dyn/cm² is given by

$$-60 \text{ dB} = 20 \log E/E_{ref}$$

$$= 20 \log E/1.0 \qquad (\because \; E_{ref} = 1 \text{ V})$$

or

$$E = 10^{-3} \text{ V}$$

Now, 6 mV output voltage of SLM $= 6 \times 10^{-3}/10^{-3} = 0.6 \text{ N/m}^2$

The sound pressure level corresponding to the acoustic pressure of 6 dyn/cm² is given by:

$$\text{SPL} = 20 \log \frac{p}{p_{ref}}$$

$$= 20 \log \frac{0.6}{2 \times 10^{-5}} \qquad (\because \; p_{ref} = 2 \times 10^{-5} \text{ N/m}^2)$$

$$= 89.54 \text{ dB}$$

Problem 14.7 *The noise signal of a certain machinery in a workshop was processed in an audio-frequency analyser which was equipped with the standard octave filters. The results obtained were as follows:*

Centre frequency (in Hz)	63	125	250	500	1000	2000	4000	8000
SPL (in dB)	84	85	84	81	72	68	66	52

Determine the overall sound pressure level in dB.

Solution Using Eq. (14.8) we add the sound pressure levels of each centre frequency in the octave band and get,

$$\text{Overall SPL} = 10 \log [a \log (8.4) + a \log (8.5) + a \log (8.4) + a \log (8.1) + a \log (7.2) + a \log (6.8) + a \log (6.6) + a \log (5.2)] = 89.87 \text{ dB}$$

14.5 ■ MICROPHONES

Microphones are transducers to convert sound pressure variations into the analogous electrical signals. The main factors in the selection of a microphone as sensitivity, frequency response, dynamic range and linearity. Microphones generally employ a thin diaphragm to convert pressure into motion. The motion is then converted into a suitable electrical output by employing a secondary transducer which may be either of the following:

1. Condenser or capacitor type microphone,
2. Electret microphone,
3. Piezo-electric microphone,
4. Electrodynamic type microphone.

14.5.1 Condenser or Capacitor Type Microphone

The capacitor or condenser type microphone is most widely used for precision measurements because of good frequency response, excellent linearity and large dynamic pressure range. A typical construction of this type microphone is shown in Fig. 14.7. The microphone cartridge consists of thin metallic diaphragm in close proximity to a rigid (stationary) back plate. The diaphragm and the back plate form the plates of the capacitor. The movement of the diaphragm caused by the impingement of sound pressure results in an output voltage given by

$$E \propto Qd \tag{14.17}$$

where E = output voltage of the capacitor,
Q = charge provided by the polarising voltage (nearly constant),
d = separation of the plates.

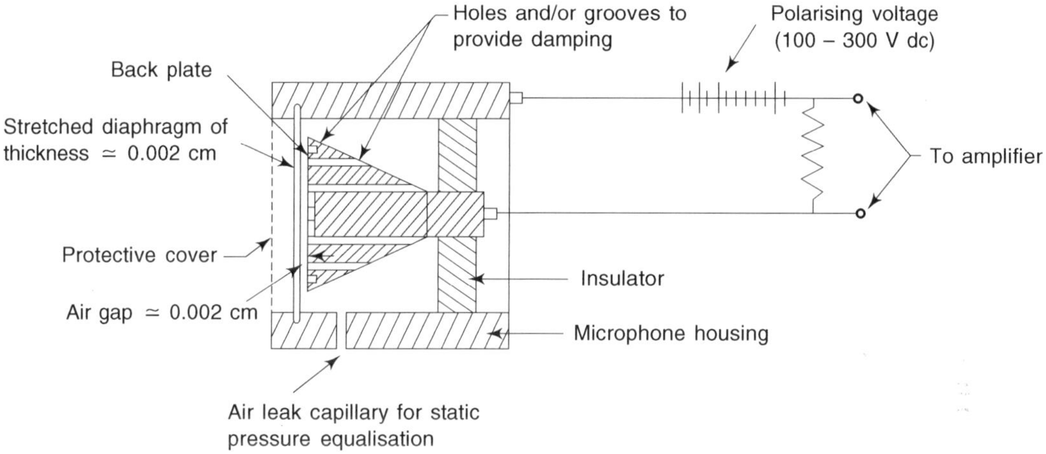

Fig. 14.7 *A Schemetic diagram of a condenser type microphone*

The microphone is provided with a capillary air-leak for equalisation of pressure on both sides of the diaphragm to prevent the diaphragm from bursting. Further, the back plate of the condenser microphone is provided with damping holes. The motion of the diaphragm causes air flow through the holes which results in energy dissipation because of fluid friction. This damping effect is utilised in the control of resonant peaks of diaphragm response.

These microphones have good frequency response even at high frequencies and also have good acoustic sensitivity. However, they are susceptible to errors caused by high humidity, which produces excessive background noise due to the leakage of the charge.

14.5.2 Electret Microphone

Electret condenser microphone is an improved version of ordinary condenser microphone. In the condenser microphone we need to supply a dc polarising voltage, whereas in electret microphone, the polarized element, namely the thin coating of the electret material supplies the bonded charge to the back plate of the electret microphone. In other words, it is a self-polarising type of condenser micorphone.

The principle of operation of the electret microphone, shown in Fig. 14.8 is precisely the same as that of the conventional condenser microphone. They have excellent frequency response and are relatively free from noise in humid environments. However, the aging of the electret material can cause errors of the order of 0.2 dB per year.

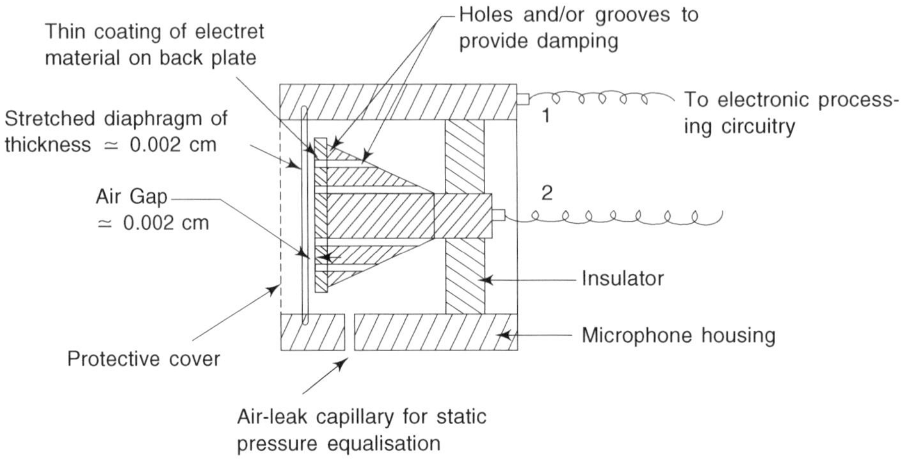

Fig. 14.8 *A schematic diagram of Electret microphone*

14.5.3 Piezo-electric Crystal Type Microphone

Certain materials possess the ability to generate an electrical potential when subjected to mechanical strain. Such materials are generally crystalline in nature and are known as piezo electric crystals. The commonly employed materials are quartz, Rochelle's salt (potassium-sodium tartarate), lead zinconate titanate, barium titanate and ammonium dihydrogen phosphate.

Rochelle's salt has strongest piezo-electric effect and is widely used as microphone transducer. However, it requires protection from moisture in the air and cannot be used above 50 °C. Further, quartz is very stable but its output is quite low. In practice, lead zirconate titanate (PZT) is commonly used in piezo-electric microphones. The force produced by the acoustic pressure on the diaphragm is used to strain the piezo-electric material which in turn produces voltage output in direct proportion to the applied force/acoustic pressure.

For the greatest sensitivity, a cantilever type of crystal element is mechanically coupled with the sensing diaphragm (Fig. 14.9). Further other constructions employ direct contact between the diaphragm and the crystal element either in bending mode or by direct contact (i.e. element is subjected to direct compression). Piezo-electric microphones are very sensitive and can measure accurately sound pressure levels below 24 dB. Further, their response at lower frequencies is also very good. However, their response at higher frequencies is not as good as of condenser microphones. In addition, they require impedance matching and are also sensitive to vibrations.

14.5.4 Electrodynamic Type of Microphone

This type of microphone is based on the principle of generation of emf when a moving conductor is placed in a magnetic field. Figure 14.10 shows the typical construction of this type of microphones. The sensing diaphragm is attached to a coil or ribbon placed within the poles of a permanent magnet. The movement of the diaphragm due to acoustic pressure generates the analogous induced voltage in the coil.

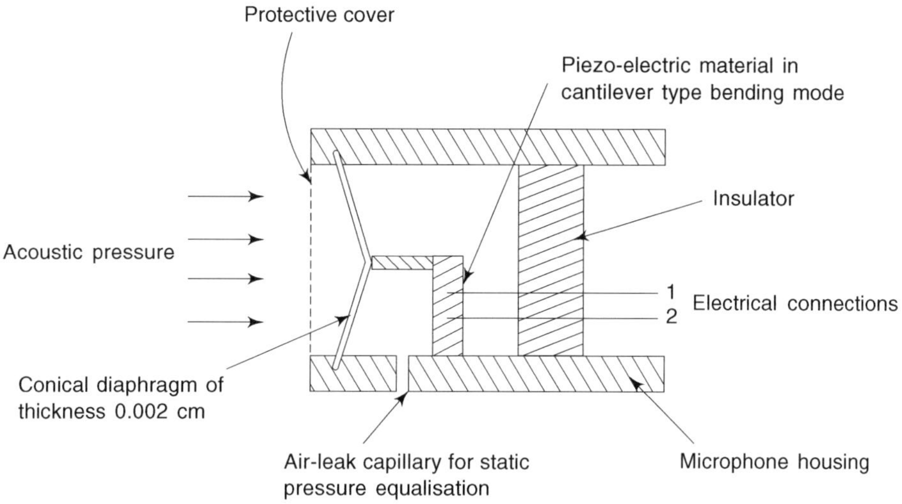

Fig. 14.9 *A schematic diagram of Piezo-electric microphone*

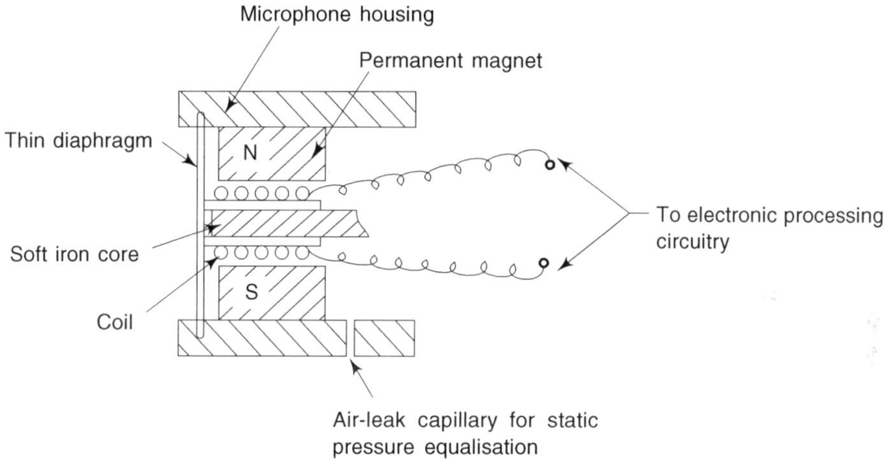

Fig. 14.10 *Schematic construction of an electrodynamic type microphone*

The output voltage V of the electrodynamic microphone is given by:

$$V = Blu \text{ volts}$$

where B = density of magnetic flux Wb/m^2

l = length of the conductor (m) and

u = velocity (m/s)

Such microphones have the advantage of being self-generating but their frequency response is poor due to the high inertia of the moving coil.

Review Questions

14.1 Indicate which of the following is correct:

(i) Decibel is a unit of
- (a) sound pressure level only
- (b) any quantity in which the ratio of two numbers is 2.0.
- (c) any quantity which is represented by the natural logarithm of the measured quantity with respect to the reference quantity.
- (d) any quantity which is represented as 10 times the logarithm of the measured quantity with respect to the reference signal.

(ii) Zero decibel SPL represents
- (a) the lowest pressure fluctuation normally discernible by human beings
- (b) zero acoustic pressure
- (c) zero acoustic power
- (d) none of the above

(iii) Two automobile whistles of 100 dB are blown at the same time. Their resultant SPL would be
- (a) 100 dB (b) 200 dB
- (c) 103 dB (d) 110 dB

(iv) The reference power for determining the sound power level is
- (a) 0.00 002 W (b) 10^{-12} W
- (c) 1 W (d) 100 W

(v) The SPL corresponding to one atmospheric pressure (1.013 bar) is
- (a) 194 dB (b) 150 dB
- (c) 174 phons (d) 150 phons

(vi) The sound pressure level of a sound source at double the distance would
- (a) become one-half (b) become one-quarter
- (c) be reduced by 6 dB (d) be reduced by 3 dB

(vii) Acoustic amplifier output is generally specified in terms of its
- (a) acoustic SPL (b) acoustic PWL
- (c) electrical wattage (d) loudness

(viii) Loudness level is the SPL at
- (a) low frequency of 100 Hz (b) frequency of 1000 Hz
- (c) high frequency of 10 000 Hz (d) none of the above

(ix) The response of the sound level meter is weighted suitably so as to provide
- (a) a response similar to that of the human ear
- (b) linearised response at different frequencies
- (c) frequency spectra at different frequencies
- (d) all of the above

(x) Audio-frequency range lies
- (a) between 20,000 and 30 000 Hz (b) between 16 and 20 000 Hz
- (c) above 40 000 Hz (d) around 1000 Hz

14.2 Fill in blanks in the following:

(i) The sound pressure level (SPL) defined as __________. The reference pressure for evaluating dB is __________.

(ii) A sound intensity of 60 dB level corresponds to ________ W/cm^2.

(iii) The sound pressure level (SPL) measured in open space (free field), at a distance of 5 m from the noise source is 80 dB. At a distance of 50 m, the SPL is ________ dB.

(iv) The high-end frequency of an octave band filter is ________ the low-end frequency of the same. The centre frequency of the octave band filter is ________ mean of the high end and low end frequencies.

(v) A noise source of 70 dB at 1 kHz has a loudness level of ________.

14.3 Convert the following sound pressure level measurements into their corresponding absolute pressure values:

(a) threshold of hearing = 0.0dB
(b) faint sound of whisper ≃ 30 dB
(c) moderate sound of normal conversational speech (1 m) ≃ 50 dB
(d) load sound of heavy duty truck (6 m) ≃ 80 dB.
(e) very loud sound of heavy duty pneumatic rivetter ≃ 130 dB
(f) threshold of pain of human hearing mechanism ≃ 140 dB
(g) sonic boom of supersonic aircraft ≃ 160 dB.

14.4 A rocket engine produces an acoustic pressure of 0.7 N/m^2. Determine the sound pressure level in dB.

14.5 Calculate the equivalent SPL of two, and four 80 dB sources.

14.6 Five sound sources of 62, 59, 65, 72 and 80 dB are added together. Determine the resultant sound pressure level.

14.7 Frequency-spectrum analysis of the noise output of the machine using one-octave band pass filters gave the following results:

Centre frequency (in Hz)	125	250	500	1000	2000	4000	8000	16000
SPL (in dB)	70	71	84	86	85	77	64	66

Calculate the overall sound pressure level.

Answers

14.1 (i) (d) (ii) (a) (iii) (c) (iv) (b)
(v) (a) (vi) (c) (vii) (b) (viii) (b)
(ix) (a) (x) (b)

14.2 (i) $20 \log_{10} (p)/(p_{ref})$, 0.00002 N/m^2 (ii) 10^{-6} (iii) 60 dB
(iv) twice, geometric (v) 70 Phons

14.3 (a) 0.00 002 N/m^2 (b) 6.32×10^{-4} N/m^2 (c) 6.32×10^{-3} N/m^2 (d) 0.02 N/m^4
(e) 63.246 N/m^2 (f) 200.00 N/m^2 (g) 2000.00 N/m^2

14.4 SPL = 90.88 dB

14.5 (a) SPL of two 80 dB sources = 83.1 dB
(b) SPL of three 80 dB sources = 84.77 dB
(c) SPL of four 80 dB sources = 86.2 dB

14.6 Resultant SPL = 80.84 dB

14.7 Overall SPL = 90.19 dB

Chapter

15

Signal and Systems Analysis

■ *INTRODUCTION* ■

Signal analysis involves operations carried out on a signal in order to suitably designate it. Determination of the frequency spectrum of dynamic signal like vibrations, noise or pressure signals, for example, is an important signal analysis operation. The peaks in frequency contents of such signals gives useful information about the nature of the signal. This is applicable to signals of periodic, random or of transient types. It has already been shown in Chapter 3 that this is made possible as seen from Fourier series or Fourier transform of such signals. Frequency analysis has been conventionally carried out, using frequency analysis instruments comprising analog filters. There have been considerable developments in the recent years on the application of digital types of devices, due to increasing use of minicomputers. With the rapid use of the fast Fourier transform algorithm, the time and effort taken for frequency analysis, are considerably reduced.

Allied with signal analysis, is what is known as *system analysis,* which involves the analysis of both input and output of any system like a vibrating system, control system or any other system, in order to characterise the system. This is made possible by analysing the output response of the system to an input signal. Experimental techniques for analysis of a complex system give meaningful information about the system and are widely used in practice, especially where theoretical methods are not easy to apply.

15.1 ■ ANALOG FILTERS AND FREQUENCY ANALYSERS

A periodic signal comprising several frequency components can be analysed in terms of its components by frequency analysis as in Fig. 15.1.

A number of filters with different centre frequencies are built in the analyser, each filter passing the output corresponding to its own characteristics.

The detector squares the signal and integrates over a defined averaging time to get average or root mean square (rms) value as desired. Thus, the strength of the signal passing out of the filter, corresponding to its centre frequency, is recorded or indicated and thus the frequency spectrum of the signal is obtained.

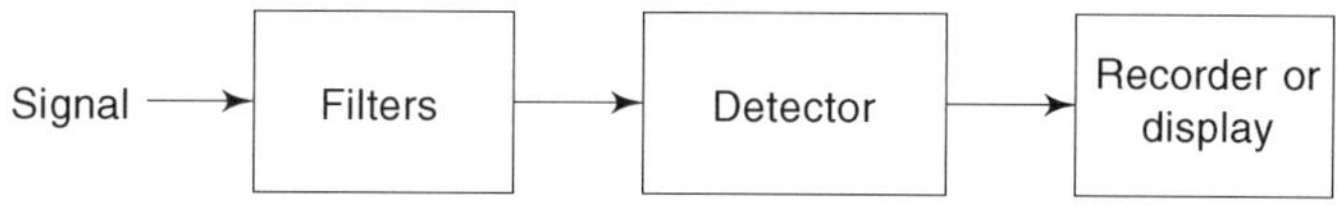

Fig. 15.1 *Frequency analyser*

Figure 15.2 shows the characteristics of an ideal filter, with nominal or centre frequency f_0, f_u and f_l are the upper and lower limits of frequencies which pass through the filter.

$$\% \text{ band width of the filter} = \frac{f_u - f_l}{f_0} \times 100$$

In case of octave filters, the band width is such that the upper limiting frequency f_u is twice the lower limiting frequency f_l. It can be easily shown that the % band width for octave filters is 70.7%, taking f_0 as the geometric mean of f_u and f_l. Also, for one-third octave filters, in which $f_u = (2)^{1/3} f_l$, % band width is seen to be 23.1%. Figure 15.3 shows a discrete stepped analyser in which multiple filters, each with a fixed centre frequency, are used. Each filter is switched in turn and

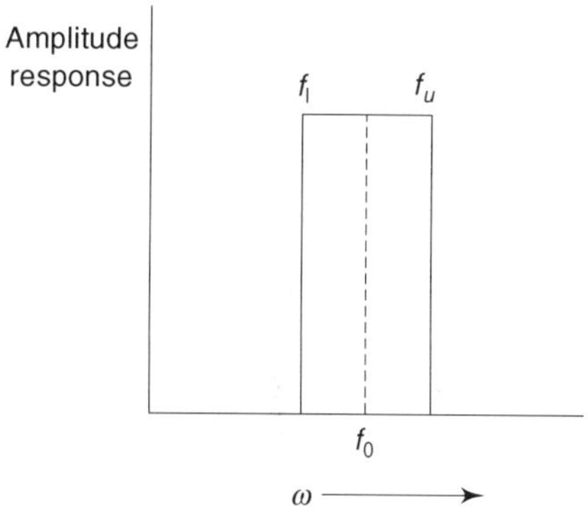

Fig. 15.2 *Filter characteristics*

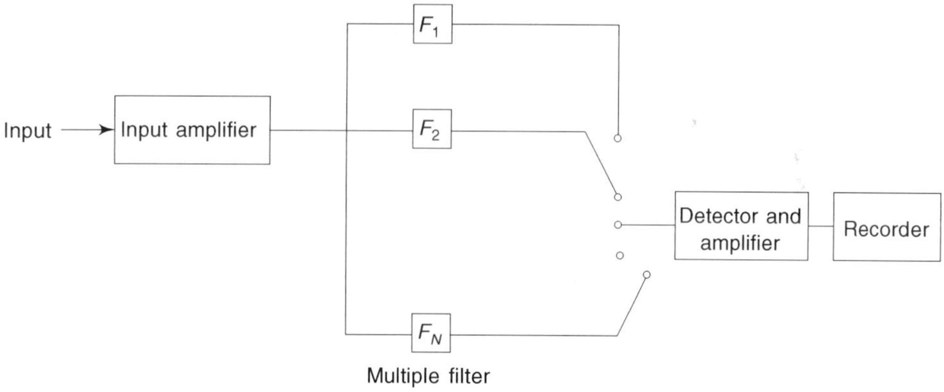

Fig. 15.3 *Discrete stepped analyser*

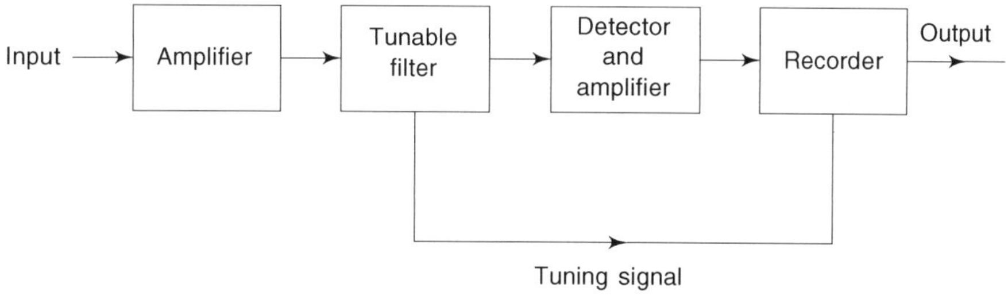

Fig. 15.4 *Sweeping filter analyser*

the output is recorded. Another type of frequency analyser is the sweeping filter type, as in Fig. 15.4, in which the filter frequency can be continuously tuned or adjusted and the frequency spectrum recorded.

In order to obtain fast analysis, the spectrum is obtained from filters in parallel as in a real time parallel analyser (Fig. 15.5) with suitable circuits, switching automatically from one filter to the other and a frequency spectrum is directly obtainable. In the case of the above analysers, the band width may be smaller than that for one-third octave analysers.

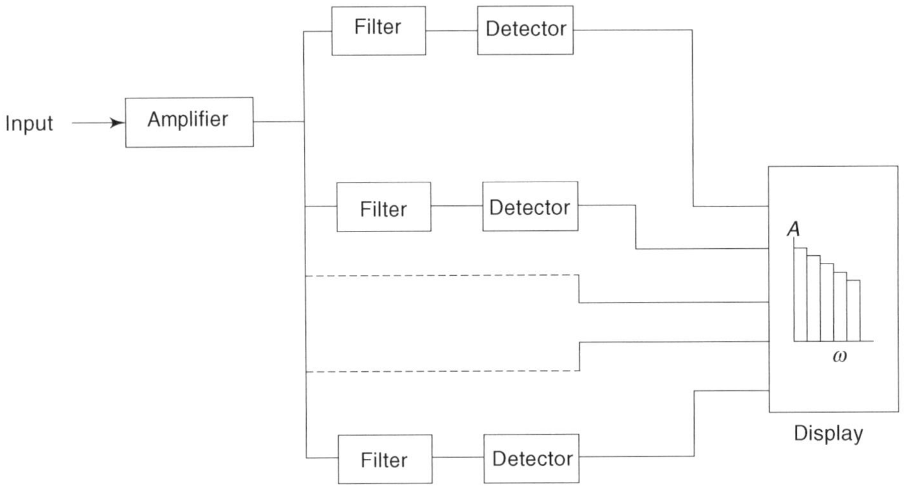

Fig. 15.5 *Real time parallel analyser*

Figure 15.6 shows a real time analyser of a different type, known as *time compression analyser.* It is based on the principle of recording the signal to be analysed at one speed and then playing it back faster, say 500 times. This results in the analysis being done at higher frequencies whereby filters with higher band widths can be employed. It is seen that the analysis time per filter is inversely proportional to the band width and thus increase in band width results in reduction in analysis time. The input signal may be of periodic, transient or random type. Filter 1 removes the spurious frequencies. It is sampled at a fast rate and converted to digital form by an *A–D* converter and recorded in the digital memory recorder. From the memory, it is read out at a fast rate and converted to analog form and the signal so obtained is time-compressed. Figure 15.6(b) shows a typical input signal and Fig. 15.6(c) shows the corresponding time-compressed signal, when the ratio of read-out rate to the recording rate is 500. Thus, the corresponding frequency contents are shifted to higher frequencies. In some commercial real-time analysers, the heterodyne technique is employed which involves the use of a single fixed band pass filter for all ranges of frequencies. The signal from the *D–A* converter is multiplied by a sine wave signal of frequency ω_c. Frequency ω_c is stepped through a range of values. If ω_s is the frequency of the signal, the multiplication shifts the frequency. Thus, the output after multiplication is

$$\text{Output} = (A_s \sin \omega_s t)\,(A_c \sin \omega_c t)$$

$$= \frac{A_s A_c}{2}\,[\cos(\omega_c - \omega_s)t - \cos(\omega_c + \omega_s)t]$$

The centre frequency of filter 2 is set to frequency $(\omega_c + \omega_s)$. After demodulation, the signal corresponding to frequency ω_s is recovered and the output indicated.

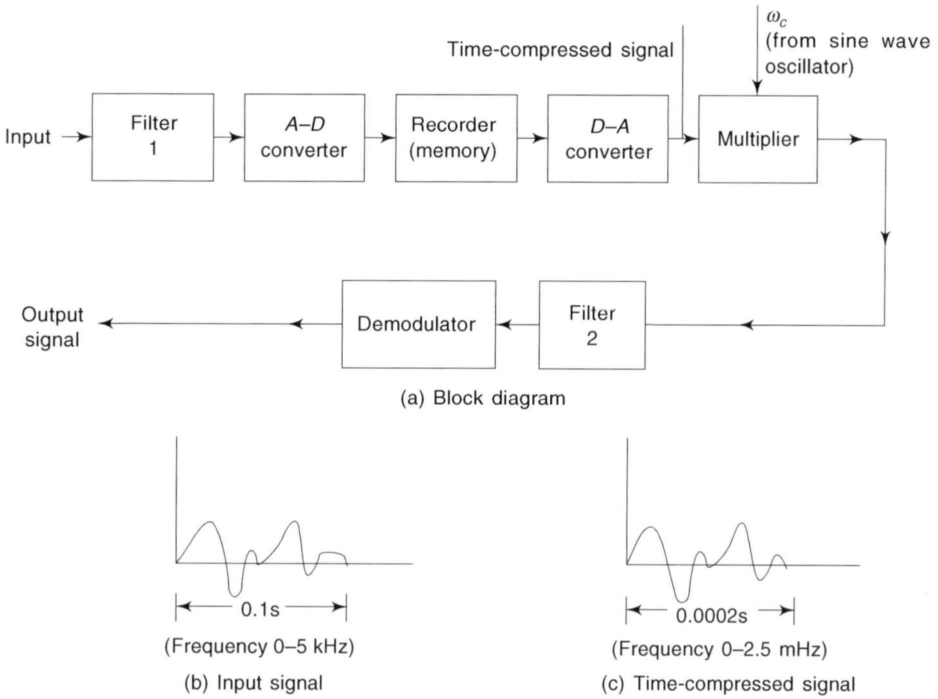

Fig. 15.6 *Time-compression analyser*

15.2 ■ FREQUENCY ANALYSIS FOR VARIOUS INPUT SIGNALS

The frequency spectrum of a periodic signal is discrete. Figure 15.7 shows the vibratory motion spectrum as in a reciprocating engine system. This can be obtained by using any of the analysers discussed in Sec. 15.2. Since the signal is continuous, the frequency analysis can be done on-line or may be recorded on a magnetic tape recorder before the analysis.

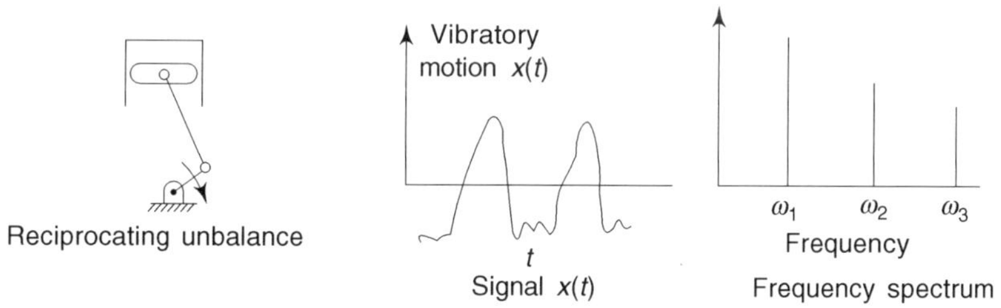

Fig. 15.7 *Frequency spectrum of a periodic signal*

For frequency analysis of transient signals, the Fourier transform of the signal $x(t)$ is desired

$$X(\omega) = \int_0^{\infty} x(t)\, e^{-j\omega t}\, dt$$

$$= \int_0^{\infty} x(t)\,(\cos \omega t + j \sin \omega t)\, dt$$

As shown in Fig. 15.8, the transient input signal $x(t)$ is multiplied by sine and cosine signals of adjustable ω, with the product terms being integrated later and indicated on the meters. The input $x(t)$ is recorded on a magnetic tape recorder and played back for analysis several times, once for each frequency.

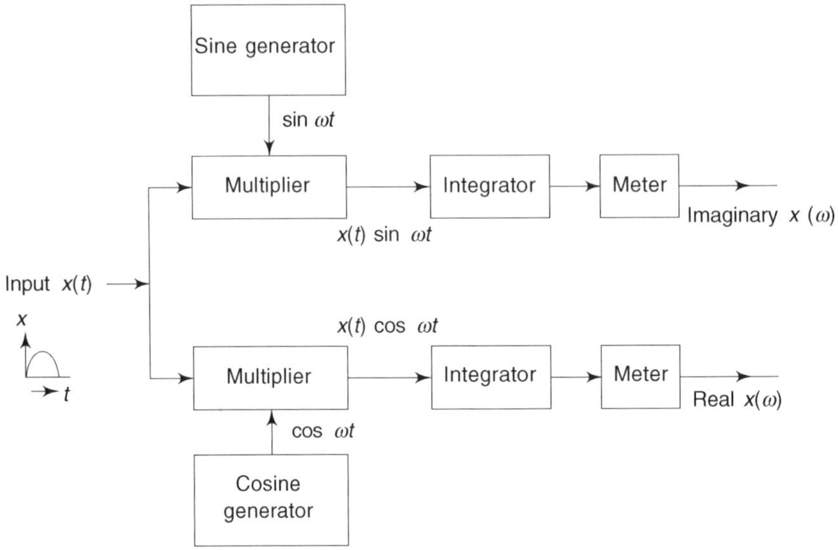

Fig. 15.8 *Fourier transform analyser*

An alternative arrangement is shown in Fig. 15.9. This uses a digital event recorder. Frequency analysis may be done for the transient signal of Fig. 15.10(a), by either having a separate playback for each frequency estimate or by analysing an equivalent periodic signal as shown in Fig. 15.10(b), by repetitive playback.

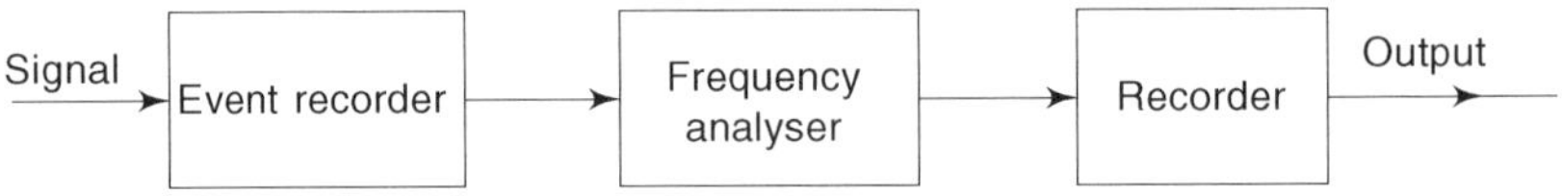

Fig. 15.9 *Arrangement for obtaining frequency spectrum*

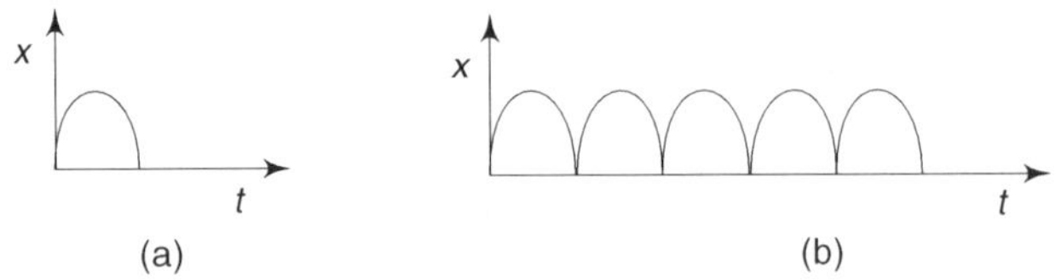

Fig. 15.10 *Transient and equivalent periodic signal*

For random signals, the frequency analysis is carried out to determine the spectral density function $S(\omega)$, which is defined as the mean square value of that part of the signal lying in a chosen frequency band or

$$S(\omega) = \frac{x_\omega^2}{\Delta\omega}$$

$$S(\omega) = \frac{x_\omega^2}{\Delta\omega}$$

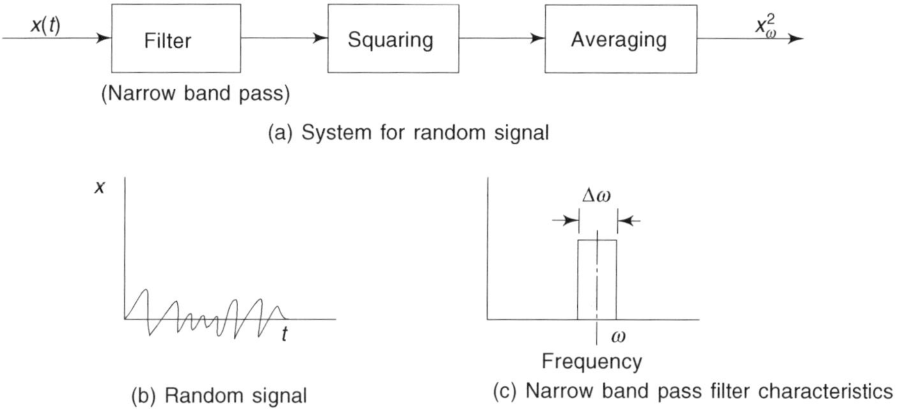

Fig. 15.11 *Random signal analysis*

As shown in Fig. 15.11, a narrow band pass filter with adjustable centre frequency and band width $\Delta\omega$ passes the output which is squared and averaged giving x_ω^2. The random signal, as shown in Fig. 15.11(b) is passed through a filter with characteristics as in Fig. 15.11(c). A plot of spectral density $S(\omega)$ is shown in Fig. 15.12. It is seen that the area under the curve gives x_ω^2.

As discussed in Sec. 3.3 and Appendix A-4, random signal may be represented by $R(\tau)$, the auto-correlation function, which is related with $S(\omega)$.

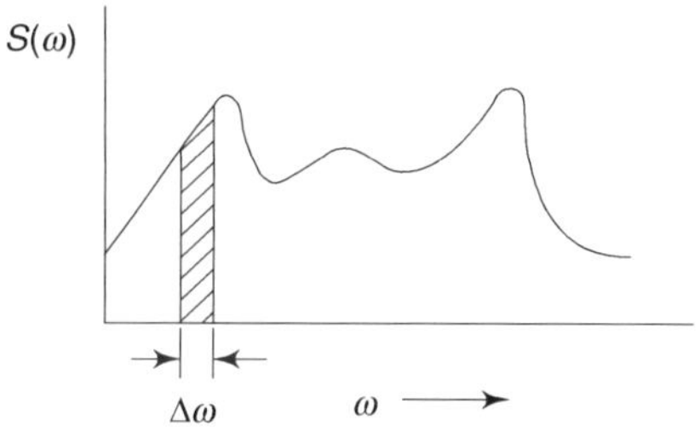

Fig. 15.12 *Spectral density $S(\omega)$ of a random signal*

15.3 ■ DIGITAL FREQUENCY ANALYSERS

With the rapid use of digital computers, frequency analysis techniques have also undergone considerable changes. Digital filters are basically different from analog ones in that the former are essentially digital processors which receive a sequence of sampled input data and carry out some digital operations on them such that the output is filtered in the desired way. Similarly, FFT (fast Fourier transform) is an algorithm for rapid calculations of the frequency spectrum from a whole block of data and has found wide applicability in recent times.

15.3.1 Sampling

The data to be analysed is sampled, usually at equally spaced time intervals, so that discrete samples are produced and fed to a digital computer. The samples are taken at a rate that is at least twice the highest frequency present in the waveform.

Figure 15.13(a) shows the sampling of a waveform, of period T at intervals of $T/2$. This appears to be the largest sampling interval that will retain the nature of the oscillatory character. In Fig. 15.13(b), the sampling is done at intervals of $0.8T$. It is seen that one of the lobes is missed. Thus, sampling interval $\Delta T \leq 1/2T_{min}$, where T_{min} is the period of the fastest oscillatory component, to be analysed.

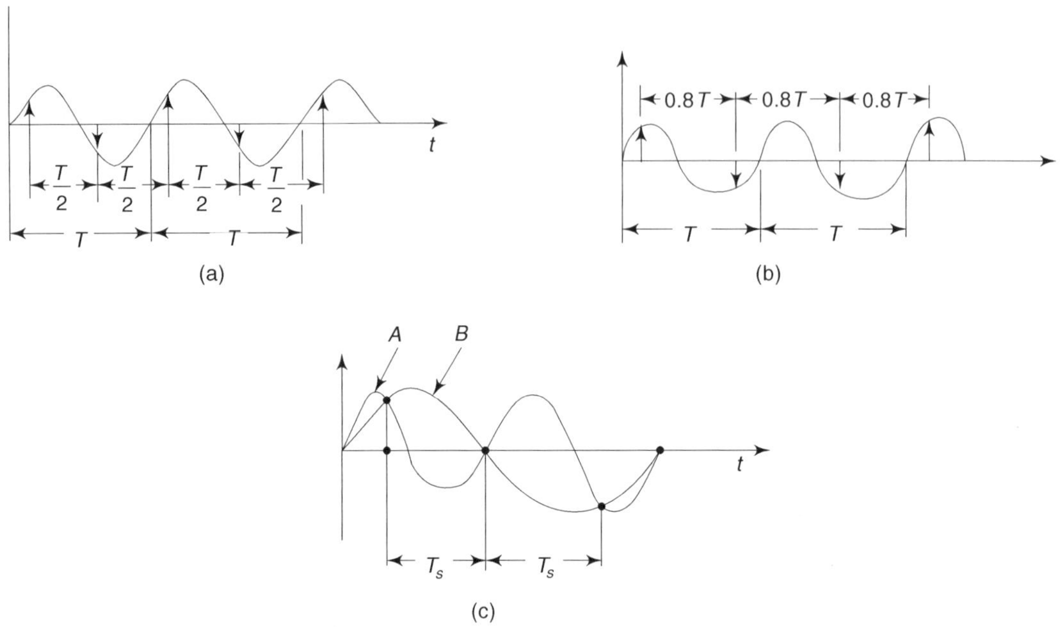

Fig. 15.13 *Sampling of continuous data*

The frequency equivalent to half the sampling rate is called the Nyquist frequency. Figure 15.13(c) shows that two sine waves A and B if sampled at interval T_s as shown, would give the same impression as far as the nature of wave form is concerned, even though A has higher frequency than B. The frequency of change of the faster varying function is known as the *alias* of the slower varying function. This feature is kown as aliasing and occurs due to slow sampling rate.

Figure 15.14(a) shows spectrum A of a signal. If the sampling frequency is ω_s, the highest frequency which is possible in the transform or spectrum is $\omega_s/2$. A signal of frequency ω and that of $(\omega_s - \omega)$ appear the same, due to aliasing. Thus, the spectrum is distorted as B as is shown in Fig. 15.14(b).

A technique to eliminate distortion is to use an anti-aliasing filter (LP type) which is used in the analysers.

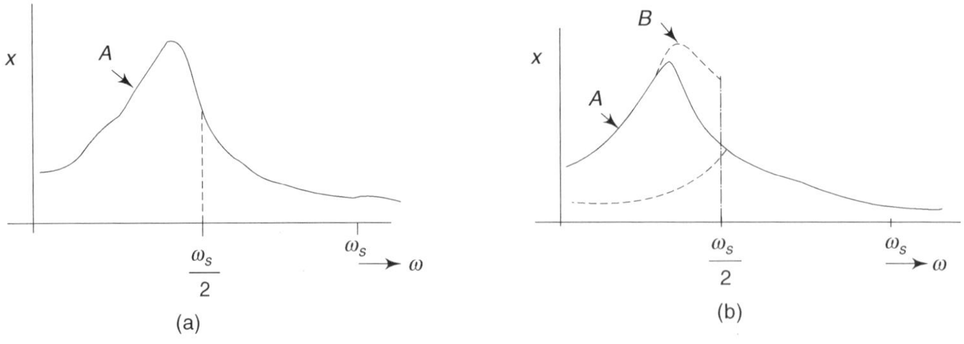

Fig. 15.14 *Distortion of signal due to aliasing*

15.3.2 Digital Filters

In the case of digital filters, the sampled data obtained from the continuous spectrum is used for carrying out digital operations on the various samples. Figure 15.15 shows the schematic diagram of a simple filter. It comprises digital components like multiplier A, adder represented by a summing junction in Fig. 15.15, a delay unit storing the previous output value viz. $y_{n-1} \cdot x_n$ and y_n are the current input and output values, respectively. Another multiplier B is also used to multiply y_{n-1} and add to Ax_n. Thus,

$$y_n = Ax_n + By_{n-1} \tag{15.1}$$

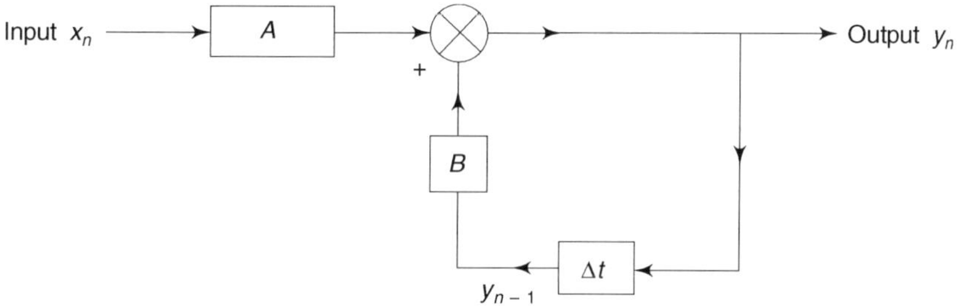

Fig. 15.15 *Schematic diagram of a simple filter*

A and B are constants, stored in memory for a given filter. Usually one independent parameter C is used, i.e.,

$$A = 1 - C$$

and $$B = C$$

With the above, Eq. (15.1) becomes

$$y_n = (1 - C)x_n + Cy_{n-1} \tag{15.2}$$

These types of filters are called recursive digital filters since the output results from input as well as previous output. It can be shown that a simple filter of the type shown in Fig. 15.14 acts like a low pass RC analog filter.

The Fourier transform of Eq. (15.2) gives

$$Y(f) = (1 - C)\, X(f) + CY(f)e^{-j2\pi f\Delta t} \tag{15.3}$$

where f stands for frequency, and the frequency response function is given as:

$$G(f) = \frac{Y(f)}{X(f)} = \frac{1-C}{1-Ce^{-j2\omega f\Delta t}} \tag{15.4}$$

Putting $2\pi f\Delta t = \theta$, it is seen that

$$|G(f)|^2 = \frac{(1-C)^2}{(1+C^2)-2C\cos\theta} \tag{15.5}$$

For the case when $\theta << 1$,

$$e^{-je} \approx 1 - j\theta$$

Thus, Eq. (15.4) becomes

$$G(f) = \frac{1-C}{1-C(1-j\theta)} \tag{15.6}$$

If C is taken $= e^{-\Delta t/\tau}$, τ being a constant and $>> \Delta t$

$$C \approx 1 - \frac{\Delta t}{\tau}$$

With the above substitution, Eq. (15.6) becomes

$$G(f) = \frac{1}{1 + j\theta \dfrac{\tau}{\Delta t}} = \frac{1}{1 + j\,2\pi f\tau}$$

Also,

$$|G(f)|^2 = \frac{1}{1 + (2\pi f\tau)^2} \tag{15.7}$$

Equation (15.7) corresponds to the frequency response of a resistance capacitance (RC) type low pass filter with $RC = \tau$.

The schematic diagram for a general recursive digital filter is shown in Fig. 15.16. The governing relation for the above is

$$y_n = Ax_n + \sum_{k=1}^{M} E_k y_{n-k} \tag{15.8}$$

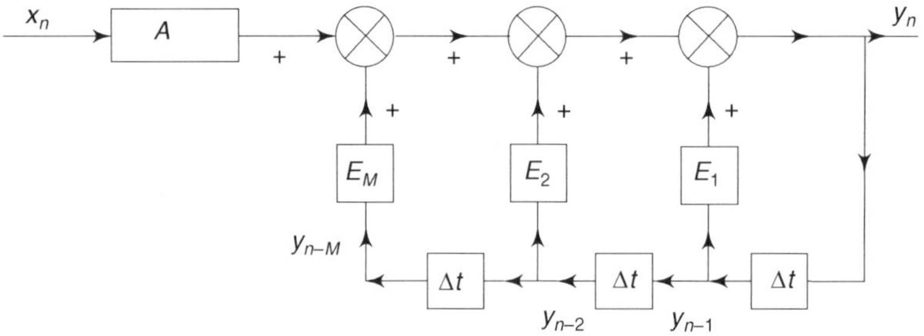

Fig. 15.16 *Schematic diagram of a general recursive digital filter*

The performance of the above filters can be analysed for different values of the constant A and E_k.

A digital frequency analyser would thus have digital filters, similar to those described and carry out digital or numerical operations on the input data, for frequency analysis.

15.3.3 Fast Fourier Transform Analyser

The fast Fourier transform (FFT) technique involves the use of a numerical algorithm for fast computation of discrete Fourier transform of a signal (which is a function of time) in the frequency domain. Fourier transform $X(f)$ of a time function $x(t)$ is given by

$$X(f) = \int_{-\infty}^{\infty} x(t)e^{-j2\pi ft}\,dt \tag{15.9}$$

f being the frequency and t the time. In practice, only finite limits of integration are used in the range $0 \rightarrow T$ and thus, finite range Fourier transform is given by

$$X(f, T) = \int_{0}^{T} x(t)e^{-j2\pi ft}\,dt \tag{15.10}$$

The above applies to a continuous function. For digital analysis, the function $x(t)$ is sampled at discrete points, normally at N intervals, with equal spacing ΔT, starting from $t = 0$.

Thus, $$x_n = x(n\Delta T) \quad n = 0, 1, 2, \ldots, N-1 \tag{15.11}$$

Thus, discrete Fourier transform would be given by

$$X(f, T) = \Delta T \sum_{n=0}^{N-1} x(n\Delta T)\, e^{-j2\pi f n \Delta T} \tag{15.12}$$

The discrete frequency values for finding $X(f, T)$ are usually taken as

$$f_d = df = \frac{d}{T}$$

Since $$T = N\Delta T$$

$$f_d = \frac{d}{N\Delta T}$$

Substituting this value in Eq. (15.12), for frequency, the discrete Fourier transform components are given by:

$$X_d = \frac{X(f_d, T)}{\Delta T} = \sum_{n=0}^{N-1} x(n\Delta T) e^{-j2\pi dn/N} \quad d = 0, 1, 2, \ldots, N-1 \tag{15.13}$$

ΔT has been taken away from the RHS for convenience. FFT methods, developed recently, are meant to find X_d at discrete frequencies f_d, by suitable numerical algorithms from Eq. (15.3). This is required when x_n is expressed as a series of N terms. In the FFT procedure, the N terms are taken as a composite of smaller number of terms and X_d is computed in an iterative way, involving much less computation time. Full details of the numerical algorithm are given by Cooley and Tukey*.

Thus, the entire process of frequency analysis of a signal involves the digital sampling and numerical evaluation of its discrete Fourier transform, which has been considerably facilitated with the wide spread use of mini computers.

15.3.4 Leakage and Windowing

For finding Fourier transform of a signal, a finite length of signal in the time domain is taken. Assuming the signal to be harmonic as in Fig. 15.17(a), the signal is periodic in the time window T. Thus, the spectrum is a line at signal frequency as shown alongside.

Figure 15.17(b) shows a signal which is not quite periodic since at the end of time T, the number of signals is not an integer and thus there is a discontinuity. This results in the frequency spectrum being a number of lines. This is known as 'leakage' as the spectrum is said to leak over a number of lines. Thus, the spectrum in Fig. 15.17(b) is inaccurate. The problem is solved by using windows. Windowing involves the use of a defined window profile before taking Fourier Transform. The signal $x(t)$ is multiplied by $y(t)$, viz. the window profile. A number of windows are used in an analyser as shown in Fig. 15.18 which shows rectangular, Hanning and exponential windows.

For rectangular window $y(t) = 1$. Hanning window is used for continuous signal, reducing the effect of start and end of the signal. The exponential window is employed for transient signal and it changes the signal mostly at the start where much the signal information in a transient signal exists. The use of proper windows reduces the inaccuracies due to leakage.

---

* J.W. Cooley and J.W. Tukey: "An Algorithm for the Machine Calculation of Complex Fourier Series", *Math. Comput.*, vol. 19, pp 297–301, April 1965.

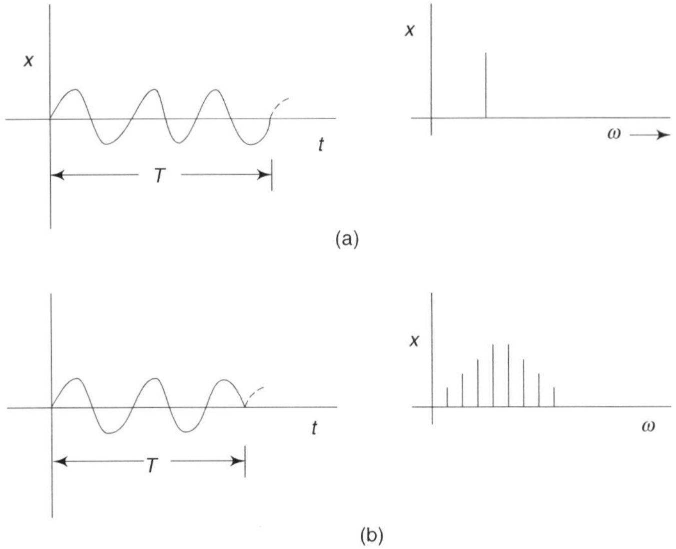

Fig. 15.17 *(a) No leakage effect (b) Leakage effect*

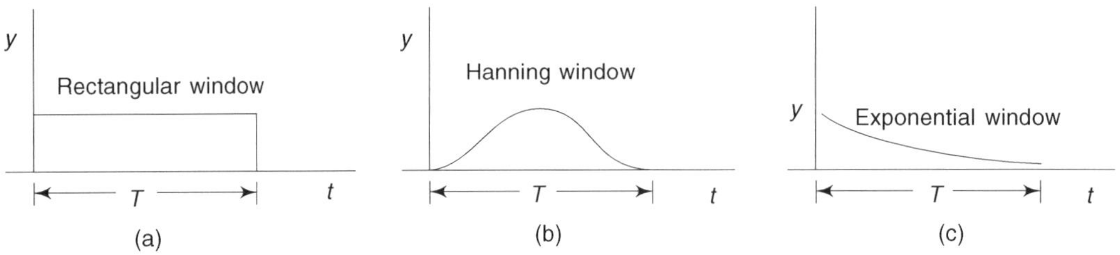

Fig. 15.18 *Different types of windows for reducing leakage effect*

15.4 ■ SYSTEM ANALYSIS BY HARMONIC TESTING

System analysis involves the application of an input signal to the system and determining its output signal characteristics relative to the input and thus form an idea about the nature of the system. The simplest technique is to apply a harmonic input signal to the system. As discussed in Chapter 3, for a linear system, the output would also be harmonic, though there may exist a phase difference between the two signals. The frequency of the input signal can be varied in order to obtain frequency response or transfer function characteristics, the transfer function being the ratio of the output to input signals.

As an example, in order to obtain vibration characteristics of a typical structure, a harmonic force of frequency ω is applied to the structure, as shown in Fig. 15.19.

The frequency of the vibrator can be varied by the oscillator. The dynamic response of the structure is measured by vibration transducers. The frequencies at which the system resonates, can be determined. In such a case, the following two parameters are needed at each frequency ω:

1. ratio of amplitude of the output response to that of the input force (also known as mechanical impedance), and
2. phase angle between the output response and input force signals.

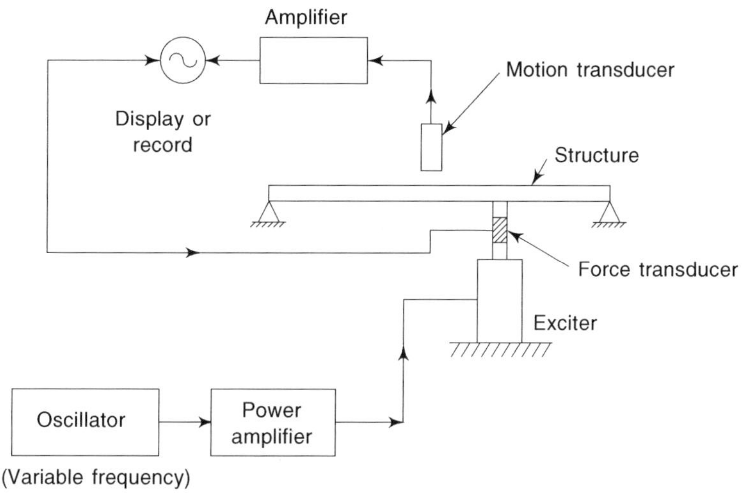

Fig. 15.19 *Testing for obtaining frequency response characteristics of a vibrating structure*

A single degree vibrating system has a governing equation corresponding to the second order system of Ch. 3 and thus the natural frequencies and damping ratio of the system can be determined from the measured data. For a complex system, a number of such parameters have to be determined for each mode of vibration, in a given frequency range.

Where a large quantity of data has to be measured and stored, the use of an automatic system analsis facility becomes essential. Figure 15.20 shows such a system using a mini computer.

The oscillator frequencies are specified and at each frequency, the response of a number of motion transducers and the output of the force transducer are analysed. The computer computes the ratio of response to the force and the phase difference between the two and plots or prints the results as desired. The whole process is completely automated and a large amount of data can be analysed. Figure 15.21 shows the first three modes of vibration of a fixed-fixed beam in bending, obtained experimentally.

15.5 ■ SYSTEM ANALYSIS BY TRANSIENT TESTING

The frequency response function $G(f)$ of a linear system may be obtained from the response $y(t)$ to any transient input $x(t)$. It requires the determination of Fourier transforms $X(f)$ and (f) or $x(t)$ and $y(t)$, respectively.

$$G(f) = \frac{Y(f)}{X(f)} \tag{15.14}$$

where

$$X(f) = \int_{-\infty}^{\infty} x(t)e^{-j2\pi ft}\, dt$$

$$Y(f) = \int_{-\infty}^{\infty} y(t)e^{-j2\pi ft}\, dt$$

Since both $X(f)$ and $Y(f)$ are complex, the frequency response function $G(f)$ of the system is also complex and is the response that would be obtained under harmonic input conditions.

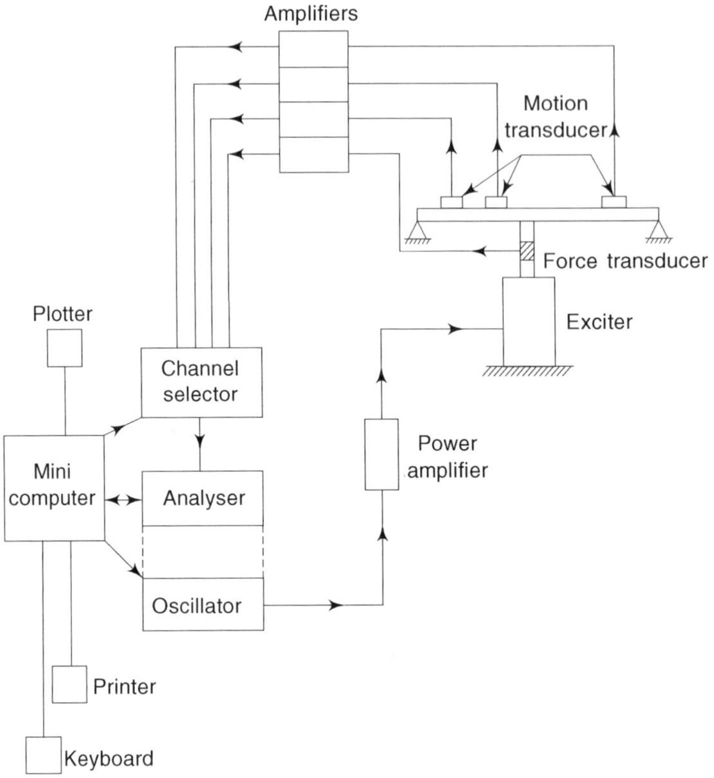

Fig. 15.20 *Computer controlled measurement system*

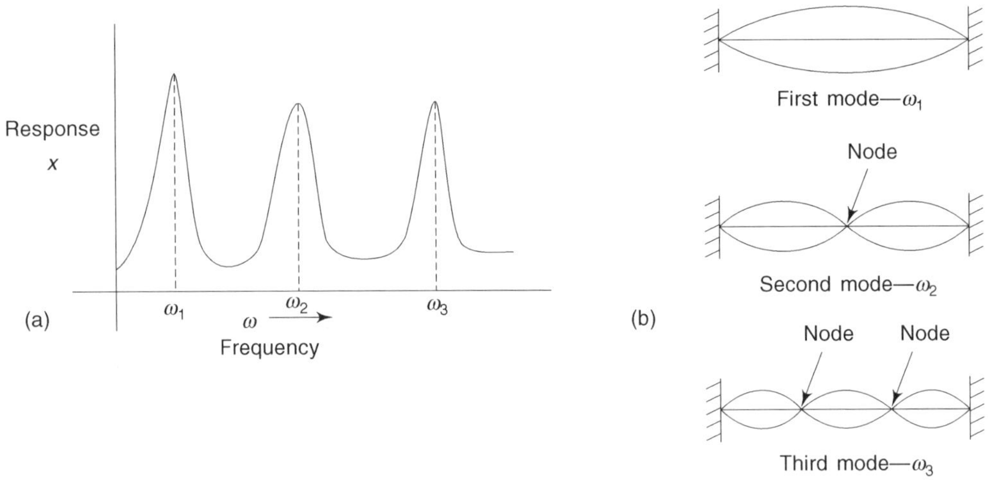

Fig. 15.21 *Frequency–response characteristics (a) and first three mode shapes and corresponding frequencies (b) of a fixed–fixed beam*

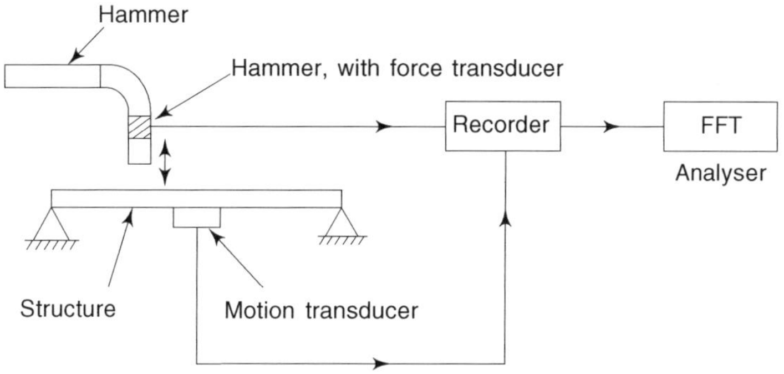

Fig. 15.22 *Frequency response characteristics by transient method*

Short duration transient inputs can be used for quicker analysis of a system, provided the Fourier transform of the input and output signals can be computed by a facility like FFT analyser. Figure 15.22 shows the arrangement for system analysis using a transient input like that obtained by hitting with a hammer. The purpose is to determine the vibration characteristics of the structure, as in Sec. 15.5. Fourier transforms of the force signal and the output signal of the motion transducer are obtained using an FFT analyser and thus frequency response function FRF of the system is obtained. From FRF's obtained by excitation at different locations on the structure, the mode-shapes corresponding to different modes of vibration can be obtained using modal identification technique.

Figure 15.23 gives another application for the determination of loudspeaker characteristics. The input excitation is in the form of a square pulse. The pulse is triggered by an interface unit of the FFT analyser. The output of the microphone is analysed by the FFT analyser. Further, the square pulse is directly fed via *B* (instead of *A* earlier) and the Fourier transform of the transient pulse determined. Thus, from the Fourier transforms of the transient input as well as the output, frequency response characteristics are determined conveniently over a wide frequency range, the upper frequency limit being determined by the duration of the pulse. This type of measurement procedure is faster and more convenient than the conventional method of harmonic testing at each frequency separately.

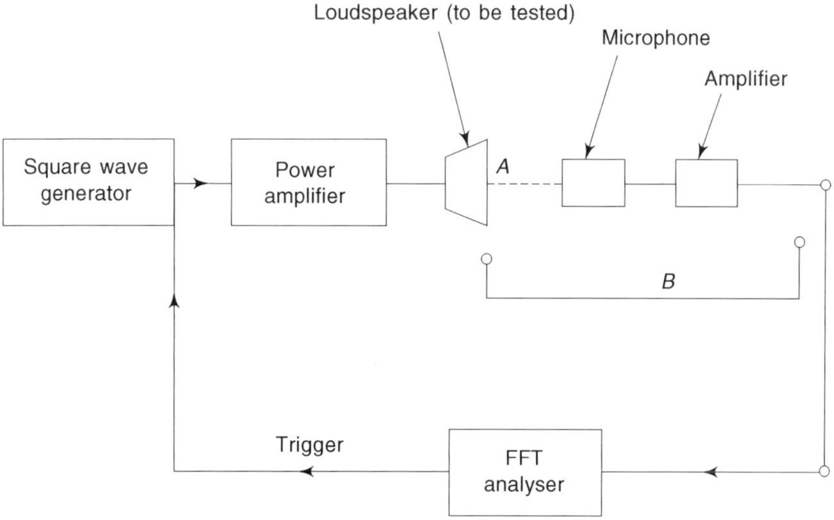

Fig. 15.23 *Measurement system for loudspeaker characteristics*

In addition to the above techniques, rapid sinewave sweep and random-input test techniques have also been developed for the purpose of system analysis.

15.6 ■ RANDOM FORCE TESTING

The dynamic characteristics of a vibratory system can also be found by using a random force excitation from a vibrator, as shown in Fig. 15.24. The random signal generator is used where the signal after amplification is fed to an electrodynamic vibrator. The signal is usually of a white noise type, in which the spectral density of the signal is constant with frequency. The various resonance of a vibrating system can be easily identified in view of excitation frequencies being spread over a wide range.

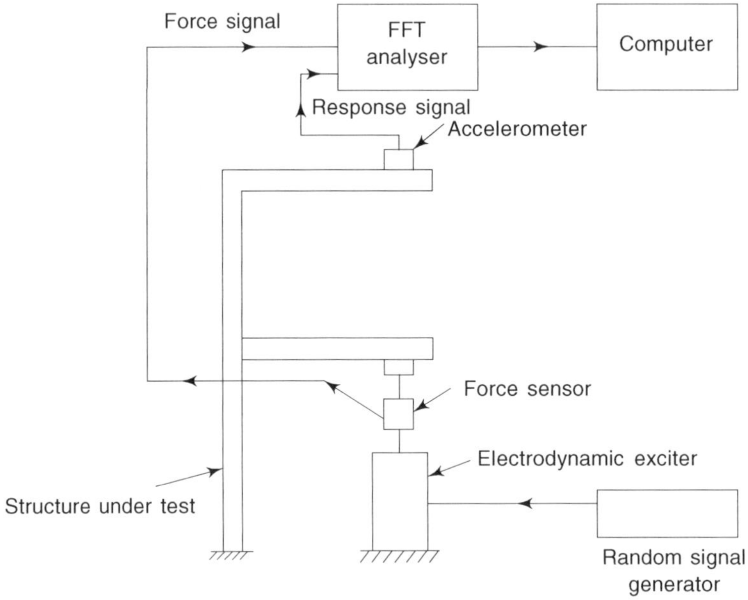

Fig. 15.24 *Experimental dynamic testing of a structure using random signals*

However, the random signals do not satisfy the Dirichlet condition $\left(\text{viz. } \int_{-\infty}^{\infty} |x(t)|\, \mathrm{d}t < \infty\right)$, which is necessary to find Fourier transforms, for advanced system analysis, as is possible with transient signals.

For a pair of random signals, the use of auto-correlation and cross-correlation functions is useful since the Fourier transforms of the correlation functions give the corresponding spectral density functions as shown in Appendix A5. This is possible in an FFT analyser, to get direct and cross-spectral density functions of the pair of signals, viz. force and response signals, and also obtain coherence function from the same. As shown in Appendix A5, the presence of noise in a signal, makes the value of coherence to be less than unity. This is a useful concept in signal and system analysis.

From the FRF's obtained for various locations of the accelerometer, it is possible to obtain the various mode-shapes by modal identification as mentioned in the previous section.

Review Questions

15.1 Indicate True or False against the following:

(a) Signal analysis means analysing the signal for finding its frequency contents.
(b) Real time analysis involves recording of the signal against time.
(c) Transient signals cannot be analysed by frequency analysis techniques.
(d) The centre frequencies of analog filters are usually arranged in geometric progression.
(e) One-third octave filter means its band width is one-third of the centre frequency.
(f) A detector in a frequency analyser squares and averages the signal.
(g) In a time-compressed signal, the frequency contents are shifted to lower frequencies.
(h) Time for analysis is inversely proportional to percentage band width of the filters.
(i) From the spectral density function of a random signal, we can find the rms value of the signal.
(j) For digital frequency analysis, sampling frequency should be at least twice the maximum frequency of the signal.
(k) Digital filtering is essentially a numerical computation process.
(l) Transfer function is the difference in phase angle between input and output signals.
(m) Transient testing for systems analysis is much faster than harmonic testing.
(n) An FFT analyser connected to vibration sensor, can be used for analysis of periodic, transient and random signals.
(o) The value of coherence equal to 1.0 indicates absence of noise in the signal.
(p) Leakage occurs in incorrect frequency spectrum due to lack of periodicity in a dynamic signal.
(q) Rectangular window is used to reduce leakage.
(r) Hanning window is preferred for transient signals.
(s) The higher number of nodes usually occur for higher modes of a vibrating system.
(t) Fourier transform is not possible for a random signal directly.
(u) Cross-correlation function relates two different signals in the frequency domain.

15.2 A sine wave signal of frequency 6 Hz is sampled with a frequency of 10 Hz. Find the alias frequency which appears corresponding to 6 Hz.

Answers

15.1	(a) T	(b) F	(c) F	(d) T	(e) F	(f) T	(g) F	(h) T
	(i) T	(j) T	(k) T	(l) F	(m) T	(n) T	(o) T	(p) T
	(q) F	(r) F	(s) T	(t) T	(u) F			

15.2 4 Hz

Chapter

16

Condition Monitoring and Signature Analysis Applications

INTRODUCTION

Condition monitoring implies determination of the condition of a machine or device and its change with time in order to determine its condition at any given time. The condition of the machines may be determined by physical parameters like vibration, noise, temperature, oil contamination, wear debris, etc. A change in any of these parameters, called signatures, would thus indicate a change in the condition or health of the machine. If properly analysed, this thus becomes a valuable tool to determine when the machine needs maintenance and in the prevention of machinery failures, which can be catastrophic and result in unscheduled breakdowns.

The parameters mentioned above may be measured or monitored continuously or at regular intervals, depending on the application. It has been seen that a modest investment on instrumentation, for measurement of these physical parameters, would ultimately result in considerable savings due to the timely diagnosis of impending failures in plant machinery.

In case of process machinery like pumps, turbines, compressors, etc. the rotating components are subjected to fatigue and wear, and maintenance problems are fairly acute needing serious consideration. The situation is further aggravated due to the present day emphasis on large machines rather than a number of smaller machines with the same total output. Thus, in such cases, information regarding the parameters like vibration, noise, temperature, wear debris, corrosion initiation and propagation of cracks and other faults, in addition to process parameters like pressure drop, flow, power, etc. which indicate the performance trend, is useful. In addition, the relation between the possible faults and the monitored parameters has to be understood for meaningful use of the data compiled.

16.1 ■ VIBRATION AND NOISE MONITORING

Vibrations or noise signature of a machine are seen to be very much related to the health of a machine. Thus, measurement of vibration levels on bearing housings, relative movement between shaft and bearings, noise emitted by a machine, etc. can provide useful information regarding faults like unbalance, misalignment, lack of oil, wear, etc. Figure 16.1, for example, indicates the frequency spectrum of vibration velocity in a ball bearing, in original new condition and also when the balls became defective. The increased level of vibrations and introduction of additional peaks is an indication of the defect.

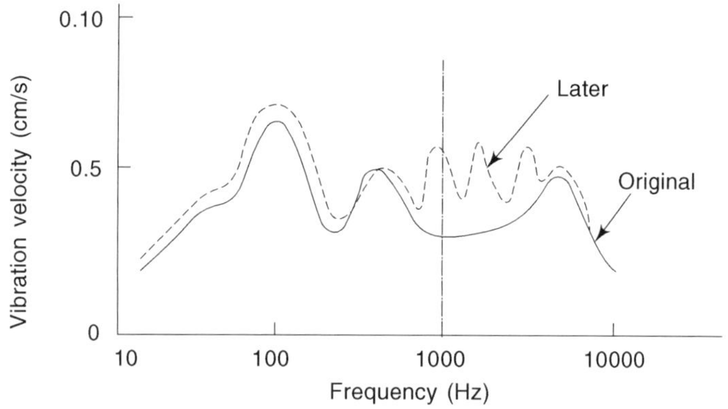

Fig. 16.1 *Vibration due to defective ball in a bearing*

16.1.1 Instrumentation for Monitoring Vibration and Noise Signals

In Chapters 7 and 14, the instruments for measurement of vibrations and noise have already been discussed. In practice, piezo-electric accelerometers are commonly employed for vibration monitoring and microphones for noise monitoring. Frequency analysis instruments, discussed in Chapter 15, are also used for obtaining the frequency spectrum of the vibration or noise signals.

Figure 16.2 shows a measuring and analysis system that may be used for monitoring the vibration signals from a number of machines. In case the data cannot be analysed in the field, it may be recorded on a magnetic tape recorder and analysed in the laboratory, as shown in Fig. 16.3.

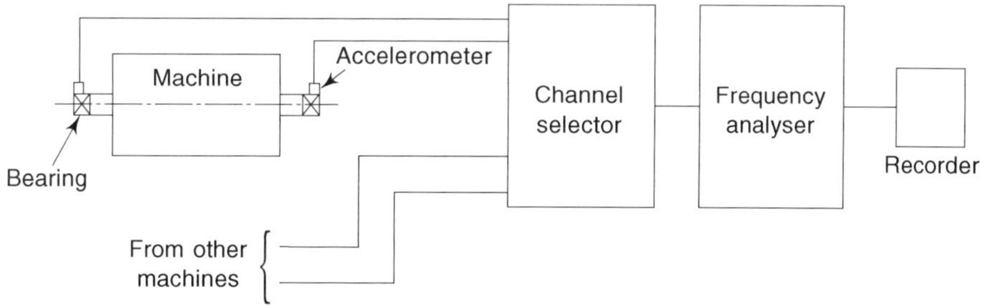

Fig. 16.2 *Vibration measuring and analysis system*

Figure 16.4 shows a typical record of the frequency spectrum from one of the transducers. The peaks in the levels have to be identified and checked whether these are below the specified limits.

The problem of identification of peak levels and specification of the limits has been discussed in Section 16.1.2.

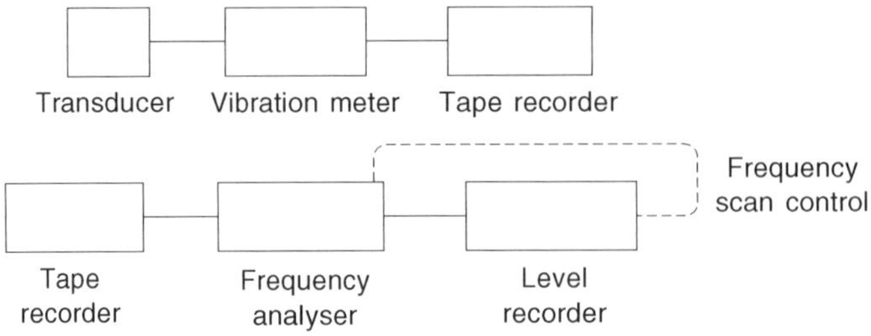

Fig. 16.3 *Field measuring system*

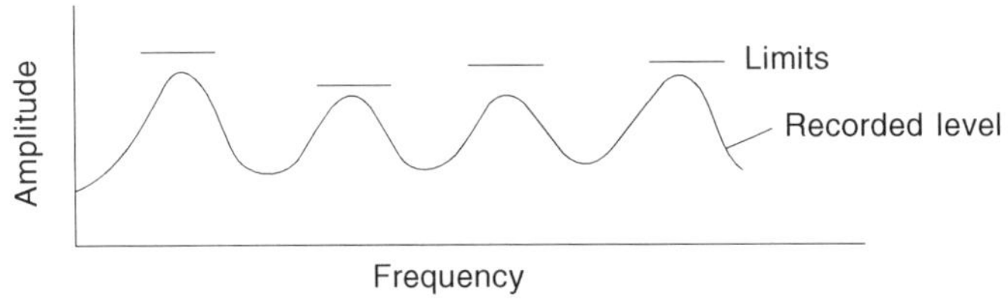

Fig. 16.4 *Typical record of vibration levels of a machine under test*

In Fig. 16.5, a system using a mini computer is shown for analysis and storage of data from a large number of machines. A real-time analyser would give an instantaneous picture of the frequency spectrum, the levels of important frequency components present in the signals are stored in the computer and compared with permissible levels, data of which is available in the memory bank. In case any of the levels is exceeded, a control signal is sent to the machine concerned. The results can also be printed on a tele-typewriter. Overall vibration levels may be indicated by a simple vibration meter, which would indicate the overall rms levels. For monitoring the frequency spectrum, the system shown in Fig. 16.6 may be used for manual operation and the levels in the various frequency bands, according to the filters chosen, may be recorded.

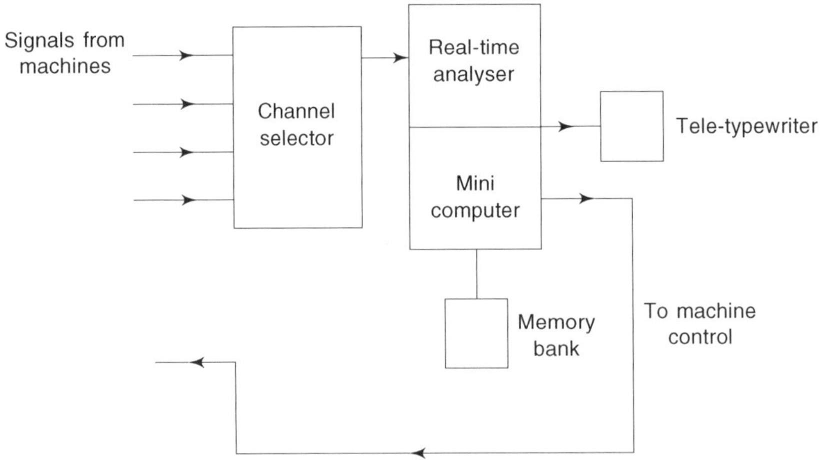

Fig. 16.5 *Use of mini computer for vibration signature analysis*

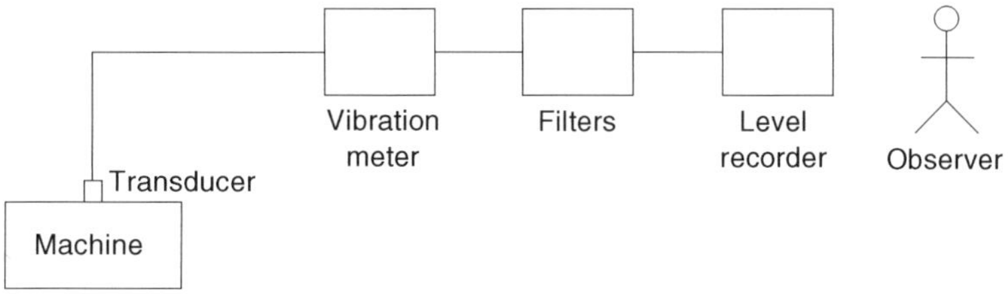

Fig. 16.6 *Vibration monitoring by observer*

The use of permanent vibration monitoring systems is necessary in applications like continuous process plants. The principle of such systems is indicated in Fig. 16.7. At a fixed number of frequency settings, the levels have to be monitored and if they exceed beyond the alarm levels, an indication or warning is to be given. Figure 16.8 shows a permanent monitoring system, based on the above principle. The centre frequency for each filter is set according to the identified predominant frequency. An alarm is given if any level is exceeded.

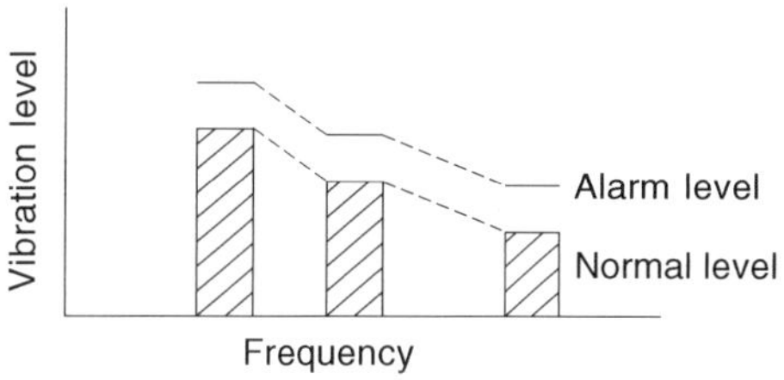

Fig. 16.7 *Principle of monitoring system*

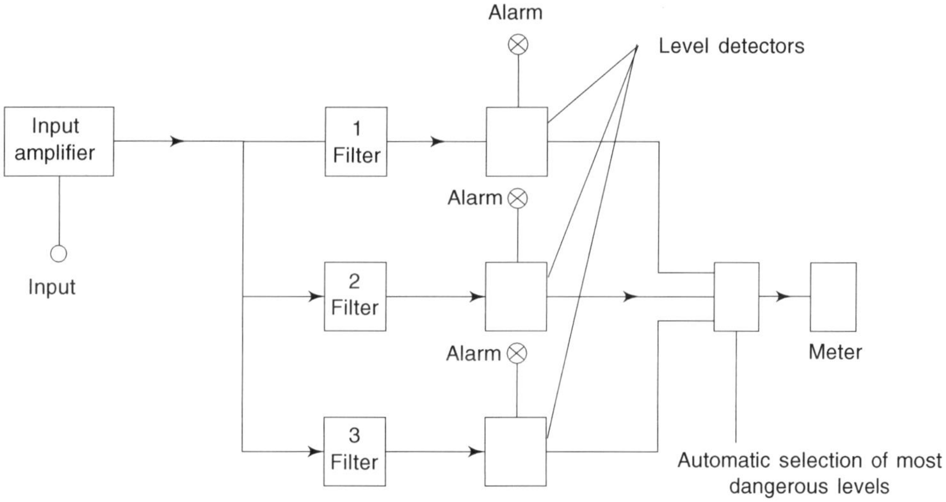

Fig. 16.8 *Permanent monitoring system*

Figure 16.9 shows a multiplexer unit which can automatically switch on a large number of channels. The settings on the attenuators are made so that only one monitor unit is necessary.

A desirable location for placing vibration transducers is on the bearing housing of any rotating machine. The vibration levels may be measured in radial and axial directions.

In Fig. 16.10, the proper location of the transducer for measuring radial vibrations is at A while that for axial vibrations at B. Locations A' or B' would not be proper locations, as the flexibility of the cover plates on which these transducers are mounted would affect transmitted vibration levels.

A proximity transducer of eddy current or capacitive or inductive type may be used for recording the motion of the shaft relative to the bearing. Two such units H and V, as shown in Fig. 16.11, would give a picture of the motion of the shaft on an indicating oscilloscope.

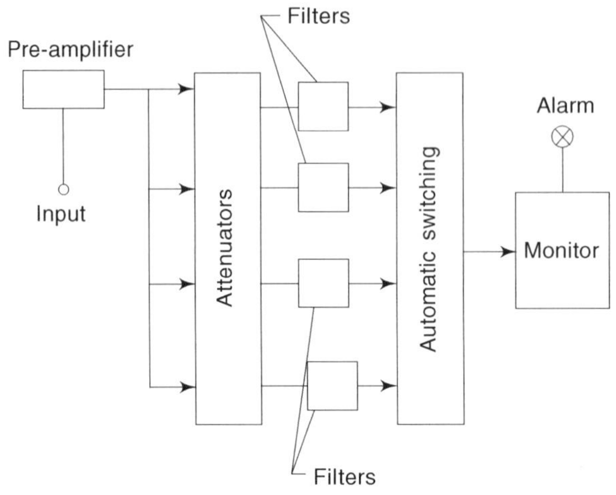

Fig. 16.9 *Multiplexer unit for automatic switching*

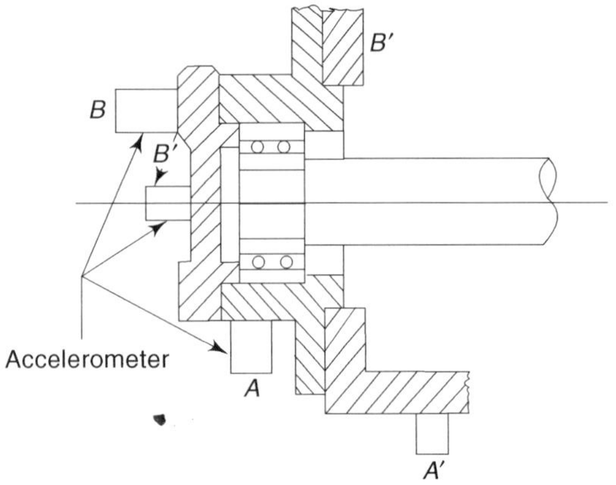

Fig. 16.10 *Proper locations for accelerometers*

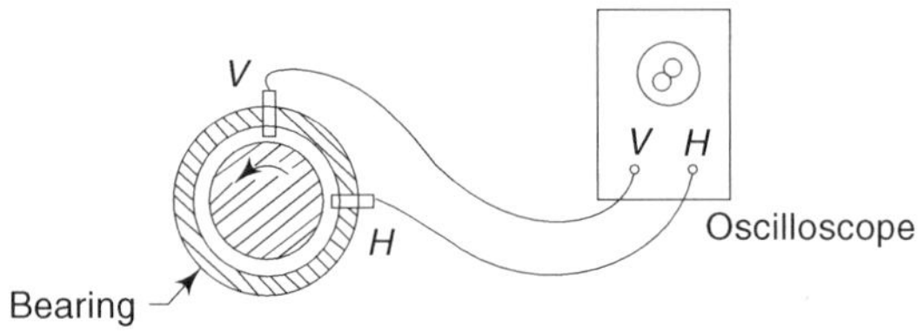

Fig. 16.11 *Horizontal-Vertical (H–V) analysis of shaft motion*

The monitoring systems for noise are identical to those discussed for vibrations. A damaged bearing would emit a sound signal different from that of a normal bearing. Sometimes, it is seen that by the time a bearing emits noise levels sufficient enough to be felt as abnormal, the damage is already considerable. It has, however, been found that prior to considerable damage, the bearings generally radiate sound which is in the ultrasonic range, the frequency of which may be around 40 kHz for ball bearings. In such cases, ultrasonic detectors and other special instrumentation are employed.

16.1.2 Analysis of Results

Table 16.1 gives general guidelines for identifying the causes of vibrations. The relations between a fault and frequency, amplitude and direction of vibrations, are given. This is a useful guide for pin-pointing the cause in case vibration levels at a certain frequency are seen to increase.

In order to determine if the vibration levels, as measured, are satisfactory or not, some of the known standards may be used.

Table 16.1 *Vibration Cause Identification*

Sl.No.	*Fault*	*Frequency*	*Direction of vibration*	*Remarks*
1.	Unbalance (rotating)	Same as running speed	Radial	Vibration amplitude is proportional to unbalance
2.	Misaligned couplings, bearings and bent shaft	Usually (2 × speed) and (1 × speed)	Radial and axial	Axial vibrations fairly large
3.	Oil film whirl	Approximately $\left(\frac{1}{2}\times\text{speed}\right)$	Radial	Occurring in high-speed turbo-machines
4.	Rolling bearing defect	Mostly at ball or roller speed. Also at ultrasonic frequencies (20–60 kHz)	Radial and axial	
5.	Damaged or worn gears	Tooth meshing frequencies = no. of teeth × rpm	Radial and axial	Side bands near tooth meshing frequencies
6.	Unbalanced reciprocating forces and couples	Speed and multiples	Usually radial	
7.	Aerodynamic and hydraulic forces	Flow pulsation frequency = no. of blades or vanes × speed	Radial and axial	

Figure 16.12 gives the rms values of overall vibration levels according to standards VD1 2056 (1964) and BS 4675 (1971), according to which the machines are classified in four groups:

1. Group K—consisting of small machines up to 15 kW,
2. Group M—consisting of medium machines 15–75 kW or up to 300 kW on special foundations,
3. Group G—consisting of large machines with rigid and heavy foundations, whose natural frequency exceeds machine speed, and
4. Group T—consisting of large machines, operating at speeds above the foundation natural frequency.

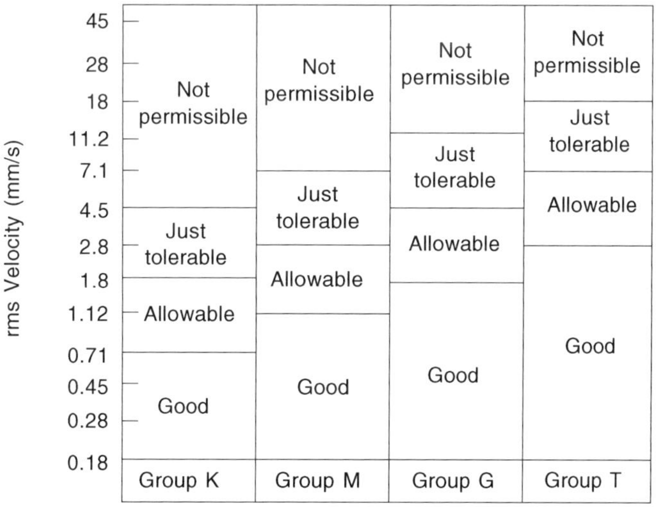

Fig. 16.12 *Permissible rms levels (from VDI 2056 (1964),* Courtesy: *VDI Verlag GMBH, Dusseldorf)*

This chart gives a rough indication of the condition of a machine. The overall levels used are not absolute indicators of machine faults as any increase in one frequency component may not get reflected in the overall levels. For this reason, it is preferable to rely on levels corresponding to various frequency components, as shown in the chart given in Fig. 16.13. For various groups of machines, as mentioned before, divide the measured values of vibration levels by a factor given below, before using Fig. 16.13.

Group	K	M	G	T
Factor	1.0	1.6	2.5	4.0

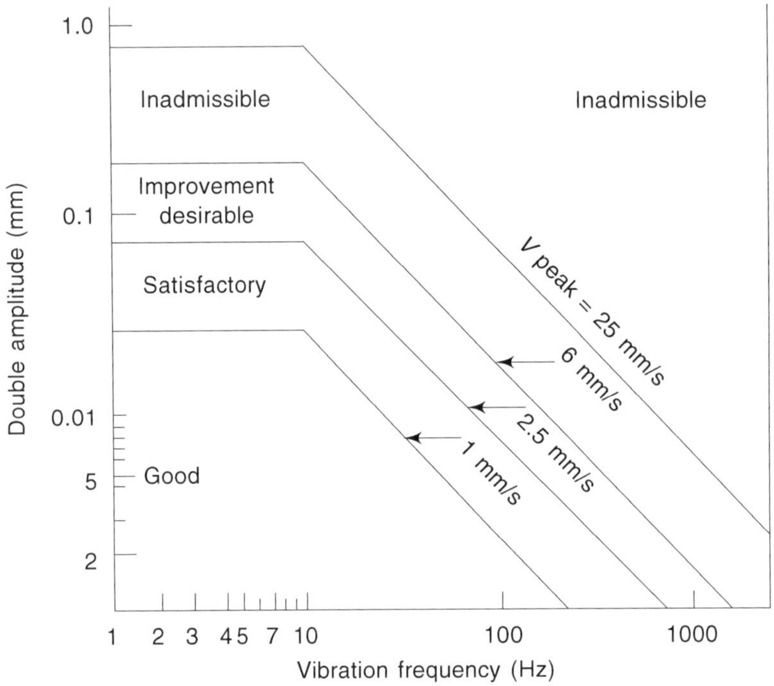

Fig. 16.13 *Chart for determining permissible levels at various frequencies (from* Tribology Handbook, *Ed., M.J. Neale,* Courtesy: *Butterworths and Co., U.K.)*

A common approach for signature analysis is to use the vibration levels when the machine is known to be in good condition as the standard or base-line levels, and compare the measured levels subsequently with the same.

The transducers for measuring vibrations may be used to measure displacement, velocity or acceleration. At high frequencies it is convenient to measure velocity or acceleration levels since the displacements are usually very small. It is also usual to indicate velocity levels for the purpose of condition monitoring as the stresses induced in machine components are often related with the same.

Table 16.2 gives the noise levels for various machine elements, which may be classified as good, acceptable or poor. These are overall levels in dB, on A scale of a sound-level meter and can only be used as a rough guideline.

Table 16.2 *Noise Levels Classification*

Elements	*Noise Level (dB A),30 cm from housing or casing*		
	Good	*Acceptable*	*Poor*
Roller/ball bearing	75	85	90
Spur gears	95	105	110
Single helical gears	85	100	105
Double helical gears	80	95	100
Epicyclic gears	75	90	95

It is found convenient to express the vibration or noise levels in dB (decibels), rather than the actual values of these parameters. As is well known, the value in dB of any parameter x is 20 log x/x_0, where x_0 is the reference value. Values of x_0, commonly used for various parameters, are as below:

Parameter	x_0
Sound pressure	2×10^5 Pa
Velocity	10^{-8} m/s
Displacement	10^{-11} m
Acceleration	10^{-5} m/s^2

16.2 ■ TEMPERATURE MONITORING

Temperature is also a useful guide to indicate the health or condition of a machine. In the handbook* on tribology, maximum permissible contact temperatures for typical machine elements are given.

The temperature, in such applications, may be measured by any of the following methods:

1. thermometer,
2. thermocouple, or
3. infrared pyrometer.

These methods have been extensively covered in Chapter 12.

It is advisable to make continuous records of temperature to indicate any deterioration of machine condition as early as possible.

Thermography is one of the latest techniques to get a thermal image of a component. This is done by an infrared camera which is used to monitor the temperature pattern in applications like turbines, bearings, piping, furnace linings, pressure vessels, etc.

The surface whose temperature is to be monitored, is scanned by a camera. For detection, a crystal of indium antimonide is used, producing an electrical signal proportional to the thermal radiations received from the object. A thermal image is obtained on a TV screen, by this system, the intensity of the electron beam is determined by the signal from the detector and the positioning in the vertical and horizontal directions is done by the signals. Thus, the shades of grey on the surface indicate target temperature. With a commercial infrared camera, it is possible to detect temperature difference as small as 0.1°C, for a target temperature of 30°C. The working distance can also be varied over a wide range. Thus, any abnormal condition like reduced lining thickness or damaged insulation or abnormal localised temperature build-up in a bearing, can be easily detected.

* M.J. Neale, *Tribology Hand Book*, Butterworths, 1973, London.

16.3 ■ WEAR BEHAVIOUR MONITORING

Wear in machines can give an indication about the condition of a machine. For this purpose, the geometrical changes in the moving components may be monitored or alternatively the wear debris produced may be analysed for size, loss rate, composition, shape, etc.

Figure 16.14 shows a system for monitoring the piston ring condition in an engine or a compressor. The sensing is done by a proximity transducer. Any broken piston rings or those not in contact with the cylinder due to excessive wear can be easily detected.

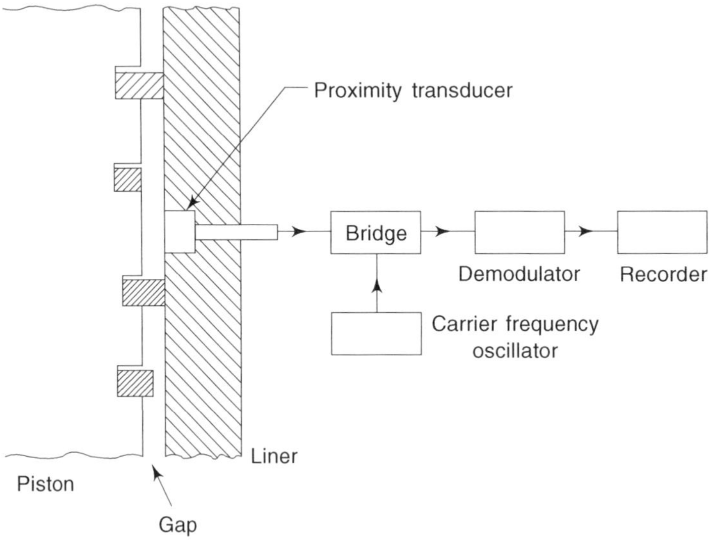

Fig. 16.14 *Piston ring wear monitoring*

For analysis of wear debris generated due to relative motion in components like bearings, gears, seals, pistons, cams, etc. a number of methods are available. These include filters and chip detectors, ferrography, spectrometric analysers, particle counters, etc.

16.3.1 Filters and Chip Detectors

Machines with recirculatory systems, can have filters to collect wear debris. Similarly, ferrous particles may be collected by magnetic plugs that collect ferrous chips. From the shape of the particles, it may be possible to have an indication of what is going on in the machine. Figure 16.15 shows typical debris shapes.

The debris from balls of ball bearings is of rounded rose petal type, from track is rounded with criss-cross scratches, from rollers of a rolling contact bearings it is rectangular with parallel lines across the width and from gear tooth debris is irregular in shape. The formation of spherical particles in large quantities, indicates pitting of rolling elements and may lead to fatigue failure. Particle size, which is normally small, may become large at the onset of a fatigue failure. Thus, the origin of the particles and a possible onset of failure can indicate failure of a component.

16.3.2 Ferrography

This involves the separation of wear debris of ferrous type magnetically and arranging it in the order of size for examination. Figure 16.16 gives a direct reading ferrograph, which can be used in the field. A

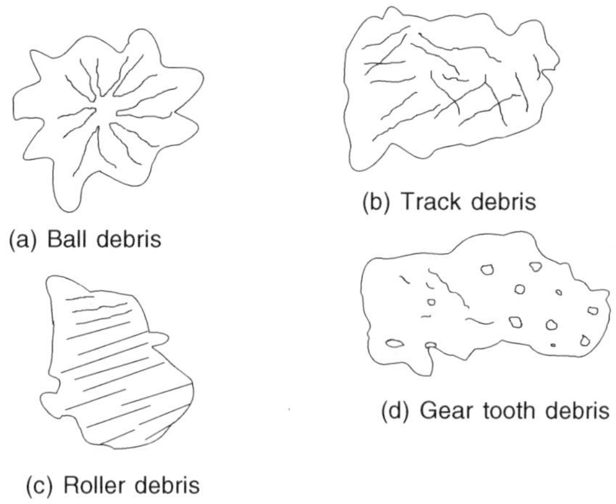

Fig. 16.15 *Typical wear debris shapes*

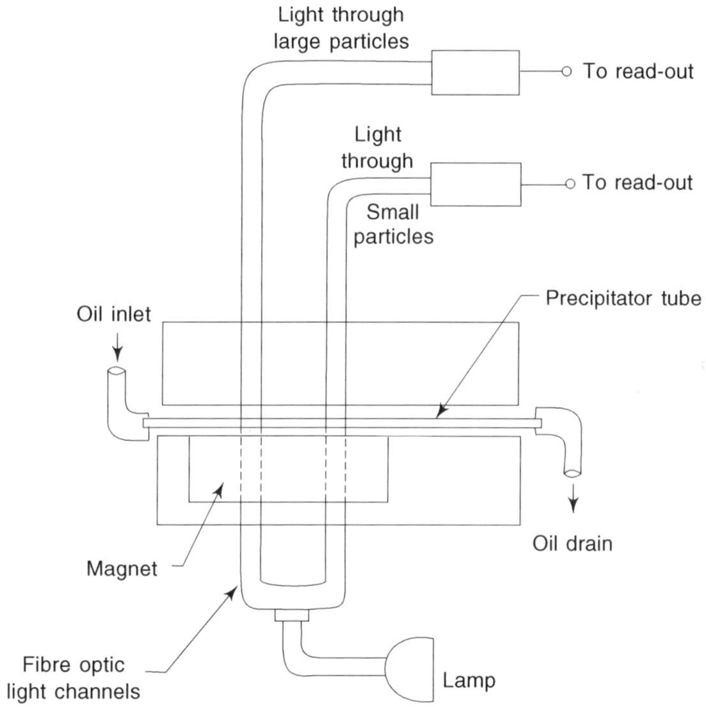

Fig. 16.16 *Direct reading ferrograph*

sample of oil containing wear debris is diluted with a solvent which breaks down any gel, which may be around the wear particles. The sample is fed to a transparent precipitator tube, on either side of which are the poles of a magnet. The magnetic force attracting the particles is proportional to their size and magnetic susceptibility. Thus, the larger particles are deposited at the entry region of the tube and the smaller particles and oxides of iron get deposited later along the tube.

The amount of material deposited is measured by the attenuation of light from a light source placed below the tube as shown, the light passing being detected by photo-electric transducers. The information about the condition of a machine is indicated by the distribution of particles along the tube, and thus the attenuation of light is measured at two points at a distance nearly 5 mm along the tube. Wear severity would be indicated by the difference between the optical densities at the two points. In normal rubbing wear, majority of particles are small. If the wear is severe, the number of bigger particles increases.

In order to make a further detailed study of the wear debris, the debris is collected on a slide as shown in Fig. 16.17, viz. a ferrographic analyser. The oil containing wear debris is sorted by size, down an inclined glass slide. The magnetic field gradient increases as the particles move downwards. The larger particles get deposited first. The slide is analysed by an optical microscope, which is arranged with the reflecting and transmitting modes of illumination. Red light is directed through the objective of the microscope. Opaque objects appear red. Green light is transmitted through transparent objects which appear green. Other objects appear in various shades of yellow-green. This helps in distinguishing metallic from oxide particles. Further, a heating of the slide to 330°C or so, turns low carbon and alloy steels to temper blue colour, cast iron to light brown colour and bronze to dark brown, chromium and aluminium remaining bright white colours. Thus, particle identification is also possible.

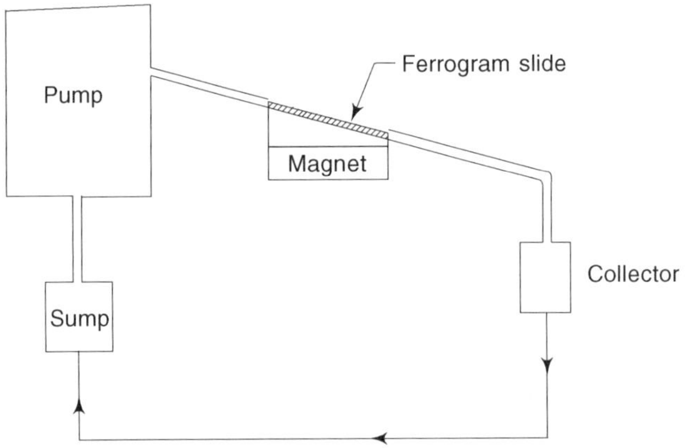

Fig. 16.17 *Ferrographic analyser*

16.3.3 Spectrometric Analysers

Oil samples containing wear debris are taken from the sump and analysed by any of the following methods:

1. atomic absorption spectrometer, or
2. emission spectrometer.

The technique is called (spectrometric oil analysis programme) SOAP. Figure 16.18 shows an atomic absorption spectrometer, which is based on the principle that every atom absorbs only light of its own specific wavelength. From the combustion of the oil sample, the metallic elements are atomised. In a light beam passing through the flame, certain wavelengths get extinguished due to absorption by the free atoms of the metals. Since each element has its characteristic wavelength, it is possible to measure the amount of light absorbed and indicate the concentration of an element in (parts per million) ppm. In Fig. 16.18, a chopper is used to produce an ac signal for the detector. The monochromator is turned to accept light of a certain wavelength. A meter indicates the difference between the light emitted by the light source and that received by the monochromator. The procedure has to be repeated for each element to be detected.

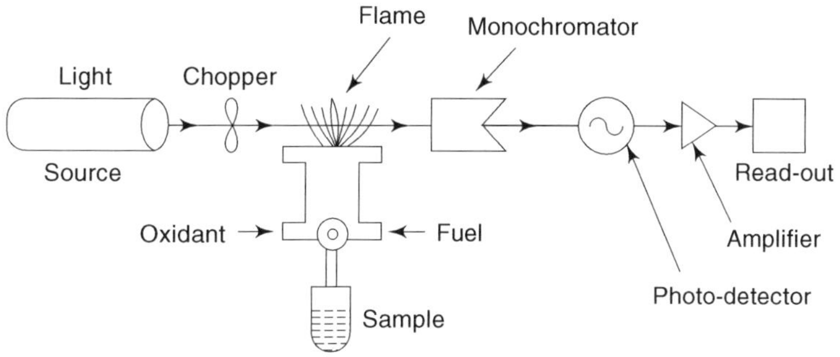

Fig. 16.18 *Atomic absorption spectrometer*

Figure 16.19 shows an emission spectrometer with which it is possible to analyse simultaneously a number of elements. A rotating graphite disc carries the oil to be subjected to high-voltage excitation. A spark excites the various metallic elements after which they emit their characteristic wave lengths, the intensity of which is measured by various detectors.

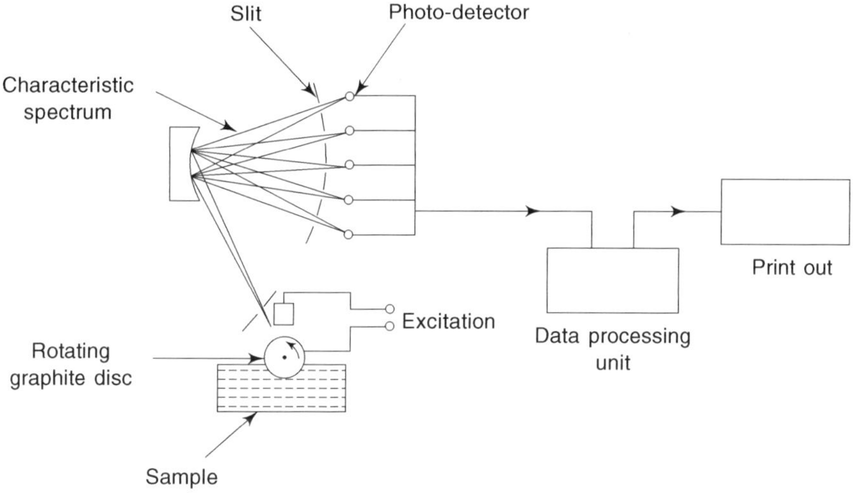

Fig. 16.19 *Emission spectrometer*

Several elements can be analysed using this method. These include copper, iron, chromium, nickel, lead, sodium, aluminium, magnesium, silicon and silver. With a knowledge of the composition of each machine element, it is possible to identify the parts wearing out fast.

16.3.4 Particle Counters

The size distribution and concentration of wear particles in oil, can be determined by various types of particle counters that have been developed. Figure 16.20 shows one such particle counter. This is based on the light blockage principle. The oil is passed through a sensor consisting of a light source and a photo-diode. A particle passing through the window blocks light proportional to its size. Accordingly, an electrical signal from the photo-diode is sent to the counter. Thus, counting and sizing of particles is possible. The turbulent flow in the narrow tube results in rotation of the particles. This is helpful in that the largest cross-section is likely to face the detector at least once, before leaving. A vacuum pump (not

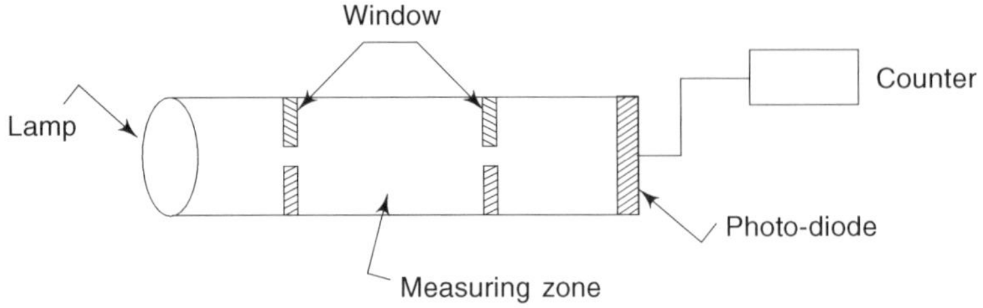

Fig. 16.20 *Particle counter*

shown in the figure) is used to suck the oil from its container through the device. Concentration levels in several ranges of particle sizes are displayed.

16.4 ■ CORROSION MONITORING

Corrosion monitoring can be a useful aid for the efficient operation of a plant or equipment. This finds application in the petroleum, chemical and power generating industries. The parameters to be measured are corrosion rate, total depth of corrosion or amount of material remaining.

The following are the basic methods of corrosion monitoring:

1. direct observation,
2. conventional chemical and electro-chemical methods,
3. NDT methods, and
4. specially developed methods.

Direct observation methods are the oldest methods, based on visual or optical examination through holes provided for the purpose on the equipment.

Conventional chemical and electro-chemical methods include analytical, potential or galvanic measurements. These measurements are generally used to monitor water quality in boiler feed water. The ratio of metallic ions in a process liquor may indicate which components are corroding.

NDT (Non-destructive testing) methods include radiography, ultrasonic, eddy current, thermography and acoustic emission methods. Ultrasonic measurements are used for flaw detection and thickness measurement in pressure vessels and piping. The major disadvantage of this method is that only a small surface can be tested and localised pitting may be missed. Radiography is quite useful as permanent records can be kept. The disadvantage is that it is slow and hazardous. Eddy current technique is primarily a surface technique. Penetrations deeper than 5–6 mm are hard to obtain. It finds use in rapid routine examination of heat exchanger tubes in nuclear and conventional power stations for monitoring corrosion.

Thermography is used to locate increased surface temperatures due to corrosion damage in furnaces, stacks and power transmission lines and acoustic emission may be used in the detection of cracks and leaks.

Specially developed methods include the electrical resistance technique, which is based on the fact that as a sample corrodes, it decreases in cross-section and so its electrical resistance increases. It is possible to find corrosion rates from recordings over a short period of time and this method can be extensively employed for plant monitoring and control.

16.5 ■ MATERIAL DEFECT MONITORING

A number of defects like cracks may be originally present in a material or may develop subsequently. It is important to monitor their state by one of the N.D.T. (Non Destructive Techniques) periodically or on-line especially during maintenance of the system. The defects may be of following types:

(a) Inherent defects These defects generally get introduced during the various production stages of the raw materials. The commonly observed defects of this type are for example, segregations, inclusions, porosity, cavities, voids, surface and sub-surface cracks.

(b) Processing defects These defects creep in during the various processing stages of the components/systems. Common examples are:

- surface and sub-surface cracks may be developed during the mechanical, thermal/heat treatment processes,
- lack of penetration of weld material may result in improper weldment,
- porosity may be caused by a poor planning procedure,
- shrinkage and porosity defects may be introduced due to poor quality of casting,
- laps and folds may be formed in a typical forging process.

(c) Service defects These defects get produced during the operating life of the components/systems. Some common examples are defects such as surface and sub-surface cracks caused due to fatigue, stress corrosion, hydrogen imbrittlement, intergranular corrosion, pitting, etc.

A number of NDT (Non-Destructive techniques) are available and the selection of a particular test method depends on its suitability for a given situation and the experience of the test personnel. Generally, these test techniques are classified in two broad categories. The first types are surface inspection/test techniques, which are primarily suitable for the examination of the flaws on the surface of the materials. The second types are the techniques used to estimate the defects by examining the internal features of the materials. Visual inspection, physical inspection, dye penetration, magnetic flux and eddy current techniques belong to the first category, whereas ultrasonic and radiographic techniques fall in the second category. Only eddy current, ultrasonic and radiographic are discussed here.

16.5.1 Eddy Current Method

This technique employs electromagnetism principle for assessing the material properties as well as the presence of defects in the materials, which are electrically conducting in nature. In this technique, a time-varying magnetic field is produced in the probe, which consists of a coil. Herein, a probe consisting of an energized coil with a high excitation frequency, usually in the range of 100 kHz to 10 MHz, is brought near the surface of the component under test. The probe induces weak electrical currents in the test sample which are sensitive to the test material conductivity and permeability. The changes in the conductivity of the material are caused by the changes in the material composition as well as the structural changes like crystal imperfections caused by voids, stress conditions or work hardening, etc. Further, presence of any discontinuity in the form of crack, etc. would disturb the eddy current flow patterns and would in turn cause changes in the permeability of the test sample.

The eddy current probe, which has an ac current excitation, produces an ac magnetic field within the materials. This in turn produces eddy currents in the material, which have its own magnetic field. This eddy current generated magnetic field cuts across the coil of the probe and induces a small current to flow through the coil in a direction opposite to that of the applied current, thus weakening the applied current. This weakening in current is caused by the change in the inductive component (due to the change in the permeability produced due to the presence of a crack, etc.) as well as in the resistive component (due to changes in the composition/structural changes, etc.) of the coil. This in turn is related to the eddy current losses. It may be emphasized again that the conductivity changes are reflected in the changes of coil resistance whereas the permeability changes affect the coil inductance. We normally begin flaw detection by calibrating the equipment. This simply means that we make our instrument insensitive to conductivity by setting a Wheatstone bridge circuit output as zero on its meter, using a known sound specimen made of the same material as the piece to be tested. The variation of the current in the test coil due to the change in its impedance caused by the presence of cracks, etc., as well as changes in the composition/structure generates a deflection in the Wheatstone bridge circuit as shown in Fig. 16.21.

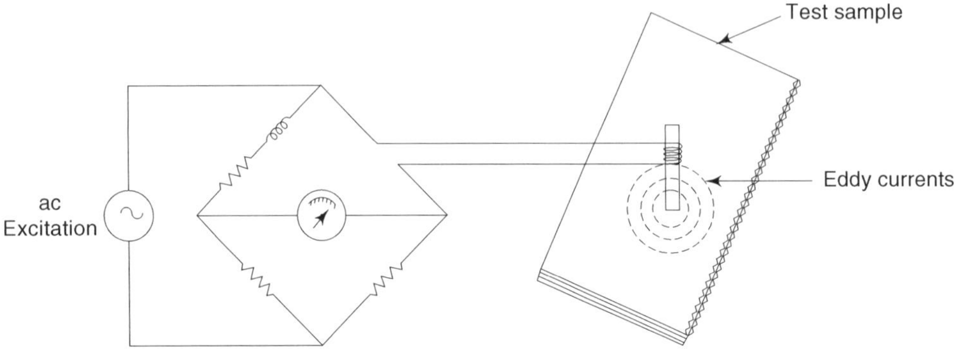

Fig. 16.21 *A schematic diagram of the eddy current flaw detection method*

The major advantages of this technique are that it is fast and it involves moderate cost. Further, it has the feasibility of very high rates of inspection. The inspection can be automated so that defective parts are identified by the test equipment. It is portable or semi-portable type. Lastly, it is free from the use of consumable material, personnel danger or hazard and contamination of parts.

The main disadvantage of this technique is that it is applicable only to electrically conductive materials and has limited penetration depth.

The applications of this technique are sorting of the components to ensure proper composition, heat treatment and hardness or dimensions, and the detection of surface or near-surface defects during high-speed inspection of rods, bars, extruded tubing and welded tubing or pipes (especially, the heat exchanger tubes). This technique is also very useful in the measuring of the thickness of non-conducting coatings on metallic substrates and metallic sheet materials. In addition, this technique is used for the inspection of rods, bars, extruded tubes and spherical ball bearings for the detection of laps, seams, radial cracks, alloy and dimensional variations as well as non-uniformity of heat treatment, etc.

16.5.2 Ultrasonic Method

The ultrasonic method makes use of low-energy sound waves of short wavelengths (which are associated with high frequencies) to measure wall thickness and also to detect defects in the materials. Ultrasonic waves are beamed into the test material and these are reflected back from the geometric boundaries as well as from the defect boundaries. Based on the time of return of these echoes, metal thickness can be measured and also the location of the defects can be identified.

This method employs the piezo-electric crystal type of transducer for the generation of sound signals of high frequency that is far in excess of the upper limit of audibility of the human ear, i.e., more than 20 kHz. The testing frequencies in this technique lie in the range of 0.5–25 MHz. When an oscillating voltage is applied to a piezo-electric crystal, it vibrates with the same frequency and vice versa. An ultrasonic flaw detector is shown in Fig. 16.22(a). The transmitter has a built-in pulse generator, which emits ultrasonic pulsed wave and this simultaneously triggers the transmitter transducer and the time base generator. Generally both the transmitter and the receiver are housed in a single probe. The pulse of the ultrasound generated from the transducer travels through the test specimen. When a flaw is encountered, a part of the ultrasound is reflected. The reflected sound travels back and reaches the receiver crystal, which converts it into electrical voltage. The voltage passing through the amplifier is fed to the CRT and a signal is observed on its screen. The CRT screen would display three echo signals, which are shown in the Fig. 16.22(b). These signals are namely, one from the top of the surface of the specimen, another from its bottom, and, the third in between these two which is known as the defect echo signal.

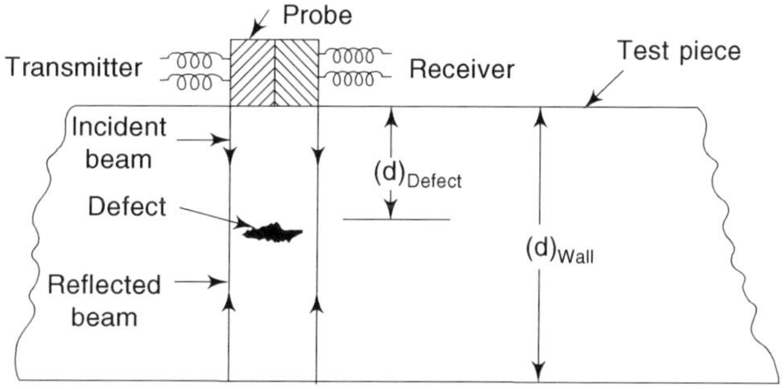

(a) Schematic diagram of pulse–echo system (Transmitter–Receiver method) for detecting depth of flaw

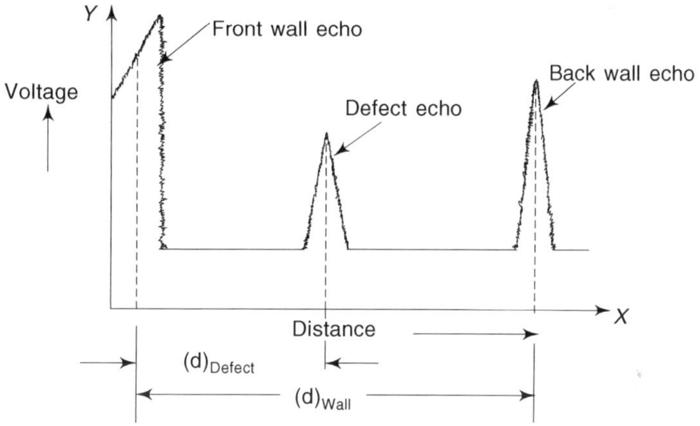

(b) Typical display on CRT screen

Fig. 16.22 *A schematic diagram of ultrasonic flaw detection method*

Ultrasonic technique is very widely used in the areas of detection, location and measurements of surface and internal voids such as cracks, slag inclusions, laminations, lack of fusion, lack of weld penetration, inclusions in welded and brazed structures, etc. In addition, it is employed in the determination of bond integrity of adhesive assemblies such as honeycomb and composite structures. The main advantages of this method are high sensitivity, fast response and its capability of automation. Further, very small defects and their sizes can be detected and their location can be determined accurately. In addition, this technique has great penetration power and defects up to 10 m thickness in steel can be inspected.

The limitation of this technique is that it requires coupling of the test material either by contact to the surface or immersion in a fluid such as water. Further, the micro-level defects of the order of half the wavelength of the ultrasonic beam cannot be detected by this method.

The technique has been successfully employed in the studies of stress corrosion cracking and fatigue failures in aircraft structures, monitoring of flaws in the weldments, pressure vessels, railway rolling stock, gas turbine blades, etc.

16.5.3 X-ray/Radiographic Technique

In this technique, the objet is exposed to X-rays on one side and a sensitive photographic plate or film is placed on the other side of it. The intensity of the X-rays is reduced by the passage through the material and also by the defects/imperfections that may be present in the material. The absorption of the X-rays is dependent on the changes in density, material thickness and chemical composition of the material. More radiation would pass through the voids than through the surrounding area. The photographic film measures the amount of radiations absorbed by, or conversely, passed through the material under test. Darker spots due to increased radiation reaching the film would indicate voids or thinner sections of materials. Thus the radiograph is really a shadowgraph. The darker areas on the film represent more penetrable and less dense areas of the material and on the other hand, lighter areas on the film represent more absorptive or denser areas of the sample. The radiograph is often used for the interpretation of the results.

The advantages of this technique are that it can be used for a wide range of materials and the exact nature of the defects can be determined from the visual images. Further, a permanent film is obtained which shows the defects in relation to significant features of the component. For example, defects in a weld such as cracks, porosity and lack of penetration have their distinct images and therefore, are easily recognizable. Furthermore, it is the best and the most reliable method for determining the extent of internal porosity in that part.

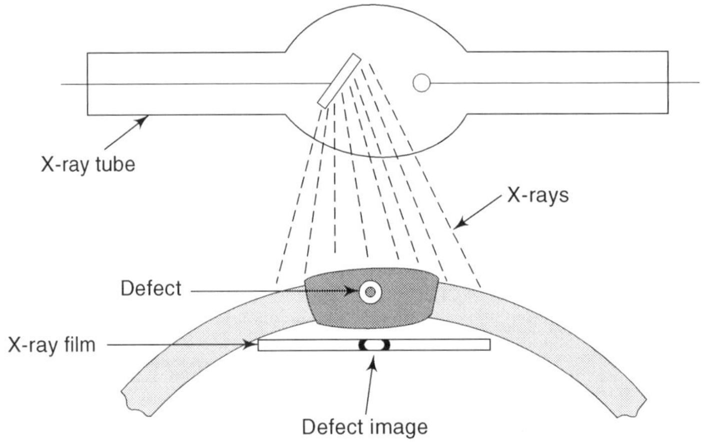

Fig. 16.23 *A schematic diagram of radiography technique*

The main limitations of this technique are the need to incorporate safety measures as well as to take proper precautions in respect of harmful radiations to the operating personnel. More over high investment in equipment and facilities are involved. Lastly radiography is inaccessible to detect small, tight, randomly oriented cracks.

The radiography technique is applied for the inspection of the incoming material, manufactured components, final products and maintenance. Further, it is widely used for the inspection of castings for shrinkage, micro-shrinkage, gas pockets, inclusions and blowholes. In addition, it is also used in forging for tear, bursts, internal cracking, welded and brazed joints for porosity, lack of fusion and/or lack of penetration and the detection of the presence of the inclusions

16.6 ■ ACOUSTIC EMISSION (AE) MONITORING TECHNIQUE

The Acoustic Emission (AE) monitoring technique is a highly sensitive type of Non-Destructive Technique for detecting a variety of defects and even the dynamic movement of defects like, for example, a typical initiation and propagation of a crack growth in a loaded sample. This way it provides useful information about the "State of Health" of the system on-line i.e., in the real time. This technique basically involves the analysis of sound signals emitted by the materials, structures or machines which are in use in the fully loaded working conditions. Further, these sound signals are typically in the range of 0.1–0.4 MHz and are beyond the range of human hearing.

Acoustic emissions are generated by high-frequency stress waves in solids which are caused by sudden defect-related movements in the stressed materials. For example, the initiation and growth of crack in a stressed material brings about rapid release of strain energy giving rise to stress induced acoustic waves in the loaded materials, structures or machines. These waves radiate into the structures and are conveniently picked by a piezo-electric type of transducer in the form of acoustic emissions. If may be noted that the source of acoustic emission energy is the elastic stress field in the loaded machines/components. In other words, without stress in the material, there are no acoustic emissions. Hence, the acoustic emission inspection technique can be used only in the loaded structure, so that we have a stress field for the generation of stress waves caused due to the defect-related deformations.

The common sources of acoustic emissions due to system defects in the stressed fields are as follows:

- Initiation and propagation of crack growth caused due to fatigue, creep or complex loading. Such a defect causes slippage/sudden changes in the orientation of grain boundaries leading to large plastic deformations.
- Large plastic deformation caused by impact loads, shock waves caused by explosions or sudden loadings on fluid flow structures due to water hammer phenomenon caused by the sudden closure of valves.
- Turbulence noise caused in fluid flow leakages past gaskets, valves, screwed fittings or high-pressure vessels developing cracks.
- Cavitation phenomenon in hydraulic machines or stalling and surge phenomenon in the aero-compressors.
- Chatter phenomenon caused by loose parts like mismatched gears or loose particles in the loaded machines like a broken ball debris in the ball bearings.
- Material property changes due to corrosion, chemical reactions, liquid–solid transformations or phase transformations.

The following are the advantages of the AE inspection technique:

1. The whole structure can be monitored on-line (even at high temperatures) from few locations both reliably and economically and without taking it out of service. Further, continuous monitoring with suitable alarms is possible.
2. Microscopic changes in the defects can be detected as sufficient energy in the form of stress waves is released from the defect source.
3. The exact location of the defect can be located by using multiple sensors and employing triangulation technique.
4. The signature analysis of signal is employed to identify the nature and type of defects in the initial stages before they become serious. Thus appropriate corrective action can be taken at that stage, to reduce the growth of that particular defect.
5. The evaluation of the nature, type and quantum of defect in AE inspection technique is based on the objective type of criteria based on the signature analysis of the defect signal. Therefore, it is less prone to subjectivity or operator's interpretation as is common in the other NDT methods.
6. Acoustic emissions is a non-intrusive method and its piezo-electric transducer is also self-generating type i.e., it requires no external power supply as is the case in ultrasonics and radiography.

The following are the limitations of the AE inspection method:

1. Non-stressed areas in a test sample do not emit any AE signal. Therefore, acoustic emission tests require an increase in stress during a test. Further, the test stresses must reach the maximum operating values and preferably stresses should be exceeding the maximum operating values so that stronger AE signals are generated by the defect-related stress waves in the test materials.
2. The AE signal for detecting a particular type of defects has cross-sensitivity to other defects simultaneously present in the stressed test sample, which constitute unwanted noise. For example AE monitoring technique when used for detecting cracks or corrosion would simultaneously pick up unwanted AE signals due to the presence of friction in some bearings as well as defects of loose parts like mismatched gears, etc. Therefore, precise discrimination of the information of the desired defects becomes feasible only after accounting for the effects of noise signals.
3. Background noise caused due to steam trap, fluid flow in pumps, rain and wind blown cables striking the structures tend to obscure the genuine AE signals. In addition, radio and electromagnetic interferences may also introduce errors in the acoustic emissions.

AE Signal Processing Parameters The following are the AE signal processing parameters of the signals of the test sample which are shown in Fig. 16.24. These may be either of continuous type or may be in the form of discrete pulses or bursts.

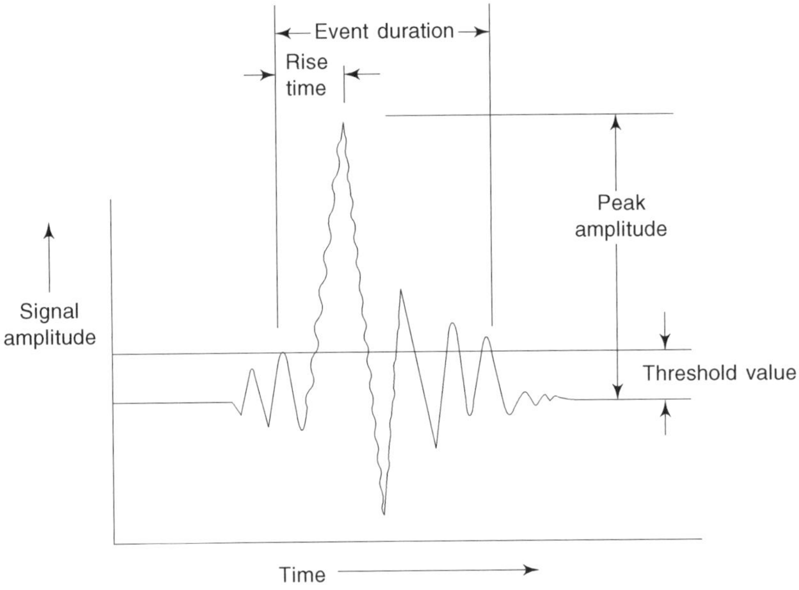

Fig. 16.24 *A typical acoustic emission signal showing the signal processing parameters*

(a) Peak Amplitude It is the highest value of the AE signal and is directly related to the magnitude of the source event.

(b) Ring Down Count (RDC) It is the number of times the AE signal crosses the threshold value. The threshold value is set to account for the background noise in the signal. RDC is obtained by incorporating a comparator circuit and pulse counter circuit in the AE signal. The value of this parameter is an indicator of the magnitude of the source event and the acoustical properties of the test sample.

(c) Integrated Event Energy Integrated energy is the area under the curve in the rectified signal envelope i.e., the energy associated with the ring down count. This parameter is dependent on the amplitude and duration of the defect in the test sample and is less dependent on the threshold settings.

(d) Event Duration It is the time elapsed form the first threshold crossing to the last. It is measured in micro-seconds. This parameter depends on the source magnitude, structural acoustics, etc.

(e) Rise Time It represents the time duration from the first peak crossing to the signal peak. This depends on the dynamic characteristics between the source and the sensor.

(f) Spectral Amplitude Analysis The frequency and amplitude content of the AE signal in the ultrasonic range i.e., between 0.1 and 0.4 MHz when considered along with other AE parameters identifies the type of defect and its severity/magnitude.

Results of a study on the application of acoustic emission (A.E.) technique to new and old ball bearings of same size, are shown in Fig. 16.25. The plots in each case give distribution of A.E. event counts against peak amplitudes and A.E. even counts against ring down counts. It is seen that both peak amplitudes and ring down counts are in higher ranges for the old used bearings, compared to those for the new ones, even though the event counts are in the lower ranges for the former cases of bearings viz. the old ones.

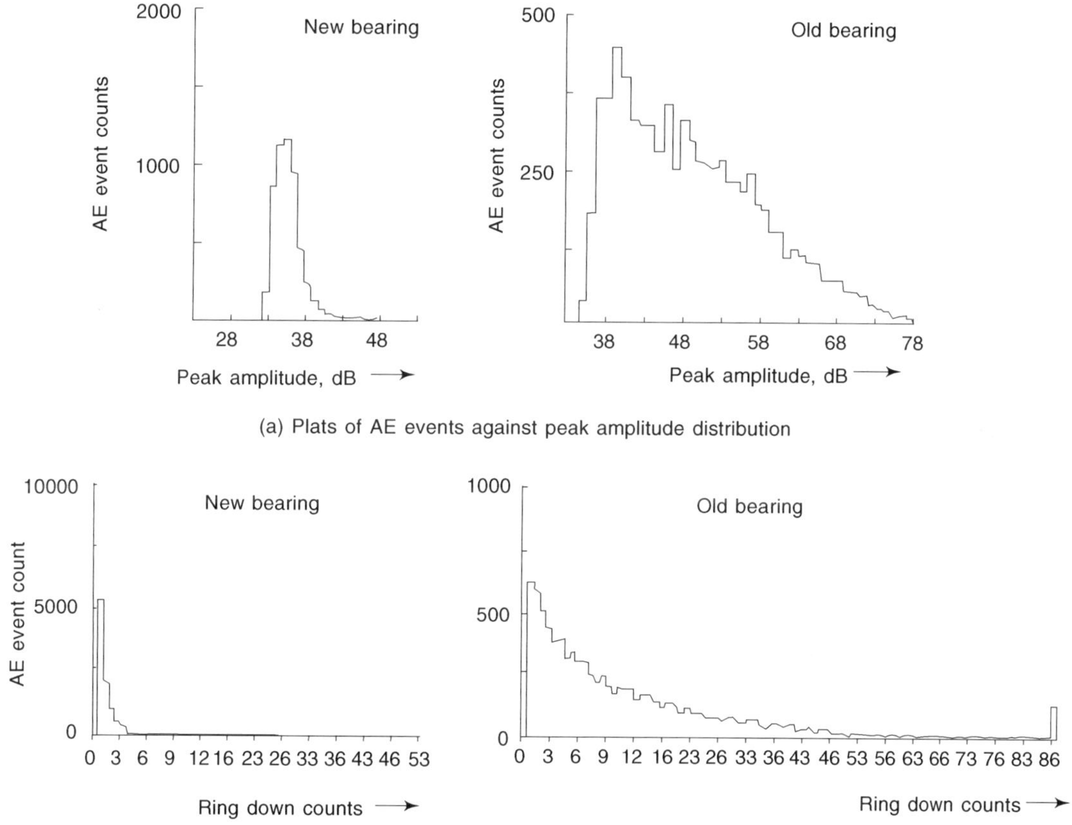

(a) Plats of AE events against peak amplitude distribution

(b) Plots of AE events against ring down-count distribution

Fig. 16.25 *Acoustic emission studies on rolling element bearings*

Applications of AE Monitoring Technique

The following are some of the proven applications where AE method is routinely used:

- Periodic and continuous monitoring of pressure vessels, oil pipelines and storage vessels (both above and below the ground) to detect and locate active discontinuities.
- In service and on-line detection of initiation and propagation of fatigue cracks in the rotating machinery, aerospace and engineering structures.
- Evaluation of material behaviour with respect to different types of failure mechanisms.
- Monitoring of fusion or resistance weldments during welding and during the cooling period.
- Detection of stress corrosion cracking and hydrogen embrittlement susceptibility of test samples.
- Condition monitoring of machinery to detect faults like loss of lubrication, presence of cavitation, bearing failure, seal failure, gear discontinuities etc.
- Detection of loose particles impact noise generated in the integrated circuits used for critical applications.
- Detection of defects like cracking of fibers/matrix or delaminations between the fibre and matrix in real time for providing indications of structural problems before a critical damage occurs.
- Condition monitoring of machinery to detect faults like loss of lubrication, cavitation bearing failure, seal failure etc:
- Detection of loose particles, impact noise in integrated circuits in critical application
- Detection of defects like cracking of fibres/matrix or delaminatious between fibres and matrix in real time for providing indications of structural problems before a critical damage occurs.

16.7 ■ PERFORMANCE TREND MONITORING

Monitoring of the performance of a system involves the measurement and recording of relevant parameters and to use the same for detecting any change in performance, e.g. in a power plant, parameters like fuel consumption rate, exhaust gas analysis, lubricating oil consumption rate, pressure, temperature, etc. are of importance and are related to the condition of the power plant.

Systems for performance trend monitoring are widely used for aero-space, marine and power generation machinery. Deterioration in combustion efficiency is detected by the measurement of fuel consumption rate or by changes in carbon dioxide and oxygen concentrations in the exhaust gases of the engines. Similarly, excessive wear in piston rings can be detected by abnormal consumption of the lubricating oil. In the case of steam turbines, failure of control valves or blade shrouds may result in low efficiency or pressure drops at certain places.

Performance trend monitoring systems are extensively used for aircraft and marine applications. Digital computer systems are also finding increasing use in such applications, whereby data analysis and computations are considerably facilitated.

Figure 16.26 shows an example of a gas turbine driving a process compressor of centrifugal type. The condition monitoring is done by vibration and performance monitoring. The locations of the vibration monitoring sensors are shown in Fig. 16.26(a), Seismic accelerometers are used at locations 1 to 5 and proximity vibration sensors are used at locations 6 and 7. Figure 16.26(b) shows the monitoring of performance parameters like pressure P, temperature T and flow rate m at various locations shown. It is possible to calculate the efficiency η of the various units. Figure 16.26(c) shows the trends in increase

(↑) or decrease (↓) of various vibration and performance parameters of the air-compressor of the gas turbine. Similar diagnostic tables are drawn for various other sub-systems and are useful in the diagnosis of the system.

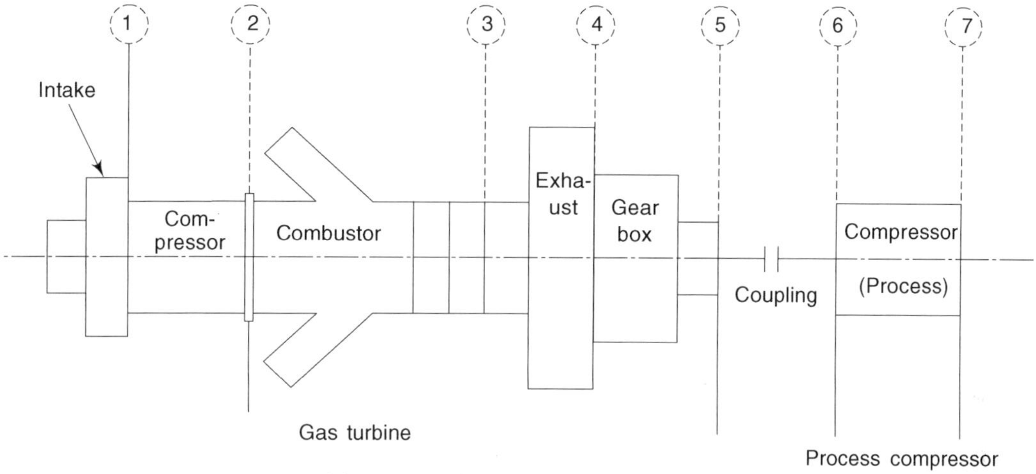

(a) Locations for vibration monitoring sensors

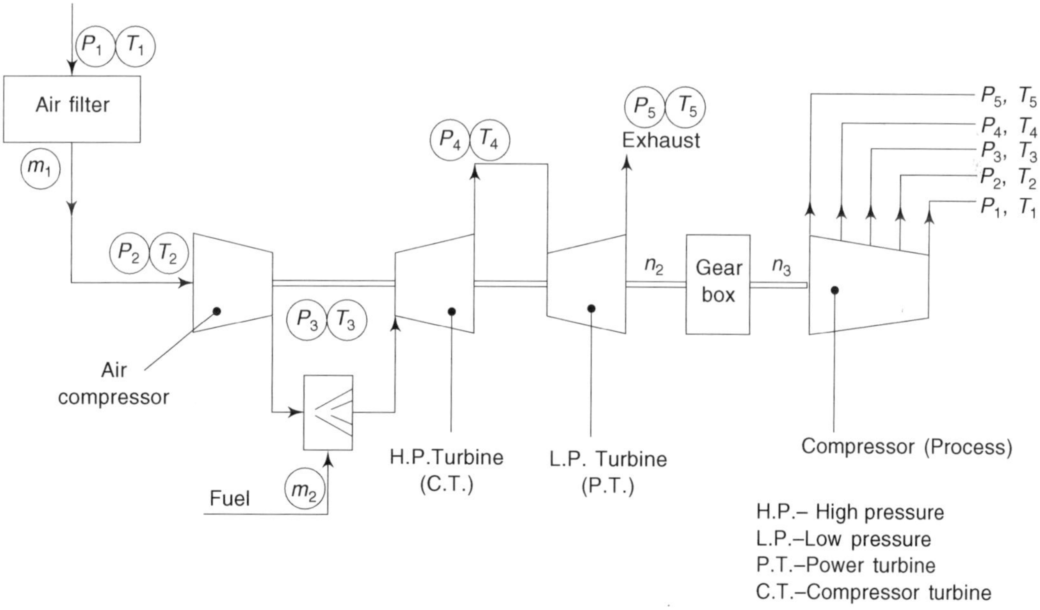

(b) Parameters detailing on the system

Faults	η_c	P_2/P_1	T_2/T_1	Compressor fluid mass flow	Vibration	ΔT Bearing	Bearing pressure	Bleed chamber pressure
Clogged fliter	—	↓	—	↓	—	—	—	—
Surge	↓	Variable	—	↓	Highly fluctuating	↑	↑	Highly fluctuating
Fouling	↓	↓	↑	↓	↑	—	—	—
Damaged blade	↓	↓	↑	↓	↑	—	—	Highly fluctuating
Bearing failure	—	—	—	—	↑	↑	↓	—

(c) Air Compressor diagnosis

Fig. 16.26 *Gas turbine vibration and performance monitoring*

16.8 ■ SELECTION OF CONDITION MONITORING TECHNIQUES

Table 16.3 gives general guidelines about the selection of suitable techniques for various applications. For rotating parts, vibration and noise monitoring as well as wear debris analysis are used. Performance trend is applicable in case of seals and power plants incorporating engines or turbines. In case of pressure vessels and piping systems, thermography, corrosion monitoring and acoustic emission are recommended.

Table 16.3 *Selection of Condition Monitoring Techniques*

Application	*Suitable Technique*
Gear drives	Vibration and noise monitoring, oil analysis, wear debris analysis
Rolling bearings	Vibration and shock pulse monitoring
Journal bearings	Temperature, oil analysis
Seals	Particle analysis, pressure reduction
Reciprocating engines and compressors	Vibration monitoring, wear debris analysis, performance trend monitoring
Turbines	Vibration monitoring, temperature and wear debris analysis, performance trend monitoring
Pressure vessels and process plants	Acoustic emission, thermography, corrosion monitoring

Various monitoring techniques are in existence. The main factor to be considered is the question of investment vis-à-vis the loss due to a sudden breakdown of machinery which may result in considerable down-time. Studies have shown that a small investment in instrumentation for condition monitoring is likely to bring high benefits, as failures can be avoided and corrective action can be taken well in time.

16.9 ■ DIAGNOSIS

The analysis of various signatures, as monitored periodically or continuously, can diagnose an impending problem or fault in a machine or plant through trend monitoring. If a number of signatures are monitored, the reasoning has to be done with care, based on experience and theoretical knowledge of the system.

With the use of computers, the analysis and diagnostic techniques, have been considerably facilitated and in the case of abnormal signatures, alarm and/or shut down is possible automatically, in order to avoid serious operational problems and failures, ensuring safety as well. Such systems, where condition monitoring and signature analysis are extensively carried out, are called 'smart systems'.

The advances in some of the intelligent diagnostic techniques like expert systems, fuzzy logic and artificial neural networks and combinations of such methods using appropriate softwares, have been currently taking place.

Figure 16.27 shows the main elements of an expert system, in which a computer program is able to produce decisions like an expert would do. There are three main components:

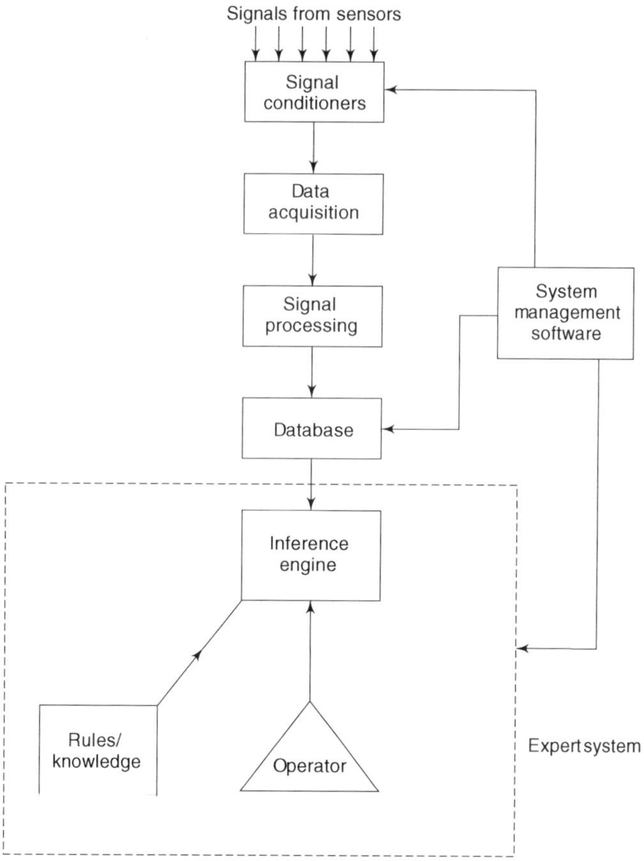

Fig. 16.27 *Main elements of an expert system*

- Knowledge base, consisting of a set of rules of the IF . . . THEN type.
- Inference engine, giving reasoning.
- Database, involving acquisition of data from sensors.

The inference engine uses the knowledge base to process the information in the database. The management software interfaces with the operator, interrogating the real-time parameters e.g. *IF* frequency is 1X (i.e. same as running speed) and sensor is radial

THEN: fault is unbalance

IF: frequency is 2X, and sensor direction is axial

THEN: fault is misalignment

The fuzzy logic systems allow processing of data that may be partially correct or incomplete and where rules cannot be exactly generated. It corresponds to human thinking where labels of fuzzy sets or logic with fuzzy rules, can be used.

The Artificial Neural Networks are meant to recognize patterns in incomplete data. These are based on human nervous systems, with several neurons or nodes having one or more hidden layers between input and output neurons, which are connected as shown in Fig. 16.28 for a simple neural network, with two hidden layers. The input–output relation for each node is determined by a set of connection weights W_i, a threshold parameter W_0 and a node activation function A, such that

$$Y = A\left(\sum_{i=1}^{n} W_i\ X_i + W_0\right),$$

where Y is the output and X_i are the inputs. The activation function 'A' which introduces a non-linear transformation between input and output at each neuron, is suitably chosen. The network is trained with available data and tested to verify the damage or fault.

The combined use of fuzzy logic and artificial neural network techniques, is quite effective for fault diagnosis and a number of studies for its application, are in progress.

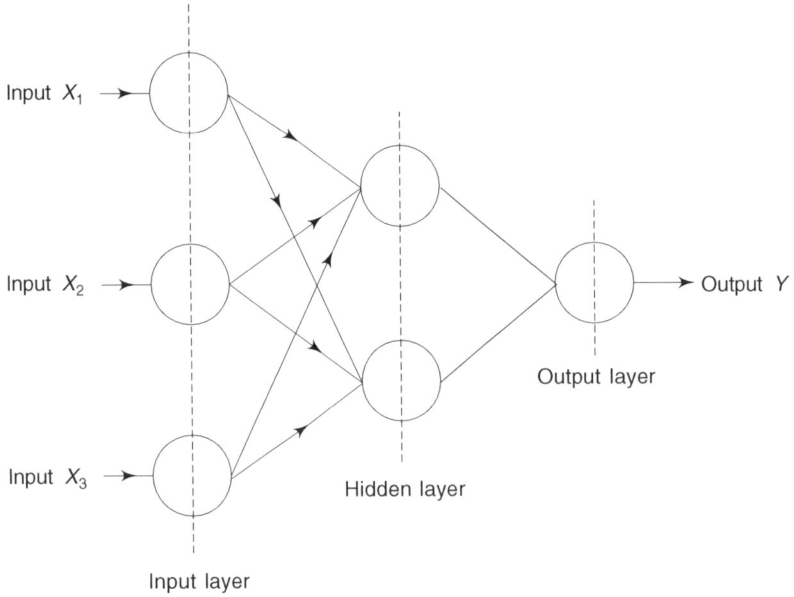

Fig. 16.28 *A typical artificial neural network*

Review Questions

16.1 Indicate True or False against the following:

(a) Condition monitoring means monitoring the condition of a machine when it is not running.

(b) Vibration, noise, temperature, wear particles, etc. are known as signatures of a machine.

(c) Vibrations of a machine reduce considerably when a fault develops.

(d) Unbalance of a machine gives vibrations at a frequency equal to the running speed.

(e) In order to measure the vibrations of a rotor, attach a piezo-electric accelerometer to the rotor.

(f) In order to have complete information, it is necessary to measure vibrations in both radial and axial directions.

(g) Frequency analysis of vibrations or noise signals from a machine is not important for signature analysis.

(h) Acceptable vibration levels of a small-power machine would be higher compared to that of a high-power machine.

(i) Bearing damage may be detected by thermography.

(j) In the case of a new machine, wear is excessive in the beginning.

(k) The size, composition, shape and loss rate of all parts can be determined by ferrography.

(l) SOAP is applicable to ferrous as well as non-ferrous particles.

(m) Size of wear particles increases before the onset of fatigue failure.

(n) Proximity sensors are used for measuring shaft motion in large machines.

(o) Acoustic emission is at frequencies in the audible range.

(p) Performance trend monitoring of a machine cannot predict a fault.

(q) Vibration monitoring may be combined with performance trend monitoring for good results in condition based maintenance.

(r) Expert systems do not need the use of a digital computer.

(s) Neurons in the artificial neural networks are essentially nodes with inputs and outputs.

(t) Fuzzy logic systems process data which may be partially correct.

Answers

16.1	(a) F	(b) T	(c) F	(d) T	(e) F
	(f) T	(g) F	(h) F	(i) T	(j) T
	(k) F	(l) T	(m) T	(n) T	(o) F
	(p) F	(q) T	(r) F	(s) T	(t) T

Chapter

17

Miscellaneous Instruments in Industrial, Biomedical and Environmental Applications

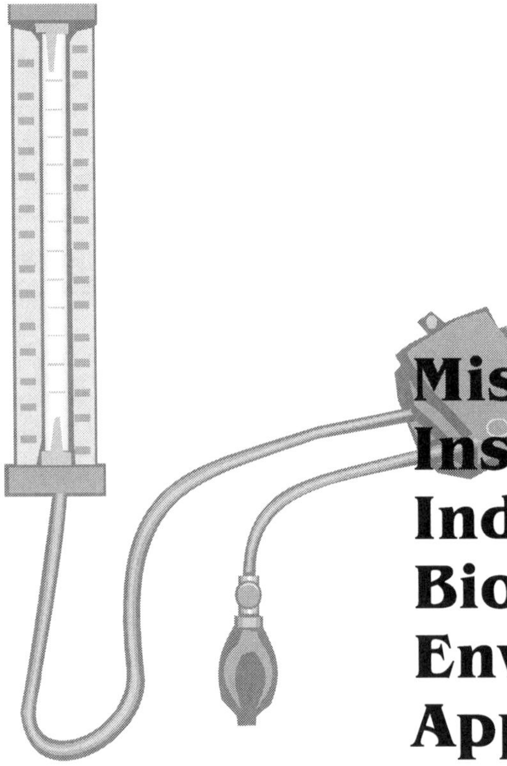

■ INTRODUCTION ■

In this chapter, some of the commonly employed miscellaneous measurement techniques in the areas of on-line measurement in industrial processes, biomedical applications and environmental air pollution studies have been discussed. The instruments for such applications are generally designed and developed to meet the real-life requirements of a measurement situation. For example, the instrument may be subjected to hostile conditions, i.e. these may be exposed to extreme temperatures, high pressures, dangerous radiations, gusty air flows, considerable shocks, large magnitude vibrations, noisy conditions, corrosive fumes, fire or explosion-prone installations, etc. Further, the measured medium may be muddy or sticky like various industrial slurries or various biomedical fluids which tend to affect the transducer functioning. Additionally, certain type of instruments together with the measurement situation may have to comply with statutory governmental/legal requirements of health/safety. These may include ensuring protection from possible hazards like electric shocks, chemical and gaseous poisoning, body damage or burn caused by direct contact with radiations like X-rays, ultraviolet rays, etc.

17.1 ■ SPECIFIC GRAVITY MEASUREMENTS

In a number of process control situations in industry, the measurement of specific gravity/density provides the best method for determining and controlling the concentration of a solution or mixture. Further, these measurements also provide a good means for direct estimation of product quality.

Specific gravity is the ratio of mass of a given volume of substance to the mass of equal volume of a certain standard substance. For liquids, the specific gravity is the ratio of the mass of a given volume of liquid to that of an equal volume of distilled water at the standard reference temperature of 4°C. For gases, the specific gravity determinations are made with reference to air at (normal temperature and pressure NTP, i.e. 20°C and 76 cm of Hg pressure) conditions.

The absolute specific gravity is not used commercially, specially in oil industry because all readings have to be expressed in decimals. Instead, the following arbitrary scales are used in practice and the instruments, namely the hydrometers, are then calibrated in terms of these arbitrary specific gravity standards.

Baumé Scale It is arbitrary scale in which the gravity of water at 10°C instead of 4°C is used as standard for specific gravity determinations. Liquids heavier than water are designated in degrees below 10° and liquids lighter than water are designated in degrees above 10°. This scale with correction for 15.5°C is defined as:

$$\text{Degrees Baumé} = \frac{140}{\text{sp. gr. } 15.5^\circ\text{C}/15.5^\circ\text{C}} - 130 \tag{17.1}$$

The denominator of the fraction, i.e. sp. gr. 15.5°C/15.5°C means that specific gravity is obtained by dividing the weight of certain volume of oil at 15.5°C by the weight of equal volume of water at 15.5°C.

API Scale It is similar to the Baumé scale and differs very slightly from it. It is defined as:

$$\text{Degrees API} = \frac{141.5}{\text{sp. gr. } 15.5^\circ\text{C}/15.5^\circ\text{C}} - 131.5 \tag{17.2}$$

Both Baumé and API scales are commonly used for petroleum products such as gasoline, kerosene, light distillates, lubricating oils, etc.

Twaddle Scale It is another arbitrary scale used for liquids heavier than water. It is defined as:

$$\text{Degrees Twaddle} = 200\,(\text{sp. gr.} - 1) \tag{17.3}$$

In this the range of specific gravity from 1.0 to 2.0 is divided into 200 equal parts, so that one degree Twaddle represents 0.005 sp. gr.

The following are some of the commonly employed intruments for *on-line* specific gravity measurements.

17.1.1 Bubbler System

This is a static pressure operated mechanism. It operates on the principle that the pressure at a point due to a liquid column is equal to g times the product of depth of the point below the surface and the density of the liquid. The pressure due to a fixed depth of liquid is therefore a measure of the density of the liquid. Figure 17.1 shows a typical arrangement of a bubbler system. In this a bubbler tube is connected to a compressed air supply through a pressure regulator. The bubbler tube has an opening at the bottom of the sampling containers. The sampling container is connected with a constantly flowing overflow tube to maintain a reference level in the container at a constant height.

The pressure regulator is adjusted so that air pressure in the bubbler tube is very slightly more than the static pressure due to the liquid column, so that some air bubbles start coming out of water. At this point the air pressure in the bubble tube is a function of the specific gravity of the liquid. The pressure is measured using either the Bourdon gauge or a suitable manometric device and is directly calibrated in terms of density/specific gravity or other arbitrary units like API, Baumé, Twaddle, etc. To improve the accuracy of the system, a suitable flow measuring device like a rotameter may be installed in the system so that a constant air flow through the bubbler system may be maintained during calibration of the instrument as well as when used in actual practice.

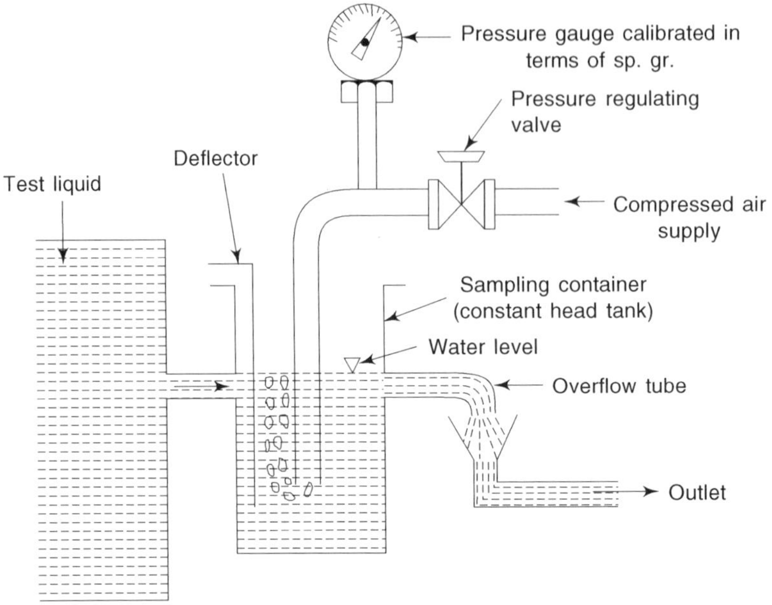

Fig. 17.1 *Bubbler system of specific gravity measurement*

In the measurement of small variations of specific gravity, the pressure in the bubbler system is amplified by means of a pneumatic amplifier. The resulting amplified pressure is then used to operate an indicator, a recorder or a controller.

17.1.2 Hydrometer Method

An ordinary hydrometer shown in Fig 17.2(a) is essentially a freely floating body which adjusts itself so that the weight of liquid displaced is equal to the weight of the hydrometer. When the density of the liquid in which it floats is reduced, the hydrometer would sink deeper into the liquid until the equilibrium between the buoyant force and the hydrometer weight is attained. Further, when the liquid density increases, the hydrometer rises. Thus, the depth of submergence of the hydrometer is a measure of the density or specific gravity of the liquid in which it floats.

Remote reading electrical signals can be obtained by making the float stem to actuate the soft iron core of the linear variable differential transducer [Fig. 17.2(b)]. The level of the liquid in the sampling container is usually kept constant by means of an overflow tube. The output of the LVDT determines the displacement of the hydrometer stem and hence the liquid density.

17.1.3 Totally Immersed Float Method

The operating principle of this device is similar to hydrometer method which is a partial immmersion device. In the totally immersed float method, the upthrust due to buoyant effect of the fluids is directly proportional to the density of the liquid.

In the device shown in Fig. 17.3, the upthrust of the glass or metal float or plummet is counterbalanced by the weight of the plummet together with the weight of a platinum-iridium chain. The plummet is so designed that at the middle of the specific gravity indicating range, it would support half the weight of the total chain length and the other half would lie at the supports provided at the lower end of the device. When the density of the fluid increases, it increases the buoyant force. The increase in buoyant

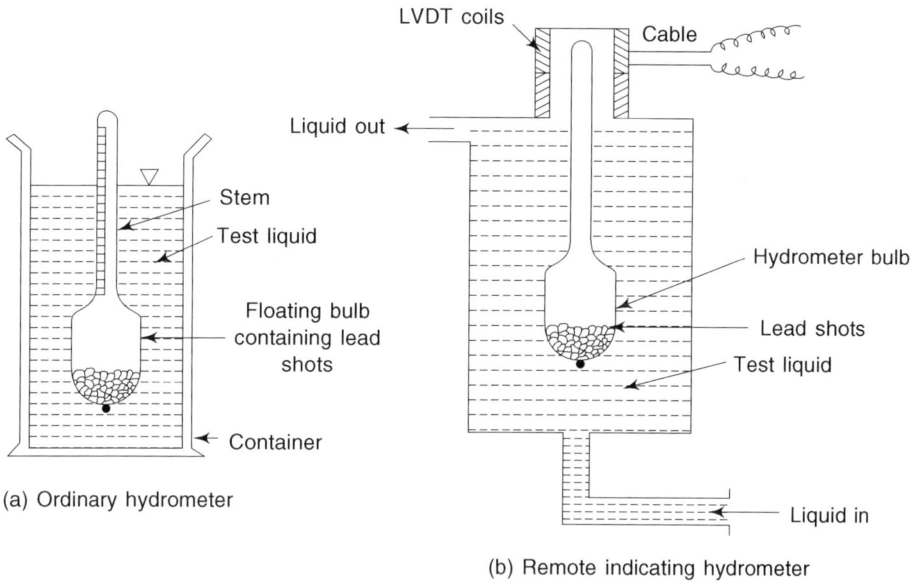

Fig. 17.2 *Hydrometer method of density measurement*

force would cause the plummet to rise. While rising, the plummet would take up a greater portion of the weight of chain. The plummet would continue to rise till the added buoyant force becomes equal to the increased proportion of the weight of the chain. When the density of the test fluid decreases, the reduced upthrust would bring the plummet down and a greater portion of the chain weight would be taken up by the supports. Ultimately, the plummet would come to a standstill when equilibrium between the upward buoyant force (surface forces) and downward body forces is achieved.

The movement of the plummet is sensed by a suitable displacement measuring device such as an LVDT. The plummet contains a ferromagnetic core which alters the inductance between the primary winding

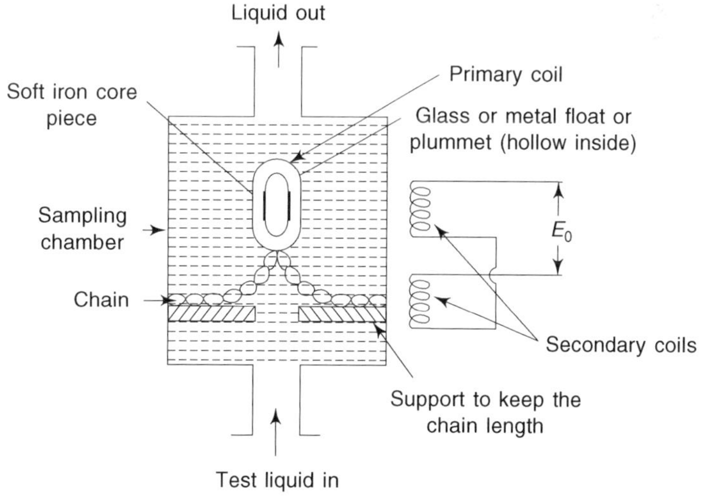

Fig. 17.3 *Schematic of totally immersed float type of density measuring device*

and two oppposed halves of the secondary winding of the differential transformer as the plummet rises or falls. The output of the transformer thus indicates the position of the plummet, which in turn indicates the density of the test fluid.

The main advantage of this device is that the instrument scale is linear. The accuracy of the device is ±2% of full-scale deflection.

17.1.4 Nuclear Absorption Method

This technique is quite versatile and is applicable for continuous and accurate density measurements of liquids including slurries or granular materials in pipes, bins, tanks, hoppers, etc. In such a system, the density gauge consists of two units, namely a source and a detector which can be clamped on a process pipe on two suitable locations usually the opposite sides of the pipe with material flow. A source containing the isotope of either cesium-137 or americium-241 or cobalt-60 is employed which emits gamma rays directed on the material medium (Fig. 17.4). The radioactive rays get partially absorbed in the material medium and the remainder are detected by detectors such as the ionisation chamber that produces a proportional output signal. The output signal in such devices for a fixed thickness of material is a function of the density of the medium. This type of instrument can measure the density over a wide range, i.e. from 500 to 4000 kg/m^3, with an accuracy ±0.1% of the measured value. However, the output of the unit is non-linear and needs to be linearised.

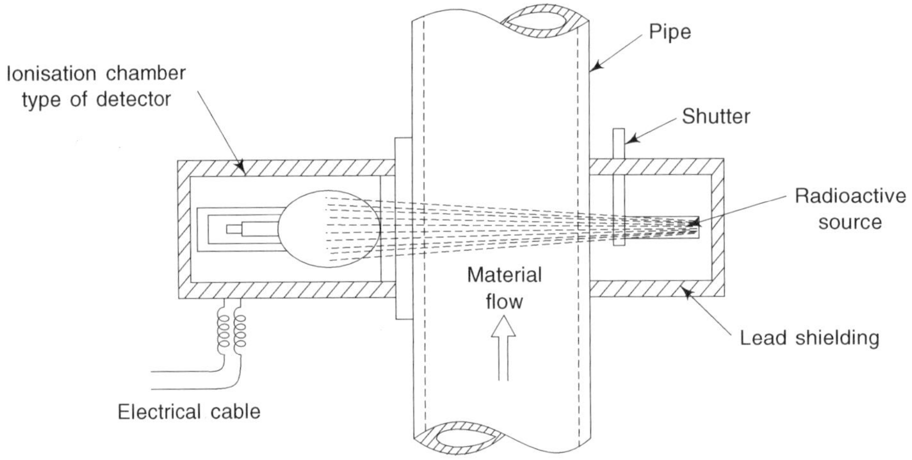

Fig. 17.4 *Nuclear absorption method of density measurement*

17.1.5 Fixed Volume Method

Measurements of specific weights can be made by continuously weighing a fixed volume of liquid enclosed in a container through which the process liquid flows in and out at a constant rate. Weighing may be carried out either by a mechanical balance or by an electrical load cell.

Another highly accurate approach in the fixed volume method consists of measuring the acoustical resonant frequency of an oscillating system such as a tube filled with a fixed volume of the process fluid. The frequency of oscillation of the tube is dependent on its mass, which in turn is dependent on the density of the liquid contained in it. Thus, a change in density produces a change in the acoustic resonant frequency.

In practice, the acoustic density meter consists of twin tubes of a high-nickel alloy which give stable frequency response. The shape of the twin tubes resembles the typical tuning fork filled with the test fluid (Fig. 17.5). A magnetically coupled vibration exciter maintains the tubes in vibration. The vibratory response of the tubes can be obtained by any typical accelerometer type of vibration pick-up. The output of the pick-up is a frequency modulated signal which is converted to a direct density read-out which enables the density to be monitored and controlled.

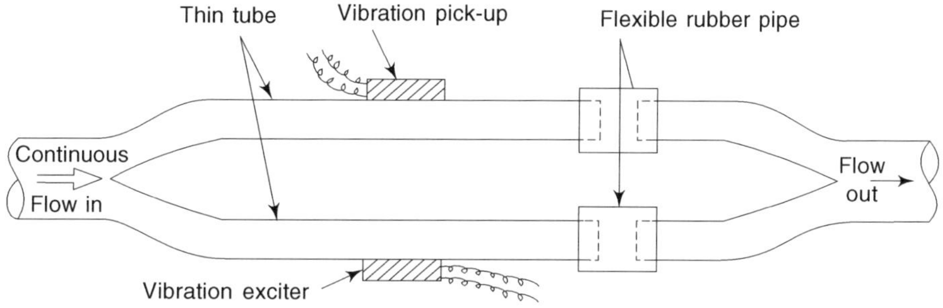

Fig. 17.5 *A typical acoustical density meter*

The device is quite robust and easy to maintain. Usually, vertical mounting is preferred, as this prevents the solids from sticking to the walls of the tube which may cause errors in the density read-outs. The range of the instrument is quite large and has an accuracy of ±0.01% in the range of 600–1600 kg/m^3. Further, the device has an excellent repeatability wih practically no hysteresis effect.

17.2 ■ MEASUREMENT OF LIQUID LEVEL

In industry, usually vast quantities of liquids such as water, solvents, chemicals, etc. are used in a number of industrial processes. Liquid level measurements are made to ascertain the quantity of liquid held in a container or vessel. The liquid level affects both pressure and rate of flow in and out of the container and therefore its measurement and/or control becomes quite important in a variety of processes encountered in modern manufacturing plants. Liquid level measurements can be broadly classified as:

1. direct methods and
2. indirect methods

17.2.1 Direct Liquid Level Measurements

Some methods by which the actual liquid level is directly indicated by means of a simple mechanical type of device are now discussed.

Dip-stick Method A commonly used method for determining the liquid level is dipping a graduated rod in a liquid. Boatmen usually dip the oars in the canal/river to know the depth of water at a particular place. Similarly, we use a dip-stick to measure the level of oil in a car engine or the height of fuel oil in a uniformly shaped storage tank. This method, though quite economical, is not very accurate specially for moving fluids. Further, it is not possible to get continuous on-line observations in industrial processes.

Sight Glass Method The sight glass or piezo-meter tube is graduated glass tube mounted on the side of the liquid containing vessel for providing a visual indication of the liquid level (Fig. 17.6). Since the liquids keep level, therefore the rise or fall of the liquid level in a tank/vessel results in a corresponding change in the level indicated by the sight tube.

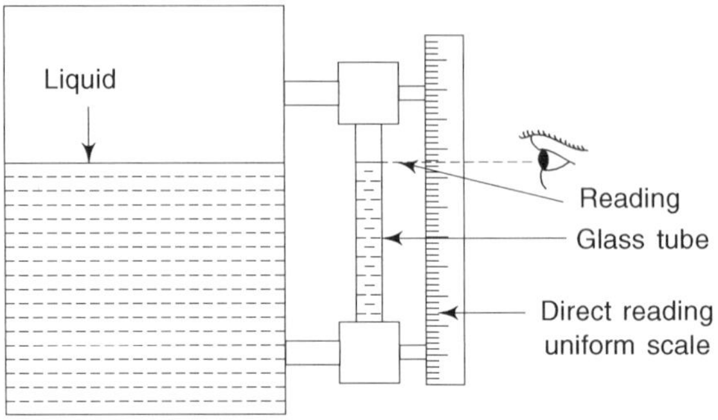

Fig. 17.6 *Sight glass level gate*

Sight tubes are usually made of toughened glass and are provided with metallic protecting covers around them. Further, the diameter of such tubes is neither too large to change the tank/vessel level, nor too small to cause capillary action in the tube.

The measurement of liquid level with this device is simple and direct for clean and coloured liquids. However, it is rather unsuitable for dirty, viscous and corrosive liquids. Further, an operator is required to record the liquid levels with this device.

Hook Gauge Sometimes it becomes necessary to accurately measure very small changes in liquid level in open tanks/containers. In a large tank/reservoir, a small change in level would mean large volumetric changes. For such applications, a simple hook gage is quite suitable. The schematic arrangement of this gauge is shown in Fig. 17.7. In this device, a vertical tubular rod is provided with a vernier scale to be clamped at a suitable height at the upper end and a U-shaped hook at the lower end. This rod moves in a guide bracket fixed to a rigid body at the datum or reference level and has a main graduated scale in it. The movable rod is brought downwards so that the hook is first pushed below the surface of the liquid. It is then gradually raised until the top of the hook breaks through the surface of the liquid. The movable rod is then clamped and the level is read off the scale. The device is accurate up to ±0.1 mm, the least count of the instrument. Further, the device is manually operated and does not lend itself to automatic reading.

Float Gauge A floating body, because of its buoyancy, would always follow the varying liquid level. Therefore, float-operated devices are capable of giving continuous, direct liquid level measurements. The floats generally used are hollow metal spheres, cylindrical ceramic floats or/disc shaped floats of synthetic materials. The top of the float is usually made sloping so that any solid suspensions in the liquid do not settle on the float and change its weight. Float gauges are sufficiently accurate when properly calibrated after installation. Further, a proper correction is required if there is a change in the liquid density due to a change in temperature.

Figure 17.8 illustrates a typical float-and-chain liquid level gauge generally used for directly measuring the liquid level in open tanks. The instrument consists of a float, a counter weight and a flexible connection that may be a chain or a thin metallic perforated tape. The counter weight keeps the chain/tape taut as the liquid rises or falls with any changes in the liquid level. The chain/perforated tape link is wound on a gear or sprocket wheel to which the pointer is attached. Any movement of this wheel would indicate on a suitably calibrated scale the level of the liquid in the tank.

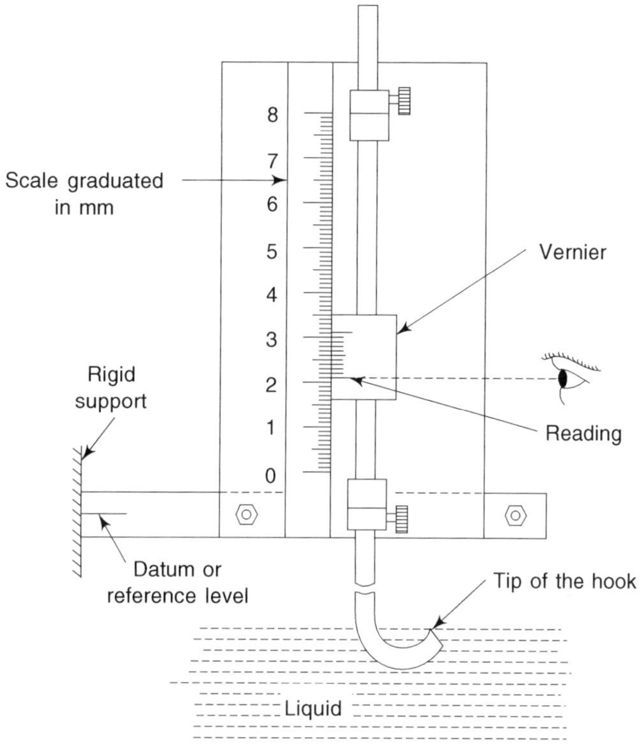

Fig. 17.7 *Hook type level indicator*

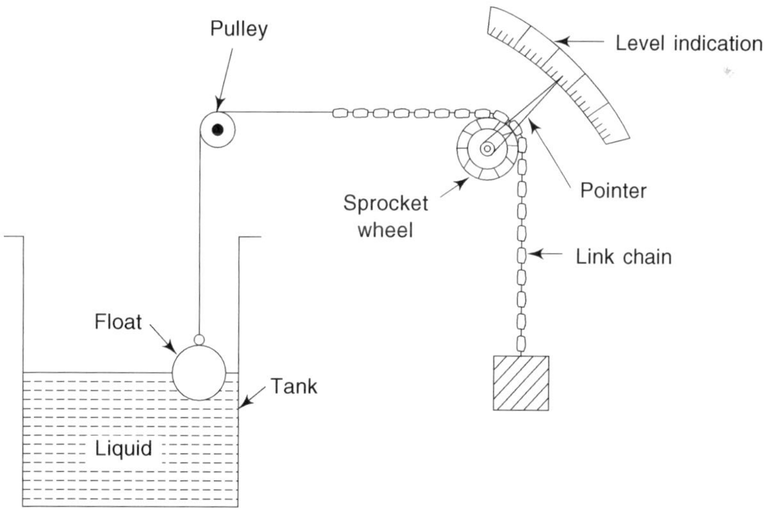

Fig. 17.8 *Float-and-chain liquid level gauge*

Another version of the float-actuated instrument is the float-and-shaft liquid level gauge (Fig. 17.9). In this unit, the motion of the float on the surface of the liquid is transferred to the shaft and the level is indicated by the pointer on the dial.

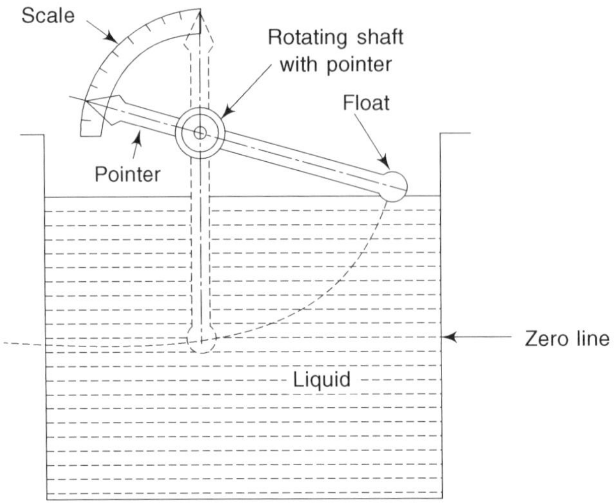

Fig. 17.9 *Float-and-shaft liquid level gauge*

Further, there are a number of float-operated schemes with electrical read-outs. In these, the float acts as a primary transducer that converts liquid level variation into a suitable displacement. This displacement is sensed by the secondary transducer such as a resistive type of potentiometric device, inductive type of LVDT, etc. Figure 17.10 shows the schematic of the float-actuated rheostatic (resistive) device. The float displacement acutates the arm which causes the slider to move over the resistive element of a rheostat. The circuit resistance changes and this resistance change is directly proportional to the liquid level in the tank.

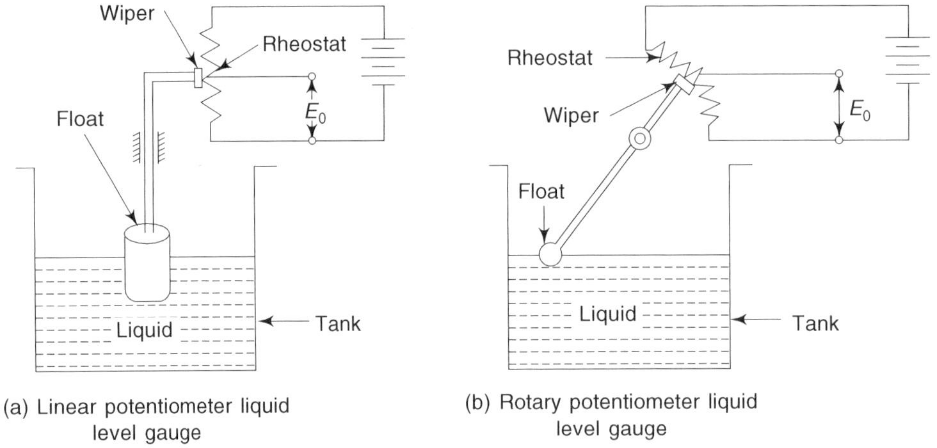

Fig. 17.10 *Typcial float-operated rheostatic liquid level gauge*

17.2.2 Indirect Liquid Level Measurements

Hydrostatic Pressure Level Measurement Device The hydrostatic pressure created by a liquid is directly related to the height of the liquid column ($p = \rho gh$). Therefore, a pressure gauge is installed at the bottom or on the side of the tank containing the liquid (Fig. 17.11). The rise and fall of the liquid level causes a corresponding increase or decrease in the pressure p which is directly proportional to the liquid

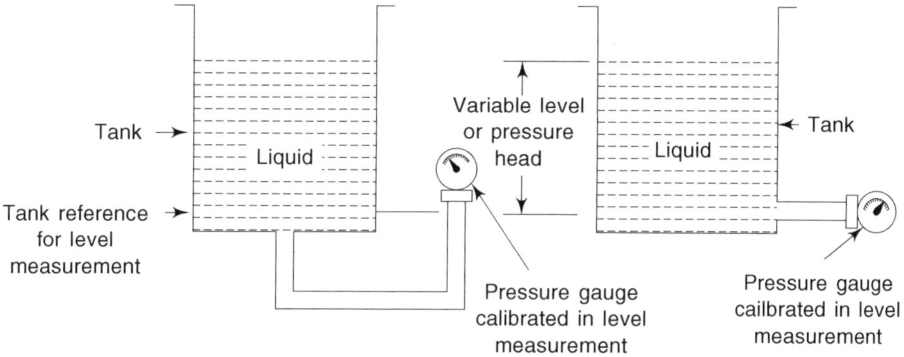

Fig. 17.11 *Typical arrangements of hydrostatic pressure type level measuring devices*

level h. The dial or scale of the pressure gauge is calibrated in the units of level measurement. These gauges function smoothly when the liquids are clean and non-corrosive. For corrosive liquids with solid suspensions, diaphragm seals between the fluid and the pressure gauge are generally employed.

Bubbler or Purge Technique for Level Measurement In this method, the air pressure in a pneumatic pipeline is so regulated that the air pressure in the bubbler tube, shown in Fig. 17.12, is very slightly in excess over that of the hydrostatic pressure at the lowermost end of the bubbler tube. The bubbler tube is installed vertically in the tank with its lowermost open end at zero level. The other end of the tube is connected to a regulated air supply and a pressure gauge. The air supply in the bubbler tube is so adjusted that the pressure is just greater than the pressure exerted by the liquid column in the tank. This is achieved by adjusting the air pressure regulator until bubbles can be seen slowly leaving the open end of the tube. Sometimes a small air flow meter is fitted in the line to control/check the excessive flow of air. When the air flow is small and the density of the fluid is uniform, then gauge pressure is directly proportional to the height of the liquid level in the open tank. In practice, the gauge is directly calibrated in the units of liquid level and if the tank is uniformly shaped, then the calibration may be in the units of volume.

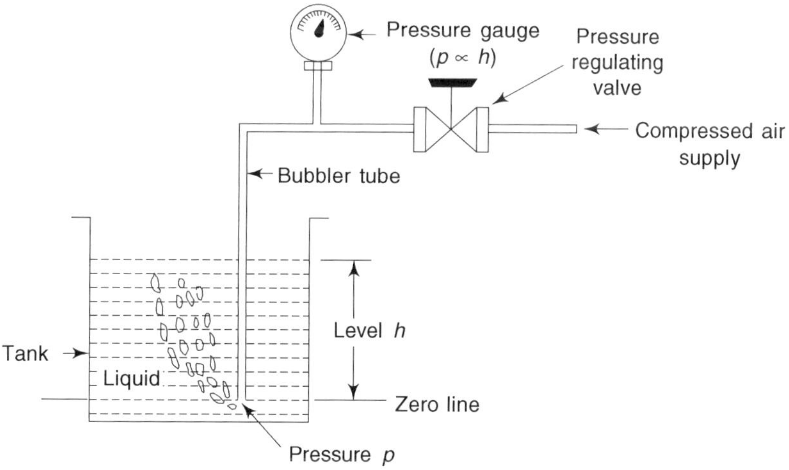

Fig. 17.12 *Bubbler or purge type of liquid level meter*

Capacitance Level Gauge A simple condenser/capacitor consists of two electrode plates separated by a small thickness of an insulator (which can be solid, liquid, gas or vacuum) called the dielectric. The change in liquid level causes a variation in the dielectric between the two plates, which in turn causes a corresponding change in the value of the capacitance of the condenser. Therefore, such a gauge is also termed a *dielectric level gauge.*

The magnitude of the capacitance depends on the nature of the dielectric, varies directly with the area of the plate and inversely with the distance between them. The capacitance can be changed by any of these factors.

In a parallel plate condenser which has identical plates each of area A (cm^2) separated by a distance d (cm) and an insulating medium with dielectric constant K ($K = 1$ for air) between them, the expression for the capacitance is given by

$$C\,(\text{in } \mu\mu F) = 0.0885\,\frac{A}{d}\,K \tag{17.4}$$

From Eq. (17.4) we observe that the capacitance varies directly with the dielectric constant which in turn varies directly with the liquid level between the plates. Figure 17.13 shows the schematic arrangement of a capacitance level gauge. The capacitance would be at a minimum when the tubes contain only air and at a maximum when the liquid fills the entire space between the electrodes. The change in capacitance can be measured by a suitable measuring unit such as a capacitive Wheatstone bridge by either manual null balancing or automatic null balancing using the null detecting circuit with a servo-motor that indicates the level reading.

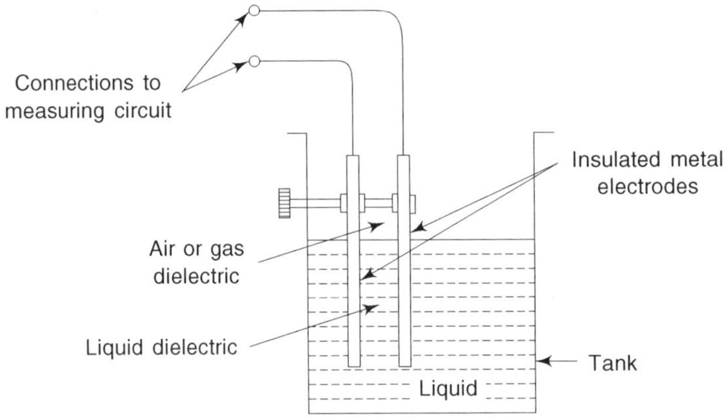

Fig. 17.13 *Dielectric liquid level gauge*

For the measurement of level in the case of non-conducting liquids, the bare probe arrangement may be satisfactory since the liquid resistance is sufficiently high. For conducting liquids, the probe plates are insulated using a thin coating of glass or plastic.

The capacitance type level gauge is relatively inexpensive, versatile, reliable and requires minimal maintenance. These units have no moving parts, are easy to install and adaptable to large and small vessels. Further, such devices have a good range of liquid level measurement, viz. from a few cm to more than 100 m. In addition, apart from sensing the level of the common liquids, these gases find wide use in other important applications such as determining the level of powdered or granular solids, liquid metals (high temperatures), liquified gases (low temperatures), corrosive materials (like hydrofluoric acid) and in very high pressure industrial processes.

Ultrasonic Level Gauge A schematic diagram of the ultrasonic level gauge is shown in Fig. 17.14. Sound waves are directed towards the free surface of the liquid under test from an ultrasound transmitter. These waves get reflected from the surface of the liquid and are received by the receiver. In this technique, liquid level variations are quite accurately determined by detecting the total time taken by the wave to travel to the liquid surface and then back to the receiver The longer this time interval, the farther away is the liquid surface, which in turn is a measure/indication of the liquid level.

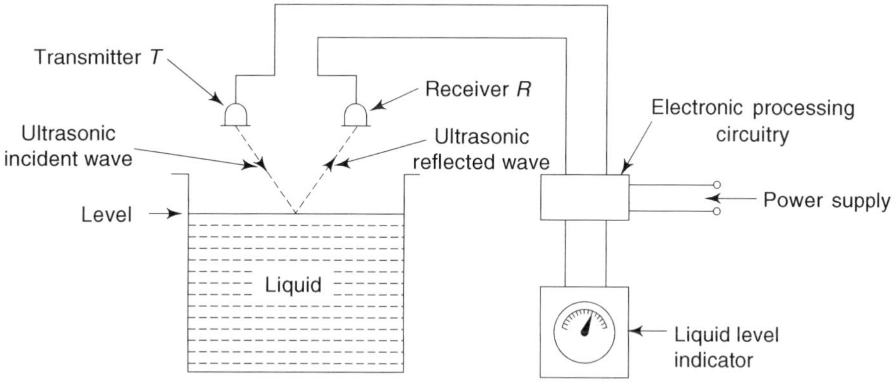

Fig. 17.14 *Schematic of ultrasonic liquid level gauge*

It may be noted that the operating principle of this instrument is quite simple. But the actual instrument is expensive and requires a high degree of experience and skill in operation. However, its main advantage is a wide range of applications in level measurement for different types of liquid and solid substances.

Nucleonic Gauge The working principle of the nucleonic gauge or gamma ray liquid level sensor is that the absorption of gamma rays varies with the thickness of the absorbing material (i.e. height of liquid column) between the source and the detector. The higher the height of the liquid column, greater is the absorption of gamma rays and consequently lower is the detector output. The output is measured and correlated with the level of liquid in the tank using the following exponential type of expression applicable in such an arrangement:

$$I = I_0 e^{-\mu \rho x} \qquad (17.5)$$

where I is the intensity of radiation falling on the detector
I_0 is the intensity of radiation at the detector with absorbing material not present
e is base of natural logarithm = 2.71
μ is the mass absorption coefficient in m^3/kg (constant for a given source and absorbing material)
ρ is the mass density of the test material in kg/m^3
x is the thickness of absorbing material in m (i.e. height of liquid column in the present case).

The schematic of the liquid level gauge is shown in Fig. 17.15. The instrument consists of a radioactive source (which may be either Ce-137, Am-241 or Co-60), a radiation detector (of ionisation chamber type) and electronic circuits incorporating the amplifiers and read-out instrument or recorder-controller.

Nuclear gauges cover a wide range of applications for recording the level of a wide variety of liquid as well as solid substances. They are quite suitable for large reservoirs of 30–40 m diameter and can give continuous measurements of heights of 20 m or more with repeatability of ±1%. Like the ultrasonic gauge, this is also a non-contact device and the level measurement is not affected by conditions of high or low temperatures, pressure, viscosity, corrosion, abrasion, etc. Further, these gauges are quite rugged

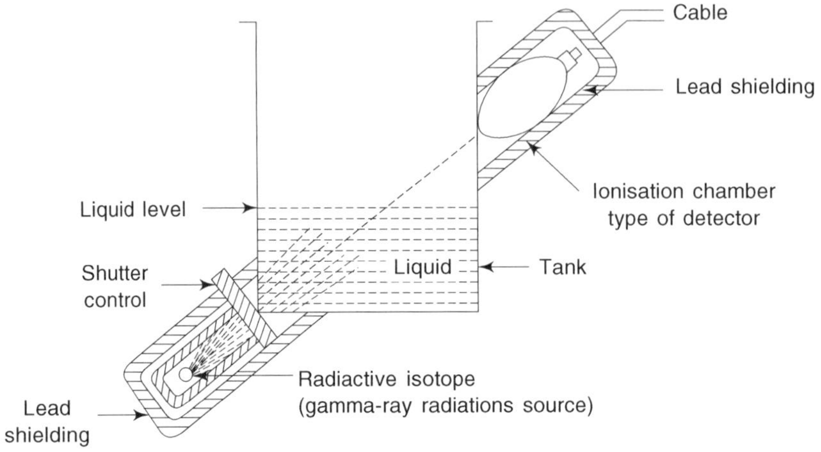

Fig. 17.15 *Schematic of gamma ray liquid level gauge*

and can withstand severe operating conditions. The main drawback in these gauges is the risks involved due to radiation effects. Therefore, adequate shielding to limit the radiation field intensity well below the Atomic Energy Commission (AEC) tolerances has to be provided for.

17.3 ■ VISCOSITY MEASUREMENTS

The internal friction or the resistance to flow within a fluid (liquid or gas) is due to the viscosity of the fluid. In Chapter 13 no direct reference was made to the effects of viscosity in the fluid flow measurements. This is because of the fact that most industrial flow measuring devices usually are employed for low-viscosity fluids in which viscosity variations (which occur due to change in temperature) are not sufficient to affect the flow measurements. However, in theoretical computations of the rate of flow, viscosity effects are accounted for by multiplying $Q_{\text{theoretical}}$ by C_d the coefficient of discharge of the flow measuring device. In fact, viscosity is one of the major factors responsible for the dissipation of energy in transporting liquids and gases in pipelines.

Further, in certain industrial machines and processes, the property of viscosity is particularly significant in the study of their tribological aspects, i.e., their lubrication and wear characteristics. In industry, oils with higher viscosity are termed as 'heavier' oils. This nomenclature does not bear any relation to their densities. The effectiveness of any lubricating oil/fluid mainly depends, among other factors, on its viscosity. Generally, lubricants should be sufficiently viscous so that they are not squeezed out of the bearings and also not too viscous so as to increase the resistance to the motion between the moving parts being lubricated. Additionally, viscosity sometimes plays an important role in modifying the physical behaviour of the fluid. For example, the proper atomisation of fuel for fuel injection or of paint spraying depends on maintaining the viscosity within certain prescribed limits.

Thus, the study of viscosity and its measurement is quite important in a number of industrial processes/applications. In a layaman's language, viscosity is a comparative measure of the resistance/internal friction in fluid flows. For example, a thick liquid like honey offers considerably higher resistance to the flow than does water. In fact, the viscosity of a liquid or gas is determined from the magnitude of resistance of the fluid when an external shearing force is applied.

Newton postulated that for certain fluids known as Newtonian fluids, the shearing stress experienced by two layers is directly proportional to the velocity gradient between them. Thus the defining equation for the dynamic or absolute viscosity, commonly known as Newton's law of viscosity is as follows:

$$\tau = \mu \frac{du}{dy} \tag{17.6}$$

where τ is the shearing stress between the fluid layers

$\frac{du}{dy}$ is the velocity gradient between the fluid layers

μ is the proportionality constant of the equation, known as the coefficient of dynamic viscosity.

It may be noted that for Newtonian fluids, the value of μ is independent of the velocity gradient du/dy. However, for non-Newtonian fluids its value varies with the velocity gradient.

From Eq. (17.6) it can be seen that the coefficient of dynamic viscosity is the ratio of shear stress to the velocity gradient and thus its dimensions are force time per unit area, i.e. Ns/m^2. Alternatively the dimensions of viscosity could be stated as mass per unit length per unit time, i.e. kg/m s. Another commonly used unit is the cgs unit 'poise' (dyn s/cm^2 or g/cm s). Smaller values of viscosity are generally expressed in terms of centipoise (1 cp = 0.01 poise). Further, the conversion factor between SI units and cgs units of dynamic viscosity is as follows:

$$1 \text{ Ns/m}^2 = 10 \text{ dyn s/cm}^2 = 10 \text{ poise} \tag{17.7}$$

Another unit of viscosity is termed as kinematic viscosity v, which is defined as the ratio of dynamic viscosity to the density of fluid, i.e., μ/ρ. The SI units of v are m^2/s and cgs units are cm^2/s (or stoke). Smaller values of viscosity are generally expressed in terms of centistokes (1 c St = 0.01 stoke). Further, the conversion factor between SI units and cgs units of kinematic viscosity is as follows:

$$1 \text{ m}^2\text{/s} = 10^4 \text{ St} \tag{17.8}$$

Occasionally, the term fluidity is used, particularly in the study of colloidal solutions. This is defined as the reciprocal of the dynamic viscosity in poise.

The majority of the viscosity measurement of oils and petroleum products are made on laboratory samples. Therefore, in addition to on-line viscosity measuring instruments, some commonly used laboratory instruments are also discussed here.

17.3.1 Capillary Tube Viscometer (Constant Head Type)

Figure 17.16 shows a commonly used laboratory instrument, viz. capillary tube viscometer (constant head type) which is based on the Hagen-Poiseuelle equation for laminar flow in a tube which is:

$$Q = \frac{\pi D^4 \Delta p}{128 \mu L} \tag{17.9}$$

where Q is volume of liquid passing through the tube per second

L is length of the capillary tube

D is diameter of the capillary tube

μ is coefficient of dynamic viscosity

Δp is pressure drop across the ends of the tube

In Eq. (17.9) the diameter is raised to the fourth power and therefore, it is essential to measure it as accurately as possible. It is usually measured in the laboratory by means of a travelling microscope. The discharge rate Q can be easily determined by using a graduated cylinder and a stop watch. The coefficient of dynamic viscosity can be computed by substituting the following in Eq. (17.9):

$$Q = \frac{\text{volume collected in the graduate jar}}{\text{time taken}}$$

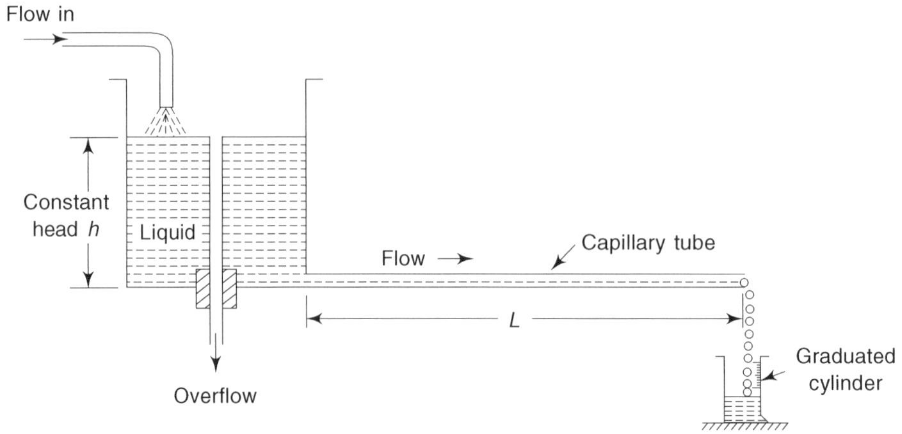

Fig. 17.16 *Capillary tube viscometer*

and $$\Delta p = \rho g h \text{ (for constant head)}$$

Therefore, Eq. (17.9) can be written as:

$$\mu = \pi D^4 h \rho g/(128QL) \tag{17.10}$$

Capillary tube viscometers have been used as secondary standards for the calibration of other type of viscometers. Further, they are commonly used in refineries to measure the viscosities of petroleum products, as such instruments are simple in construction, need practically on maintenance and are easy to use. However, they are unsuitable for unclean fluids as the dirt or grit tends to clog the capillary tube.

17.3.2 Efflux Type Viscometers

The maintenance of uniform diameter as well as the accurate measurement of pressure and the diameter are the difficulties experienced in the constant pressure type capillary tube viscometer. Therefore, another adaptation of the capillary tube principle is used for routine industrial measurements of viscosity. Such instruments are called efflux type of viscometers. In these instruments, there is no provision for applying constant head/pressure to the capillary tube. In use, the time of efflux for a fixed quantity of liquid through a standardised short capillary is determined, the liquid initially being at some specified level in the container. The measured time is taken as a direct measurement of the viscosity for the efflux type viscometer used for the test (Fig. 17.17).

Commercially available viscometers based on the efflux principle are namely Saybolt, Redwood and Engler viscometers In the Saybolt viscometer, the time in seconds for 60 ml of oil to flow through the orifice gives a measure of the viscosity of the oil, which is expressed in Saybolt universal seconds. The specified volume of efflux in the Redwood viscometer is 50 ml. Further, in the Engler viscometer, the specified volume of efflux is 200 ml. However, in this case, the measure of viscosity is expressed in terms of Engler degrees which is the ratio of the time of flow of 200 ml of the test fluid at the given temperature to the time taken for the same volume of water at 20°C.

The measure of viscosity in terms of Saybolt universal seconds, Redwood seconds and Engler degrees can be converted into the kinematic viscostiy v by the following empirical relation:

$$v = At + \frac{B}{t} \tag{17.11}$$

where v is the kinematic viscosity in centistokes

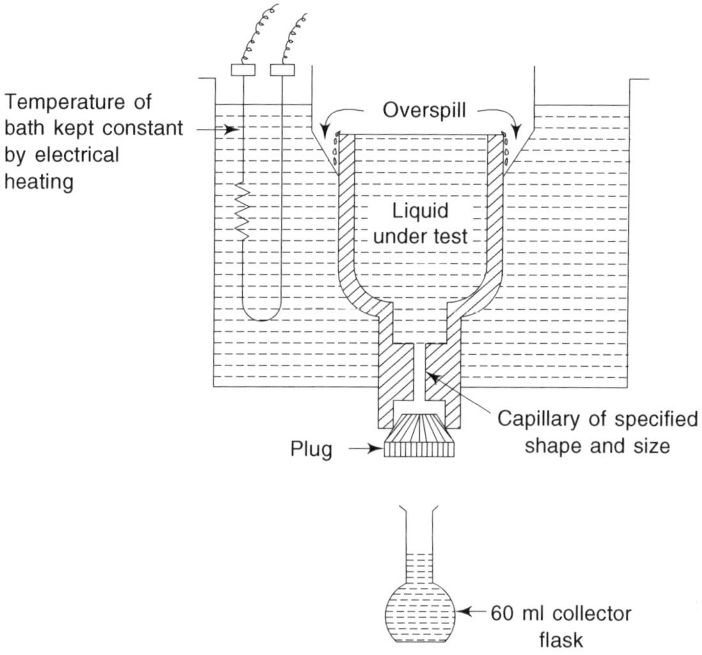

Fig. 17.17 *Schematic of Saybolt viscometer*

t the time of efflux in seconds (or degrees for Engler viscometers)

A and B are constants applicable to the type of viscometer

Commonly accepted values of A and B for different viscometers are as follows:

Viscometer type	*A*	*B*
Saybolt viscometer	0.22	180
Redwood viscometer	0.26	172
Engler viscometer	1.47	374

These viscometers are widely used in petroleum and other allied industries. However, they are unsuitable for continuous measurements which may be required in some industrial processes.

17.3.3 Rotating Concentric Cylinder Viscometer

The unit basically consists of two concentric cylinders and the small intervening annular space contains the test fluid whose viscosity is to be determined (Fig. 17.18). The outer cylinder is rotated at a constant angular speed. The viscous drag due to the liquid between the cylinders produces a torque on the inner cylinder, which would rotate if it was not restrained by an equal and opposite torque developed by the torsion wire. As the spring torque is proportional to the angle through which it turns, therefore the angular movement of the pointer on a fixed disc is used as a measure of viscosity.

If the annular space $(r_2 - r_1)$ is sufficiently small in comparison to the radius of the inner cylinder, then the following relation can be established between the torque produced on the inner cylinder due to viscous drag and the viscosity of the fluid as

$$\mu = T(r_2 - r_1)/(2\pi\omega r_1^2 r_2 L) \tag{17.12}$$

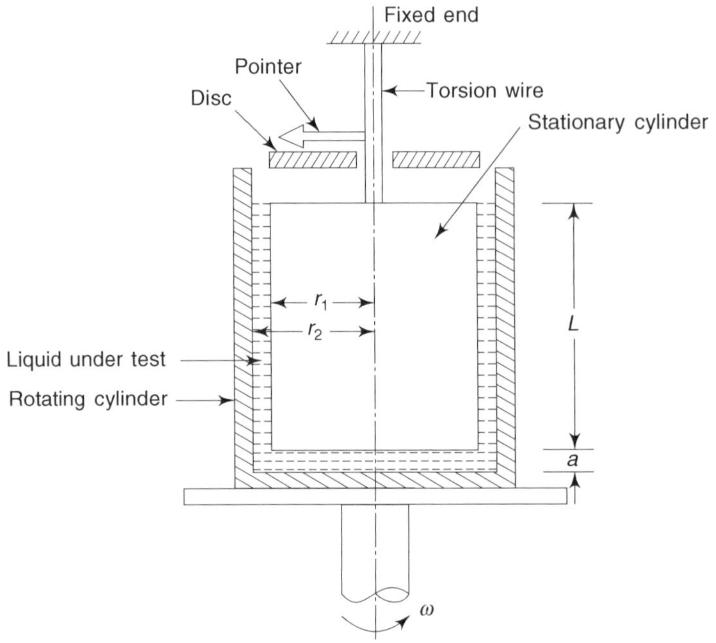

Fig. 17.18 *Schematic of rotating concentric cylinder viscometer*

When the gap a is small, then a small additional viscous drag torque is produced on the inner cylinder due to the bottom disc. Then Eq. (17.12) takes the following form

$$\mu = \frac{T}{\pi \omega r_1^2 \left[\dfrac{r_1^2}{2a} + \dfrac{2Lr_2}{r_2 - r_1} \right]} \tag{17.13}$$

This instrument can be used for continuous reading, if suitable arrangements are made for a continuous sample of the liquid to flow through the apparatus. In certain applications, it may be possible to overcome this sampling difficulty by immersing the viscometer directly in the main liquid line. The instrument is quite useful in studying the anomalous behaviour of non-Newtonian liquids as a wide range of shear rates are obtainable when a three or five speed gear box is used for the outer cylinder and three to four interchangeable inner cylinders are used. Further, with such an arrangement, the measuring range of the instrument becomes from a few centipoises to several thousand poises.

17.3.4 Variable Area Viscometer

The device is based on the principle that for a constant flow-rate, the position of a viscosity-sensitive floating bob in a tapered tube is governed by the viscous forces which are a function of the viscosity of the test fluid. Figure 17.19 shows a variable area/tapered tube (similar to the rotameter discussed in Ch. 13) mounted in a bypass line. A sampling or meter pump draws some of the liquid from the main process stream and pumps it through the unit which has a flow-rate setting float and viscosity measuring float. The viscosity measuring float is designed to be sensitive to the viscosity variations. Further the flow-rate sensitive float is designed to be sensitive to the flow variations and immune to viscosity effects. The flow rate is controlled and the flow-rate setting float is set at the fixed index mark. For this constant

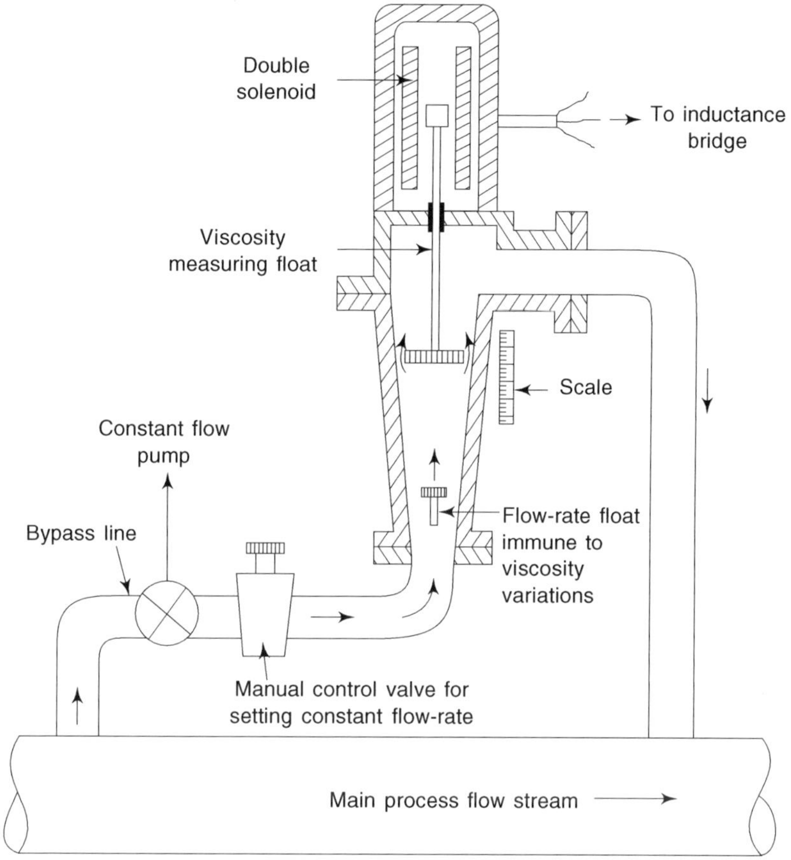

Fig. 17.19 *Schematic of variable area viscometer*

flow rate, the position of the viscosity-sensitive float is a measure of liquid viscosity. The output of the instrument can be made electrical by incorporating a linear variable differential transducer (LVDT) or any other suitable transducer.

The main advantage of this device is its capability to measure continuously the viscosity of liquids having viscosities as high as 10,000 poise. Further, the instrument can be installed in the main process stream or in a bypass so that viscosities can be measured under the prevailing temperature and pressure conditions.

17.4 ■ MEASUREMENT OF HUMIDITY AND MOISTURE

The amount of water vapour content in the atmosphere, i.e., humidity, is an important process variable in a number of industrial processes. This is because humidity/moisture content affects the behaviour of many commercial materials such as paper, textiles, tobacco, soap powders, fertilizers, paints, lacquers, leather, celluloid, artificial resins, glues, films, wood products, etc. Humidity measurement and control are necessary during many industrial processes as well as in heating and air-conditioning systems. For example, environmental heat and cold can be endured without much discomfort if the humidity of air is carefully controlled so that it establishes human comfort conditions.

Humidity of the atmosphere represents the measure of water vapour present in the air. Alternatively, it could be considered as the degree of dampness of the air. Humidity is generally expressed in terms of 'absolute humidity' or 'relative humidity'.

The absolute humidity of a gas is defined as the mass of water vapour present in a unit volume of gas and is usually expressed in g/m^3. The other term, relative humidity (RH) compares the humidity of air with humidity of saturated air at the same temperature and pressure.

Relative humidity is defined as the ratio of mass of water vapour present in a given volume of gas to the mass of water vapour necessary to saturate the same volume of gas at the same temperature. Thus, dry air (i.e., with no moisture content) has RH = 0.0% and saturated air has RH = 100%. Saturated air means that the air contains all the moisture content that it can hold at that particular temperature and pressure.

Traditionally, the relative humidity of the air is determined by a psychrometer, a device containing two thermometers. The temperature sensing bulb of one measures the environmental temperature (dry bulb temperature) and around the bulb of the other is a wick from which water is evaporated to produce cooling which indicates the wet bulb temperature. Relative humidity is related to the difference between the dry and wet bulb temperatures. For remote indication of these temperatures, suitably waterproofed thermistors may be used.

Relative humidity can also be determined by the dew-point technique in which the temperature of a polished metal surface is reduced (usually by evaporation of volatile liquid) until there is a visible condensation of water vapour. This temperature of condensation is called the dew point. Further, a relationship exists between the dew point, environmental temperature and relative humidity. The relationship between various humidity variables such as wet and dry bulb temperatures, dew point and percent RH and grams of moisture per cubic metre of air, etc. at specified atmospheric pressures are given in the form of a chart known as the psychrometric chart. The measurement of dry bulb and wet bulb temperatures or alternatively the determinations of dew point can be automated to a certain extent, but the complexity of calculations involving the psychrometric chart hinders the development of this technique into a continuous reading instrument. The following are some of the instruments which are generally used for on-line humidity measurements.

17.4.1 Hair Hygrometer

Certain materials such as human hair, animal membranes, wood and paper undergo changes in linear dimensions when they absorb moisture from the atmosphere. Human hair becomes longer as the humidity of the surrounding air increases, and shortens when the air becomes drier. This property of hair can be used to operate a pointer or recording pen through a system of mechanical linkage. The indicator scale can be calibrated to give a direct indication of the humidity. Figure 17.20 illustrates the schematic diagram of the hair hygrometer. The transducer element consists of strands of hair to give it increased mechanical strength. The hair strands are generally arranged parallel to each other with sufficient space between them for giving free access to the air sample under test. Further, for proper functioning, the element is maintained under light tension by a spring.

This instrument is quite cheap and is available in convenient pocket-size shape. It is not a precision instrument and is generally used for industrial purposes where readings of high precision are not required. Such instruments are recommended for use in the temperature range 5–35°C and in the relative humidity range 40–95%. In the specified range of temperature and humidity, the accuracy of a well-made instrument is within 3–4%. The calibration of these instruments should be checked periodically, and adjusted in necessary, using the precision wet and dry bulb hygrometer.

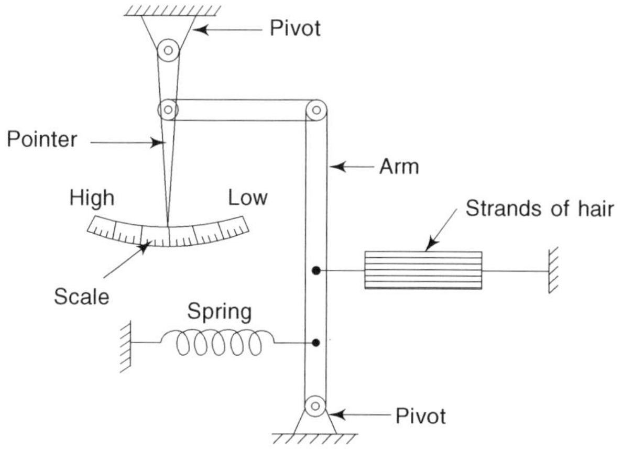

Fig. 17.20 *A schematic of hair hygrometer*

17.4.2 Humistor Hygrometer

Humistor is the short form of humidity sensitive resistor. The electrical resistance of such resistors is found to vary reproducibly with the changes in relative humidity of the surrounding air and therefore these are conveniently used as sensing elements for humidity measurements.

The sensing element, i.e., the humistor consists of two metal grids bonded to a sheet of plastic. This arrangement is given a coating of moisture-sensitive, i.e. hygroscopic salt such as lithium chloride. Alternatively, other salts like barium fluoride, potassium hydrogen phosphate or aluminium oxide can also be used. As the relative humidity rises, the film becomes more conductive and the electrical resistance of the grid is lowered. The variation in resistance is calibrated in terms of RH units. It has been found that a single transducer generally can cover only a small range of the order of 11%. When a large range, of the order of 5–99%, of relative humidity is required then a combination of seven or eight transducer elements are used, each designed for a specific part of the total range.

Since the humistor is an electrical sensor, it meets the industrial need of speed, versatility, accuracy and high sensitivity. Such instruments are capable of detecting a fraction of 1% change in relative humidity. However, its resistance/relative humidity relationship is non-linear and therefore the scale calibration is not uniform in these instruments.

17.5 ■ MEASUREMENT OF pH VALUE

In a number of industrial applications involving chemical, medical or bacteriological processes, etc. it is important to accurately measure and control the acidity or alkalinity of the solutions. The degree of acidity and/or alkalinity of a solution is denoted by using a scale, termed as the pH scale. The pH value is based on the actual measurement of the hydrogen concentration in g/l of the solution. For example, the hydrogen ion concentration for pure water, which is a neutral solution at 25°C is 0.0 000 001 or 1.0 $\times$ 10^{-7} g/l. To avoid handling such small numbers, the pH scale is suitably defined in logarithmic form as follows, to obtain convenient numerical number:

$$\text{pH} = \log \frac{1}{h} \tag{17.14}$$

where h is the hydrogen ion concentration in g/l.

Now substituting the value of hydrogen ion concentration for neutral solution of pure water we get,

$$(\text{pH})_{\text{neutral solution}} = \log \frac{1}{(10^{-7})}$$

$$= \log (10)^7 = 7.0 \qquad (17.15)$$

It may be noted that acid solutions increase in strength as the pH value decreases from 7.0 to 0.00 and alkaline solutions increase in strength as the pH value increases in value above 7.0. Further, a change of pH of one unit represents a change of concentration by a ratio of 10. For example, a solution with a pH of 5.0 is 10 times more acidic than a solution with a pH of 6.0.

The practical range of the pH scale is between 0 and 14. A pH of 0.0 corresponds to an acid soultion of unit strength and a pH of 14 corresponds to a basic solution of unit strength.

17.5.1 pH Meter

The need for quantitative measurement of acidity or alkalinity, i.e., the pH value, has led to the development of special hydrogen ion detectors and associated circuitry. The most common industrial method of pH determination is by a glass electrode and a salt bridge connecting the calomel cell electrode (Fig. 17.21) to the sample solution whose pH is to be determined. In this, the glass electrode acts as a measuring electrode because the thin glass membrane allows the passage of hydrogen ions in the form of H_3O^+. Inside the glass bulb is a highly acidic buffer solution (usually pH = 1.0). This electrode, which forms an electrolytic half cell, operates on the principle that an electrolytic potential is observed between two solutions of different ion concentrations when they are separated by a thin glass membrane. Another half cell known as calomel (mercury and mercurous chloride) pH refernce cell consists of a glass bulb filled with a saturated solution of potassium chloride in contact with the measured solution through the salt bridge tube through a porous asbestos fibre plug. In some cases, a capillary is arranged to provide

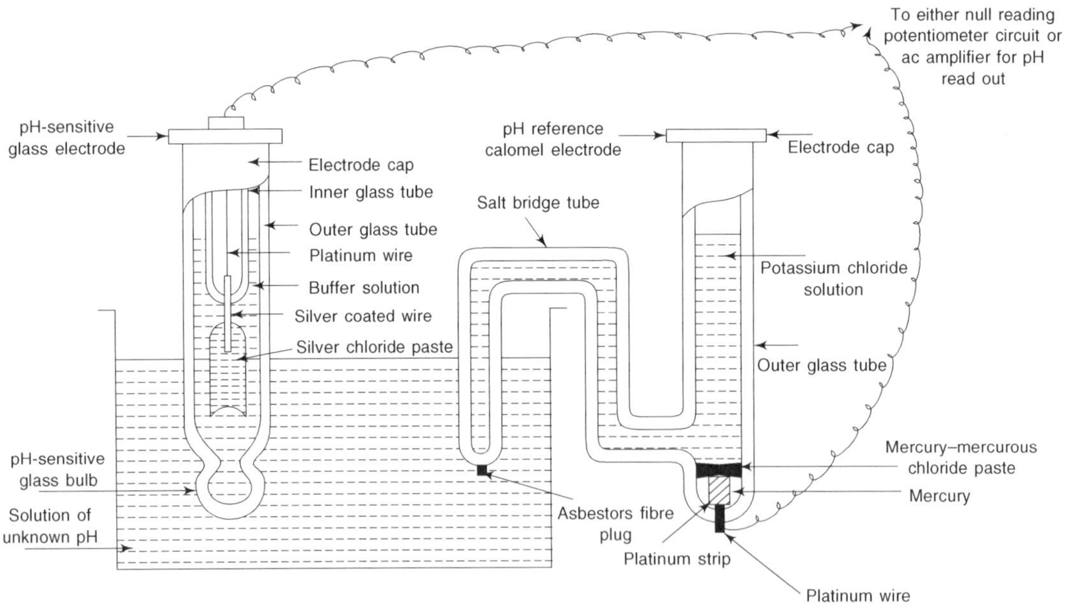

Fig. 17.21 *A typical pH measuring cell which includes pH measuring glass bulb half cell and calomel (mercury and mercurous chloride) pH reference half cell*

the liquid junction between the potassium chloride solution and the measured solution. The electrolyte (potassium chloride) is kept at a slightly higher pressure so that it slowly escapes from the electrode into the measured liquid. Further, the electrolytic potential at this reference electrode is constant.

The algebraic sum of the potentials of the two half cells is proportional to the concentration of hydrogen ions in the solution and this becomes the measure of pH value. Thus, the voltage produced by the two electrodes is applied at the input of a measuring instrument such as the null-balance millivolt potentiometer which incorporates a standard cell for calibration of the slide wire of the potentiometer. Alternatively, a high-impedance dc amplifier is used so that the measuring instrument does not draw appreciable current from the electrodes. This is because any current flow tends to cause polarisation of electrode which disturb the accuracy of pH measurement.

The pH-sensitive glass electrode produces a voltage change of 59.1 mV for 1.0 pH solution at 25°C. The composition of the inner cell solution of the glass electrode is so adjusted that for a pH value of 7.0 (neutral solution), the voltage of the glass electrode and the reference electrode is zero.

A glass electrode is quite adequate for pH measurements up to 9.0. For pH values more than 9.0, a negative static charge error is produced due to sodium or potassium ions present in the alkaline solutions. In such applications, it is necessary to select a glass electrode made up of a special type of glass. Further, these electrodes can operate in the temperature range 0–100°C at pressure from 1 to 10 atmospheres. Also, these electrodes are not affected by oxidation-reduction potentials. The accuracy of these electrodes is of the order of ±0.02 pH units.

Other measuring electrodes besides glass electrodes are not commonly used. The hydrogen electrode is a laboratory standard but is prone to contamination and requires a continuous supply of pure hydrogen. The quinhydrone electrode is limited in usefulness because it is suitable for a small range. Further, the antimony electrode has the same advantages but solutions containing copper, silver, mercury or lead ions poison the electrode resulting in serious errors.

17.6 ■ BIOMEDICAL MEASUREMENTS/BIOMETRICS

The use of engineering measurement techniques have made substantial contributions in the accurate determinations of various physiological parameters, that give useful information to clinicians to detect the malfunction of the system being measured and also aid in proper diagnosis and patient care. Some of the biomedical instruments may be designed and developed for a particular specialised requirement but a majority of them are adaptations of the widely used physical measuring devices. For example, the electrical resistance of a thermistor changes with temperature, regardless of whether the temperature of an engine or a human body is being measured. The principle of the device is the same. Only the shape/size of the device may be different. For monitoring the temperature of biomedical subjects, a thermistor device is generally used along with a suitable bridge circuit in which the terminal device may be a voltmeter, calibrated in terms of degrees Celsius or degrees Fahrenheit. Similarly, a strain gauge is commonly used for measuring stresses in the various structural components. Its principle of operation is that the electrical resistance is changed by the stretching of a wire of semiconductor material of which it is made. With suitable bridge configuration, an electrical output that is proportional to the amount of strain can be obtained. Since the pressure can also be translated into strain by various means, therefore blood pressure can be determined by suitable adaptation of this device. Some of the commonly used biomedical instruments are described here.

17.6.1 Blood Pressure Measurement

Blood pressure is one of the most important physiological variables and is considered as one of the important parameters of the state of the cardio-vascular system. Dangerously high blood pressure (hypertension) and extremely low blood pressure (hypotension) indications helps the doctors to treat their patients suitably to avoid their untimely death. Blood pressure measurements are carried out by the following methods.

Indirect Method The indirect method routinely used for measuring blood pressure uses a device known as the sphygmomanometer (*Sphygmos* is a Greek word meaning pulse). The instrument consists of an inflatable rubber bladder enclosed in an inelastic fabric called the cuff, a rubber squeeze pump-and-valve assembly and a manometer. The manometer is either mercury column type or an aneroid type pressure measuring assembly similar to the Bourdon gauge.

The cuff is wrapped and secured with either hooks or a velcro fastener, around the patient's upper arm at a point mid-way between the elbow and the shoulder. The stethoscope is placed over the branchial artery at the elbow level where it is close to the skin surface [Fig. 17.22(a)]. The cuff is normally inflated manually with a rubber bulb so that the pressure inside the inflated bladder is more than arterial pressure. This pressure compresses the artery against the underlying bone and causes a complete stoppage of blood flow in the blood vessel (artery). The pressure in the cuff is gradually reduced by operating the needle type pressure release valve so that there is gradual fall of mercury in the manometer, of the order of 3 mm/s (which is usually considered best).

When the systolic pressure in the pulsatile blood pressure pulse [Fig. 17.22(b)] first just exceeds the cuff pressure, the branchial artery takes the shape of a venturimeter with extremely small throat area. Turbulence is generated as the blood flow experiences a sudden expansion downstream of the tiny throat opening of the venturi. The sounds generated by the turbulence are known as Korotkoff sounds and these can be heard through the stethoscope placed over the branchial artery downstream of the cuff. The pressure of the cuff that is indicated on the manometer when the first Korotkoff sound is heard is known as systolic blood pressure.

As the cuff pressure continues to drop, the throat area of the venturi formed in artery also starts increasing and this severity of sudden expansion downstream of the throat continues to decrease. Consequently, the turbulence level down stream of the venturi and hence the Korotkoff sounds start becoming less loud and muffled. Ultimately, these sounds disappear when the cuff pressure drops just below the diastolic pressure. In short, to read the blood pressure one has to note the gauge pressure at the onset of Korotkoff sounds (systolic blood pressure) and also when the sound just disappears altogether (diastolic blood pressure). These pressure are usually expressed in the form of ratio of systolic pressure over diastolic pressure (i.e., 120/80).

The routinely used method of locating the systolic and diastolic pressure values by listening to the Korotkoff sounds is called the *auscultation* method. This method is quite simple in operation and is considered accurate enough for ordinary clinical use.

An alternative technique called the palpation method for the blood pressure measurement, uses the sense of touch to detect the patient's pulse in the radial artery at the wrist level. The cuff is inflated until the palpation activity of the radial pulse disappears due to stoppage of blood flow in the branchial artery. The pressure in the cuff is slowly released and when the pulse becomes palpable, then the systolic value is recorded. Palpation cannot detect the diastolic pressure because there is no known palpable change at that pressure value.

Both ausculation and palpation method are subjective and often fail when the blood pressure is very low (i.e. when the patient is in a state of 'shock'). Therefore, for the measurement of low blood pressures, the stethoscope is usually replaced by a microphone placed beneath the cuff over the branchial artery. The output of the microphone is suitably conditioned to operate a beeper or lamp when the systolic and diastolic features are recognised. Such instruments are useful in the emergency rooms, intensive care

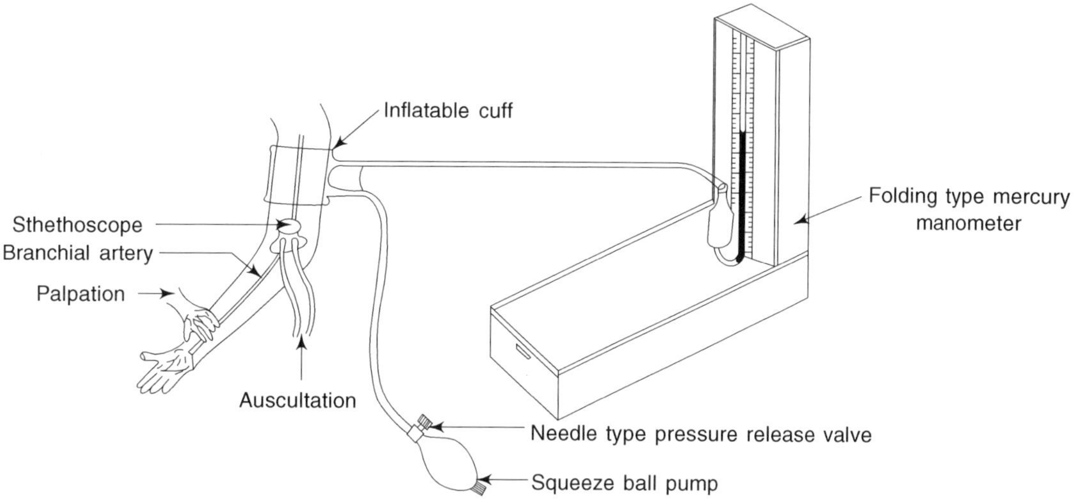

Fig. 17.22(a) *Schematic of arterial blood pressure measurement using auscultation and palpation method*

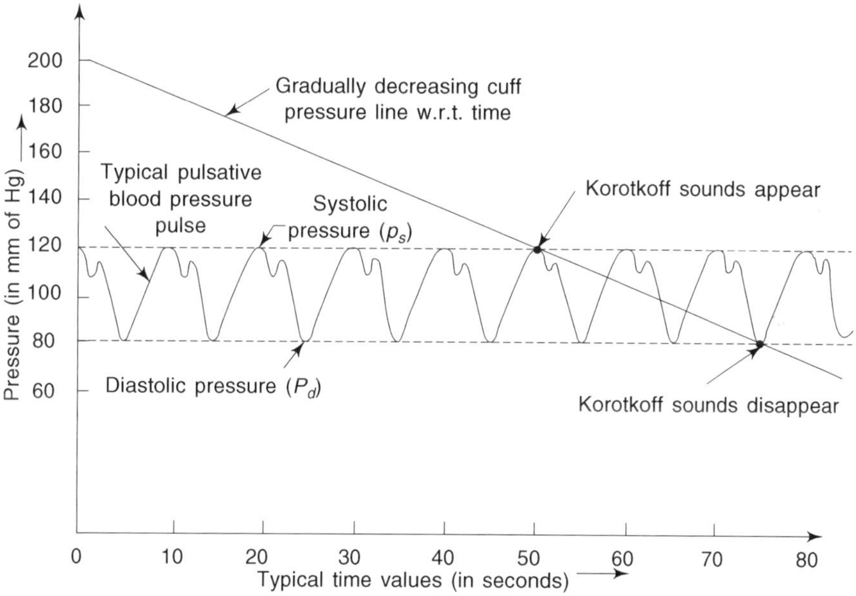

Fig. 17.22(b) *Typical variations of cuff pressure of sphygmomonometer and blood pressure which produce Korotkoff sounds*

units (ICU) and coronary care units (CCU) of hospitals where ambient high noise levels tend to obscure the perception of Korotkoff sounds.

These days semi-automatic sphymomanometers have been developed. In such instruments, the pressure cuff is automatically inflated up to about 200 mm and is allowed to deflate slowly at the desired rate of about 3 mm/s [Fig. 17.22(b)]. Further, the pressure transducer is incorporated in the sphygmomanometer cuff and a microphone beneath the cuff. The pressure and the Korotkoff sound signals are recorded simultaneously verus time on a two-channel recorder. From these records, the values of systolic and diastolic pressures are interpreted.

Sphymomanometry is a simple and useful noninvasive (i.e. does not involve any traumatic surgical operation for insertion of transducer/catheter, etc.) technique. Further, it is quite accurate for ordinary cinical use. This is because a variation of less than 10 mm exists between the present indirect method and the direct method of blood pressure measurement. However, only systolic and diastolic arterial pressure readings can be obtained, with no indication of the shape of the waveform.

Direct Method This method is usually of invasive type and gives a continuous and permanent record of the blood pressure variations at any desired site of measurement in the blood stream (e.g. in the aorta). The pressure measurements are usually made by means of a suitable pressure transducer connected to the patient through a thin piece of tubing called the *catheter*. The pressure measuring systems used should have good frequency response so as to faithfully reproduce the pulsatile nature of blood pressure variations between the maximum value systolic pressure P_s and lowest value diastolic pressure P_d. The pressure transducer is sometimes made of miniature size so as to be accommodated on the arterial catheter tip. Alternatively, sterile saline solution is first introduced into the catheter so that fluid pressure is transmitted to the transducer outside the body (i.e. extracorporeal procedure). The catheter end usually terminates on a specified terminal of a three-way stop cock which is either secured to the patient's body or to a convenient support (Fig. 17.23) near his bedside. The second terminal of the stop cock is connected to a syringe that is used to administer medicines directly to the patient through the catheter or to withdraw arterial blood samples for laboratory analysis. The third terminal is connected to the transducer which is usually a diaphragm pressure gauge. Some commercially available instruments use strain gagues, whereas others incorporate the linear variable differential change to transducer as secondary displacement transducer to the diaphragm which converts the pressures to the corresponding displacements.

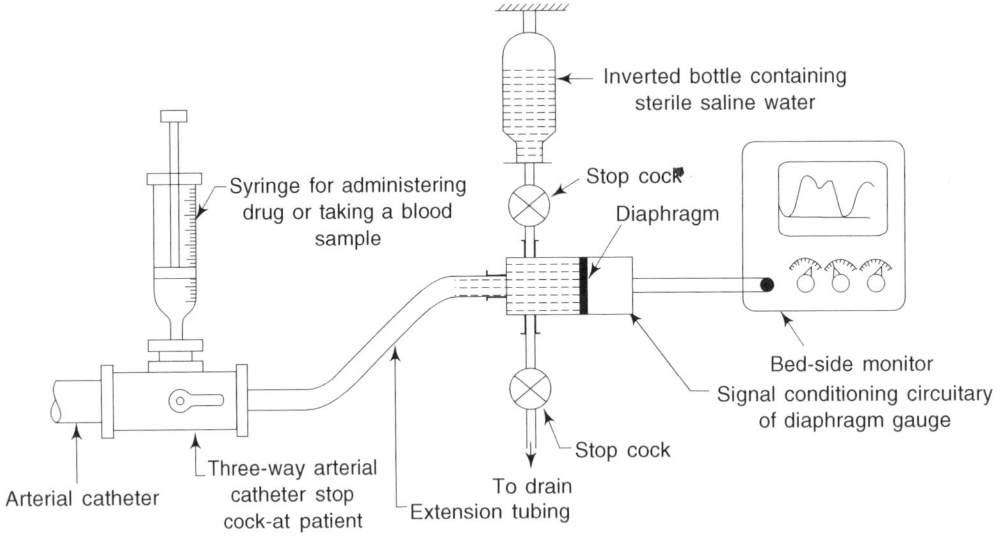

Fig. 17.23 *A schematic layout of direct blood pressure measurement system*

Currently, in such instruments a very sensitive type of transducer namely bonded silicon element bridge is being employed. The transducer is available in the form of a chip which is directly used as the pressure sensing diaphragm giving a sizeable output voltage for extremely small displacements of the order of a few micrometers. For example, even with a low voltage excitation of 10 V, a range of 300 mm of Hg is obtained with 3 μm deflection, producing 30 mV signal.

It may be noted that in certain applications, the mean arterial pressure (MAP) may be required. The value of MAP can be obtained by integrating the pressure wave form by using a suitable integrating circuit. However, the empirical relation commonly used in practice is as follows:

$$\text{MAP} = P_d + \frac{1}{3}\,[P_s - P_d] \tag{17.16}$$

17.6.2 Measurements of Bioelectric Potentials

The electrochemical activity in certain types of body cells gives rise to ionic voltages usually termed as *bioelectric potentials.* Generally, such cells in the resting state have a lower base value of ionic potential, the *resting potential.* Further, on account of some applied stimulus, the cells becomes excited and the ionic potential rises to a certain higher value. The magnitude of the bioelectric potential in the excited state is designated as the action potential. Now, a cell that has been excited and displays action potential is said to be *depolarised.* In other words, the process of changing the resting potential state to the action potential state due to an applied stimulus is termed depolarisation. Similarly, the reverse process of returning from the action potential state to the resting potential state due to the removal of stimulus is known as *repolarisation.* Further, when the action potential is generated, there is a brief period of time during which the cell does not respond to any stimulus. This period is called the *refractory period* and is of the order of 1 ms in a typical nerve fibre.

Certain systems in the body generate the bioelectric potentials and give their own characteristic depolarisations and repolarisation wave forms which convey useful information about the function they represent. Through the use of suitable transducers (body electrodes) and other associated circuitry, these natural monitoring bioelectric potentials are converted into corresponding electrical voltages which are measured/recorded. In fact, the study of the bioelectric events is nowadays being extensively employed by the clinicians in evaluating/quantifying the nature and degree of various diseases associated with nerve condition, brain activity, heart functions, muscle activity and so on.

One of the routinely used techniques for diagnosing various diseases and conditions associated with the heart is by means of an electro-cardiogram abbreviated as ECG (sometimes EKG, from the German spelling electro-kardiogram). ECG is a graphic recording or display of the electrical activity of the heart muscle myocardium, which is represented in the form of time varying voltages in the cardiac cycle. In standard ECG recording there are five electrodes connected to the patient. There are two electordes on the upper extremities, i.e. right arm (RA) and left arm (LA); two electrodes on the lower extremities, i.e. right leg (RL) and left leg (LL) and one electrode on the chest (C). These electrodes are connected to the differential bioelectric amplifier through a lead selector switch. The recording obtained across different pairs of electrodes (commonly called leads) give different waveform shapes and amplitudes. Each lead conveys a certain amount of unique information that is not available in other leads.

Figure 17.24 shows a typical waveform of lead *I* which is between the left and right arms. The different features of ECG wave form represent rhythmic electrical depolarisation and repolarisation of myocardium associated with the contractions of auricles and ventricles and are designated by a letter system. The *P*-wave indicates auricle contraction while ventricle contraction occurs immediately following the P–R interval represented by the QRS complex and a refractory period (resting for repolarisation) is indicated by the *T*-wave. QRS complex is considered the most important part of the ECG as it represents the depolarisation of both the ventricles. It has been observed that the time duration of ECG features is generally of the same order over a wide range of heart rates. For example, the QRS complex requires approximately 90 ms, *P–R* interval is roughly 150–200 ms and *S–T* interval is about 50–150 ms.

Looking at the ECG, the clinician would typically look at the heart rate. The normal value lies between 60 and 100 beats/min. Rates faster than this (i.e. 100 beats/min) are called *tachecardia* (fast heart). The

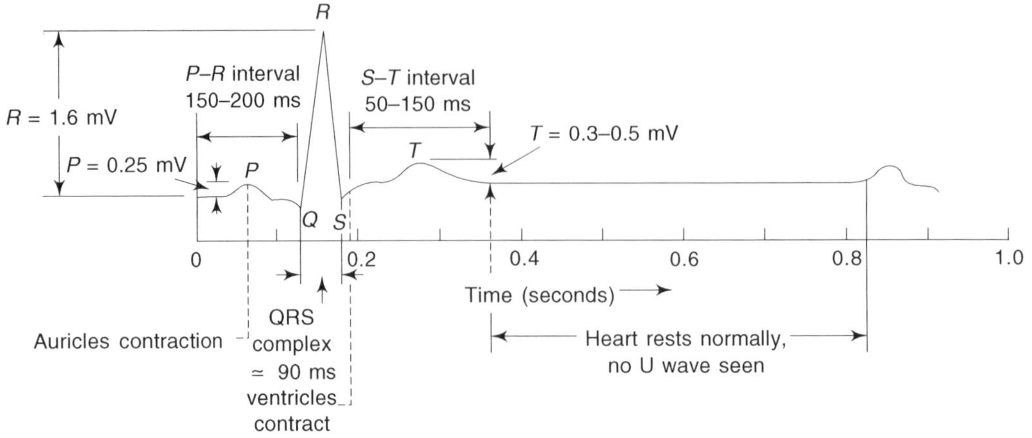

Fig. 17.24 *A typical ECG waveform showing the time and amplitudes in various segments*

opposite condition, i.e. too slow a heart rate of less than 60 beats/min is called *bradycardia*. Further, he would see if the cycles are evenly spaced. If not, an *arrhythmia* may be indicated. Arrhythmia may be caused due to premature auricle contraction, premature ventricle contractions, heart block, etc.

It should be noted that the electrocardiograph (device for recording ECG) should be capable of measuring accurately electrical potentials in the range of 0.1–4 mV (peak amplitude and should be designed to have good frequency response in the range of dc to 100 Hz.

The other physiological measurements based on the bioelectric phenomenon are as follows:

Electro-encephalogrom (EEG) EEG represents the record of electrical activity of the brain which is usually in the range of 10–100 μV Peak amplitude with band width range from dc to 100 Hz. EEG helps to detect brain disorders such as cerebral brain lesions (from the irregularity/asymmetry of EEG tracings), epilepsy (recurrent attacks of disturbed brain function resulting in convulsions, etc.) and other mental diseases, etc. The bioelectrical potentials are usually measured with surface electrodes on the scalp. But, sometimes needle electrodes are also used for recording the brain activity at specific locations.

Electro-myogram (EMG) EMG represents a record of the electrical activity of the skeletal muslces, which is usually in the range of 50 μV to 1 mV peak amplitude with bandwidth range from 10 to 3000 Hz. EMG is used to determine muscle action, muscle fatigue or to detect muscle diseases such as paralysis of muscle/muscle injury, etc.

Electro-retinogram (ERG) ERG represents the record of potential from the retina. It is used to detect retinal activity with variables like colour, duration and intensity of light, effects of drugs/anesthesia, etc. Further, it is also used to diagnose retinal detachment, chloroid vision, etc.

Electro-oculogram (EOG) EOG represents the record of corneal-retinal potentials associated with the eye movements. It is used to study the effect of intensity of light on the eye movement to determine the nature and degree of optical nerve injury, etc.

Electro-gastrogram (EGM) EGM represents the record to potentials associated with muscle activity in the gastro-intestinal tract. It is a useful tool for the clinicians to determine the lesions/injuries of the various gastro-intestinal muscles.

17.7 ■ MEASUREMENT OF ENVIRONMENTAL AIR POLLUTION PARAMETERS

Environmental air pollution can be defined as the addition to our atmosphere of any material(s) having a harmful effect on our lives. Typical gaseous pollutants include gases such as carbon monoxide (CO), carbon dioxide (CO_2): compounds of sulphur like sulphur dioxide (SO_2); hydrogen sulphide (H_2S); oxides of nitrogen like nitrous oxide (N_2O), nitric oxide (NO), nitrogen dioxide (NO_2); ozone (O_3); hydrocarbons, organic para-oxides, aldehydes, organic acids and other partially oxidised organic compounds including nitrogen containing esters such as peroxyacylnitrates (PANs). In addition, the other pollutants are vapours of water and other solvents present in smoke; aerosols such as inorganic sulphates, nitrates, chloride and ammonium salts and particulates which are small particles usually of solids such as atmospheric dust, coal dust, fly ash, insecticide dust, pollen, metallic foundry dust, milled flour, etc.

Air pollutants can be classfied as follows:

1. *A primary pollutant* It is an emission of high concentration which originates from the source. It is usually lethal in nature.
2. *A secondary pollutant* It is an emission of considerably lower concentration which is formed through the reaction of primary pollutants. This reaction can occur at the emission point or at a remote location.

Air pollution is mainly caused by the exhaust emissions of various process industries, combustion type vehicles, thermal power stations, domestic heating, forest and agricultural fires, etc. The various components of air pollution can have quite a harmful effect on human health. For example, CO is known to cause heart disease. SO_2, NO, NO_2 and the particulates are irritants to breathing and create discomfort to the eyes. Further, inhalation of O_3 in large quantities may lead to a severe headache.

The quantities of pollutants can be expressed either on a volumetric or mass basis. The commonly used volumetric unit for gaseous pollutants is parts per million which is abbreviated as ppm and is defined as:

$$1 \text{ ppm} = \frac{1 \text{ volume of gaseous pollutant}}{10^6 \text{ volumes of air (containing pollutants)}}$$

$$= 0.0001 \text{ per cent by volume} \qquad (17.17)$$

Alternatively, the amount of pollutant in the air can be expressed in mass units, i.e. in the form of $\mu g/m^3$. Further, it is possible to convert the quantities in ppm to $\mu g/m^3$ knowing the molecular weight of the particular gaseous pollutant and assuming the ideal gas equation to hold good at the ambient temperature and pressure conditions.

The following are the commonly used devices empolyed in the environmental air pollution studies.

17.7.1 Orsat Apparatus for Exhaust Gas Analysis

The analysis of boiler flue gases or the emissions of power generating combustion type engines is important not only to the air pollution studies but also to optimising the combustion phenomenon for obtaining the most efficient burning rates resulting in better energy utilisation.

A simple and commonly used laboratory device, namely Orsat apparatus shown in Fig. 17.25, is employed to analyse the products of combustion, i.e., to determine the percentage of CO_2, O_2, CO and N_2. A sample of the products of combustion is taken into the measuring burrette using a hand-operated or hydraulic aspirator or a small electric pump. The sampling inlet valve is shut off and the sample is forced into the first reagent pipette containing caustic potash (KOH) solution to absorb only CO_2. The sample is then brought back to the measuring burette and the decrease in volume from the original volume is recorded to determine the percentage of CO_2. This procedure is successively repeated with the other two reagent pipettes containing pyrogallic acid and cuprous chloride solutions for the absorption of O_2 and

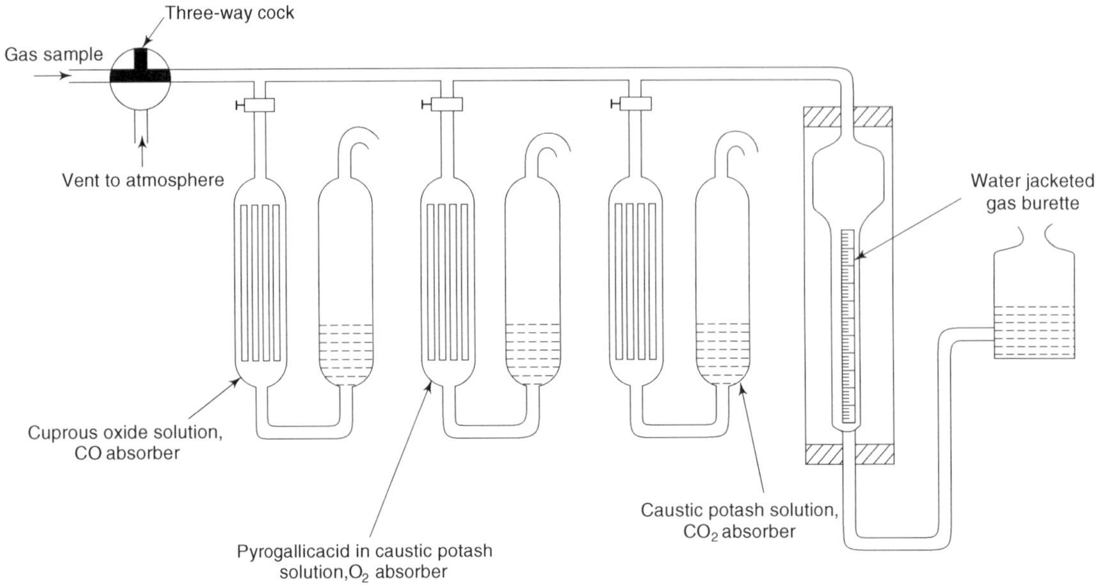

Fig. 17.25 *Orsat gas analysis apparatus*

CO, respectively. Finally, the proportion of N_2 is determined by difference. From the volumetric analysis, the dry molecular weight of the sampled gas can be determined as follows:

$$M_D = 0.44\ (\%\text{ of } CO_2) + 0.25\ (\%\text{ of } CO) + 0.32\ (\%\text{ of } O_2) + 0.28\ (\%\text{ of } N_2) \quad (17.18)$$

where M_D = dry molecular weight of the sampled gas.

17.7.2 Gas Chromatography

Gas chromatograph technique is probably the best method for *continuous sampling simulation* for the volumetric analysing of the products of combustion and other gas samples as well. The term 'continuous sampling simulation' has been used because the chromatorgraphy technique involves sampling and selective separation of the various componetns of volumetric determinations.

The basic block diagram of the gas chromatograph is shown in Fig. 17.26. It consists of four major components, namely sample introduction unit, the fractionating column, a detector and a recording device. The fractionating column is usually considered as the heart of the system. It is made of a copper or stainless steel tube 0.25–50 mm in diameter and 1–20 m in length. The column is usually packed with a fine mesh of solid adsorbing materials such as charcoal, granular silica jel or activated alumina. Alternatively, it may be filled with a suitable adsorbant liquid depending on the particular type of gas component to be sampled.

The sample gas for analysis in a gas chromatograph is usually kept small. It is typically collected in a small glass syringe and introduced into the chromatographic column so that its various components are selectively adsorbed by the suitable solid/liquid adsorbants. After this, an inert carrier gas, usually, ultra-pure nitrogen or helium, fed at constant flow rate transports the sample through the column to the detector. Different components of the sample gas are retained in the column for different lengths of time so that the dectector senses different components at different times. The device usually is calibrated for the retention times of the various components present in the sampled gas. Chromatographic devices are equipped with suitable closely controlled heating or cooling systems to achieve proper separation times for different components.

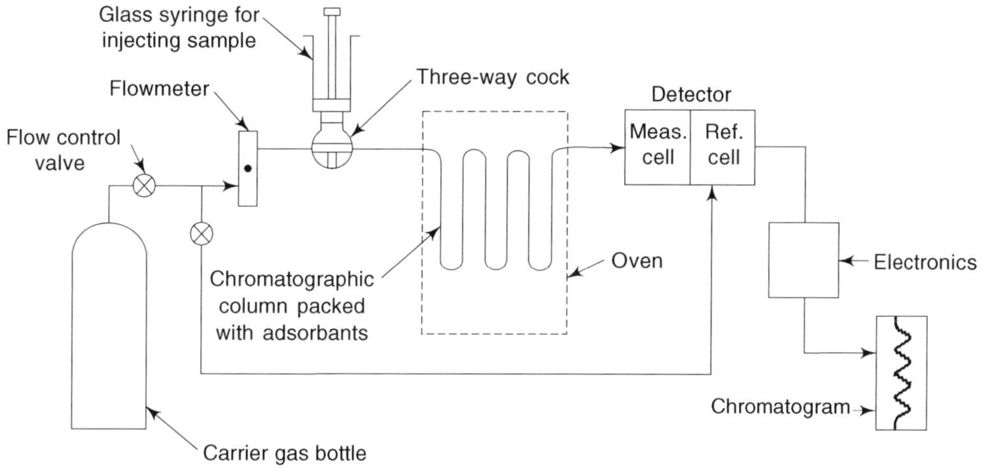

Fig. 17.26 *Block diagram of gas chromatograph*

Detectors used in the gas chromatograph are of several types. Differential thermal conductivity cells are quite popular as they are simple in construction, relatively inexpensive, sensitive and stable. They contain two cells of either matched resistance wires or thermistors arranged in the Wheatstone bridge circuit. The principle of thermal conductivity detectors depends on the cooling effect of gas flowing over a heated element. There is a different cooling effect when a constituent of the sampled gas is present in the carrier gas than when the carrier gas has no constituent in it. By passing pure carrier gas through one cell and the carrier gas plus the constituent in the other cell, a differential signal proportional to the amount/concentration of the constituent is generated. Other types of detectors are hydrogen flame ionisation detectors, argon ionisation detectors, etc. In addition, special detectors are available for specific elements.

The output of the detector is processed electronically and displayed on a chart recorder. The display indicates the concentration curve or peak of each component of the mixture. The qualitative identification of a particular component is made by the retention time which is the time elapsed from injection of the sample to detection. The quantitative or concentration measurement is based on the height of the peak.

Chromatographic units are available for 100 or more constituents in the mixtures with minimum sensitivities of the order of 0.005 ppm within the accuracy limits of ±2% or so.

17.7.3 Non-dispersive Infrared and Non-dispersive Ultraviolet Gas Analysers

Non-dispersive infrared (IR) gas analyser (also known as infrared absorption spectrometer) is shown schematically in Fig. 17.27. It is used to determine the concentration of gaseous pollutants in the test gas samples. The principle of operation of the instrument is based on the absorption of IR radiations in narrow wavelength bands, with each gas exhibiting its own peculiar characteristics. For example, each hydrocarbon has a characteristics absorption spectrum, so that the graph of wavelength versus percentage absorption (transmission) enables the hydrocarbon to be identified. The fundamental law which governs the relationship between the various variables is known as Beer's law and is as follows:

$$C = \frac{1}{ax} \log (I_0/I_x) \tag{17.19}$$

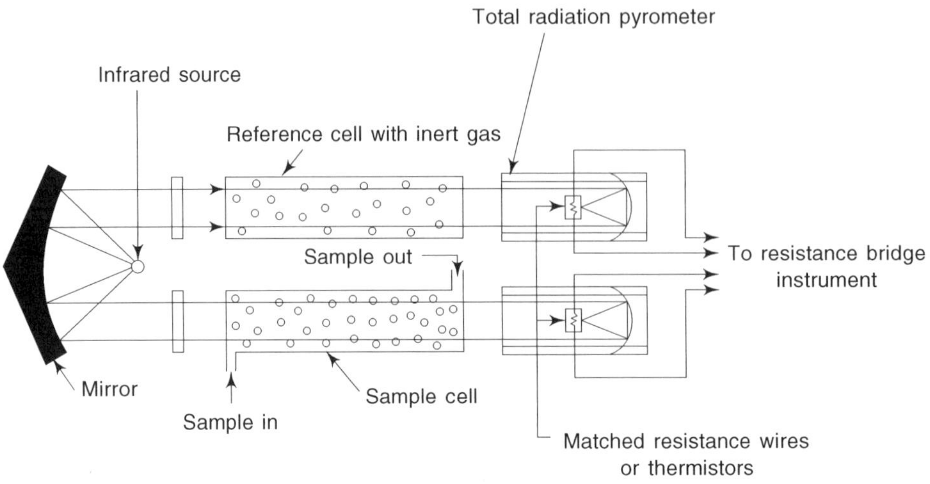

Fig. 17.27 *Schematic of non-dispersive infrared gas analyser*

where C = concentration of substance
a = absorption factor of substance
x = thickness of sample (along the optical path)
I_0 = intensity of beam before sample
I_x = intensity of beam after sample

In operation, different types of filters the placed in the path of infrared rays so that only the absorption wavelength band for the particular constituents in the gas sample to be studied are successively maintained. These are then passed through the reference cell containing an inert gas (usually nitrogen) as well as the sample cell which admits the passage of sample gas containing the pollutants under investigation. No IR absorption takes place in the reference cell, while the absorption in the sample cell is proportional to the number of molecules per unit volume (concentration) of the component of interest.

The transmitted IR radiations of the reference and sample cells are detected by two total radiation pyrometers containing either matched resistance wires or thermistors. The absorption of IR radiations causes the temperature of the resistance wires/thermistors to rise. This is sensed by a suitable bridge circuit and is further amplified to give a suitable display on an appropriate read out/recording device.

Infrared spectrometry is useful in the qualitative and quantitative determination of a number of organic gases and liquids. Further, this type of device has in general higher minimum sensitivity than the chromatograph system. However, certain gases like oxygen, nitrogen, hydrogen, chlorine and all other elemental diatomic gases as well as the inert gases do not absorb the infrared radiations and therefore cannot be studied using this method.

The oxides of nitrogen i.e., NO and NO_2 do not have the absorption spectrum in the IR range. Therefore, IR lamp in the spectrophotometer is replaced by a UV lamp and then with suitable filter, we can detect the oxides of nitrogen by using NDUV method.

17.7.4 Smoke Density Measurement

The discharge of large amounts of thick fumes and heavy smoke in the environment of a typical posh city (with a lot of automobiles) or an industrial township destroys the natural transparancy of clean atmospheric air. This phenomenon is commonly known as generation of *smog* which means accumulation of vapours (such as fog) aerosols (particles of very fine size) and particulates such as dirt, dust, bacteria and other air-borne solids. The presence of smog is evidenced by decreased visibility on a typically still day while driving on a highway.

The simplest method of smoke density measurement is based on the degree of opacity caused by the fumes/smoke issuing from a stack. It is usually determined by visual comparison of the discharge gas stream with the Ringleman scale. This scale consists of a set of six Ringleman smoke charts ranging from white (0) to increasing darker shades with even increments to grey shades and finally to complete black (5). Since the method of comparison is based on manual observation, therefore, it is not suitable for continuous measurements.

However, it is possible to determine the opacity of smoke within $\pm 1/2$ Ringleman number by an experienced 'smoke reader'.

For large plants, an electronic visiometer is usually employed for continuous sampling of smoke density of the exhaust stack. The simplest system consists of a lamp and a photo cell located on opposite sides of the stack. As smoke issues from the stack, it obstructs light proportional to its density. The intensity of the light incident on the photo cell produces a proportional electrical signal which can be recorded or used to operate an alarm device.

Alternatively, the sample of stack is passed through a standardised glass container. It is then illuminated by a pulsed xenon lamp. The light scattered by the sample volume is proportional to its smoke density which is measured by a photo-multiplier tube and associated solid state electronic circuitry. This system is quite versatile and can be made to interface with any type of data handling equipment using the conventional voltage outputs to drive a recorder or data logger, etc.

Review Questions

17.1 Indicate if the following statements are true or false. In case of false statements write the correct statement.

(i) In the nucleonic liquid level/density measuring gauge, the instrument output is directly proportional to the measurement (i.e. liquid level/density).

(ii) Capillary tube viscometer can be used as a flow metering device if the viscosity of the liquid is known.

(iii) The accuracy of the bubbler system for measuring liquid level/density is improved if the rate of feed to the bubbler pipe is kept constant.

(iv) Ultrasonic method for liquid level measurement is usually not preferred in industrial applications, even though it is very accurate.

(v) Dip-stick method of level indication is a popular technique in industrial applications because it is simple and inexpensive.

(vi) Dew point temperature and wet bulb temperature are always same.

(vii) pH value is defined as the logarithm to the base 10 of the reciprocal of hydrogen ion concentration in grams per litre in a given solution.

(viii) In a pH meter, the glass electrode is the reference electrode.

(ix) Chromatography is a term used for separating and identifying the components in a given mixture. The name implies that colour is involved in detection but it is not mandatory.

(x) Hydrocarbons can be detected quantitatively by infrared spectrometry but it is not possible even to sense the presence of hydrogen by the same technique.

17.2 Indicate the correct statement in the following:

(i) The calibration curve of the dielectric liquid level gauge is usually
(a) continuously linear (b) continuously non-linear
(c) partly linear and partly non-linear (d) none of the above

(ii) The general equation with usual notation for efflux type of viscometers is
(a) $v = k_1 t + k_2/t$ (b) $t = k_1/v + k_2 v$ (c) $v = k_1 e^t + k_2 e^{-t}$ (d) $v = e^{k_1} + e^{k_2/t}$

(iii) The relationship between kinematic viscosity (v) and dynamic viscosity (μ) is
(a) v (centistokes) = μ (centipoise)/ρ (g/cm^3)
(b) μ (centistokes) = v (centipoise)/ρ (g/cm^3)
(c) μ (centipoise) = v (centistokes)/ρ (kg/m^3)
(d) v (stokes) = μ (poise) × ρ (kg/m^3).

(iv) The instrument which is unsuitable for continuous viscosity measurements is
(a) variable area viscometer (b) rotating cylinder viscometer
(c) efflux type viscometer (d) none of the above

(v) Two fluids A and B are flowing in two indentical pipes at the same Reynolds number (in the laminar range). The densities of A and B are same but the pressure drop per unit length is 4 times in the case of A than B. The ratio of viscosities of A and B is
(a) 1/4 (b) 4 (c) 1/2 (d) 2

(vi) The temperature to which a vapour gas mixture must be cooled (at varying humidity) to saturate it is
(a) dew point (b) wet bulb temperature
(c) atmospheric temperature (d) none of the above

(vii) The pH value of a solution having a hydrogen ion concentration of 2.3×10^{-11} g/l is
(a) acidic in nature (b) alkaline in nature
(c) neutral solution (d) none of the above

(viii) The reference electrode in pH measurements is
(a) glass electrode (b) hydrogen electrode
(c) antimony electrode (d) Hg-calomel electrode

(ix) In infrared absorption spectrometry, the concentration of substance is proportional to
(a) ratio of intensities of beams without the sample (I_0) and after it passes through the sample (I_x)
(b) square of the ratio of (a)
(c) log of the ratio of (a)
(d) in of the ratio of (a)

(x) The production of Korotkoff sounds in the indirect blood pressure measurement is due to
(a) pressure pulses produced due to the difference in systolic and diastolic blood pressures
(b) turbulence and eddying flows downstream of throat of the venturi-shaped branchial artery which is formed due to cuff pressure
(c) heart sounds transmitted in the blood flow
(d) pulsatile nature of blood flow

(xi) ECG represents
(a) the rhythmic curve of depolarisation and repolarisation of the myocardium, the heart muscle
(b) the waveform of the voltage produced in the heart muscle due to an externally applied electrical stimulus
(c) the record of electrical currents flowing in the cardiac muscle
(d) the record of intensity of heart sounds at different intervals sensed by an electro-acoustic sensor.

(xii) Orsat apparatus is used for
(a) gravimetric analysis of flue gas
(b) volumetric analysis of flue gas
(c) mass flow of flue gas
(d) measuring smoke density of a flue gas

(xiii) The carrier gas employed in gas chromatography is
(a) methane (b) oxygen (c) ammonia (d) nitrogen

(xiv) Non-dispersive infrared spectrometry is suitable for detecting
(a) carbon monoxide (b) oxygen (c) helium (d) nitrogen

(xv) A chromatograph is used for
(a) measuring flow rate of gas (b) measuring the temperature of gas
(c) analysing the composition of a gas (d) measuring the pressure of a gas

(xvi) Environmental pollution is not caused by
(a) air pollution (b) water pollution
(c) noise pollution (d) radioactive fall out
(e) high humidity content in atmospheric air.

(xvii) When the reading of pH meter changes from 7 to 9, the hydrogen ion concentration of the solution
(a) is halved (b) is doubled
(c) increases 100 times (d) decreases 100 times

(xviii) A capillary tube viscometer with known dimensions, is used for measuring the viscosity of oil. In order to obtain viscosity, it is necessary and sufficient to measure
(a) volume of fluid collected in a given time
(b) pressure drop across the capillary
(c) both (i) and (ii)
(d) not only (i) and (ii) but also the flow should be in the laminar range of Reynold number.

(xix) The oxides of nitrogen in the automobile emissions can be detected by
(a) Orsat analysis (b) Flame ionisation detector
(c) Non-dispersive ultraviolet technique (d) Non-dispersive infrared technique

(xx) Korotkoff sounds are used
(a) as a reference for sound level measurements
(b) for studying heart muscle functioning
(c) for blood pressure measurements
(d) for studying heart valve functioning

17.3 Fill in the blanks in the following.

(i) In the aircraft, the fuel tanks are located in the hollow spaces of wings and the fuel level gauge is of ________ type.

(ii) The presence of hydrogen ions in water at 25°C is experimentally found to be ________ g/l and pH value corresponds to ________.

(iii) In the capillary tube viscometer, the type of flow in the capillary should invariably be ______.

(iv) The volumetric unit of pollutant measurement is ________ whereas the corresponding gravimetric unit is ________.

(v) In the measurement of pH, a reference electrode made of ________ is used.

(vi) Flame ionisation detector in the gas chromatography is used for the detection of ________.

(vii) The viscometer which gives direct and readable indication of the viscosity is ________.

(viii) The practical range of scale of pH meter varies from ________ to ________.

(ix) Orsat gas analyser used for __________ of the products of combustion.

(x) The instrument which gives on-line indication of concentration of CO in the emissions of IC engine is __________.

17.4 In the air bubble level gauge, the bubble tube was immersed 650 mm below the water surface. If the air pressure was measured by means of a Bourdon gauge, what would be the pressure indicated by the gauge in kN/m^2.

17.5 The viscosity of an oil of specific gravity 0.90 was measured with a capillary tube viscometer consisting of a glass pipe of an accurate bore of 10 mm with a length of 0.6 m under a head of 2 m of oil. A quantity of 840 cm^3 was found to flow in 60 s. Calculate the dynamic viscosity of oil in poise. Further, check if the flow is laminar.

17.6 The viscosity of a polymer solution was measured using the rotating cylinder viscometer in which the inner cylinder of diameter 20 cm was stationary. The outer cylinder of diameter 20.4 cm contained the polymer solution up to a height of 40 cm. The clearance at the bottom of two cylinders was 4 mm. The outer cylinder was rotated at 500 rpm and the torque registered by the torsion meter attached to the inner cylinder was found to be 9.8 Nm. Determine the viscosity of the liquid under test in poise.

Answers

17.1	(i) False	(ii) True	(iii) True	(iv) True	(v) False
	(vi) False	(vii) True	(viii) False	(ix) True	(x) True.
17.2	(i) b	(ii) (a)	(iii) (a)	(iv) (c)	(v) (b)
	(vi) (a)	(vii) (b)	(viii) (d)	(ix) (c)	(x) (b)
	(xi) (a)	(xii) (b)	(xiii) (d)	(xiv) (a)	(xv) (c)
	(xvi) (e)	(xvii) (d)	(xviii) (d)	(xix) (c)	(xx) (c)

17.3 (i) dielectric level gauge (ii) 10^{-7}g/l, 7
(iii) laminar (iv) ppm, $\mu g/m^3$
(v) calomel cell (vi) hydrocarbons
(vii) rotating cylinder viscometer (viiii) 0 to 14
(ix) volumetric analysis (x) Non-dispersive infrared technique

17.4 6.377 kN/m^2

17.5 Coefficient of dynamic viscosity μ = 5.16 poise, Reynold's number = 3.10

17.6 μ = 1.43 poise

Chapter

18

Recent Developments in Instrumentation and Measurements

■ *INTRODUCTION* ■

The recent developments in instrumentation and measurements are based on the use of digital computers and development of new types of sensors. The data acquisition using computer-based systems, storage, analysis and processing of data is now widely used. The development of silicon microsensors and inclusion of microcontrollers on a single chip are the recent developments. The transducers are tending to become smart or intelligent with compensation, linearisation self-test, calibration, etc. being carried out using the in-built intelligent microprocessors. Further developments include communication of large amount of data from several sensors and actuators using the field bus, reducing the cost of wiring and making the data available at several levels in a distributed computer controlled system. Some IEEE standards on smart sensors are available and others are expected soon. This would result in interchangeable modules. The field bus, based entirely on digital signal transmission need to be standardised. Presently, signal transmission from sensors and actuators is being done in the plants using both analog and digital signals. These developments have been triggered due to the wide availability of digital computers for measurement and control at reasonable prices.

18.1 ■ COMPUTER-AIDED MEASUREMENTS

18.1.1 Introduction

In applications involving measurements, analysis, transmission and storage of data from a number of sensors, it is useful to employ a digital computer. The present-day microcomputers or PCs are fairly versatile and can be interfaced to an experimental set-up, involving a number of sensors, where large amount of data is required to be stored. Also, the data can be transmitted over a distant location or from the PC to the mini or mainframe computer.

A computer would need data in digital form to be fed into it and the output data would also be in the same form. Since a large number of transducers are of analog type, A–D converters have to be employed between the sensor and the computer. Also, the data transfer and communication hardware should be of standard form, for compatibility and to permit interchangeability. The data may be transmitted in serial or parallel mode. On this basis, the bus configurations have been standardised.

A number of analysis operations like averaging, determination of Fourier transforms, etc. may be carried out with data stored in the computer. In cases where the digital computer is used for the purpose of control, the digital data stored is converted to analog form by a D–A converter and the signal is applied to the final control elements of the system, which are essentially of analog type.

This section gives brief details of the data acquisition system, using digital devices like computers. Some of the configurations of interface units used along with the microprocessor unit and the data transfer and communication devices and techniques are briefly covered.

18.1.2 Microcomputers for Data Acquisition Purposes

A microprocessor-based measurement system or a data acquisition system is shown in Fig. 18.1.

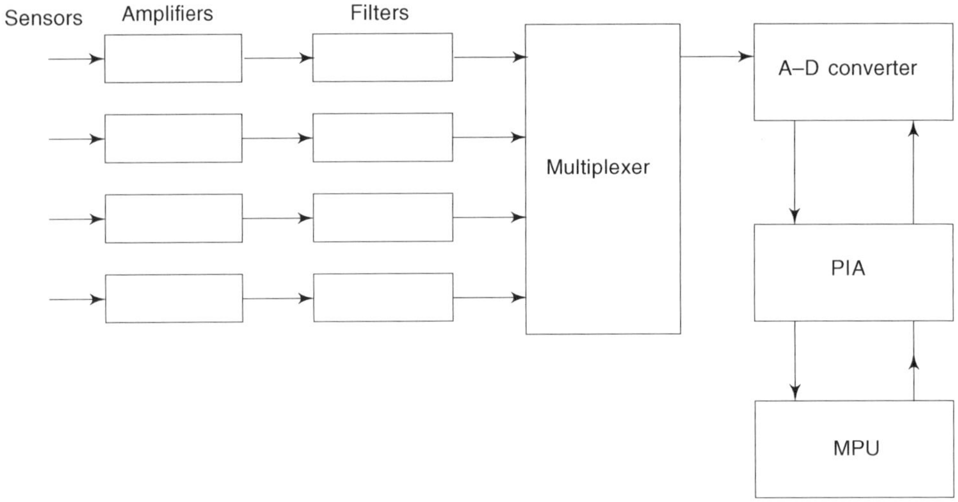

Fig. 18.1 *Microprocessors based measurement system*

Signals from a number of sensors like temperature, pressure, vibration etc., after amplification and filtering of noise or undesirable signals, are fed to a multiplexor unit, which is like a switching unit and takes up one signal at a time for being digitised by an A–D converter. The output of the A–D converter is fed through a Peripheral Interface Adapter (PIA) to a Microprocessor Unit (MPU) for storage and analysis operations. A PIA is like a bridge between outside world and the computer and transmits data essentially in parallel mode. Figure 18.2 shows the two modes of data transmission, viz. parallel and serial modes.

In parallel mode, bits comprising a character are transmitted simultaneously over parallel lines, one line being used for each bit. The speed of transmission is high but the cost is also high because of increased number of wires. However, in the case of serial mode of data transmission, only one wire is used for transmission of all the bits and the position and duration of each is suitably controlled and kept track of, through an operation called *handshaking* through which it is possible to regulate when the transmitter is

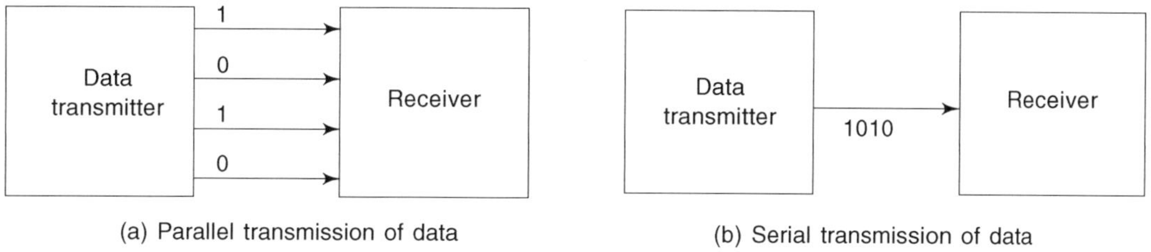

Fig. 18.2 *Parallel and serial data transmission*

to start transmission or whether the receiver is ready to receive it. In the case of a synchronous signal interface, the rate of data transmission is determined by clocks at a fixed rate in the transmission device while in an asynchronous type of interface, data is proceeded and followed by start and stop framing bits.

Figure 18.3 shows a microcomputer with same interface units. The clock generates pulses, which controls the rate at which the microcomputer works. The microcomputer essentially comprises MPU (or Central Processing Unit (CPU)) of memory and input/output I/O unit. The MPU or CPU consists of central logic circuitry logic circuitry for interpreting and carrying out logical and mathematical operations

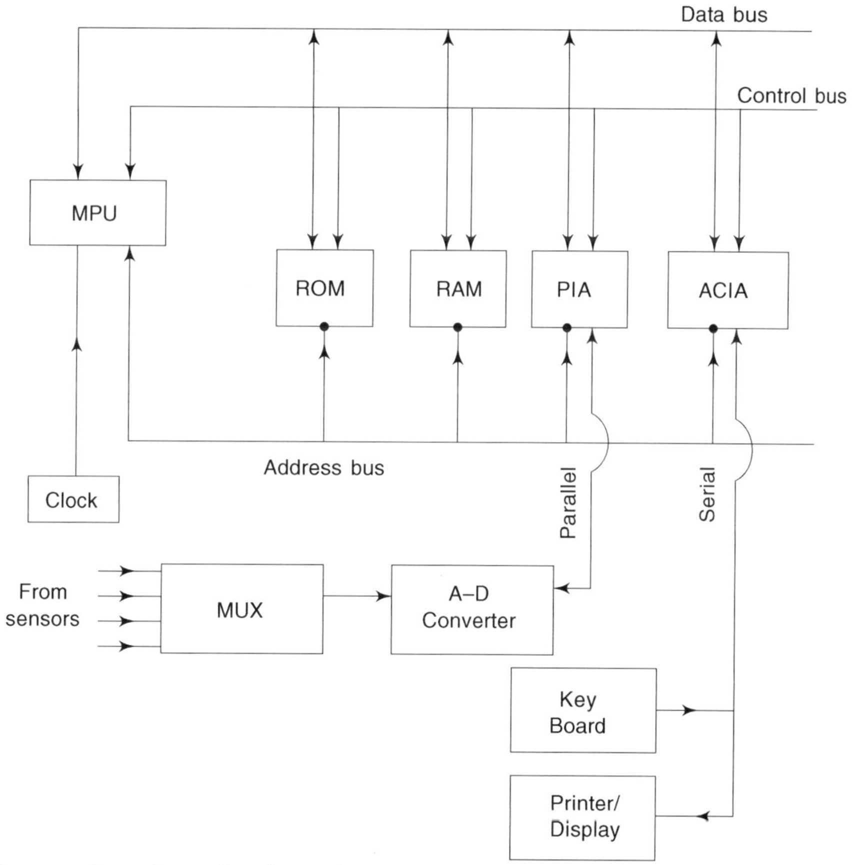

Fig. 18.3 *Microcomputer and some interface units*

as per the program. It serves as one control centre for directing flow of information of digitised types. It accepts inputs from keyboard or transducers through high pathways (called buses) and directs the data to memories, displays, etc.

The memories shown in Fig. 18.3 are of the following types:

1. Read Only Memory (ROM) which operates instructions for system organisation and operation.
2. Random Access Memory (RAM) which is essentially a bank for temporary deposit and withdrawal of data or information form programs.

The CPU interacts both ways with disk or magnetic tape type memory, in addition to I/O units which comprises of printer or plotter terminals, etc. There are three types of buses or pathways as shown in Fig. 18.3.

— Address bus, which is for accessing different memory locations.

— Data bus, which is for handling bits of data. There is a two-way flow control of data by the use of gating.

— Control bus, for exercising control or synchronisation function.

Figure 18.3 also shows output, from various sensors through MUX (Multiplexer Unit), A–D converters and interface unit PIA. It handles data in parallel form. Another unit Asynchronisation Interface Adapter (ACIA) which handles data serially is shown connected to keyboard and printer/display units.

Currently, microcontrollers using microprocessors with input/output devices including A–D and D–A converters and other peripherals are available on a chip for use.

18.1.3 Data Transfer and Communication

As pointed out earlier, PIA is a parallel type of data transfer and communication device consisting of data bus buffers, data registers and peripheral interface as shown in Fig. 18.4. Such units are also called PPI (Programmable Peripheral Interface).

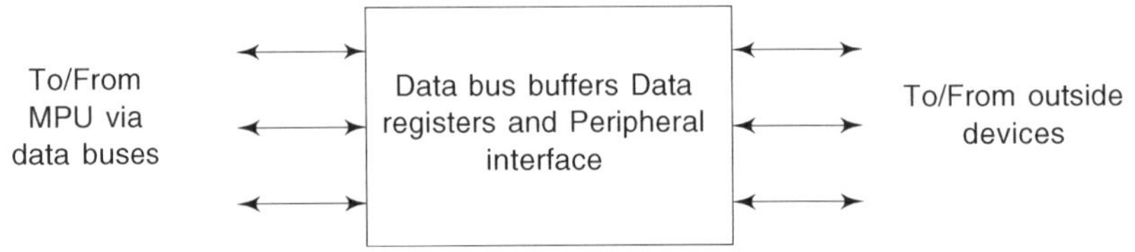

Fig. 18.4 *Peripheral interface adapter*

Figure 18.5 shows the use of telephone lines for transfer of data from a microcomputer to a mainframe system. A signal type of standard interface RS-232C is used, in addition to Modem, which is a modulator/demodulator type device, which codes and decodes signal on to a carrier wave and vice versa.

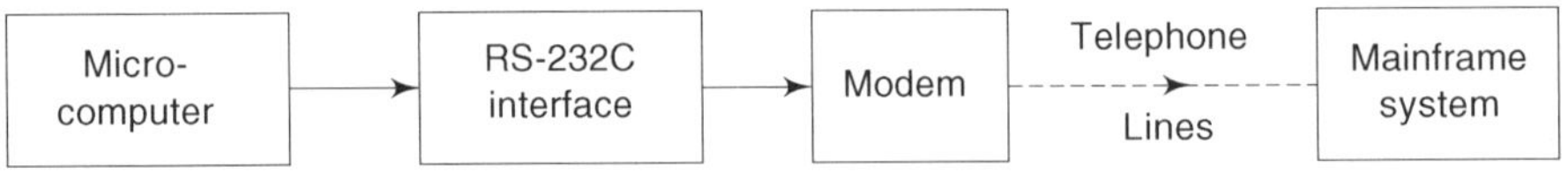

Fig. 18.5 *Serial transfer of data from microcomputer to mainframe system*

The outline of general-purpose interface bus structure (GPIBS) is shown in Fig. 18.6. It is adopted by IEEE 488-HP for parallel bit and byte serial mode of data transmission. There are three buses—for data transmission, transfer control or handshaking and general bus activity management. The communication between devices may be for the following:

— Listening, as in a printer or plotter

— Talking, as in the case of transducers or counters.

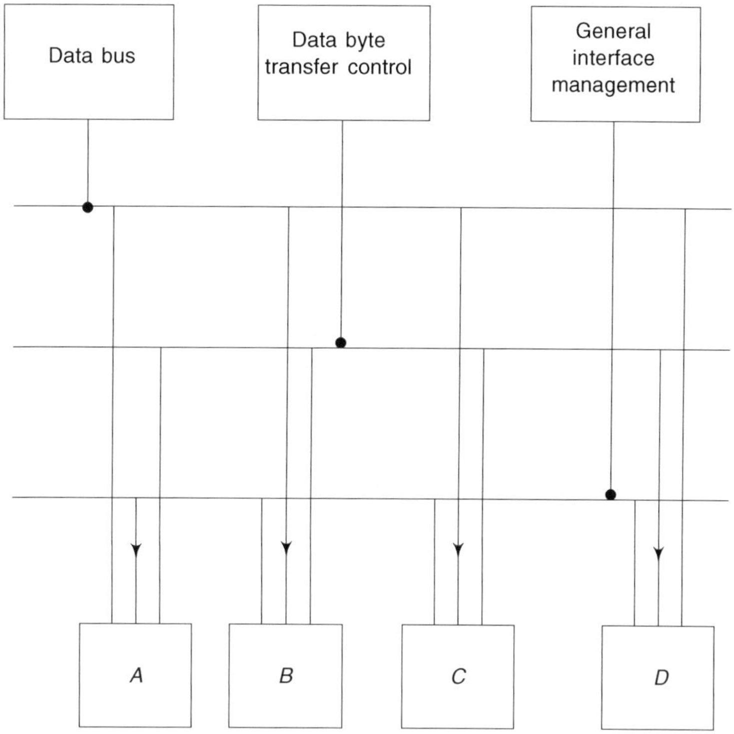

A. Talk, listen and control: Computer
B. Talk and Listen: *A–D* Converter, Tape drive spectrum analyzer.
C. Talk: Transducer, Counter
D. Listen: Printer/plotter

Fig. 18.6 *General-purpose interface bus structure*

— Talking and Listening, as for *A–D* converters, tape drive, spectrum analysers, etc.
— Talking, Listening and Control, as for a computer. The interface has been used for data transfer and communication between various devices in a measurement system.

18.1.4 Data Acquisition System

In practice, data is acquired from the sensors through the signal conditioning elements and transferred to a computer for the analysis and storage as shown in Fig. 18.7. Usually, a data acquisition (DAQ) board is plugged in a slot at the back of a computer. The board includes a mutlitplexer,

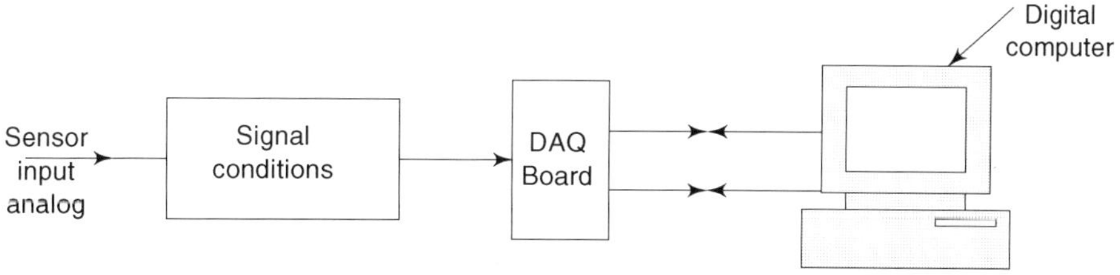

Fig. 18.7 *A typical data acquisition system using a computer*

A–D and D–A converters, registers, control elements, etc., as shown in Fig. 18.8. A computer software is used for controlling data acquisition through the board. The data may be acquired through an interrupt or direct memory access (DMA).

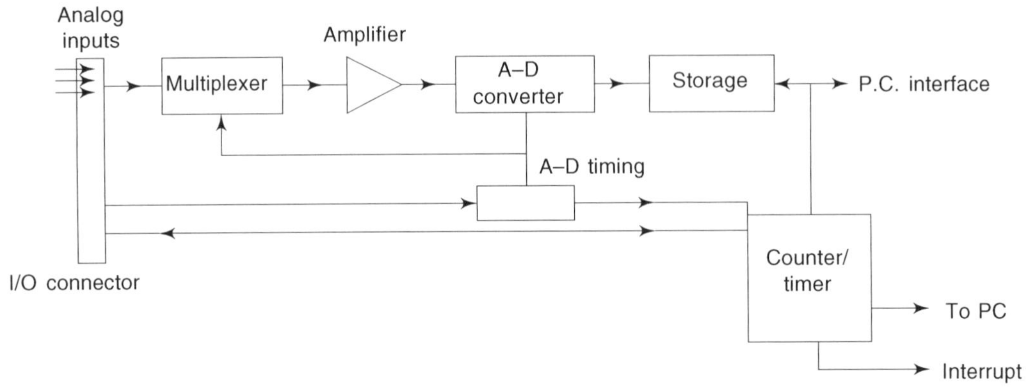

Fig. 18.8 *A typical schematic diagram of Data Acquisition (DAQ) board*

The current trend is to use graphical programming using Virtual Instruments (VI), for acquisition, processing, storage and presentation of measurement data. A VI consists of software and hardware for transferring the data from an instrument to a computer. 'Lab View' is one of the graphical programming softwares that can be used with a PC for the above purpose.

18.2 ■ FIBRE OPTIC TRANSDUCERS

18.1.1 Introduction

Optical Fibre Sensors (OFS) are a relatively new type of sensors which work on the principle of alteration of light by external variables like motion, pressure, temperature, flow, strain, pH, particle size, turbidity, smoke density, etc. The fibres act as light conduits to conduct light and consist of thin strands of dielectric materials like glass, quartz, and plastics, in which some dopants may be introduced for certain special applications.

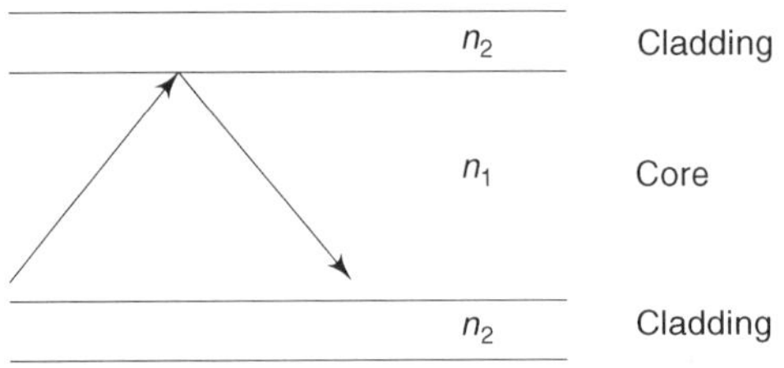

Fig. 18.9 *A typical optical fibre*

A light ray propagating in the fibre is trapped by total reflection at the boundary of the core and the cladding material as shown in the Fig. 18.9. This is possible only if refractive index n_2 of the material is less than the refractive index n_1 of the core. The core diameter is very small, the average value being 50 μm.

18.2.2 Advantages and Disadvantages

Some of the advantages of the optical fibre sensors are the following:

1. Enhanced safety due to its non-metallic construction and absence of high voltages, sparks, etc.
2. Capability of their use in harsh environments involving corrosive gases due to chemical inertness
3. Low weight and small size, and hence small sensor size
4. High sensitivity and fast responses of measured quantities because of low thermal and inertial mass

5. Immunity to electromagnetic interference (EMI) as optical fibre are non-conducting. They neither generate nor receive EMI's
6. Capability with fibre optics telemetry systems
7. Their compatibility of coupling with the modern miniaturised VLSI electronic circuits

Some of the disadvantages of Optical Fibre Sensors (OFS) are

1. Fragility, requiring careful packaging
2. Alignment problems due to small size of optical components
3. Increased costs

18.2.3 Types of Configurations

There are basically two types of optical fibre sensor configurations as shown in Fig. 18.10. These are

(a) Extrinsic sensors (or incoherent sensors)
(b) Intrinsic sensors (or coherent sensors)

The emitter, which may be a light emitting diode or a laser source, emits the light rays through the fibre, which get modulated due to the outer signal, which is to be measured. The output fibre is connected to the detector, which converts the optical energy into electrical energy. These detectors work on the principle of creation of an electron–hole pair in semiconductors or the release of electrons from the cathode of the photomultiplier tube. In the case of the extrinsic sensor, as in Fig. 18.10(a) the intensity modulation of light takes place outside the fibre, while in the case of Fig 18.10(b) viz. intrinsic sensor, it takes place within the fibre. Examples of the two types of sensors are given in Fig. 18.11.

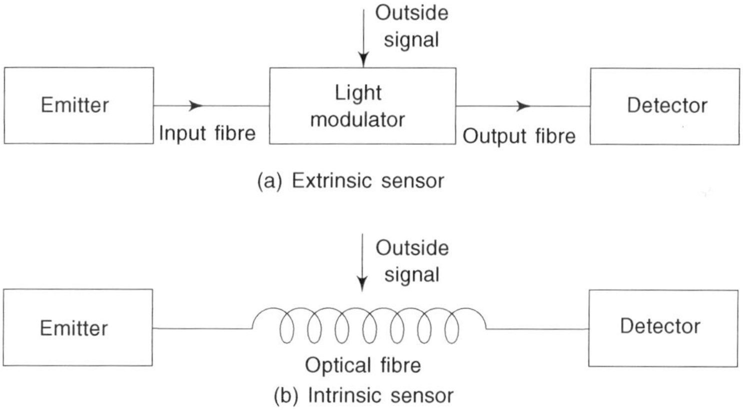

Fig. 18.10 *Types of Optical Fibre Sensor configurations*

Figure 18.11(a) shows an extrinsic or external intensity modulator type of sensor in which the position of the reflector due to motion to be measured may change, thus changing the light intensity in the output fibre. This is detected by a detector.

In Fig. 18.11(b), the fibre bending due to the pressure, which is the input variable, to be measured, induces radiation losses, changing the intensity at the output. This happens within the fibre itself and hence the configuration is called as intrinsic type. This causes radiation of light even at small deformation and is also called micro-bend sensor. In this type of sensor configuration, apart from the intensity modulator, there may be phase or frequency modulation, which after detection may be used for several applications.

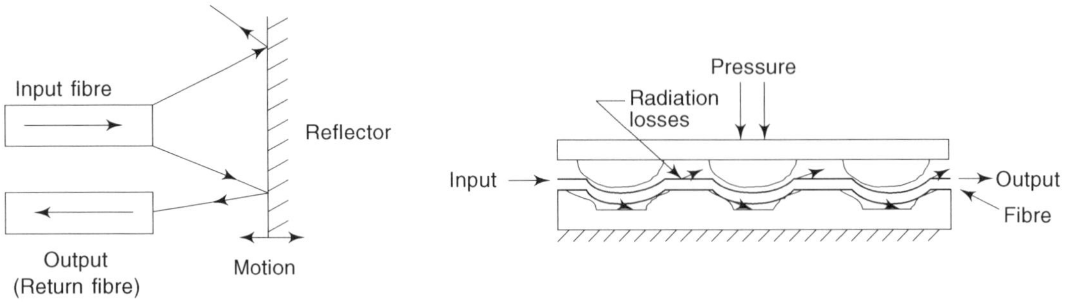

(a) Extrinsic type (b) Intrinsic type

Fig. 18.11 *Examples of extrinsic and intrinsic type sensors*

18.2.4 Applications

A number of applications for the measurement of various physical variables are described as follows:

Figure 18.12 shows an arrangement for the measurement of the pressure involving an extrinsic type of sensor in which the light from the input fibre is reflected from a diaphragm and picked up by the output fibres. The application of pressure deflects the diaphragm and the intensity of the reflected light depends on the diaphragm deflection.

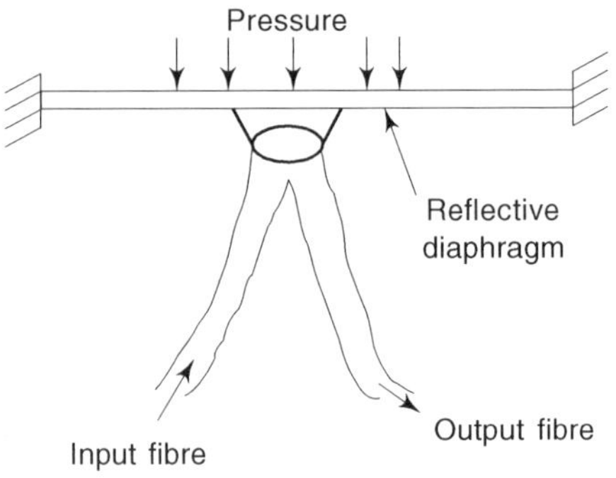

Fig. 18.12 *A typical fibre optic pressure transducer*

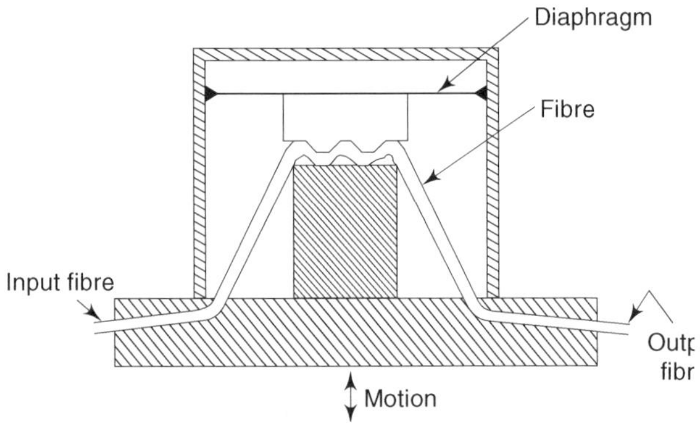

Fig. 18.13 *A typical fibre optic accelerometer*

Figure 18.13 shows a seismic accelerometer for measuring vibrations of the object to which it is attached. It uses a micro-bend type sensor, which bends due to the inertial force of the mass. The diaphragm acts as a spring of the seismic transducer and the relative motion of the mass relative to the frame changes the intensity of the light from the output fibre since the fibre bending due to relative motion induces radiation losses. The measurement of intensity in the output fibre gives an idea of the acceleration of the vibrating effect. It is an intrinsic type of optical fibre sensor arrangement, since the changes in radiation losses take place in the fibre itself.

Liquid level sensors which work on the principle of variation in refractive index are also available. If the cladding is removed from a portion of the fibre, the bare core loses more light when it is in the liquid than when it is in the air. A number of such sensors at various levels can indicate the level of the liquid. A continuous level sensor is also possible by measurement on the reflected light from the air–liquid surface.

Figure 18.14 shows the configuration of a temperature sensor. It is seen that the phase of light gets changed due to the change in the length of the fibre as a result of application of longitudinal strain due to thermal expansion. The change in phase is converted to intensity variations using a Mach Zehnder interferometer as shown in the figure. The intensities at the two detectors are seen to be proportional to $1 \pm \cos \delta$, when δ is the phase change due to the phase modulation. The value of δ for glass fibre is about 100 rad/°C/m of length and thus the device is highly sensitive. This type of device is of intrinsic or of coherent type.

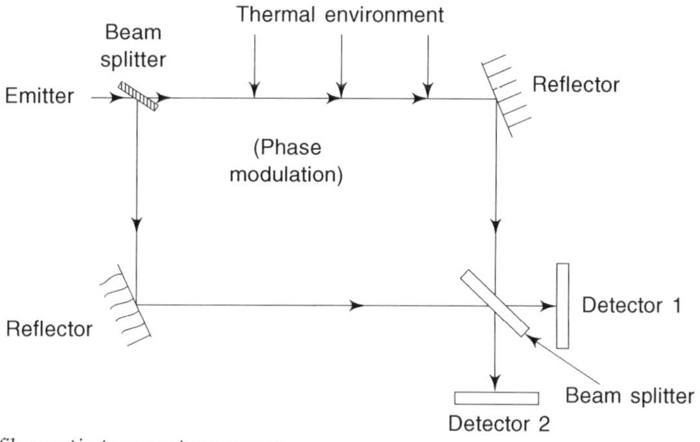

Fig. 18.14 *A typical fibre optic temperature sensor*

In the flow sensor, as shown in Fig 18.15, the frequency of light waves scattered from the particles in the moving fluid is Doppler shifted, the frequency shift being proportional to velocity. The input optical fibre carries light, which is focussed on the fluid flow. The output fibre carries the light scattered by the particles in the flowing fluid and particle velocity is found from the modulated frequency spectrum. This is an extrinsic or incoherent type of fibre optic sensor since the sensing takes place outside the fibre.

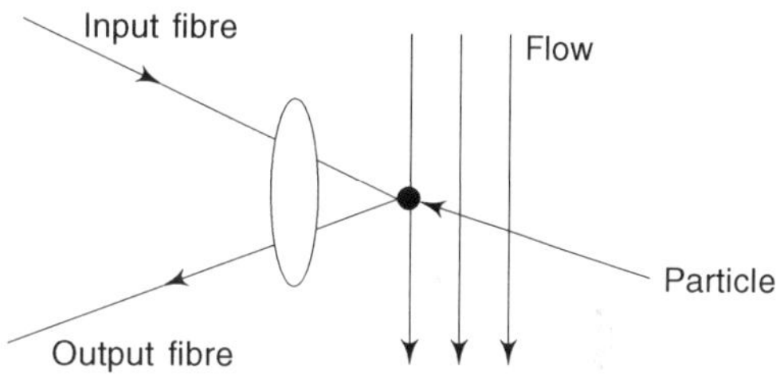

Fig. 18.15 *A typical fibre optic flow sensor*

18.3 ■ MICROSENSORS

There is a trend towards miniaturisation of sensors, since a large number of them are used in smart machines and structures. Also, integrated sensors and signal conditioning devices are used in mechatronic applications. Different types of microsensors including piezoresistive, Hall effect, thermal, chemical etc. are available. These are based on the use of silicon and other semiconductor materials. A few of them are described as follows.

18.3.1 Silicon Microsensors

Silicon is used as a base material for making microsensors. This is due to the fact that silicon has got good physical properties and can be shaped to the desired shape using micro-machining by etchants. Therefore, it is a good substrate for incorporating the associated signal conditioning circuitry.

These sensors are micromachined or etched from a single crystal of silicon and are small in size and quite rugged. The sensors along with the signal conditioning electronic components are mounted on the same silicon substrate.

Usually, a piezoresistive material or a semiconductor, is diffused on to the silicon substrate at the region of maximum strain, as shown in Fig. 18.16. The resistance of the piezoresistive material changes with strain, which with a suitable circuitry like a Wheatstone bridge, can be made to give an output which depends on the strain. The strain may be due to any of the physical variables to be measured like pressure, force, vibrations, etc.

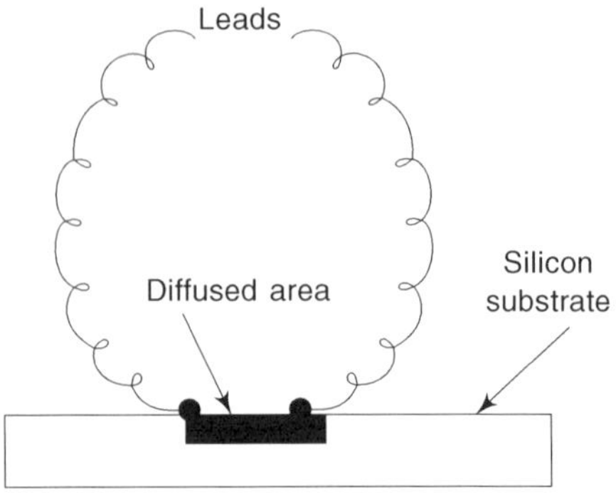

Fig. 18.16 *Diffused type semiconductor strain gauge*

The micromachining of silicon is usually done by etchants, which dissolve the exposed portion of silicon, producing the desired shape. The diffusion of piezoresistor is carried out first on the silicon wafer, after which a SiO_2 mask on silicon surface is applied to the surface not to be etched. Then the wafer is exposed to the etchant, which produces the desired surface and does not affect the mask. Lastly, the protection mask may be removed with a chemical, which does not affect the silicon.

Figure 18.17 shows a silicon type micropressure transducer using a diaphragm integral with the silicon substrate. Then a piezoresistive sensor is diffused at the appropriate locations so that the strain and resistance changes are sufficient enough to produce an output from the bridge formed by the sensors. The principle is similar to that of a diaphragm type pressure transducer.

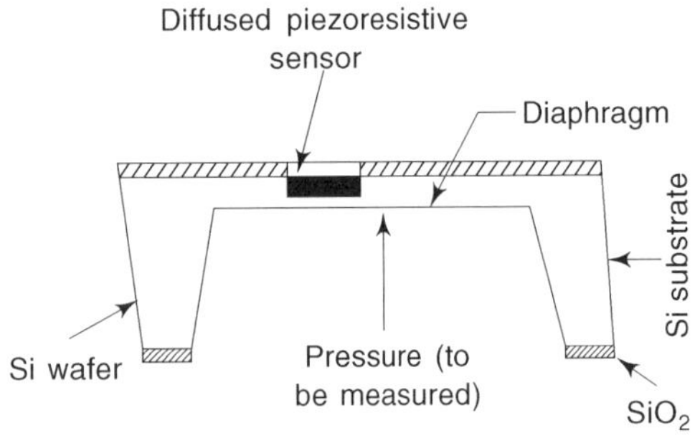

Fig. 18.17 *Diaphragm type silicon micropressure transducer*

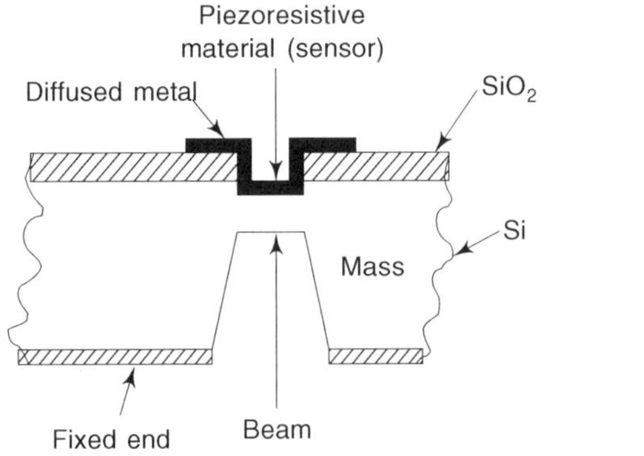

Fig. 18.18 *Cantilever type silicon micro-accelerometer*

Figure 18.18 shows a silicon type micro-accelerometer, which is used for measuring vibrations. The fixed end is attached to a casing, which in turn is attached to the vibrating object. The above is a seismic type sensor. The beam and the mass are integral and are of silicon. The piezoresistive material is diffused on the beam, which acts like a spring. The diffused metal is for connection of the sensor to the amplifier.

The relative motion of the mass with respect to the fixed end cause strains in the beam, which results in the change in resistance of the piezoresistive material. As before, the piezoresistive material forms a part of a Wheatstone bridge which produces an output proportional to the strain in the beam, the latter being designed in such a way that the strain is proportional to the acceleration of the object to which the fixed end is attached.

Problem 18.1 *Figure 18.19 shows a part of a micro-accelerometer of capacitance type for measuring the vibrations of an object to which its base is attached. Its seismic mass m is 10^{-5} kg, mounted at the free end of a silicon cantilever beam with a stiffness of 0.5 N/mm at the mounting position. The maximum amplitude of the harmonic acceleration signal to be measured in 20 g. What should be the minimum gap between the mass and the base? Also, find the maximum frequency of vibration for which the sensor can be used for measuring acceleration.*

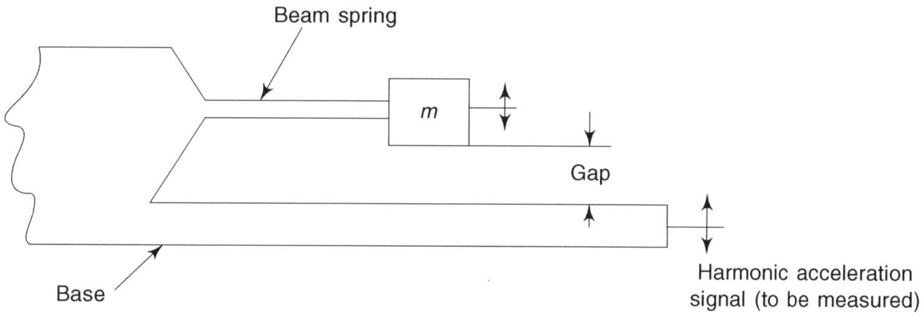

Fig. 18.19 *Arrangement for Problem 18.1*

Solution

$$\text{Mass } m = 10^{-5} \text{ kg}$$
$$\text{stiffness } k = 0.5 \text{ N/mm} = 500 \text{ N/m}$$
$$\text{Natural frequency } \omega_n = \sqrt{\frac{k}{m}} = \sqrt{\frac{500}{10^{-5}}}$$
$$= 7100 \text{ rad/s}$$

Maximum value of ω to be measured

$$= 0.4\ W_n$$
$$= 2840 \text{ rad/s} = 452 \text{ Hz}$$

As seen from Chapter 7, for an accelerometer

$\dfrac{\omega_n^2 z_0}{A} \approx 1$, where z is the motion amplitude of mass relative to the base, and A is the amplitude of acceleration of the base

or
$$\frac{(7100)^2 z_0}{20 \times 9.81} = 1$$
$$z_0 = 3.9 \times 10^6 \text{ m} = 3.9\ \mu\text{m}.$$

18.4 ■ SMART SENSORS

The sensors having decision-making and communication logic added to the basic sensors, are called smart or intelligent sensors. Some of the other features included in such sensors are: compensation for interfering inputs, linearisation, self-test and calibration facility. There are usually microcomputers and other elements on the same chip, whenever possible.

For data acquisition using the smart pressure sensor, the functional diagram is as shown in Fig. 18.20. If a piezoresistive sensor is used, as primary sensor for pressure, the main interfering input is temperature and so a secondary temperature sensor is used to compensate for the effect of temperature on resistance change. The memory of the microcontroller has calibration data stored in it. The microcontroller is essentially a microprocessor with input/output (I/O) facility.

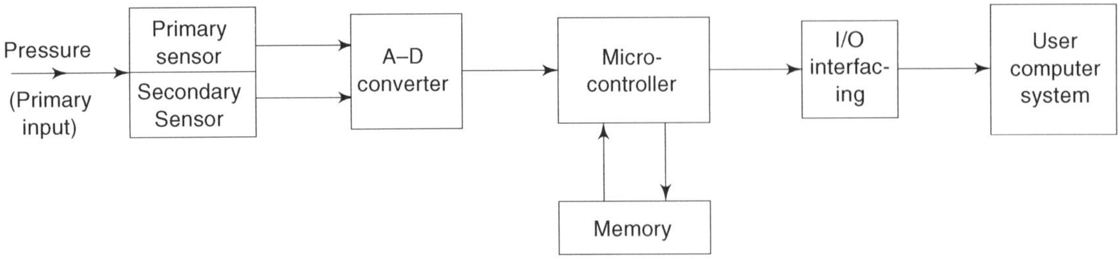

Fig. 18.20 *Data acquisition system using smart pressure sensor*

The IEEE 1451.2 standard for smart sensors is as shown in Fig. 18.21.

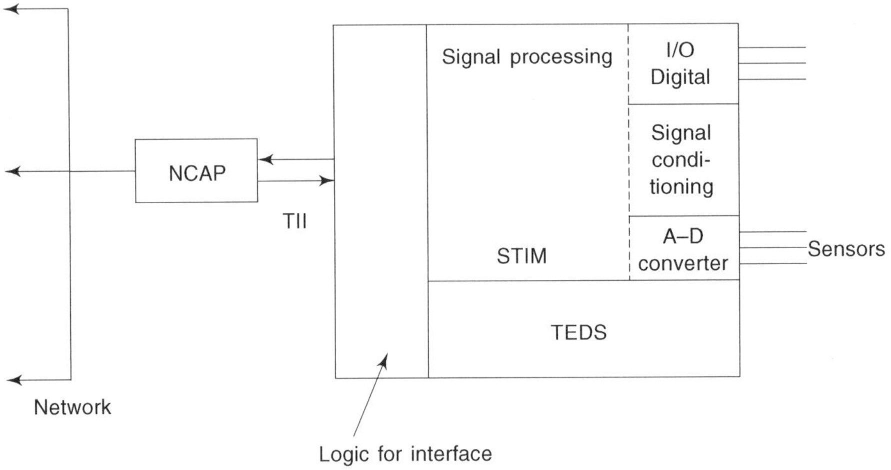

Fig. 18.21 *Smart sensor modules*

The various modulers are:

STIM — Smart Transducer Interface Module
NCAP — Network Application Processor
TEDS — Transducer Electronic Data Sheet
TII — Transducer Independent Interface

The TEDS contains information of the transducer's identification, viz. make, model, specification, calibration, location etc. Any transducer can be interfaced through STIM, performing the functions of signal conditioning, linearisation, signal processing, pattern recognition of monitored signals etc. The TII interfaces STIM with NCAP. It is a 10-wire serial I/O bus that defines data transport from STIM to NCAP, address and data transport from NCAP to STIM, latching data, performing triggering functions, data transport acknowledge, interrupt used by STIM to request services of NCAP etc. NCAP is a device used for supporting a network interface, and may provide local intelligence including the possible use of a microcontroller. It is the smart sensor's window to the outside world. Any control network may be connected to a group of sensors, with an appropriate NCAP.

One of the useful applications of the above unit is for a condition monitoring system for a plant, to monitor the health of machines and process, to detect possible damage. Instead of using a dedicated computer with spectrum analyser or data acquisition system or signal processing and pattern recognition, a smart unit on the lines of IEEE 1451.2 can accept inputs from various sensors, process the signals and diagnose fault if any and communicate. This would be quite economical for the plant.

18.5 ■ SMART TRANSMITTERS AND FIELD BUS

A conventional analog transmitter transmits the sensor signals for various physical parameters using 4–20 mA current, as discussed in Chapter 5. The signal is communicated to any access point in the measurement loop or control room in a plant. The use of computers has resulted in several techniques involving the use of field buses for transmitting digital or hybrid (both analog and digital) signals.

One of the common transmitters, of the smart type, uses the Highway Addressable Remote Transducer (HART) protocol. It employs the use of Frequency Shifting Key (FSK), for digital signals on the existing 4–20 mA lines for analog signals as shown in Fig 18.22. The FSK arrangement involves the use of two frequencies–1200 and 2200 Hz for bits 1 and 0, respectively.

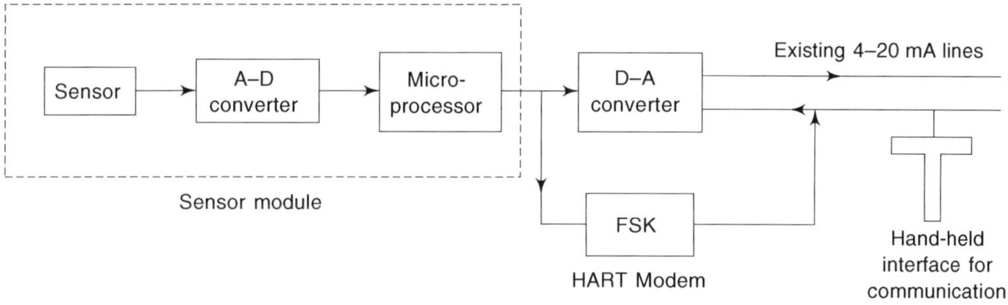

Fig. 18.22 *Data transmission using Hart protocol*

Thus, there is simultaneous communication for both analog and digital signals.

The current developments include the development of various 'field buses' which link various sensors and actuators at the field or process levels with the main system communication devices or distributed control systems. This includes linking of sensors, controllers, PC and Programmable Logic Controllers (PLC) so that the information is available at various levels in a plant, and results in reduction of wiring in the plant.

Figure 18.23 shows a typical Distributed Control System (DCS) with the desired information communicated at various levels in a factory or a plant, using computer networking.

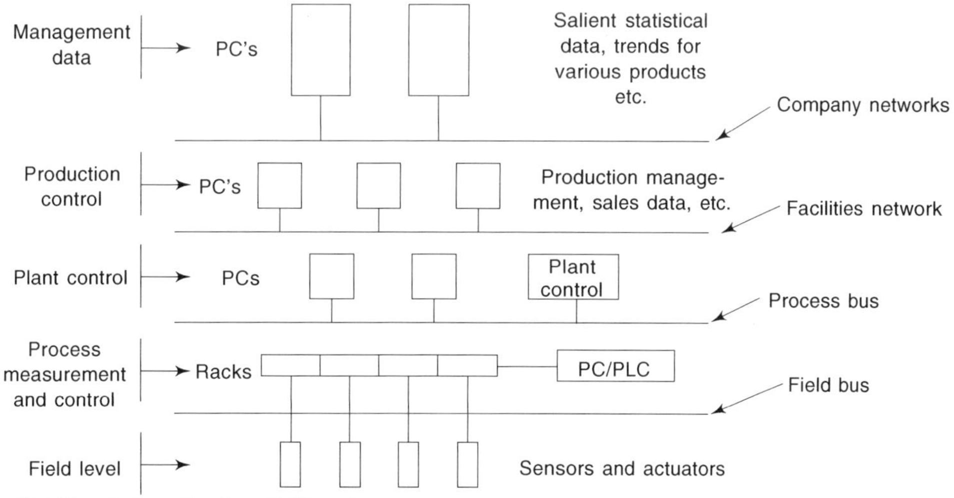

Fig. 18.23 *Field bus integration in a DCS system*

18.6 ■ VIRTUAL INSTRUMENTATION

It is defined as a layer of software and/or hardware, which is added to a digital computer so that the user can use the same as if it were like a conventional instrument.

The standard architecture for such an instrument consists of

- data acquisition
- analysis involving calculations, e.g., Fourier transform from time to frequency domain, which is done by software
- presentation including user interface.

Virtual Instrumentation (VI) presents a new way of doing testing and analysis. With the use of software, the VI functions just like an actual instrument.

With V.I., only the front end, in software form, needs to be specific to the analysis, since a computer can handle most of the tasks involving data acquisition, analysis and presentation.

Softwares that emulate instruments like oscilloscope, digital multimeter, spectrum analyser, etc., can be programmed as the front end instead of using separate hardware front ends for different experiments.

Advances of V.Is are low cost, flexibility and ability to customize various instruments.

Graphical programming language (e.g. LABVIEW) can be easily used for building a V.I. The functional blocks can be selected from a pallete menu. The blocks may vary from arithmetic operations to data acquisition and instrument control.

A set of drivers is used to provide a series of high level calls or icons so that the user can initialize and control the instruments. The drivers are a set of software routines that control a programmable instrument, configuring, initialising, reducing from/writing to and triggering.

DSP (Digital Signal Processing) software can carry out numerical analysis like Fast Fourier Transform analysis on joint time frequency analysis.

*In Fig. 18.24, dynamic force and acceleration signals are shown being transmitted to a desktop computer. An electrodynamic exciter has been employed for providing the harmonic force input. An adjustable frequency signal of desired frequency for driving the exciter. A power amplifier amplifies the signal of the oscillator and is then fed to the exciter via a power amplifier. The wiring diagram of the complete test set-up, including the instrumentation and connection of various equipment, is shown in Fig. 18.24. It shows the charge amplifier and DC filters that route the acquired force and acceleration signals from the exciter to the data acquisition card in the computer through the connector block.

A virtual block is developed in Labview for acquiring and digitizing the data and then for further analysis, in order to find the complex shear modulus of a polymeric viscoelastic material. The front panel is shown in Fig. 18.25. The DC filtered analog signals are first digitized using a data acquisition card. The signals are represented on the front panel as they are acquired. The signals are then filtered that noise and some other frequency components are removed by a bandpass filter. The filtered signals are then displayed separately as also the first versus acceleration signal, in order to find the complex shear modulus of a polymeric viscoelastic material. Figure 18.25 shows various settings, analysis and output display/indicators.

* *Further details are reported in:*
"Passive and active control of vibration and noise in machines and structures", by T.K. Kundra, B.C. Nakra, K. Gupta, N. Tandon & S.P. Singh, DST sponsored project report, October, 2003. The work was carried out at Indian Institute of Technology, Delhi.
Contribution of Dr. A.K. Darpe of I.I.T. Delhi is also gratefully acknowledged for this section by the authors of this book.

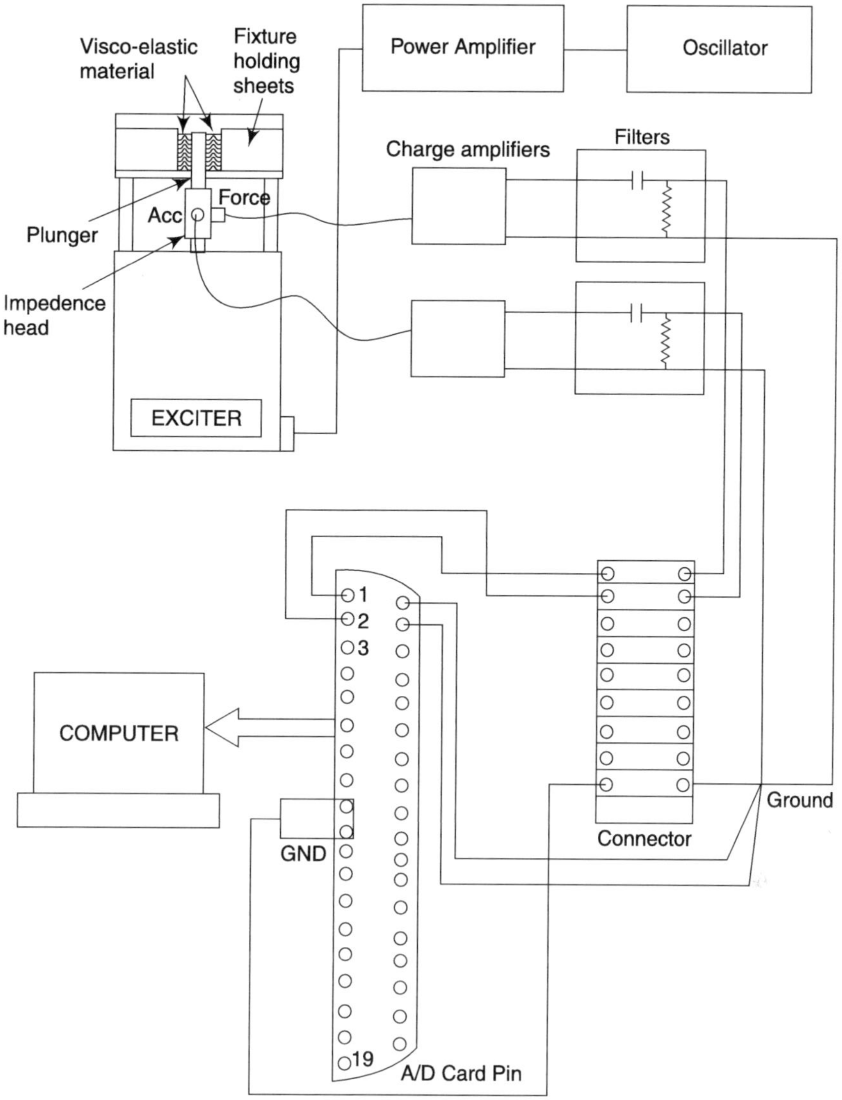

Fig. 18.24 *Connection diagram showing instruments and equipment for data acquisition of force and acceleration signals from the impedance head mounted on the exciter.*

For the purpose of computation of the complex shear modulus, some parameters can be set on the front panel. They include the calibration constant for the acceleration and force signals, surface area and thickness of the VEM sample being tested. For the signal acquisition, one can set the frequency of excitation. The upper and lower cut-off for bandpass filter are automatically set depending on the range parameter set on the front panel as shown in Fig. 18.25. A total of 4000 samples at 25 kHz sampling frequency are acquired from two analog input channels of the A/D card for acquiring force and acceleration signals from the test setup.

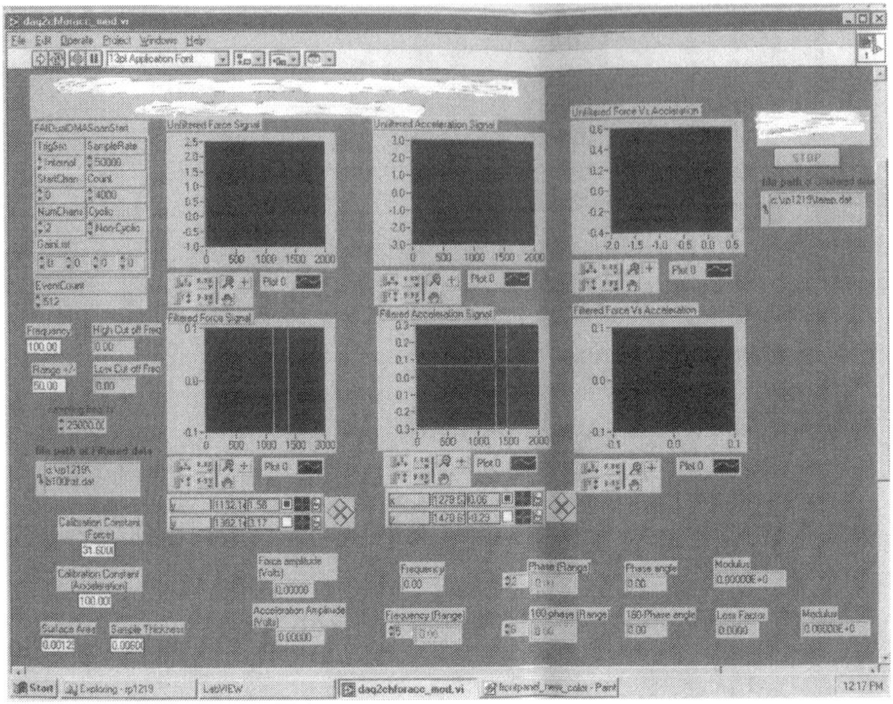

Fig. 18.25 *Top part of the virtual instrument for acquiring and analysing force and acceleration data from a test set-up.*

Review Questions

18.1 Indicate True (T) or False (F) against each of the following:

- (i) ROM in a computer is for temporary storage.
- (ii) PIA is a parallel type of data transfer device.
- (iii) A DAQ board used with a PC has a memory of its own.
- (iv) An optical fibre sensor gives a lot of noise with the signal due to electromagnetic interference.
- (v) In an intrinsic type of OFS, the light intensity modulation takes place inside the fibre.
- (vi) An OFS type sensor is usually miniature in size.
- (vii) Micromachining of silicon sensors is done on a small milling machine.
- (viii) Silicon sensors are usually miniature types.
- (ix) Field bus is for transfer of data using a standard cable.
- (x) A transmitter can be used to transmit signal from a sensor to a control room.
- (xi) Telephone wires can be used to transmit a signal from a sensor.
- (xii) HART-based transmitters transmit only digital signals.
- (xiii) Distributed control systems do not use computers.
- (xiv) Hall effect sensors can be of analog or digital types.
- (xv) Silicon is more sensitive than indium antimonide when used as a plate in a Hall effect sensor.

18.2 A capacitive type micro-accelerometer has a seismic mass of 10^{-6} kg mounted at the free end of a silicon cantilever, with the cantilever stiffness being 250 N/m at the free end. The mass has a square surface 1 mm × 1 mm, with a thickness of 400 μm, and forms the movable plate of the capacitive sensor, while the base of the accelerometer forms the fixed plate. The air gap between the two plates is of thickness 1 μm. The above variable capacitance sensor forms a part of the linearising circuit using an operational amplifier (as in Chapter 5). The input supply to the circuit is 10 V and fixed capacitance at the input of the circuit is 20 pF while the variable capacitance sensor is in the feedback path from the output to the input side of the operational amplifier.

Find the output sensitivity of the measuring system (mV/g of acceleration) and also find the highest frequency till which it can be used as an accelerometer.

Answers

18.1 (i) F (ii) T (iii) T (iv) F (v) T (vi) T (vii) F
(viii)T (ix) T (x) T (xi) T (xii) F (xiii) F (xiv) T
(xv)T

18.2 880 mV/g, 1000 Hz

Chapter

19

Control Engineering Applications

■ *INTRODUCTION* ■

Automatic control finds a wide range of applications in our day-to-day life as well as in several engineering disciplines. Typical applications include chemical process plants, production machines, power plants, transportation systems, environmental control techniques, automated aeronautical and space systems, consumer goods, etc. The term 'control' implies to regulate, direct or command.

For obtaining the desired performance characteristics, control engineers study the dynamic characteristics of the total system, measuring instrument, controller, as well as the process. Further, they employ both accurate and precise measurements of the process variables to be manipulated. After this, they design, develop and implement the enabling technologies for powering the actuators based on the design procedures, mathematical skills and analytical tools. The actuator components involved may be of electrical mechanical, hydraulic or pneumatic type.

Hydraulic devices are used in machine tools and aerospace systems where a high power-to-size ratio is needed. Pneumatic devices are preferred in production plants and chemical processes from safety considerations and due to easy availability of compressed air in the plants. In addition, electrical devices are fairly common in a variety of small and compact type of control systems in industries and also in household appliances.

A control system's dynamic performance analysis is important so that it responds to commands and disturbances, as desired, during transient and steady state. Also, it should be stable under operating conditions. Thus the control parameters have to be properly chosen. In addition, the instrument used for measuring the value of the control variable should be accurate under dynamic conditions and should involve minimum lag or delay in measurement. The above requirements need mathematical modelling and solution of the governing equations, which have been introduced in the present chapter.

19.1 ■ TYPES OF CONTROL SYSTEMS

19.1.1 Classification of Control Systems

The classification may be done in several ways given as follows:

Open loop or closed loop type In the *open loop type* control system, there is no comparison of the controlled variable with the desired input. Herein, for each reference input, there is a fixed output, which has no effect on the control action. For any typical input, there may be large variations in the controlled variable because of the effect of the external disturbances. The functional elements of the open loop type of control system are shown in Fig. 19.1 (a).

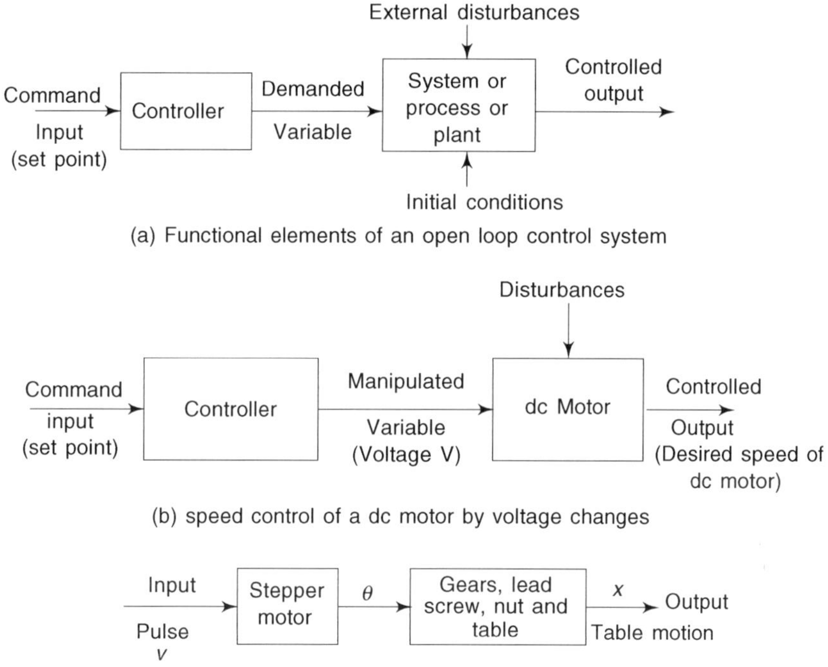

(a) Functional elements of an open loop control system

(b) speed control of a dc motor by voltage changes

(c) Open loop control of machine tool table

Fig. 19.1 *Open loop control systems*

In this system, the command input to the controller i.e., the set point is calibrated (depending on the initial conditions, transfer function of the system and the behavioural characteristics of the external disturbance on the system), to give the desired value of the controlled output. For example, the traffic lights at the road intersections for controlling the vehicular traffic is an open type of control system. Herein, the red, amber and green light timings (input to the control action) are pre-set values which are determined on the basis of average vehicular traffic at any particular road intersection.

However, in certain processes, a number of set points are calibrated on a control knob (controller) to give a controlled output corresponding to any selected value of set point. The common example of such a system is the speed control of a dc motor by voltage changes, as shown in Fig. 19.1(b). In a such a system, the variable speed of the motor is controlled by means of a position of calibrated knob (controller) which is marked with the desired dc motor speeds. Fig 19.1(c) shows the use of a stepper motor for controlling the motion of a machine tool. It can also be used for robot motion control.

The main advantages of the open loop control systems are that they are simple in construction, easy to maintain and economic in operation. In addition, there is no stability problem involved. However, the disadvantage of these systems is that they need careful calibrations and adjustments for obtaining accurate results. Further, the external disturbances or load may affect adversely the controlled output, which may differ from the desired value.

In the *closed loop control system*, the output signal has a direct effect on the control action. In this system, the difference between the measured value of the output response and the desired or reference input or the error is utilised for controlling the manipulated variables. The output signal is fed back and the feed back loop is completed as shown in Fig. 19.2. The closure of the loop permits the comparison of the output signal with the reference input. The error signal suitably actuates the control element to minimise the error. This would then result in bringing the output of the system to the desired input value.

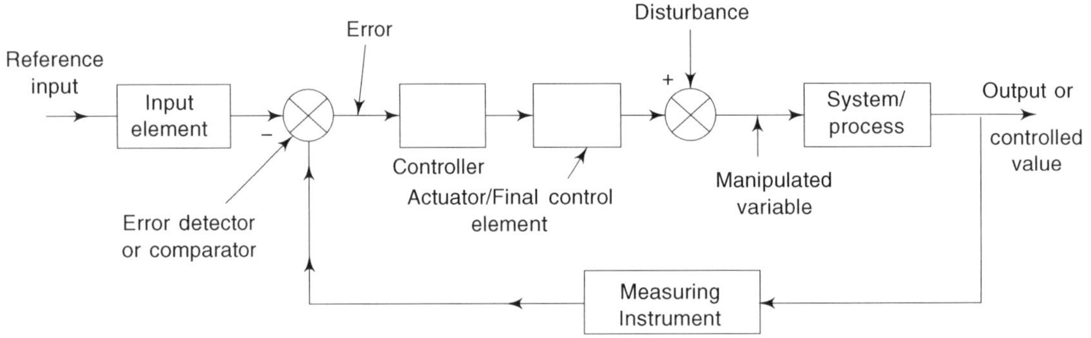

Fig. 19.2 *Block diagram of a typical closed loop system*

The closed loop control systems are also termed as feedback type or error actuated type of control systems. They may be of either manual or automatic type. For example, the control of traffic at the roadways intersection becomes a manual type of closed loop control system when a traffic policeman controls the traffic. The policeman estimates (measures) the amount of traffic volume coming from a particular direction and then accordingly adjusts the command input i.e., the time interval to be allowed for traffic flow in that direction.

Currently, the complex type of control systems with very fast response do not employ the time consuming manual operations. They incorporate suitable systems which would perform the desired functions more rapidly and automatically. The main advantages of such automatic systems are their ability to provide fast and precise control in the face of complex external disturbances. Since they are of automatic operating types, therefore they reduce the drudgery of the manually operating repetitive and continuous job of adjustments of the manipulated variables. Further, the feedback loop components are relatively inexpensive. Therefore, these systems are very cost effective in terms of providing economy in plant operation, increased productivity and improved quality. However, these systems sometimes have the tendency to overcorrect the error signal and this may cause instabilities in the system, if not properly designed.

Figure 19.2 shows the main elements of a typical closed loop system. The desired input is compared with the value of the output or the controlled variable, after its measurement by an instrument and the error signal is further processed by a controller and applied to an actuator like a motor or a valve, called the final control element. Disturbances like environmental temperature in a thermal system or load in a rotating system like an engine-generator system, may also change the output values like temperature or speed. The manipulated variables applied to the system or process, controls the output value.

Figure 19.3(a) shows the diagram of a speed control system of a turbine-driven generator, regardless of changes in load. The manipulated variable viz. steam flow rate in the case of a steam turbine or the fuel flow rate in the case of a gas turbine, controls the speed of the system. The speed is sensed by the governor, which also acts as a comparator (or error detector) and controller. The sleeve motion controls an amplifier (hydraulic servo in several cases), which controls the motion of the valve stem, controlling the flow rate of fuel. Figure 19.3(b) shows the block diagram of the system, representing various functional elements.

Regulator and servomechanism When the controlled or desired value of the control system is relatively fixed, the control system is known as a regulator e.g. in a voltage regulator the desired voltage value is fixed. A servomechanism (often used in motion control), on the other hand, may have any position, to be desired by the control system, e.g. a gun control system is expected to attain any desired position, according to the target position which changes as in the case of an aircraft.

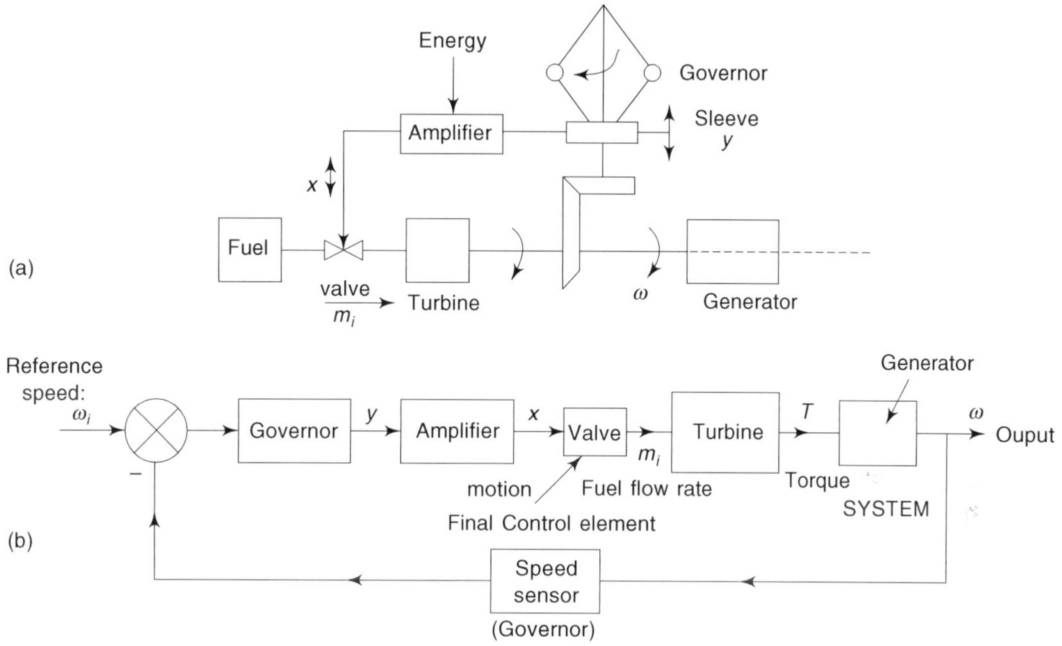

Fig. 19.3 *Representation of a speed control system*

Analog and digital control systems The variables or signals in an analog system may vary continuously and can acquire any value, while those in a digital or discrete or sampled data system, vary in a discrete fashion. A computer controlled system has discrete data of input and output to the computer which usually does the functions of a comparator and controller. The final control element like a valve is usually of analog type since the process or the system may have variations in the system parameters, in a continuous fashion. A computer controlled system is used in practice when a number of controlled variables are involved (like in a boiler system). In such a system, the controlled variables are: steam temperature, steam pressure and water level, while the manipulated variables are: fuel flow rate, air flow rate and feed water flow rate. The steam flow rate from the boiler is controlled depending on the load on the system (like in a turbine). In the case of a multi-variable system like the above, the variables interact in a complex way, and modelling and control are also quite complicated. Thus, computer control becomes necessary.

Sequence Control There are applications like chemical processes, production systems and consumer products which involve control actions of discrete nature, one action followed by the other in a sequence, e.g. in a cloth washing machine, the sequential control may involve loading of cloths, rotation of the drum, supply of water at a desired temperature for a fixed time and adding of detergent, draining of water, followed by clean wash, rinsing, drying and stopping of the drum. Each action is often time based and requires logic control actions, e.g. the machine door may close before rotation and supply of water start, and the machine may stop after all operation are completed, with provision of an emergency stop. The system in such cases may use logic devices and controllers like programmable logic controllers or microcontrollers.

19.2 ■ EXAMPLES OF FEEDBACK CONTROL SYSTEM AND THEIR BLOCK DIAGRAMS

Figure 19.4(a) shows a level control system in which the control is of on/off type. Any change in level from the desired position, would change the state of the switch. If the level rises, the switch would be off and vice versa. When the switch is on the solenoid winding is energized and the solenoid rod moves up, opening the valve and flow q_i is supplied to the tank. When the level rises above the set point, the switch is opened (or is off) and the solenoid is de-energised, closing the valve, and the inflow q_i is cut off so that the level falls to the set point. In such a case, the level is always fluctuating. Figure 19.4(b)

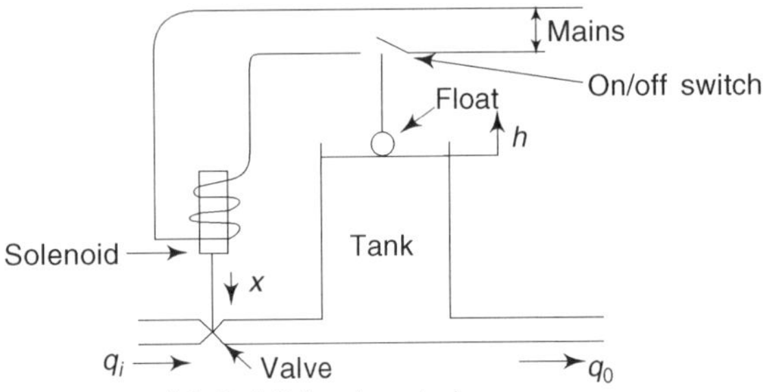

(a) On/off level control process

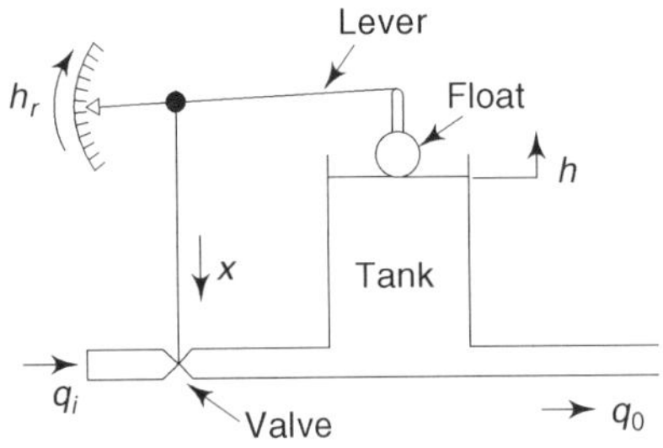

(b)

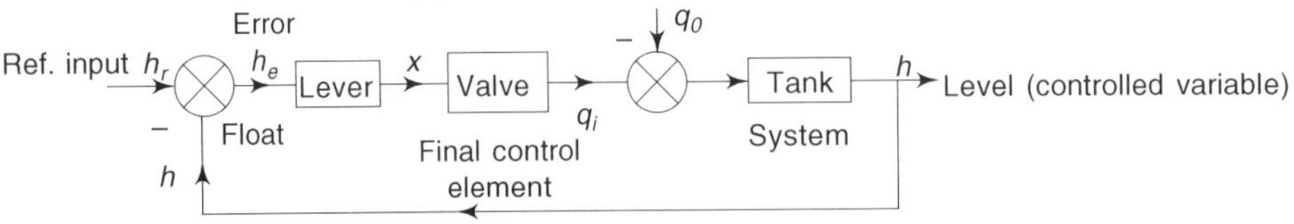

(c)

Fig. 19.4 *Level control process*

shows a proportional type of level control process or system in which the level sensed by a float closes or opens the valve, regulating the flow q_i proportional to the change in level h from the set point or desired value h_r. The block diagram of the system in Fig. 19.4(c) shows various functional elements of the system of Fig. 19.4(b). The outflow rate is q_o which acts as the load or disturbance.

Figure 19.5(a) shows a heat exchanger system, in which the heat from the hot fluid is transferred to water. The controlled variables is the temperature of the fluid at the outlet which is sensed by a sensor like a thermocouple and compared with the set point. The error is amplified/controlled and the electrical signal is converted to a pneumatic pressure signal using an *I–P* (current to pressure) converter. The pressure signal is applied to an elastic element like a diaphragm and the deflection of the element controls the motion of the final control value, controlling the inflow rate, q_{in}, of water going into the exchange. A block diagram of the system showing the various functional elements and also the variables, is shown in Fig. 19.5(b).

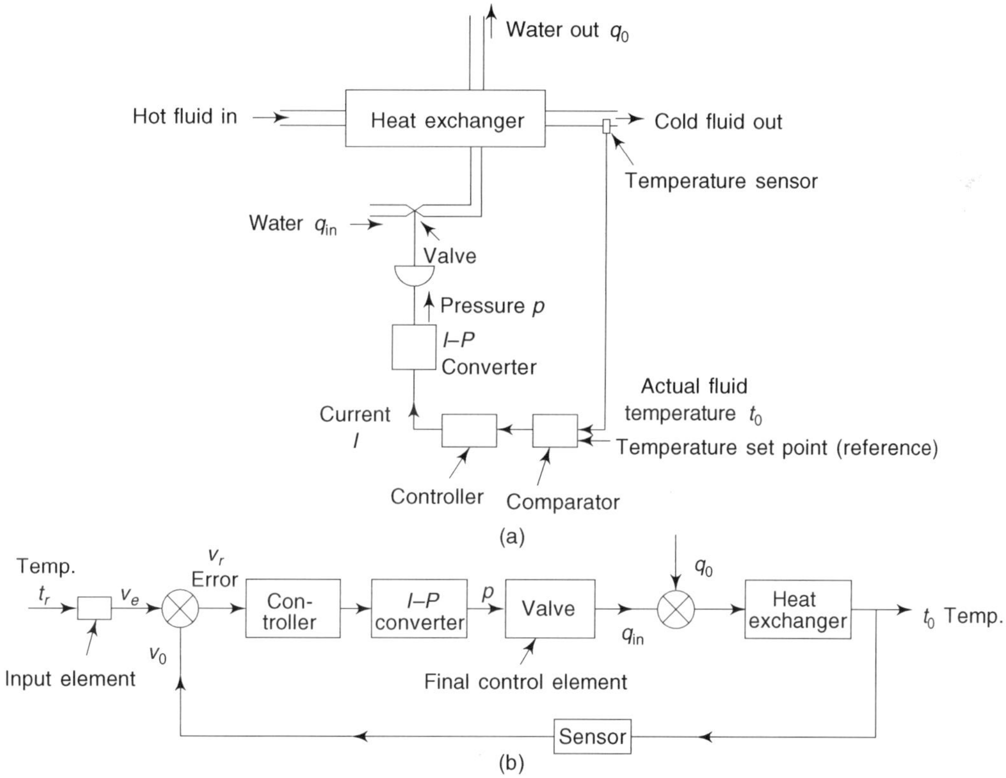

Fig. 19.5 *Fluid temperature control system*

The above system can be computer controlled as shown in the block diagram Fig. 19.6, in which the computer acts as the comparator and the controller. The temperature sensor and Analog-to-Digital-Converter (ADC) are in the feedback loop while Digital-to-Analog Converter (DAC) is in the forward loop as shown in Fig. 19.6. The starred variables are the sampled data or digital variables.

The block diagram in Fig. 19.7 shows a control system for controlling the motion of a robot arm. This is a computer controlled servo-system for providing a desired rotation θ_r to robotic arm. The system uses a digital encoder for sensing the motion θ_0 and has a DAC in the forward loop. The computer can control the various other arms of the robot as desired, by similar servo-systems. The starred variables are the digital variables.

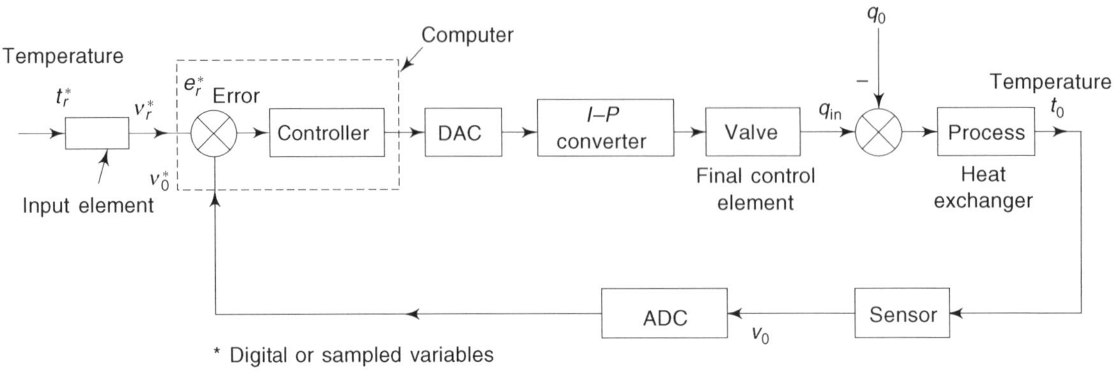

Fig. 19.6 *A schematic diagram of computer controlled fluid temperature system*

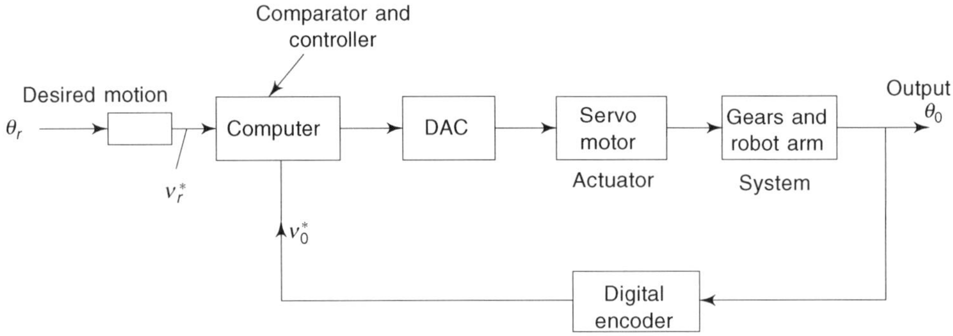

Fig. 19.7 *A schematic diagram of robot arm control*

19.3 ■ TRANSFER FUNCTIONS OF ELEMENTS, SYSTEM AND PROCESSES

19.3.1 Definition and Representation of Transfer Function in 't' and 's' Domains

Transfer function (TF) of an element is given by:

$$\text{TF} = \frac{\text{Output variable}}{\text{Input variable}}$$

This is represented by a block diagram as shown in Fig. 19.8 and gives the expression relating the two variables, from the governing equation of the element.

For example, for a rotating system with mass moment of inertia I and viscous damping coefficient B due to friction at the bearing, the equation relating the torque $u(t)$ and angular speed $\omega(t)$ is given as

Fig. 19.8 *A block diagram of control element showing 'transfer function'*

$$u(t) = I\,\frac{d\omega(t)}{dt} + B\omega\,(t) \tag{19.1}$$

In the above equation, the first term on the RHS is the inertia torque and the second term is the damping torque, with damping assumed as viscous or proportional to angular velocity.

Taking time derivative operator $D = \frac{d}{dt}$, Eq. (19.1) gives transfer function (TF) in time or t domain as:

$$\text{TF} = \frac{\omega(t)}{u(t)} = \frac{1}{ID + B} \tag{19.2}$$

Similarly, transfer function (TF) between rotation $\theta(t)$ and torque $u(t)$ is:

$$\text{TF} = \frac{\theta(t)}{u(t)} = \frac{1}{ID^2 + BD} \tag{19.3}$$

where $D^2 = \frac{\mathrm{d}^2}{\mathrm{d}t^2}$

Referring to Fig. 19.9(a), if the drive torque is $u(t)$ and load torque is $u_L(t)$, the governing equation is:

$$u(t) = u_L(t) + \text{ID}\ \omega(t) + B\omega(t).$$

or $u(t) - u_L(t) = (ID + B)\ \omega(t)$

or $$\omega(t) = \frac{u(t) - u_L(t)}{ID + B} \tag{19.4}$$

The graphical representation is as shown in Fig. 19.9. For convenience, $u(t)$ and other variable which are dynamic or function of time t are shown without the brackets as u, etc. Addition or subtraction is shown by a comparator, the output of which is $u_1 = (u - u_L)$, which in turn is the input to the block diagram giving Fig. 19.9 (b) as the representation of Eq. (19.4)

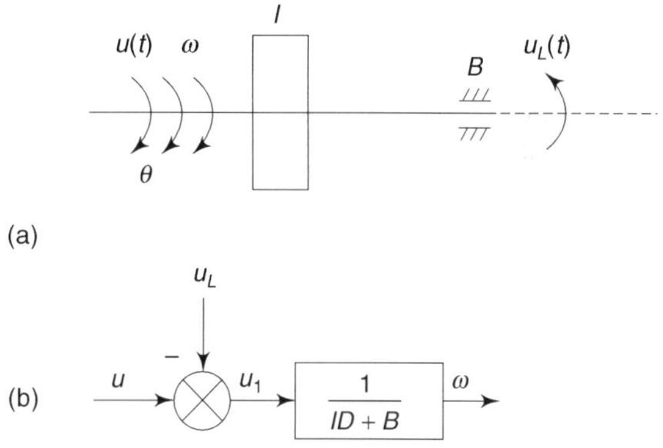

Fig. 19.9 *A schematic diagram of rotating system representation*

or $$\frac{\omega(t)}{u_1} = \frac{1}{ID + B}$$

Graphical representation like that in Fig. 19.9 (b) would form a part of the block diagram of the total system or process being controlled along with the representations for controllers, amplifiers, measuring elements, etc.

An alternate representation often used is in Laplace transform domain or s domain. It is shown below that D may be replaced by s, D^2 by s^2, and so on.

Laplace transform of any function $x(t)$ is $X(s)$, i.e.

$X(s) = \int_0^\infty e^{-st}\, x(t)\, \mathrm{d}t$, where s is the Laplace transform operator,

or $X(s) = L\, x(t)$, implying that $X(s)$ is the Laplace transform of $x(t)$.

It is known that Laplace transform of $\frac{\mathrm{d}x}{\mathrm{d}t}$ or of Dx is given by

$$L\,[D\,x(t)] = s\,X(s) - x(0),$$

where $x(0)$ is the value of $x(t)$ at $t = 0$ or its initial value. Since the system parameters like $x(t)$ vary from steady state, $x(0) = 0$, giving

$$L\,[Dx\,(t)] = s\,X\,(s)$$

or conversion form t domain to s domain of the system variables may be done by putting $D = s$ and taking $X(s)$ in place of $x(t)$.

Similarly,
$$L\,[D^2\,x(t)] = s^2\,X(s) - s\,x(0) - \dot{x}\,(0),$$

where $x(0)$ and $\dot{x}(0)$ are the values of $x(t)$ and $\frac{dx}{dt}(t)$ at $t = 0$.

Since all initial values are zero at $t = 0$, for reasons given earlier, D^2 may be substituted by s^2 and $x(t)$ by $X(s)$ for conversion from t domain to s domain. Further, D^n may be replaced by s^n, where n is an integer.

In the present chapter, the solution would be carried out either in t or s domain.

19.3.2 Comparator or Error Detector

The value of reference input $r(t)$ or $R(s)$ is compared with that of the controlled variable $c(t)$ or $C(s)$ by a comparator or error detector, which may be a potentiometric element, or a synchro or a lever etc., the representation of the equation $e(t) = r(t) - c(t)$ is as shown in Fig. 19.10. Its representation in the 's' domain is also shown.

Fig. 19.10 *Representation of comparator is t and s domains*

19.3.3 Controllers

The transfer functions of controller of a proportional type is K, as seen from the equation:

$$b(t) = Ke(t),$$

where K is a constant relating $e(t)$ viz. error signal being the controller input signal and the controller output signal $b(t)$. An amplifier is essentially a proportional controller.

Other types of controllers are:

- $P - D$ controller with transfer function
$$K + K_d D = K(1 + T_d\,D)x,$$
where $$T_d = \frac{K_d}{K}$$
- $P - I$ controller, with transfer function
$$K + \frac{Ki}{D} = K\left[1 + \frac{1}{T_i D}\right],$$
where $$T_i = \frac{K}{K_i}$$
- $P–I–D$ controller (or 3 term controller), with transfer function
$$K\left(1 + T_d\,D + \frac{1}{T_i D}\right).$$

These controllers may be of electrical, hydraulic or pneumatic types. In the alone transfer functions, T_d and T_i have units of time and are known as the derivative and the integral of time, respectively.

19.3.4 Actuators or Final Control Elements

Field controlled dc motor The error signal from the comparator after amplification is supplied to an actuator like a motor to control the system. In a field controlled dc motor, the voltage v_f is fed to the field, with armature current i_a, remaining constant. Change in $v(t)$, changes the current $i_f(t)$ and hence the speed $\omega(t)$ and position $\theta(t)$. The governing equations are:

$$v(t) = (R_f + L_f D)\, i_f(t) \tag{19.5}$$

where R_f and L_f are the field resistance and inductance, respectively, the latter being used since the variables are time dependent. $D = \dfrac{d}{dt}$ is the time derivative operator.

Motor torque $u(t) = K\, i_f(t)$, (19.6)

K being the motor torque constant (torque per unit field current),

$$u(t) = u_L(t) + (ID + B)\, \omega(t), \tag{19.7}$$

with $u_L(t)$ being load torque. I the mass moment of inertia of the rotating system, B is the viscous damping constant of the system and $\omega(t)$ is the angular speed (rad/s) of the motor.

The representation of Eqs. (19.5) – (19.7) is shown by transfer functions and comparator as in Fig. 19.11(b), where t in brackets is not shown for convenience. This practice will be followed also in other sections in this chapter. Similarly in Fig. 19.11(c), the representation of equations relating $v(t)$ and $\theta(t)$ is shown in time domain.

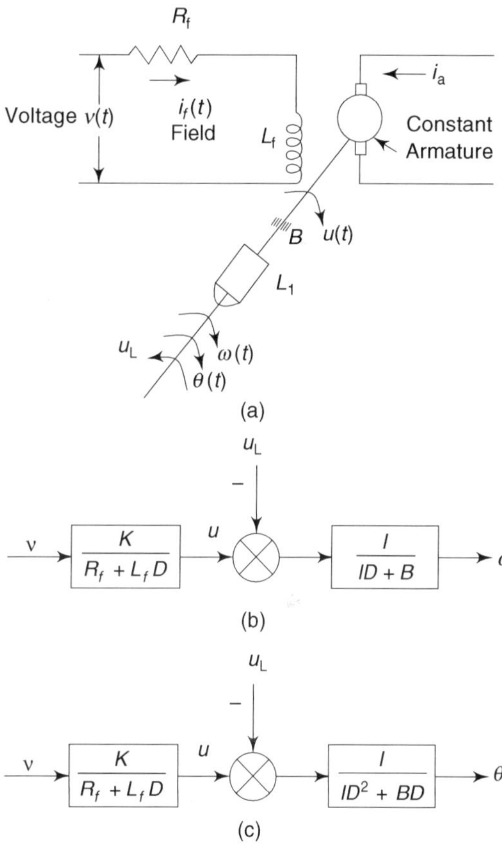

Fig. 19.11 *A schematic diagram of field controlled dc motor*

Armature Controlled Motor The outline diagram is shown in Fig. 19.12(a). The field current is kept constant, v, is the input voltage to the armature winding and is related with the error voltage from the comparator, after amplification/modification by the controller.

The transfer functions can be obtained from the governing equations as below.

Back emf across the armature = $k_b\, \omega$, where K_b is the back emf constant and ω is the angular speed of the armature.

$$v = k_b\omega + R_a I_a + L_a\, DI_a, \tag{19.8}$$

where R_a and L_a are the resistance and inductance of the armature winding, respectively and I_a is the armature current.

Motor Torque $u = K_a\, i_a$, (19.9)

where K_a is the armature constant

$$u = u_L + (ID + B)\,\omega \tag{19.10}$$

where u_L is the load torque, I the mass moment of inertia and B is the viscous damping constant of the rotating system.

The block diagram relating input v and output ω is the time domain, corresponding to Eqs. (19.8) – (19.10) is shown in Fig. 19.12(b).

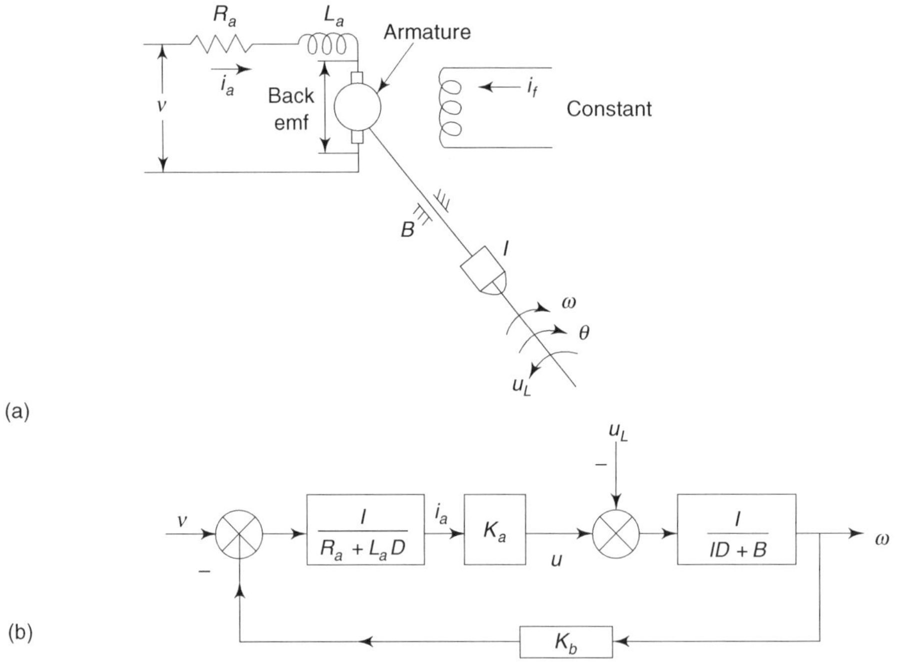

Fig. 19.12 *A schematic diagram of armature controlled dc motor*

Hydraulic Servomotor or Actuator The final control element in a feedback control system may be of hydraulic type (instead of an electric one like a motor) where loads are high. One such element is shown in Fig. 19.13(a). The input is motion x and output in y. Motion x is applied at the end E of a lever. It instantaneously pivots about G as in Fig. 19.13(b). Assuming F to be located at the middle of lever GF, the motion e of point F moves the piston valve up (as shown) and high pressure oil moves through the upper port of the power cylinder. (Under steady state conditions, the piston valve covers both the ports.) This moves the power piston down, as shown in Fig. 19.13(a). The lever pivots about point F as shown in Fig. 19.13(c). The motions are small, so the resulting motion

$$e = \frac{x}{2} - \frac{y}{2} \tag{19.11}$$

The flow rate of oil through a port, may be assumed to be proportional to the amount by which it is uncovered or equal to be, where b is called the port constant. Assuming oil to be incompressible,

Flow rate of oil through the port = be

$$= A\,\frac{dy}{dt}$$

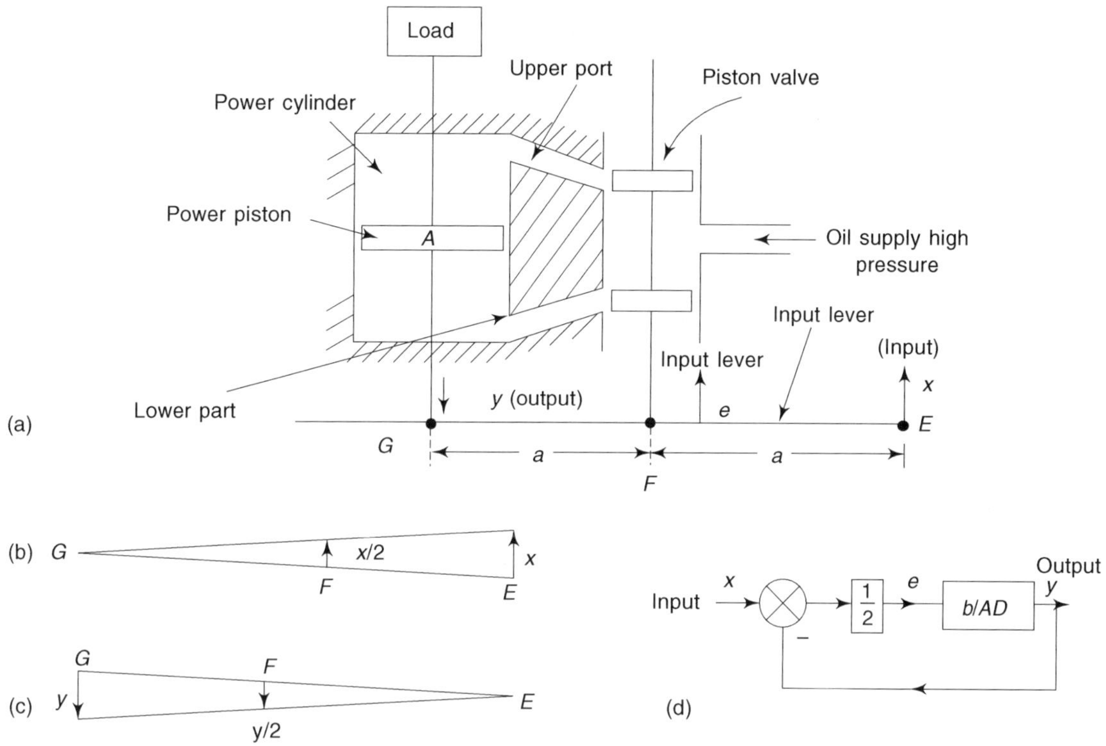

Fig. 19.13 *A schematic diagram of hydraulic servomotor or actuator*

or $$be = ADy, \tag{19.12}$$

D being $\frac{d}{dt}$ and A the area of power piston.

Using Eqs. (19.11) and (19.12), the transfer functions relating various variables are shown in Fig. 19.13(d) in the form of a block diagram.

19.3.5 Process Systems

The fluid and thermal systems can be represented in a fashion similar to that of the electrical systems.

Liquid Systems Figure 19.14 shows a liquid level process or system, in which the level h is the controlled variable, A the area of tank. Inflow and outflow rates are q_i and q_o, respectively, while the pressure at input is p_i and that in the tank at the bottom is p.

$$p = \rho gh, \tag{19.13}$$

Where ρ is mass density of the liquid.

The governing equations are

$$q_i = \frac{p_i - p}{R} \tag{19.14}$$

where R is the fluid resistance. Inflow rate q_i through the obstruction like a valve is the ratio of pressure drop to the value of the resistance R. (This is analogous to the electrical system where current is the ratio of the potential difference to the resistance.)

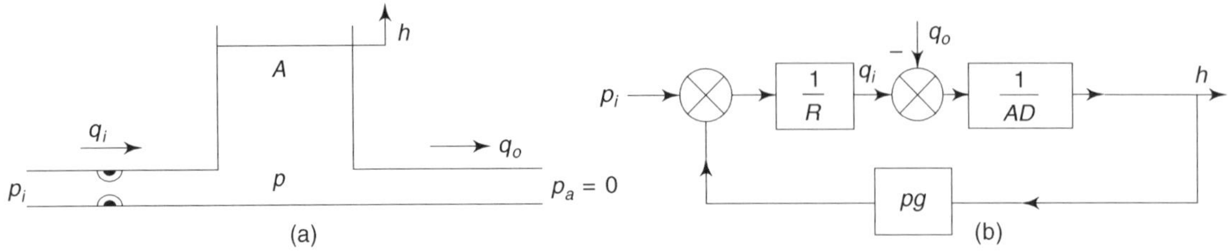

Fig. 19.14 *A schematic diagram of liquid level process*

Also,
$$C \frac{dp}{dt} = q_i - q_o, \tag{19.15}$$

where C is the fluid capacitance of the tank and $(q_i - q_o)$ is the net inflow rate increase in the tank. Taking the flow rates as volumetric ones for the liquid, and A the area of the tank, we may also write.

$$A \frac{dh}{dt} = q_i - q_o \tag{19.16}$$

From Eqs. (19.13)–(19.16), we get tank capacitance $C = \dfrac{A}{\rho g}$. This is analogous to the case of electrical system.

Equations (19.13) – (19.16) may be represented graphically by a block diagram as shown in Fig. 19.14(b).

Gas systems A gas pressure system with chamber capacitance C and flow resistance R is shown in Fig. 19.15.

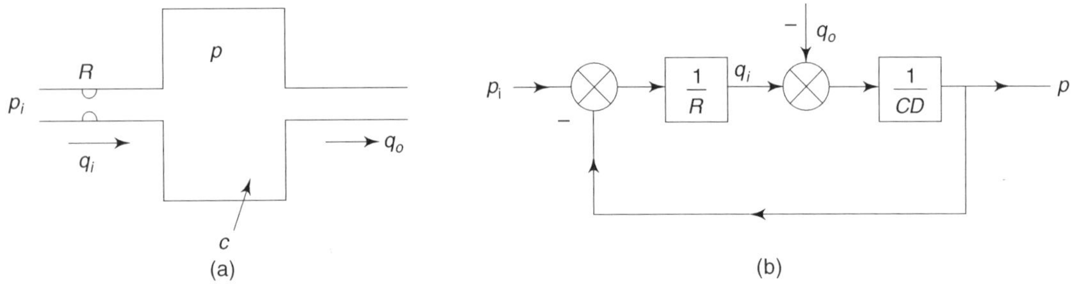

Fig. 19.15 *A schematic diagram of gas pressure process*

Here, q_i and q_o are used to denote mass flow rates of the gas at the inlet and the outlet, respectively, since the gas as a fluid is compressible. The governing equations are similar to Eqs. (19.14) and (19.15) and the corresponding block diagram is shown in Fig. 19.15. Gas capacitance C and resistance R can be found for any system using the gas laws and the flow equations.

Thermal Systems A thermal system shown in Fig. 19.16(a) can be represented by similar equations except that C is the thermal capacitance of the chamber and R is the resistance to the heat flow rate from the chamber to outside, the temperatures being 'θ' in the chamber and θ_o, outside. Inflow heat rate is q_i and outflow heat rate is q_o. The governing equations are:

$$C \frac{d\theta}{dt} = q_i - q_o \tag{19.17}$$

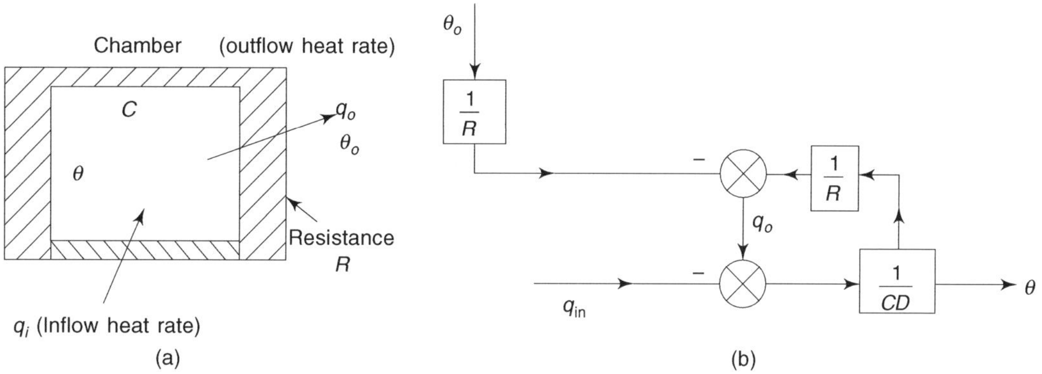

Fig. 19.16 *A schematic diagram of a typical thermal process.*

$$\frac{\theta - \theta_0}{R} = q_0 \tag{19.18}$$

The equations may be represented as in Fig. 19.16(b). The values of C and R may be found for any system using heat transfer and thermodynamics equations. All the equations written in the time domain can be easily converted to Laplace domain if required.

19.4 ■ BLOCK DIAGRAMS OF FEEDBACK CONTROL SYSTEM

Figure 19.17 shows the block diagram of a typical control system. The input signal r and feedback signal b are normally of electrical type and so is the error signal e.

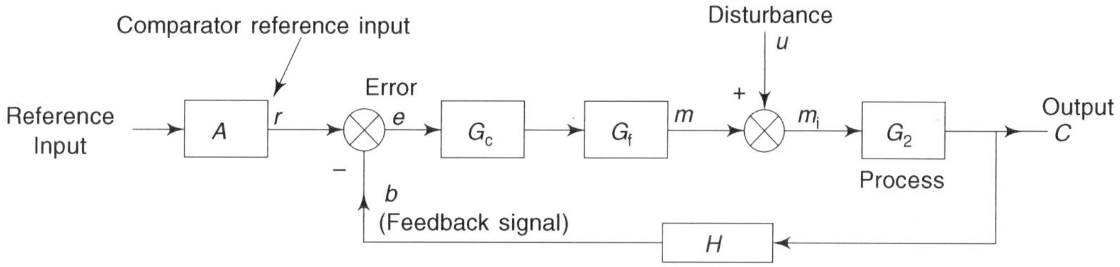

Fig. 19.17 *Typical block diagram of feedback control system*

The transfer function H in the Feedback loop is that of the measuring system or instrument used to measure the value of the controlled variable or the output.

G_c is the transfer function of the controller, which may be of proportional type, or proportional cum derivative type (*PD* type), or proportional cum integral type (*PI* type), or three-term controller, viz. *PID* type as mentioned earlier.

G_f is the transfer function of the final control element or actuator like electric motor, hydraulic or pneumatic type actuator or a value; G_2 is the transfer function of the process/system being controlled; 'm' and m_1 are the manipulated variables and u is the load or disturbance.

Figures 19.18 and 19.19 show the block diagrams in t and s domains, respectively. Any of these diagrams may be used for analysis. Since A, the transfer function of the reference input in Fig. 19.17 is normally a scaling factor, it has not been included in Figs. 19.18 and 19.19 for simplicity.

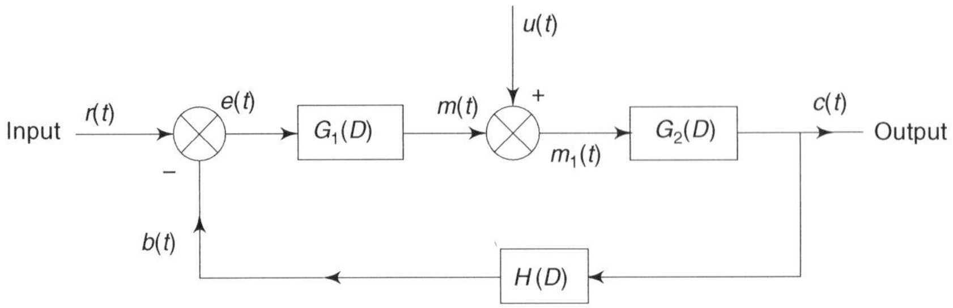

Fig. 19.18 *Block diagram used for analyses in t (time) domain.*

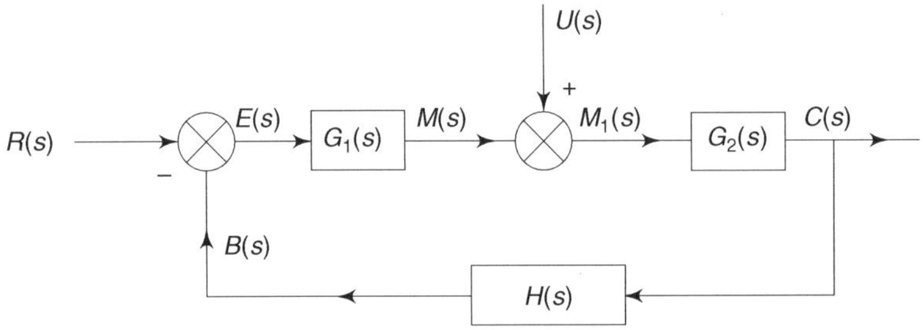

Fig. 19.19 *Block diagram used for analysis in s domain.*

Taking the system shown in Fig. 19.19, the relevant equations are:

$$E(s) = R(s) - B(s)$$

$$= R(s) - H(s)\, C(s) \tag{19.19}$$

$$M_1(s) = M(s) + U(s)$$

$$= G_1(s)\, E(s) + U(s) \tag{19.20}$$

$$C(s) = G_2(s)\, M_1(s) \tag{19.21}$$

From Eqs. (19.19)–(19.21), we get the relation between the output $C(s)$ and the two inputs $R(s)$ and $U(s)$ as below:

$$C(s) = \frac{G_1(s)G_2(s)[R(s)]}{1+G_1(s)G_2(s)H(s)} + \frac{G_2(s)[U(s)]}{1+G_1(s)G_2(s)H(s)} \tag{19.22}$$

It may be seen that transfer function $H(s)$ of the measuring system influences the output of the system. Considering one input at a time,

If $U(s) = 0$

$$C(s) = \frac{G_1(s)G_2(s)[R(s)]}{1+G_1(s)G_2(s)H(s)}$$

Taking $G(s) = G_1(s)\, G_2(s)$, viz.

Transfer function of the forward loop.

$$C(s) = \frac{G(s)}{1+G(s)H(s)}[R(s)] \tag{19.23}$$

$$\frac{C(s)}{R(s)} = \frac{G(s)}{1+G(s)H(s)} \tag{19.24}$$

$G(s)/R(s)$ is called the closed loop transfer function of the system. $G(s)\ H(s)$ is called the open loop transfer function of the system. It is the transfer function of the system relating $b(t)$ and $e(t)$ if the loop was opened at the comparator.

The denominator of $\frac{C(s)}{R(s)}$ put to zero is called the characteristic equation. From Eq. (19.24), $1 + G(s)\ H(s) = 0$ is derived as the characteristics equation. It is seen to influence the transient response of the system, as will be shown later. From Eq. (19.24), we get a differential equation.

$$[1 + G(s)H(s)]\ C(s) = G(s)\ R(s) \tag{19.25}$$

The L.H.S. is a polynomial in s normally. Since $s = D = \frac{d}{dt}$, it is an ordinary differential equation, viz.

$$[1 + G(D)\ H(D)]\ c(t) = G(D)\ r(t) \tag{19.26}$$

The classical solution of the above differential equation for a given input $r(t)$ is the sum of the complementary solution (with R.H.S. = 0) plus the particular solution. The initial conditions are applied to find the unknown constants.

Similarly, using Eq. (19.22), the output $C(s)$ for input $U(s)$ with $R(s) = 0$ is given by:

$$C(s) = \frac{G_2(s)U(s)}{1+G(s)H(s)}$$

where $G(s) = G_1(s)\ G_2(s)$, as taken in Eq. (19.23).

19.5 ■ TRANSIENT AND STEADY STATE RESPONSE OF CONTROL SYSTEMS

All variables in a block diagram of a control system represent changes from the steady state values, when a reference input or disturbance is applied. For analysis purpose, to determine the transient and steady response to a given input, the input is taken to be either step type or ramp type.

For example, a step type reference input would be $r(t) = f$, say, where f is a constant. If $f = 1$, it is called unit step type. Similarly, a ramp type input is of type $r(t) = Ct$, where C is a constant and $r(t) = t$ for unit ramp input. The same applies to load or disturbance $u(t)$.

The transient response may be determined in the t domain by finding the sum of the classical and particular solutions for a given input (as mentioned earlier) or it may be determined in the s domain by using Laplace transform method. It is important that the transient response of a control system is not too slow and also is not too much oscillatory, reaching high values of output $C(t)$ which may be damaging. Further, the steady state value of the output should be as close to the desired or reference input as possible or error $e\ (t)$ should be minimum, preferably tending to zero, at steady state.

The steady state error can be found from the transient solution using $t \to \infty$. It is somewhat easier to find the steady state value of error by using the final value theorem of Laplace transform of error function $E(s)$ i.e.

Steady state error
$$\underset{t\to\infty}{e(t)} = \underset{s\to 0}{sE(s)} \tag{19.27}$$

where $E(s)$ is the Laplace transform of $e(t)$. For example, if

$$\text{error } e(t) = \left[\frac{ID^2 + aD}{ID^2 + aD + K}\right] r(t),$$

$$\text{steady state error} = \underset{s \to 0}{sE(s)} = \left[\frac{Is^2 + as}{Is^2 + as + k}\right] R(s),$$

where $R(s)$ is Laplace transform of $r(t)$.

For unit step input $r(t) = 1$, $R(s) = \frac{1}{s}$ and for $r(t) = t$, the Laplace transform $R(s) = \frac{1}{s^2}$, as may be seen from the table of Laplace transforms.

The following examples give a technique for finding transient response and steady state errors for a first-order system or process, with a proportional controller when subjected to inputs $r(t)$ or $u(t)$.

Problem 19.1 *For the system with a block diagram as in Fig. 19.20, taking K = 200, find*

(i) closed loop transfer function
(ii) transient response for unit step reference input r(t)
(iii) steady state error for unit ramp reference input r(t)
(iv) steady state error for unit step disturbance input u(t)

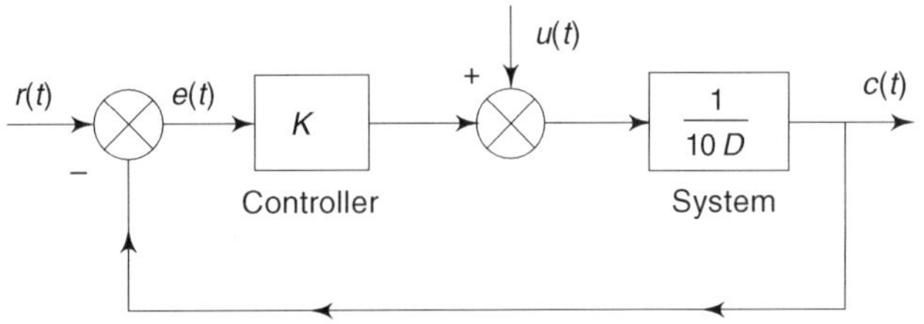

Fig. 19.20 *Figure for Problem 19.1*

Solution K is constant or it is a proportional controller. Using Eq. (19.22) in time domain for the above problem,

$$G_1(t) = K,\ G_2(t) = \frac{1}{10D},\quad D = \frac{d}{dt}$$

$$H(t) = 1$$

$$c(t) = \frac{KG_2(t)\,r(t)}{1 + KG_2(t)} + \frac{u(t)\,G_2(t)}{1 + KG_2(t)}$$

(i) If $u(t) = 0$ i.e. load input does not change, using equation $c(t) = \dfrac{Kr(t)}{\dfrac{1}{G_2(t)} + K}$

$$\text{closed loop transfer function } \frac{c(t)}{r(t)} = \frac{K}{K + \dfrac{1}{G_2(t)}}$$

(ii) For $r(t) = 1$, viz. unit step input,

$$c(t) = \frac{K}{K + 10KD} = \frac{200}{200 + 20D}$$

$$(10D + 200)\,c(t) = 200$$

$c(t)$ = complementary solution + particular solution

$$= Ae^{-20t} + 1$$

Using initial condition, $c = 0$ at $t = 0$

$$A = -1$$

or $$c(t) = [1 - e^{-20t}]$$

This is plotted in Fig. 19.21.

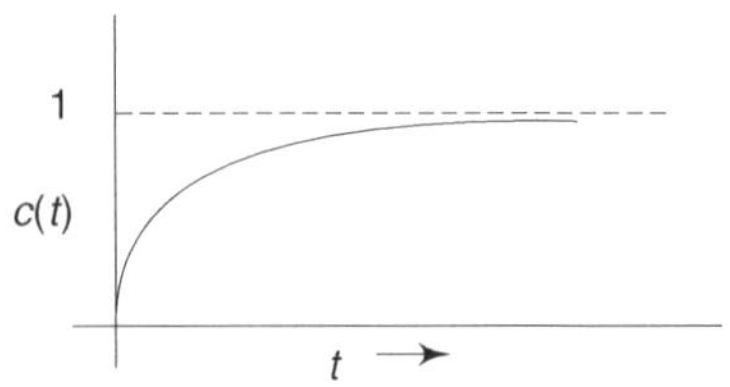

Fig. 19.21

The steady state value of

$$c(t) = 1$$

Thus there is no steady state error since $r(t) = 1$

(iii) For $r(t) = t$, viz. unit ramp input,

$$U(t) = 0,$$

Using $$c(t) = \frac{Kr(t)}{(10D + K)}$$

Error $$e(t) = r(t) - c(t)$$

$$= \left[1 - \frac{K}{10D + K}\right] r(t) = \left(\frac{10D}{10D + K}\right) r(t)$$

In Laplace transform domain

$$E(s) = \left(\frac{10s}{10s + K}\right) R(s)$$

$$\text{Steady state error} = \underset{t \to \infty}{e(t)} = \underset{s \to 0}{sE(s)}$$

$$= \underset{s \to 0}{s} \left(\frac{10s}{10s + 200}\right) \frac{1}{s^2}$$

$$= \frac{10}{200} = \frac{1}{20}$$

(iv) For $u(t) = 1$, $r(t) = 0$

using equation $$c(t) = \frac{G_2(t)}{1 + KG_2(t)} = \frac{1}{\frac{1}{G_2(t)} + K}$$

$$c(t) = \frac{u(t)}{10D + K} = \frac{u(t)}{10D + 200}$$

$\therefore$ $$e(t) = r(t) - c(t)$$

$$= -c(t) = -\frac{u(t)}{10D + 200}$$

For $$u(t) = 1,\ u(s) = \frac{1}{s},$$

or $$\text{steady state error} = \underset{s \to 0}{sE(s)}$$

$$= -\frac{s\frac{1}{s}}{10s + 200} = -\frac{1}{200}$$

19.6 ■ EFFECT OF VARIOUS TYPES OF CONTROL ACTIONS ON DYNAMIC PERFORMANCE

19.6.1 System with Reference Input

As in Eq. (19.23), output $c(t)$ is related with input $r(t)$ and disturbance $u(t)$.

Considering the case of the only input $r(t)$, with

$u(t) = 0$, we get

$$c(t) = \frac{G_1(t)\,G_2(t)\,r(t)}{1 + G_1(t)\,G_2(t)\,H(t)} \tag{19.28}$$

Proportional Control Taking a second order system with

$$G_2(t) = \frac{1}{ID^2 + aD}$$

and proportional controller with $G_1(t) = K$, K being constant. Also, $H = 1$ for unity feedback system. Substituting in Eq. (19.28),

$$c(t) = \frac{kr(t)}{ID^2 + aD + K}$$

or

$$(ID^2 + aD + K)\,c(t) = Kr(t) \tag{19.29}$$

Equation (19.29) is a second order differential equation. For step input $r(t) = f$, f being a constant,

$$(ID^2 + aD + K)\,c(t) = Kf$$

The solution $c(t)$ of the above equation may be obtained by classical method or by Laplace transform. Classical method will be used here. Dividing both sides of the above equation by K, non-dimensional parameters can be used as done in Section (3.3) for a second order instrument system.

$\omega_n = \sqrt{\dfrac{K}{I}}$, ω_n being undamped natural frequency.

$$\text{Damping ratio } \xi = \frac{a}{2\sqrt{KI}},$$

we get

$$\left(\frac{D^2}{\omega_n^2} + \frac{2\xi}{\omega_n} D + 1\right) c(t) = f \tag{19.30}$$

The above equation is similar to Eq. (3.60), for which the solution is obtained as in Eq. (3.66) for the case $\xi < 1$, or the underdamped case.

$$\frac{c(t)}{f} = \left[1 - e^{-\xi\omega_n t}\left(\cos \omega_n \sqrt{1-\xi^2}\, t + \frac{\xi}{\sqrt{1-\xi^2}} \sin \omega_n \sqrt{1-\xi^2}\, t\right)\right] \tag{19.31}$$

Equation (19.31) is obtained from Eq. (3.66) by putting $f = Kx_s$ in the latter equation. This can also be written as

$$\frac{c(t)}{f} = \left[1 + \frac{e^{-\xi\omega_n t}}{\sqrt{1-\xi^2}} \sin(\omega_n \sqrt{1-\xi^2}\, t + \phi)\right]$$

where

$$\sin \phi = \sqrt{1-\xi^2} \tag{19.32}$$

Equation (19.32) is similar to Eq. (3.67).

A plot of the transient output response $c(t)$ for step input $r(t) = f$ is shown in Fig. 19.22 for values of ξ equal to 0.1 and 0.2.

If may be seen that the response in oscillatory with circular frequency $\omega_n \sqrt{1-\xi^2}$ and decays with time, the decay rate increasing with increase in ξ as expected. It is seen that the first value of the peak output is the highest and has to be limited to prevent possible damage to the system. The difference between the first peak value and the input value is called peak overshoot. Its value can be found using Eq. (19.31). Expressed as percentage peak overshoot, it is given as:

$$\% \text{ peak overshoot} = 100 \times \frac{-\pi\xi}{e^{\sqrt{1-\xi^2}}}$$

It is seen to occur at time t_p which equals $\dfrac{\pi}{\omega_n\sqrt{1-\xi^2}}$.

The peak overshoot is 53% for $\xi = 0.2$ and 25.4% for $\xi = 0.4$.

The transient response is shown in Fig. 19.22, it should be fast and reach the steady state in the shortest possible time. For this purpose, the following timings are specified, as in Fig. 19.22:

— rise time (t_r)
— settling time (t_s).

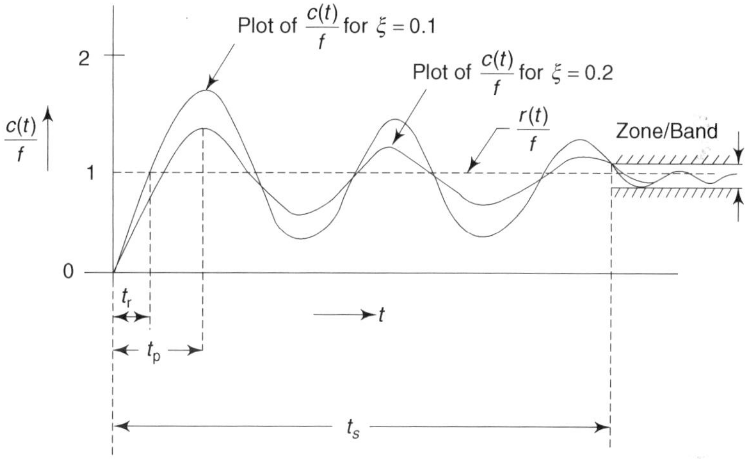

Fig. 19.22 *Transient response to step input*

The rise time t_r is the time taken for the output to reach the input step value and is labelled in Fig. 19.22 for $\xi = 0.1$. Its expression can be found from Eq. (19.31) and is seen to be

$$t_r = \begin{cases} \dfrac{1 \cdot 8}{\omega_n}, & \text{for } \xi = 0.2 \\ \dfrac{2 \cdot 2}{\omega_n}, & \text{for } \xi = 0.4 \end{cases}$$

Another useful feature is 'settling time t_s', which is defined as the time taken for the output to reach within a zone or band of $\pm x\%$ about the steady state value and later remain within that zone, as shown in Fig. 19.22. It can also be found from Eq. (19.22), and the value is seen to be:

$$t_s = \begin{cases} \dfrac{3 \cdot 9}{\xi\omega_n}, & \text{for } x = 2\% \\ \dfrac{3}{\xi\omega_n}, & \text{for } x = 5\% \end{cases}$$

The steady error $e(t) = r(t) - c(t)$ is seen to be zero as $t \to \infty$ in Eq. (19.22). Alternatively, it may be found by the final value theorem in Laplace transform domain as follows.

Using Eq. (19.29) in the s domains,

$$C(s) = \frac{KR(s)}{Is^2 + as + K} \tag{19.33}$$

Error

$$E(s) = R(s) - C(s)$$

$$= R(s) - \frac{KR(s)}{Is^2 + as + k}$$

$$= \frac{(Is^2 + as)\,R(s)}{Is^2 + as + K}$$

For step input

$$r(t) = f,\ R(s) = \frac{f}{s}$$

Thus,

$$E(s) = \frac{(Is^2 + as)\,f}{s(Is^2 + as + K)}$$

Steady state error

$$= \underset{s \to 0}{sE(s)} = 0$$

Thus, the steady state error is seen to be zero, due to step input.

If the reference input were unit ramp type, or $r(t) = t$,

Laplace transform $R(s) = \frac{1}{s^2}$, and for the above case, as in Eq. (19.33), steady state error, by application of the final value theorem is seen to be $\frac{a}{K}$. This error thus increases with increase of damping and decreases with increase of gain K.

Thus, if the controlled variables were position and the input were constant ramp type, viz. constant velocity, there would be a steady state error in the value of position.

Proportional Plus Derivative Control (P–D Control) The transfer function of the controller for P–D control is: $K(1 + T_d D)$, T_d being derivative time and K, the proportional constant. Derivative control cannot be used alone as the controller output for a constant input would be zero.

Using the above transfer function, instead of K in Eq. 19.28, we get:

$$[ID^2 + (a + KT_d)\,D + K]\,c(t) = (K + KT_d)\,r(t) \tag{19.34}$$

Thus, damping ratio $\xi = \frac{a + KT_d}{2\sqrt{KI}}$,

which shows that the use of derivative control increases the effective damping of the system and thus improves the transient response by decreasing the peak overshoot. It can be checked using the final value theorem that the steady state error for the above case would be zero and $\frac{a}{K}$ for unit step and unit ramp input, respectively, and thus remain unaffected by adding derivative control to the proportional one.

Proportional Plus Integral Control (P–I Control) The transfer function of the controller for *P–I* control is: $K\left(1 + \frac{1}{T_i D}\right)$, T_i being integral constant and K, the proportional constant. Using the above transfer function instead of K in Eq. (19.28), we get:

$$\left[ID^3 + aD^2 + KD + \frac{K}{T_i}\right] c(t) = \left(KD + \frac{K}{T_i}\right) r(t) \tag{19.35}$$

$$e(t) = r(t) - c(t)$$

Substituting for $c(t)$ from Eq. (19.34), we get

$$e(t) = \left[\frac{ID^3 + aD^2}{ID^3 + aD^2 + KD + \dfrac{K}{T_i}} \right] r(t) \tag{19.36}$$

To find the steady state error using the final value theorem for unit step input or $r(t) = 1$ i.e. $R(s) = \frac{1}{s}$ and using Eq. (19.36) in s domain, steady state error $= \underset{s \to 0}{sE(s)} = 0$

The steady state error can also be shown to be zero for unit ramp reference input.

Since the governing Eq. (19.35) is of third order, it may result in instability of the system, as discussed in the next section, unless the parameters are carefully chosen.

19.6.2 System with Disturbance or Load Input

Considering the case of only load input $u(t)$, with $r(t) = 0$, we get using Eq. (19.23),

$$c(t) = \frac{G_2(t)\,u(t)}{1 + G_1(t)\,G_2(t)\,H(t)} \tag{19.37}$$

Considering a second order system with unity feedback, viz. $G_2(t) = \dfrac{1}{ID^2 + aD}$, $H(t) = 1$
Taking the case of proportional control or
$G_1(t) = K$, we get from Eq. (19.28)

$$(ID^2 + aD + K)\; c(t) = u(t)$$

Error

$$e(t) = r(t) - c(t)$$
$$= -\,c(t) \qquad [\text{since } r(t) = 0]$$
$$= -\frac{u(t)}{ID^2 + aD + K}$$

In Laplace transform domain,

$$E(s) = \frac{-U(s)}{Is^2 + as + K}$$

The steady state error due to unit step load input $U(s) = \frac{1}{s}$ gives on application of the final value theorem,

Steady state error $= \underset{s \to 0}{sE(s)} = -\dfrac{1}{K}$.

Unlike in the similar case with unit step reference input in the previous section where steady state error is zero, the load step change gives a steady state error. This may be explained from the block diagram given in Fig. (19.23), by the fact that $r(t) = 0$ and to respond to the change in $u(t)$, an error would be present, under steady state equilibrium conditions.

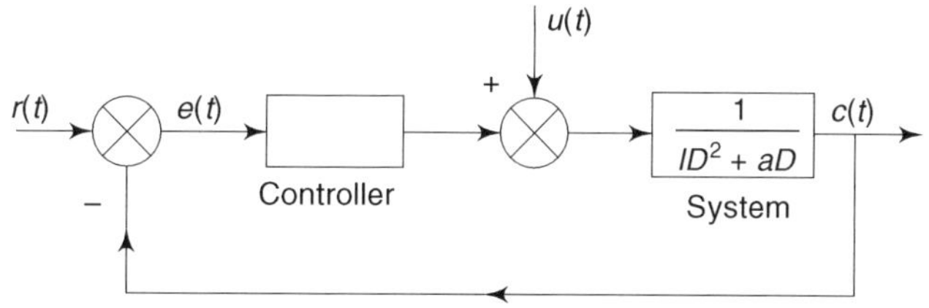

Fig. 19.23 *Block diagram of the system*

Problem 19.2 *A position control feedback system is shown in Fig. 19.24. The potentiometers unit acts as both sensor and comparator with a transfer function of 0.628 V/revolution. Neglect the motor field inductance and armature inertia. Load inertia I = 1kg-m². Speed ratio of mortor to load = 10*

Bearing viscous damping constant B = 0.4 Nm s/rad

$$R_f = 20\ \Omega$$

Motor torque constant = 5 Nm/A of field current

Difference amplifier gain K = 20.

Draw a block diagram, find the closed loop transfer function $\frac{\theta_o}{\theta_r}$ *and* $\theta_o(t)$ *for unit step input* $\theta_r(t)$*. Also, determine the peak overshoot of the system.*

Solution The governing equations are

Error voltage $e = (\theta_r - ,\theta_0)\, C_1$, where C_1 is the potentiometer constant.

$$e_f = Ke$$

$$i_f = \frac{e}{R_f}$$

Motor torque $T_m = C\, i_f$, where C is the motor torque constant, $T_0 = 10\ T_m$

$$T_0 = I\ddot{\theta}_0 + B\dot{\theta}_0,$$

where I is the mass moment of inertia of load, and B is the viscous damping constant of the load rotating system and dot . is differentiation with respect to t.

or $$T_0 = (ID^2 + BD)\ \theta_0$$

where $$D = \frac{d}{dt}$$

The block diagram for the system of Fig. 19.24 is shown in Fig. 19.25 as below.

Forward loop transfer function

$$G = \frac{C_1 C}{R_f\ (ID^2 + BD)}$$

Substituting $C_1 = 0.628/2\pi$ V/rad $= 0.1$ V/rad

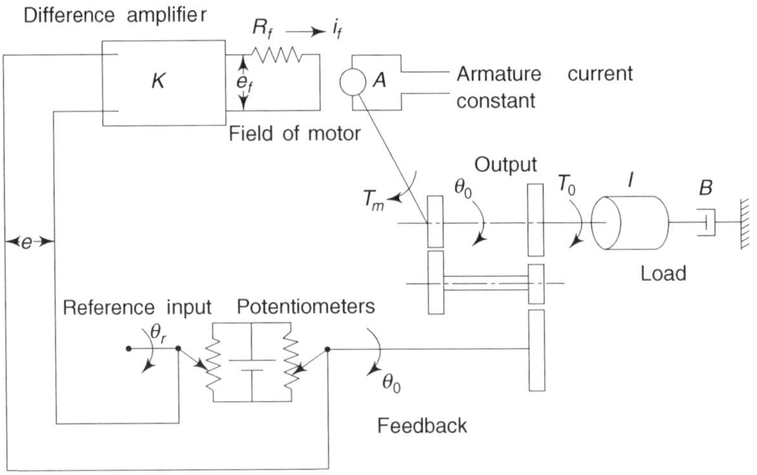

Fig. 19.24 *Figure for Problem 19.2*

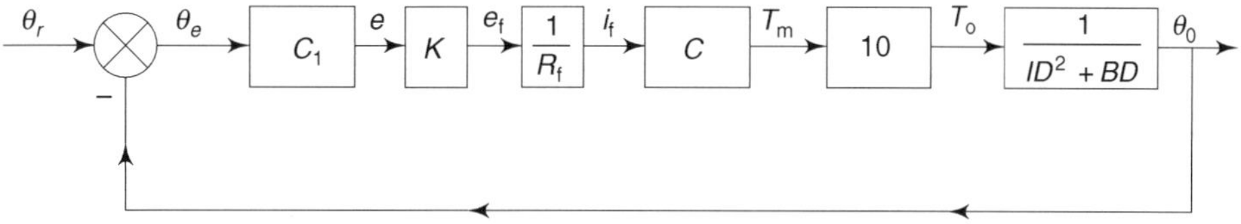

Fig. 19.25 *Block diagram of system of Fig. 19.24*

$$C = 5 \text{ Nm/A}$$
$$R_f = 20\ \Omega$$
$$K = 20,\ I = 1 \text{ kg m}^2$$

and
$$B = 0.4 \text{ Nm s/rad}$$

$$G = \frac{0.5 \times 20\,(10)}{20\,(D^2 + 0.4D)} = \frac{5}{(D^2 + 0.4D)}$$

Closed loop transfer function $= \dfrac{G}{1+G}$, for $H = 1$,

H being the feedback element transfer function,

or
$$\frac{\theta_0}{\theta_r} = \frac{1}{\frac{1}{G} + 1} = \frac{1}{\frac{D^2 + 0.4D}{5} + 1}$$

or
$$\frac{\theta_0}{\theta_r} = \frac{5}{D^2 + 0.4D + 5}$$

For
$$\theta_r = 1$$

$$(D^2 + 0.4\,D + 5)\ \theta_0 = 5\ \theta_r\,(t) = 5 \text{ for } \theta_r(t) = 1$$

Using Eq. (19.31),

$$\theta_0(t) = \left[1 - e^{-\xi\omega_n t}\left(\cos \omega_n \sqrt{1-\xi^2}\ t + \frac{\xi}{\sqrt{1-\xi^2}} \sin \omega_n \sqrt{1-\xi^2}\ t\right)\right]$$

where undamped natural frequency $\quad \omega_n = \sqrt{5} = 2.24$ rad/sec

Damping ratio $\quad \xi = \dfrac{0.4}{2\sqrt{5}} = 0.089$

% Peak overshoot $\quad = 100 \times e^{\frac{-\pi\xi}{\sqrt{1-\xi^2}}} = 74.6\%$

Problem 19.3 *For the temperature control system of Fig. 19.26, draw the block diagram and derive the expression relating input θ_r output θ and disturbance θ_a, in Laplace transform notation. Find the characteristic equation and damping ratio of the system.*

Temperature θ is the controlled variable in the chamber of thermal capacitance C with outside temperature θ_a, R is the thermal resistance to heat flow from the chamber to outside.

If motion $z = C_1\theta_r$, where θ_r is the reference or desired input, heat inflow rate from heater $q_{in} = k$ y. The temperature sensing system consisting of bulb, capillary and liquid filled bellows has transfer function $\dfrac{x}{\theta} = \dfrac{C_2}{\tau s + 1}$, τ being the time constant representing measurement delay and s is the Laplace transform operator

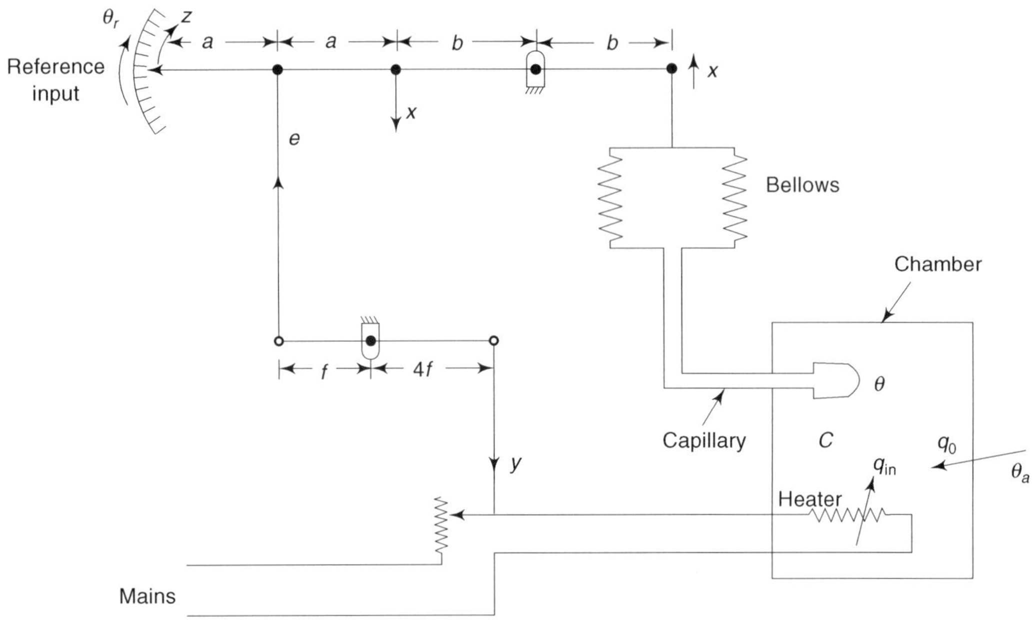

Fig. 19.26 *Figure for Problem 19.3*

$$C_1 = C_2 = 1 mm/°C,$$
$$\tau_m = 10 s$$
$$\frac{q_{in}}{y} = 100 W/mm$$
$$R = 0.1 °C/W$$
$$C = 400 J/C°$$
$$RC = 40 s$$

Solution The governing equations are

$$q_{in} - q_o = CD\theta, \quad R = \frac{\theta - \theta_a}{q_0}$$
$$y = 4e$$
$$e = z - x$$
$$z = 0.1\ \theta_r, \ z \text{ being in mm and } \theta_r \text{ in °C},$$
$$\frac{x}{\theta} = \frac{C_2}{1 + T_m s} = \frac{1}{1 + 10s}$$

Eliminating q_0 from the first two equations:

$$q_{in} = \left(\frac{RCD + 1}{R}\right)\theta - \frac{\theta_a}{R}$$

The block diagram is shown in Fig. 19.27.

In the above, $C_1 = C_2 = 1$ mm/C°, $\tau_m = 10$ s

$$\frac{R}{1 + RCs} = \frac{0.1}{1 + 40s}$$

Figure 19.28 is the simplified block diagram

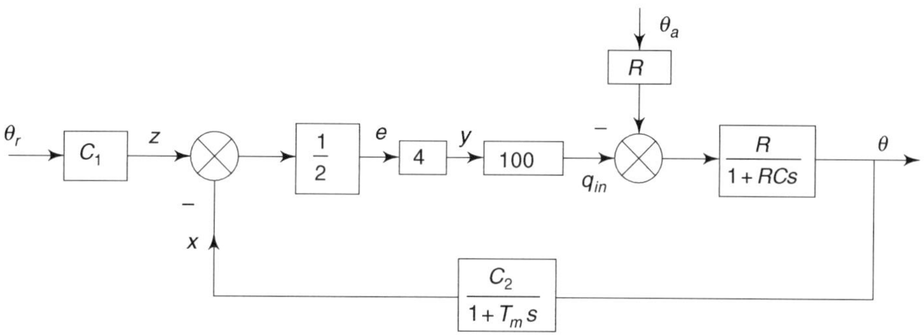

Fig. 19.27 *Block diagram of Fig. 19.26.*

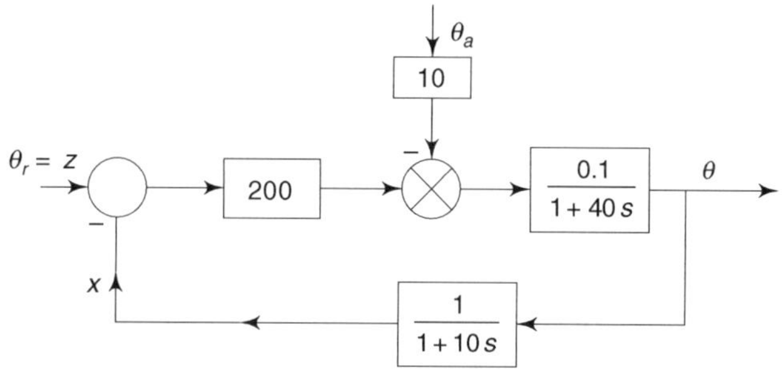

Fig. 19.28 *Simplified block diagram of Fig. 19.27*

From Fig. 19.28, we can write

$$\left\{\left[\theta_r - \frac{\theta}{1+10s}\right]200 - 10\,\theta_a\right\}\frac{0.1}{1+40s} = \theta$$

or
$$200\,\theta_r - \frac{200\,\theta}{1+10s} - 10\theta_a = 10\,(1+40s)\,\theta$$

Simplifying, we get the following relation between output θ and inputs θ_r and θ_a,

$$(400\,s^2 + 50\,s + 21)\,\theta = (200\,s + 20)\,\theta_r - (10\,s + 1)\,\theta_a$$

$$\text{Damping ratio } \xi = \frac{50}{2\sqrt{400\,(21)}} = 0.27$$

19.7 ■ STABILITY OF CONTROL SYSTEMS

A control system is stable if it returns to equilibrium position after it is disturbed. The transient response after application of reference input or disturbance, settles down to a steady state in the case of first and second order control systems discussed. This is due to the fact that the characteristic equation in each of the cases discussed, had roots with negative real parts and thus there is an exponential decay of the complementary part in the transient solution. If, however, any root of the characteristics equation has a positive real part, the output will keep on increasing with time and the system will become unstable, as in Fig. 19.29 for both the cases.

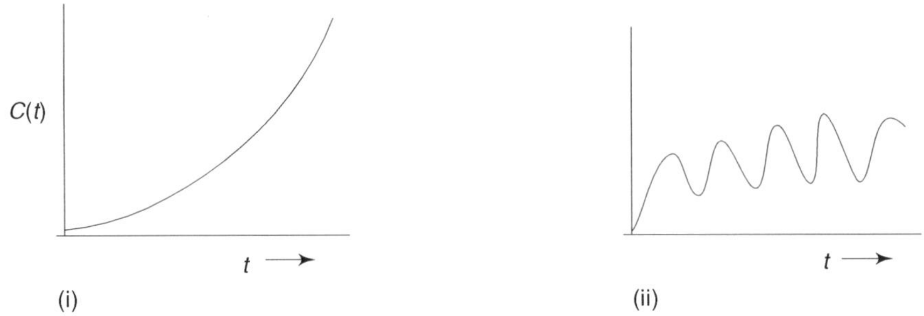

Fig. 19.29 *Response of unstable systems*

One of the major causes of instability in a control system is the presence of time lags and friction e.g. in a position control system, the motor torque controlled by the error signal, may vary in such a way due to the above mentioned reasons, that the former does not become zero but remains finite when error signal is zero. Due to the system inertia and friction (varying with speed in a rather complex fashion), the torque may keep on increasing the response at the null position, causing instability.

To check for instability, it needs to be checked if any root of its characteristic equation has a positive real part. If so, the parameters need to be changed till all roots of the characteristic equation have negative real parts.

Rouths' criterion gives the number of roots of a characteristic equation which have positive real part and is stated as below. Consider a characteristic equation of a general form, given as:

$$a_0 s^n + a_1 s^{n-1} + a_2 s^{n-2} + \dots + a_{n-1} s + a_n = 0 \tag{19.38}$$

where n is an integer.

Arrange the coefficients in an array, as below and calculate additional terms, as shown.

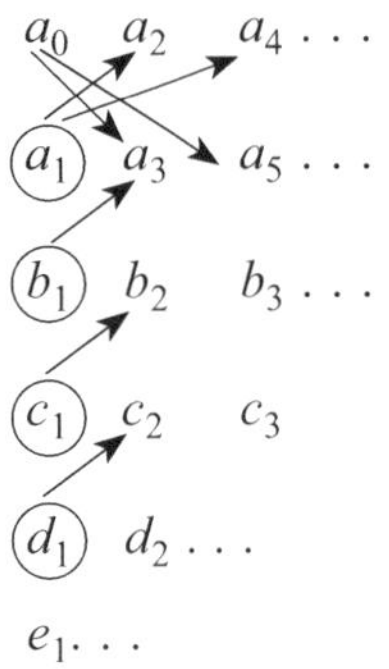

where

$$b_1 = \frac{a_1 a_2 - a_0 a_3}{a_1}, \quad b_2 = \frac{a_1 a_4 - a_0 a_5}{a_1}$$

$$c_1 = \frac{b_1 a_3 - a_1 b_2}{b_1}, \quad c_2 = \frac{b_1 a_5 - a_1 b_3}{b_1}$$

$$d_1 = \frac{c_1 b_2 - b_1 c_2}{c_1}, \quad d_2 = \frac{c_1 b_3 - b_1 c_3}{c_1}$$

$$e_1 = \frac{d_1 c_2 - c_1 d_2}{d_1}, \dots$$

This is continued in each row and column, until only 0's are obtained, for additional terms. According to Rouths' criterion, the number of changes of sign in the left hand column, of the array, starting from, say, top to bottom, equals the number of roots with positive real part.

Thus, there should be no change of sign is the left hand column, for the system to be stable.

For example, for the system, with block diagram as in Fig. 19.30,

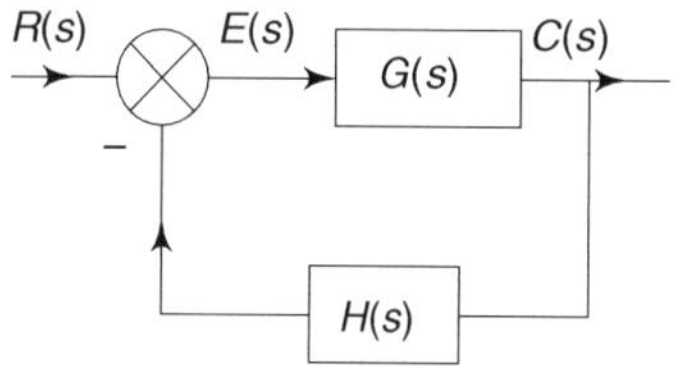

Fig. 19.30 *A typical block diagram of the system*

$$\frac{C(s)}{R(s)} = \frac{G(s)}{1 + G(s)H(s)}$$

Characteristics equation is

$$1 + G(s)\,H(s) = 0$$

If
$$G(s) = \frac{10}{5s^2 + 10s} \quad \text{and} \quad H(s) = \frac{5}{s+1},$$

Characteristic equation is

$$1 + \frac{50}{(5s^2 + 10s)(s+1)} = 0$$

or
$$5s^3 + 15s^2 + 10s + 50 = 0$$

Arranging as array,

5	10
15	50
–6.67	0
50	0
0	

$$b_1 = \frac{15 \times 10 - 5 \times 50}{15} = -6.67$$

$$b_2 = 0$$

$$c_1 = \frac{-6.67\,(50) - 0}{-6.67} = 50$$

$$c_2 = 0$$

$$d_1 = 0$$

The left column shows two changes of sign, going from say top to bottom. Thus, the system has two roots with positive real part, according to Routh's criterion and the system is unstable.

Problem 19.4 *In problem (19.2), find transient response $\theta_o(t)$ due to unit ramp input $\theta_i(t) = t$ and calculate the steady state error.*

If the transfer function of the difference amplifier is changed to 20 $\left(1 + \frac{K_i}{D}\right)$, find the value of K_i at which the system become unstable.

Solution

$$\text{Amplifier gain} = 0.5\left(20 + \frac{20\,K_i}{D}\right) = 10\left(1 + \frac{K_i}{D}\right)$$

Transfer function

$$G(D) = \frac{10}{20}\,\frac{\left(1 + \frac{K_i}{D}\right)}{(D^2 + 0.4\,D)}$$

$$= 0.5\left(\frac{D + K_i}{D^3 + 0.4\,D^2}\right)$$

Characteristics equation is $1 + G(D) = 0$, for unity feedback system

or $$D^3 + 0.4\,D^2 + 0.5\,D + 0.5\,K_i = 0$$

Using Routh's criterion, the array is:

1	0.5
(0.4)	$0.5\,K_i$
$0.2 - 0.5\,K_i$	0
0	

For no change of sign in the left column, K_i is to be $< \dfrac{0.2}{0.5} < 0.4$

Review Questions

19.1 Indicate true (T) or false (F) against each of the following:

(i) A feedback control system is less accurate than an open loop system.

(ii) A feedback control system is always stable.

(iii) Open loop systems are also used in some applications in practice.

(iv) The transfer function of an element is the ratio of input variable to output variable.

(v) A block diagram of a system relates changes in output and input variables from their steady state values.

(vi) Characteristic equations of a control system is related with an open loop transfer function of a system.

(vii) Closed loop transfer function of a system with forward loop transfer function G and feedback loop transfer function H is given by $\dfrac{G}{1+GH}$

(viii) For a second order system, peak overshoot due to step input is related with both damping ratio and undamped natural frequency of the system.

(ix) Steady state error in a first order system due to unit ramp reference input depends on the damping in the system.

(x) Steady state error in a second order system due to step input of load depends on the gain of its proportional controller and damping of the system.

(xi) For stability of a feedback control system, no root of its characteristics equation, should have a positive real part.

(xii) The introduction of integral control to a system with proportional control, eliminates its steady state error.

(xiii) Derivative control adds to the overall damping of a feedback system.

(xiv) A robotic system may use either an open loop control system or a closed loop control system.

19.2 Draw a block diagram showing various functional elements of a flow control system for a fluid flowing is a pipe. The actual flow rate is measured by a differential pressure type device.

19.3 In a boiler system used in a power plant, the controlled variables are: steam pressure, its temperature and water level in the boiler. The manipulated variables are: fuel flow, air flow and feed water flow rates. Use a computer controlled system and draw the block diagram.

19.4 Draw a block diagram showing various functional elements of a speed control system, as shown in Fig. 19.31.

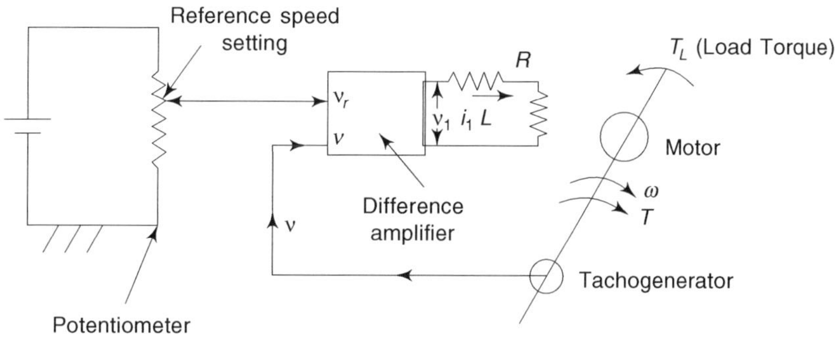

Fig. 19.31

Torque T is proportional to current i_1, Gain of difference amplifier = K. Tachogenerator measures ω, the angular speed, giving voltage v proportional to ω.

19.5 For the system with block diagram as in Fig. 19.32.

(i) Derive an expression for the closed loop transfer function.

(ii) Find the transient response due to unit step input $r(t) = 1$ and also find the steady state error.

(iii) Find the steady state error due to unit ramp input.

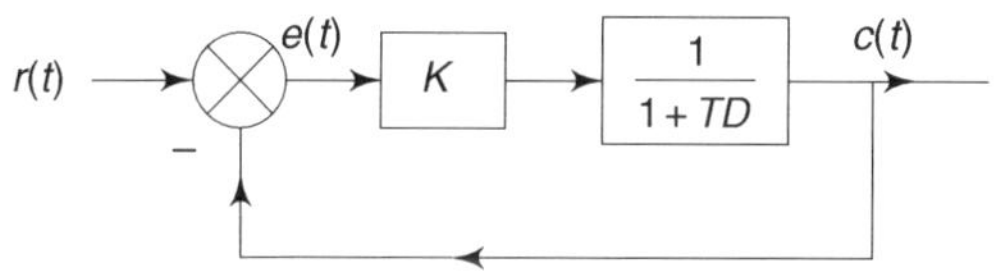

Fig. 19.32

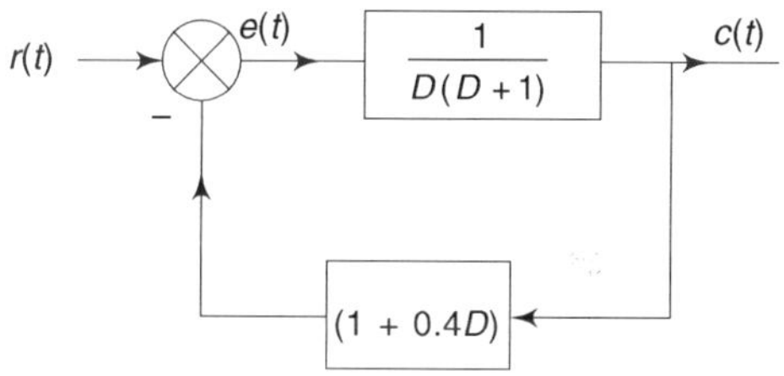

Fig. 19.33

19.6 For the system with the block diagram as shown in Fig. 19.33.

(i) Derive an expression for the closed loop transfer function and also for characteristic equation.

(ii) Calculate the damping ratio of the system from its characteristic equation.

(iii) If the transfer function of the feedback element is 1, what would be the corresponding damping ratio?

19.7 For the governor speed control system of a turbine-driven generator, shown in Fig. 19.34 draw the block diagram relating the reference input r_i and output speed ω_0, and find the transfer function $\dfrac{\omega_0}{r_i}$.

Spring stiffness = 2N/cm, force at the sleeve due to centrifugal face on the governor balls = 0.4 N/rad/s. Turbine acceleration = 0.02 rad/s^2 for 1 mm of motion y. For the hydraulic servo, area of power piston = 15 cm^2. Part constant = 1 cm^2/s.

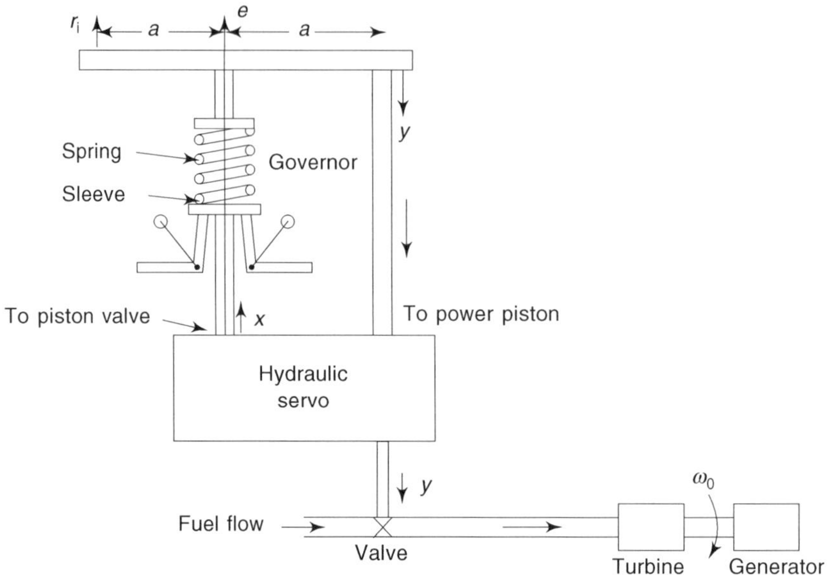

Fig. 19.34

19.8 For the level control system shown in Fig. 19.35 reference input is h_r and output is h_c. Area A_T of the tank = 1.5 m^2. Resistance $R = 4 \times 10^4$ Ns/m^5. Mass density $\rho = 1000$ kg/m^3

Inflow rate $q_{in} = Ky$, where $K = 2$ m^2/s.

Draw the block diagram and find the relation between h_r and h_c. For unit step input h_r, find the steady state error.

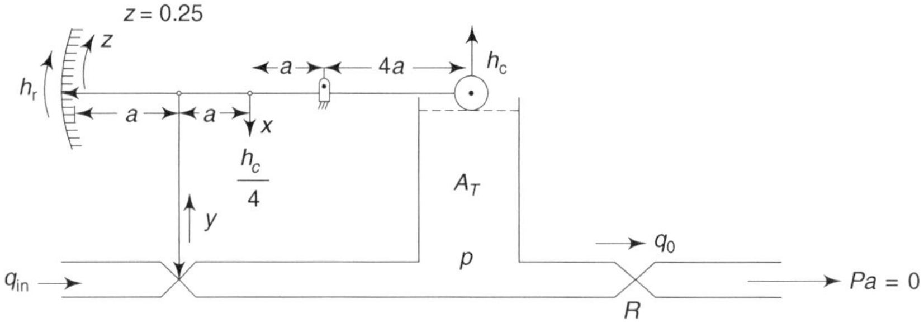

Fig. 19.35

19.9 For the system with block diagram as shown in Fig. 19.36, $\tau = 2$,

(i) find if the system is stable or not.

(ii) If $\tau = 0.5$ instead of 2, other parameters remaining unchanged, check the system for stability.

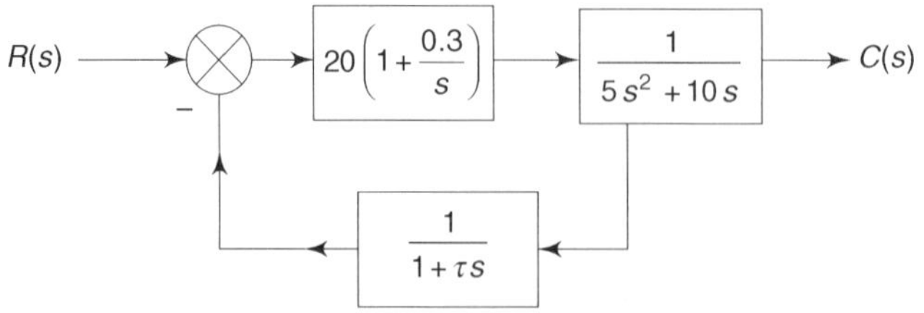

Fig. 19.36

Answers

19.1	(i) F	(ii) F	(iii) T	(iv) F	(v) T
	(vi) T	(vii) T	(viii) F	(ix) T	(x) F
	(xi) T	(xii) T	(xiii) T	(xiv) T	

19.5 $\dfrac{K}{TD + (K+1)}, \dfrac{1}{K+1}$

19.6 $\dfrac{1}{D^2 + 1.4\,D + 1}$, 0.7, 0.5

19.7 $(450\,D^2 + 15\,D + 1.2)\,\omega_0 = 300\,x_i(t)$

19.8 $(6D + 1.23)\,h_c = h_r(t)$; 0.81

19.9 (i) unstable (ii) stable.

Chapter

20

Electrical Measurements

■ *INTRODUCTION* ■

There are a large number of machines or their components used in industrial and household applications in which the actuating forces and the resulting motions are derived from the input electrical power. In fact, an extremely large measurement range is encountered in respect of the electrical power measurements. For example, the typical current signals received on earth from sensors fitted in *Mars* spacecraft are of the order of *nano-amperes* or less. On the other hand, the current output of 100,000 MW power plant is of the order of several *mega-amperes*. Therefore, the measurements of electrical quantities form an important area of instrumentation engineering.

20.1 ■ ADVANTAGES OF ELECTRICAL MEASURING INSTRUMENTS

There is a widespread use of electrical measuring devices in modern-day instrumentation systems as compared to their counterparts in mechanical devices. This is because of their following notable capabilities.

1. The frequency and transient response of the signals are excellent because of low inertia effects which is due to the extremely low mass of electrons moving in the circuits.
2. Modern instruments are generally incorporated with VLSI (Very Large Scale Integration) fabrication techniques. This way, miniaturisation and micro-miniaturisation of the devices are possible.
3. Because of small size of the instruments, they need comparatively less power and also, in general, have less loading effects on the measured quantities.
4. Being economical in power consumption along with associated light weight, these systems can be made battery-operated and portable.

5. Quite often non-contact measurements are possible. Especially, in biometrics, many non-invasive techniques are currently being employed.
6. Complex signal manipulation modules can be incorporated conveniently in these instruments. These are, for example,
   (i) amplification/attenuation
   (ii) summing/differencing/multiplication/division of signals or alternatively averaging of signals.
   (iii) integration/differentiation or double integration/differentiation of signals
   (iv) Fourier analysis/frequency analysis of signals
   (v) A to D or D to A conversion
   (vi) Complex manipulations like *Z-transform*, multi-point *correlations* and *speech recognition*
7. These systems can be integrated with modern high-speed digital computer systems along with software like Artificial Neural Networks (ANNs) and fuzzy logic giving the desired intelligent information in *real time* (i.e., online processed information).
8. These systems can be easily connected to telemetry systems, and thereby remote indication and recording of measured data becomes feasible.
9. Lastly, a new area of *Mechantronics* has currently emerged which combines the use of mechanical, electrical and electronic devices. These are quite often coupled with microprocessor systems also. This finds a vast number of applications in instrumentation, control engineering and robotic systems.

20.2 ■ MEASUREMENT OF RESISTANCE, INDUCTANCE AND CAPACITANCE

Bridge circuits are most widely used nowadays and are most accurate in the measurement of resistances, capacitances and inductances. In addition, bridge circuits are used in control applications. A common example is the use of the fully automatic type of bridge which carries out null condition to achieve precise control in a host of industrial applications.

In this section, the use of bridge circuits in the measurement of resistance (R), inductance (L) and capacitance (C) has been discussed. A bridge circuit consists of a simple network of four arms forming a closed circuit with four nodes. Two opposite nodes are connected to an excitation source which may be either ac or dc depending on the type of bridge. The other two opposite nodes are connected to a current detector in the form of a galvanometer. In ac bridges, the galvanometer is replaced by a pair of head phones. A simple ac bridge consisting of four arms of R, L and Cs has been discussed. It may be noted that when all the four arms are purely resistive, it is termed as a Wheatstone bridge.

20.2.1 General Description of a Typical ac Bridge Circuit

A general form of an ac bridge is shown in Fig. 20.1. The four bridge arms have impedances consisting of R, L and Cs.

This bridge enables to determine the impedances at audio as well as radio frequencies. It may be noted that for audio frequencies ear phones are quite suitable as null detectors. The best sensitivity of the human ear is around 1000 Hz. For low frequencies of 50 Hz a *vibration* galvanometer is most convenient. However, for ordinary laboratory testing, ac galvanometer may be employed.

The inductive and capacitive impedances of the various arms may be determined as follows:

$$\text{Capacitive impedance} = 1/(j\omega C) \tag{20.1}$$

and

$$\text{Inductive impedance} = j\omega L \tag{20.2}$$

where ω is the circular frequency of excitation voltage in rad/s,
C is the capacitance in farads and

L is the inductance in henrys.

For a series circuit of resistance (R), capacitance (C) and inductance (H), the total impedance Z in j-operator notation is given by

$$Z = R + [j\omega L + 1/(j\omega C)]$$

or

$$= R + j\,[\omega L - 1/(\omega C)] \quad (20.3)$$

where $j = \sqrt{(-1)}$, which represents the imaginary axis.

The condition for bridge balance is that the potential difference between junction B and D should be zero. Therefore, the voltage drop from A to B and from A to D should be equal both in magnitude and phase. In other words,

$$E_{AB} = E_{AC} \quad (20.4)$$

or

$$I_1 Z_1 = I_2 Z_2 \quad (20.5)$$

where

$$I_1 = \frac{E}{Z_1 + Z_3}$$

and

$$I_2 = \frac{E}{Z_2 + Z_4}$$

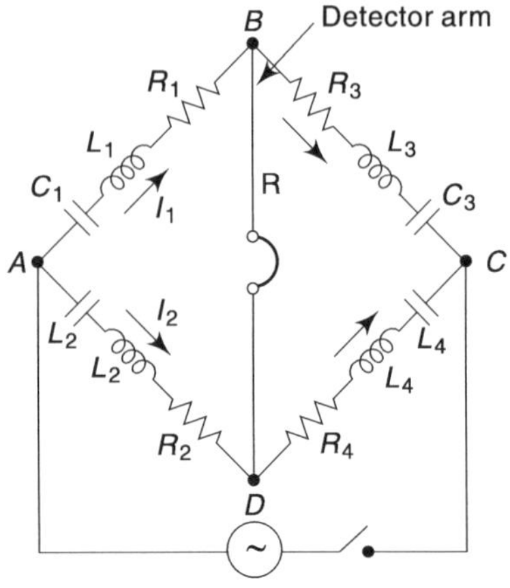

Fig. 20.1 *A typical ac Bridge circuit*

By substituting the values of I_1 and I_2 in Eq. (20.5) and on simplification, we get

$$Z_1 Z_4 - Z_2 Z_3 = 0 \quad (20.6)$$

or

$$\frac{Z_1}{Z_2} = \frac{Z_3}{Z_4} \quad (20.7)$$

or

$$Z_2 Z_3 = Z_1 Z_4 \quad (20.8)$$

Equation (20.8) represents the general equation of ac bridge balance. In other words, it implies that product of impedances in one pair of opposite arms is equal to the product of impedances of the other pair of opposite arms in the case of ac bridge balance condition.

An impedance is generally written as

$$Z = |Z| \angle\theta \quad (20.9)$$

where $|Z|$ is the magnitude of the impedance, and θ is the phase angle.

Therefore, Eq. (20.8) can be written as

$$(|Z_2| \angle\theta_2)\,(|Z_3|\ \theta_3) = (|Z_1|\ \theta_1)\,(|Z_4|\ \theta_4)$$

or

$$(|Z_2|)\,|Z_3| \angle(\theta_2 + \theta_3) = (|Z_1|)\,(|Z_4|) \angle(\theta_1 + \theta_4) \quad (20.10)$$

It is obvious from Eq. (20.10) that for an ac bridge balance, the following two conditions must be satisfied simultaneously. They are

(a) The product of the magnitude of impedances of the opposite arms must be equal, i.e.,

$$(|Z_2|)\,(|Z_3|) = (|Z_1|)\,(|Z_4|) \quad \text{and} \quad (20.11)$$

(b) The sum of the phase angle of the opposite arms must be equal, i.e.,

$$\angle\theta_2 + \angle\theta_3 = \angle\theta_1 + \angle\theta_4 \quad (20.12)$$

It may be noted that for an ac bridge balance, all the four impedances need not be identical in magnitude and phase angles. Further, the individual arms of the ac bridge may have series or parallel combinations of R, L and Cs. This results in a large number of possible combinations. However, all of them may not be of practical importance. Some useful bridges have been obtained by making two of the four arms of an ac bridges purely resistive. The measurement of unknown capacitance or inductance can be conveniently carried out by standard capacitor/inductances, using the ac bridges with two purely resistive arms.

Problem 20.1 *The impedances of an AC bridge hawing an excitation voltage of 1 kHz are as follows:*
Arm AB with impedance $Z_1 = 100\ \Omega \angle 60°$ (inductive impedance)
Arm AD with impedance $Z_1 = 300\ \Omega \angle 0°$ (purely resistive)
Arm BC with impedance $Z_3 = 50\ \Omega \angle 30°$ (inductive impedance)
and Arm DC with impedance Z_4 = unknown impedance
Determine the R, L or C components of the unknown impedance considering it as series circuit.

Solution For bridge balance, we get,

$$Z_1 Z_4 = Z_2 Z_3$$

Writing the impedances in polar form, we get

$$[100\ \Omega \angle 60°]\ [Z_4] = [300\ \Omega \angle 0°]\ [50\ \Omega \angle 30°].$$

Using Eqs 20.11 and 20.12, we get :

$$Z_4 = (300)\ (50)/(100) = 150\ \Omega \text{ and}$$

$$\theta_4 = 0° + (30°) - (60°) = -30°$$

Therefore, the unknown impedance

$$Z_4 = 150\ \Omega \angle -30°.$$

Further, the negative angle of impedance indicates that Z_4 consists of a series $R - C$ circuit.

Now resistance $R_4 = 150 \cos 30° = 129.9\ \Omega$

and capacitive impedance

$$X_{C_4} = 150° \sin 30° = 75\ \Omega$$

$$= 1/(2\pi \times 1000 \times C_4)$$

or

$$C_4 = 2.12 \times 10^{-6} \text{ F}$$

$$= 2.12\ \mu\text{F}$$

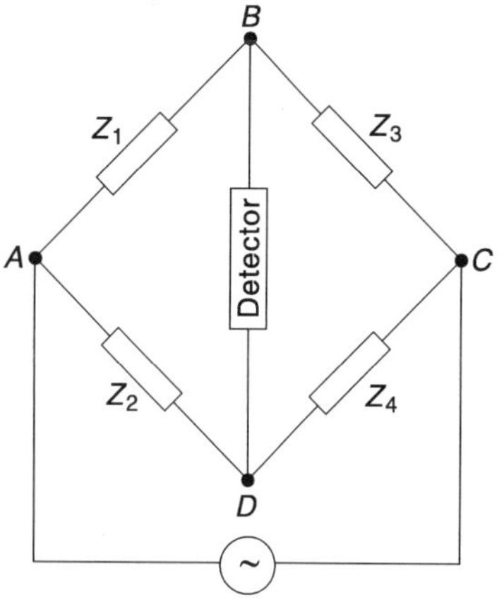

Fig. Prob. 20.1

20.2.2 Measurement of Resistance

Wheatstone-bridge Method A Wheatstone bridge is commonly used for both accuracy and precise measurements of resistance in the range of 1 Ω to 100 kΩ. The ac bridge discussed earlier takes the shape of the Wheatstone bridge if all the arms are purely resistive. The excitation voltage to the bridge may be either ac or dc type. This has been discussed in detail in chapter 4.

Advantages

1. It is a low-cost device and does not require skilled operation.
2. The accuracy of measurement of resistance depends on the accuracy of adjustable, standard resistor which provides null condition for the determination of the unknown resistance. With the use of high-quality standard resistors, accuracies of ± 0.5 % can be achieved.
3. It is used extensively in industrial applications like quality control of resistance wires, determination of resistance of transformers, motor windings, relay coils and solenoids.

Disadvantages

1. It is not possible to measure with reasonable accuracy low values of resistances below 1 Ω, as well as high values of resistances above 100 kΩ.
2. Small errors are caused due to the resistance of connecting wires and contact resistances of the binding posts.

Ammeter—voltmeter method The ammeter–voltmeter method employs the simple Ohm's law to determine the value of an unknown resistance. The procedure is automated by the use of an AVO meter which has been discussed later in this chapter.

Figure 20.2 shows a simple ohmic circuit for determining the unknown resistance value R_x by using an ammeter and a voltmeter.

Current passing through Ammeter

$$I = I_V + I_R$$

or
$$I_R = I - I_V \tag{20.13}$$

True value of unknown resistance

$$R_x = \frac{V}{I_R}$$

or
$$R_x = \frac{V}{I - I_V} = \frac{V}{I\left[1 - \frac{V}{IR_V}\right]} \tag{20.14}$$

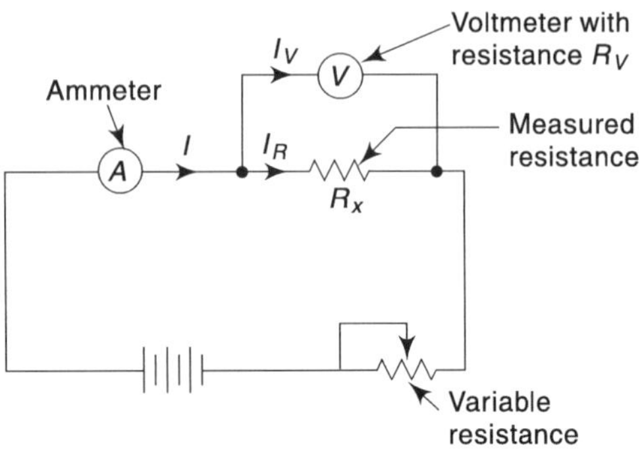

Fig. 20.2 *A typical Ohmic circuit*

Now, the value of measured resistance

$$R_m = \frac{V}{I} \tag{20.15}$$

From Eqs (20.14) and (20.15), we get,

$$R_x = R_m\left[\frac{1}{1 - R_m / R_v}\right] \tag{20.16}$$

where $\left[\frac{1}{1 - R_m / R_v}\right]$ is the correction factor.

Advantages

1. It is a commonly employed low-cost procedure which does not require skilled operation.
2. Accuracies of the order of ±1% can be achieved if the accuracies of the ammeter and voltmeter are of the order of ± 0.5%.
3. For a constant voltage dc source, the procedure could be automated and the scale of the moving coil galvanometer may be calibrated in ohms, and the instrument could be termed as an ohm-meter.

Disadvantages

1. A correction factor needs to be applied on the measured value to obtain the true value of the resistance.
2. The low values of resistances invariably have a high percentage of error because of the inherent, lower accuracy level in the lower half scales of both the ammeter and the voltmeter.

20.2.3 Measurement of Inductance

Measurement of an unknown inductance say, L_x and its associated internal resistance say, R_x, can be conveniently carried out by the use of the following bridge circuits.

Inductance Measuring ac Bridge Figure 20.3 shows a typical ac bridge for the measurement of an unknown inductance L_x having an internal resistance R_x. This unknown inductance with its internal resistance forms one arm of the bridge circuit. Its corresponding ratio arm consists of a standard inductor L_4 and a standard resistor R_4.

The other ratio arms R_1 and R_3 are purely resistive.

For the ac bridge balance condition, we have,

$$Z_x Z_3 = Z_1 Z_4$$

or $$[R_x + j\omega L_x] R_3 = R_1 [R_4 + j\omega L_4] \quad (20.16a)$$

Equating the real and imaginary parts, we get,

$$R_x R_4 = R_1 R_4 \quad (20.17)$$

and $$L_x R_3 = R_1 L_4 \quad (20.18)$$

Using Eq. (20.18), the value of the unknown inductance L_x can be evaluated.

Further, Q-factor of the inductance $= \dfrac{\omega L_x}{R_x}$ (20.19)

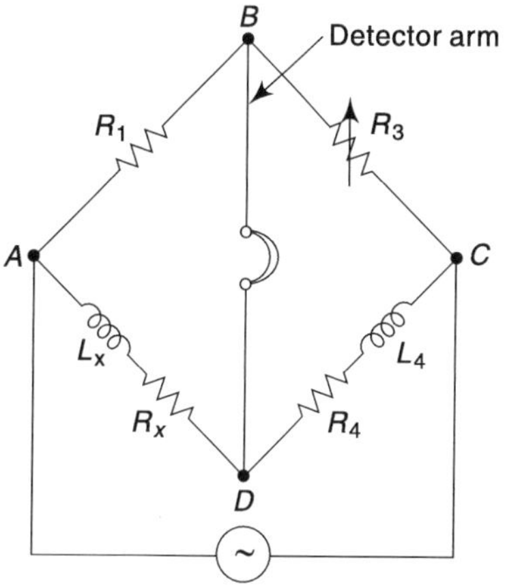

Fig. 20.3 *A typical inductance measuring ac bridge*

Advantages

1. The bridge circuit is a simple modification of the Wheatstone bridge in which one of the resistive ratio arms is replaced by an inductive ratio arms.
2. The bridge equations are independent of the frequency of the input voltage to the bridge circuit.
3. The bridge yields simple equations for unknown values of inductances and associated internal resistances.
4. The Q-factor of any coil can be determined conveniently using this bridge.
5. The bridge can be used to measure the values of an unknown inductance within the accuracy of $\pm$ 2% for low values of Q-factor, i.e., Q in the range of 1 to 10.

Disadvantages

1. The bridge balance becomes diffcult to achieve with large as well as very low Q-factors. Therefore, accuracy of measurement reduces considerably.
2. If the Q of the unknown reactance is more than the Q of the standard inductor then a variable resistance in series, with unknown resistance is necessary to obtain bridge balance.

Maxwell's Bridge Maxwell's bridge, shown in Fig. 20.4, employs a standard capacitor to determine the value of an unknown inductance. The opposite arm of the bridge containing unknown inductor has a standard resistor and a standard capacitor in parallel. It may be noted that R_x is the internal resistance of the inductor.

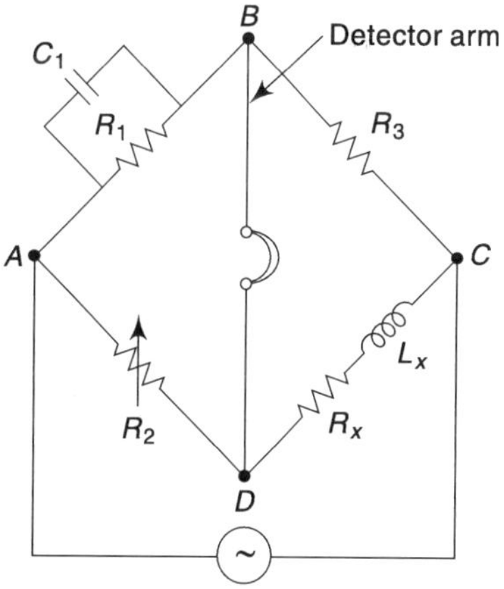

Fig. 20.4 *Maxwell's Bridge for determining unknown inductance*

For the bridge balance condition, we have

$$Z_1 Z_x = Z_2 Z_3 \quad (20.20)$$

or $$\frac{1}{\left[\dfrac{1}{R_1} + j\omega C_1\right]}\left[R_x + j\omega L_x\right] = R_2 R_3$$

or $$R_x + j\omega L_x = \frac{R_2 R_3}{R_1} + j\omega C_1 R_2 R_3 \quad (20.21)$$

Equating the real and imaginary parts, we get,

$$R_x = \frac{R_2 R_3}{R_1} \quad (20.22)$$

and $$L_x = C_1 R_2 R_3 \quad (20.23)$$

Further, Q-factor of inductance $= \dfrac{\omega L_x}{R_x}$

$$= \omega C_1 R_1 \tag{20.24}$$

Advantages

1. The bridge balance equations are independent of frequency.
2. The bridge yields simple equations for unknown values of inductances and its associated internal resistances.
3. The Q-factor of any coil can be determined conveniently using this bridge.
4. The bridge is suitable for the measurement of an unknown inductane with an accuracy of ± 2% in the low values range of Q-factor ranging between 1 and 10.

Disadvantages

1. The bridge balance is difficult to achieve with very large Q factor (greater than 10) and for very low Q factors (less than 1).
2. The bridge requires a standard capacitor of high accuracy and this is quite expensive.
3. The resistance in parallel to capacitance is generally of high value and resistance boxes of high values are costly.

Problem 20.2 *Determine the inductive impedance connected in the balanced Maxwell's bridge circuit shown in Fig. Prob. 20.2. Also, determine the Q-factor of the coil if the frequency of the exitation voltage is 1000 Hz.*

Solution Using Eqs (20.18) and (20.19), we get

$$R_x = \frac{R_2\, R_3}{R_1}$$

$$= \frac{(600)(1000)}{2000}$$

$$= 300\ \Omega \qquad \textbf{Ans.}$$

Further, $L = C_1\, C_2\, R_3$

$= (0.2 \times 10^{-6}) \times 180 \times 1000$

$= 120$ mH.

Using Eq. (20.20), we get

Q-factor of coil $= \omega L_x/(R_x)$

$= 2\pi \times (1000) \times 0.12/(300)$

$= 2.51$

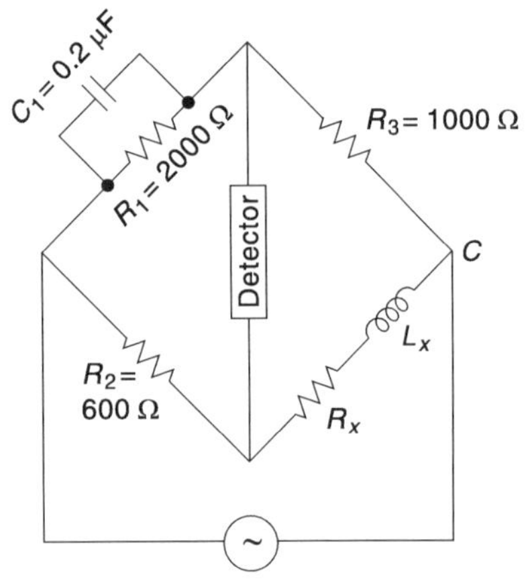

Fig. Prob. 20.2

Hay's Bridge Hay's bridge, shown in Fig. 20.5, is used to determine the value of unknown inductance. It differs from Maxwell's bridge by having a standard capacitor in series with a standard resistor in place of the parallel combination used in Maxwell's bridge. This bridge is suitable for measuring inductors of high Q-factors.

For bridge balance condition, we get,

$$Z_1\, Z_x = Z_2\, Z_3$$

or

$$\left[R_1 - \frac{j}{\omega C_1}\right]\left[R_x + j\omega L_x\right] = R_2\, R_3$$

or

$$R_1\, R_x + \frac{L_x}{C_1} + j\omega L_x\, R_1 - \frac{jR_x}{\omega C_1} = R_2\, R_3 \tag{20.25}$$

Equating the real and imaginary parts and solving for R_x and L_x, we get,

$$R_x = \frac{\omega^2 C^2 R_1 R_2 R_3}{1+\omega^2 R_1^2 C_1^2} \quad (20.26)$$

$$L_x = \frac{R_2 R_3 C_1}{1+\omega^2 R_1^2 C_1^2} \quad (20.27)$$

Using Eqs (20.26) and (20.27), the Q-factor becomes

$$Q = \frac{\omega L_x}{R_x}$$

$$= \frac{1}{\omega R_1 C_1} \quad (20.28)$$

Substituting the value of Q in Eq. (20.27), we get,

$$L_x = \frac{C_1 R_2 R_3}{\left[1+\left(\frac{1}{Q}\right)^2\right]} \quad (20.29)$$

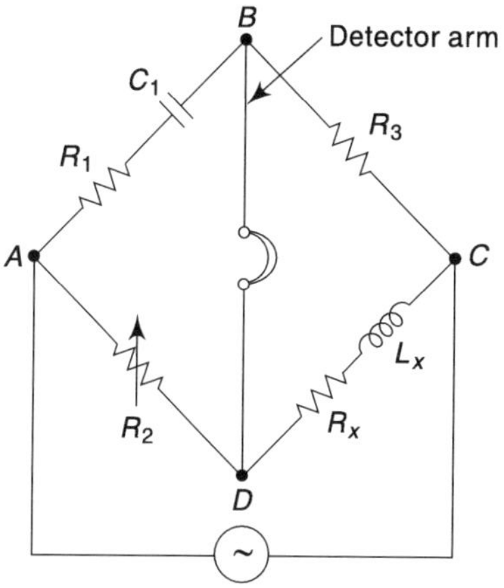

Fig. 20.5 *Hay's bridge for determining unknown inductance*

For $Q \geq 10$, the term $\left(\frac{1}{Q}\right)^2$ is of the order of 0.01 and can be neglected.

Thus $L_x = C_1 R_2 R_3$ for $Q > 10$ (20.30)

This expression is the same as the one derived in Maxwell's bridge (see Eq. 20.23).

Advantages

1. The bridge is mainly suitable for measurements of unknown inductances for high Q coils, especially for those coils having Q greater than 10.
2. This bridge yields a simple expression for the value of unknown inductances for high Q coils. Further, the expression of the Q-factor is also very simple.
3. This bridge requires relatively low value of resistance in series with the capacitor, and a small value resistance box which is relatively less expensive can be used.
4. A commercial bridge gives the accuracy of the order of ± 2% for high Q coils.

Disadvantages

1. The bridge is not suitable for unknown inductances of low Q-factors (< 10). This is because the value of $1/Q^2$ in the denominator becomes important and cannot be neglected.
2. For low Q-coils bridge balance is obtained with difficulty.

20.2.4 Measurement of Capacitance

Measurement of an unknown capacitance, say C_x, and its associated leakage resistance R_x can be conveniently carried out by the use of the following bridge circuits.

Capacitance Measuring Bridge Figure 20.6 shows a typical ac bridge for the measurement of an unknown capacitance C_x, which may have a leakage resistance R_x. It has ratio arms R_1 and R_3 which are purely resistive. The known standard capacitor C_4 has been connected in series with R_4, which also includes the leakage resistance of the standard capacitor.

For the ac bridge balance condition, we have,

$$Z_x Z_3 = Z_1 Z_4$$

or $$\left[R_x - \frac{j}{\omega C_x}\right] R_3 = R_1 \left[R_4 - \frac{j}{\omega C_4}\right] \quad (20.31)$$

Equating the real and imaginary parts, we get,

$$R_x R_3 = R_1 R_4 \quad (20.32)$$

$$R_3/(\omega C_x) = R_1/(\omega C_4)$$

or $$C_x = \frac{R_3}{R_1} C_4 \quad (20.33)$$

Using Eqs (20.32) and (20.33), the values of unknown capacitance C_x and unknown leakage resistance R_x can be evaluated.

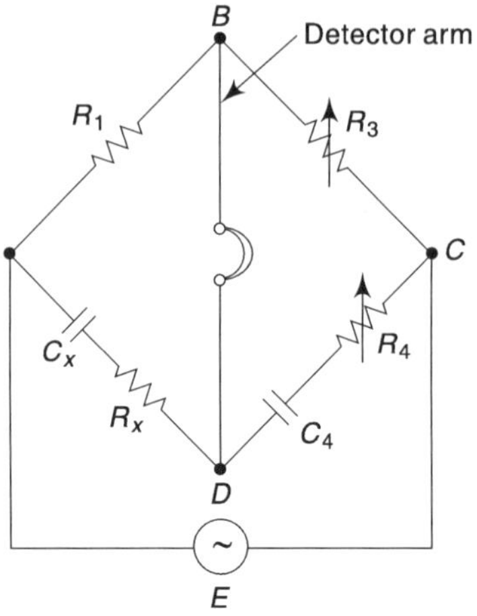

Fig. 20.6 *A typical capacitance measuring ac bridge*

Advantages

1. It is a simple bridge for comparing an unknown capacitance with a known capacitance.
2. The bridge equations are of simple type for determining the values of unknown capacitance and leakage resistance of the capacitor.

Disadvantages

1. The bridge balance is difficult to obtain with an imperfect capacitor which is commonly associated with dielectric losses.
2. The bridge is suitable for measurement of high quality mica-capacitors or air capacitors which are expensive.
3. The bridge yields relatively less accurate values of Q-factor.

Schering Bridge This bridge is used for the precise measurements of unknown capacitance and dielectric loss of capacitors. Figure 20.7 shows a typical circuit arrangement of the Schering bridge. In this bridge, C_2 is a standard capacitor of high quality. Generally either mica capacitor of low loss type or air capacitor which is loss free type is employed.

For the bridge balance condition, we have

$$Z_1 Z_x = Z_2 Z_3$$

or $$\frac{1}{\left[\frac{1}{R_1} + j\omega C_1\right] \times \left[R_x - \frac{j}{\omega C_x}\right]} = [R_3] \times \left[-\frac{j}{\omega C_2}\right]$$

or $$\left[R_x - \frac{j}{\omega C_x}\right] = [R_3] \times \left[-\frac{j}{\omega C_2}\right] \times \left[\frac{1}{R_1} + j\omega C_1\right] \quad (20.34)$$

Equating the real and imaginary parts, we get,

$$R_x = R_3 \frac{C_1}{C_2} \quad (20.35)$$

and $$C_x = \frac{R_1}{R_3} C_2 \quad (20.36)$$

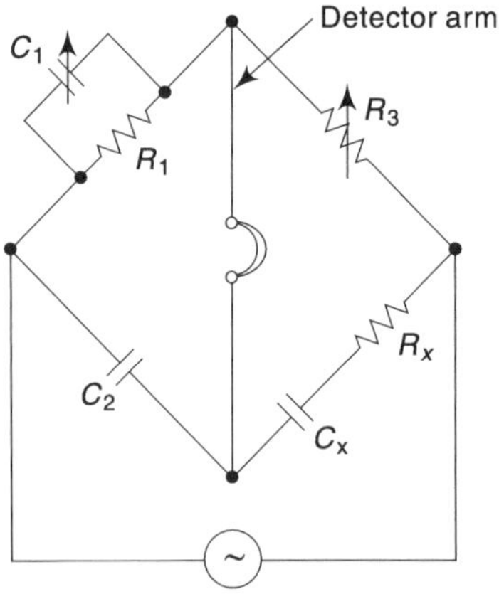

Fig. 20.7 *A typical circuit diagram of Schering bridge*

Using Eqs (20.35) and (20.36) the dissipation factor D which is the ratio R_x and X_x can be determined.

$$D = \frac{R_x}{L_x} = \frac{R_x}{1/(\omega C_x)} = \omega R_x\, C_x$$

$$= \omega \times \left[R_3 \frac{C_1}{C_2}\right] \times \left[\frac{R_1}{R_3} C_2\right] = \omega R_1\, C_1 \qquad (20.37)$$

It may be noted that dissipation factor D is the reciprocal of the Q-factor i.e.,

$$D = 1/Q$$

Advantages

1. The equation of capacitance $C_x = \frac{R_1}{R_3} C_2$ and since C_2 and R_1 are fixed at the bridge balance, the dial of R_3 can be directly calibrated in terms of measured unknown capacitance.
2. The dissipation factor $D = \omega R_1 C_1$ and in case the frequency is fixed then the dial of C_1 can be directly calibrated in terms of the dissipation factor D at that particular frequency.
3. The bridge is commonly used for testing small capacitors at low voltages as it yields very precise measurements. Further, the commercial units are used to measure capacitances in the range of 100 pF to 10 μF with accuracy of ± 2 %.

Disadvantages

1. In the high-voltage Schering bridge circuit, stray capacitances in the bridge elements may introduce errors.
2. The bridge requires the standard capacitor of high accuracy and this is quite expensive.

20.3 ■ MEASUREMENT OF VOLTAGE AND CURRENT

20.3.1 Principle of Operation of Ammeter and Voltmeter

Ammeters and voltmeter are basic electrical measuring instruments and their operating ranges are specified in terms of safe operating currents and voltages respectively. However, there is no basic difference in their principle of operation. The deflection torque produced in all types of ammeters and voltmeters except those of electrostatic type depends on the current flowing through the coil of the instrument. Basically, an ammeter has a low input resistance. Therefore, when this instrument is connected *in series* with any circuit, it does not cause any appreciable voltage drop, and consequently the power absorbed from the circuit is minimum. On the other hand, a voltmeter has high input resistance. Therefore, when it is connected *in parallel* with any circuit, it does not take appreciable current, and thereby the power absorbed from the circuit is minimum. Hence, both ammeters and voltmeters are basically current-sensing devices.

Multi-Ranging of Ammeter A typical dc voltage-measuring instrument of D'Arsonval type is termed a galvanometer. As mentioned before, it can be converted to ammeter by providing a shunt resistance R_{sh} in parallel with the galvanometer resistance R_G, as shown in Fig. 20.8.

Since the shunt resistance is in parallel with the galvanometer resistance, therefore the voltage across both the resistances is the same.

Therefore, $$V_{sh} = V_G = V$$

or $$I_{sh}\, R_{sh} = I_G\, R_G$$

or $$R_{sh} = \frac{I_G\, R_G}{I_{sh}} \qquad (20.38)$$

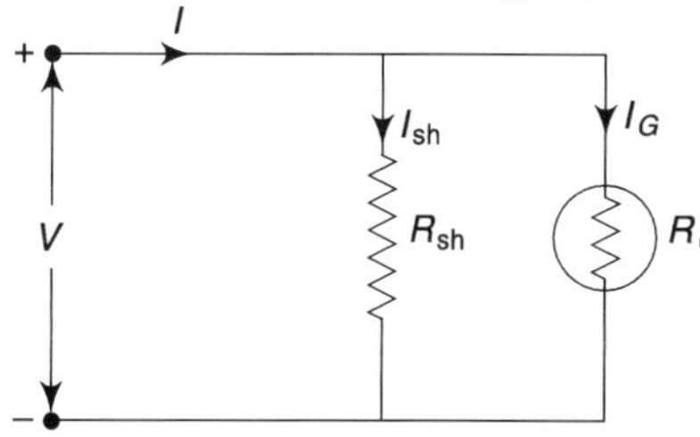

Fig. 20.8 *A typical ammeter circuit*

Herein, $I_{sh} = I - I_G$

Hence. $$R_{sh} = \frac{I_G R_G}{I - I_G} \quad (20.39)$$

Using Eq. (20.39), we can determine the shunt resistance for a full-scale galvanometer current, I_G corresponding to the maximum measured current I.

The ratio of the maximum value of the measured current I to the full-scale galvanometer current I_G, is termed the multiplying power of the shunt, M_{sh} (i.e, $M_{sh} = I/I_G$).

Using Eq. (20.39), it can be shown that

$$M_{sh} = 1 + (R_G/R_{sh}) \quad (20.40)$$

or $$R_{sh} = \frac{R_G}{(M_{sh-1})} \quad (20.41)$$

For multiranging of a galvanometer, for ranges I_1, I_2 .... I_i, we incorporate shunt resistances R_{sh_1}, R_{sh_2}, ... R_{sh_i} in the galvanometer circuit as shown in Fig 20.9. The shunt resistance corresponding to the desired measuring range in the multi-range ammeter can be selected by means of the selector switch S, shown in Fig. 20.9.

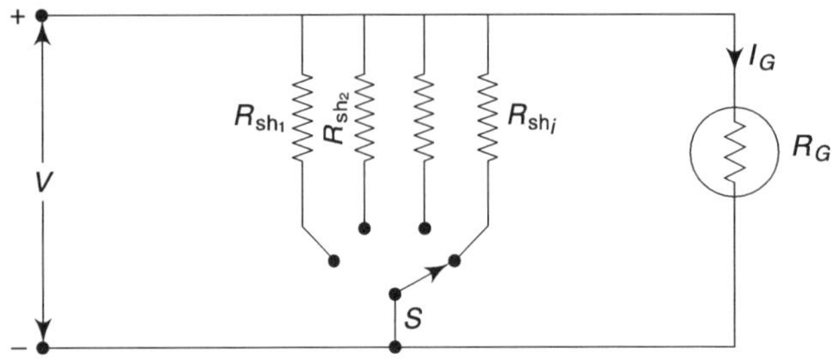

Fig. 20.9 *A typical multi-range ammeter circuit*

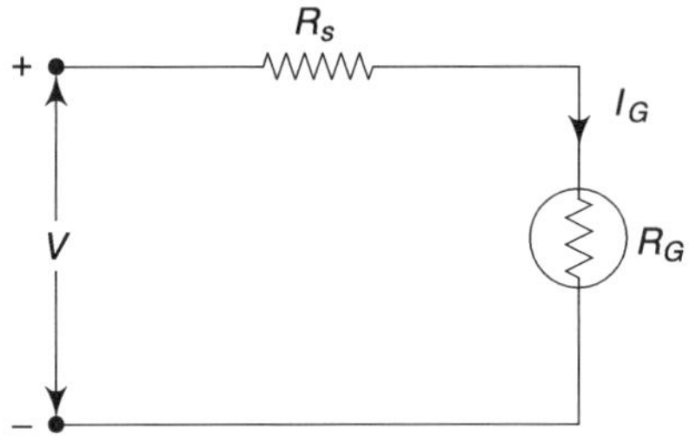

Fig. 20.10 *A typical voltmeter circuit*

Multiranging of Voltmeter The dc galvanometer can be converted to a dc voltmeter by incorporating a series resistor R_s known as a multiplier resistor. In the simple ohmic circuit shown in Fig. 20.10, the voltage drop V across the circuit is given by:

$$V = I_G (R_s + R_G)$$

Therefore, $$R_s = \frac{V}{I_G} - R_G \quad (20.42)$$

Thus, the multiplier resistance limits the current flowing through the galvanometer so that it does not exceed the value corresponding to the full-scale deflection I_{fsd}.

For multi-ranging of galvanometer, for ranges V_1, V_2, ...V_i, we incorporate, multiplier resistances R_{s_1}, R_{s_2}, ... R_{s_i} in the galvanometer circuit as shown in Fig. 20.11.

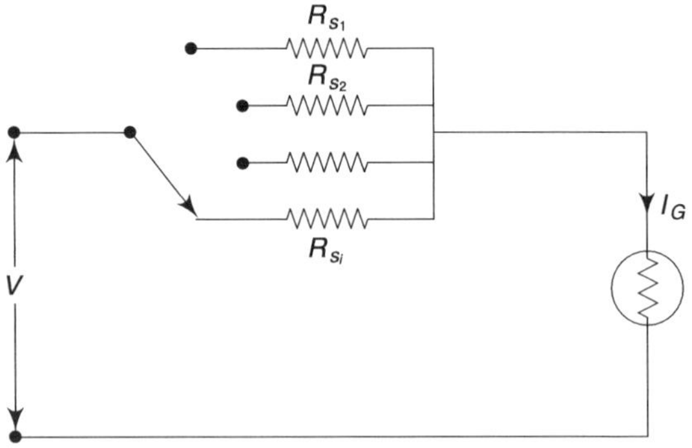

Fig. 20.11 *A typical multi-range voltmeter circuit*

The general requirements of the multiplier resistances are the same as that of the shunt resistances which have been listed earlier. Here, also the series resistances of manganin in the form of either bobbins or strips are used as multiplier resistances for multi-range type of voltmeters.

Problem 20.3 *A dc galvanometer of 5-ohm resistance reads up to 50 mA. Determine the value of the resistance*
(a) in parallel to enable the instrument to read up to 1 A
(b) in series to enable it to read 10 V

Solution:

(a) Resistance of galvanometer $R_G = 5\Omega$

Full-scale deflection current $I_G = 50 \times 10^{-3}$ A

Current to be measured $I = 1$A

Multiplying factor of shunt $M_{\text{sh}} = \dfrac{I}{I_G} = \dfrac{1}{50 \times 10^{-3}} = 20$

Using Eq. (20.41), we can determine the shunt resistance R_{sh} to be connected in parallel to enable to read up to 1A:

$$R_{\text{sh}} = \frac{R_G}{M_{\text{sh}-1}}$$

$$= \frac{5}{20 - 1} = 0.26316\ \Omega$$

(b) Voltage to be measured = 10 V

Using Eq. (20.42), we can determine the series resistance R_s required for reading upto 10 V is:

$$R_s = \frac{V}{I_G} - R_G$$

$$= \frac{10}{50 \times 10^{-3}} - 5$$

$$= 195\ \Omega$$

20.3.2 Permanent Magnet Moving Coil (PMMC) Instrument

It is one of the most useful and accurate instruments for measurement of dc voltages and currents. Its principle of operation is based on the motor action produced by the flow of a small current through a moving coil suspended in the magnetic field of a horse-shoe type permanent magnet. This basic deflecting-moment producing mechanism is also known as D'Arsonval moving coil movement. Herein, the electromagnetic deflecting torque is directly proportional to the current flowing in the coil. This deflecting torque is balanced by the mechanical torque of a torsion-type control spring attached to the moving coil. The balancing of these torques determines the angular position θ of the moving coil.

In Fig. 20.12, a typical taut suspension-type of D'Arsonval movement is shown, in which the attached mirror also moves by an angle θ, equal to the angular movement of the moving coil. This mirror is then part of the lamp-and-scale display device. Herein, the incident ray from the lamp striking the mirror rotates by the angle θ and causes the reflected ray to deflect by an angle 2θ. This way, the optically amplified light signal is displaced as a movement of light spot on a ground glass-type calibrated scale.

It may be noted that the output of the galvanometer can be conveniently modified with the help of suitable shunts or multiplier resistances to cover a wide range of measurements, either as ammeter or voltmeter respectively.

In actual practice, the taut suspension type of mechanism is replaced by a V-shaped jewel-type pivot suspension at the top and bottom supports (see Fig. 20.13). Further the permanent type of U-shaped

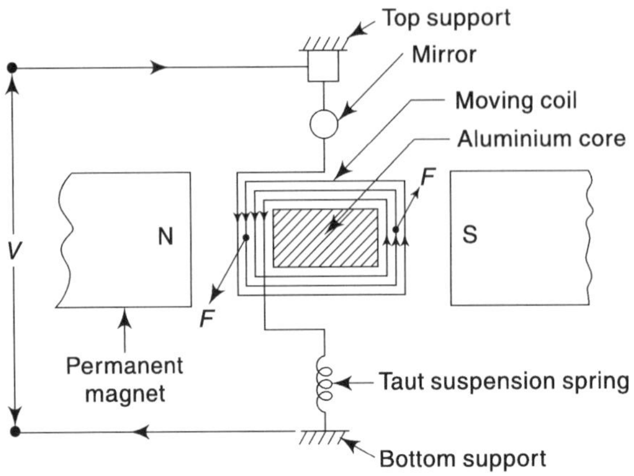

Fig. 20.12 *A typical D'Arsonval Type PMMC Galvanometer*

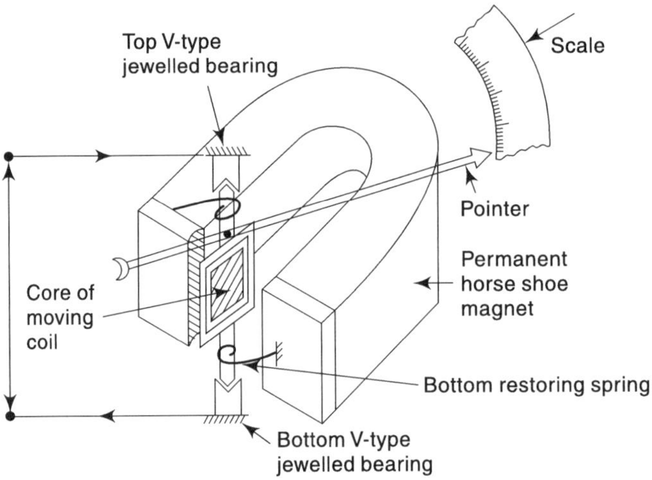

Fig. 20.13 *Permanent Magnet Moving Coil (PMMC) instrument*

horse-shoe magnet is generally employed. This is made of Alnico material, and gives a flux density of the order of 2000 to 5000 Wb/m^2. The moving coil area generally ranges from 0.5 to 4 cm^2.

The controlling torque is provided by two phosphor–bronze hair springs which may be either helical or spiral. These are coiled in opposite directions. The damping of the instrument is provided by means of eddy current damping. This method of damping is based on the principal that when a conducting type of non-magnetic core of the coil is moved in the magnetic field, it generates eddy currents. According to Lenz's law, the eddy currents generates a force in opposition to the forces that cause rotation of the conducting material.

The governing equation of the instrument is:

$$T_d = BIN\,(l \times b) = BINA \qquad (20.43)$$

where T_d = deflecting torque in N.m

B = flux density in Wb/m^2

I = current in the moving coil in Amperes

$l =$ length of coil in m
$b =$ breadth of coil in m
$A =$ Area of coil $= (l \times b)$ in m^2
and $N =$ number turns of wire of the coil.

From Eq. (22.43), it can be seen that

deflection torque $T_d \propto I$ (22.44)

In this instrument, the deflection torque T_d is equal and opposite to spring-controlled torque T_c. In other words,

$$T_d = T_c \text{ (in magnitude)} \quad (20.45)$$

Further, the controlling torque T_c is proportional to the coil deflection/rotation θ. In other words,

$$T_c \propto \theta \quad (20.46)$$

Using Eqs. (22.44) to (22.46) we get,

$$\theta \propto I \quad (20.47)$$

Since the instrument deflection θ is directly proportional to the current in the coil I, the scale markings of the basic PMMC instrument are uniformly spaced. In other words, PMMC is a linear type of measuring device.

The advantages and limitation of the PMMC are as follows.

Advantages

1. The power consumption of the instrument is quite low, i.e., it typically ranges between 25 to 150 μW.
2. It is a linear instrument, i.e., its scale markings are evenly spaced.
3. The instrument is free of hysteresis error.
4. The instrument has relatively high torque/weight ratio.
5. No effect of stray magnetic fields because of strong unidirectional magnetic field of the instrument.
6. The instrument has built-in eddy current damping because of non-magnetic conducting core consisting of either copper or aluminium.
7. The instrument has the capability of measuring voltages and currents by providing suitable high resistances or shunts respectively. In addition, these resistances could be modified to extend the measuring range of the instrument.

Limitations

1. It can be used only for dc measurements.
2. The instrument is relatively costlier as compared to moving-iron instruments.
3. With time, the jewelled bearings wear out and cause a change in calibration.

Problem 20.4 *The resistance of a moving-coil voltmeter is 15×10^3 ohms. Its coil has 100 turns and is 30 mm long and 20 mm wide. The flux density in the air gap is 0.1 Wb/m^2. Calculate the deflection torque produced by the application of 240 volts to the instrument.*

Also, calculate the deflection θ of the moving coil if control springs give a deflection of one degree for an applied torque of 16×10^{-7} N.m.

Solution Given data in the problem are

Number of turns in the coil, $N = 100$

Area of the coil $A = (l \times b)\ (30 \times 10^{-3} \times 20 \times 10^{-3})\ m^2$

$= 6 \times 10^{-4}\ m^2$

Voltage, $V = 240$ V

and Coil resistance $R = 15000\ \Omega$

$\therefore$ Current flowing through the coil, $I = \frac{240}{15000} = 0.016\text{A}$

Using Eq. (20.43) we get,

$$\text{Deflection torque } T_d = BINA$$
$$= 0.1 \times 0.016 \times 100 \times (6 \times 10^{-4})$$
$$= 9.6 \times 10^{-5} \text{ N.m}$$

The control springs provide the restoring torque T_c which is given by

$$T_c = k\theta$$

where K is the spring constant (given as 16×10^{-7}/degrees rotation).

For steady state deflection of moving coil, we have,

$$T_d = T_c$$
$$\therefore \quad T_c = 9.6 \times 10^{-5}$$
$$= (16 \times 10^{-7}) \times \theta$$
$$\text{or} \quad \theta = 60°$$

Problem 20.5 *A taut suspension type D'Arsonval galvanometer has been used as a micro-ammeter. The sensitivity of its moving coil, which is attached with a mirror is 0.2 rad/μA. In this instrument, the movement of the incident ray with angle θ from a light source gets amplified as the reflected ray with angle 2θ from the mirror. Determine the overall sensitivity of the movement of the light spot in mm/μA if the scale is placed at a distance of 125 mm from the mirror.*

Solution The functional element of the taut suspension-type D'Arsonval galvanometer are shown in Fig. 20.7.

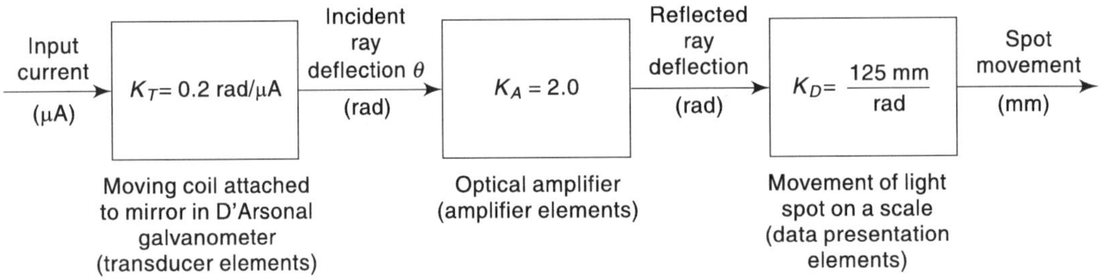

Fig. Prob. 20.5 *Block diagram of functional elements of D'Arsonval-type galvanometer with lamp-and-scale arrangement*

Given data in the problem are

(i) sensitivity of transducer element $k_T = 0.2$ rad/μA

(ii) sensitivity of lamp and scale optical type of amplifier $k_A = 2$ rad/rad.

(iii) sensitivity of the data presentation element $k_D = 125$ mm/rad.

Now the multiplication of the sensitivities of the functional elements gives the overall sensitivity K_{overall} of the instrument.

$$\therefore \quad (K)_{\text{overall}} \text{ of micro-ammeter} = 0.15 \times 2 \times 125$$
$$= 50 \text{ mm/μA}$$

20.3.3 Moving Iron Instrument

In a moving iron instrument shown in Fig. 20.14, current is passed through a fixed coil which magnetises one fixed iron and another moveable iron which in turn is attached to the pointer and restoring spring. By a suitable arrangement in the coil these irons are either made to attract or repel depending on whether they were magnetized in the opposite direction or in the same direction respectively. Generally, repulsion-type instruments are most commonly used.

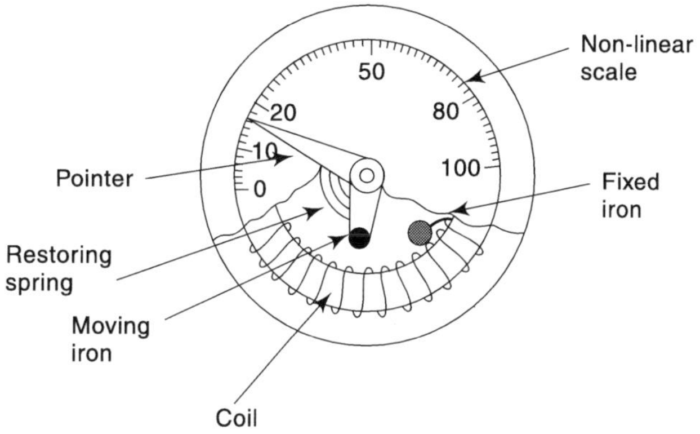

Fig. 20.14 *A typical moving iron instrument*

It may be noted that when there is no current flowing through the coil then both moving as well as fixed irons are touching each other. At this position, the pointer is set to a zero reading.

In the moving-iron instrument, the magnetic field in both fixed and moving irons it proportional to the current flowing in the coil. Therefore, the deflecting torque in the instrument is given by

$$T_D = kI^2 \tag{20.48}$$

where k is the constant of instrument, and I is the rms value of current.

As the deflecting torque is proportional to the square of the current, therefore, the instrument can measure both dc as well as ac currents. However, the scale divisions of the instrument are non-uniform.

Further, it is possible to linearise the scale by employing the moving iron in the form of a rectangular vane and fixed iron in the form of a tapered vane as shown in Fig. 20.15.

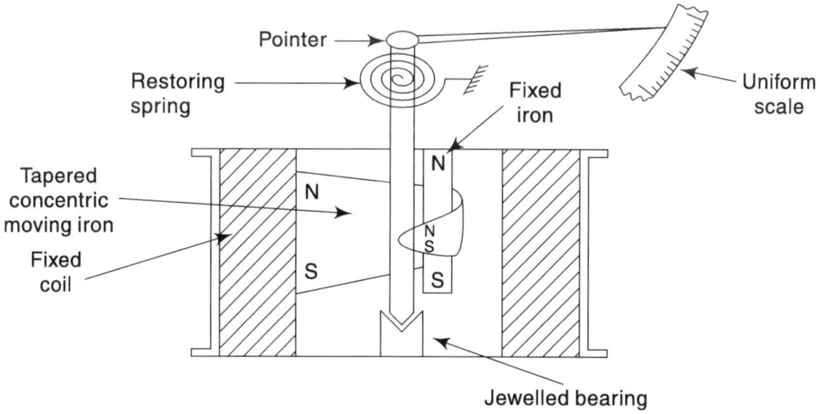

Fig. 20.15 *Concentric iron vanes in the moving-iron instrument in the repulsion-type mode*

Advantages

1. They are suitable for both dc as well as ac circuits.
2. Being simple in construction, they are quite inexpensive.

3. They generate a relatively higher operating torque.
4. These instruments are of robust type and can withstand overloads momentarily.
5. These are preferred in low frequency and high power circuits.

Disadvantages

1. They are comparatively less accurate as compared to the PMMC type instruments.
2. They possess non-uniform scale.
3. They are prone to hysteresis error produced in the iron due to stray magnetic fields.
4. In case of ac measurements the change in frequency and waveform introduce serious errors.

Problem 20.6 *The movement of a pointer on the scale of a moving-iron instrument in its calibration run was found to be 58 mm when it was fed with a constant current of 10 A at 50 Hz. It is a second-order instrument and is represented, with usual notation by the equation:*

$$J\ddot{\theta} + B\dot{\theta} + k\theta = T_d$$

The instrument has a spring stiffness k such that a torque of 11.0 Nm gives a steady-state deflection of 90°. Further, the polar moment of inertia of rotating part J is 7×10^{-4} kg.m² and the damping constant B of the instrument is 9.8×10^{-2} N.m/(rad/s). Determine the calibration constant C if the deflection torque $T_d = C I^2$. Given that the length of the pointer of the instrument is 40 mm.

Solution The stiffness of torsional spring

$$k = \frac{11.00 \text{ Nm}}{\frac{\pi}{2} \text{ rad}} = 7.0 \text{ N.m/rad}$$

The governing equation of the instrument is

$$J\ddot{\theta} + B\dot{\theta} + k\theta = T_d$$

Dividing throughout by k, we get

$$\frac{J}{k} + \frac{B}{k}\dot{\theta} + \theta = \left(\frac{1}{k}\right) T_d$$

Substituting $\sqrt{\frac{k}{J}} = \omega_n^2$, $\frac{B}{2\sqrt{kJ}} = \xi$ and $\frac{1}{k} = K$ we get,

$$\left(\frac{D^2}{\omega_n^2} + 2\frac{\xi}{\omega_n} D + 1\right)\theta = K\,T_d$$

The operation of the instrument can be represented in the form of a block diagram and is shown in Fig. Prob. 20.6.

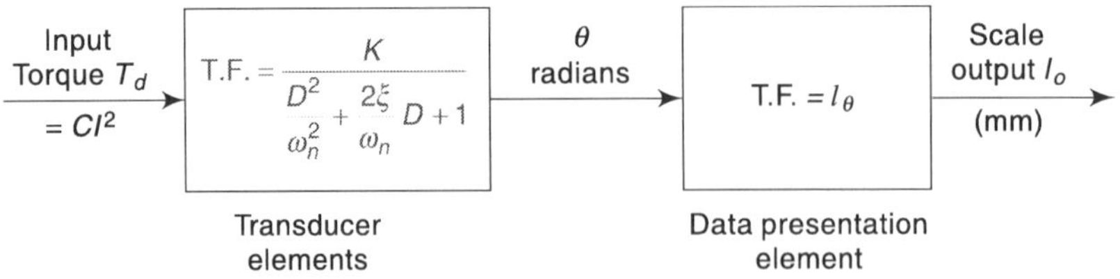

Fig. Prob. 20.6 *Block diagram of moving iron instrument*

Natural circular frequency ω_n of system $= \sqrt{\dfrac{k}{J}}$

$$= \sqrt{\frac{7.0}{7 \times 10^{-4}}} = 100 \text{ rad/s}$$

$$\text{Damping ratio } \xi = \frac{B}{2\sqrt{kJ}} = \frac{9.8 \times 10^{-2}}{2\sqrt{7.0 \times 7 \times 10^{-4}}}$$

$$= 0.7$$

and static sensitivity K of the instrument

$$= \frac{1}{k}$$

$$= \frac{1}{7} = 0.143 \text{ rad/N.m}$$

Further, frequency ratio $r = \dfrac{\omega}{\omega_n} = \dfrac{2\pi \times 50}{100} = 3.14$

Thus, the governing equation of the instrument becomes

$$\theta_o = \frac{0.143\, T_d}{\dfrac{D^2}{1.0{,}000} + \dfrac{1.4\, D}{100} + 1} \text{ rad}$$

The solution of this second-order different equation (refer/Ch. 4) gives:

$$\theta_o = \frac{K\,(CI^2)}{\sqrt{(1 - r^2)^2 + (2\xi r)^2}} = \frac{C\,(0.143) \times 10^2}{\sqrt{(1 - 3.14^2)^2 + (2 \times 0.7 \times 3.14)^2}}$$

$$= 1.45 \text{ C rad.}$$

The output of the transducer element becomes input to the data-presentation element.

It may be noted that data-presentation element is a zeroth-order element with a static sensitivity constant l = 40 mm which is the length of the pointer.

The output of the data-presentation element

$$I_o = l\, \theta_0$$

or I_o = (40) (1.45 C)

= 58 mm (given).

Therefore, C = 58/(40 × 1.45) = 1.0 N.m/A^2.

20.3.4 Electro dynamic (EDM) Type Instrument

Electrodynamic or electro-dynamometer (EDM) instruments are very similar to PMMC instruments. Herein, the horse-shoe type of permanent magnet of PMMC is replaced by two fixed coils on either side of the moving coil. These coils generate magnetic fields proportional to either the current (or definite fraction of current) to be measured or current flowing, proportional to the voltage to be measured. These coils are generally air cored to avoid errors due to hysterisis effects and eddy currents. The moving coil which is connected to the pointer of the instrument is connected in series with the fixed coils and is shown in Figs 20.16 (a) and (b) for the case of ammeter and voltmeter respectively.

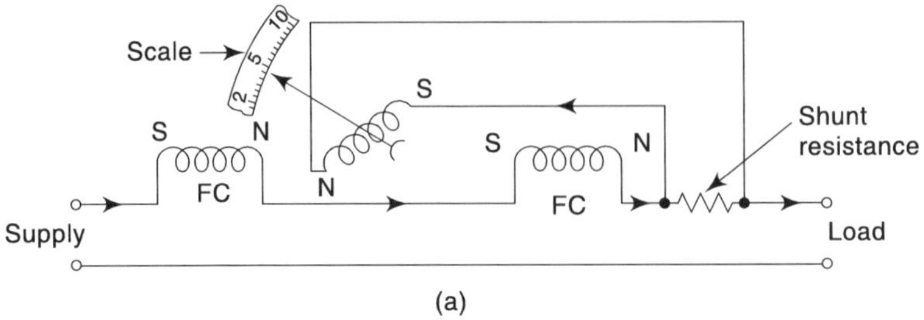

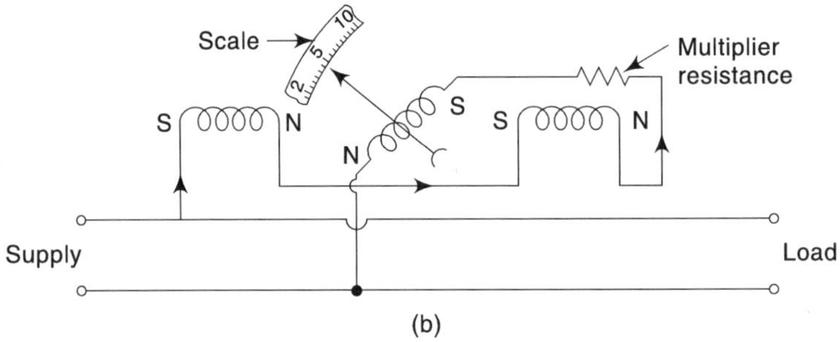

Fig. 20.16 *(a) A typical EDM ammeter (b) A typical EDM voltmeter*

The deflection torque developed in the moving coil suspended in the magnetic field of fixed coils is given by

$$T_D = BINA \tag{20.49}$$

where B is the strength of the magnetic field of the fixed coils in Wb/m^2

I is the current flowing in both the moving as well as fixed coils

N is the number of turns of wire in a coil

and A is the effective area of the moving coil in m^2

In the electrodynamic instrument the magnetic flux density B of the coil is directly proportional to the current I flowing through it, below its saturation range i.e., $T_D \propto I^2$ (20.50)

Due to the square law, the measuring scale of the instrument is clearly non-linear. Due to this there is overcrowding of scale makings at low current values and there is progressive spreading of scale divisions at higher current values.

Advantages

1. The instrument is of universal type. In other words, it can be used for both ac and dc systems.
2. Due to absence of an iron core, the instrument does not display hysterisis effects.
3. Eddy-currents losses are absent as the fixed coils have an air core.
4. The instrument can be modified to be used as wattmeter if current of the load circuit is flowing in fixed coils and the current proportional to voltage is flowing in the moving coil and vice versa.
5. The instruments can measure a range of currents and voltages up to 10 A and 600 V respectively with grade-I accuracy specification of ± 1 % and is conveniently available.
6. These instruments have the same calibration for dc instruments as well as ac measurements.

Disadvantages

1. These instruments have low torque/weight ratio as compared the D'Arsonval instruments. Hence the sensitivity of these instruments is relatively low because of production of poor deflecting torque, especially in the low measured value ranges.
2. The scale of the instrument is non-uniform because of the square-law response.
3. Power consumption of the instruments is relatively higher. This is because the current to be measured not only passes through the moving coil but also passes through the fixed coil to generate the magnetic flux necessary for generating a deflecting torque in the moving coil.
4. The instrument needs to be protected adequately by covering it with a laminated shield so as avoid the errors caused by stray magnetic fields.

20.3.5 Rectifier Instruments

Rectifier instruments are quite useful for the measurement of *ac currents and voltages* of radio and communication circuits in the audio frequency range (i.e., from 20 Hz to 20 kHz), by converting them into unidirectional dc. Then, these are measured using the high sensitivity D'Arsonval-type moving-coil movement, placed between the magnetic field of a permanent magnet. It may be noted that this type of measuring arrangement is more sensitive than the corresponding electrodynamic or moving-iron-type instruments for the measurement of ac currents and voltages.

The rectifier element of such instrument generally consists of either germanium of silicon diode because of their capabilities in terms of higher peak inverse voltages (PIV) as well as higher current ratings as compared to conventional copper oxide elements. Further, the instrument may be provided with either a half-wave rectifier or a full-wave rectifier circuit and the instrument would measure the average value of the rectified output because of mass inertia effect of the moving coil of the instrument. It may be noted that for the same input, the output of a half-wave rectifier in 50% of that of a full wave rectifier. Therefore, the scale of the instrument using a half-wave rectifier shall be multiplied by 2 if the instrument scale is calibrated for a full-wave rectifier.

Figure 20.17 shows a typical rectifier instrument with a full-wave rectifier bridge circuit. It may be noted that even though the rectifier instrument measures average value of voltage, its meter is calibrated in terms of rms voltage. Therefore, the scale is multiplied by the form factor K_f which is defined as the ratio of rms value of waveform to the average value of the waveform.

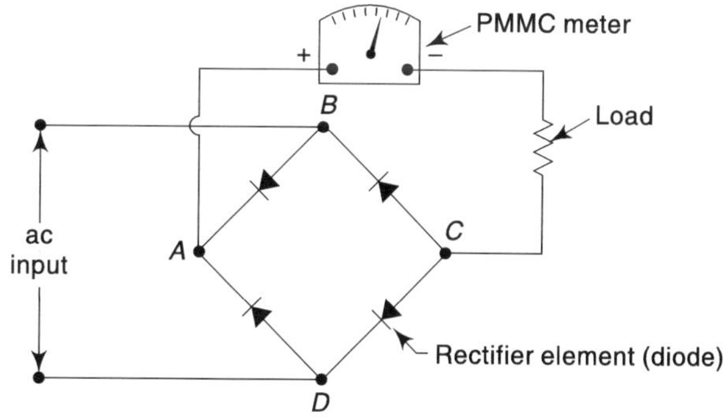

Fig. 20.17 *A typical rectifier-type voltmeter with a full-wave rectifier circuit*

As mentioned earlier, in a rectifier instrument, the PMMC meter scale is designed to indicate rms values. This is because, its moving coil moves the pointer with an average value of the input waveform. Therefore, it is important to evaluate the meter-scale multiplier value which is the form factor K_f of the waveform.

For the case of a typical sinusoidal input i.e., $v = V_m \sin \omega t$, the value of K_f for full-wave rectification is evaluated as follows:

$$v_{\text{rms}} = \left[\frac{1}{2\pi}\int_o^{2\pi} (V_m \sin \omega t)^2\, d(\omega t)\right]^{\frac{1}{2}}$$

$$= \frac{V_m}{\sqrt{2}} \tag{20.51}$$

Similarly

$$v_{av} = \frac{1}{2\pi}\int_o^{2\pi} (V_m \sin \omega t)\, d(\omega t)$$

$$= \frac{2}{\pi} V_m \tag{20.52}$$

Therefore, the value of form factor K_f for the sinusoidal type of ac waveform becomes

$$K_f = \frac{\text{rms value of ac signal}}{\text{average value of ac signal}}$$

$$= \frac{(V_m/\sqrt{2})}{\frac{2}{\pi}V_m} = 1.11 \tag{20.53}$$

Hence, in order to that a rectifier instrument shows real true rms values of sinusoidal input, its scale marking are made using a multiplying factor of 1.11. Further, since the output of the half-wave rectifier circuit is 50 % of that of the full-wave rectifier circuit, therefore, the scale markings of PMMC are multiplied by a factor of 2.22.

Problem 20.7 *A rectifier-type ac voltmeter is fed with a 10 V rms signal. Determine, the equivalent dc output for the case of: (a) half-wave rectifier circuit, and (b) full-wave rectifier circuit.*

Solution

(a) Input signal $= 10.0$ V rms

Peak voltage, $V_m = \sqrt{2}\, v_{\text{rms}} = \sqrt{2} \times 10 = 14.14$ V

Average value of voltage, v_{av} for the case of half wave rectifier circuit $= \frac{1}{\pi} V_m$

$$= \frac{(14.14 \text{ V})}{\pi} = 4.5 \text{ V}$$

Hence, $V_{\text{dc}} = 4.5$ V

(b) Since the output of the full-wave rectifier circuit is twice that of the half-wave rectifier circuit,

therefore, $V_{\text{dc}} = 2 \times 4.5$ V

$= 9$ V

Advantages

1. These instruments can be employed at frequencies up to 20 kHz which is well above the range of other ac instruments.
2. These instruments have low power consumption coupled with higher sensitivity as compared to other instruments.
3. They have a uniform scale.
4. It is used widely in the commonly used multimeters for measuring multiple ranges of ac voltages and currents as it is generally provided with half-wave rectifier circuit alongwith multiple shunts and multiplying resistances which can be suitably selected, using a selector switch.

Disadvantages

1. The measured values of rectifier type instruments are in terms of average values rather them the rms values.
2. For sinusoidal inputs, the scale reading needs to be multiplied by 1.11 for full-wave rectifier circuit for converting the average values to rms values. This becomes 2.22 if a half-wave rectifier circuit is used.
3. If the input signal has a different type of waveform such as square, triangles or saw-tooth type then instrument calibrated with a multiplying factor of the sinusoidal ac input would indicates erroneous readings.

20.3.6 Thermal Instruments

The principle of operation of thermal type of ammeters depends on the amount of heating effect caused by current which is being measured. There are two types of thermal instruments. They are

(i) Hot wire type and
(ii) Thermocouple type.

Hot-wire Instrument The principle of operation of the hot-wire instrument is based on the increase in length due to thermal expansion of the fine platinum-iridium resistance wire attached between two fixed supports. The heating effect is caused by the current flowing through the wire. Due to increase in length between fixed supports, a sag in the wire is caused. This sagging effect is suitably amplifier by means of mechanical linkages and is converted into the motion of the pointer on the graduated scale.

The schematic diagram of a hot-wire ammeter is shown in Fig. 20.18. It has fixed supports A and B in which a hot wire of platinum–iridium having a diameter of the order of 0.1 mm is attached. At ambient conditions, the sag of the wire is adjusted to zero and the pointer also indicates a zero on its scale. It is attached by means of another wire at E and another fixed support C. This phosphor–bronze wire is then attached by means of a silk thread wound round a pulley with sufficient tension and tied at the fixed point F.

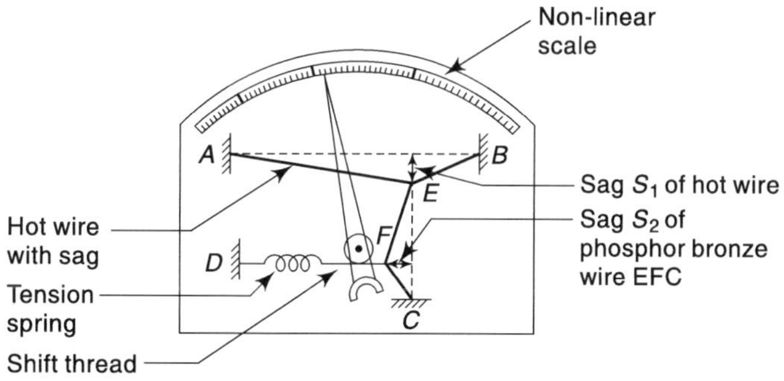

Fig. 20.18 *A typical hot-wire type of ammeter*

This instrument is a double-sag instrument. The second sag S_2 causes the movement of the silk thread. This, in turn, causes the movement of the pulley. The attached pointer with the pulley indicates the current flowing in the hot wire on the graduated scale.

It can be shown that the sag of the hot wire is proportional to the square root of change in the length dL which is the input parameter of the hot wire. Further, the change in length of the hot wire is directly proportional to (I^2R) heating effect.

Therefore, $$\text{sag } S_1 \text{ of hot wire} \propto \sqrt{dL}$$

or $$\propto \sqrt{I^2} \quad (\because \quad dL \propto I^2)$$

or $$\propto I \tag{20.54}$$

Similarly, it can be shown that the second sag S_2 which is the change in length in the phosphor–bronze wire also satisfies the square-root relationship of the input parameter I. Hence, the meter reading is proportional to $\sqrt{I}$ and consequently, the scale of the instrument is not uniform.

Advantages

1. The instrument is suitable for measurement of currents and voltages in very high frequency ranges including radio frequencies where other instruments are found to be unsuitable.
2. The instrument indicates an r.m.s value of current irrespective of its waveform. Thus, the instrument can be used for dc as well as ac.
3. The instrument is not affected by stray magnetic fields as no magnetic effect is used in the operation of instrument.
4. The calibration is same for ac and dc inputs.
5. The instrument is fairly accurate.
6. It is simple in construction and is also quite economical.
7. The instrument has fairly good sensitivity.

Disadvantages

1. The fine-diameter hot wire of the instrument is quite delicate and fragile.
2. The instrument is not capable of taking up overloads as the hot wire may melt. Therefore, as a precautionary measure, a safety fuse is incorporated in the hot-wire circuit.
3. These instruments are associated with a high power consumption.
4. The scale of the instrument is non-linear.
5. The instrument reading is associated with a small lag as it takes time to heat the wire to a steady-state value.
6. The instrument requires frequent zero adjustments because of the variations in the ambient temperature.

Thermocouple Instruments Thermocouple instruments are quite useful in the measurement of currents and voltages of high frequencies. Herein, the measured input is passed through a thin wire either of copper or constantan of diameter 0.025 mm, and this acts as a heater element. The heat generated equals to I^2R in the thin wire is thus proportional to the squared value of current. Further, this is not dependent on either the frequency or waveform of the measured input.

The heat generated in the heater element is then transferred to the hot thermal junction. The output of the thermocouple is fed to the permanent magnet moving coil (PMMC) type of instrument which is associated with a high degree of accuracy as well as sensitivity.

The instrument is available in the following configurations which are shown in Fig. 20.19.

(a) Self-heating type

In this type, the measured input is passed through the thermocouple element itself and not through the heater wire. This has a disadvantage that the thermocouple act as a shunt to the meter.

(b) Contact Type

In this type, the heater element is in direct contact with the hot junction of the thermocouple. This is however, less sensitive than the self-heating type of configuration.

(c) Non-contact Type

In this type, the thermocouple element is electrically insulated from the hot junction by means of an insulating bead. This way, the thermocouple element becomes a non-contact type of instrument. The provision of insulation between heater element and thermocouple junction becomes essential for the measurement of currents in the high voltage circuits. This arrangement makes the instrument sluggish and less sensitive.

(d) Vacuum Type

The sensitivity of the non-contact arrangement can be increased by placing the heater wire and thermocouple hot junction side by side in a vacuum chamber in which the heat loss due to conduction is avoided.

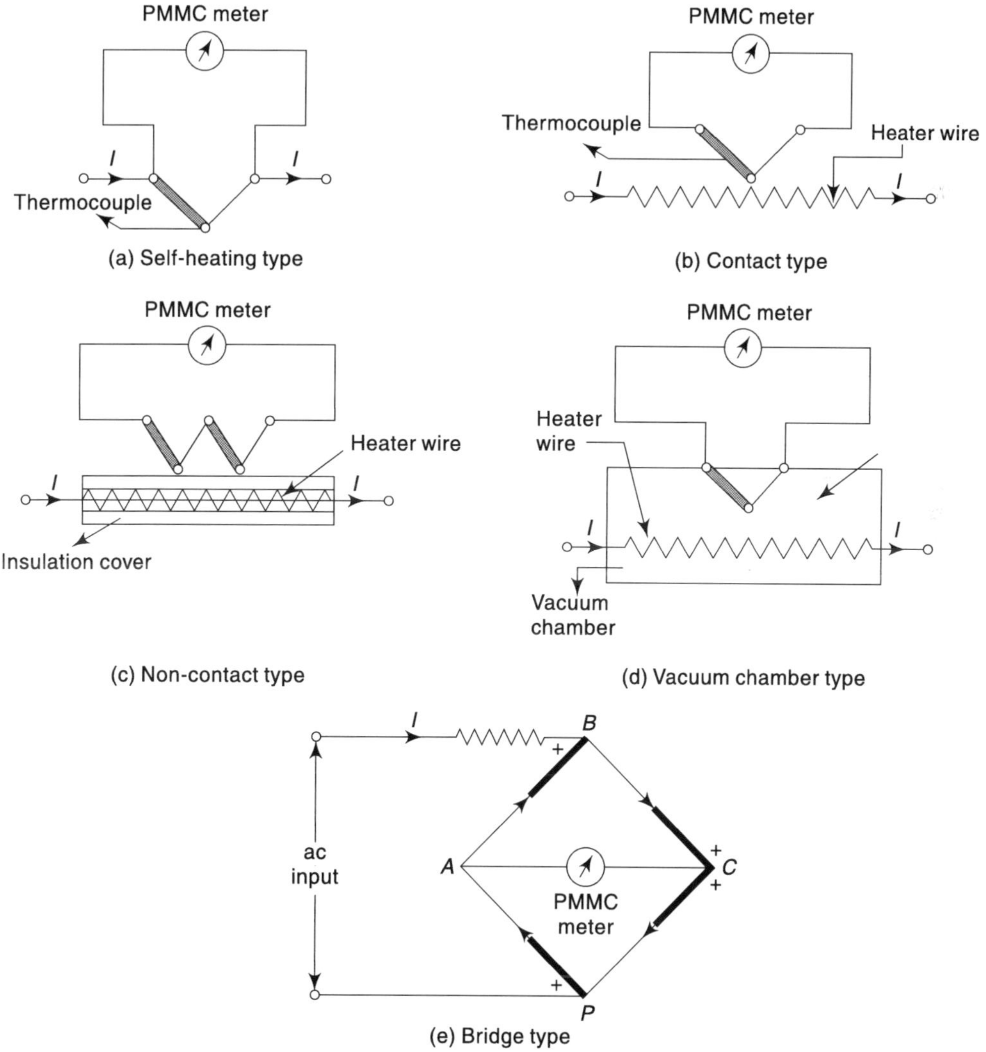

Fig. 20.19 *Typical thermocouple-type ammeter configurations*

(e) Bridge Type

In this configuration, no separate heater is used and the current to be measured flows directly through the thermocouple elements and raises their temperature by self heating effect, proportional to the square of the current. The thermal elements are arranged in the four arms of the bridge. In this configuration the shunt resistance effect of the thermal element is avoided. Further, the instrument gives greater output voltage in the meter as compared to the single, self-heating element-type arrangement.

Advantages

1. The capacitance and inductance in thermocouple instruments are extremely small. Therefore, these instruments are unaffected by the frequency and waveform of the measured inputs. Because of this, these instruments are capable of measuring input up to 100 MHz, i.e., in the RF domain.
2. The accuracy of the instruments is quite high, i.e., up to ± 1%.
3. These instruments are associated with high sensitivity.
4. They are not affected by stray magnetic fields.
5. The instruments are not expensive on they employ commonly used and conveniently available combinations of thermocouples like copper–constantan (T-type) iron constanton (J-type), chormel–constantan (E-type), chromel–alumel (K-type), Platinum–Platinum/10% rhodium (S-type).

Disadvantages

1. Heater elements of the instruments can withstand only small overloads.
2. The power consumption of the instrument in the measurement process is quite substantial
3. Rise in temperature in the heater element causes a change in the resistance of the heater.

20.3.7 Multimeter

A multimeter is a multi-purpose instrument as is evident from its name. It is one of the most versatile, general-purpose, compact and portable type of shop instruments. The instrument is alternatively termed either Ampere-Volt-Ohm (AVO) meter or Volt-Ohm-Milliammeter (VOM) meter.

Basically, it is a three-in-one multirange instrument suitable for measurement of

(a) Voltages (both dc and ac)
(b) Currents (both dc and ac) and
(c) Resistances

The usual ranges available in commonly available multimeters are as follows:

(a) dc voltage ranges of 100 mV, 3 V, 10 V 30 V, 300 V, 600 V and 1000 V
(b) dc current ranges of 100 μA, 300 μA, 1 mA, 3 mA, 10 mA, 1 A, 3 Aand 10 A
(c) ac voltage range of 3V, 10 V, 30 V, 300 V, 600 V and 100 V
(d) Resistance ranges of 2 kΩ, 200 kΩ and 20 MΩ and ranges marked as X 1, X 100 and X 10 k

It may be noted that the measurement ranges and circuit detail may vary from one instrument to another. However, the essential components of any multimeter are as follows:

(a) Balanced bridge amplifier coupled to a D'Arsonval type of PMMC indicating meter suitable for measuring dc inputs.
(b) Input attenuator or range switch to limit the input voltage to the desired value.
(c) Rectifier unit to convert the ac input voltage to proportional dc voltage for making them suitable for a PMMC meter.
(d) Internal battery which is operational in the resistance measurement circuit.
(e) Function switch to select the desired measurement range available in the instrument.

Figure 20.20 shows a typical portable type of multimeter. The rotary function switch has an 'off' marking shown herein at the top. On one side, muiltirange circuits for dc amperes and dc voltages can be selected. The position of dc (A) or dc (V) would correspond to the appropriate value of the shunt resistance or multiplier resistance respectively corresponding to the measuring range selected.

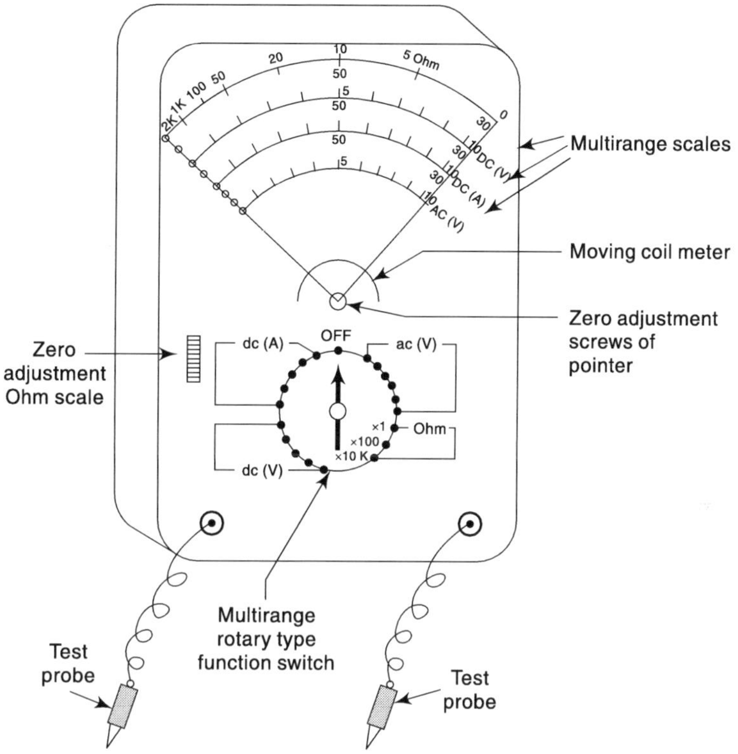

Fig. 20.20 *A typical portable type of multimeter*

When the function switch is put on ac (V), then the input of the instrument is connected to a half-wave rectifier circuit for converting ac input to dc output. This is done as the D'Arsonval instrument measures only dc input. It may be noted that the sensitivity of an ac input with half-wave rectification is only 45 % of dc sensitivity. Therefore, the multiplier resistance is determined in such a way that the output voltage gets enhanced by a factor of 2.22 so that the shortfall of the output due to a half-wave rectifier is fully compensated.

Lastly, when the function switch is put on the ohm-meter circuit, the built-in power supply is switched on in the ohmic circuit. Usually, the measured resistance is placed in series with a moving coil meter. The current through the meter is inversely proportional to the resistance being measured and the scale can be calibrated in terms of resistance (ohm). The resistance indicating scale is obviously reciprocal scale and is of non-linear type.

When the test probes are connected together, it causes a full-scale deflection of the pointer which corresponds zero resistance. The zero adjustment potentiometer can be adjusted to adjust the zero value of ohm on the measuring scale. On the other hand, when the test probes are not connected to anything, then the meter indicates infinity (∞). Hence, the resistance meter scales are in reverse direction to the voltage and current scales.

It may be noted that the ohm scales of multimeters are quite accurate but are suitable for measuring small resistances only. At the higher end, such a scale is difficult to read because of overcrowding and is less accurate because of non-linearity.

General operating instructions of a multimeter

1. To protect the meter from overloading preferably start with a higher range and move down the range successively.
2. For higher accuracy, the range selected should be such that the meter reading falls on the upper half of the meter scale.
3. The polarity of circuit should be checked for the cases of measurements of dc currents and voltages.
4. Ohm-meter batteries should be replaced frequently to ensure accuracy in resistance measurements.
5. The instrument should be used carefully and should not be exposed long to shock and vibrations, high temperatures, fumes and dusty environments.
6. Extra precautions should be taken when the meter is used for checking high voltages or high currents in the circuits.
7. In case of measurement of resistance, it should be ensured that power supply to the circuit is switched off; otherwise the voltage across the resistance may overload/damage the meter.

20.3.8 Induction-Type Wattmeter and Energy Meter

An induction-type instrument is suitable for measurements of ac inputs only. In these instruments the deflecting torque is generated by the interaction of eddy currents generated in the aluminium disc supported on jewelled bearings, by the two fluxes generated by electromagnets. Figure 20.21 shows two laminated electromagnets, namely, a shunt electromagnet and the series electromagnet. A shunt electromagnetic is excited by the current proportional to the load whereas a series magnet is excited by the line current of the load.

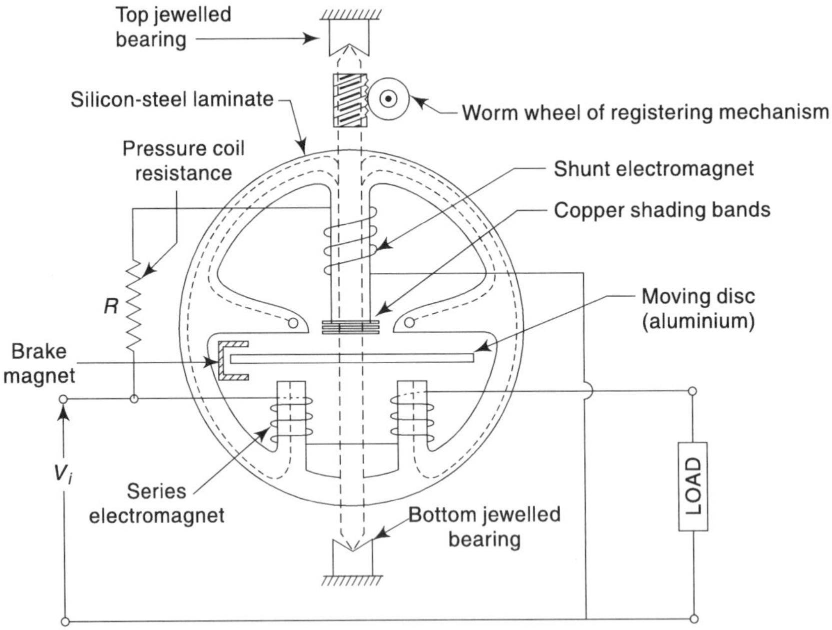

Fig. 20.21 *Induction-type energy meter*

The torque generated in the aluminium disc or drum is maximum when the phase difference ϕ between the magnetic fields of shunt and series electromagnetic, i.e., is equal to 90°. Copper shading bands are provided on the shunt magnet. With suitable adjustments of these copper shading bands, the maximum phase difference, $\phi = 90°$, is obtained between the fluxes of the said electromagnets.

An induction-type of instrument can be used in two modes.

(a) Induction-type Wattmeter The deflection torque produced on the aluminium disc is balanced by the controlling torque of a torsion spring. This way, the deflection of the pointer on the scale would indicate the wattage of the load.

(b) Induction-type Energy Meter The deflecting torque produced by the electrical load causes the aluminium disc in this case, to rotate. The rate of rotation is proportional to the wattage of the load. The rotational speed of the disc can be suitably controlled by means of a permanent magnet positioned near the edge of the disc. The magnetic field of this magnet provides a braking torque to the rotating disc. The position of the permanent magnet is adjustable and therefore, the braking torque can be suitably adjusted to obtain the required speed range of the aluminium disc.

The consumption of energy in watt-hour (Wh) is proportional to the number of rotations of the aluminium disc, in a given time. For totalising the number of rotations, the spindle of the disc is provided with a worm-screw and worm-wheel mechanism. The worm wheel, in turn is connected to a train of gears which are connected to the suitable energy consumed registers.

Advantages

1. These instruments are quite robust. They can handle currents up to 1 A and voltages up to 750 V.
2. They are not affected by stray magnetic fields.
3. The instruments develop relatively stronger torques.
4. The scale of the instruments are uniform when used as wattmeters.

Disadvantages

1. The meter cannot be used for dc inputs.
2. The instrument is comparatively less accurate as compared to EDM-type meter.
3. The power consumption of the meter is comparatively higher as compared to an EDM-type meter.
4. The instrument is accurate only at the stated frequency and temperature. The variation of frequency and ambient temperature introduces errors in the measurements.
5. The wear of the bearings with time increases the frictional torque and therefore, instruments need to be calibrated periodically.

20.3.9 Power Meter

The electrodynamic (EDM) movement discussed in Section 20.3.4 for the measurements of voltages and currents can be conveniently used to measure power for both dc inputs as well as ac inputs of any waveform of voltage and current. It may be noted that when the EDM movement is used for the measurement of only voltages or currents then the fixed coils and the moveable coil are connected is series, and the output of the instrument is proportional to I^2.

However, when the EDM movement is used as a dc wattmeter or a single-phase ac wattmeter then, generally, the fixed coil, known as current coil, is excited by the total line current flowing through the circuit, and the moving coil is not connected in series with this circuit. Instead, the moving coil located in the magnetic field of the fixed coil, is excited independently by the small current proportional to voltage by introducing a high multiplier resistance across the load circuit. The fixed coil is generally termed the pressure coil of the instrument. The layout of the electrical circuit of the Electrodynamic (EDM) type of wattmeter is shown is Fig. 20.22.

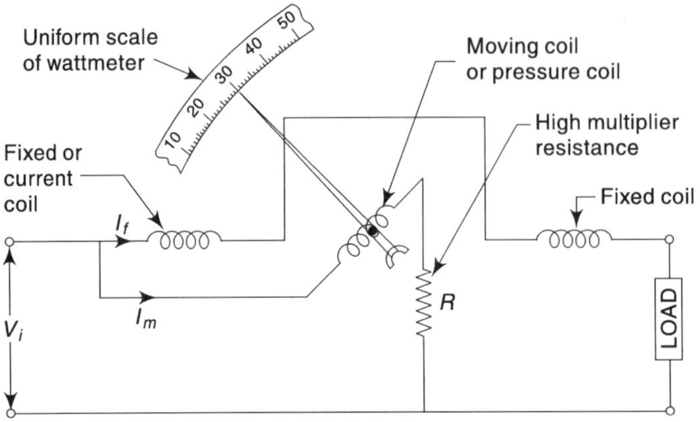

Fig. 20.22 *A typical electrodynamic (EDM) type of wattmeter*

The principle of operation of the electrodynamic (EDM) wattmeter depends on the movement of the moving coil. Its movement in turn depends on the interaction of magnetic fields of the fixed and the moving coils. Further, the intensity of magnetic fields in these coils depends on the currents I_f and I_m respectively flowing in these coils. Herein, I_f is the line current of the load and I_m is the current proportional to the voltage. Therefore, the meter can be conveniently calibrated in terms of watts or true power consumed by the load.

Say V_i = Input voltage

I_f = current in the fixed coil

= line current of load (see Fig. 20.15)

R = Resistance of the moving coil circuit

Therefore I_m = current in the moving coil

$$= \frac{V_i}{R} \text{ (by Ohms law)} \tag{20.55}$$

Now deflecting torque

$$T_d \propto I_f \times I_m$$

$$\propto I_f \frac{V_i}{R}$$

$$= K(I_f V_i) \quad \text{(where } K = \text{constant)}$$

$$= K \text{ (Wattage of load)} \tag{20.56}$$

Hence, for a dc circuit, the output of the meter is directly proportional to the power consumed or the wattage of the load. Therefore, the meter scale has uniform graduations. It may noted that there may be fluctuations in the load wattage. In that case, the meter does not indicate the instantaneous power. Due to the inertia of the moving parts it indicates the average power consumed in the load circuit.

The EDM meter is capable of measuring ac power also. In case, the inputs is of sinusoidal type then,

Say $V_i = V_m \sin\theta$ and

$I_f = I_m \sin(\theta - \phi)$

where ϕ is the phase difference between voltage and current in the load circuit.

Then, average value of deflection torque is

$$T_d = K\left[\frac{1}{2\pi}\int_0^{2\pi}(V_m \sin\theta)(I_m \sin(\theta-\phi)\,d(\theta))\right] \tag{20.57}$$

On simplification,

$$T_d = K\left[\frac{V_m\, I_m}{2}\right]\cos\phi \tag{20.58}$$

$$= \frac{V_m}{\sqrt{2}}\cdot\frac{I_m}{\sqrt{2}}\cos\phi \tag{20.59}$$

or
$$T_d = K\left[V_{\text{rms}} \times I_{\text{rms}} \times \cos\phi\right]$$

$$= K \text{ [ac power of load circuit]} \tag{20.60}$$

where
$$V_{\text{rms}} = \frac{V_m}{\sqrt{2}}$$

$$I_{\text{rms}} = \frac{I_m}{\sqrt{2}}$$

and
$\cos\phi$ = power factor of load circuit.

Advantages

1. The instrument indicates true power for both dc as well as ac inputs.
2. The scale of the instrument is linear.
3. By careful design, an instrument with a high degree of accuracy can be obtained and this in turn could be used as a standard for calibration purposes.
4. Due to absence of an iron core, the instrument does not display hysteresis effects.
5. Eddy current losses are absent as the fixed coils have air cores.

Disadvantages

1. These instruments have low torque/weight ratio. Hence, the sensitivity of the instrument is relatively small. This is because of production of poor deflecting torque, especially at low ranges of load power.
2. Power consumption of the instrument is relatively high. This is because there is some voltage drop in the fixed coils in current circuit and the voltage circuit, i.e., the moving coil also consumes current proportional to the voltage.
3. The instrument is prone to errors due to stray magnetic fields. Therefore, the instrument need to be provided with adequate magnetic shielding by enclosing it in an iron casing.

20.4 MAGNETIC FLUX MEASUREMENTS

For measurement of magnetic flux parameters, the following two types of devices are used commonly:

1. Hall-effect transducer
2. Magnetometer coil transducer

20.4.1 Hall-Effect Transducer

It is commonly used to determine magnetic flux density. Figure 20.23 shows a diagram to explain the principle of the Hall-effect transducer. A potential difference v is created due to the motion of beam of charged particles when a current passes through a plate, which is subjected to a magnetic field at right angles to the plate.

Forces act on the charged particles which are deflected on one side of the plate which is negatively charged, with the other side being positively charged. This results in the potential difference v as shown in Fig. 20.23, when

$$v = \frac{KBI}{t} \quad (20.61)$$

where I is the current (amperes)

B is the magnetic flux density of the field (Wb/m^2), t is the thickness of the plate (metre) and K being the Hall coefficient.

The units of flux density is Wb/m^2 viz., webers/m^2 or Tesla.

The value of K for *Si*, Germanium and *InSb* are 6×10^{-6}, 3.5×10^{-2} and 6×10^{-4} Vm/AT respectively where T stands for Tesla.

Thus, from Eq. (20.61), the value of B can be determined if the value of the other parameters are known.

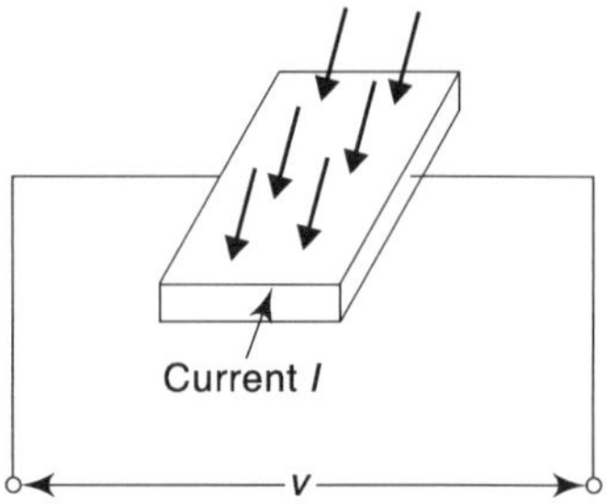

Fig. 20.23 *Hall effect transducer*

Problem 20.8 *The voltage output of a plate of Hall effect transducer is 5 mV. If the plate is made of Si whose K is 6×10^{-6} Vm/AT, plate thickness is 2 mm and the value of current through the plate is 4 amperes, find the value of flux density B.*

Solution Using Eqn. (20.61), viz

$$V = \frac{KIB}{t}$$

or

$$B = \frac{Vt}{KI} = \frac{5 \times 10^{-3} \times 2 \times 10^{-3}}{6 \times 10^{-6} \times 4}$$

$$= \frac{10}{24} = 0.42 \text{ Tesla}$$

The Hall effect transducer is used in several applications like fuel level measurement in a fuel tank, brushless motor for determining alignment of windings on its stator for switching current. The advantages of the transducer—low cost, immunity to environmental current and possibility of high switching rate.

20.4.2 Magnetometer Search Coil

This works on the principle of voltage being induced in a coil called the magnetometer search coil due to its motion, the coil being placed in a plane perpendicular to the direction of the magnetic field.

The motion may be rotation of the coil or its oscillations. The voltage induced is proportional to the rate of change of magnetic flux and depends on the number of coils N, angular velocity of motion of the coil ω and magnetic flux density B. It is seen that rms voltage v induced in the coil is given by

$$v = \frac{1}{\sqrt{2}} NAB\omega \quad (20.62)$$

where A is the area of cross section of the coil (m^2) and ω is the angular velocity of the coil in rad/sec, v is in volts and B has units of Tesla.

20.5 ■ WAVEFORM GENERATION AND MEASUREMENTS

Waveform signal generators are also called oscillators. The signals may be of harmonic or non-harmonic types. The signals are normally periodic or oscillating and are used for measurement testing of a system of mechanical, structural, electronic, control and instrument systems. The frequency of output signals of oscillators may range from a few Hz (Hertz) to several mega Hz and higher. The oscillators may usually generate harmonic, square or triangular on periodic signals as shown in Fig. 20.24.

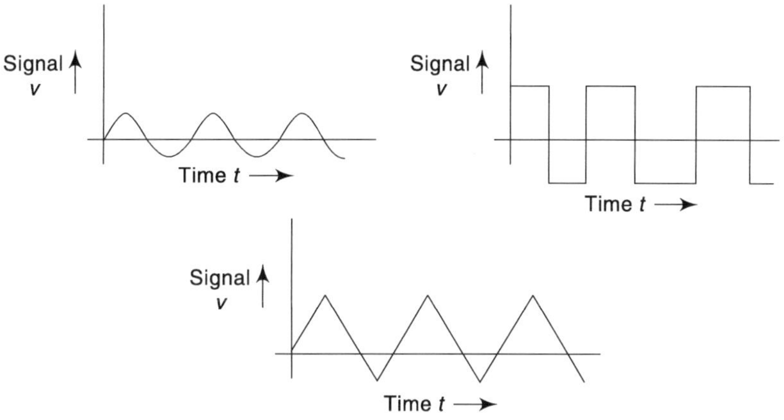

Fig. 20.24 *Type of oscillators*

20.5.1 Principle of Oscillators

The output of oscillators is produced by an oscillating circuit whose block diagram is shown in Fig. 20.25.

The three elements of an oscillator include

- Tank circuit
- Amplifier
- Feedback circuit

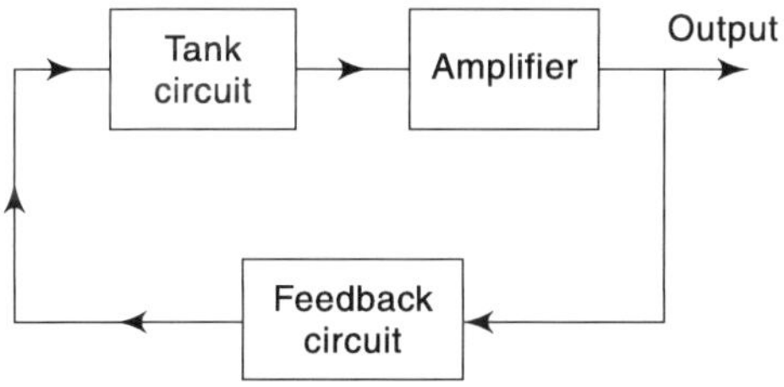

Fig. 20.25 *Block diagram of oscillator*

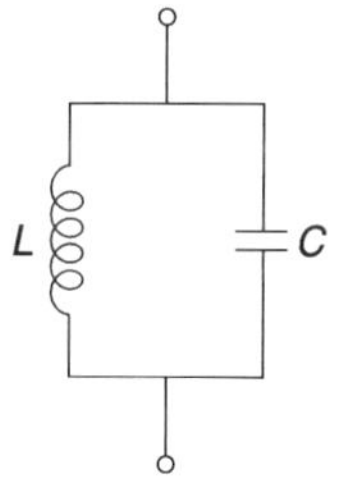

Fig. 20.26 *LC type tank circuit*

A tank circuit may be of *LC* type as shown in Fig. 20.26. The natural frequency of the circuit can be shown as

$$f(\text{Hz}) = \frac{1}{2\pi\sqrt{LC}} \tag{20.63}$$

The tank circuit may be of *RC* type or crystal type.

The signal from the tank circuit is applied to an amplifier which may be a transistor or operational amplifier type. The feedback circuit is used to apply the amplifier output to the tank and amplifier circuit as in Fig. 20.25.

If the transfer function of the forward loop is A and of feedback loop is B, as shown in Fig. 2.27, the equations relating outputs voltages v_o and input voltage v_1 is obtained as below:

$$\frac{v_o}{e} = A \tag{20.64}$$

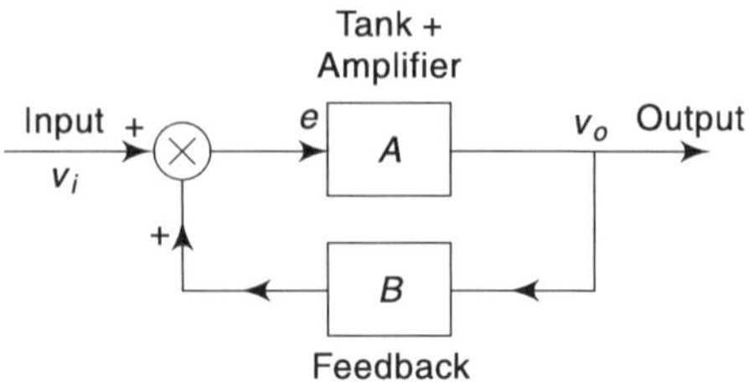

Fig. 20.27 *Principle of oscillator*

For positive feedback as shown in Fig. 20.28

Further, $$e = v_i + B\, v_o \tag{20.65}$$

Eliminating e from the above two equations, we get

$$\frac{v_o}{v_i} = \frac{A}{1 - BA}$$

$\frac{v_o}{v_i}$ is infinite or system is unstable if

$$1 - BA = 0 \tag{20.65}$$

or $$|BA| = 1$$

and $$\angle BA = 0°$$

This is called Barkhausen criterion for sustained oscillations at a frequency say ω, which is the harmonic or sinusoidal signal frequency (rad/s). In such circuits, instability with positive feedback is responsible for signal generation. This is possible with a small signal like the one with 'noise' which is always present in a signal.

There are several types of oscillators, viz., Wien bridge oscillator, phase shift oscillator, Hartley oscillator, crystal oscillator etc.

20.5.2 Wien-bridge Oscillator

A sketch of a Wien-bridge oscillator for generating sinusoidal signals using an operational amplifier (op. amp.) is shown in Fig. 20.28. This is used to generate signals in the frequency range of 1 Hz to 10 MHz.

The operational amplifier is used in the forward circuit and resistance R and capacitance C are used in the feedback circuit as shown.

Denoting voltages as v_o, v_a and v_b as in Fig. 20.28 and currents as ir, if, iR, i_c and i, it can be shown easily by circuit equations, that transfer function

$$A = \frac{v_o}{v_b} = \frac{R_a + R_f}{R_a} \tag{20.66}$$

Assume current i_R to be negligible in view of the very small current drawn by an op-amp. due to its high input impedance.

Similarly, transfer function $$B = \frac{v_b}{v_o}$$

$$= \frac{RCD}{(RC)^2\, D^2 + 3RCD + 1} \tag{20.67}$$

where operator $$D = \frac{d}{dt}$$

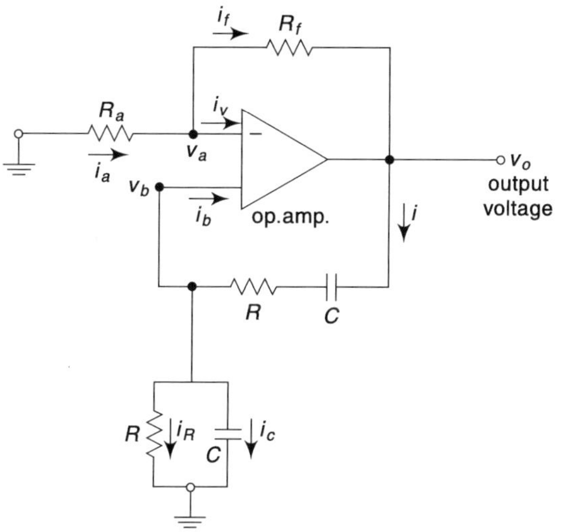

Fig. 20.28 *Wien bridge oscillator*

For satisfying Barkhausen criterion, magnitude of overall transfer function

$$\frac{A}{1 - AB} = 1$$

and phase angle = 0

Applying the above equations and taking

$$D = j\omega,$$

which is valid for harmonic oscillations, we can find the conditions for producing oscillations at a frequency ω (rad/s).

20.5.3 Phase-shift Oscillator

A phase-shift oscillator using a common emitter-amplifier configuration is shown in Fig. 20.29.

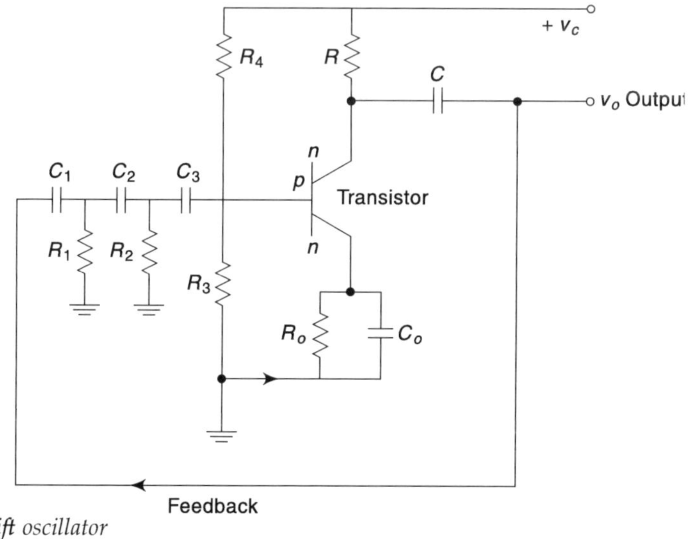

Fig. 20.29 *Phase-shift oscillator*

In the feedback circuit, three resistance and capacitance element R_1, C_1, R_2, C_2, R_3, C_3 are used which provide a phase shift of 180° while the amplifier provides another phase shift of 180°. So, the total phase shift is 360° or 0°, thus satisfying Barkhausen's criterion.

It can be shown that frequency of oscillator

$$f_{\text{osc.}} = \frac{0.065}{\bar{R}\bar{C}} \text{ (Hz)}$$

where $R_1 = R_2 = R_3 = \bar{R}$

and $C_1 = C_2 = C_3 = \bar{C}$

20.5.4 Hartley Oscillator

This type of oscillator makes use of feedback with inductances, L_1 and L_2 as shown in Fig. 20.30, which act as an auto transformer providing feedback between output and input. An *npn* transistor is used in common emitter amplifier arrangement.

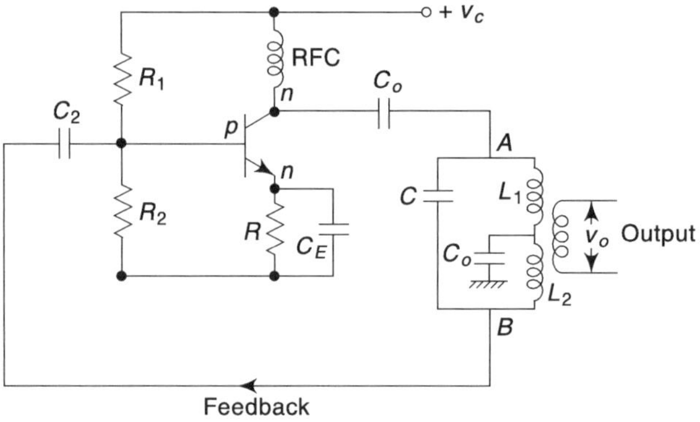

Fig. 20.30 *Hartley oscillator*

It is a positive feedback oscillator

The tank circuit between *A* and *B* comprises of L_1, L_2 and *C*. The transistor provides a phase of 180°, with *LC* tank circuit providing a phase shift of 180°, the total phase shift being 360°. RFC is radio frequency choke between + v_c and collector, while other capacitors block dc and pass ac.

Frequency of oscillations is given by Eq. (20.68) as

$$f_o = \frac{1}{2\pi\sqrt{LC}} \text{ (Hz)} \tag{20.68}$$

where $L = L_1 + L_2$

20.5.5 Colpitt's Oscillator

This type of oscillator uses capacitive feedback, with the arrangement being identical to that of a Hartley oscillator. Two capacitances C_1 and C_2 and one inductance *L* are use as shown in Fig. 20.31, which shows the tank circuit between *A* and *B*.

Frequency of oscillations in given by Eq. (20.69) as

$$f_o = \frac{1}{2\pi\sqrt{\dfrac{L C_1 C_2}{C_1 + C_2}}} \tag{20.69}$$

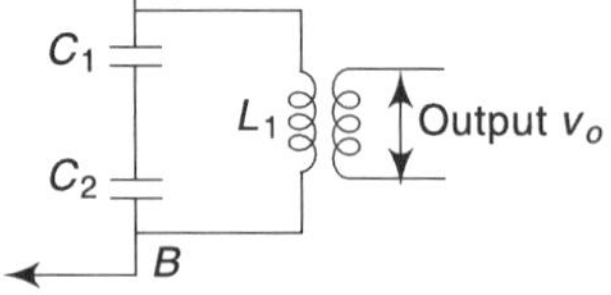

Fig. 20.31 *Tank circuit of Colpitt oscillator*

20.5.6 Crystal Oscillator

Use of piezoelectric crystals for signal generation is seen to give a high degree of stability of frequency. For this reason, they are used for time or frequency reference normally at high frequency. Application of ac voltage to a crystal results in its vibration at the frequency of the applied signal, which can be adjusted to give high vibrations. The corresponding frequency is called resonant frequency.

Figure 20.32 (a) shows a crystal oscillator with a crystal arranged for operation in series and in feedback path between collector and base of a transistor, connected in a common emitter amplifier arrangement. Resistances R_1 and R_2 are like a potential divider to provide bias to the transistor. R_3 is emitter resistance and C_3 is emitter capacitor, RFC is radio frequency of choke blocks of the ac component in the supply, C_0 is capacitance of the mounting electrodes of the crystal. The equivalent circuit of the system is shown in Fig. 20.32(b).

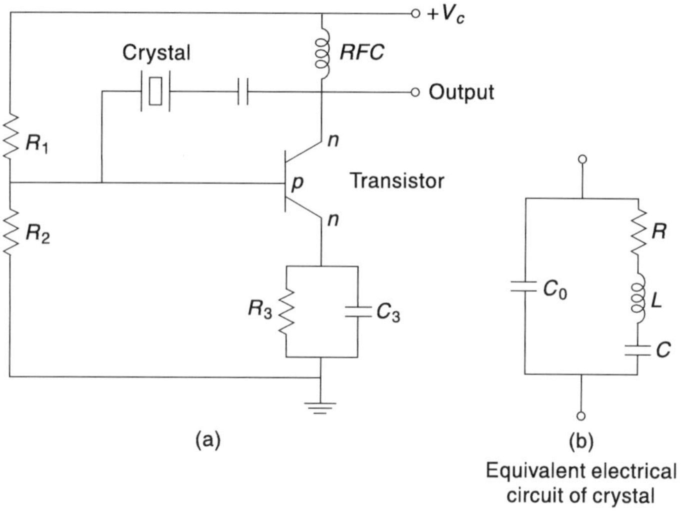

Fig. 20.32 *Crystal oscillator*

The frequency f of oscillation is seen to be as in Eq. (20.70) viz.

$$f = \frac{1}{2\pi\sqrt{LC}} \tag{20.70}$$

where L and C are equivalent inductance and capacitances of the circuit.

20.5.7 Square-wave Generator

The arrangement for a square-wave generator is shown in Fig. 20.33(a). The capacitance C charges and discharges alternately. This causes a high or low operational amplifier saturation output as shown in Fig. 20.33(b) and (c).

By writing and solving equations for branches containing R_1 and R_2, voltages v and v_b, for current i, taking current i_b as negligible, we get

$$v_b = \frac{R_2\, v}{R_1 + R_2} = gv \tag{20.71}$$

where
$$g = \frac{R_2}{R_1 + R_2}$$

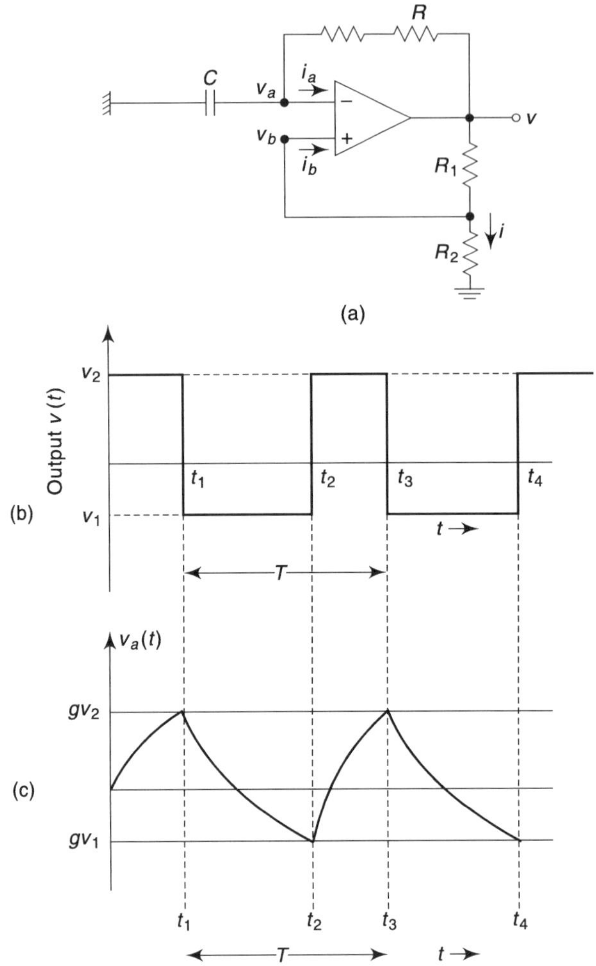

Fig. 20.33 *Square wave generator*

Similarly, for the branch containing R and C, it is possible to get the first-order differential equation for v in terms of v_a, assuming current i_a as negligible. The solution of the differential equation can be obtained during increase and decrease of v_a against time t.

The operational amplifier v (t) will be at saturation levels as in Fig. 20.33 (b), giving

For $v_b > v_a$, output $v = v_2$

For $v_b < v_a$, output $v = v_1$ (20.72)

Also, v_a (t) will vary between gv_2 and gv_1 as shown in Fig. 20.33(c),

where $$g = \frac{R_2}{R_1 + R_2}$$

Voltages $v_a(t)$ and v_a (t) are periodic and time period T is seen to be:

$$T = R_c \log_e \frac{(v_2 - g\,v_1)(v_1 - g\,v_2)}{v_1\,v_2\,(1-g)^2} \tag{20.73}$$

20.6 FREQUENCY AND PHASE MEASUREMENT

20.6.1 Frequency Measurement

There are several techniques for measurement of frequency of a periodic repetitive signal say a sinusoidal signal. These include measurements using a CRO or a use of frequency counter.

Measurement Using CRO As shown in Fig. 20.34, it can be easily done using a CRO. The voltage signal is applied to the y axis (vertical) and time base is along t (horizontal axis). As shown, time T between the two successive peaks or zero crossings can be easily measured on a calibrated *CRO*. This is called time period of the signal, with which it repeats.

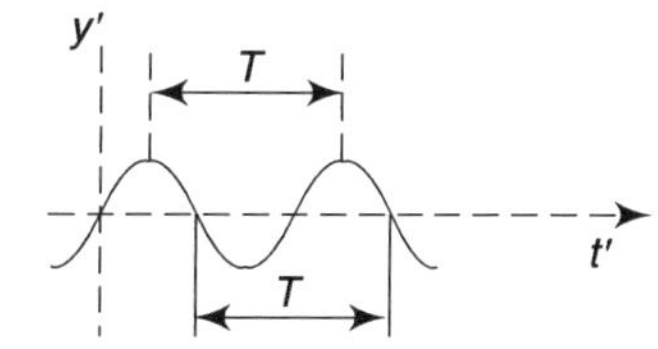

Fig. 20.34 *Frequency measurements using a CRO. (Cathode Ray Oscillascope)*

$$\text{Frequency } f \text{ (Hz)} = \frac{1}{T}, \ T \text{ being in seconds.}$$

Frequency Counter A sketch of the frequency counter is given in Fig. 20.35. The time period of the signal with an unknown frequency is compared with that of a gate-enable signal of known time period from a clock oscillator as shown in Fig. 20.35(a). A comparison of the two signals as in Fig 20.35(b) gives easily the unknown frequency of the signal, e.g.,

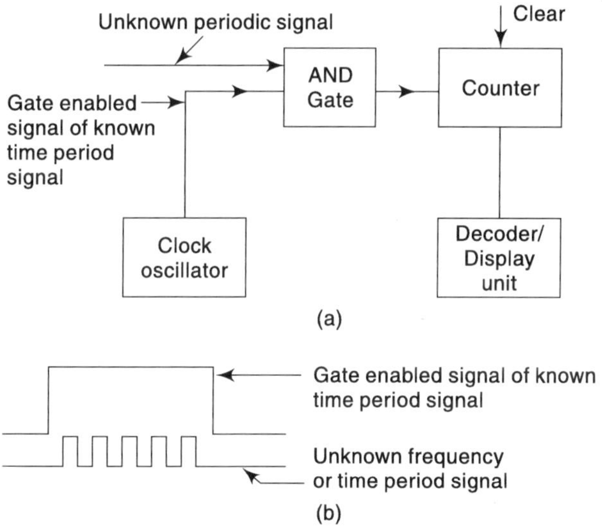

Fig. 20.35 *Frequency measurement using a frequency counter*

if the time period of the gate enabled signal from the clock is 1 sec, (or frequency 1 Hz), the unknown frequency of the second periodic signal is given by its number of cycles in 1 second.

20.6.2 Phase Measurement

The phase angle between two signals say harmonic ones can be obtained by several methods including use of a CRO or a phase meter.

Phase Measurement using CRO As explained in Chapter 6, in order to measure phase between two harmonic signals A and B as in Fig. 20.36, the time difference between the two peaks (or zero crossings) can be measured by using Lissajous figure, which is obtained by feeding one of the signals to the vertical input control and the other to the horizontal input controls of the CRO. For harmonic signals, the Lissajous figure is in the form of an ellipse as shown in Fig. 20 37.

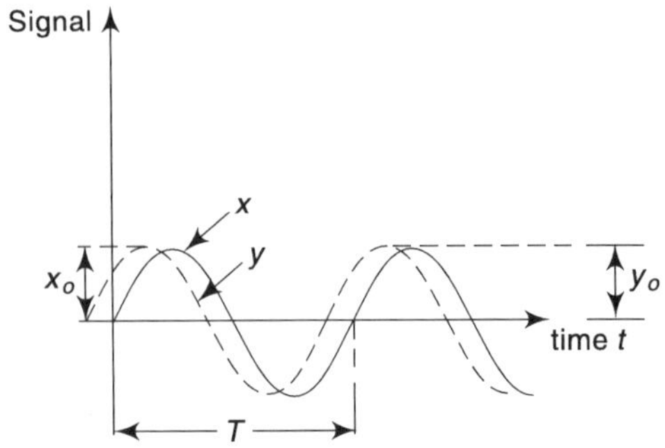

Fig. 20.36 *Voltage signals x and y as seen on the CRO*

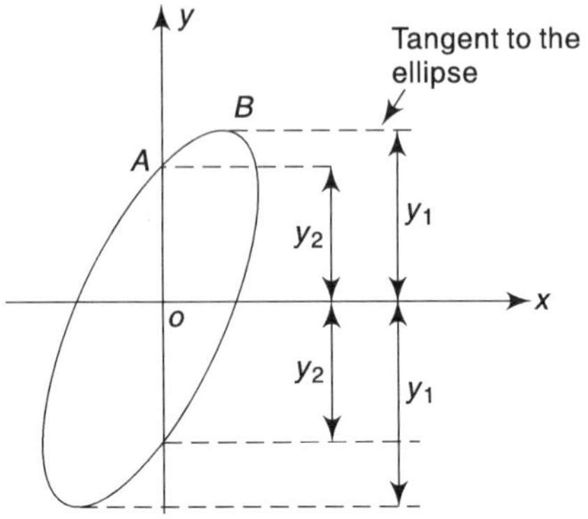

Fig. 20.37 *Ellipse as seen from CRO with signals y and x applied to vertical and horizontal input terminals*

If signal $x = x_o \sin \omega t$ (20.74)

$$y = y_o \sin (\omega t + \phi) \quad (20.75)$$

where ω is the circular frequency (rad/s), which is same for the two signals and ϕ is the phase difference (rad) between the signals, fed to the vertical and horizontal plates of the CRO

It is seen from Eqs (20.74) and (20.75) that

At $t = 0, \quad x = x_o$

Further, at $t = 0, \quad y = y_o \sin \phi = oA = y_2$ (20.76)

(from Fig. 20.37)

Thus, $y_2 = y_o \sin \phi = y_1 \sin \phi$ (20.77)

Since peak value $y_o = y_1$ from Fig. 20.37

Thus, from Eq. (20.77),

$$\sin \phi = \frac{y_2}{y_1} \quad (20.78)$$

Knowing y_2 and y_1, ϕ can be easily obtained.

Problem 20.9 *Find the phase difference between the harmonic signal x and y for which Lissajous figures are obtained as shown in Figs. 20.38 (a) or (b).*

Solution For Fig. 20.38 (a), using Eqn. (20.78),

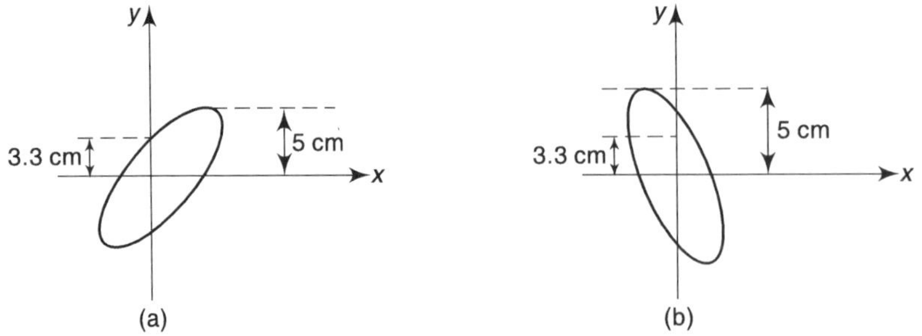

Fig. 20.38 *Calculation of phase difference*

$$\sin\phi = \frac{3.3}{5} = 0.66$$

$$\phi = 41.3 \text{ deg}$$

For Fig. 20.38 (b),

$$\phi = 180 - 41.3 = 138.7 \text{ deg.}$$

Phase meter For finding phase difference between two signals of same frequency, the principle of phase meter is as shown below:

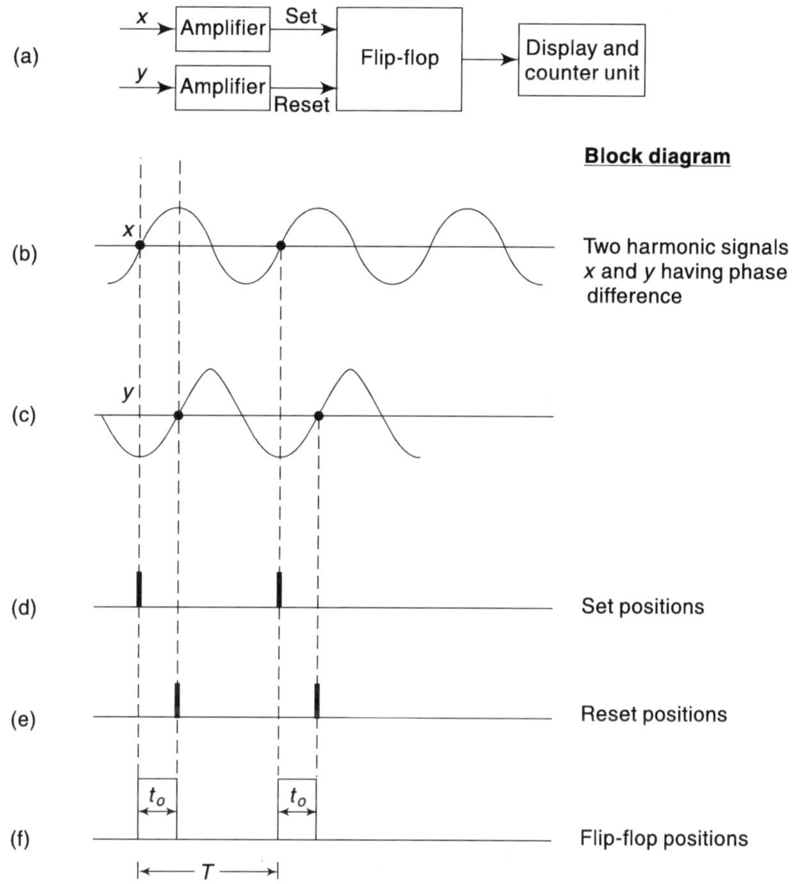

Fig. 20.39 *Principle of phase meter*

The two input signals are amplified as shown in Fig. 20.39 (a). The positive zero-crossing of x sets a flip-flop while the positive zero-crossing of y signal resets it, as shown in Fig. 20.39 (b) to (e). Thus, the output of the flip-flop is a rectangular pulse train of width t_o as shown in Fig. 20.39 (f).

If the time period of the input signals is T, the phase ϕ between x and y is given by

$$\phi = \frac{t_o}{T} \times 360 \text{ degrees.} \tag{20.79}$$

It the clock frequency is $\frac{360}{T}$, the clock signal is gated for delay period and the number of pulses can be counted, giving an indication of the phase angle ϕ.

Review Questions

20.1 Indicate if the following statements are true or false. If false, rewrite the correct statement.

(i) A moving coil galvanometer is made into a dc ammeter by connecting a high resistance in series with the meter.

(ii) The response of a moving iron instrument is of linear type.

(iii) For a sensitive type of galvanometer, a taut suspension type of support is generally used.

(iv) The instrument with square law response can be used for measuring both dc as well as ac inputs.

(v) An ac voltmeter using a full-wave rectification and having sinusoidal input has a sensitivity equal to 0.9 times the dc sensitivity.

(vi) The deflection of the hot-wire instruments depends on the averages value of current of any waveform and frequency.

(vii) Error due to hysterisis is predominant in the moving-iron instrument.

(viii) The permanent magnet moving coil-type of instrument is capable of measuring dc inputs only.

(ix) The operating frequency range of the moving iron instrument is from dc to 125 Hz.

(x) If the damping torque of a measuring instrument is more than the critical damping, the instrument is called over-critically damped.

20.2 Fill in the blanks in the following:

(i) A multi-range ammeter is a galvanometer provided with multiple, ________.

(ii) For the rectifier instrument to indicate the rms value of the measured input, its scale is multiplied by the ________ of the waveform.

(iii) Uniformity of scale of an ammeter indicates that the meter is of ________ type.

(iv) An electrodynamometer type of instrument can be used to measure both ________ inputs.

(v) The voltage range of dc voltmeter can be extended by providing ________.

(vi) Multiplying power of a shunt is defined as the ratio of maximum value of measured current to that of ________.

(vii) The torque produced by an indicating instrument by utilizing magnetic, electrodynamic, thermal and electrostatic effect is known as ________.

(viii) The force responsible for the reduction of amplitude of oscillations of an ammeter is ________.

(ix) The shunt resistance of the ammeter is ________ than the meter resistance.

(x) At the steady-state deflected position of the measuring instrument, the controlling torque developed in the instrument is ________.

20.3 Choose the appropriate answer in the following:

(i) Which of the following instruments is associated with high power consumption?

(a) permanent magnet moving coil type
(b) moving iron type
(c) electrodynamometer type
(d) hot wire type

(ii) The deflecting torque produced in the moving iron instruments is proportional to
 (a) current (b) $(\text{current})^2$
 (c) $(\text{current})^3$ (d) $\sqrt{(\text{current})}$

(iii) The main limitation of the PMMC instrument is
 (a) high power consumption
 (b) absence of effective eddy current damping
 (c) low torque/weight ratio
 (d) high cost relative to moving-iron instrument

(iv) Electrodynamometer type of instruments can be used to measure
 (a) ac input only
 (b) dc input only
 (c) both ac as well as dc inputs
 (d) very high frequency, MHz range type inputs only

(v) A rectifier-type moving-coil instrument respond to
 (a) rms values of all waveforms
 (b) average values of all waveforms
 (c) rms values of only sinusoidal waveforms
 (d) peak values of all waveforms

(vi) The controlling torque of an electrical measuring instrument is proportional to
 (a) θ (b) θ^2 (c) $\frac{1}{\theta}$ (d) $\sqrt{\theta}$

(vii) For measuring currents in the radio frequency range, which of the following instruments is used?
 (a) moving iron type (b) moving coil type
 (c) thermocouple type (d) rectifier type

(viii) In a thermocouple type of instrument, heat energy transferred to the hot junction is converted back into electrical energy due to
 (a) Faraday effect (b) piezo-electric effect
 (c) Seebeck effect (d) Joule–Thomson effect

(ix) The operating frequency range of a rectifier type of instrument is
 (a) up to 20 Hz (b) between 20 Hz to 20 kHz
 (c) from 20 kHz to 50 MHz (d) above 50 MHz

(x) Which of the following ammeters is associated with uniformity in the scale markings?
 (a) thermocouple type (b) moving-iron type
 (c) hot-wire type (d) permanent magnet moving coil type

20.4 Indicate True or False against the following
 (a) Hall effect sensor is used to measure rate of change of magnetic flux
 (b) A magnetic search coil gives an output voltage which is proportional to second time derivative of magnetic flux
 (c) In a waveform generator, negative feedback is used.
 (d) Barkhausens's criteria states that in the total phase of an oscillator circuit, total phase shift is 180°
 (e) In a Hartley oscillator, capacitances are used in the feedback circuit.
 (f) In a Colpitt's oscillator, inductances are used in the feedback circuit.
 (g) A crystal oscillator is used for variable low-frequency oscillator.
 (h) Both frequency and phase difference between two signals can be measured using a CRO.
 (i) Lissajous figure is an ellipse if the frequency of signals in horizontal and vertical inputs in a CRO are same.

20.5 The impedances in the various arms of an AC bridge (shown in Fig. Prob. 20.5) having an excitation frequency of 5 kHz are as follows:
   (i) arm AB with impedance $Z_1 = (10 - 10j)\ \Omega$
   (ii) arm AD with impedance $Z_2 = (5 - 5j)\ \Omega$
   (iii) arm BC with impedance $Z_3 = 10\ \Omega$ and
   (iv) arm DC with impedance $Z_4 = 5\ \Omega$
      (a) Check if the bridge is balanced?
      (b) If the arm *AB* has impedance $Z_2 = (10 + 10j)\ \Omega$ then check the bridge balance in this case?

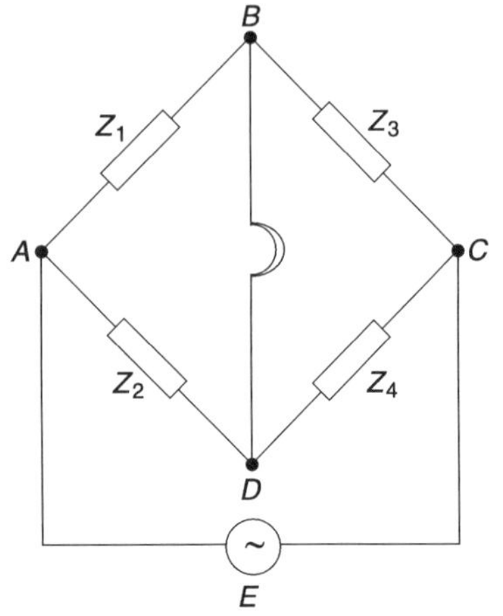

Fig. Prob. 20.5

20.6 (a) A moving coil instrument has a resistance of 100 Ω and full scale deflection current of 2 mA. Determine the values of individual multiplier resistance for each voltage range of a multi-range voltmeter for using it for the following voltage measuring ranges:
   (i) 0 to 1 V   (ii) 0 to 10 V
   (iii) 0 to 30 V   (vi) 0 to 100 V
(b) Determine the values of resistances R_1, R_2, R_3 and R_4 if multiplier resistances are connected as shown in Fig. Prob. 20.6(b).

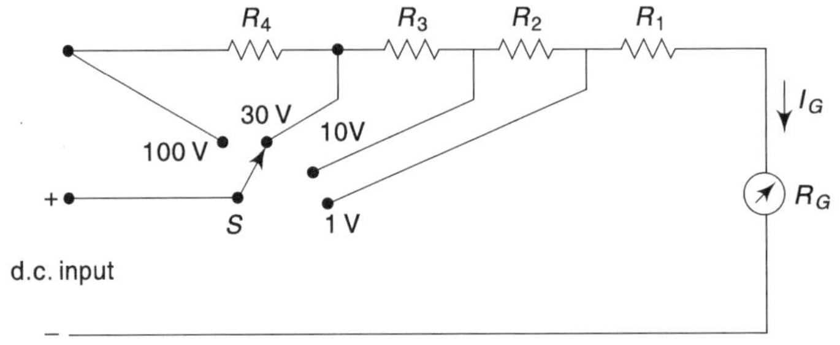

Fig. Prob. 20.6(b)

20.7 (a) A D'Arsonval movement has a resistance of 100 Ω and full scale deflection current of 2 mA. Determine the values of individual shunt resistances to make it a multi-range ammeter for using it in the following current ranges.
   (i) 0 to 100 mA   (ii) 0 to 1 A   (iii) 0 to 10 A   (iv) 0 to 100 A
(b) Determine the value of resistances R_1, R_2, R_3 and R_4 if the shunt resistances are connected as shown in Fig. Prob. 20.7(b).

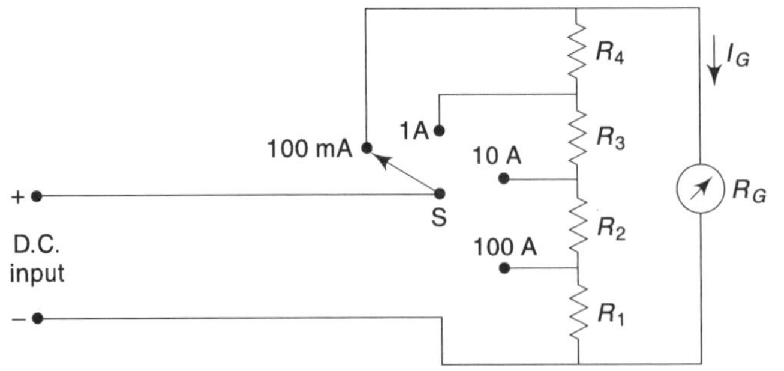

Fig. Prob. 20.7(b).

20.8 A D'Arsonval movement with an internal resistance of 500 ohms and full scale current of 50 mA is available. Show how it could be adopted as voltmeter to measure voltages up to 1000 V and as ammeter to measure currents up to 100 A.

20.9 An unknown resistance was measured by ammeter-voltmeter method. A voltmeter having a multiplying resistor of 200 kΩ gave a reading of 150 V on its 200 V scale when it was connected to an unknown resistance in series with a milliammeter. If the milliammeter reading was 12 mA, determine

(a) the apparent resistance of the unknown resistor,
(b) the actual resistance of the unknown resistor, and
(c) the percentage error in the measurement due to loading effect of the voltmeter.

20.10 A Maxwell's capacitance bridge shown in Fig. Prob. 20.10 is used to measure the unknown inductive impedance, the various values are

$$C_1 = 0.15\ \mu\text{F} \quad \text{and} \quad R_1 = 1500\ \Omega$$
$$R_2 = 800\ \Omega$$
$$R_3 = 1200\ \Omega.$$

(a) Determine the unknown impedance values L_x and R_x.
(b) Determine the Q-factor of the coil if excitation frequency is 2 kHz.

20.11 An AC Bridge was connected in Hay's bridge configuration shown in Fig. Prob. 20.11. The various of electrical components connected in the arms were as follows:

(i) Arm AB – a capacitor of 0.5 μF in series with resistance of 800 Ω
(ii) Arm BC – a resistance of 2000 Ω
(iii) Arm AD – a resistance of 1000 Ω and
(iv) Arm DC – a choke coil of unknown inductance L_x and resistance R_x.

Determine the inductive impedance of the choke coil if bridge balance was achieved at the supply frequency of 500 Hz.

20.12 An AC bridge in Schering mode configuration is used to measure the impedance of a series type of R-C circuit, which has been connected to arm DC of the bridge (shown in Fig. Prob. 20.12). The other three arms of the bridge are as follows:

(i) Arm BC = Resistance of 500 Ω
(ii) Arm AD = Capacitance (loss free) of 100 μF
(iii) Arm AB = capacitor of 0.4 μF in parallel with 1000 Ω resistance

(a) For the balanced bridge, determine the values of unknown capacitance and resistance, if the bridge was excited by 10 kHz supply voltage.
(b) Determine the dissipation factor of the R-C circuit under test.

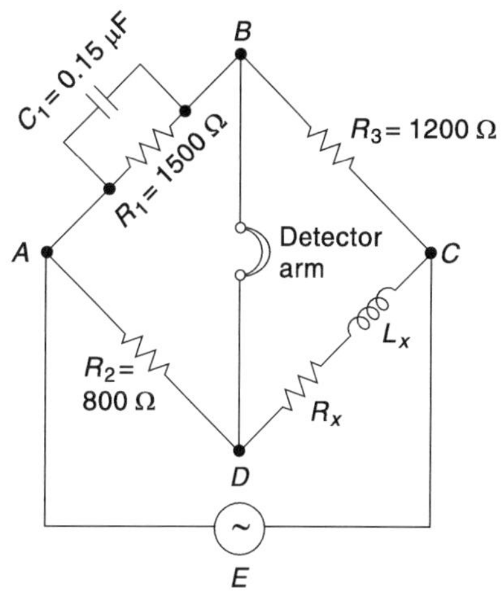

Fig. Prob. 20.10

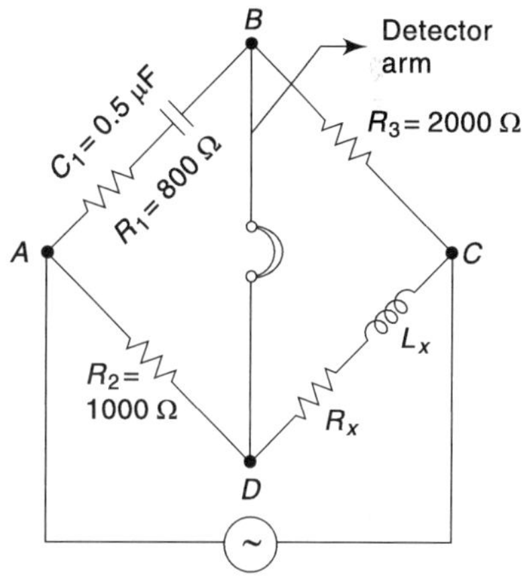

Fig. Prob. 20.11

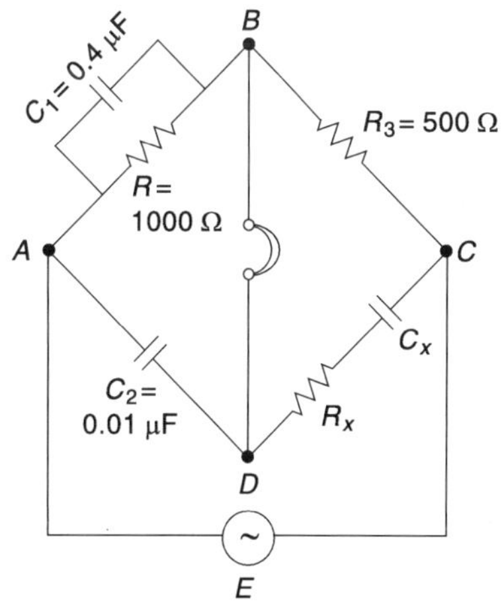

Fig. Prob. 20.12

20.13 A permanent magnet moving coil instrument has the following data:
(i) Coil dimensions = 30 mm × 25 mm
(ii) Flux density in the air gap = 2.1×10^{-3} Wb/m^2
(iii) Control spring constant = 1.0×10^{-6} Nm/radian
Determine the number of turns required to produce an angular deflection of 90° in the moving coil of the instrument when a current of 10 mA is flowing in it.

20.14 Find the desired thickness of a plate of a Hall effect transducer if the voltage output is 100 mV. The plate is made of In Sb whose Hall coefficient is 6×10^{-4} Vm/AT and flux density of the magnetic field is 0.8 Tesla. The current through the plate is 3 amperes.

20.15 For a Wien bridge oscillator, find the frequency of oscillation and other parameters, if resistance R = 200 ohms and capacitance C = 10 μF, and the bridge satisfies the Barkhausen criteria for an oscillator.

20.16 For a phase shift oscillator, find the frequency of oscillation for $R_1 = R_2 = R_3$ = 1 kΩ. ohm and $C_1 = C_2 = C_3$ = 2 μF.

20.17 For a Hartley oscillator of Fig. (20.30), find the frequency of oscillation if $L_1 = L_2 = 6 \times 10^{-3}$ H and C = 10 μF.

20.18 Draw the shape of lissajous ellipse for two harmonic signals of same frequency if phase difference is 30°, 90°, 150°, 210°, and 300°.

Answers

20.1 (i) False. A moving coil galvanometer is made into a high range ammeter by connecting a low resistance in the parallel with the meter which is known as shunt resistance.
(ii) False. The response of the moving iron instrument is of non-linear type.
(iii) True
(iv) True
(v) True
(vi) False. The deflection of the hotwire instrument depends on the rms value of current of any waveform and frequency.
(vii) true
(viii) true
(ix) true
(x) False. If the damping torque of a measuring instrument is more than the critical damping, the instrument is called over-damped on dead beat type.

20.2 (i) Shunts (ii) form factor (iii) D'Arsonval's type (iv) ac and dc
(v) multiplier resistance (vi) full scale (vii) deflecting torque
(viii) damping force (ix) less (x) equal to deflecting torque

20.3 (i) (a) (ii) (b) (iii) (a) (iv) (c)
(v) (b) (vi) (a) (vii) (c) (viii) (c)
(ix) (b) (x) (d)

20.4 (a) F (b) F (c) F (d) F
(e) F (f) F (g) F (h) T
(i) T

20.5 (a) The products of magnitudes of impedances and sum of phase angles of opposite arms of the bridge are equal, hence the bridge is balanced.

(b) In this case, the products of magnitudes of impedances of opposite arms are equal. However, the sum of phase angles of opposite arms are unequal. Since, the second condition is not satisfied, therefore, the bridge is unbalanced.

20.6 (a) (i) Series resistance R_s in 0 to 1 V range = 400 Ω
(ii) Series resistance R_s in 0 to 10 V range = 4.9 kΩ
(iii) Series resistance R_s in 0 to 30 V range = 14.9 kΩ and
(iv) Series resistance R_s in 0 to 100 V range = 49.9 kΩ
(b) (i) R_1 = 400 Ω (ii) R_2 = 4.5 kΩ
(iii) R_3 = 10.0 kΩ and (iv) R_4 = 35.0 kΩ

20.7 (a) (i) Shunt resistance R_{sh} in 0 to 100 mA range = 2.04 Ω
(ii) Shunt resistance R_{sh} in 0 to 1 A range = 0.2 Ω
(iii) Shunt resistance R_{sh} in 0 to 10 A range = 0.02 Ω
(iv) Shunt resistance R_{sh} in 0 to 100 A range = 0.002 Ω
(b) (i) R_1 = 0.002 Ω (ii) R_2 = 0.018 Ω
(iii) R_3 = 0.18 Ω (iv) R_4 = 1.836 Ω

20.8 (i) Multiplier resistance in voltmeter mode = 19500 Ω
(ii) Shunt resistance in ammeter mode = 0.25 Ω

20.9 (a) 12500 Ω (b) 13333.3 Ω
(c) –6.67 %

20.10 (a) L_x = 144 mH and R_x = 640 Ω
(b) *Q*-factor = 20.82

20.11 L_x = 732 mH and R_x = 542.3 Ω

20.12 (a) C_x = 0.02 μF and R_x = 2 kΩ (b) D = 0.628

20.13 N = 100 turns.

20.14 1.44 mm

20.15 159.2 Hr, R_f = 400 ohm

20.16 32.5 Hz

20.17 460 Hz

Part 3

DATA ANALYSIS

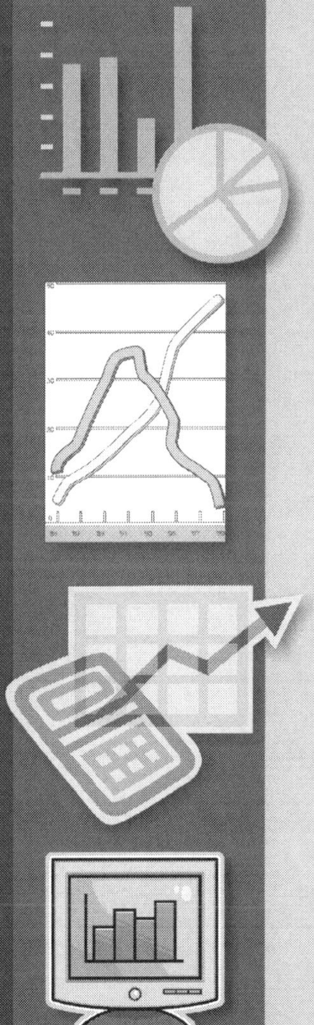

Chapter

21

Basic Statistical Concepts

■ *INTRODUCTION* ■

The main objective of any experiment is to obtain data and suitably process it to obtain useful factual information/conclusions with regard to the subject under investigation. However, the data is invariably subject to uncertainties due to a number of factors. Some of these could be minimised or eliminated, whereas the others are sometimes inherent in the instrument. Therefore we are generally faced with the problem of attempting to extract maximum amount of information from a group of data which does not give exact answer. Normally, it is a good policy to separate the chaff from the grain. In other words, we try to estimate the amount of uncertainties in the given data. To do this, as a first step, it is useful to obtain a clear perspective of the type and nature of the 'raw' data. Therefore, this raw data is grouped in various conveniently selected ranges of measured values and the frequency of occurrence in those ranges is determined. Quite often, a plot of the measured values against the frequency of occurrence or a parameter related to the frequency of occurrence is obtained. This generally gives a good visual description of the data, i.e., whether the data has a tendency to group itself in a central region, degree of scatter from the central region, whether the distribution is symmetrical or not, etc. Consequently depending on the type and nature of distribution of the given data, a suitable statistical analysis is employed to interpret the results from the given data in terms of degree of dispersion, level of confidence, etc.

21.1 ■ TYPES OF MEASURED QUANTITIES

In general, we come across two types of measured quantities, namely discrete and continuous quantities.

21.1.1 Discrete Quantities

Discrete quantities are those that are made up of elementary events that have distinct values. Very often, these are whole numbers or integers, but it is not necessary that they should always be so. For example, it is possible to have 2 to 3 children in a family but not 2.5. Similarly, when a die rolled, we expect to get an integer value between 1 and 6, i.e., it is not possible to get a value of say 4.5. The simplest way to present such data is to arrange it in order of increasing magnitude and plot the same as absolute frequency diagram (or bar and line chart) with the measured quantity as abscissa and the frequency of occurrence as ordinate (Fig. 21.1). Alternatively, the ordinate of the frequency diagram may be made non-dimensional by dividing the frequency of occurrence of the event by the total number of events. Such a diagram is then termed as the relative frequency diagram.

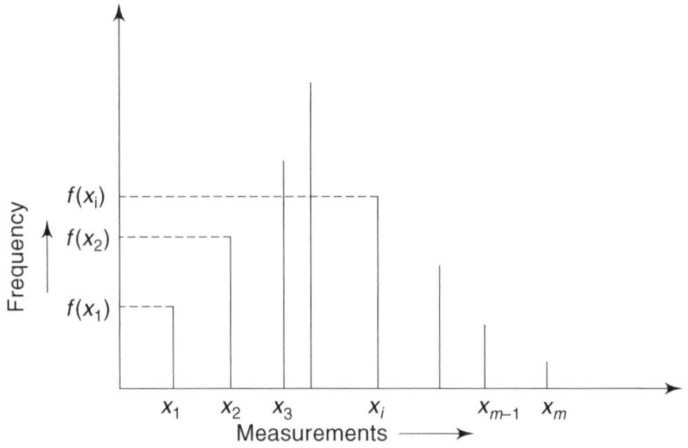

Fig. 21.1 *A typical discrete frequency distribution*

Problem 21.1 *Two dice are rolled 200 times. The following results are obtained:*

Number	*Number of occurrences*
2	6
3	10
4	16
5	22
6	28
7	38
8	30
9	24
10	14
11	8
12	4

Show suitably the distribution of frequencies diagrammatically.

Solution Since the data is discrete, it can be represented by either an absolute frequency diagram or a relative frequency diagram. For plotting a relative frequency diagram, the relative frequencies are determined as follows:

Number obtained from the dice experiment	*Absolute frequency* $(f)_i$	*Relative frequency* $(f_r)_i = (f)_i / \sum_{i=1}^{n} (f)_i$
2	6	0.03
3	10	0.05
4	16	0.08
5	22	0.11
6	28	0.14
7	38	0.19
8	30	0.15
9	24	0.12
10	14	0.07
11	8	0.04
12	4	0.02
Total	200	1.0

The relative frequency of the occurrences are plotted against the number obtained from the dice experiment to obtain the required relative frequency diagram which suitably represents the distribution the frequencies diagrammatically (Fig. 21.2).

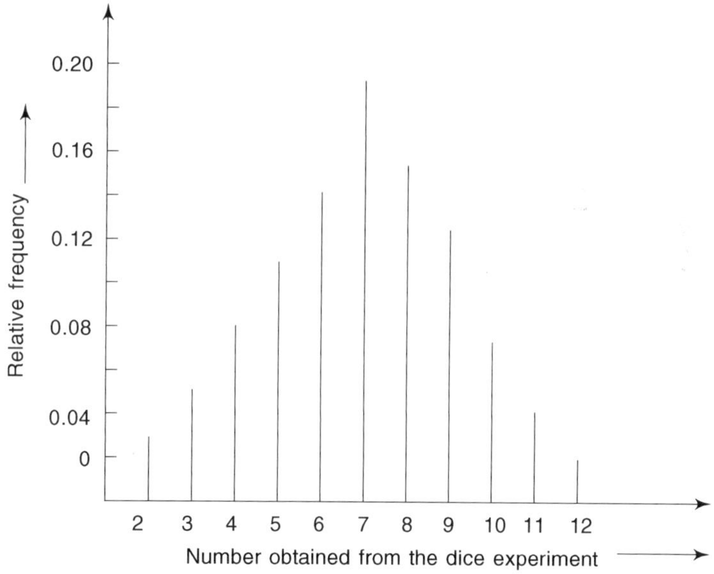

Fig. 21.2 *Relative frequency diagram of the dice experiment*

21.1.2 Continuously Distributed Quantities

Continuously distributed quantities are those in which the events may have any value between the given limits. In other words, we may say that it is always possible to have an intermediate value between any two typical values, no matter how close they may be. For example, the ball bearings produced by a certain machine may be of any diameter between the maximum and minimum limits. Similarly, the weights of all males in the population is a continuously distributed variate. No matter what the two weight selected may be within a range of given values, another possible value can always be determined between them.

Histogram A histogram is a graphic display of the sample data (which is in general continuous and in which the abscissa indicates the recorded values and the ordinate indicates the frequency of occurrence in a specified range of measured values. Therefore, an important part of making a histogram is grouping the observations into suitable groups (which are also termed as intervals, classes or cells) by selecting the proper boundaries. To do so, we first determine the range of values, $R(x)$ which is given by:

$$R(x) = (x_{\text{max}} - x_{\text{min}}) \tag{21.1}$$

where x_{max} is the highest value in the data and x_{min} is the lowest value.

Further, $$R(x) = \text{number of classes of classes} \times \text{class interval} \tag{21.2}$$

Before we decide the class interval, number of classes, class boundaries, etc., the following points should be borne in mind for drawing a histogram:

1. the class intervals must accommodate all the data,
2. it is practical to have at least 6 class intervals but not more than 16 class intervals, and
3. each piece of data must correspond to and be counted in one and only one class interval.

To satisfy the above-mentioned criteria, generally the following procedure is followed.

The value of $R(x)$ obtained from Eq. (21.1) is divided by a suitable round figure value of 'class interval' so that the 'number of classes' lies between 6 and 16. If this number obtained does not happen to be a whole number, then the integer value on the higher side is generally taken to represent the number of classes. It may be instructive to note that there is no hard and fast rule for keeping the class interval uniform. Therefore, if the situation demands a certain fixed number of classes, then the class intervals may be made unequal to adjust them in the given range of values. However, in practice these are kept same, as it facilitates comparison of the frequency of occurrence between different classes, by eliminating the necessity of allowing for differences in the class interval size.

After deciding the class interval and the number of classes, the class mid-points and class boundaries are determined. A class mid-point or class mark is the value of the variable which lies halfway between the upper and lower class boundaries. Now, an important consideration determining the various class boundaries is that no measurement should fall on the boundary so that we can readily determine into which class a particular observation belongs. Therefore, normally we fix the class boundary value to one more significant digit than the observations. This ensures that none of the data coincides with the boundary.

After this the raw data are grouped into the various chosen classes. This is best done by using a tally or score sheet. The number of observations falling within the confines of a particular class is known as the class frequency. Finally, the pictorial diagram representing the class frequency/relative frequency or percentage of frequency as ordinate with respect to various classes as abscissa, known as absolute frequency/relative frequency or the percentage frequency histogram, respectively, depending on the dimensions of the frequency values used in representing the ordinate (Fig. 21.3), is drawn.

Another way of describing the frequency distribution diagrammatically is by drawing the frequency polygon. This is obtained by plotting the various class frequencies versus the class mid-points. Alternatively, it is obtained by joining the mid-points of the frequency bars of the absolute frequency histogram by means of a smooth curve (Prob. 21.2).

Additionally, frequency distribution can be represented by cumulative frequencies versus class mid-points. Such a curve is called an *ogive.* An ogive may be drawn to represent the frequencies of

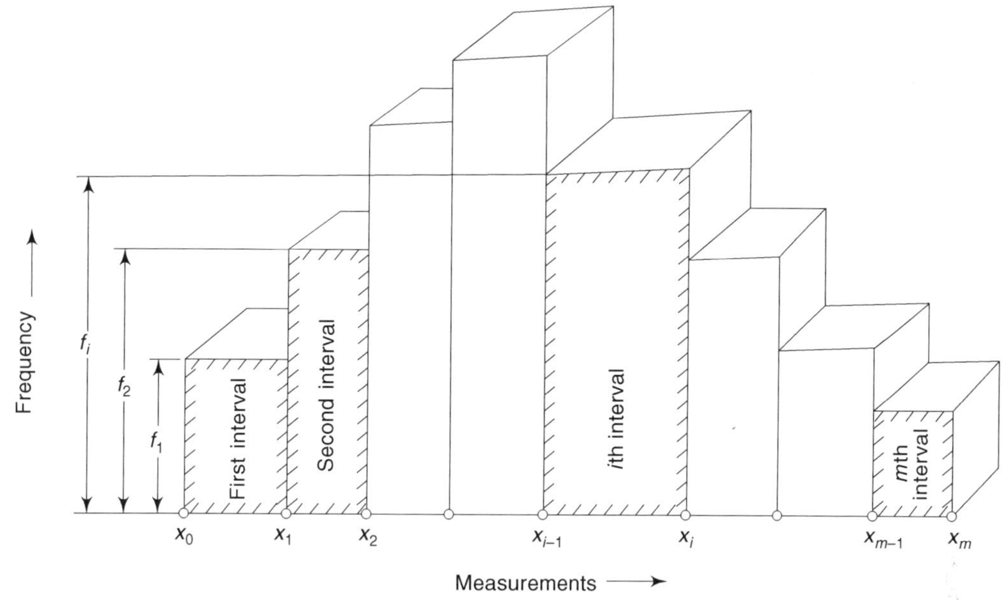

Fig. 21.3 *Pictorial representation of absolute frequency diagram of continuously distributed quantities*

occurrences either as "or more" or "less than" cumulative frequency distribution. Whenever, we refer to an ogive without qualification, the 'less than' type is implied.

It is worth mentioning here that the absolute frequency/relative frequency/percentage frequency histograms; frequency polygon as well as the ogive type representation of the data are normally used for visualising the nature of the data. However, it may be difficult to make a quantitative comparative estimate of different frequency distributions if their class intervals are not the same. Therefore, in actual practice, these frequency distributions are normalised with respect to the class width also, and are thus commonly represented in the form of *normalised* histograms.

Normalised Histogram In a normalised histogram, normalised frequency is taken as the ordinate instead of the absolute frequency or a percentage frequency. Normalised frequency is obtained as follows.

We decide to group the given data in m classes say, and choose the class boundary values as x_0, x_1, x_2, . . . x_i, . . . x_m, having the frequency of occurrence in the various class intervals as $f_1, f_2, \ldots, f_i, \ldots, f_m$, respectively. Now the total number of measurements, i.e.

$$n = \sum_{i=1}^{m} f_i \tag{21.3}$$

Class interval of ith class $= (x_i - x_{i-1})$ (21.4)

Relative frequency in ith class, $(f_r)_i = f_i \Big/ \sum_{i=1}^{m} f_i$ (21.5)

Normalised frequency in ith class, $(F_n)_i = \dfrac{(f_r)_i}{(x_i - x_{i-1})}$ (21.6)

Now, in normalised histogram, the normalised frequency $(F_n)_i$ (which is relative frequency per unit interval) is plotted against the class intervals as is shown in Fig. 21.4.

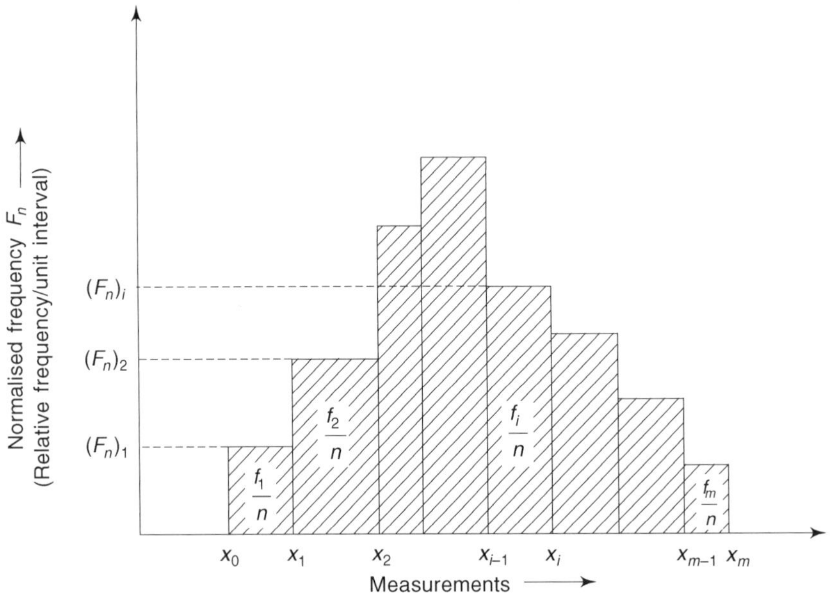

Fig. 21.4 *Normalised histogram*

Now area of the ith bar in the normalised histogram, A_i

$$= \text{ordinate} \times \text{class interval}$$

$$= \frac{1}{(x_i - x_{i-1})} \frac{f_i}{\sum_{i=1}^{m} f_i} (x_i - x_{i-1}) = \frac{f_i}{\sum_{i=1}^{m} f_i}$$

$$= \frac{f_i}{n} \tag{21.7}$$

$$= (f_r)_i \tag{21.8}$$

Therefore, the area assigned to each bar in the normalised histogram represents the relative frequency of occurrence $(f_r)_i$ of the events within the given class interval. In case all the class intervals are equal, then the ordinate of each class interval is proportional to the relative frequency. The area of the histogram is then given by

$$\sum A_i = \sum \frac{f_i}{n} = \frac{n}{n}$$

$$= 1 \tag{21.9}$$

Thus, the area of the normalised histogram is unity. It would be shown (Ch. 22) that the area of the Gaussian distribution is also unity. Therefore, if the given continuously distributed data has to be compared with a Gaussian distribution, then normalised histogram of the data becomes more appropriate as compared to the ordinary histogram, frequency polygon or an ogive curve.

Problem 21.2 *A steel fabricator has tested the strength of a particular type of a rivetted joint. The following table gives the strength in kN of a sample of 100 joints. Choose a suitable class interval/number of classes and*

(a) construct a group data table showing class boundaries, class marks, relative frequencies and normalised frequencies,

(b) represent the data in the form of a numerical frequency histogram,

(c) draw a frequency polygon with respect to the relative frequencies,

(d) draw a normalised histogram, and

(e) draw an ogive curve indicating the % less than cumulative frequencies.

1447	1478	1469	1446	1433	1469	1478	1447	1426	1469
1429.5	1479	1447.5	1434.5	1469	1479	1450	1430.5	1471.5	1479.5
1451.5	1436	1471.5	1480.5	1455.5	1436.5	1473	1484	1451	1438
1470	1484.4	1449	1436	1468	1482	1448	1441	1469.5	1481
1453	1442.5	1474	1430.5	1453	1436.5	1471	1480.5	1449	1437
1475	1480.5	1449	1474.5	1488	1417	1471	1491	1423.5	1471
1492	1439	1486	1457	1449.5	1457	1489	1457	1441	1494
1513	1458	1505.5	1495	1458.5	1464	1496	1464.5	1499	1461
1501	1464	1456	1459	1457.5	1459	1459	1461.5	1456	1461
1465	1461	1461.5	1462	1465	1463.5	1470	1466	1469	1466

Solution Maximum value in the data, $x_{max} = 1513$ kN

Maximum value in the data, $x_{min} = 1417$ kN

Range, $R(x) = 1513 - 1417$

$= 96$ kN

Taking the class interval to be say 10 kN, we get

$$\text{Number of classes} = \frac{R(x)}{\text{class interval}} = \frac{96}{10}$$

$$= 9.6 \approx 10 \text{ (the next higher integer)}$$

Now, the available data in the problem is up to one decimal place. Therefore, to avoid any data falling on the boundary, we assign class boundary values up to two decimal places. The various class boundaries, class marks, class frequencies, relative frequency, relative frequency per unit interval, etc. are shown in Table 21.1.

(a) Table 21.1 presents the given raw data in grouped data form, showing the various class boundaries, class marks with their absolute frequencies, relative frequencies and normalised frequencies.

(b) The tabular data of Table 21.1 has been plotted in Fig. 21.5(a). The absolute frequency as ordinate versus the various classes represents the numerical frequency histogram.

(c) The dotted curve shown in Fig. 21.5(a) with the ordinate as relative frequency versus the various class marks represents the frequency polygon with respect to the relative frequency.

(d) The frequency distribution with respect to normalised frequency as ordinate [shown on the right side of Fig. 21.5(a)] versus the various classes represents the normalised histogram.

(e) For drawing an ogive for indicating '% less than' frequency distribution, we make Table 21.2 of the lower class boundary versus '% less than' strength values of the samples.

Table 21.1

Sl. No.	*Class boundaries*	*Class mark*	*Observation tally*	f_{xi}	*Relative frequency* $(fr)_i$	*Relative frequency per unit interval* $(F_n)_i$
1	1415.55–1425.55	1420.55	𝍷𝍷	2	0.02	0.002
2	1425.55–1435.55	1430.55	𝍸 𝍷𝍷	7	0.07	0.007
3	1435.55–1445.55	1440.55	𝍸 𝍸 𝍷	11	0.11	0.011
4	1445.55–1455.55	1450.55	𝍸 𝍸 𝍸	15	0.15	0.015
5	1455.55–1465.55	1460.55	𝍸 𝍸 𝍸 𝍸 𝍷𝍷𝍷	23	0.23	0.023
6	1465.55–1475.55	1470.55	𝍸 𝍸 𝍸 𝍸	20	0.20	0.020
7	1475.55–1485.55	1480.55	𝍸 𝍸	10	0.10	0.010
8	1485.55–1495.55	1490.55	𝍸 𝍷𝍷	7	0.07	0.007
9	1495.55–1505.55	1500.55	𝍷𝍷𝍷𝍷	4	0.04	0.004
10	1505.55–1515.55	1510.55	𝍷	1	0.01	0.001

$$\sum_i^n fx_i = 100 \qquad \sum_i^n fr_i = 1$$

Table 21.2

Sl. No.	*Strength values of test samples (in kN)*	*Cumulative frequency*	*Cumulative relative frequency*	*% Cumulative frequency*
1	less than 1415.55	0	0.00	0.0
2	less than 1425.55	2	0.02	2.0
3	less than 1435.55	9	0.09	9.0
4	less than 1445.55	20	0.20	20.0
5	less than 1455.55	35	0.35	35.0
6	less than 1465.55	58	0.58	58.0
7	less than 1475.55	78	0.78	78.0
8	less than 1485.55	88	0.88	88.0
9	less than 1495.55	95	0.95	95.0
10	less than 1505.55	99	0.99	99.0
11	less than 1515.55	100	1.00	100.0

The tabulated data of Table 21.2 has been plotted in Fig. 21.5(b) to show the ogive curve indicating '% less than' cumulative frequency distribution.

21.2 ■ CENTRAL TENDENCY OF DATA

One of the important parameters describing the numerical information concerns the location to the central tendency of the data. There are a number of ways by which the central tendency of the data may be determined. For example, median, mode, arithmetic mean, etc. of the data may be determined. However, it my be pointed out that each of the above measures may be highly appropriate to describe only a certain type of data and completely inappropriate for another.

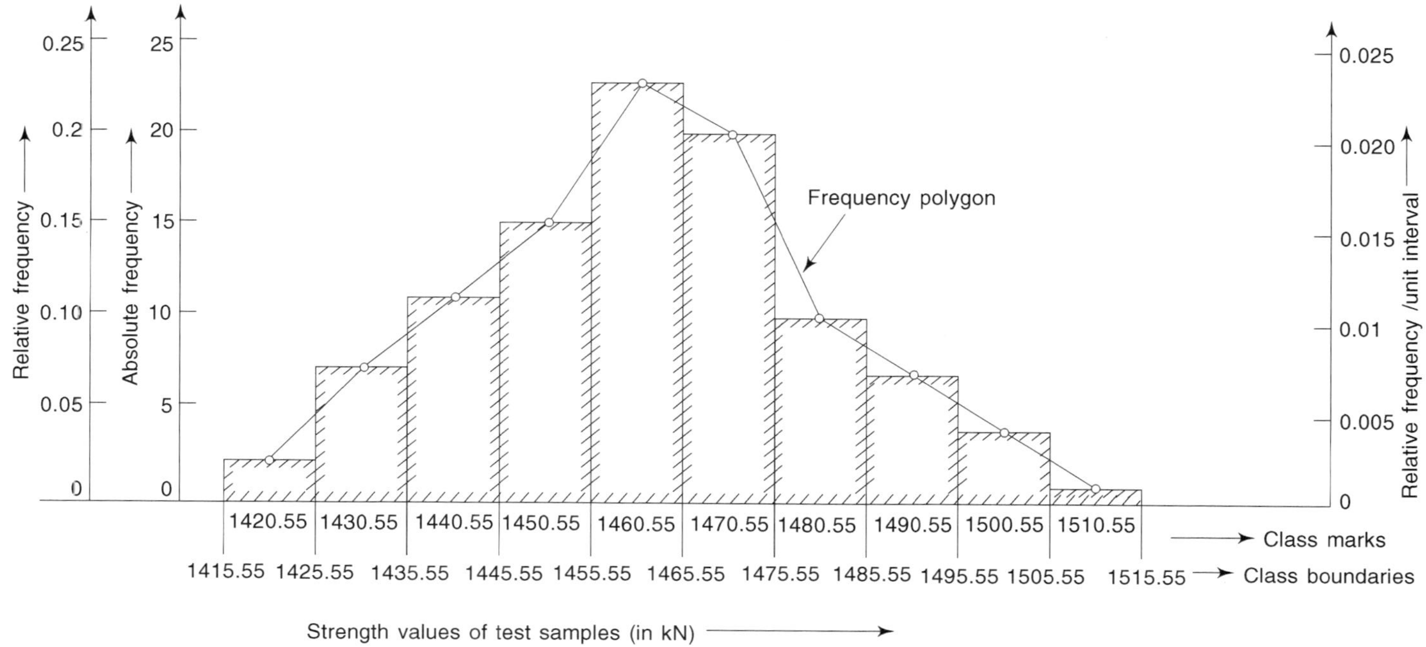

Fig. 21.5(a) *Various graphical representations of continuously distributed data of the test smaple of Problem 21.2*

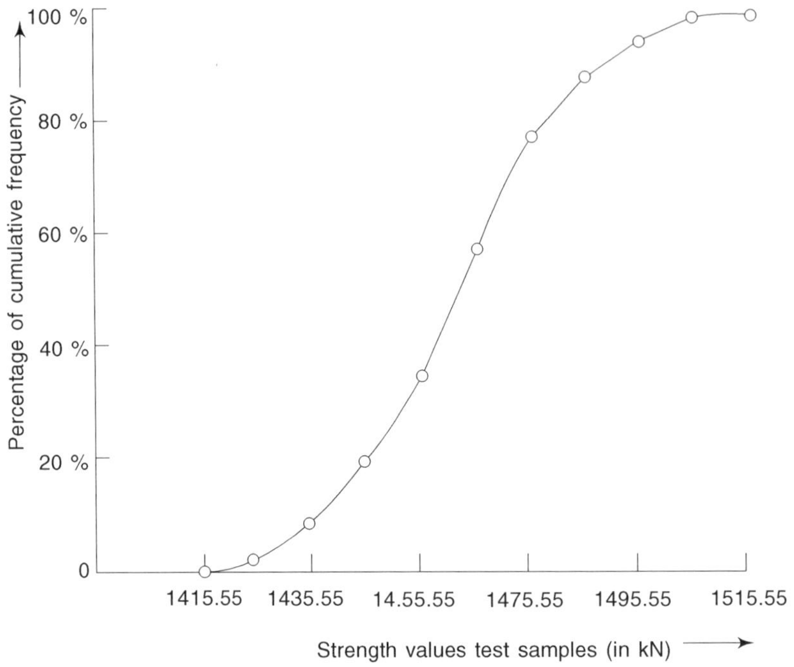

Fig. 21.5(b) *Percentage less than cumulative frequency distribution of the given data of Problem 21.2*

21.2.1 Mode

The word *mode* means fashion in French. Therefore, any quantity that is quite popular or widely occurring or in fashion is known as the mode or the modal value. In other words, we can say that *mode* is the value of the variate that occurs with greatest frequency.

For example, let us say that a set of items contains the following discrete values: 15, 18, 17, 14, 17, 13 and 16. It is obvious in this set of values that the frequency of number 17 is the highest. Therefore, this would be termed as the modal value.

It may be noted from the above example that, in a set of discrete values, the modal value can be determined quite easily. However, in the case of continuously distributed data, the modal value would be in the group of data which has the highest frequency density. In other words, it would be within the highest rectangle of the histogram. To determine the modal value, consider a part of the histogram showing the highest frequency rectangle along with the adjacent lower class and higher class rectangles (Fig. 21.6). We define the mode as the abscissa $\hat{X}$ of the point of intersection P of the lines QS and RT. It can be shown that the value of mode $\hat{X}$ is given by the formula:

$$\text{Mode } \hat{X} = L_1 + \left[\frac{\Delta_1}{\Delta_2 + \Delta_2}\right] C \tag{21.10}$$

where L_1 = lower class boundary of the modal class (i.e. class containing mode)
D_1 = excess of the modal frequency over frequency of next lower class
D_2 = excess of the modal frequency over frequency of next higher class
C = modal class interval size

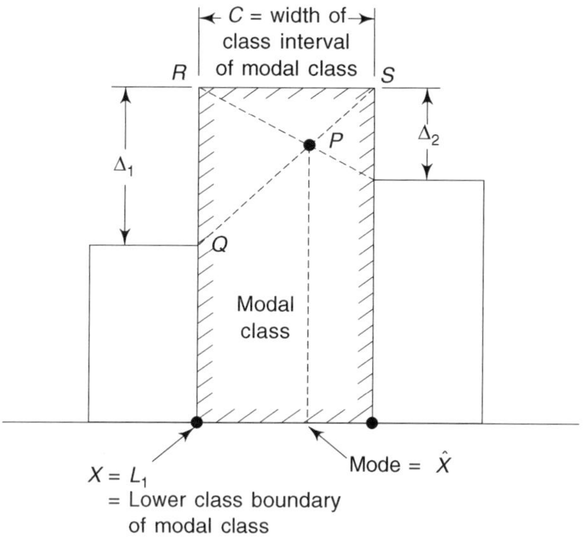

Fig. 21.6 *Part of the histogram showing modal class and the adjacent lower class and higher class rectangles*

Occasionally, a distribution may be bimodal, i.e. it contain two distinct values which may occur much more frequently than other values of the variable. When this occurs, it is usually an indication that the sample comes from two different populations.

In general, mode is considered to be a poor measure of the central tendency of the data as quite often the most frequently occurring quantity does not appear near the centre of the data. However, for descriptive analysis, mode is a useful measure to describe the most frequently occurring value. In other words it represents the variate with highest frequency in a given set of data.

21.2.2 Median

The median is defined as the middle value of a group of data arranged in the numerical order (i.e. in an array) which may be either in ascending order (i.e., from the lowest value of observation to the highest one) or vice versa. In case the sample size is odd, then it is quite easy to locate the middle value. On the other hand, if the sample size is even, we determine the median by taking the average of the two middle values. Median is also termed as 50th percentile and is denoted as P_{50}. Fiftieth percentile means a score P_{50} such that approximately 50% of the measurement in the distribution are less than P_{50} and 50% of the measurements in the distribution are greater than P_{50}.

Let us say that the n observations of the data are listed either in the ascending or in the descending order from 1 to n. Then mathematically we can express the value of median as:

$$\text{Median} = \{X\}_{(n+1)/2}, \text{ when } n \text{ is odd} \tag{21.11}$$

$$= \frac{1}{2}[\{X\}_{n/2} + \{X\}_{(n/2)+1}], \text{ when } n \text{ is even.} \tag{21.12}$$

For example, a set of items contains five values: 15, 18 – 12, 42 and 12. For finding the median, we first arrange the values in the ascending order and mark the serial number of each value and we get,

Value	–12	12	15	18	42
Serial No.	1	2	3	4	5

Median value here corresponds to serial number $(n + 1)/2 = (5 + 1)/2 = 3$, which is 15. It may be noted that median divides the set of observations equally, i.e. two being on one side and two to the other side of the median value in the aforesaid example.

Suppose another set of items contains six values: 15, 18, –12, 42, 12 and 66. For determining the median, we arrange them in the descending order and mark the serial number of the values getting

Value	66	42	18	15	12	–12
Serial No.	1	2	3	4	5	6

Median value here corresponds to the average values at serial numbers $(n/2)$ and $(n/2 + 1)$, i.e. 3 and 4. Therefore, the median value is $(15 + 18)/2 = 16.5$. Again, we find that the median value divides the set of observations equally, i.e. two being on one side and two on the other side of the median values.

The median is in fact, a positional measure, in the sense that its value is determined by the value of observation occupying a particular position in the distribution. It may be noted that the median is not affected by the extreme values of the data. For example, we are interested in determining the average national income. The arithmetic mean of incomes of all individuals will have a bias introduced by the extremely high incomes of a very few individuals. Therefore, in this case, median income value is used in practice to indicate the average national income.

In certain experimental data, the distribution of the data may not be symmetrical. In other words, the histogram of the data shows the distribution to be *skewed*. In such cases, the median is considered as the central location of the data. One common example of this measure is the fatigue failure of metals.

For grouped data the median obtained by interpolation is given by

$$\text{Median} = L + \left\{\frac{\left(\frac{n}{2}\right) - (\sum f_l)}{f_{\text{median}}}\right\} C \tag{21.13}$$

where L_1 = the lower class boundary of the median class (i.e. class containing the median)
n = the total number of observations in the data
$(\Sigma f)_l$ = sum of frequencies of all the classes lower than the median class
f_{median} = frequency of the median class
C = width of the median class interval

Geometrically, the median value is obtained by the intersection of the abscissa of the histogram with a vertical line that divide the histogram into two parts having equal areas.

However, it may be pointed out here that the value obtained for the median may be non-representative if the data does not cluster at the centre of the distribution.

21.2.3 Arithmetic Mean

Arithmetic mean is another name of the term simple average which we commonly use in our day-to-day life. It is the most common and most useful of the various estimates of the central tendency of the data. By definition, the arithmetic mean of a group of data is equal to the sum of values in the group divided by the number of values comprising the group. It is possible to determine either the mean of the sample of the data or the mean of the population. The symbol $\overline{X}$ (pronounced X bar) traditionally denotes the former, the mean of the sample and $\overline{\overline{X}}$ (pronounced X double bar) the latter. It may be noted that the *population* is the total number of values in the data from which the *sample* is taken. Whereas when we refer to the sample, we do not mean just a portion of the population, but a portion selected in such a way that the observed values have no *bias*. This process is generally termed as *random* sampling.

Let us say that a particular sample has a total of n observations in which the various observed values are: x_1 occurring f_1 times, x_2 occuring f_2 times, . . . , x_m occurring f_m times.

Now, the sum of the frequencies of occurrences of various values obviously gives us the total number of observations of the sample.

$$\therefore \qquad f_1 + f_2 + \ldots + f_i + \ldots + f_m = \sum_{i=1}^{m} f_i$$

$$= n \tag{21.14}$$

Further, the mean of the example $\bar{X}$ is given by

$$\bar{X} = \frac{\sum_{i=1}^{m} f_i x_i}{\sum_{i=1}^{m} f_i} \tag{21.15}$$

The same expression can also be applied for continuously distributed data where x_i represents the value of class mid-point of the ith class having the frequency f_i.

Similarly, the mean of the population can also be determined. Say the population has N observations and $\bar{X}_1, \bar{X}_2 \ldots, \bar{X}_j \ldots, \bar{X}_k$ are the means of the values of k samples.

The mean of the population, $\bar{\bar{X}}$, is given by

$$\bar{\bar{X}} = \frac{\sum_{j=1}^{k} \bar{X}_j}{k} \tag{21.16}$$

Weighted Arithmetic Mean Sometimes, we associate the numbers $x_1, x_2, x_3, \ldots, x_i, \ldots, x_n$ with certain weighing factors or weights $w_1, w_2, w_3, \ldots, w_i, \ldots, w_n$, respectively, depending of the significance or importance attached to these numbers. In such a case, the weighted arithmetic mean is defined as:

$$\bar{X} = \frac{w_1 x_1 + w_2 x_2 + \ldots + w_n x_n}{w_1 + w_2 + \ldots + w_n}$$

$$= \frac{\sum_{i=1}^{n} w_i x_i}{\sum_{i=1}^{n} w_i} \tag{21.17}$$

It is interesting to note the similarity between Eqs. (21.15) and (21.17). In fact, Eq. (21.15) can be considered a weighted arithmetic mean with weights $f_1, f_2, f_3, \ldots, f_i, \ldots f_m$.

Problem 21.3 *A student's marks in the written examination, oral examination and laboratory examination in 'Experimental Methods and Analysis' course were 87, 71 and 61, respectively.*

(a) If the weighing factors accorded to these marks are 5, 3 and 2, respectively, determine the average marks obtained by the student.

(b) Determine the average marks if equal weightage for all the examinations is used.

Solution

(a) Using Eq. (21.17), the weighted arithmetic mean of the students marks are given by

$$(\bar{X})_{\text{weighted}} = \frac{(5)(87)+(3)(71)+2(61)}{5+3+2} = 77 \text{ marks}$$

(b) The simple arithmetic means (with equal weighing factors) of the students' examination marks are given by

$$(\bar{X})_{\text{simple}} = \frac{87+71+61}{3} = 73 \text{ marks}$$

21.3 BEST ESTIMATE OF TRUE VALUE OF DATA

We have so far described three parameters namely mode, median and arithmetic mean that can be used to describe the central tendency of the data. For unimodal curves which are moderately skewed (asymmetrical), there is an empirical relation correlating these parameters:

$$\text{mean} - \text{mode} = 3\ (\text{mean} - \text{median}) \tag{21.18}$$

Figure 21.7 shows the relative position of the mode, median and mean for the frequency curves skewed to the right as well as to the left. However, for symmetrical curves, the mode, median and mean all coincide.

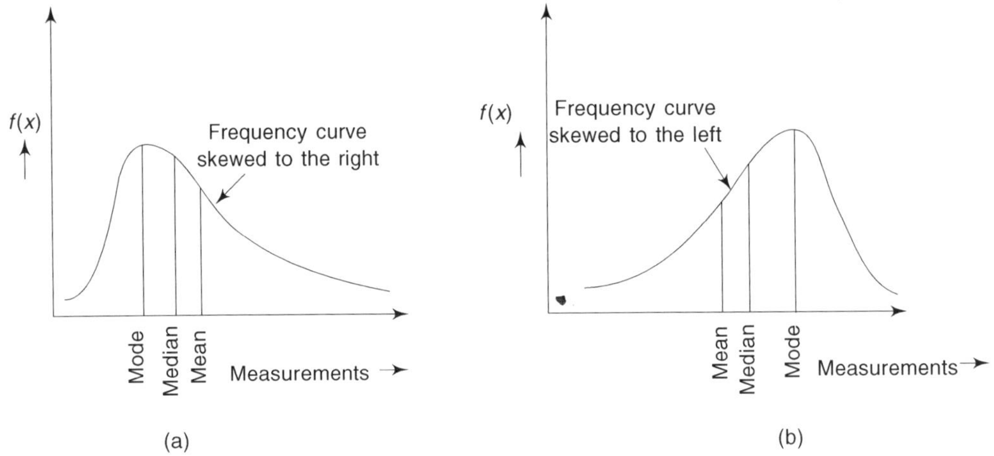

Fig. 21.7 *Relative positions of mean, median and mode of moderately skewed frequency curves*

In experimental data analysis, one is usually faced with the problem of determining the best estimate of the true value in the given data. Let us suppose that the measurements $x_1, x_2, x_3, \ldots, x_i, \ldots, x_n$ are made of a quantity whose true value is X. Then the measurement would have the following errors:

$$\begin{aligned} \text{Error in } x_1,\ e_1 &= x_1 - X \\ \text{Error in } x_2,\ e_2 &= x_2 - X \\ &\vdots \\ \text{Error in } x_i,\ e_i &= x_i - X \end{aligned} \tag{21.19}$$

In Eq. (21.19), some of the errors would be positive while the others would be negative. Now, if we give equal weightage to negative and positive errors, then the best way is to square all the errors.

Now, the most plausible criteria for the best choice of X would be that errors in the various measured quantities $e_1, e_2, \ldots, e_i, \ldots, e_n$ should be minimum. Alternatively, the sum of the squares of the errors should be minimum.

The sum of the squares of the errors can be written as:

$$S_e = e_1^2 + e_2^2 + \ldots + e_i^2 + \ldots + e_n^2$$

$$= (x_1 - X)^2 + (x_2 - X)^2 + \ldots + (x_i - X)^2 + \ldots + (x_n - X)^2 \quad (21.20)$$

In Eq. (21.20) S_e, the sum of the squares of the errors, is a function of X, the best estimate of the true value. Now, we wish to determine the value of X that would make S_e minimum. At the minimum $X = X_0$ (say), then

$$\frac{dS_e}{dX} = -2(x_1 - X) -2(x_2 - X) \ldots -2(x_i - X) \ldots -2(x_n - X) = 0 \quad (21.21)$$

Hence, $X_0 = (x_1 + x_2 + \ldots + x_i + \ldots + x_n)/n$

$$= \overline{X}_n$$

$$= \text{arithmetic mean} \quad (21.22)$$

Therefore, in general, arithmetic mean is considered the best estimate of the true value of any given data.

Problem 21.4 *The table below shows the frequency distribution of the resistance of resistor manufactured by a company.*

Resistance (Ω)	*Frequency*
93–95	4
96–98	15
99–101	33
102–104	21
105–107	7

Determine the following:

(a) Arithmetic mean

(b) Median value

(c) Modal value

(d) Use the empirical formula, mean — mode = 3 (mean — median), to calculate the modal value and compare this value with that obtained in part (c).

Solution

(a) In the problem, the resistance values in the ranges of 93–95, 96–98, etc. are given. For finding the arithmetic mean, we would assume that 4 resistors have a resistance of 94 Ω, 15 resistors have a resistance of 97 Ω, etc.

$$\therefore \quad \text{Arithmetic mean } \overline{X} \text{ becomes} = \frac{(4)(94) + 15(97) + 33(100) + 21(103) + 7(106)}{4 + 15 + 33 + 21 + 7}$$

$$= \frac{8036}{80} = 100.45\ \Omega$$

(b) The median value can be determined either by the method of interpolation or by directly using Eq. (21.13). By definition, median is that value for which half the total frequency (80/2 = 40) lies above it and the other half below it.

Now the sum of the first two class frequencies is 4 + 15 = 19. Thus, to give the desired 40, we require 21 more of the 33 cases in the third class. The median value would now lie at 21/33 of the width between 99 and 101.

$$\text{Median} = 99 + \frac{21}{33}(101 - 99)$$
$$= 99 + 1.273 = 100.273\ \Omega$$

Alternatively using Eq. (21.12), we get,

$$\text{Median} = L_1 + \left\{\frac{\frac{n}{2} - (\sum f_i)}{f_{\text{median}}}\right\} C$$
$$= 99 + \frac{40 - (4 + 15)}{33}(101 - 99)$$
$$= 99 + 1.273 = 100.273\ \Omega$$

(c) Equation (21.10) gives the values of the modal value for the continuously distributed data. Here,

$$L_1 = 99$$
$$\Delta_1 = 33 - 15 = 18$$
$$\Delta_2 = 33 - 21 = 12$$

$$\text{Modal value } X = L_1 + \left\{\frac{\Delta_1}{\Delta_1 + \Delta_2}\right\} C$$
$$= 99 + \left\{\frac{18}{12 + 18}\right\}(101 - 99) = 100.2\ \Omega$$

(d) Using the given empirical relation we get the value of mode as:

$$\text{Mode} = \text{mean} - 3\ (\text{mean} - \text{median})$$
$$= 100.45 - 3(100.45 - 10.273) = 99.92\ \Omega$$

It may be noted that there is a good agreement between the modal value calculated in part (c) of the problem and the modal value obtained using the empirical relation for moderately skewed distributions.

21.4 ■ MEASURES OF DISPERSION (SPREAD OR VARIABILITY)

Measures of central location alone usually do not give an adequate description of the experimental data. In fact, additionally the variability or the spread of the data should also be taken into account. For example, if a person is having his head in a refrigerator and his feet in an oven, then on the average he can be said to be feeling fine. But in practice, the person would be most uncomfortable because of the extreme difference/spread between the two temperatures. Therefore, another important parameter of experimental data is the amount of scatter or the degree of the clustering of the data.

The various measures of spread or dispersion of the data are as follows.

21.4.1 Range

The simplest way to represent dispersion is by the range which is the difference between the maximum and minimum values of the given data. The major disadvantage associated with the range as a measure of dispersion is that it is based solely on the dispersion of the extreme values. Further, another drawback is that the range fails to provide information about the clustering or the lack of clustering of the observed values within the two extreme values. Therefore, it is hardly employed as a measure of dispersion. However, it is employed in practice to have an approximate idea of the extent of scatter values only in the available data.

21.4.2 Average or Mean Deviation

The average deviation is an improvement over the range as the values of all the observations are included in its computation. In this, the deviation of each data point from the arithmetic mean is evaluated, the absolute value taken and the average of these calculated. Thus,

$$\text{Mean deviation (MD)} = \frac{\sum_{i=1}^{n} |(x_i - \bar{X})|}{n} = \frac{\sum_{i=1}^{n} |d_i|}{n} = \overline{\sum_{i=1}^{n} |x_i - \bar{X})|} \tag{21.23}$$

where d_i is the deviation of the ith observation from the mean.

In Eq. (21.23) $|(x_i - \bar{X})|$ represents the absolute value of the deviation of x_i from $\bar{X}$ (the absolute value of a number is the number without the associated sign and is indicated by two vertical lines placed around the number). For example $|-5| = 5$, $|+4| = 4$, $|7| = 7$ and $|-0.52| = 0.52$).

If $x_1, x_2, \ldots, x_i, \ldots, x_n$ occur with frequencies $f_1, f_2, \ldots, f_i, \ldots, f_n$, respectively, then the mean deviation is written as:

$$\text{Mean deviation (MD)} = \frac{\sum_{i=1}^{n} f_i |x_i - \bar{X}|}{\sum_{i=1}^{n} f_i} = \frac{\sum_{i=1}^{n} f_i |d_i|}{n}$$

$$= \frac{\sum f_i |(x_i - \bar{X})|}{n}$$

$$= \sum_{i=1}^{n} f_i \overline{(x_i - \bar{X})} \tag{21.24}$$

where $\sum_{i=1}^{n} f_i = n$, total number of observations in the data.

Equation (21.24) is also useful for grouped data where x_i represent the class mark and f_i the corresponding class frequencies.

Occasionally, the mean deviation may also be defined in terms of absolute deviations from the median value or any other average instead of mean. An interesting property of the sum $\sum_{i=1}^{n} |(x_i - a)|$ is that it is a minimum when a is the median. In other words, the mean deviation about the median is minimum.

It may be noted that it would be more appropriate to use the terminology mean absolute deviation rather than mean deviation. In other words, strictly speaking, the average of absolute deviations is the index of dispersion usually termed as mean deviation. In fact, the average of actual deviations from the mean values can be shown to be zero, which is as follows:

$$\frac{\sum_{i=1}^{n} d_i}{n} = \sum_{i=1}^{n} \frac{(x_i - \bar{X})}{n}$$

$$= \frac{\sum x_i}{n} - \frac{n\bar{X}}{n}$$

$$= \bar{X} - \bar{X}$$

$$= 0 \tag{21.25}$$

In the computation of the mean deviation, we come across some deviations that are positive while the rest are negative. That is where in Eq. (21.25), they balance each other about the mean. Therefore, for each deviation, the absolute modulus has to be computed which is rather awkward from the point of view of computation. Hence, the mean deviation is less commonly employed as a measure of the dispersion of the data.

21.4.3 Variance

Now, instead of considering the mean of absolute deviation about the mean value, we determine the mean square deviations about the mean value. The squaring operation avoids the inherent problem of taking absolute deviations, because all squares are positive in sign. The mean square deviation (MSD) also termed as the *variance* is traditionally a very commonly used measure of the variability of the data. This is because of its following notable features:

1. it employs all the values of the data and is sensitive to any change in the value of any data,
2. it is independent of the central location, i.e. the mean value, because it uses the deviations about the mean value, and
3. its computation is reasonably simple because the squared values of all deviations from the mean are positive.

Variance is usually denoted by the symbol σ^2 which is the square of the Greek letter *sigma.* It is computed as follows:

$$\sigma^2 \equiv \frac{1}{n}\sum_{i=1}^{n}(x_i - \bar{X})^2 \tag{21.26}$$

Expanding the RHS we get,

$$\sigma^2 = \frac{1}{n}\sum_{i=1}^{n}(x_i^2 - 2x_i\bar{X} + \bar{X}^2)$$

$$= \frac{1}{n}\sum_{i=1}^{n}x_i^2 - \frac{2\bar{X}}{n}\sum_{i=1}^{n}x_1 + \frac{n\bar{X}^2}{n}$$

$$= (\overline{x^2}) - 2(\bar{X})^2 + (\bar{X})^2$$

$$= (\overline{x^2}) - (\bar{X})^2 \tag{21.27}$$

Thus, we may express the variance as the difference of the mean of squared values minus the square of the mean value of the set of data.

21.4.4 Standard Deviation

Quite often the square root of variance denoted σ and termed standard deviation or root mean square deviation is used to represent the measure of dispersion. The main advantage of standard deviation as compared to the variance is that the unit of standard deviation is the same as that of the measured quantity. It is defined as:

$$\sigma = \sqrt{\frac{\sum_{i=1}^{n} d_i^2}{n}} = \sqrt{\frac{\sum_{i=1}^{n}(x_i - \bar{X})^2}{n}} \tag{21.28}$$

For the case of grouped data it may be written as

$$\sigma = \sqrt{\frac{\sum_{i=1}^{n} f_i (x_i - \bar{X})^2}{n}} \tag{21.29}$$

where $\sum_{i=1}^{n} f_i = n$ = total number of observations

It may be noted that for moderately skewed frequency distributions, the empirical relation between mean deviation and standard deviation is as follows:

$$\text{Mean deviation} = \frac{4}{5}\ (\text{standard deviation}) \tag{21.30}$$

21.4.5 Coefficient of Variation

The coefficient of variation or coefficient of dispersion is a non-dimensional number representing the degree of scatter of the data with respect to the mean value. It is obtained by dividing the standard deviation by the mean value, i.e.,

$$\text{Coefficient of variation (C.V.)} = \frac{\text{standard deviation}}{\text{mean value}}$$

$$= \frac{\sigma}{\bar{X}} \tag{21.31}$$

It is generally expressed as a percentage and in such a case it is also termed *percentage variation.* It may be noted that the coefficient of variation fails to be useful when $\bar{X}$ is close to zero.

21.4.6 Adjusted Standard Deviation

The measure of dispersion of the data σ, suffers from the lacuna that it is biased when the number of observations is small. For example, when only one observation of a physical quantity is considered, the value of σ reduces to zero. This implies that the measurement has no dispersion and consequently does not suffer from any measurement error at all. Obviously, this is a highly biased result when only one observation is used in the calculations. However, when two or more observations are used, the bias in the parameter progressively decreases till it becomes negligible for large values of n. Therefore, the value of rms deviation σ is adjusted so that it is more realistic to give an unbiased estimate of the precision. This is achieved by dividing the sum of squares of the deviations from the mean by $(n - 1)$, which represents the degree of freedom; in place of n which is the total number of observations. The number of degrees of freedom refers to the number of independent pieces of information generated from the given set of data. This can be determined by using inductive reasoning as follows:

A simple case of sample size 2 provide only one useful input with regard to estimating dispersion around the population mean, because the sample set must not only provide information on dispersion but also the variable from which dispersion is measured, namely $\bar{X}$. Thus, a sample size of 2 provides only one independent observation with respect to dispersion. Similarly, for a sample size of 10, we can find 10 deviations. However, we would say that effectively, there are 9 independent deviations as the tenth could be found out from the fact that the sum of the deviations is equal to zero. Therefore, in general a sample size of n provides $(n - 1)$ independent observations with respect to the population standard deviation. In fact, the general expression for the degrees of freedom is $(n - m)$ where n is the number of observations and m the number of constants that have to be estimated from the original data in order to generate values from which deviations have to be computed. In the present case $m = 1$, i.e. only one constant, the arithmetic mean has to be computed. Hence, the adjusted standard deviation (denoted by s) also termed the unbiased or the best estimate of precision of the apparatus, can now be written as

$$s = \sqrt{\frac{\sum_{i=1}^{n}(x_i - \bar{X})^2}{(n-1)}}$$

$$= \sqrt{\frac{\sum_{i=1}^{n} d_i^2}{(n-1)}} \tag{21.32}$$

Comparing Eqs. (21.28) and (21.32) the relationship between s and σ can be written as:

$$s = \sqrt{\frac{n}{(n-1)}}\,\sigma \tag{21.33}$$

The quantity $\sqrt{\frac{n}{(n-1)}}$ is known as Bessel's correction factor.

Equation (21.33) shows that as n increases, the difference between s and σ decreases. Thus, as the sample gets larger the standard deviation of the sample becomes as less biased estimate of the precision of the apparatus. In practice, for $n > 25$ Bessel's correction factor is quite small and is usually neglected.

21.5 ■ STANDARD DEVIATION OF THE SAMPLE MEANS

The term adjusted standard deviation or the best estimate of the precision of apparatus denotes the magnitude of random error of any single measurement. In other words, we can say that

$$X = (\bar{x})_{\text{measured}} \pm s \tag{21.34}$$

Similarly, the precision in the sample means would indicate how close is the measured value of a sample mean to the true value. In other words, the internal standard error (denoted as S_n by some authors) or the best estimate of uncertainty U_n is given by $\sigma(\bar{X}_n)$.

Therefore,

$$X = (\bar{X}_n)_{\text{measured}} \pm \sigma(\bar{X}_n)$$

$$= (\bar{X}_n)_{\text{measured}} \pm U_n \text{ (when } n\text{, the sample size, is large)} \tag{21.35}$$

21.5.1 Best Estimate of Uncertainty

As explained before, the best estimate of uncertainty represents the extent of random error in the measured values. This is determined by considering m samples of size n in a population of $(m \times n)$ observations. In this, a typical sample mean can be represented as:

$$\bar{X}_i = \frac{\sum_{j=1}^{n} x_{ij}}{n} \tag{21.36}$$

where i represents the ith sample and j the observation in the ith sample.

Further, the mean of the means, also known as population mean can be defined as:

$$\bar{\bar{X}} = \frac{\sum_{i=1}^{m} \bar{X}_i}{m} \simeq \frac{m\bar{X}}{m} \simeq \bar{X} \tag{21.37}$$

where $\bar{X} \simeq \bar{X}_1 \simeq \bar{X}_2 = \ldots = \bar{X}_m$ = constant
If the deviation in the sample is denoted by d_{ij} and that in the sample mean by D_i then,

$$\sigma^2(x) = \frac{1}{nm} \sum_{j=1}^{n} \sum_{i=1}^{m} d_{ij}^2 \tag{21.38}$$

and

$$\sigma^2(\bar{X}_n) = \frac{1}{m} \sum_{i=1}^{m} D_i^2 \tag{21.39}$$

where $\sigma^2(x)$ and $\sigma^2(\bar{X}_n)$ represents the variation of individual measurement and the sample means, respectively.

Now

$$\begin{aligned} D_i &= (\bar{X}_1 - \bar{\bar{X}}) \\ &\simeq \frac{1}{n} \sum_{j=1}^{n} x_{ij} - \frac{n\bar{X}}{n} \quad (\because \bar{\bar{X}} \simeq \bar{X}) \\ &\simeq \frac{1}{n} \sum_{j=1}^{n} (x_{ij} - \bar{X}) \\ &\simeq \frac{1}{n} \sum_{j=1}^{n} d_{ij} \end{aligned} \tag{21.40}$$

Substituting the value of D_i from Eq. (21.40) in Eq. (21.39) we get,

$$\begin{aligned} \sigma^2(\bar{X}) &= \frac{1}{m} \sum_{i=1}^{m} \left\{ \frac{1}{n} \sum_{j=1}^{n} d_{ij} \right\}^2 \\ &= \frac{1}{mn^2} \sum_{i=1}^{m} \left\{ \sum_{j=1}^{n} d_{ij} \right\}^2 \end{aligned} \tag{21.41}$$

The double sum on the RHS of Eq. (21.41) when evaluated contains two different types of terms. There are some terms in which d_{ij} is squared while the other terms contain the product of two different d_{ij}. Because, there is equal likelihood of the terms being negative and positive, therefore, in the limit, when we take a very large number of observations mn, the sum of product terms of two different d_{ij}, tends to zero.

$\therefore$

$$\sum_{i=1}^{m} \left\{ \sum_{j=1}^{n} d_{ij} \right\}^2 \simeq \sum_{i=1}^{m} \sum_{j=1}^{n} d_{ij}^2 \tag{21.42}$$

New Eq. (21.41) becomes,

$$\begin{aligned} \sigma^2(\bar{X}_n) &= \frac{1}{mn^2} \sum_{i=1}^{m} \sum_{j=1}^{n} d_{ij}^2 \\ &= \frac{1}{n} \left\{ \frac{1}{mn} \sum_{i=1}^{m} \sum_{j=1}^{n} d_{ij}^2 \right\} \\ &= \frac{\sigma^2(x)}{n} \end{aligned} \tag{21.43}$$

Now if $\sigma(x)$ in Eq. (21.43) is replaced by s_n the unbiased estimate of the best precision, then $\sigma(\bar{X}_n)$ is replaced by U_n the best estimate of internal uncertainty (also known as the internal standard error). Hence,

$$U_n = \pm \frac{s_n}{\sqrt{n}}$$

$$= \pm \frac{\sigma_n(x)}{\sqrt{(n-1)}} \tag{21.44}$$

Hence, the results of the experiment may be summarised in the form:

$$X = \bar{X}_n \pm U_n \tag{21.45}$$

where $\bar{X}_n$ is the mean of the sample data and U_n the best estimate of internal uncertainty, also termed the internal standard error.

Problem 21.5 *Ten samples of a steel wire were tested on a universal testing machine. The breaking strengths in tonnes (t) of the samples were:*

4.3, 4.5, 4.7, 4.2, 4.5
4.6, 4.4, 4.6, 4.9, 4.5

Compute the following:
(a) the mean value of the breaking strength,
(b) mean deviation of the data
(c) standard deviation,
(d) best estimate of precision of the apparatus, and
(e) best estimate of the uncertainty in the data.

Solution The mean value, mean deviation and standard deviation can be evaluated from the following:

I	x_i*(in t)*	d_i*(in t)*	d_i^2 *(in t^2)*
1	4.3	–0.22	0.0484
2	4.5	–0.02	0.0004
3	4.7	0.18	0.0324
4	4.2	–0.32	0.1024
5	4.5	–0.02	0.0004
6	4.6	0.08	0.0064
7	4.4	–0.12	0.0144
8	4.6	0.08	0.0064
9	4.9	0.38	0.1444
10	4.5	–0.02	0.0004
	$\Sigma x_i = 45.2$	$\Sigma\lvert d_i\rvert = 1.44$	$\Sigma d_i^2 = 0.356$
	$\bar{X} = \dfrac{45.2}{10}$	$\text{MD} = \dfrac{1.44}{10}$	$\sigma = \sqrt{\dfrac{0.356}{10}}$
	$= 4.52\ t$	$= 0.144\ t$	$= 0.189\ t$

Thus we have,

(a) Mean value of breaking strength = 4.52 t

(b) Mean deviation = 0.144 t

(c) Standard deviation = 0.189 t

It may be noted that for a moderately skewed distribution, mean deviation is 4/5 of the standard deviation. In this example, (4/5) (0.189) = 0.151 is very nearly equal to the calculated value of MD = 0.144. This shows that the results of the problems are in good agreement with the empirical relation (Eq. (21.30)).

(d) The best estimate of precision of data is given by the adjusted or unbiased estimate of standard deviation

$$s_n = \sqrt{\frac{n}{(n-1)}}\,\sigma_n$$

$$= \sqrt{\frac{10}{9}}\; 0.189 = 0.199\ t$$

(e) The best estimate of the uncertainty or the internal standard error is given by

$$U_n = \frac{s_n}{\sqrt{n}}$$

$$= \frac{0.199}{\sqrt{10}} = 0.063\ t$$

Thus, in this problem, the experimental result of the given data can be expressed as follows:
Breaking strength of steel wire = 4.52 ± 0.063 t

21.6 ■ EVALUATION OF SAMPLE MEAN AND STANDARD DEVIATION BY METHOD OF CODING

In evaluating the mean value and the standard deviation, sometimes we come across an extensive data with considerable number of digits. This results in time-consuming arithmetic manipulations and additionally there are chances of errors, especially if the calculations are done by hand. Therefore, we employ a method of *coding* that simplifies the calculations to a large extent. The data is coded by:

1. Changing the origin of the data,
2. changing the units of the data, and
3. changing both the origin and the units of the data.

21.6.1 Evaluation of Mean and Standard Deviation by Changing the Origin of the Data

The procedure for coding by changing the origin of the data is also called the method of assumed mean. Sometimes we intuitively employ this method of calculation in our day-to-day life. For instance, if 4 observations are given as say 100.20, 100.30, 100.15 and 100.25 and the mean value is to be determined. Instead of adding them up, getting the total of 400.80 and then dividing by 4 to get 100.20; it is much easier to just find the mean of 0.20, 0.30, 0.15, and 0.25, i.e. 0.20 and add it to 100.

In effect, we are working with the values of x' where $x' = x - 100$. In other words, we have shifted the origin of data to $X_A = 100$. Now, it is obvious that for a general case, we can write the observed values in the following from:

$$X_i = x'_i + X_A \tag{21.46}$$

where X_A is the assumed mean (a constant value)

The mean value in Eq. (21.46) can be obtained by taking the summation and dividing by the total number of observations

$$\Sigma \frac{x_i}{n} = \frac{1}{n} \Sigma (x_i' + X_A)$$

or

$$\bar{X}_n = \bar{X}_n' + X_A \tag{21.47}$$

where $\bar{X}_n'$ is the mean value of quantities with shift of origin of the data.

Similarly, the variance in x, i.e. $\sigma^2(x)$ can be obtained by substituting the values of x_i and $\bar{X}_n$ from Eqs. (21.46) and (21.47) in the following equation:

$$\sigma^2(x) = \frac{1}{n}\Sigma \ (x_i - \bar{X}_n)^2$$

$$= \frac{1}{n}\Sigma \{(x'_i + X_A) - (\bar{X}_n' + X_A)\}^2$$

$$= \frac{1}{n}\Sigma (x'_i - \bar{X}_n')^2$$

$$= \sigma^2(x') \tag{21.48}$$

or

$$\sigma(x) = \sigma(x') \tag{21.49}$$

21.6.2 Evaluation of Mean and Standard Deviation by Changing the Units of the Data

Sometimes, a change in units facilitates in reducing the number of digits required in arithmetic manipulations. For instance, if the values of 4 observations were 400, 600, 200 and 500, and you were required to find their mean, you would probably disregard the zeros at the first instance and find the mean of 4, 6, 2 and 5 to be 4.25. After that you would then put back the zeros and present the mean to be 425. What you are really doing in this example is that you are changing the units of the original data by a suitable multiplying factor which in the present case is 1/100. Thus, we can write the observed values in the following general from:

$$x_i = ax_i'$$

where a is a suitable multiplying factor for changing the units of the data.

The mean and standard deviations in Eq. (21.50) can be found in a similar manner as in Sec. 21.7.1 and the results would be:

$$\bar{X}_n = a\bar{X}_n' \tag{21.51}$$

and

$$\sigma_n(x) = a\sigma_n(x') \tag{21.52}$$

21.6.3 Evaluation of Mean and Standard Deviation by Changing Both Origin and Units

Equation (21.46) is used in changing the origin of the system by subtracting/adding a value X_A known as the assumed mean. However, the resulting values sometimes do not simplify the calculations to the desired extent. Therefore, in addition, it is sometimes profitable to combine the effect of changing the units also as explained in Eq. (21.50). Thus, the observed values may be written in the following form:

$$x'_i = x_i - X_A$$

and $$x''_i = \frac{x'_i}{h}$$

or $$x_i = h x''_i + X_A \tag{21.53}$$

Here h is a multiplying scale factor, whereas a is the scale factor in Eq. (21.50).

The mean value and variance in Eq. (21.53) can be derived from the first principle. The results obtained are as follows:

$$\bar{X}_n = h\ \bar{X}''_n + X_a \tag{21.54}$$

and $$\sigma_n(x) = h\sigma_n(x'') \tag{21.55}$$

Problem 21.6 *The breaking loads of 80 specimens of a certain material were measured and the results were arranged in the grouped frequency table as follows:*

Breaking load x_i in kN (mid-interval value)	*Number of specimens f_i*
942.5	3
962.5	10
982.5	14
1002.5	25
1022.5	17
1042.5	9
1062.5	2

Determine the mean and standard deviation of the data using the method of coding.

Solution The calculation of the mean and standard deviation of the given frequency distribution can be easily done by shifting the origin of the data by subtracting the middle value of the data, i.e. X_A = 1002.5. In addition, further simplification can be done by dividing shifted origin values by the class interval, i.e. h = 20.0. Now, the change of origin in the data as well as that in units of the same along with the necessary computations for the calculations of mean value and the standard deviation can be conveniently carried out in the form of the following table:

x_i	f_i	$x'_i = x_i - X_A$	$x'' = \frac{xi}{h}$	$f_i x''$	$f_i x''^2$
942.5	3	–60	–3	–9	27
962.5	10	–40	–2	–20	40
982.5	14	–20	–1	–14	14
1002.5	25	0	0	0	0
1022.5	17	20	1	17	17
1042.5	9	40	2	18	36
1062.5	2	60	3	6	18
Total	80			–2	152

Now the relationship between x_i and x_i'' is

$$x_i'' = \frac{x_i - X_A}{h}$$

or

$$x_i = hx_i'' + X_A$$

$$\bar{X}_n = \frac{h\sum f_i x_i''}{\sum f_i} + X_A$$

$$= \frac{(20)(-2)}{80} + 1002.5 = 1002.0 \text{ kN}$$

Similarly, the standard deviation of the data can be calculated as:

$$\sigma_n(x) = h\sigma_n(x'')$$

$$= 20 \times \sqrt{\frac{\sum f_i (x_i'')^2}{N} - (\bar{X}_n'')^2}$$

$$= 20 \times \sqrt{\frac{152}{80} - \left(\frac{-2}{80}\right)^2} = 27.56 \text{ kN}$$

21.7 ■ EVALUATION OF BEST ESTIMATE MEAN VALUE AND LEAST ERROR IN A MULTIPLE SET OF DATA

When repeat measurements of a particular quantity are made by different persons, using different instruments and employing different sample sizes, then we obtain the mean value and standard deviation for each set of data. From these measured values, the evaluation of mean values and the internal estimate of uncertainties in each set of data can be represented as: $(\bar{X}_n \pm U_n)$, $(\bar{X}_m \pm U_m)$, . . . $(\bar{X}_l \pm U_l)$.

Now it is possible to evaluate the combined best estimate of true value of data by the minimising combined internal estimate of uncertainty. *We initially consider two sets of data* i.e., $(\bar{X}_n \pm U_n)$ and $(\bar{X}_m \pm U_m)$. If U_n is very much smaller than U_m, then the first experiment would be considered accurate and we would neglect the second data. On the other hand, if U_m is vary much smaller than U_n, then we shall believe the second data as compared to the first one. Further, if two errors U_n and U_m are equal then, it would be reasonable to take the mean of the combined measurement as

$$\bar{X}_{n,m} = \frac{1}{2}(\bar{X}_n + \bar{X}_m) \tag{21.56}$$

It may be noted that in Eq. (21.56), equal importance has been given to two experimental values of equal accuracy. In general, the simplest way to combine the results of two experiments is:

$$\bar{X}_{n,m} = \alpha\,\bar{X}_n + (1 - \alpha)\,\bar{X}_m \tag{21.57}$$

where $0 \leq \alpha \leq 1$.

In Eq. (21.57), when $\frac{U_n}{U_m} << 1$, then $\alpha = 1$ and the first equation predominates and vice versa.

Using the Eq. (2.7) of Ch. 2 i.e., propogation of uncertainties for the combined quantities in Eq. (21.57), we get,

$$U_{n,m}^2 = \left(\frac{\partial \bar{X}_{n,m}}{\partial \bar{X}_n}\right)^2 U_n^2 + \left(\frac{\partial \bar{X}_{n,m}}{\partial \bar{X}_m}\right)^2 U_m^2$$

$$= \alpha^2 U_n^2 + (1 - \alpha)^2 U_m^2 \tag{21.58}$$

Equation (21.58) would have a minimum when

$$\frac{d}{d\alpha}(U^2_{n,m}) = 0 \qquad (21.59)$$

This gives

$$2\alpha\, U_n^2 + 2(1-\alpha)(-1)\, U_m^2 = 0 \qquad (21.60)$$

or

$$\alpha = U_m^2/(U_n^2 + U_m^2) \text{ and} \qquad (21.61)$$

$$(1-\alpha) = U_n^2/(U_n^2 + U_m^2) \qquad (21.62)$$

Dividing the numerator and denominators by $U_n^2\ U_m^2)$ in Eqs. (21.61) and (21.62) we get,

$$\alpha = U_n^{-2}/(U_n^{-2} + U_m^{-2}) \text{ and} \qquad (21.63)$$

$$(1-\alpha) = U_m^{-2}/(U_n^{-2} + U_m^{-2}) \qquad (21.64)$$

Substituting the values of (α) and $(1-\alpha)$ in Eqs. (21.57) and (21.58) we get,

$$\bar{X}_{n,m} = \frac{1}{U_n^{-2} + U_m^{-2}}\left(\frac{\bar{X}_n}{U_n^2} + \frac{\bar{X}_m}{U_m^2}\right) \text{ and} \qquad (21.65)$$

$$U^{-2}_{n,m} = U_n^{-2} + U_m^{-2} \qquad (21.66)$$

Equations (21.65) and (21.66) give the best of the mean value and the least error respectively in the twin data sets i.e., $(\bar{X}_n \pm U_n)$ and $(\bar{X}_m + U_m)$. Further, Eq. (21.65) shows that the weighing factor for $\bar{X}_n$ and $\bar{X}_m$ are proportional to $1/(U_n^2)$ and $1/(U_m^2)$, respectively.

Now, for a general case, the multiple sets of data consist of the following:

$$\left.\begin{aligned} X_n &= \bar{X}_n \pm U_n, \\ X_m &= \bar{X}_m \pm U_m, \\ &\cdots\cdots\cdots \\ &\cdots\cdots\cdots \\ X_l &= \bar{X}_l \pm U_l \end{aligned}\right\} \qquad (21.67)$$

For such a case, the best estimate of mean value can be obtained from Eq. (21.65) by using the weighing factors proportional to $1/(U_n^2)$, $1/(U_m^2)$ . . . . . . $1/(U_l^2)$ for $\bar{X}_n$, $\bar{X}_m$, ..., $\bar{X}_l$, respectively.

Hence,

$$\bar{X}_{n,m,\ldots,l} = \frac{1}{(U_n^{-2} + U_m^{-2} + \ldots U_l^{-2})}\left[\frac{\bar{X}_n}{U_n^2} + \frac{\bar{X}_m}{U_m^2} + \ldots \frac{\bar{X}_e}{U_l^2}\right] \qquad (21.68)$$

and similarly,

$$U_n^{-2}{}_{,m,\ldots,l} = U_n^{-2} + U_m^{-2} + \ldots + U_l^{-2} \qquad (21.69)$$

Problem 21.7 *Three experiments were conducted to measure the speed of light.*

(i) In experiment A, 10 measurements were taken and the mean velocity was determined as 2.9985×10^{10} *cm/s with a mean square deviation of* 4.41×10^{15} *(cm/s)*^{2}.

(ii) In experiment B, 17 measurements were taken and the mean velocity was determined as 2.9972 with a mean square deviation of 2.25×10^{14} *(cm/s)*^{2}.

(iii) experiment C, 10 measurements were taken and the mean velocity was determined as 2.9979 with a mean square deviation of 1.02×10^{15} *(cm/s)*^{2}.

Determine the best estimate of the velocity of light and find the least error by combining the data of the all the sets.

Solution The uncertainty U_A in the experiment A is given by

$$U_A = \sqrt{\frac{\sigma_A^2}{(n_A - 1)}} = \pm\sqrt{\frac{4.41\times 10^{15}}{9}} = \pm\ 0.007 \times 10^{10}\ \text{cm/s.}$$

Therefore, velocity of light in experiment A is given by

$$V_A = (2.9985 \pm 0.0007) \times 10^{10}\ \text{cm/s}$$

Similarly, V_B and V_C can be evaluated which are as follows:

$$V_B = (2.9972 \pm 0.0005) \times 10^{10}\ \text{cm/s and}$$

$$V_C = (2.9979 \pm 0.0008) \times 10^{10}\ \text{cm/s}$$

Now using Eq. (21.68) best combined estimate of the velocity of light is given by

$$V_{A,\,B,\,C} = \frac{1}{(0.0007)^{-2} + (0.0005)^{-2} + (0.0008)^{-2}}$$

$$\times \left\{\frac{2.9985}{(0.0007)^2} + \frac{2.9972}{(0.0005)^2} + \frac{2.9979}{(0.0008)^2}\right\} \times 10^{10}$$

$$= 2.99769 \times 10^{10}\ \text{cm/s}$$

Further, using Eq. (21.69) the best combined estimate of the least error is given by

$$U_{A,B,C} = [(0.0007)^{-2} + (0.0005)^{-2} + (0.0008)^{-2}]^{-1/2} \times 10^{10}$$

$$= 0.00036 \times 10^{10}\ \text{cm/s}$$

Hence, the combined best estimate of the velocity of light based on the three sets of data is $V_{A,B,C} = (2.99769 \pm 0.00036) \times 10^{10}$ cm/s.

Review Questions

21.1 Fill in the blanks in the following:

(i) It is seldom profitable to have less than ________ classes and more than ______ classes in graphical representation of data.

(ii) Area of a normalised histogram is ________.

(iii) Ogive is a representation of ________ data.

(iv) In statistics, there are several averages in common use. They are:
(a) ________, (b) ________, (c) ________.

(v) A sample is a portion of ________.

(vi) If $\overline{X}$ is the mean of n values of data, then

$$\sum_{i=1}^{n} (x_i - \overline{X}) = \_\_\_\_\_\_\_\_.$$

(vii) For a discrete variate, the modal value corresponds to the highest ________.

(viii) If X_a has any value other than $\overline{X}$, then

$$\sum_{i=1}^{n} (x_i - \overline{X})^2 \text{ is } \_\_\_\_\_\_\_\_ \text{ than } \sum_{i=1}^{n} (x_i - X_a)^2.$$

(ix) $n\left(\sum_{i=1}^{n} x_i^2\right)$ is ________ than $\left(\sum_{i=1}^{n} x_i\right)^2$

(x) Coefficient of variation = ________ /mean value.

(xi) $\sqrt{\dfrac{\sum_{i=1}^{n}(x_i - \bar{X}_n)^2}{(n-1)}}$ is the ________ estimate of ________ of apparatus.

(xii) In experimental analysis ________ is considered the best estimate of the true value.

(xiii) $\sigma(\bar{X}_n)$ is given by ________.

(xiv) Standard deviation of the deviations from the assumed mean is equal to the ________ of the data.

(xv) Frequency polygon is obtained by joining the ________ frequency bars of the absolute frequency histogram.

21.2 Indicate if the following statements are true or false. If false, rewrite the correct statement:

(i) A normalised histogram represents the pictorial representation of discrete type of data.

(ii) Coefficient of variance = $\dfrac{\text{variance of data}}{\text{mean value}}$

(iii) Mean of samples is approximately equal to the mean of sample means. Similarly, the standard deviation of samples is also the standard deviation of the sample means.

(iv) An experiment is considered consistent if its internal estimate of uncertainty is the lowest possible.

(v) The most reliable measure of the variability of data is the variance of the data.

(vi) Sum of the relative frequencies per unit interval of all the classes is unity.

(vii) For moderately skewed distributions, the following relation holds good:
mean – median = 3 (mean – mode)

(viii) The best estimate of true value can be written as

$$x = \bar{X}_n \pm \frac{\sigma}{n}$$

(ix) Standard deviation represents the best estimate of the precision of data.

(x) It is possible to change the origin as well as the units of the data simultaneously.

(xi) A graph of cumulative frequency distribution is known as the normalised histogram.

(xii) A relative frequency distribution represents frequency in terms of percentages/fractions.

(xiii) A less than type of ogive curve is S-shaped which slopes upwards and to the right.

(xiv) A histogram consists of a series of rectangles in which the width of each class represents the number of items falling with in that class.

(xv) The measure of the average for the income of a worker in a factory is the median value and the measure for representing the number of petals of a typical flower is the modal value.

21.3 The following table shows the diameters in millimeters of a sample of 60 ball bearings. Choosing suitable class interval and number of classes, construct a table showing:

(i) class boundaries

(ii) class marks

(iii) class relative frequency, and

(iv) normalised frequency.

Also draw the normalised histogram and ogive of the data.

6.35	6.39	6.36	6.42	6.33	6.24	6.35	6.36	6.37	6.28
6.44	6.36	6.41	6.36	6.32	6.34	6.27	6.35	6.32	6.35
6.37	6.26	6.31	6.35	6.41	6.36	6.40	6.42	6.29	6.38
6.32	6.34	6.33	6.35	6.38	6.28	6.40	6.42	6.36	6.44
6.40	6.30	6.34	6.29	6.35	6.35	6.46	6.31	6.37	6.32
6.35	6.37	6.29	6.38	6.34	6.39	6.30	6.32	6.30	6.33

21.4 The following two sets of data are obtained for the shear strength of stainless steel spot welds of 0.041 cm thickness in kg per weld:

(i)	140	160	144	148	144	148
	146	162	150	152	150	158
	148	146	152	154	152	160
	146	156	154	152	156	143
	150	150	138	153	158	165
(ii)	146	148	154	154	152	146
	142	146	160	153	148	154
	152	154	152	158	160	150
	146	150	150	156	150	152
	148	148	160	160	152	154

Calculate the mean, mode, median, variance and standard for the first set both by the standard method and the method of coding.

Plot the discrete value relative frequency distribution curves for the first set. Superimpose on these the corresponding curve for the second set and comment. Also calculate the mean and the standard deviation of the first set by grouping the data into a suitable number of intervals, again using both the standard method and the method of coding.

21.5 The following measurements of a particular dimension of a mass produced automobile component were made and tabulated by an engineer.

Class boundaries (cm)	*Frequency* f_i
4.35–4.45	1
4.45–4.55	4
4.55–4.65	8
4.65–4.75	15
4.75–4.85	11
4.85–4.95	9
4.95–5.05	2

Find the mean, median, mode, range, mean deviation and standard deviation of the data.

21.6 The following table lists a sample of experimental data of the repair times of a certain aircraft system along with certain calculations that have been made from it.

X_i *(h)*	f_i	v'	$f_i v'$	$f_i v'^2$
5.0	5	–2.5	–12.5	31.25
5.5	9	–2.0	–18.0	36.0

6.0	13	–1.5	–19.5	29.25
6.5	19	–1.0	–19.0	19.0
7.0	31	–0.5	–15.5	7.75
7.5	28	0	0	0
8.0	17	0.5	8.5	4.25
8.5	12	1.0	12.0	12.0
9.0	8	1.5	12.0	18.0
9.5	5	2.0	10.0	20.0
10.0	3	2.5	7.5	18.75

Find the mean, standard deviation, adjusted standard deviation and internal standard error of the given data.

21.7 Ten measurements (in mm) of the diameter of a ball bearing are

24.0, 24.2, 23.8, 24.5, 23.6, 23.8, 24.4, 24.0, 23.8, 24.1

Determine the mean, median and mode which represent the central tendency of the data.

21.8 The change of origin and units of a given data x'_i gives

$$x'_i = \frac{x_i - X_A}{h}$$

where X_A is assumed as mean and h is a constant.

Show that (i) mean value $\bar{X}_n = h \dfrac{\sum f_i x'_i}{\sum f_i} + X_A$ and

(ii) standard deviation $\sigma_x(x) = h \sqrt{\dfrac{\sum f_i x_i'^2}{\sum f_i} - \left(\dfrac{\sum f_i x_i'^2}{\sum f_i}\right)^2}$

21.9 101 observations of daily emissions of sulphur dioxide (in tonnes (t)) in a thermal power plant were arranged in the form of a cumulative frequency distribution which was as follows:

Sulphur-dioxide (in t)	*Cumulative frequency*
less than 3.05	0
less than 6.05	3
less than 9.05	10
less than 12.05	22
less than 15.05	38
less than 18.05	63
less than 21.05	80
less than 24.05	90
less than 27.05	99
less than 30.05	101

Determine: (i) mean, mode and median of the data,
(ii) best estimate of precision of the data,
(iii) accuracy of mean $\sigma(\bar{X})$ of the data.

21.10 A sample consisting of $x_1, x_2, \ldots, x_i, \ldots, x_n$ has a mean value of $\bar{X}_n$ and a variance of s_n^2. Now, an additional observation x_{n+1} becomes available, and the mean and variance of $(n+1)$ observations is $\bar{X}_{n+1}$ and s_{n+1}^2. Show that:

$$(n+1)s_{n+1}^2 = n s_n^2 + \frac{n}{n+1}(x_{n+1} - \bar{X}_n)^2.$$

21.11 An electronic component was subjected to life test programme under high temperature to accelerate the failure mechanism. The failure time (in hours) for 32 randomly selected components is shown below.

102.3	178.3	68.3	126.5	93.6	150.7	52.3	112.4
82.3	24.5	156.8	6.7	134.8	199.7	75.3	124.2
75.3	160.8	145.3	112.5	57.8	148.5	105.9	92.8
188.1	201.3	94.1	84.3	107.3	39.8	133.8	115.6

(i) Arrange the data in the less than type of cumulative distribution.
(ii) Determine the mean, mode median, mean deviation, standard deviation and internal standard error of the data.

21.12 An industrial township has an availability of a maximum water supply of 500 million litres per day. An estimate of the projected daily requirements 5 years hence is as follows:

Month	*Consumption per day (in million litres)*
January	327
February	422
March	421
April	478
May	521
June	558
July	576
August	588
September	582
October	472
November	452
December	421

(i) Draw the histogram of the data and determine that by what percentage of time, the demand would exceed the supply.
(ii) Determine the arithmetic mean, standard deviation and the adjusted standard deviation.
(iii) Discuss what needs to be done to ensure a smooth supply of water.

21.13 The internal standard error of mean of 25 measurements is 2.4 cm. If the measurement procedure and instruments are unchanged, how many measurements should one make to reduce the standard error to 1.5 cm.

21.14 The cold start times (in seconds) of a diesel engine were found to be as follows:

3.3, 3.5, 3.6, 3.7, 3.0
3.5, 3.6, 3.5, 3.7, 3.5
3.6, 3.5, 3.6, 3.4, 3.6
3.9, 3.6, 3.5, 3.6, 3.2

(i) Plot the normalised histogram.
(ii) Evaluate the mean, standard deviation and internal standard error of the data.

21.15 A laboratory study of the time delay (in seconds) or relay switches gave the following results:

Time delay (in seconds)	1.2	1.25	1.30	1.35	1.40	1.45	1.50	1.55	1.60
frequency	1	3	8	15	23	15	8	3	1

Determine the mean value of time delay, standard deviation and adjusted standard deviation.

21.16 An incomplete frequency distribution is given as follows:

Variable	*Frequency*
10–20	14
20–30	30
30–40	f_1
40–50	65
50–60	f_2
60–70	25
70–80	18
Total	230

If the median value is 46, determine the missing frequencies f_1 and f_2.

21.17 Determine the class interval if the arithmetic mean of the following distribution is 40.2 and the assumed mean is 40.

Step deviation	–3	–2	–1	0	1	2	3
Frequency	5	10	21	26	22	12	4

21.18 A room thermostat was set to switch-on at 23°C. To check the accuracy of the thermostat setting, an accurate thermometer was used to record the temperature everytime the light went on. In one day, the following 30 readings were taken:

24, 22, 23, 21, 22, 20, 24, 23, 22, 21 25, 23, 22, 24, 21
23, 20, 24, 22, 21, 25, 21, 22, 24, 22, 24, 23, 25, 21, 24

What is the average temperature of the room? How much is the uncertainty in the measured temperature.

21.19 The marks obtained by 200 candidates in an examination were grouped in the intervals 0–5, 5–10, . . ., 95–100. The mean and the standard deviations were found to be 40 and 15. Later, it was discovered that a score of 43 was misread as 53 in obtaining the frequency distribution. Determine the corrected mean and standard deviation corresponding to the corrected frequency.

21.20 Means and standard deviations of two sample sizes n_1 and n_2 are a_1, s_1 and a_2, s_2, respectively. The two samples are combined to form a single composite sample of size N (= $n_1 + n_2$) and standard deviation S. Show that:

$$S^2 = \frac{n_1 s_1^2 + n_2 s_2^2}{N} + \frac{n_1 n_2 (a_1 - a_2)^2}{N^2}$$

21.21 In a right-angled triangle, the hypotenuse C was measured as (5.02 ± 0.02) cm. The other two sides A and B were measured as (3.0 ± 0.01) cm and (4.0 ± 0.01) cm, respectively. Determine:
(i) the length of the hypotenuse and its uncertainty from the measurements of A and B, and
(ii) combined best estimate of the length of the hypotenuse using the direct measurement as well as that estimated from the measurements of the sides A and B.

21.22 Three experiments were set up to measure the charge of an electron in 10^{-19} coulombs. The results obtained were:
(i) 1.604 ± 0.003
(ii) 1.606 ± 0.006
(iii) 1.602 ± 0.004
Determine the combined best estimate of the charge of an electron and its uncertainty.

Answers

21.1 (i) 5, 20
(ii) unity
(iii) cumulative frequency of
(iv) mean, mode and median
(v) given data
(vi) zero
(vii) frequency of the data
(viii) less than
(ix) greater than
(x) standard deviation
(xi) best, precision
(xii) arithmetic mean
(xiii) $\dfrac{\sigma}{\sqrt{(n-1)}}$
(xiv) standard deviation
(xv) middle points of the

21.2 (i) false, a normalised histogram represents the pictorial representation of *continuous* type of data.
(ii) false, RHS of the expression should be standard deviation/mean value
(iii) false, mean of samples is approximately equal to the mean of sample means, but the standard deviation of samples is not equal to the standard deviation of sample means.
(iv) false, an experiment is considered consistent if the internal estimate of uncertainty is equal to the external estimate of uncertainty.
(v) true
(vi) false, sum of relative frquencies of all classes is unity.
(vii) false, mean–mode = 3 (mean–median).
(viii) false, $x = \bar{X}_n + \dfrac{\sigma}{\sqrt{(n-1)}}$.
(ix) false, adjusted standard deviation represents the best estimate of the precision of data.
(x) true.
(xi) false, it is known as ogive curve.
(xii) true.
(xiii) true.
(xiv) false, frequency in the histogram is represented by the height of the rectangle.
(xv) true.

21.3 Choosing a class interval of 0.2 and class boundaries up to three decimal places, the following is a typical distribution showing class boundaries, class marks, absolute frequencies, relative frequencies and normalised frequencies in the various class intervals.

Class boundaries	*Class marks*	*Absolute freq uency*	*Relative frequency*	*Normalised frequency*
6.225–6.245	6.235	1	1/60	5/6
6.245–6.265	6.255	1	1/60	5/6
6.265–6.285	6.275	3	1/20	5/2
6.285–6.305	6.295	6	1/10	5
6.305–6.325	6.315	7	7/60	35/6
6.325–6.345	6.335	7	7/60	35/6
6.345–6.365	6.355	15	1/4	25/2
6.365–6.385	6.375	7	7/60	35/6
6.385–6.405	9.395	5	1/12	25/6
6.405–6.425	6.415	5	1/12	25/6
6.425–6.445	6.335	2	1/30	5/3
6.445–6.465	6.455	1	1/60	5/6

21.4 mean value (first set) = 151.56
mode (first set) = 152.0
median (first set) = 151.0
standard daviation (first set) = 6.35
variance (first set) = 40.32

21.5 mean value = 4.732 cm
median = 4.730 cm
mode = 4.714 cm
range = 0.7 cm
mean deviation = 0.1118 cm
standard deviation = 0.136 cm

21.6 mean value = 7.27 h
standard deviation = 1.12 h
adjusted S.D. = 1.124 h
internal standard error U_I = 0.092 h

21.7 X_{mean} = 24.02 mm, X_{median} = 24.0 mm,
X_{mode} = 23.8 mm

21.9 $\bar{X}$ = 16.47 t, σ_{n-1} = 5.55 t
X_{mode} = 16.64 t, X_{median} = 16.55 t

$\sigma(\bar{X})$ = 0.555 t

21.11 $\bar{X}$ = 110.63 h, X_{mode} = 106.67 h
X_{median} = 110 h, MD = 38.24 h
σ = 47.69 h, U_I = 8.57 h

21.12 (i) Demand exceeds supply during May, June, July, August and September

(ii) $\bar{X} = \dfrac{1}{365}$ [327 × 31 + 422 × 28 + . . . + 421 × 31] = 484.74 million litres/per day

σ_n = 78.16 million litres per day

(iii) Since, the demand exceeds supply over 5 months and the mean is < 500 million litres per day; therefore, the problem can be handled by suitable storage tanks.

21.13 $n = 64$

21.14 (i) Range = 3.0–3.9 and the distribution of the data is

2.95–3.15	315–3.35	3.35–3.55	3.55–3.75	3.75–395
1	2	7	7	1

(ii) $\bar{X}$ = 3.52s, σ = 0.182s U_n = 0.042s

21.15 $\bar{X}$ = 1.4s σ_n = 0.076s σ_{n-1} = 0.077s

21.16 $f_1 = 32$ and $f_2 = 46$

21.17 class interval = 10

21.18 $\bar{X}$ = 22.6°C σ_n = 1.45°C U_1 = 0.26°C

21.19 $\bar{X}$ = 39.95 σ = 14.974

21.21 (i) Length of the hypotenuse = (5.0 ± 0.01) cm

(ii) Combined best estimate of hypotenuse = (5.004 ± 0.009) cm

21.22 combined best estimate of the charge of the electron based on three sets of data is: (1.6037 ± 0.0022) × 10^{-19} coulombs.

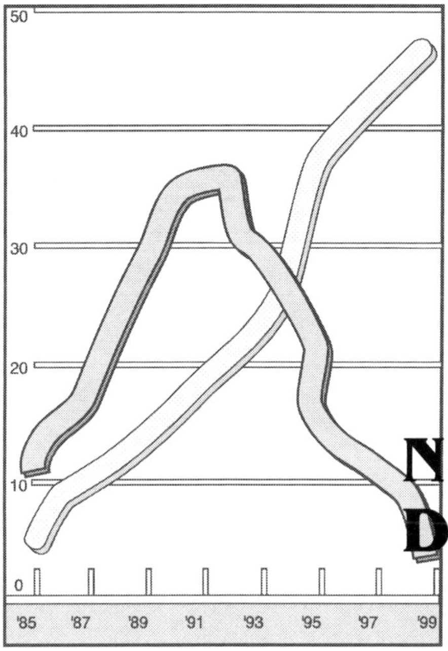

Chapter

22 Normal Distribution

INTRODUCTION

In general, measuring instruments are associated with a number of factors causing random errors. Therefore, the instrument readings exhibit a dispersion/scatter in the data. However, the magnitude of the individual errors is usually small. Now, if the measured value is observed for a very large number of times, the data exhibit a continuous distribution. This could be easily represented by a normalised histogram, i.e., by representing the relative frequency for unit class interval as ordinate and the measured value as abscissa. If the magnitude of the class interval is kept small and the ordinates of the various class mid-points are joined by a smooth curve, the resulting distribution is called the *limiting frequency distribution*. This distribution has a characteristic bell shape (Fig. 22.1) and is commonly termed *normal* or *Gaussian* distribution.

The normal distribution is by far the most commonly occurring distribution. It serves as a model for a number of variates like experimental random errors, dimensions of mass produced articles, and the measurable biological characteristics such as man's weight, height, etc. This distribution can be represented mathematically as

$$p(x) = \frac{1}{\sqrt{2\pi\sigma}} \exp\left\{-\frac{(x-\bar{X})^2}{2\sigma^2}\right\} \tag{22.1}$$

where $p(x)$ is the probability density function which, for a given interval, represents the relative frequency of occurrence of the measured value in that interval

σ the standard deviation of the measured values

$\bar{X}$ the mean of the measured values

The normal distribution model is employed in decision-making processes like the determination of the probability that the measured value lies within a given range. Alternatively, if the level of probability or the confidence level is a certain fixed value (which may be a requirement for a certain situation) then it is possible to determine the allowable scatter or dispersion from a given mean value.

Also, the properties of normal distribution are used for comparing various normally distributed samples using statistical criteria known as *significance tests.* Further, it is possible to determine the 'goodness of fit' of the measured values with those of the expected normally distributed values in the different ranges of measurements using the criteria known as χ^2-test (pronounced as chi-square test). This criteria of χ^2-test is also applicable for determining whether any non-normal distribution conforms to any other known theoretical distribution or not.

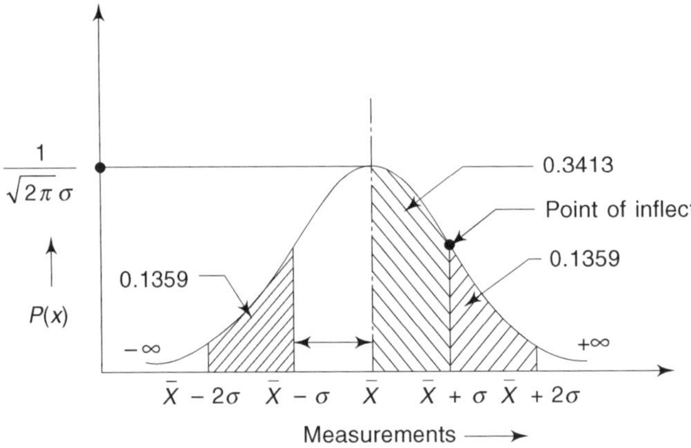

Fig. 22.1 *Typical Gaussian distribution*

22.1 ■ PROPERTIES OF GAUSSIAN DISTRIBUTION

Any typical Gaussian curve has the following features:

1. It has a maxima at $x = \bar{X}$, i.e., at the mean value.
2. Points of inflexion of the curve are at $x = \bar{X} \pm \sigma$.
3. It is symmetrical about the ordinate at $x = \bar{X}$ and divides the curve into two equal parts.
4. Because of the symmetry of the curve, the median is equal to the mean. Further, since the mean value occurs at the peak probability density, it also represents the mode. Thus, for a normal distribution mean = mode = median.
5. The x-axis is an asymptote of the curve.
6. The area under the normal distribution curve is unity, i.e.

$$\int_{-\infty}^{\infty} P(x)\,dx = \frac{1}{\sqrt{2\pi}\,\sigma}\int_{-\infty}^{\infty} \exp\left\{-\frac{(x-\bar{X})^2}{2\sigma^2}\right\}dx = 1 \tag{22.2}$$

7. For the same mean value, the distribution has a sharp peak for smaller values of σ and is flatter for higher values of σ. If σ is small, that means that scatter in the data is small and consequently more values are concentrated near the mean value. Since the area of the normal distribution curve is unity, therefore the ordinate becomes higher making the peaky shaped curve. Hence, smaller the value of σ, the larger is the maxima of the curve. In fact, the equation of the maximum value of the probability density function is as follows (Fig. 22.2).

$$\{p(x)\}_{\max} = \frac{1}{\sqrt{2\pi}\,\sigma} = \frac{0.399}{\sigma} \tag{22.3}$$

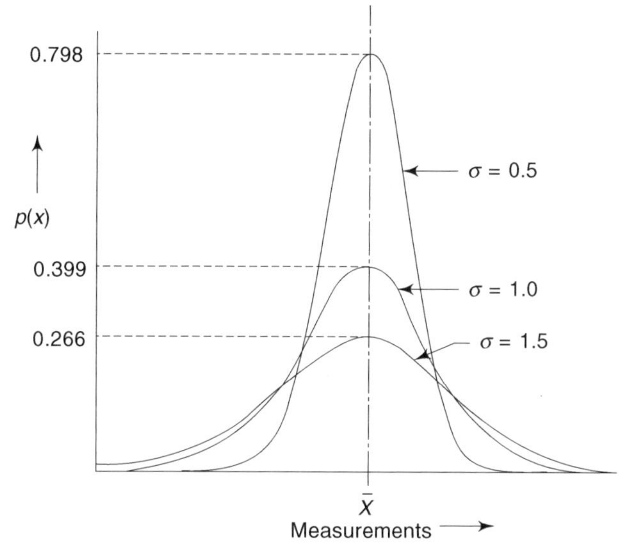

Fig. 22.2 *Effect of σ on the shape of normal distribution curve*

8. The probability that the mean value $\bar{X}$ takes the value between x_1 and x_2 is the area of the normal distribution curve between x_1 and x_2 as shown in Fig. 22.3. It is known as the integral Gaussian probability in the range between x_1 and x_2 and is represented as

$$\left|P(x)\right|_{x_1}^{x_2}$$

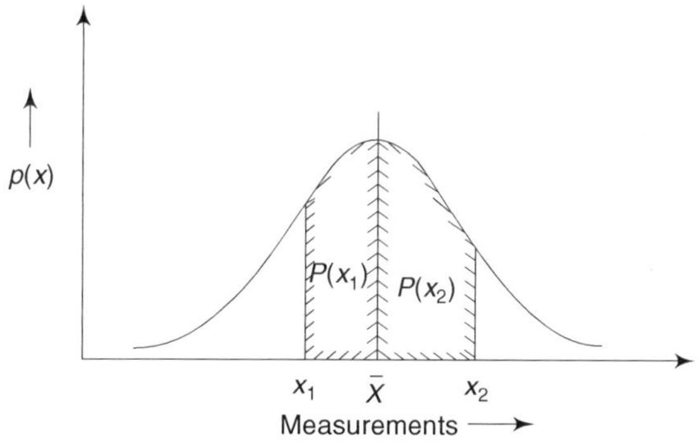

Fig. 22.3 *Integral Gaussian probability of occurrence in a specified range of limiting values*

To determine the integral Gaussian probability in the specified range, we proceed as follows:

$$\left|P(x)\right|_{x_1}^{x_2} = \int_{x_1}^{x_2} \frac{1}{\sqrt{2\pi}\,\sigma} \exp\left\{\frac{-(x-\bar{X})^2}{2\sigma^2}\right\} dx$$

$$= \int_{x_1}^{\bar{X}} \frac{1}{\sqrt{2\pi}\,\sigma} \exp\left\{\frac{-(x-\bar{X})^2}{2\sigma^2}\right\} dx + \int_{\bar{X}}^{x_2} \frac{1}{\sqrt{2\pi}\,\sigma} \exp\left\{\frac{-(x-\bar{X})^2}{2\sigma^2}\right\} dx$$

$$= |p(x)|_{x_1}^{\bar{X}} + |P(x)|_{\bar{X}}^{x_2}$$

$$= P(x_1) + P(x_2) \tag{22.4}$$

where $P(x_1)$ and $P(x_2)$ are the integral Gaussian probabilities between x_1 and $\bar{X}$ and $\bar{X}$ and x_2, respectively. The procedure of determining the values of integral Gaussian probabilities is explained in Sec. 22.4.

It may be noted that $|P(x)|_{x_1}^{x_2}$ would be the difference between $P(x_1)$ and $P(x_2)$ if both x_1 and x_2 lie on the same side of $\bar{X}$.

22.2 ■ AREA UNDER THE NORMAL DISTRIBUTION CURVE

The area under a normal distribution curve between the limits $-\infty$ to ∞ is the integral Gaussian probability of occurrence of the measured value in the very large range between $-\infty$ to ∞. Obviously, the integral Gaussian probability would be unity as all the possible measured values have been taken into account, i.e.,

$$|P(x)|_{-\infty}^{\infty} = 1$$

To obtain this result, we integrate the Gaussian distribution equation [Eq. (22.1)] as follows:

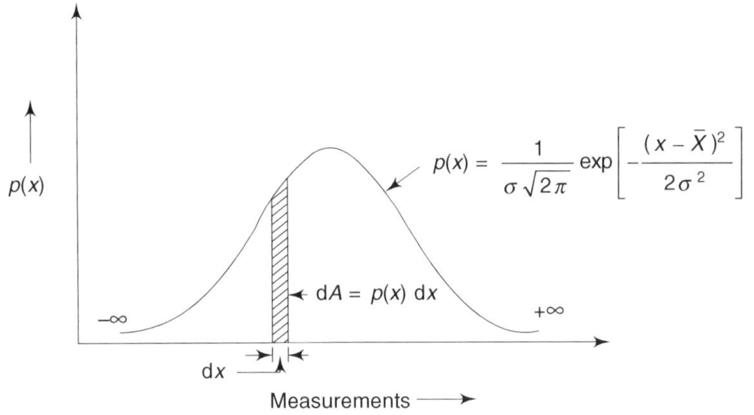

Fig. 22.4 *Typical Gaussian distribution showing elemental area*

The elemental area dA of the normal distribution

$$= p(x)\,dx$$

$$= \frac{1}{\sigma\sqrt{2\pi}} \exp\left\{\frac{-(x-\bar{X})^2}{2\sigma^2}\right\} dx \tag{22.5}$$

Integrating Eq. (22.5) between the limits $-\infty$ and ∞ we get,

$$[A]_{\text{normal}} = \int_{-\infty}^{\infty} \frac{1}{\sigma\sqrt{2\pi}} \exp\left\{\frac{-(x-\bar{X})^2}{2\sigma^2}\right\} dx \tag{22.6}$$

The expression on the RHS of Eq. (22.6) can be simplified by substituting $z = (x - \bar{X})/\sigma$ and correspondingly $dz = dx/\sigma$

$$\therefore \qquad [A]_{\text{normal}} = \frac{1}{\sqrt{2\pi}} \int_{-\infty}^{\infty} \exp\left(\frac{-z^2}{2}\right) dz \qquad (22.7)$$

The term $\int_{-\infty}^{\infty} \exp\,(-z^2/2)\, dz$ is known as the normal error function and its value can be shown by numerical integration or otherwise to be $\sqrt{2\pi}$.

Substituting the value of the normal error function in Eq. (22.7) we get the result that the area under the normal distribution curve is unity.

22.3 ■ DETERMINATION OF MEAN VALUE AND STANDARD DEVIATION OF THE CONTINUOUS DISTRIBUTION OF GAUSSIAN TYPE

The Gaussian distribution equation (Eq. (22.1)) involves two parameters namely, standard deviation σ and mean value $\bar{X}$. Using the standard procedure of determining the mean value and the standard deviation of a continuous distribution, it is possible to show that these parameters come out to be the same. This would show that the use of the parameters σ and $\bar{X}$ in the Gaussian distribution equation is in conformity with their definitions.

22.3.1 Determination of Mean Value for a Gaussian Distribution

Mean value of the continuous variate is defined by

$$\bar{X} = \Sigma x \cdot f_r\,(x) \qquad (22.8)$$

where $f_r(x)$ is the relative frequency of occurrence of the measured value x.

Now, for the Gaussian distribution, $p(x)$ the probability density function is the relative frequency per unit interval for the value x.

∴ The relative frequency $f_r(x)$ of occurrence of x in the interval dx is,

$$f_r(x) = p(x)\, dx \qquad (22.9)$$

Substituting this value in Eq. (22.8) we get

$$\bar{X} = \Sigma x \cdot p(x)\, dx \qquad (22.10)$$

Replacing the summation sign by integral sign with limits $-\infty$ to ∞ and substituting the expression for Gaussian distribution in place of $p(x)$ in Eq. (22.10), we get

$$X = \int_{-\infty}^{\infty} x \cdot p(x)\, dx$$

$$= \int_{-\infty}^{\infty} x \cdot \frac{1}{\sigma\sqrt{2\pi}} \exp\left\{\frac{-(x-\bar{X})^2}{2\sigma^2}\right\} dx \qquad (22.11)$$

This integral can be easily evaluated if we make the following substitution in Eq. (22.11):

$$z = \frac{x - \bar{X}}{\sigma} \quad \text{and} \quad dz = \frac{dx}{\sigma}$$

$$\therefore \qquad X = \int_{-\infty}^{\infty} \frac{1}{\sqrt{2\pi}} (z\sigma + \bar{X}) \exp\left(\frac{-z^2}{2}\right) dz \qquad (22.12)$$

$$= \int_{-\infty}^{\infty} \frac{\sigma}{\sqrt{2\pi}} z \exp(-z^2/2)\, dz + \int_{-\infty}^{\infty} \frac{\bar{X}}{\sqrt{2\pi}} \exp(-z^2/2)\, dz$$

$$= \frac{\sigma}{\sqrt{2\pi}} \left| -\exp\left(-\frac{z^2}{2}\right) \right|_{-\infty}^{\infty} + \bar{X} \int_{-\infty}^{\infty} \frac{1}{\sqrt{2\pi}} \exp\left(-\frac{z^2}{2}\right) dz \tag{22.13}$$

$$= 0 + (\bar{X})\,(1) \qquad \left[\because \int_{-\infty}^{\infty} \frac{1}{\sqrt{2\pi}} \exp[(-z^2/2)]dz = 1\right]$$

$$= \bar{X}$$

22.3.2 Determination of Standard Deviation for Gaussian Distribution

Variance of a continuous variate is defined by

$$\sigma^2 = \Sigma(x - \bar{X})^2 \cdot f_r\,(x) \tag{22.14}$$

For the normal destribution this expression becomes

$$\text{Variance} = \int_{-\infty}^{\infty} (x - X)^2 \cdot p(x)\, dx \tag{22.15}$$

$$= \int_{-\infty}^{\infty} (x - \bar{X})^2 \cdot \frac{1}{\sqrt{2\pi}\,\sigma} \exp\left\{\frac{-(x - X)^2}{2\sigma^2}\right\} dx \tag{22.16}$$

Substituting $z = (x - \bar{X})/\sigma$ and $dz = dx/\sigma$ we get

$$\text{Variance} = \frac{1}{\sqrt{2\pi}} \int_{-\infty}^{\infty} \frac{\sigma^2 z^2}{2} \exp\left(\frac{-z^2}{2}\right) dz$$

Treating $z^2 \exp(-z^2/2)$ as $z \cdot z \exp(-z^2/2)$ and integrating by parts, we get

$$\int z^2 \exp(-z^2/2)\, dz = -z \exp(-z^2/2) + 1 \cdot \exp(-z^2/2)\, dz \tag{22.18}$$

Substituting the value of the integral in Eq. (22.17) we get,

$$\text{Variance} = \frac{\sigma^2}{\sqrt{2\pi}} \left| -z \exp(-z^2/2) \right|_{-\infty}^{\infty} + \frac{\sigma^2}{\sqrt{2\pi}} \int_{-\infty}^{\infty} \exp(-z^2/2)\, dz$$

$$= 0 + (\sigma^2) \cdot (1) \qquad \left[\because \frac{1}{\sqrt{2\pi}} \int_{-\infty}^{\infty} \exp(-z^2/2)\, dz = 1\right]$$

$$= \sigma^2$$

22.4 ■ STANDARDISED NORMAL DISTRIBUTION

In order to reduce the different normal distributions to a general form, it is common to standardise them by moving the origin of the coordinates of the mean value as well as to choose a scale on the x-axis in terms of σ. The suitable variable selected is $z = (x - \bar{X})/\sigma$, termed the *standardised normal variate.*

The probability of occurrence $P(x)$ in the range x_1 to x_2 given by the normal distribution equation is

$$P(x)\ \{x_1 \le x \le x_2\} = \int_{x_1}^{x_2} \frac{1}{\sqrt{2\pi}\,\sigma} \exp\left\{-\frac{(x - \bar{X})^2}{2\sigma^2}\right\} dx \tag{22.19}$$

This equation can be transformed in the form of $P(x)$ in the range of z_1 and z_2 as follows:

Since $\quad z \text{ (by definition)} = \dfrac{x - \overline{X}}{\sigma}$

$$\therefore \qquad z_1 = \frac{x_1 - \overline{X}}{\sigma}$$

$$z_2 = \frac{x_2 - \overline{X}}{\sigma} \tag{22.20}$$

$$\text{and} \qquad dz = \frac{dx}{\sigma}$$

Substituting these values in Eq. (22.19) we get,

$$P(z)\ \{z_1 \le z \le z_2\} = \int_{z_1}^{z_2} \frac{1}{\sqrt{2\pi}} \exp\left\{-\frac{z^2}{2}\right\} dz \tag{22.21}$$

This equation is termed the *standardised normal equation.*

Comparing Eqs. (22.19) and (22.21), it is obvious that every time the probability of occurrence between x_1 and x_2 is required, the evaluation of integral in Eq. (22.19) is required, which is rather complex as compared to that in Eq. (22.21). This is due to the fact that in the latter case, the use of tabulated values of the normal error function, i.e. $\int \exp(-z^2/2)\, dz$, simplifies the calculations considerably.

Secondly, the shape of the curve in Eq. (22.19) depends on the values of $\overline{X}$ and σ and will be different for different cases. Whereas the equivalent Eq. (22.21) shows that all normal distributions in x, with whatever values of $\overline{X}$ and σ, reduce to the same shape of the standardised normal distribution in terms of the standard normal variate z with mean value zero and standard deviation unity. In other words, substituting $\overline{X} = 0$ and $\sigma = 1$ in Eq. (22.19) the equation reduces to the form of Eq. (22.21). It is because of this generality of Eq. (22.21) that the probability density function $p(z)$ (i.e. the ordinate of the standard normal curve) as well as the integral Gaussian probability $P(z) = \int p(z)\, dz$ (i.e. the area of the standardised normal curve between the given limits) are tabulated versus z (Tables 22.1 and 22.2).

It may be noted that while using the tables to evaluate the probabilities of normally distributed variates, it is advisable to sketch the area that corresponds to the probability required. Further, one should be careful in the use of integral Gaussian tables, as not all the tables give the same area. Some tables give the area from 0 to z, while others may give from $-z$ to z, from z to ∞ or from $-\infty$ to z, as shown in the Figs 22.5(a), (b), (c) and (d). The shaded areas represent the probability that the observation falls in the corresponding interval. However the tabulated values of the integral Gaussian probability presented in the text (Table 22.2) are as in Fig. 22.5(a).

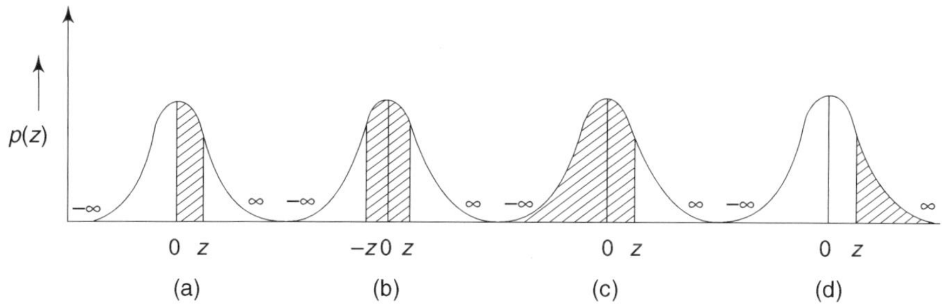

Fig. 22.5 *Integral Gaussian probability in different ranges*

Table 22.1 *Normal Probability Density Function p(z)*

$$p(z) = \frac{1}{\sqrt{2\pi}} \exp(-z^2/2)$$

Each entry in the table indicates normal probability density function $p(z)$ corresponding to $\pm z$

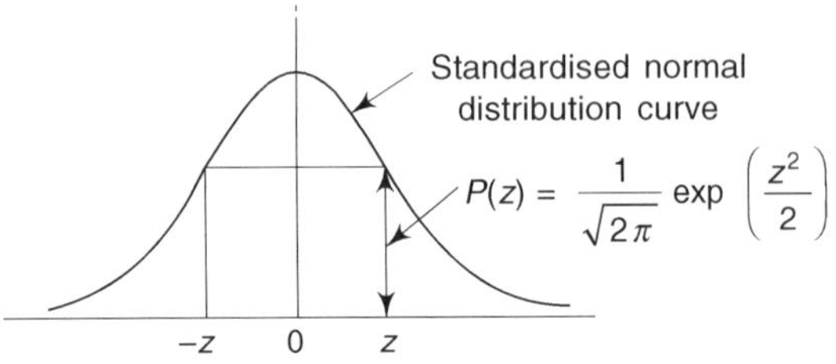

To illustrate: the ordinate $p(z)$ of the standardised normal distribution curve corresponding to $z = \pm 1.0$ is 0.2420

± z	*0.00*	*0.02*	*0.04*	*0.06*	*0.08*
0.0	0.3989	0.3989	0.3986	0.3982	0.3977
0.1	0.3970	0.3961	0.3951	0.3939	0.3925
0.2	0.3910	0.3894	0.3876	0.3857	0.3836
0.3	0.3814	0.3790	0.3765	0.3739	0.3712
0.4	0.3683	0.3653	0.3621	0.3589	0.3555
0.5	0.3521	0.3485	0.3448	0.3410	0.3372
0.6	0.3332	0.3292	0.3251	0.3209	0.3166
0.7	0.3123	0.3079	0.3034	0.2989	0.2943
0.8	0.2897	0.2850	0.2803	0.2756	0.2709
0.9	0.2661	0.2613	0.2565	0.2516	0.2468
1.0	0.2420	0.2371	0.2323	0.2275	0.2227
1.1	0.2179	0.2131	0.2083	0.2036	0.1989
1.2	0.1942	0.1895	0.1849	0.1804	0.1758
1.3	0.1714	0.1669	0.1626	0.1582	0.1539
1.4	0.1497	0.1456	0.1415	0.1374	0.1334
1.5	0.1295	0.1257	0.1219	0.1182	0.1145
1.6	0.1109	0.1074	0.1040	0.1006	0.0973
1.7	0.0940	0.0909	0.0878	0.0848	0.0818
1.8	0.0790	0.0761	0.0734	0.0707	0.0681
1.9	0.0656	0.0632	0.0608	0.0584	0.0562
2.0	0.0540	0.0519	0.0498	0.0478	0.0459
2.1	0.0440	0.0422	0.0404	0.0387	0.0371
2.2	0.0355	0.0339	0.0325	0.0310	0.0297
2.3	0.0283	0.0270	0.0258	0.0246	0.0235
2.4	0.0224	0.0213	0.0203	0.0194	0.0184
2.5	0.0175	0.0167	0.0158	0.0151	0.0143
2.6	0.0136	0.0129	0.0122	0.0116	0.0110
2.7	0.0104	0.0099	0.0093	0.0088	0.0084
2.8	0.0079	0.0075	0.0071	0.0067	0.0063
2.9	0.0060	0.0056	0.0053	0.0050	0.0047
3.0	0.0044	0.0042	0.0039	0.0037	0.0035

Table 22.2 *Integral Gaussian Probability P(z)*

Each entry in the table indicates the area under the standard normal curve from 0 to z

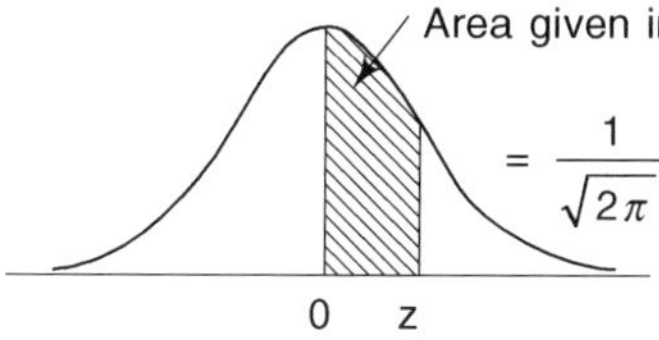

$$\text{Area given in the table} = |p(z)|_0^z = \frac{1}{\sqrt{2\pi}} \int_0^z \exp\left(-\frac{z^2}{2}\right) dz$$

To illustrate: area under the standard normal curve between the maximum ordinates and a point 1.96 standard deviations away is 0.4750

z	*0.00*	*0.01*	*0.02*	*0.03*	*0.04*	*0.05*	*0.06*	*0.07*	*0.08*	*0.09*
0.0	0.0000	0.0040	0.0080	0.0120	0.0160	0.0199	0.0239	0.0279	0.0319	0.0359
0.1	0.0398	0.0438	0.0478	0.0517	0.0557	0.0596	0.0636	0.0675	0.0714	0.0753
0.2	0.0793	0.0832	0.0871	0.0910	0.0948	0.0987	0.1026	0.1064	0.1103	0.1141
0.3	0.1179	0.1217	0.1255	0.1293	0.1331	0.1368	0.1406	0.1443	0.1480	0.1517
0.4	0.1554	0.1591	0.1628	0.1664	0.1700	0.1736	0.1772	0.1808	0.1844	0.1879
0.5	0.1915	0.1950	0.1985	0.2019	0.2054	0.2088	0.2123	0.2157	0.2190	0.2224
0.6	0.2258	0.2291	0.2324	0.2357	0.2389	0.2422	0.2454	0.2486	0.2518	0.2549
0.7	0.2580	0.2612	0.2642	0.2673	0.2704	0.2734	0.2764	0.2794	0.2823	0.2852
0.8	0.2881	0.2910	0.2939	0.2967	0.2996	0.3023	0.3051	0.3078	0.3106	0.3133
0.9	0.3159	0.3186	0.3212	0.3238	0.3264	0.3289	0.3315	0.3340	0.3365	0.3389
1.0	0.3413	0.3438	0.3461	0.3485	0.3508	0.3531	0.3554	0.3577	0.3599	0.3621
1.1	0.3643	0.3665	0.3686	0.3708	0.3729	0.3749	0.3770	0.3790	0.3810	0.3830
1.2	0.3849	0.3869	0.3888	0.3907	0.3925	0.3944	0.3962	0.3980	0.3997	0.4015
1.3	0.4032	0.4049	0.4066	0.4082	0.4099	0.4115	0.4131	0.4147	0.4162	0.4177
1.4	0.4192	0.4207	0.4222	0.4236	0.4251	0.4265	0.4279	0.4292	0.4306	0.4319
1.5	0.4332	0.4345	0.4357	0.4370	0.4382	0.4394	0.4406	0.4418	0.4429	0.4441
1.6	0.4452	0.4463	0.4474	0.4484	0.4495	0.4505	0.4515	0.4525	0.4535	0.4545
1.7	0.4554	0.4564	0.4573	0.4582	0.4591	0.4599	0.4608	0.4616	0.4625	0.4633
1.8	0.4641	0.4649	0.4656	0.4664	0.4671	0.4678	0.4686	0.4693	0.4699	0.4706
1.9	0.4713	0.4719	0.4726	0.4732	0.4738	0.4744	0.4750	0.4756	0.4761	0.4767
2.0	0.4772	0.4778	0.4783	0.4788	0.4793	0.4798	0.4803	0.4808	0.4812	0.4817
2.1	0.4821	0.4826	0.4830	0.4834	0.4838	0.4842	0.4846	0.4850	0.4854	0.4857
2.2	0.4861	0.4864	0.4868	0.4871	0.4875	0.4878	0.4881	0.4884	0.4887	0.4890
2.3	0.4893	0.4896	0.4898	0.4901	0.4904	0.4906	0.4909	0.4911	0.4913	0.4916
2.4	0.4918	0.4920	0.4922	0.4925	0.4927	0.4929	0.4931	0.4932	0.4934	0.4936
2.5	0.4938	0.4940	0.4941	0.4943	0.4945	0.4946	0.4948	0.4949	0.4951	0.4952
2.6	0.4953	0.4955	0.4956	0.4957	0.4959	0.4960	0.4961	0.4962	0.4963	0.4964
2.7	0.4965	0.4966	0.4967	0.4968	0.4969	0.4970	0.4971	0.4972	0.4973	0.4974
2.8	0.4974	0.4975	0.4976	0.4977	0.4977	0.4978	0.4979	0.4979	0.4980	0.4981
2.9	0.4981	0.4982	0.4982	0.4983	0.4984	0.4984	0.4985	0.4985	0.4986	0.4986
3.0	0.4987	0.4987	0.4987	0.4988	0.4988	0.4989	0.4989	0.4989	0.4990	0.4990
3.1	0.4990	0.4991	0.4991	0.4991	0.4992	0.4992	0.4992	0.4992	0.4993	0.4993
3.2	0.4993	0.4993	0.4994	0.4994	0.4994	0.4994	0.4994	0.4995	0.4995	0.4995
3.3	0.4995	0.4995	0.4995	0.4996	0.4996	0.4996	0.4996	0.4996	0.4996	0.4997
3.4	0.4997	0.4997	0.4997	0.4997	0.4997	0.4997	0.4997	0.4997	0.4998	0.4998
3.5	0.4998	0.4998	0.4998	0.4998	0.4998	0.4998	0.4998	0.4998	0.4998	0.4998
3.6	0.4998	0.4998	0.4999	0.4999	0.4999	0.4999	0.4999	0.4999	0.4999	0.4999
3.7	0.4999	0.4999	0.4999	0.4999	0.4999	0.4999	0.4999	0.4999	0.4999	0.4999
3.8	0.4999	0.4999	0.4999	0.4999	0.4999	0.4999	0.4999	0.5000	0.5000	0.5000
3.9	0.5000	0.5000	0.5000	0.5000	0.5000	0.5000	0.5000	0.5000	0.5000	0.5000

22.5 CONFIDENCE LEVEL

As mentioned earlier, the area under the normal distribution curve between $-\infty$ and ∞ is unity. This should indeed be so because the probability of all possible measured values lying between $-\infty$ and ∞ has to be unity. In actual practice, we generally specify a certain range of permissible values of scatter or dispersion from the mean value and determine the probability that the measured value lies in that range. This probability can be evaluated by finding the area under the normal distribution curve in the specified range. When this probability is expressed as a percentage, it is termed as *confidence level.*

For example, we wish to predict how much is the probability of occurrence of the measured value x in the range ($\bar{X} + \sigma$) to ($\bar{X} - \sigma$). Obviously, it is the area under the normal distribution curve in the specified range of ($\bar{X} + \sigma$) to ($\bar{X} - \sigma$). Using Table 22.2 we find that the integral Gaussian probability in the range of ($\bar{X} \pm \sigma$) (alternatively between $z = \pm 1$) is 0.6826. Therefore, we can say that chances are better than 2 : 1 (actually 68.26 : 31.74) that the measured value lies between $\bar{X} \pm \sigma$. Alternatively we may say that the confidence level in such a case is 68.26%.

Using the integral Gaussian table, the confidence levels for other cases can also be determined. Some typical values of confidence levels are shown in Fig. 22.6.

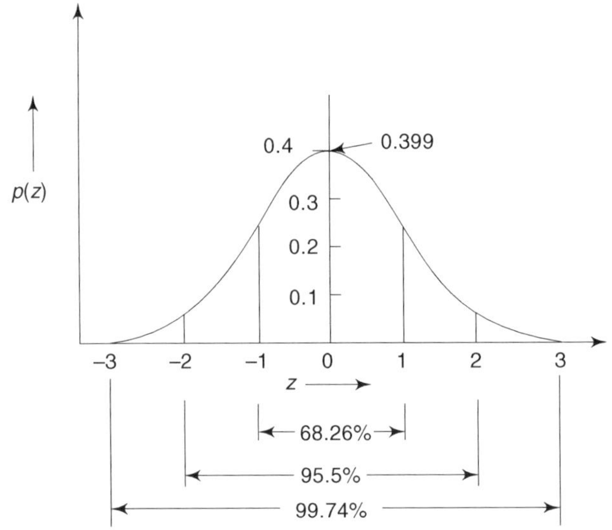

Fig. 22.6 *Confidence levels for different ranges of measured values*

In any experiment, we generally try to set the confidence limits within which we expect the measured value to lie within a given probability [Figs. 22.7(a), 22.7(b) and 22.7(c)]. To do this, we must decide the probability of error that we are willing to accept. The percentage probability of error is defined as 100 minus the confidence level. In general, we accept errors up to 5%, i.e. 95% confidence level for which $z = \pm 1.96$ (from Table 22.2) for two-sided confidence and $z = 1.645$ for one-sided confidence. However, where human life is involved, we insist on low probability of error, of the order of 1%. This gives a confidence level of 99% for which $z = \pm 2.58$ for two-sided confidence and 2.326 for one-sided confidence.

In practice, we come across two types of problems. Some are *direct* in which the limits of x are given and the probability of occurrence in this range is required to be determined. Whereas the other problems are of inverse type in which the probability of occurrence, i.e. confidence level is usually taken as 95% and the limits of x are required to be determined.

Problem 22.1 *Analysis of the data of the machine components having completed their anticipated useful lives due to wear and tear usually follow the normal distribution pattern. If the components of a particular type, in one of the samples have a mean wear out life $\bar{X}$ of 1000 h with a standard deviation σ of 25 h, determine the proportion of components that would have wear out life in hours (a) greater than 1050 and (b) between 950 and 1025.*

Solution In such problems, we basically determine the area under the normal distribution curve pertaining to the particular region indicated by the problem. Therefore, it is usually preferable to represent the data pictorially by sketching a normal distribution curve.

(a) *Greater than $x = 1050$ h*

In a given sample, the mean wear out life, $\bar{X} = 1000$ h and standard deviation $\sigma = 25$ h.

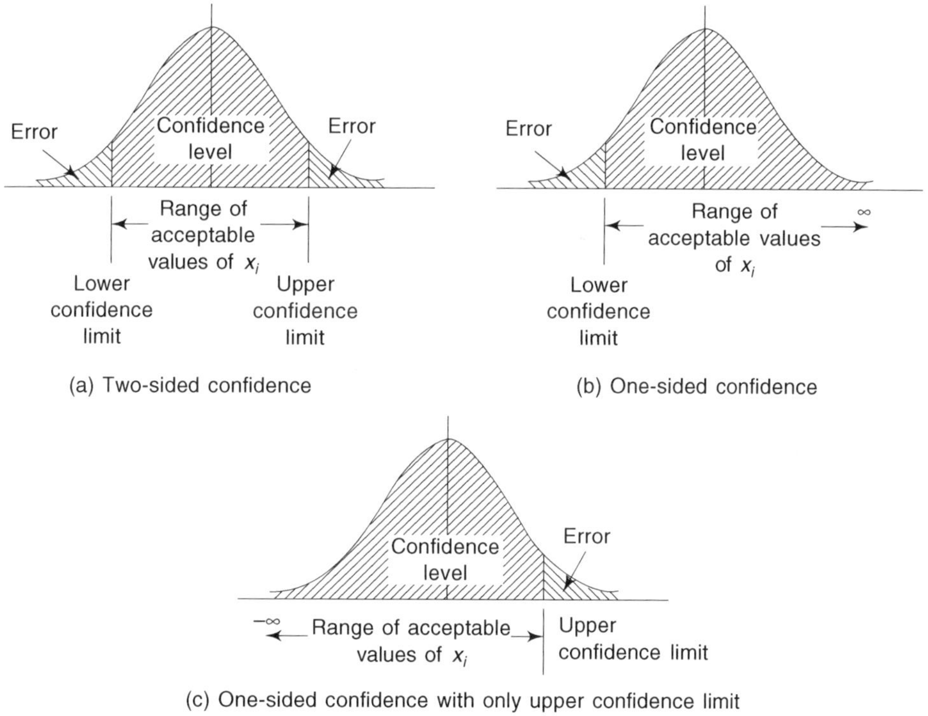

Fig. 22.7 *Pictorial representation of Gaussian distribution with given confidence limits*

Standard normal variate z_1 at the cut-off value of x_1 equal to 1050 h

$$= \frac{x - \bar{X}}{\sigma}$$

$$= \frac{1050 - 1000}{25} = 2.0$$

It is now clear from Fig. 22.8(a) that the proportion of components having a life greater than 1050 h corresponds to the area under the normal distribution curve from $z_1 = 2.0$ to $z_2 = \infty$. This area is equal to the area of the normal distribution curve between $z = 0$ and $z_2 = \infty$ minus the area between $z = 0$ and $z_1 = 2.0$.

In other words, the required integral Gaussian probability is:

$$|P(Z)|_{2.0}^{\infty} = |P(Z)|_{0}^{\infty} - |P(Z)|_{0}^{2}$$

$$= 0.50 - 0.4772 \qquad \text{(From Table 22.2)}$$

$$= 0.0228$$

Hence, we can say that 2.28% of the machine components have life greater than 1050 h.

(b) *Between 950 and 1025 h*

Here, $x_3 = 950$ h and $x_4 = 1025$ h

$\therefore$ Corresponding $z_3 = \dfrac{950 - 1000}{25} = -2.0$

and $z_4 = \dfrac{1025 - 1000}{25} = 1.0$

The shaded area in Fig. 22.8(b) represents the proportion of components having a life between 950 and 1025 h.

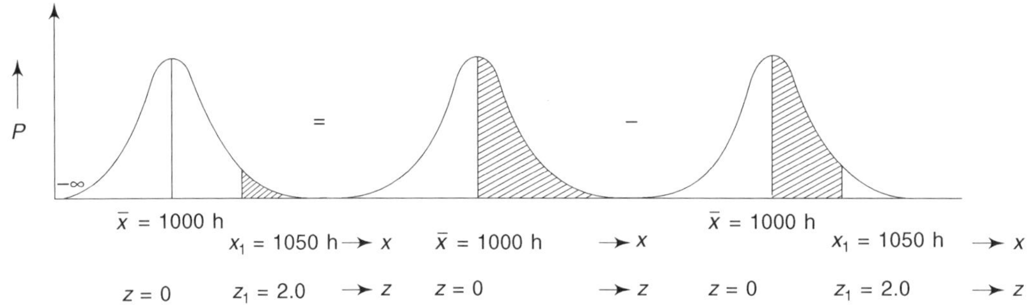

Fig. 22.8(a) *Figure for Problem 22.1*

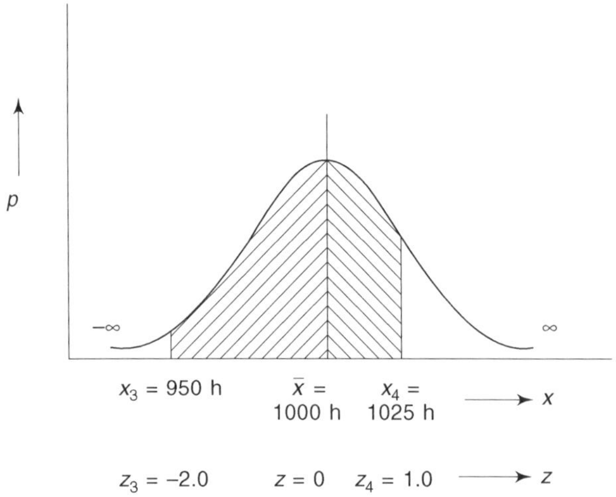

Fig. 22.8(b) *Figure for Problem 22.1*

Therefore, the required integral Gaussian probability is:

$$\left|P(Z)\right|_{-2.0}^{1.0} = \left|P(Z)\right|_{-2.0}^{0} + \left|P(Z)\right|_{0}^{1.0}$$

$$= 0.4772 + 0.3413 \text{ (From Table 22.2)} = 0.8185$$

Hence, the proportion of components having life between 950 and 1025 h is 81.85%.

Problem 21.2 *A study has indicated that the life of TV picture tubes manufactured by a certain firm is normally distributed with a mean life of 5 years (1 year = 365 days) and a standard deviation of 500 days. The manufacturer gives a guarantee of 1 year. Determine (a) what percentage of picture tubes will he have to replace in 1 year? (b) If the manufacturer wishes to replace the same amount of picture tubes with 2 years guarantee, what should he do?*

Solution (a) The mean life of the TV picture tube,

$$\bar{X} = 5 \times 365 = 1825 \text{ days}$$

Standards deviation, $\sigma = 500$ days

Guarantee period, $x_1 = 365$ days

The standard normal variate for the cut-off value of 365 days, z_1

$$= \frac{365 - 1825}{500} = -2.92$$

The given data can be represented pictorially on the normal distribution curve as shown in Fig. 22.9.

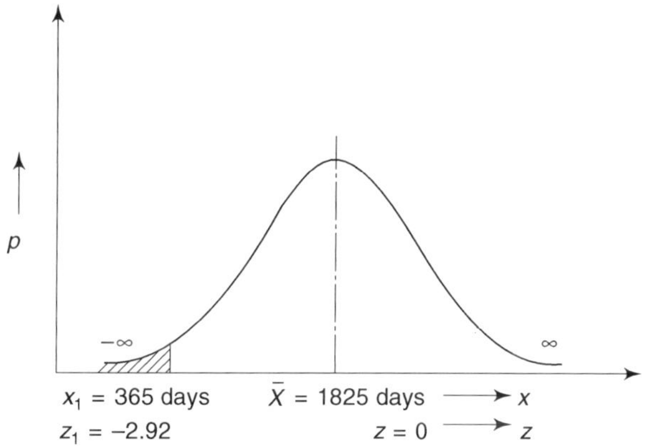

Fig. 22.9 *Figure for Problem 22.2*

The proportion of TV picture tubes the manufacturer has to replace is indicated by the shaded area on the normal distribution diagram.

Therefore, the required integral Gaussian probability between $z = -2.92$ to $z = -\infty$ is determined from Table 22.2 as follows:

$$\left|P(Z)\right|_{-\infty}^{-2.92} = \left|P(Z)\right|_{0}^{-\infty} - \left|P(Z)\right|_{0}^{-2.92}$$

$$= 0.5 - 0.4982 = 1.8 \times 10^{-3}$$

Hence, the percentage of picture tubes to be replaced with 1 year guarantee is 0.18.

(b) If the manufacturer wants to replace the same percentage of picture tubes, then he can do one of the following:

(i) Improve the quality, i.e.

$$-\frac{\bar{X} - 730}{500} = -2.92$$

$$\bar{X} = 2190 \text{ days} = 6 \text{ years}$$

(ii) Increase the precision with which he manufacturers the picture tubes, i.e.

$$-\frac{730 - 1825}{\sigma} = -2.92$$

$$\sigma = 375 \text{ days}$$

(iii) Any combination of the above two factors satisfying the relation:

$$-\frac{\bar{X} - 730}{\sigma} = -2.92$$

Problem 22.3 *An assembly of shaft and hub is used in many engineering situations. They are selected at random from a large supply with manufacturer's specifications of mean shaft diameter of 31.0 mm and standard deviation σ_1 of 0.04 mm, and with mean hub diameter of 31.1 mm and standard deviation σ_2 of 0.03 mm. Determine the number of times we shall have a satisfactory fit out of 250 cases, if a satisfactory fit is defined as one having a diameteral clearance of at least 0.03 mm and at the most 0.18 mm.*

Solution Mean diameteral clearance

$$\varepsilon = (\text{mean hub diameter}) - (\text{mean shaft diameter}) = 31.1 - 31 = 0.1 \text{ mm}$$

Overall standard deviation in diameteral clearance

$$\sigma(\varepsilon) = \sqrt{\sigma^2_{\text{hub diameter}} + \sigma^2_{\text{shaft diameter}}}$$

$$= \sqrt{0.04^2 + 0.03^2} = 0.05 \text{ mm}$$

The mean, maximum and minimum clearance are shown in normal distribution curve in Fig. 22.10.

Now, the proportion of satisfactory fits is given by the area of the normalised Gaussian curve between the minimum and maximum clearance.

$$Z_{\max} = \frac{\varepsilon_{\max} - \bar{\varepsilon}}{\sigma(\varepsilon)} = \frac{0.18 - 0.1}{0.05} = 1.6$$

and

$$Z_{\min} = \frac{\varepsilon_{\min} - \bar{\varepsilon}}{\sigma(\varepsilon)} = \frac{0.03 - 0.1}{0.05} = -1.4$$

From Table 22.2 of the integral Gaussian probability we get,

$$\left|P(Z)\right|_0^{1.6} = 0.4452$$

and

$$\left|P(Z)\right|_{-1.4}^{0} = 0.4192$$

$\therefore$ % of satisfactory clearance = 100 {0.4452 + 0.4192} = 86.44

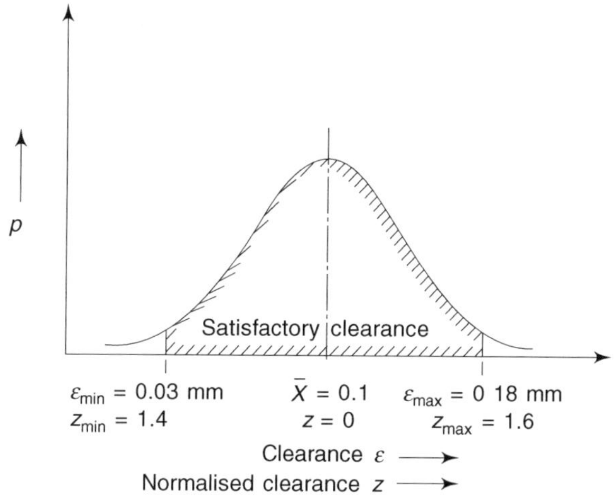

Fig. 22.10 *Figure for Problem 22.3*

Hence, the number of times we have a satisfactory fit out of 250 cases

$$= \frac{86.44}{100} \times 250$$

$$= 216.1 \simeq 216$$

Problem 22.4 *Structural engineers sometimes apply a factor of safety to the statistical minimum failing stress. They usually employ the criteria that 95% of the tests should exceed the minimum failing stress. For a certain specimen of timber, 100 tests for strength properties were carried out which were found to be normally distributed with a mean failing stress of 30.25 N/mm² and standard deviation of 3.95 N/mm². Determine the value of the minimum failing stress.*

Solution The strength properties of the timber can be shown on the normal distribution curve as in Fig. 22.11.

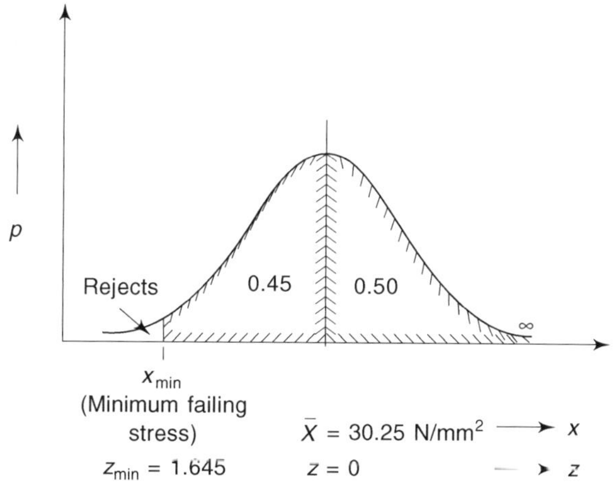

Fig. 22.11 *Figure for Problem 22.4*

In this case one-sided confidence level is 95% which $z = -1.645$ (Table 22.2).

$$\frac{x_{\min} - \bar{X}}{\sigma} = -1.645$$

or
$$\frac{x_{\min} - 30.25}{3.95} = -1.645$$

which gives $x_{\min}$, the minimum failing stress as 23.75 N/mm^2.

22.6 ■ CENTRAL LIMIT THEOREM

The *central limit theorem* is an important statistical theorem. It states that 'the sample means of a population do follow Gaussian distribution, whatever the distribution of the individual measurements'. Therefore, the standard normal variate z for the sample means population can be defined as:

$$z = \frac{\bar{X} - \bar{\bar{X}}}{\sigma_m} \tag{22.22}$$

where $\bar{X}$ is the mean value of any sample

$\bar{\bar{X}}$ is the population mean

and σ_m is the standard deviation of means which is very nearly equal to the internal estimate of uncertainty (or internal standard error) U_i, when the number of observations is large

It may be noted that the normal distribution of measurements of a sample has lesser precision as compared to the corresponding normal distribution of the sample means of the population. In other words, the latter distribution has a sharper peak than the former (Fig. 22.12). This is because of the fact that

$$\sigma_m = \frac{\sigma}{\sqrt{n}} \tag{22.23}$$

In most casts, we are interested in the analysis of internal standard error of the data and employ the standard normal variate z with respect to the population mean and the standard deviation of the means. However, in some problems, we are interested in the standard deviation of the sample data. In such cases, the value of z is calculated using the sample mean and the sample standard deviation.

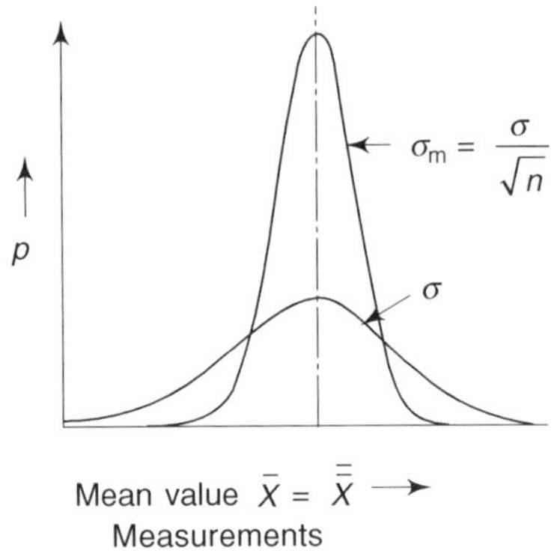

Fig. 22.12 *A typical distribution showing the measurements of the sample means of a population*

Problem 22.5 *In a manufacturing process, the time required to complete a certain electronic component was to be studied. The time needed had a mean of 75 min. and a standard deviation of 10 min. for the case of 25 randomly selected components in a sample. Determine (a) population mean, (b) standard deviation of the means, and (c) the size of the sample if the internal standard error (or internal estimate of uncertainty) is not to exceed 1 min.*

Solution (a) The population mean in nearly equal to the sample mean:

$$\therefore \quad \bar{\bar{X}} = 75 \text{ min.}$$

(b) Standard deviation of mean $\sigma_m = \dfrac{\sigma}{\sqrt{n-1}}$

$$= \frac{10}{\sqrt{25-1}} \simeq 2 \text{ min.}$$

(c) Internal standard error $U_i = \dfrac{\sigma}{\sqrt{n-1}}$

If $U_i = 1$ min.

then $1 = \dfrac{10}{\sqrt{(n-1)}}$

which gives n, the size of sample = 101.

Problem 22.6 *A firm manufactures ball bearings for a certain application. Several samples of size n were taken at random for a population of a large number of balls with mean $\bar{X}$ and standard deviation σ. What is the range within which the sample mean $\bar{X}$ can lie with 95% confidence level?*

Solution Since the samples have been taken from a large population, their means follow the normal distribution (as per the central limit theorem) with mean $\bar{\bar{X}} = \bar{X}$ and standard deviation $\sigma_m = \sigma/\sqrt{n}$. For two-sided confidence level of 95%, the value of standard normal variate is

$$Z = \pm 1.96$$

or $$\frac{\bar{X} - \bar{\bar{X}}}{\sigma_m} = \pm 1.96$$

or $$\bar{X} = \bar{\bar{X}} \pm 1.96 \frac{\sigma}{\sqrt{n}}$$

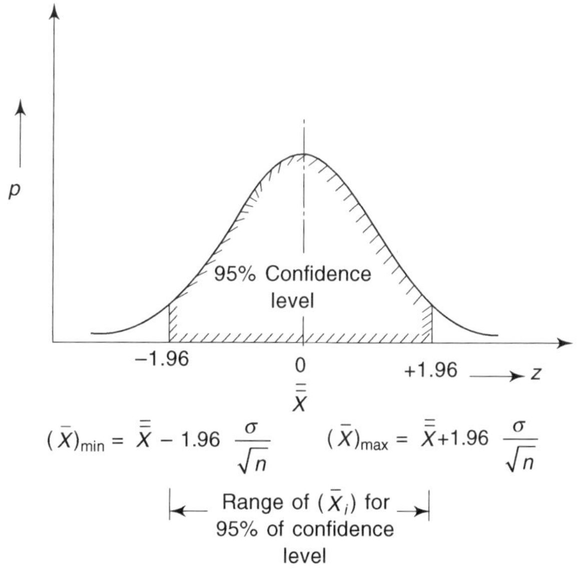

Fig. 22.13 *Figure for Problem 22.6*

which gives

$$(\bar{X})_{max} = \bar{\bar{X}} - 1.96 \frac{\sigma}{\sqrt{n}}$$

and

$$(\bar{X})_{min} = \bar{\bar{X}} - 1.96 \frac{\sigma}{\sqrt{n}}$$

Hence, the length of abscissa shown in Fig. 22.13 between $(\bar{X})_{max}$ and $(\bar{X})_{min}$ gives the required range within which the sample mean $\bar{X}$ would be with 95% confidence level.

22.7 ■ SIGNIFICANCE TEST

If the normal distribution of a particular quantity is known and the quantity is measured again under somewhat changed conditions, the mean value is unlikely to be the mean of the original distribution. If the difference in the means is small, it would be reasonable to assume that the distribution is from the same population. On the other hand, if the difference is considerable, then it would be reasonable to assume that the changed circumstances have altered the values and the result is significant. In other words, we can say that the original data and the subsequent data taken under somewhat changed conditions are not from the same population.

To test whether there is a significant change in the original and subsequent data, we use a *significance* test based on the difference of means. The statistician's criterion of the significance test is that if the difference in the mean values of the samples deviates 1.96 times the internal standard error of the difference of means, then the change is significant.

Let us say that a particular sample has n_1 quantities which gives the mean value as $\bar{X}_1$ with a standard deviation of σ_1. Similarly, another sample under somewhat changed conditions has n_2 quantities which give the mean value as $\bar{X}_2$ with a standard deviation σ_2.

It should be recalled here that the internal standard error (or the internal estimate of uncertainty) of a sample of n quantities with standard deviation σ is given by

$$U = \frac{\sigma}{\sqrt{(n-1)}} \tag{22.24}$$

Keeping this in mind, we proceed as follows to determine whether the change is significant or not.

1. Find the difference in mean values which gives the range of variation of the mean values, i.e.,

$$R(\bar{X}) = \bar{X}_1 - \bar{X}_2 \tag{22.25}$$

2. Determine the internal standard error of $R(\bar{X})$, i.e.,

$$\{(U^2)\}_{R(\bar{X})} = \left[\frac{\partial R(\bar{X})}{\partial(\bar{\bar{X}}_1)}\right]^2 \cdot U_1^2 + \left[\frac{\partial R(\bar{X})}{\partial \bar{X}_2}\right]^2 \cdot U_2^2$$

$$= U_1^2 + U_2^2 \tag{22.26}$$

Substituting the values of U_1 and U_2 in terms of the corresponding standard deviations of the samples we get,

$$(U)_{R(\bar{X})} = \sqrt{U_1^2 + U_2^2}$$

$$= \left[\left\{ \frac{\sigma_1}{\sqrt{(n_1 - 1)}} \right\}^2 + \left\{ \frac{\sigma_2}{\sqrt{(n_2 - 1)}} \right\}^2 \right]^{1/2} \quad (22.27)$$

3. Knowing the values of $R(\bar{X})$ and $(U)_{R(\bar{X})}$ from Eqs. (22.25) and (22.27), respectively, we now employ the significance test criterion which is as follows:

   If $$R(\bar{X}) > |1.96(U)_{R(\bar{X})}| \quad (22.28)$$

   then the difference of means is significant, otherwise, we can assume that the original and subsequent distributions are from the same population.

In other words, we can say that up to 95% confidence level, the means of the two samples are not from the same population if the significant test given by Eq. (22.28) is positive. Conversely, if the difference in means $R(\bar{X})$ lies within 1.96 times the internal standard error in $R(\bar{X})$, then the result is insignificant and we can say that probability of error is less than 5% in considering both the samples to be from the same population. Further, if the difference of means deviates more than 2.58 times the combined internal standard error in $R(\bar{X})$, then the result is highly significant and in this case, the confidence level considered is 99%.

Problem 22.7 *Ten specimens of mild steel are chemically analysed for carbon content in two different laboratories. The percentage of carbon content contained are as follows:*

Laboratory A	0.23	0.23	0.21	0.25	0.28
	0.24	0.20	0.18	0.23	0.24
Laboratory B	0.21	0.19	0.24	0.20	0.22
	0.18	0.20	0.22	0.25	0.17

Test the hypothesis that there is no significant difference in the two laboratories in their determination of percentage of carbon.

Solution From the given data we find that:

The mean value of percentage of carbon found in Laboratory A, $\bar{X}_A = 0.229$ and the standard deviation in the values in Laboratory A, $\sigma_A = 0.02625$.

Similarly in Laboratory B, $\bar{X}_B = 0.208$ and the standard deviation in the percentage of carbon $\sigma_B = 0.024$.

The internal estimate of uncertainty $$U_A = \frac{0.02625}{\sqrt{(10-1)}} = 0.00875$$

and the internal estimate of uncertainty, $$U_B = \frac{0.024}{\sqrt{(10-1)}} = 0.008$$

Combined internal estimate of uncertainty of difference in means,

$$U_{R(\bar{X})} = \sqrt{(0.00875^2 + 0.008^2)}$$
$$= 0.011857$$

The range of mean values $$R(\bar{X}) = \bar{X}_A - \bar{X}_B$$
$$= 0.229 - 0.204 = 0.021$$

Assuming 95% confidence level for the significance test, we get,

the allowable range of combined internal standard error in $R(\bar{X}) = 1.96\ U_{R(\bar{X})}$

$$= 1.96 \times 0.011857 = 0.0232$$

Now, it is obvious that the difference in means $R(\bar{X})$ lies within 1.96 times the combined internal estimate of uncertainty $U_{R(\bar{X})}$. Hence, we can say that there is no significant difference in the determination of percentage of carbon in the specimens of mild steel, carried out in the two laboratories.

22.8 ▮ CHI-SQUARE TEST FOR GOODNESS OF FIT

When a set of measurements is obtained, it is believed that the measurements are a sample of some known theoretical distribution; say normal frequency distribution which is generally hypothesised in cases involving experimental statistics. For comparing the different parts of the observed distribution, we subdivide the data into a number of classes say n and determine the observed frequency in each class. Then we estimate the expected frequency of each class by assuming that the distribution conforms to a particular theoretical distribution.

For example, if the assumed distribution considered is Gaussian, then the following procedure may be adopted to calculate the expected values of frequencies for a given set of data

1. Calculate the mean value and standard deviation of the data.
2. For each class interval, calculate the standard normal variates z_u and z_l for the upper and lower boundary values, respectively.
3. From the integral Gaussian table, determine the integral Gaussian probability between 0 and z_u and 0 and z_l.
4. The difference in the above values gives the integral Gaussian probability in the given interval if both the upper and lower boundaries lie either between 0 and ∞ or 0 and $-\infty$. The sum of these values gives the integral Gaussian probability if the upper boundary lies between 0 and ∞ and the lower boundary lies between 0 and $-\infty$ and vice versa.
5. Multiplying the integral Gaussian probability in a given class interval by the total number of observations gives the expected frequency of occurrence of the variable in that interval.
6. The summation of the expected frequencies in all classes sometimes does not equal the total number of observations. The slight difference is caused by small rounding-off errors due to interpolations in the integral Gaussian table. Therefore, the expected frequencies in step (5) are multiplied by a suitable correction factor so as to make the sum of expected frequencies equal to the number of observations.

After determining the expected frequencies in the various classes, we determine the χ^2 (pronounced chi-square)-parameter as follows:

Let us say that there are n classes ($n > 1$) and the expected and observed frequencies in the various classes are denoted by

$$f_{e_1},\ f_{e_2},\ f_{e_3},\ \ldots f_{e_n} \quad \text{and} \quad f_{o_1},\ f_{o_2},\ f_{o_3},\ \ldots f_{o_n}$$

Now, our aim is to determine whether the observed frequencies and the expected frequencies are close enough for us to conclude that they come from the same probability distribution. To do so, we define the χ^2-parameter as:

$$\chi^2_{(n-m)\text{d.f.}} = \sum_{i=1}^{n} \left\{ \frac{(f_{oi} - f_{ei})^2}{f_{ei}} \right\} \tag{22.29}$$

where n is the number of values that are summed up to produce the values of χ^2
m is the number of constants used in the calculation of expected frequencies
$n - m$ is the degrees of freedom and
subscript d.f. stands for the degrees of freedom

The values of the numerator in the χ^2 (chi-square) expression represent the squares of deviations between the expected and observed frequencies in various classes which is always positive. These values are normalised in each class by dividing them by the respective expected frequency of each class. It may be noted that the same order of deviation in the expected and observed frequencies causes relatively larger contribution in the χ^2-parameter at the tail portions of the normally distributed data, as compared to the values close to the mean value of the data. This is because of relatively large values of the expected frequencies near the mean value of the data which happens to be in the denominator of the χ^2-parameter. In order to restrict the unusually large contributions in χ^2-parameter when the expected frequencies are small, the empirical criterion commonly used in practice is to regroup the various classes in such a way so that the expected frequency in each class is not less than 5.0.

Further, a correction is sometimes applied to the chi-square values when the degree of freedom F, i.e. $(n - m)$ is of the order of 1.0. This is termed as *Yate's correction* and accounts for the inaccuracies involved when the results of continuous distributions are applied to discrete data. The correction consists of writing Eq. (22.29) in the following form:

$$\chi^2_{F=1.0} = \sum_{i=1}^{n}\left[\frac{\{|(f_{oi} - f_{ei})| - 0.5\}^2}{f_{ei}}\right] \tag{22.30}$$

If the sample distribution agrees with the assumed theoretical distribution then $\chi^2 = 0$. This is of course very unlikely because even if the sample is taken from the parent distribution, one would not expect *exact* agreement in every interval. But, larger the value of χ^2, the more is the disagreement between the assumed distribution and the observed values. In other words, in such a case, the smaller is the probability that the observed distribution matches the expected distribution. Thus, the chi-square parameter is quite useful in statistical analysis of data as it helps to test a particular hypothesis in the given data.

In applying the chi-square test, we first determine the value of χ^2 for the given data. Then, we determine the values of degrees of freedom F which is equal to $(n - m)$. Knowing the values of χ^2 and F, we determine the probability that the actual measurements match the expected distribution from either the chi-square tables (Table 22.3) or from the χ^2-F diagram (Fig. 22.14) which gives cross-plots of chi-square probability, $P(\chi^2)$, for various values of χ^2 and F.

22.9 ■ CRITERIA FOR GOODNESS OF FIT

The statistical criteria for the goodness of fit, i.e., how well as set of observed data fit the assumed theoretical distributions are as follows:

1. If the value of probability in the χ^2-test lies between 0.1 and 0.9, then the observed distribution is considered to follow the assumed distribution. In other words, there is no reason to suspect the hypothesis. In certain cases, the lower limit of chi-square probability (also termed *significance level* or simply *level*) may be reduced to 0.05.
2. If the value of the probability in the χ^2-test is below the lower prescribed limit, then the result is significant and the sample data is considered to be entirely different from the assumed distribution. In such cases, the value of the χ^2-parameter is usually quite large.
3. If the value of the χ^2-parameter is nearly zero or very small, then the probability may exceed the upper limit of 0.9. Such cases are hardly encountered in practice. If it is so, then we normally consider the data to be *suspiciously good*.

Table 22.3 *χ^2–F Table indicating the Probability $P(\chi^2)$*

This table gives the values of χ^2 which have various probabilities of being exceeded by a sample taken from the given parent distribution. The number of degrees of freedom is *F*. to illustrate: for a sample with 6 degrees of freedom, the probability, $P(\chi^2)$ is 0.95 if χ^2 = 1.635 and 0.1 if χ^2 = 10.645.

$F\backslash P(\chi^2)$	0.99	0.95	0.90	0.80	0.70	0.50	0.30	0.20	0.10	0.05	0.01	0.001
1	0.000157	0.00393	0.0158	0.0642	0.148	0.455	1.074	1.642	2.706	3.841	6.635	10.827
2	0.0201	0.103	0.211	0.446	0.713	1.386	2.408	3.219	4.605	5.991	9.210	13.815
3	0.115	0.352	0.584	1.005	1.424	2.366	3.665	4.642	6.251	7.815	11.341	16.268
4	0.297	0.711	1.064	1.649	2.195	3.357	4.878	5.989	7.779	9.488	13.277	18.465
5	0.554	1.145	1.610	2.343	3.000	4.351	6.064	7.289	9.236	11.070	15.086	20.517
6	0.872	1.635	2.204	3.070	3.828	5.348	7.231	8.558	10.645	12.592	16.812	22.457
7	1.239	2.167	2.833	3.822	4.671	6.346	8.383	9.803	12.017	14.067	18.475	24.322
8	1.646	2.733	3.490	4.594	5.527	7.344	9.524	11.030	13.362	15.507	20.090	26.125
9	2.088	3.325	4.168	5.380	6.393	8.343	10.656	12.242	14.684	16.919	21.666	27.877
10	2.558	3.940	4.865	6.179	7.267	9.342	11.781	13.442	15.987	18.307	23.209	29.588
11	3.053	4.575	5.578	6.989	8.148	10.341	12.899	14.631	17.275	19.675	24.725	31.264
12	3.571	5.226	6.304	7.807	9.034	11.340	14.011	15.812	18.549	21.026	26.217	32.909
13	4.107	5.892	7.042	8.634	9.926	12.340	15.119	16.985	19.812	22.362	27.688	34.528
14	4.660	6.571	7.790	9.467	10.821	13.339	16.222	18.151	21.064	23.685	29.141	36.123
15	5.229	7.261	8.547	10.307	11.721	14.339	17.322	19.311	23.307	24.996	30.578	37.697
16	5.812	7.962	9.312	11.152	12.624	15.338	18.418	20.465	23.542	26.296	32.000	39.252
17	6.408	8.672	10.085	12.002	13.531	16.338	19.511	21.615	24.769	27.587	33.409	40.790
18	7.015	9.390	10.865	12.857	14.440	17.338	20.601	22.760	25.989	28.869	34.805	42.312
19	7.633	10.117	11.651	13.716	15.352	18.338	21.689	23.900	27.204	30.144	36.191	43.820
20	8.260	10.851	12.443	14.578	16.266	19.337	22.775	25.038	28.412	31.410	37.566	45.315
21	8.897	11.601	13.240	15.445	17.182	20.337	23.858	26.171	29.615	32.671	38.932	46.797
22	9.542	12.338	14.041	16.314	18.101	21.337	24.939	27.301	30.813	33.924	40.289	48.268
23	10.196	13.091	14.848	17.187	19.021	22.337	26.018	28.429	32.007	35.172	41.638	49.728
24	10.856	13.848	15.659	18.062	19.943	23.337	27.096	29.553	33.196	36.415	42.980	51.179
25	11.524	14.611	16.473	18.940	20.867	24.337	28.172	30.675	34.382	37.652	44.314	52.620
26	12.198	15.379	17.292	19.820	21.792	25.336	29.246	31.795	35.563	38.885	45.642	54.052
27	12.879	16.151	18.114	20.703	22.719	26.336	30.319	32.912	36.741	40.113	46.963	55.476
28	13.565	16.928	18.939	21.588	23.647	27.336	31.391	34.027	37.916	41.337	48.278	56.893
29	14.256	17.708	19.768	22.475	24.577	28.336	32.461	35.139	39.087	42.557	49.588	58.302
30	14.953	18.493	20.599	23.364	25.508	29.336	33.530	36.250	40.256	43.773	50.892	59.703

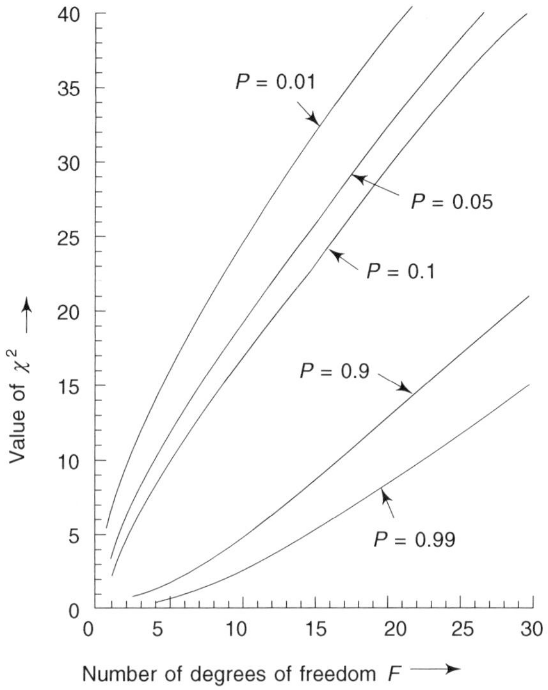

Fig. 22.14 *χ^2–F diagram indicating various values of χ^2 probability*

Problem 22.8 *The coefficient of friction between glass sheet and wood was measured in the laboratory by a technique that is free from systemic errors. The data obtained was as follows:*

Coefficient of friction values μ	*Observed frequency f_{o_i}*
0.44–0.46	3
0.46–0.48	10
0.48–0.50	12
0.50–0.52	16
0.52–0.54	10
0.54–0.56	6
0.56–0.58	3

Determine if the values of the coefficient of friction follow the Gaussian distribution or not. Test χ^2-values up to 10% level.

Solution We first determine the mean value and the standard deviation of the given data using the method of assumed mean or otherwise.

The mean value X of the coefficient of friction μ of the given data is 0.5067 and the standard deviation σ in the given data is 0.03062.

Using the integral Gaussian tables, we determine the integral Gaussian probabilities between 0 and z_l and 0 and z_u for different classes. From these values the integral Gaussian probability of each class $P(\Delta z)$ is determined. It may be noted that the summation of the integral Gaussian probabilities of all the classes is found to be generally very slightly less than unity. Therefore, the expected frequencies in the various classes are calculated by multiplying $P(\Delta z)$ by the total number of observations and a correction factor which is the reciprocal of $\Sigma P(\Delta z)$.

Sl. No.	Classes	f_{ol}	z_l	z_u	$P(z_l)$	$P(z_u)$	$P(\Delta z)$	f_{el}
1	0.44–0.46	3	–2.178	–1.525	0.4853	0.4364	0.0489	2.99
2	0.46–0.48	10	–1.525	–0.872	0.4364	0.3084	0.1280	7.83
3	0.48–0.50	12	–0.872	0.219	0.3084	0.0867	0.2217	13.57
4	0.50–0.52	16	–0.219	0.434	0.0867	0.1678	0.2545	15.57
5	0.52–0.54	10	0.434	1.088	0.1678	0.3617	0.1939	11.87
6	0.54–0.56	6	1.088	1.741	0.3617	0.4592	0.0975	5.97
7	0.56–0.58	3	1.741	2.394	0.4592	0.4952	0.0360	2.20
	Total	60					0.9805	60

In the above table, the expected frequencies of the first and last classes are less than 5. Therefore, these are combined with the adjacent classes to make them more than 5 and then the various ratios $(f_{oi} - f_{ei})^2/f_{el}$ are calculated as follows:

Sl. No.	f_{oi}	f_{ei}	$f_{oi} - f_{ei}$		$(f_{ol} - f_{ei})^2/f_{ei}$
1	13	10.82	2.18		0.439
2	12	13.57	–1.57		0.182
3	16	15.57	0.43		0.012
4	10	11.87	–1.87		0.295
5	9	8.17	0.83		0.084
				Total	1.012

$\therefore \quad \chi^2 = 1.012$

The number of degrees of freedom F are given by $F = n - m$

In this problem the number of terms which are summed to give χ^2, i.e. $n = 5$. Further, the number of constraints m is equal to the number of quantities obtained from the observations which are used in the calculation of expected frequencies. In the present problem $m = 3$ because there are three quantities namely, the total number of observations, the mean value and standard deviation of the data which have been used in the calculation of expected frequencies.

$$\therefore \quad F = 5 - 3$$
$$= 2$$

For 2 degrees of freedom, the value of χ^2 at 10% level of chi-square probability is 4.605 (from Table 22.3). Now, the value of χ^2 obtained from the given data, i.e. 1.012 is far from being large enough to justify the rejection of the normal distribution model. Alternatively, we can also state that there is no reason to suspect that the data of the coefficient of friction values obtained in the experiment follow the assumed normal distribution. This is because the value of $P(\chi^2)$ corresponding to $\chi^2 = 1.012$ and $F = 2$ is 0.62 (from Table 22.3) which lies between required range of 0.1 to 0.9.

22.10 ■ CONTINGENCY TABLES

There are certain experiments in which the observed frequencies occupy a single row or column. Such data is obtained due to the variation of a single variable and is therefore arranged in the form of *one-way classification table*. If the data has n number of columns, then this type of data has the form of $1 \times n$ (pronounced 1 by n) table.

Sometimes each data point of two or more similar experiments is sampled and classified with respect to certain stipulated variations of the conditions. Alternatively, the data points of a single experiment may be classified with respect to multiple variations of conditions. This results in a multi-way classification table or $m \times n$ table in which observed frequencies occupy m rows and n columns. Such tables are generally called *contingency* tables.

In the solution of contingency table problems, we generally put forward a hypothesis (called the *null* hypothesis) that helps us to determine the expected frequencies corresponding to the various observed frequencies according to the rules of probability. After this, we investigate the goodness of *fit* between the expected and observed frequencies as per the null hypothesis using the χ^2-test. We compute the χ^2-parameter and determine the degree of freedom in the given data. If the χ^2-parameter is greater than $(\chi^2)_{0.10\ \text{level}}$, then the result is considered significant, i.e. the null hypothesis (N.H.) is rejected in favour of an *alternative* hypothesis) (A.H.).

It may be noted that the number of degrees of freedom F in an $m \times n$ contingency table when $m > 1$ and $n > 1$ is given by

1. $$F = (m-1)(n-1) \qquad (22.31)$$
   if the expected frequencies can be computed without determining the population parameters such as mean value, standard deviation, etc.
2. $$F = \{(m-1)(n-1) - k\} \qquad (22.32)$$
   if the expected frequencies calculations involve the use of k number of population parameters.

Problem 22.9 *A test was conducted to investigate the effect of a new vaccine against a particular infectious disease on laboratory animals and the following data was obtained:*

	Got disease	*Did not get disease*	*Total*
Vaccinated	8	40	48
Not vaccinated	15	27	42
Total	23	67	90

(a) Using significance levels of 0.05, test whether the new vaccine is effective in combating the particular infectious disease under investigation.
(b) Repeat the problem using Yate's correction.

Solution It is possible to determine the expected frequencies if we assume a null hypothesis that there is no difference between the vaccinated and unvaccinated groups, i.e., the vaccination and this disease are independent. Now, it is clear that in a sample of 90, the proportion which got the disease is 23/90. Taking this as an estimate of overall proportion which got the disease, $23/90 \times 48$, i.e. 12.27 is the expected number of laboratory animals out of total the sample of 48 which were vaccinated. By similar reasoning, the expected number of laboratory animals that were vaccinated and did not get the disease would be $67/90 \times 48$, i.e. 35.73. Continuing this way, the other expected frequencies have been calculated which are shown in the following table in parentheses below the observed frequencies.

	Got disease	*Did not get disease*	*Total*
Vaccinated	8	40	48
	$\left(\frac{23}{90}\times 48 = 12.27\right)$	$\left(\frac{67}{90}\times 48 = 35.73\right)$	
Not vaccinated	15	27	42
	$\left(\frac{23}{90}\times 42 = 10.73\right)$	$\left(\frac{67}{90}\times 42 = 31.27\right)$	
Total	23	67	90

Knowing the observed and expected frequencies, the χ^2-parameter without and with Yate's correction can be evaluated as follows:

Sl.No.	f_{oi}	f_{ei}	$(f_{oi} - f_{ei})^2/f_{ei}$	$\{\lvert f_{oi} - f_{ei}\rvert - 0.5\}^2/f_{ei}$
1	8	12.27	1.486	1.158
2	40	35.73	0.510	0.398
3	15	10.73	1.699	1.325
4	27	31.27	0.583	0.455
		Total	4.278	3.336

$$\chi^2 = 4.278$$

$$(\chi^2)_{\text{with Yate's correction}} = 3.366$$

We observe there are three restrictions in the data: (i) the number vaccinated, (ii) the number not vaccinated, and (iii) the additional restriction involved in the calculation of expected frequencies is those that contracted the disease and those that did not. The number of degrees of freedom is thus:

$$F = 4 - 3 = 1$$

Alternatively, the number of degrees of freedom could also be obtained by using Eq. (22.31)

$$F = (m - 1)(n - 1)$$
$$= (2 - 1)(2 - 1) = 1$$

From chi-square tables (Table 22.3), the critical value of χ^2 at 0.05 significance level corresponding to one degree of freedom is 3.841. This is less than χ^2-value obtained in the experiment. Therefore, the result is significant and we reject the null hypothesis (N.H.) in favour of the alternative hypothesis (A.H.). Hence we can say that up to 0.05 level, there is a significant difference between the vaccinated and not vaccinated groups, i.e., the new vaccine is effective in combating the particular disease under investigation.

Since the degrees of freedom in the given problem is 1, it is desirable to incorporate Yate's correction in the χ^2-value. The χ^2-value with Yate's correction is 3.336. It is less than $(\chi^2)_{0.05\ \text{level}}$, i.e. 3.841. Therefore, the result is insignificant. Hence, the test favours the null hypothesis (N.H.) and we can say up to 5% significance level that the vaccination and this particular infectious disease are independent.

Review Questions

22.1 Fill in the blanks in the following:

(i) The probability that a measured value would lie between $\bar{X} + 1.645\sigma$ and $\bar{X} - 1.645\sigma$ is __________.

(ii) The tails of the normal distribution curves are __________ to the abscissa (base line).

(iii) A confidence interval is a range within which the population __________ would most likely fall.

(iv) (1–confidence level) represents the __________ in any typical experiment.

(v) The shape of normal distribution is __________ (choose any one of the shapes mentioned, i.e. parabolic, rectangular hyperbolic, half sine curve, bell-shaped curve).

(vi) Area under the normal distribution curve is __________.

(vii) "No matter what the shape of original distribution, a notable event occurs with the distribution of means of sample values. They approach a normal distribution." The previous sentences are the statement of __________ theorem.

(viii) The points of inflection of the normal curve are at $x_i = \bar{X} \pm b$ where $b =$ __________.

(ix) If a student received a score of 90 in a test with standard error of 3, he could feel 90% confident that his true score lay between __________ and __________.

(x) Null hypothesis states that no population __________ exists.

(xi) The data is suspiciously good if the χ^2 probability __________.

(xii) In a normally distributed data, the constraints which affect the degrees of freedom are: (a) __________, (b) __________, and (c) __________.

(xiii) The value of χ^2 for 9 : 3 : 3 : 1 hypothesis for 66, 22, 30 10 is __________.

(xiv) In a contingency table with m rows and n columns, the degrees of freedom, when population parameters are not used is given by __________.

(xv) A test is highly significant if the range of means of the sample is __________.

22.2 Indicate whether the following statements are true or false. If false rewrite the correct statement.

(i) The maximum height of the standard form of normal distribution is always $1/\sigma\sqrt{2\pi}$.

(ii) If the results from experiments repeated for many times follow the normal frequency distribution, it does not confirm that the error is only due to randomness.

(iii) The sum of the relative frequencies of all classes is always 1.0.

(iv) In a mass produced sample of balls, usually, less than 50% of the balls have diameter in the limits of 1 standard deviation on either side of the mean diameter.

(v) The area of normal distribution curve between two points represents the confidence level/relative frequency of occurrence/probability of occurrence of any event in that region.

(vi) The normal distribution of sample means is flatter as compared to the normally distributed sample curves.

(vii) The distribution of sample means of skewed (non-normal) distributions is always skewed.

(viii) Normalised histogram represents the normal distribution of experimental data.

(ix) The results of two experiments will be significantly different if the combined internal standard error of the difference of means at 95% confidence level is more than the range of means.

(x) Lower values of χ^2 of the data than the critical value at a given significance level ensure rejection of the null hypothesis in favour of the alternative hypothesis.

22.3 In a certain manufacturing process, the diameter of shafts produced has a mean diameter of 20 cm and a standard deviation of 0.5 mm. If the shaft diameters range from 19.9 to 20.1 cm are acceptable, how many rejects would you expect in a random lot of 100 shafts.

22.4 A mass produced electronic gadget had $\overline{X}$ = 163 h. If the probability of survival of the gadget is at least 90% for at least 130 h, determine the standard deviation of the test sample.

22.5 An accelerated test programme for determining the time of failure of an electronic component obeys a normal probability distribution with $\overline{X}$ = 2 h and σ = 2 h. Out of a sample of 600 test components, how many are likely to survive for at least 3 h.

22.6 A manufacturer agrees to supply precision machine screws with length specification as:

$$l = (10.05 \pm 0.12) \text{ mm}$$

His first lot of screws consisted of:

$$l_1 = (10.00 \pm 0.1) \text{ mm},$$

and the second lot had:

$$l_2 = (10.1 \pm 0.1) \text{ mm}$$

what is the probability that a screw selected randomly in each lot would be defective?

22.7 A random sample was taken of the internal mean diameters of nuts produced by a manufacturer. The mean was 7.5 mm with a standard deviation of 0.1 mm. If nuts having a diameter of 7.375 and 7.583 are acceptable, what percentage are rejected?

22.8 A producer has to control the process very carefully when making the packets of dry peas or any other commodity sold by weight. He does not want to risk having under-weight packets, leading to less customers. However, he does not want to give too much away in over-weight packets. A machine used for filling 0.5 kg packets of dry peas, produces packets with weights which are normally distributed having a mean of 0.52 kg and a standard deviation of 0.01 kg. Find: (a) what proportion of packets have less than 0.5 kg peas, and (b) more than 0.53 kg.

22.9 A wire type strain gauge produced by a certain firm has a mean resistance of 75 Ω with a standard deviation of 0.3 Ω. They are used in a certain application where the requirements are 75 ± 0.42 Ω.

(a) What proportion of gauges will be defective?

(b) What should the precision be if the manufacturer wants to have 90% within the required range? Assume the resistance of gauges manufactured to be normally distributed.

22.10 A firm manufactures machines that are used to test the gripping power of the men from the age group 18–28. It is found that the grip strength is normally distributed with a mean value of 120 kg with a standard deviation of 20 kg. A prize equal to the fee is guaranteed if the needle on the dial shows 130 kg and double the fee if the reading is 150 kg. If a sample of 500 men test their gripping power, what would be the earning of a person owning such a machine if the fee charged per man is Re. 1.

22.11 Specifications for concrete used in civil engineering jobs may require specimens to be made and tested on site. These specimens are of stated size and shape (usually cubes). The normal model is judged suitable for analysing the results of strength tests on the cubes. Further, for a set of conditions, the coefficient of variation of cube strengths is used to describe the variability. If a research laboratory report gave a figure of 0.8 for the ratio of minimum strength to mean strength taking the definition of minimum strength as the strength below which a certain percentage of specimens may be expected to fail, determine the coefficient of variation if the percentage given is: (a) 5% and (b) 1%.

22.12 A cosmetic firm uses a machine to pack cream in bottles. The statistical analysis carried out gives the mean weight of the packed bottles and empty bottles as 200 g and 50 g with standard deviation of 10 g and 8 g, respectively. Can a sample of 10 bottles have a mean cream content of 140 g? State with 95% confidence level.

22.13 A firm manufacturers resistors for a certain application that requires the resistors to be of 100 Ω with a maximum allowable deviation of 1.2 Ω. The manufacturer takes several samples of 50 resistors and finds the mean of means to be 101 Ω with a standard deviation of means as 0.5 Ω.

(a) With how much confidence can we say that the product should be acceptable?

(b) To have 95% confidence, what should be the standard deviation of a sample of 50 randomly selected products?

22.14 Five automobile tyres, each of two different brands, were tested for wear (in grams) after driving for 2000 km under the same conditions. The data obtained is as follows:

Brand A	13.5	13.1	12.4	12.6	13.9
Brand B	12.5	12.7	12.0	13.7	12.6

Can we conclude that brand A is significantly superior to brand B?

22.15 50 bulbs of brand A were tested and it was found that 68.3% had a life of 1160 – 1240 h with a normal distribution. The manufacturer of brand B claimed that his product had a longer life. Subsequently, 100 bulbs of brand B were tested and found to follow normal distribution with 50% having a life of 1190–1230 h. Are the bulbs of brand B significantly better than brand A?

22.16 Of two similar groups of patients A and B consisting of 50 and 101 individuals, respectively, the first was given a new type of sleeping pill and the second a conventional type. For patients in group A, the mean number of hours of sleep was 7.88 h with a standard deviation of 0.24 h. For patients in group B, the mean number of hours of sleep was 7.75 h with a standard deviation of 0.32 h. Test whether the new type of sleeping pill is significantly better than the conventional pill. Use 95% confidence level.

22.17 100 bearings of brand X were tested and found to have a mean operating life of 30,000 h with a standard deviation (adjusted) of 2000 h; while 64 bearings of brand B showed a mean operating life of 32,000 h with a standard deviation (adjusted) of 4000 h. Is brand Y better? You may assume that both brands follows a Gaussian distribution.

22.18 A speed-breaker in the form of a mild obstruction on the road was put up in front of a roadside school gate. The means and standard deviations of the vehicle speeds, respectively before and after the putting up of the speed breaker were:

$$\bar{X}_1 = 40.5 \text{ km/h} \qquad \sigma_1 = 5.82 \text{ km/h} \qquad n_1 = 100$$

$$\bar{X}_2 = 31.8 \text{ km/h} \qquad \sigma_2 = 8.51 \text{ km/h} \qquad n_2 = 80$$

Indicate whether the installation of the speed-breaker has influenced the psychology of the automobile drivers or not.

22.19 Two engineering classes independently measured the static coefficient of friction of wood on glass by a technique known to be free from systematic errors. Their results were as follows:

	Class I	*Class II*
X	0.412	0.423
σ	0.0254	0.0241
n	50	64

On the basis of the above data alone, can it be stated that there is a significant difference in the results obtained by the two classes? Use a confidence level of 95%.

22.20 Average shear strength of 100 steel specimens grouped in the form of a frequency table are:

Shear strength of specimen (in N/mm^2)	*Observed frequency f_{o_i}*
1500.5	1
1550.5	0
1600.5	3
1650.5	16
1700.5	18
1750.5	24
1800.5	21
1850.5	14
1900.5	2
1950.5	0
2000.5	1

22.21 Test whether the data obeys the normal distribution. If so, compute the chi-square probability.

The volume of an aluminium block was determined by immersing it 100 times in water and noting the amount of volume displaced. The measurements yielded the following distribution:

Volume (in cm^3)	120.2	120.4	120.6	120.8	121.0	121.2	121.4	121.6	121.8
Frequency	3	6	13	18	20	18	13	6	3

Test if the data supports a Guassian distribution of errors. Use 10% significance level.

22.22 A die was rolled 60 times and the distribution of numbers on the uppermost die face was found to be as follows:

Number on the uppermost die face	1	2	3	4	5	6
Number of occurrences	16	7	11	6	14	6

Test the hypothesis that all faces of the die have equal probability of landing uppermost, i.e. the die is not loaded. Use 10% significance level using x^2 test.

22.23 While reading a scale where the last figure is estimated, some observers show a marked preference for particular digits. The following table shows the distribution of the last figure in 200 randomly chosen routine readings made by one observer

Last figure	0	1	2	3	4	5	6	7	8	9
Observed frequency	40	15	17	16	19	30	13	15	21	14

Test whether there is evidence of such a preference, if the chance/risk of wrongly accusing the observer regarding his bias is 5%.

22.24 A study was conducted on a set of patients who did not sleep well. Some were given a newly developed sleeping pill, while the others were given identical looking sugar pill and all patients thought that they were being given the sleeping pills. They were later asked whether the pill helped them in sleeping well or not. The results of their responses is given in the 2 × 2 contingency table. Assuming that all the patients told the truth, test the null hypothesis that there is no difference between the newly developed sleeping pill and the sugar pill at a significance level of 0.05. Further, also examine whether Yate's correction makes any difference in the acceptance or rejection of the null hypothesis.

	Patients who slept well	*Patients who did not sleep well*
Patients who took sleeping pills	33	17
Patients who took sugar pills	19	31

22.25 A study was conducted to determine the effectiveness of a vaccine in combating an infectious disease. In an area where the disease was prevalent, a total of 200 people were tested, 80 of whom were vaccinated in the last 12 months and 120 were not. In each case, it was noted whether they contracted the disease and, if so, whether it was in severe or mild form. Test if the vaccine is effective. Use 5% significance level.

	No disease	*Mild*	*Severe*
Vaccinated	45	25	10
Not vaccinated	30	50	40

22.26 Two treatments were tried out to control a certain type of plant infestation with the following results.

Treatment Type	*Number of plants examined*	*Number of plants found infected*
A	150	15
B	150	6

Can we conclude that treatment B is superior to treatment A in controlling this type of infestation? Use 5% significance level.

22.27 The following table shows the number of defective and satisfactory articles in two samples, one taken before and the other taken after the introduction of a modification in the process of manufacture.

	Defective articles	*Satisfactory articles*	*Total*
Before	20	120	140
After	5	75	80
Total	25	195	220

(a) Determine whether the data provides a strong evidence that the modification affects the quality of articles produced. Use 10% significance level.

(b) Repeat this problem using Yate's correction.

22.28 The observed frequencies in a 2 × 2 contingency table are given below:

	1	*2*	*Total*
A	a_1	a_2	N_A
B	b_1	b_2	N_B
Total	N_1	N_2	N

Assuming null hypothesis, determine the expected frequencies in the various elements of the contingency table. Further, show that the x^2-parameter without Yate's correction is given by:

$$N(a_1b_2 - a_2b_1)^2/(N_1N_2N_AN_B).$$

22.29 For the following 3 × 2 contingency table, show that:

$$\chi^2 = \frac{N}{N_A}\left[\frac{a_1^2}{N_1} + \frac{a_2^2}{N_2} + \frac{a_3^2}{N_3}\right] + \frac{N}{N_B}\left[\frac{b_1^2}{N_1} + \frac{b_2^2}{N_2} + \frac{b_3^2}{N_3}\right] - N$$

	1	*2*	*3*	*Total*
A	a_1	a_2	a_3	N_A
B	b_1	b_2	b_3	N_B
Total	N_1	N_2	N_3	N

22.30 A study was conducted on a particular dimension of the articles produced by a set of four machines by the use of 'go' and 'no-go' gauges. The articles were categorised as oversize, within tolerances and undersize. The table below shows the number of articles in each of these three categories produced by machines. A, B, C and D during a certain period of time.

	Oversize	*Within tolerances*	*Undersize*	*Total*
Machine A	6	120	4	130
Machine B	7	102	11	120
Machine C	4	168	8	180
Machine D	5	151	14	170
Total	22	541	37	600

Using 10% level, check if there is a significant difference in the performance of the given set of machines.

Answers

22.1 (i) 0.90 (ii) asymptotic (iii) data
(iv) error (v) bell-shaped curve (vi) unity
(vii) central limit (viii) σ (ix) 94.935 and 85.065
(x) difference (xi) is greater than 90%
(xii) mean value; standard deviation and total number of data
(xiii) 2.667 (xiv) $(m-1)(n-1)$
(xv) greater than 2.58 times the combined internal standard error of the range of means

22.2 (i) true (ii) false (iii) true (iv) false (v) true
(vi) false (vii) false (viii) false (ix) false (x) false

22.3 number of rejects ≈ 5

22.4 σ = 25.78 h

22.5 number of samples surviving at least 3 h = 185

22.6 0.2860 for both

22.7 30.89%

22.8 (a) 2.28% (b) 15.87%

22.9 (a) 16.16% (b) standard deviation in resistance = 0.255 Ω

22.10 312 rupees (round figure)

22.11 (a) 0.122 (b) 0.085

22.12 No, the mean cream content of 140 g in 10 bottles is less than the required cream content of 142.98 g at 95% confidence level.

22.13 (a) confidence level = 65.54% (b) $\sigma = 0.854\ \Omega$

22.14 No, tyres of brand A and brand B are from the same population.

22.15 No, there is no significant difference between bulbs of brand A and brand B.

22.16 New type of sleeping pill is significantly better than the conventional pill.

22.17 Yes, bearing of brand Y is significantly superior to brand X.

22.18 At 95% confidence level, the result is significant and therefore, the installation of speed-breaker has brought down the speeds of vehicles.

22.19 Difference of means = 0.011 and 1.96 times in internal estimate of uncertainty of the difference of means = 0.00927. Since the former is greater than the later, hence the result is significant.

22.20 The data follows Gaussian distribution and $P(x^2) = 0.35$.

22.21 The data follows Gaussian distribution at 10% significance level.

22.22 For 5 d.f. at 10% level, the critical value of χ^2 from $\chi^2 - F$ table is 9.24. The value of χ^2 obtained from the data using null hypothesis is 9.4 and is significant being greater than χ^2_{crit} . Hence, the die is loaded.

22.23 For 9. d.f. at 10% level, the critical value of χ^2 from $\chi^2 - F$ table is 16.919. The value of χ^2 obtained from the data using the null hypothesis is 32.1. Hence the result is significant and we can say that observer has a preference of 0 as the last figure.

22.24 For 1 d.f. system at 5% significance level $\chi^2 = 3.841$ and χ^2 observed = 7.85. Since $\chi^2_{0.05}$ is less than tabular χ^2 observed, the result is significant and hence the sleeping pill is effective. However, χ^2 observed = 6.84 with Yate's correction. Now, with Yate's correction, the result becomes insignificant and therefore, the sleeping pill is effective.

22.25 $\chi^2_{\text{observed}} = 22.2$ and $\chi^2_{0.05}$ for a 2 d.f. system = 5.99. Since $\chi^2_{0.05} < \chi^2_{\text{observed}}$, hence the result is significant and the vaccine is effective.

22.26 For 1 d.f. system at 5% significance level $\chi^2_{\text{tabular}} = 3.841$ and $\chi^2_{\text{observed}} = 4.174$. Since $(\chi^2)_{0.05}$ is less than χ^2_{observed}, hence the result is significant and the treatment A is superior to that of B. However, χ^2_{observed} with Yate's correction becomes 3.28. Now $\chi^2_{0.05}$ is greater than χ^2_{observed}. Therefore, the result with Yate's correction becomes insignificant. Hence, there is no difference between treatments A and B.

22.27 (a) For 1 d.f. at 10% level, the critical value of χ^2 from $\chi^2 - F$ tables is 2.71. The value of χ^2 obtained from the data using N.H. is 3.264 which is greater than χ^2_{crit} . Hence the result is significant and we can say that the introduction of modification in the process of manufacture improves the quality of the articles produced.

(b) The modified values of χ^2 using the Yate's correction obtained from the data with N.H. is 1.60. This is less than $\chi^2_{\text{crit}} = 2.71$. Hence the result is insignificant and we can say that the introduction of modification in the process of manufacture does not affect the quality of the articles produced.

22.30 $[\chi^2_{\text{date}}]_{\text{using N.H.}} = 9.64$.

$[\chi^2_{10\%\ \text{level}}]_{\text{d.f.} = 6} = 10.645$ (from $\chi^2 - F$ table)

Hence there is no significant difference in the performance of given set of machines.

Chapter

23

Graphical Representation and Curve Fitting of Data

INTRODUCTION

Very often, we observe that a relationship is found to exist between two or more variables. For example, the weight of a person depends on his age, height, etc. Similarly, the wear of a car tyre depends to a large extent on the mileage it has run. It is frequently desirable to express the relationship governing the various variables in a physical phenomenon in a mathematical form.

Now, in order to determine the governing relations of the variables, the first step becomes the collection of the data showing the corresponding values of variables under consideration. For example, if x and y denote respectively the load and extension in a typical tension test of a specimen on universal testing a machine, then a simple data of n observations would consist of extensions $y_1, y_2, y_3, \ldots, y_n$ corresponding to the loads $x_1, x_2, x_3, \ldots x_n$. The next step would be to plot the various data points $(x_1, y_1), (x_2, y_2), \ldots, (x_n, y_n)$ on a rectangular coordinate system so as to obtain the scatter diagram of the tabulated data. The drawing of the scatter diagram helps the experimenter to visualise the smooth curve approximating the data. It also helps him to note especially the peculiarities such as maxima, minima, inflections, hysteresis, etc. which become more apparent than when the information is in tabular form.

After drawing the scatter diagram we try to pass mostly by free hand the best possible smooth curve through the various data points. Such a curve is much more likely to represent the truth than the scatter of the individual points and is called the *approximating curve* of the given data. For example, in Fig. 23.1(a) the data appears to be approximated well by a straight line and we say that a linear relationship exists between the variables. Further, in Fig. 23.1(b), the best approximating curve is not a straight line and so we can say that a non-linear relationship exists between the variables.

It may be noted that it is often wasteful to empoly the ranges of the coordinate axes such that zero origin is included in the graphical representation of the data. However, it should be shown only if the data point happens to be there,data occur on both sides of it or if there is a theoretical reason for expecting the curve to pass through it.

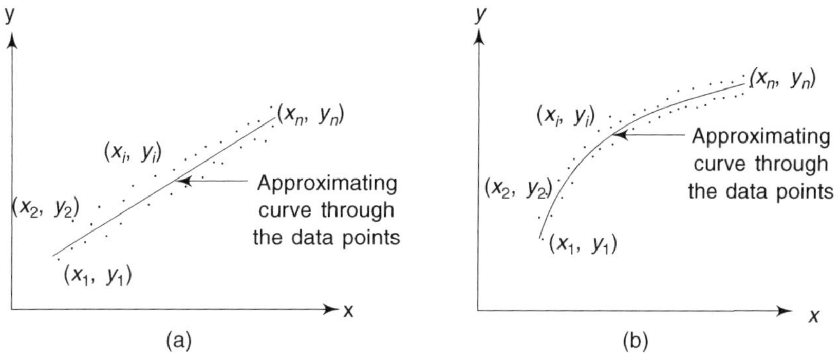

Fig. 23.1 *Typical scatter diagrams and the approximating curves through the data points*

23.1 ■ EQUATIONS OF APPROXIMATING CURVES

For the purpose of reference, the governing equations/functional relationships of the various common types of approximating curves are listed in Table 23.1. In the various equations, as a matter of standard practice, abscissa gives the values of x which represents the 'cause', i.e., the independent variable and the ordinate gives the values of varible y which represents the corresponding 'effect', i.e., the dependent variable. However, in certain exceptional cases, the roles of x and y can be interchanged.

Table 23.1 *Equations of Common Type of Approximating Curves*

(x and y are the independent and dependent parameters, respectively. All other letters in the equations represent constants.)

Sl. No.	*Type of Curve*	*Governing equation/functional relationships*
1	Stratight line or linear relationship	$y = a_0 + a_1x$
2	Parabolic or quadratic curve or polynomial of second degree	$y = a_0 + a_1x + a_2x^2$
3	Cubic curve or polynomial of third degree	$y = a_0 + a_1x + a_2x^2 + a_3x^3$
4	Quartic curve or polynomial of fourh degree	$y = a_0 + a_1x + a_2x^2 + a_3x^3 + a_4x^4$
5	Polynomial of nth degree	$y = a_0 + a_1x + a_2x^2 + \ldots + a_nx^n$
6	Hyperbolic curve	$y = \frac{1}{a_0 x}$
7	Modified hyperbolic curve	$y = \frac{1}{a_0 + a_1 x}$ or $\frac{1}{y} = a_0 + a_1x$
8	Exponential curve	$y = ab^x$ or $\log y = \log a + x (\log b) = a_0 + a_1x$ Also $y = ae^{bx}$ or $\ln y = \ln a + bx = a_0 + a_1x$
9	Power law or Geometric curve	$y = ax^b$ or $\log y = \log a + b \log x$
10	Modified exponential curve	$y = ab^x + c$
11	Modified power law curve	$y = ax^b + c$
12	Gompertz curve	$y = pq^{bx}$ or $\log y = \log p + (\log q)b^x = ab^x + c$
13	Modified Gompertz curve	$y = pq^{bx} + d$
14	Logistic curve	$y = \frac{1}{ab^x + c}$ or $\frac{1}{y} = ab^x + c$
15	Logarithmic function curve	$y = a_0 + a_1 (\log x) + a_2 (\log x)^2$
16	Trigonometric curve	$y = a_0 + a_1 \cos \omega x + a_2 \sin \omega x$

23.2 ■ GRAPHICAL REPRESENTATION OF FUNCTIONAL RELATIONSHIPS

The simplest type of approximating curve is a straight line. Further, a straight line is easily recognisable, but it is far less easy to identify polynomial, exponential or logarithmic functions. To facilitate the nature and type of functional relationship of the approximating curve from the scatter diagram, some of the commonly occurring graphical shapes of the curves corresponding to the various functional relationships are shown in Fig. 23.2.

It has been found that certain types of functional relationships can be conveniently plotted in the form of a straight line by choosing suitable scales for the ordinate as well as the abscissa (Table 23.2). In such cases, it becomes quite easy to determine the various constants as well as to estimate the extent of uncertainties in the various parameters of the functional relationship. The vairous scales empolyed for linearising the functional relationships may be linear versus linear, polar, logarithmic versus logarithmic or logarithmic-linear, etc. depending on the nature of the non-linear curve. For example, a power law relationship appears as a straight line on a log versus log graph paper whereas an exponential type curve becomes the same on a semi-log graph paper.

It may be noted that logarithmic plotting is specially useful when a variable covers a wide range. Another advantage is that the percentage changes or the fractional changes in the variables correspond to the same length on the logarithmic scale, whatever the value of the variable.

23.3 ■ DETERMINATION OF PARAMETERS IN LINEAR RELATIONSHIPS

If the scatter diagram of the data happens to be linear, then the governing equation has two constants that are given by

$$y = a_0 + a_1 x \tag{23.1}$$

where a_1 and a_0 are the slope and the intercept of the line, respectively.

To determine a_1 and a_0, we employ one of the following methods.

Table 23.2 *Plotting of various functional relationships in the form of a straight line*

(x and y are the dependent and independent parameters; a, b and c are constants; x_1 and y_1 are the coordinates of points obeying a particular relationship)

Sl. No.	*Type of curve and its governing equation*	*Ordinate*	*Abscissa*	*Recommended graph paper*	*Graphical representation showing slope and intercept parameter*
1.	Linear or straight line $y = ax + b$	y	x	Liner versus linear	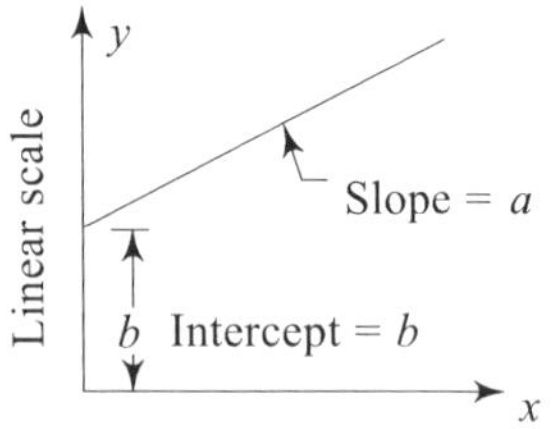
2.	Power law curve $y = ax^b$	$\log y$	$\log x$	log versus log	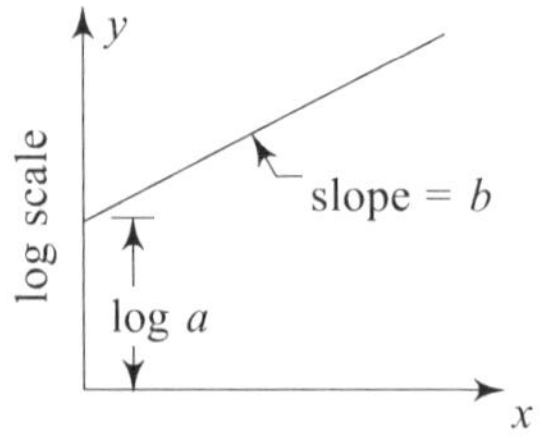

3.	Modified power law curve $y = ax^b + c$	$\log (y - c)$	$\log x$	log versus log	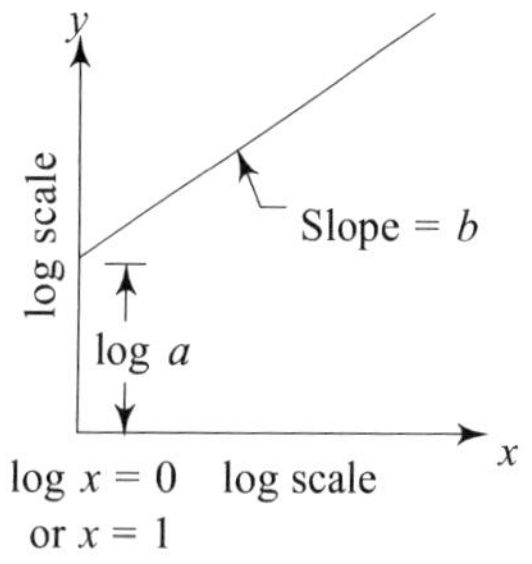
4.	Exponential type curve $y = ae^{bx}$	$\log y$	x	log versus linear	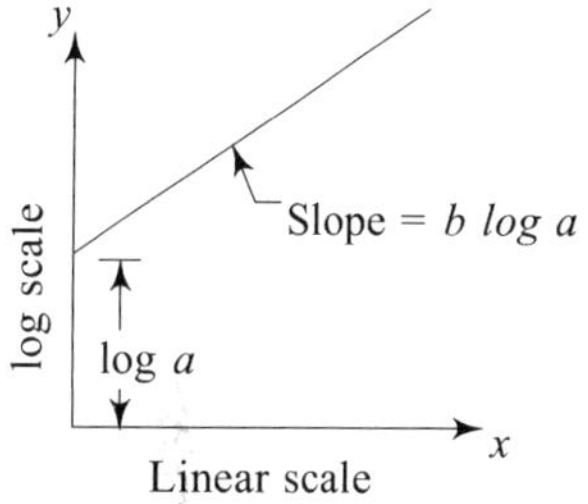
5.	Exponential type curve $y = ab^x$	$\log y$	x	log versus linear	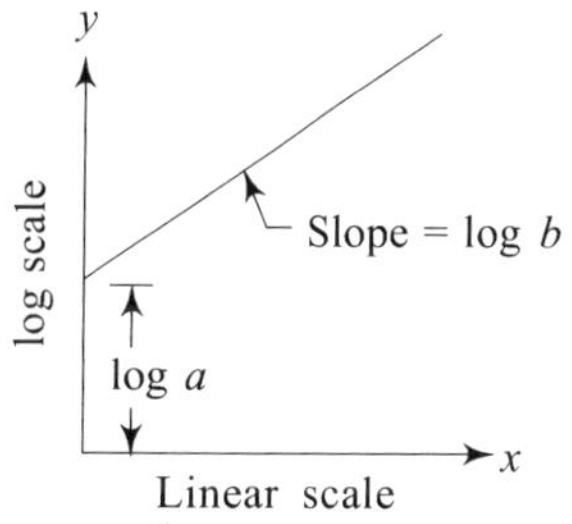
6.	Hyperbolic type curve $y = \frac{1}{ax + b}$	$\frac{1}{y}$	x	Reciprocal versus	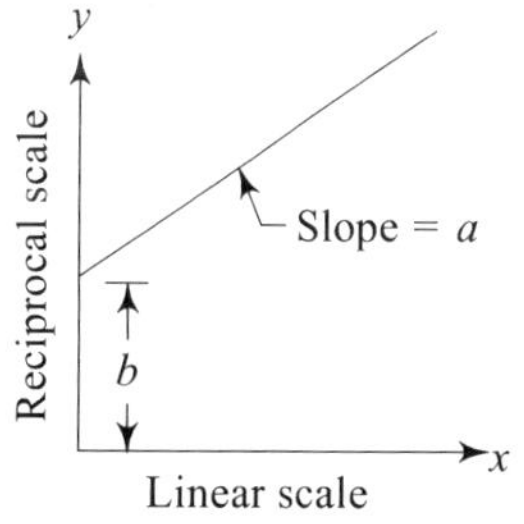
7.	Parabolic type curve $y = a + bx + cx^2$	$\frac{y - y_1}{x - x_1}$	x	Linear versus linear	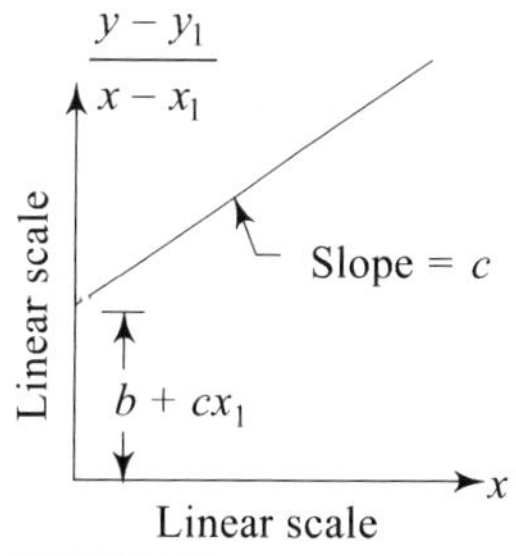

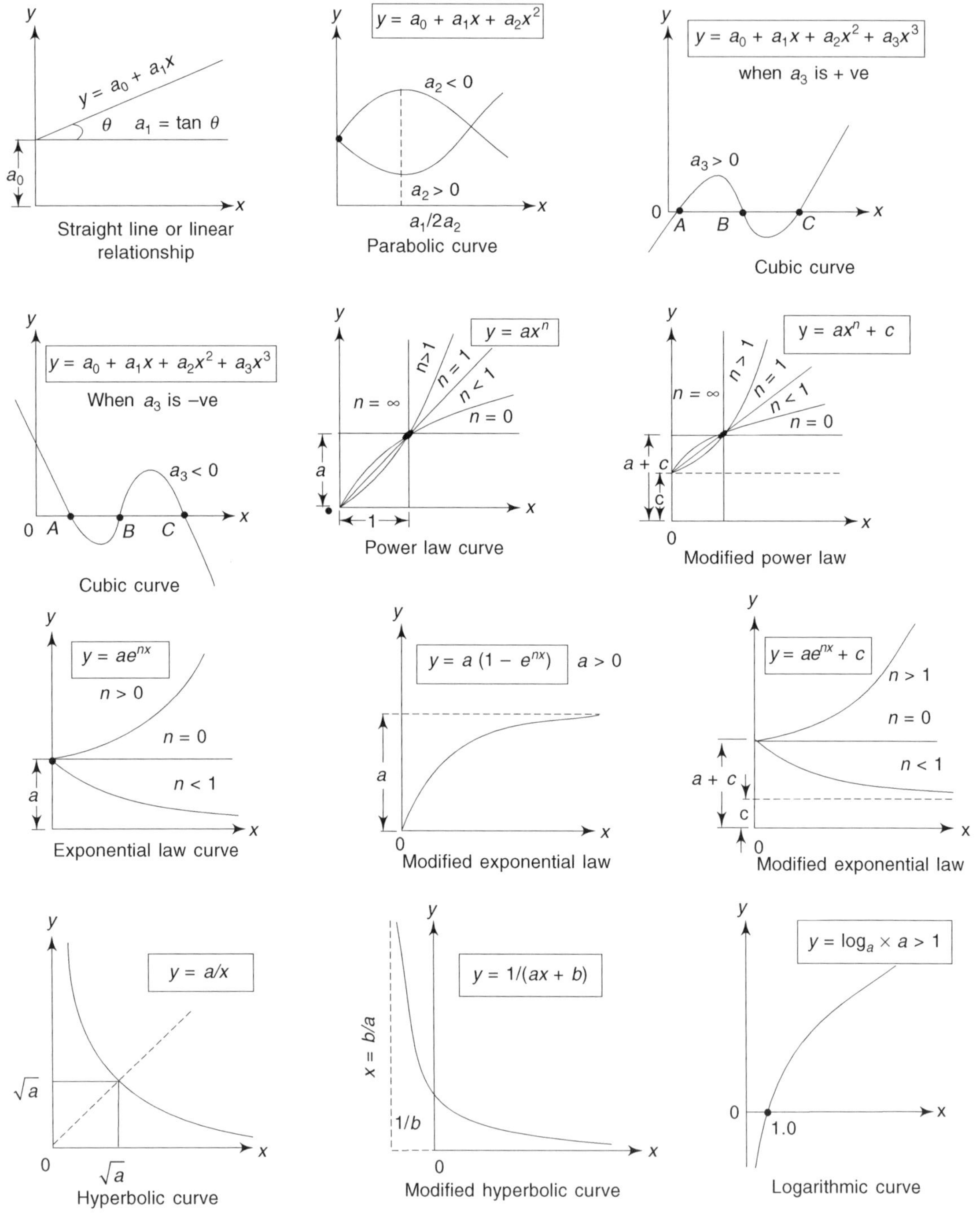

Fig. 23.2 *Graphical representation of the curves on linear versus linear graph paper corresponding to the various functional relationships*

23.3.1 Graphical Method

Generally, it is possible to draw a fairly good straight line through a given set of data points using a transparent ruler, in such a way that the experimental data points lie uniformly about the line. Although this method is quite convenient, yet for accurate work it has a number of shortcomings, viz.

1. Different observers draw the line differently.
2. An observer would tend to draw the line appreciably differently if the data is plotted to a different scale, say if the scale of the ordinate is elongated.
3. The graphical procedure does not provide any idea of how good the straight line is, i.e., no estimate of uncertainty in the slope is available.

23.3.2 Method of Sequential Differences

Suppose there are n pairs of observations, i.e., (x_1, y_1), (x_2, y_2), . . ., (x_n, y_n) whose scatter diagram is of linear type [Fig. 23.1(a)]. To determine the slope of the line that best represents these n points, we determine the slopes of each pair of adjacent points which gives $(n - 1)$ values of the slope. The best estimate of the slope is then given by the average of these values. Further, it is possible to calculate the best estimate of uncertainty in the mean value from the deviations in the individual slopes. Now, the various slopes for the various pairs of points are

$$\left.\begin{aligned} a_1 &= \frac{(y_3 - y_2)}{(x_3 - x_2)} \\ a_2 &= \frac{(y_3 - y_2)}{(x_3 - x_2)} \\ a_{n-1} &= \frac{(y_n - y_{n-1})}{(x_n - x_{n-1})} \end{aligned}\right\} \tag{23.2}$$

The mean of all slopes $\bar{a} = \sum_{i=1}^{(n-1)} a_i/(n-1)$ (23.3)

After determining the mean slope $\bar{a}$, we now determine the intercept of the linear relationship. This is determined by assuming that the centroid of the data lies on the straight line.

The centroid of the given data points is given by

$$\bar{X} = \sum_{i=1}^{n} \frac{x_i}{n} \tag{23.4}$$

and

$$\bar{Y} = \sum_{i=1}^{n} \frac{y_i}{n}$$

Substituting the values of mean slope, $\bar{X}$ and $\bar{Y}$ in the straight line relation of Eq. (23.1), we get the intercept $\bar{a}_0$ as

$$\bar{Y} = \bar{a}\bar{X} + \bar{a}_0$$

or

$$\bar{a}_0 = \bar{Y} - \bar{a}\bar{X} \tag{23.5}$$

However, this method suffers from certain limitations that are as follows:

1. The straight line fit determined by this method is generally not very accurate if the range of a_i is large.
2. If the x_i's are at equal intervals, say Δx, then,

$$\bar{a} = \frac{(y_n - y_1)}{(\Delta x)} \frac{1}{(n-1)} \tag{23.6}$$

i.e., $\bar{a}$ is a function of the first and the last points of the data and no weightage is given at all to the intermediate points in the determination of mean slope. Thus, this is serious flaw in this method as unfortunately the first and the last points of a straight line plot are usually most suspect.

23.3.3 Method of Extended Differences

In this method, the given data points are divided into two equal groups, namely high values and low values of x and the corresponding points in the two groups are differenced. Let us say that we have an even number of data points (in case of odd number of data points, the most suspect point or alternatively any one point is discarded in the calculations), then, two groups of data are

low values

$$\text{(group 1)} = (x_1, y_1), (x_2, y_2), \ldots, (x_n, y_n) \tag{23.7}$$

and high values

$$\text{(group 2)} = (x_{n+1}, y_{n+1}), (x_{n+2}, y_{n+2}), \ldots, (x_{2n}, y_{2n}) \tag{23.8}$$

Now, the values of slopes between the corresponding high and low values of groups 1 and 2 from Eqs. (23.7) and (23.8) are:

$$\left.\begin{aligned} a_1 &= \frac{y_{n+1} - y_1}{x_{n+1} - x_1} \\ a_2 &= \frac{y_{n+2} - y_2}{x_{n+2} - x_2} \\ &\vdots \\ a_n &= \frac{y_{2n} - y_n}{x_{2n} - x_n} \end{aligned}\right\} \tag{23.9}$$

Thus, the mean slope is given by

$$\bar{a} = \sum_{i=1}^{n} \frac{a_i}{n} \tag{23.10}$$

Knowing the mean slope, the value of the mean intercept is calculated using the procedure explained in Eqs. (23.4) and (23.5). Further, the magnitude of the uncertainty or the internal standard error may be calculated from the deviations of the individual slopes from the mean value.

It may be noted that the procedure of extended differences gives the mean slope which is the same as the slope of the centroids of the high (group 2) and low values (group 1).

To sum up, the method of extended differences is considered better than the method of sequential differences because it takes into account all the data points even if the intervals between different data points are the same. Further, it has the advantages of ease and simplicity and gives good results whether the errors arise in x or y coordinates. Thus, it is usually recommended for ordinary calculations. However, in situations where we know which coordinate is giving rise to experimental errors, the well-known method of least squares is usually preferred over this method.

23.3.4 Method of Least Squares

This is general method for determining the best fitting curve, say a best fitting line or best fitting parabola or best fitting any other approximating curve for a given set of data. The main advantage in this procedure is that it does not depend at all on the judgement of an individual in determining the best fitting curve.

Let us first consider a general case of a set of data points in which the scatter diagram of the experimental points represents the curve C (Fig. 23.3). In this method, we assume that the scatter in the points is due to errors in measuring y and the best fitting curve is that which minimises the sum of the squares of the errors in the y-direction. In other words, of all the approximating curves for a given set of data points, the curve having the property that the sum of error squares S_e given by $(e_{y1}^2 + e_{y2}^2 + \ldots + e_{yn}^2)$ is minimum, is called the best fitting curve. Further, a curve having this property is said to fit the data in the least square sense and is called the least square curve. Thus, a line satisfying this property is a least square line, a parabola the least square parabola, etc.

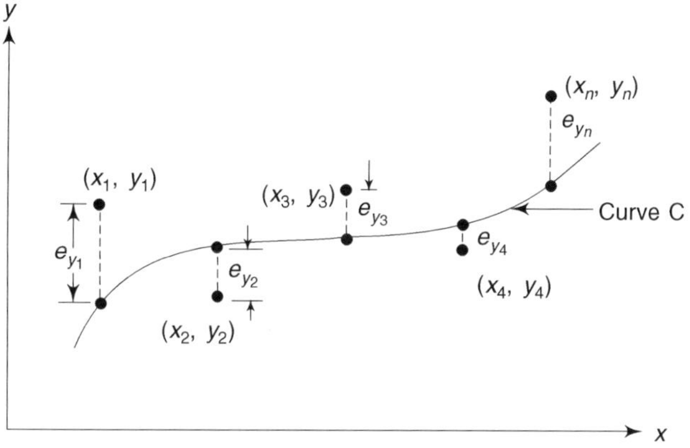

Fig. 23.3 *A typical scatter diagram of data showing errors in measuring y with respect to curve C*

It may be noted that if values of x, the independent vairable are considered more accurate than y, then the sum of error squares in the y-direction are minimised with respect to the approximating curve and the resulting curve is called the least square regression curve of y on x. Conversely, if the values of y are more accurate than x, then the sum of squares in the x-direction are minimised and the resulting curve would be the least square regression curve of x on y.

23.3.5 Linear Least Square Curve Fitting

Figure 23.4 shows the scatter diagram of the data in which a straight line curve, often called linear regression equation, could possibly be fitted. The general form of the linear equation is

$$y = a_0 + a_1 x \tag{23.11}$$

where a_0 and a_1 represent the intercept and the slope of the line, respectively.

If the values of x are more accurate than y (i.e., regression of y on x) then we estimate the values of ordinate y^* from the more accurately observed values of x; that is,

$$y_i^* = a_0 + a_i x_i \tag{23.12}$$

where y_i is the estimated value of the dependent variable y_i. Thus, the error e_{yi} is given by

$$e_{yi} = y_i - y_i^* \tag{23.13}$$

The least square principle states that for the best fitting straight line, the sum of the squares of the errors should be minimum. That is,

$$S_e = \sum_{i=1}^{n} e_{yi}^2 \tag{23.14}$$

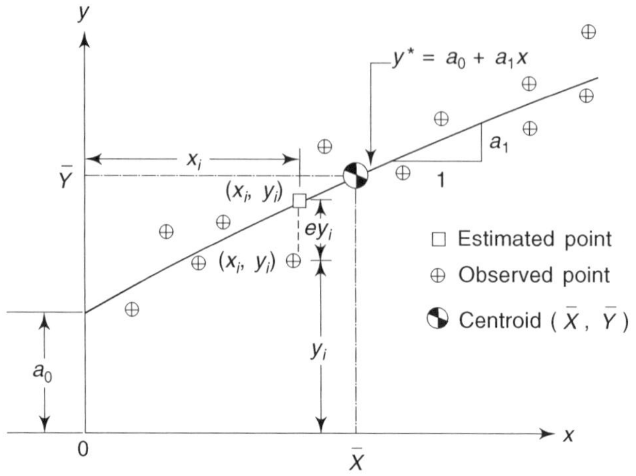

Fig. 23.4 *Ralationship of terms in a linear regression equation*

is minimised. Substituting the value of e_{yi} from Eq. (23.13) we get the sum of error squares as:

$$S_e = \sum_{i=1}^{n} (y_i - y_i^*)^2 \tag{23.15}$$

Substituting the value of y_i^* from Eq (23.13) in Eq. (23.15) we get,

$$S_e = \sum_{i=1}^{n} (y_i - a_0 - a_1x_i)^2 \tag{23.16}$$

In order to minimise S_e, we determine the partial derivatives of S_e with respect to a_0 and a_1 in Eq. (23.16). Each derivative must be zero at its minimum. Thus, this process gives two equations with two unknowns. Therefore,

$$\frac{\partial S_e}{\partial a_0} = -2 \sum_{i=1}^{n} (y_i - a_0 - a_1x_i) = 0 \tag{23.17}$$

and
$$\frac{\partial S_e}{\partial a_1} = -2 \sum_{i=1}^{n} x_i(y_i - a_0 - a_1x_i) = 0 \tag{23.18}$$

Equations (23.17) and (23.18) can be simplified to

$$\sum_{i=1}^{n} y_i - a_0n - a_1 \sum_{i=1}^{n} x_i = 0 \tag{23.19}$$

and
$$\sum_{i=1}^{n} x_iy_i - a_0 \sum_{i=1}^{n} x_i - a_1 \sum_{i=1}^{n} x_i^2 = 0 \tag{23.20}$$

Solving Eqs. (23.19) and 23.20 simultaneously we get,

$$a_0 = \frac{\sum_{i=1}^{n} y_i \sum_{i=1}^{n} x_i^2 - \sum_{i=1}^{n} x_i \sum_{i=1}^{n} x_i y_i}{n \sum_{i=1}^{n} x_i^2 - \left(\sum_{i=1}^{n} x_i\right)^2} \tag{23.21}$$

and
$$a_1 = \frac{n \sum_{i=1}^{n} x_i y_i - \sum_{i=1}^{n} x_i \sum_{i=1}^{n} y_i}{n \sum_{i=1}^{n} x_i^2 - \left(\sum_{i=1}^{n} x_i\right)^2} \tag{23.22}$$

The denominator in Eqs. (23.21) and (23.22) is the same and could be replaced by a quantity say Δ and further short notation $\sum x_i$, $\sum x_i y_i$, etc. in place of $\sum_{i=1}^{n} x_i$, $\sum_{i=1}^{n} x_i y_i$, etc. could be employed to present the above equations as:

$$a_0 = \frac{\left(\sum y_i \sum x_i^2 - \sum x_i \sum x_i y_i\right)}{\Delta} \tag{23.23}$$

and
$$a_1 = \frac{\left(n \sum x_i y_i - \sum x_i \sum y_i\right)}{\Delta} \tag{23.24}$$

where
$$\Delta = \left\{n \sum x_i^2 - \left(\sum x_i\right)^2\right\}$$

It may be noted that by dividing Eq. (23.19) by n we get

$$\frac{\sum y_i}{n} = a_0 + a_1 \frac{\sum x_i}{n} \tag{23.25}$$

or
$$\bar{Y} = a_0 + a_1 \bar{X} \tag{23.26}$$

This shows that the least square line would pass through the centroid ($\bar{X}$, $\bar{Y}$).

In practice, in numerical problem solutions, we determine the value of slope a_1 by employing Eq. (23.24) in which we have to determine the values of $\sum x_i$ and $\sum y_i$. Then, it becomes easier to determine the value of the intercept a_0 by employing Eq. (23.25) in place of Eq. (23.23).

In case the values of y happen to be more accurate than x, then we regress x on y (i.e. $x = b_0 + b_1 y$) and the expressions for the slope b_1 intercept b_0 of the line become

$$b_1 = \frac{\left(n \sum x_i y_i - \sum x_i \sum y_i\right)}{\Delta'} \tag{23.27}$$

where
$$\Delta' = \left\{n \sum y_i^2 - \left(\sum y_i\right)^2\right\}$$

and
$$b_0 = \frac{\sum x_i}{n} - b_i \frac{\sum y_i}{n} \tag{23.28}$$

23.3.6 Determination of Uncertainties (Internal Standard Errors) in the Slope and Intercept Values for Linear Regression

In the least square regression of y on x, we have assumed that the values of x are accurate and have minimised the deviations in y using the least squares criterion. Therefore, initially we determine the standard deviations and then the uncertainty (internal standard error) in y values. Subsequently, the expressions for uncertainties in a_0 and a_1 which are functions of uncertainties in y are determined.

$$\sigma_n^2(y) = \frac{\sum_{i=1}^{n}(y_{\text{measured}} - y_{\text{estimated}})^2}{n}$$

$$= \frac{1}{n}\left\{\sum_{i=1}^{n}(y_i - a_1 x_i - a_0)^2\right\}$$

$$= \frac{1}{n}\{\sum y_i (y_i - a_1x_i - a_0) - a_1$$

$$x_i (y_i - a_1x_i - a_0) - a_0 \quad (y_i - a_1x_i - a_0)\} \tag{23.29}$$

In may be noted that the second and third terms of Eq. (23.29) are the same as Eqs. (23.19) and (23.20), i.e., the normal equations of the linear least square regression. Since both these equations are equal to zero, therefore, Eq. (23.29) becomes

$$\sigma_n^2(y) = \frac{1}{n}\left(\sum y_i^2 - a_1 \sum x_i y_i - a_0 \sum y_i\right) \tag{23.30}$$

From Eq. (23.30), the internal estimate of uncertainty $U_n(y)$ can be calculated as

$$U_n(y) = \frac{\sigma_n(y)}{(n-2)^{1/2}} \tag{23.31}$$

It may be noted that in drawing a straight line, two points are required and therefore one extra degree of freedom is lost. Thus, the denominator in the expression (23.31) is $(n - 2)^{1/2}$ in place of $(n - 1)^{1/2}$.

Now, we determine the internal estimate of the uncertainties in the slope and the intercept of the line.

From Eq. (23.24) we get,

$$a_1 = \frac{1}{\Delta}\{n \sum x_i y_i - \sum x_i \sum y_i\}$$

$$= \frac{1}{\Delta}\{n(x_1y_1 + x_2y_2 + \ldots + x_ny_n) - n\bar{X}(y_1 + y_2 + \ldots + y_n)\}$$

$$= \frac{n}{\Delta}\{(x_1 - \bar{X})y_1 + (x_2 - \bar{X})y_2 + \ldots + (x_n - \bar{X})y_n\} \tag{23.32}$$

Employing the formula of propagation of error for a linear combination we get,

$$\sigma^2(a_1) = \frac{n^2}{\Delta^2}\{(x_1 - \bar{X})^2 \cdot \sigma^2(y_1) + \ldots + (x_n - \bar{X})^2 \cdot \sigma^2(y_n)\} \tag{23.33}$$

Substituting $\sigma(y_1) \simeq \sigma(y_2) \simeq \ldots \simeq \sigma(y_n) \simeq \sigma(n)$ in Eq. (23.33) we get,

$$\sigma^2(a_1) = \frac{n^2}{\Delta^2}\{\sum (x_i - \bar{X})^2 \cdot \sigma^2(y)\}$$

$$= \frac{n^2}{\Delta^2}\left\{\sum (x_i^2 - 2\bar{X}x_i + \bar{X}^2)\right\} \cdot \sigma^2(y)$$

$$= \frac{n^2}{\Delta^2}\left\{\sum x_i^2 - 2\bar{X}\sum x_i + n\bar{X}^2\right\} \cdot \sigma^2(y)$$

$$= \frac{n^2}{\Delta^2}\left\{\sum x_i^2 - \frac{\left(\sum x_i\right)^2}{n}\right\} \cdot \sigma^2(y) \quad \left[\because \bar{X} = \frac{1}{n}\sum x_i\right]$$

$$= \frac{n^2}{\Delta^2}\left\{\frac{\Delta}{n}\right\} \sigma^2(y) \qquad \left[\because \sum x_i^2 - \frac{\left(\sum x_i\right)^2}{n} = \frac{\Delta}{n}\right]$$

$$= \frac{n}{\Delta}\sigma^2(y)$$

$$\sigma(a_1) = \frac{\sqrt{n}}{\sqrt{\Delta}}\sigma(y) \tag{23.34}$$

Similarly, it can be shown that

$$U_n(a_1) = \frac{\sqrt{n}}{\sqrt{\Delta}} \cdot U_n(y) \tag{23.35}$$

Thus, Eq. (23.35) gives the expression for determining the internal estimate of uncertainty in the slope of the least square line in terms of uncertainty in the values of y.

Now, the expression for the intercept of the least square line is given by Eq. (23.23) as

$$a_0 = \frac{1}{\Delta}\left(\sum y_i \sum x_i^2 - \sum x_i \sum x_i y_i\right)$$

$$= \frac{1}{\Delta}\{\sum x_i^2 \, (y_1 + y_2 + \ldots + y_n) - n\,\bar{X}\,(x_1 y_1 + x_2 y_2 + \ldots + x_n y_n)\}$$

$$= \frac{1}{\Delta}\left\{\left(\sum x_i^2 - n\bar{X}\,x_i\right)_i + \ldots + \left(\sum x_i^2 - n\bar{X}x_n\right)y_n\right\} \tag{23.36}$$

Again, using the linear combination formula for the propagation of errors we get,

$$\sigma^2(a_0) = \frac{1}{\Delta^2}\sum\left\{\sum x_i^2 - n\bar{X}x_i\right\}^2 \cdot \sigma^2(y_i) \tag{23.37}$$

Substituting $\sigma(y_1) \simeq \sigma(y_2) \simeq \ldots \simeq \sigma(y_n) \simeq \sigma(y)$ in Eq. (23.37) we get,

$$\sigma^2(a_0) = \frac{1}{\Delta^2}\sum\left\{\sum x_i^2 - n\bar{X}x_i\right\}^2 \cdot \sigma^2(y)$$

$$= \frac{1}{\Delta^2}\sum\left\{\left(\sum x_i^2\right)^2 - 2\,n\bar{X}x_i \cdot \sum x_i^2 + n^2\,\bar{X}^2\,x_i^2\right\} \cdot \sigma^2(y)$$

$$= \frac{1}{\Delta^2}\left\{n\left(\sum x_i^2\right) - 2\,n\bar{X}\sum x_i \cdot \sum x_i^2 + n^2\,X^2\,\sum x_i^2\right\} \cdot \sigma^2(y)$$

$$= \frac{n\sum x_i^2}{\Delta^2}\left\{\sum x_i^2 - 2\,\bar{X}\sum x_i + n\bar{X}^2\right\} \cdot \sigma^2(y)$$

$$= \frac{n\sum x_i^2}{\Delta^2}\left\{\sum x_i^2 - \frac{\left(\sum x_i\right)^2}{n}\right\} \cdot \sigma^2(y) \qquad \left[\because \bar{X} = \frac{\sum x_i}{n}\right]$$

$$= \frac{n\sum x_i^2}{\Delta^2}\left\{\frac{\Delta}{n}\right\} \cdot \sigma^2(y) \qquad \left[\because \left\{n\sum x_i^2 - \left(\sum x_i\right)^2\right\} = \Delta\right]$$

$$= \frac{\sum x_i^2}{\Delta}\,\sigma^2(y) \tag{23.38}$$

$$\therefore \qquad \sigma(a_0) = \frac{\left(\sum x_i^2\right)^{1/2}}{(\Delta)^{1/2}}\,\sigma(y) \tag{23.39}$$

Now, it can be shown that

$$U_n(a_0) = \frac{\left(\sum x_i^2\right)^{1/2}}{(\Delta)^{1/2}}\,U_n(y) \tag{23.40}$$

Thus, Eq. (23.40) gives the expression for determining the internal estimate of uncertainty in the intercept of the least square line in terms of uncertainty in the values of y.

Problem 23.1 *Successive masses of 1 kg each (of high accuracy) were added at the hook at the lower end of a vertically hanging wire. The position of a mark at the lower end was measured using an ordinary scale. The following results were obtained:*

Load x (kg)	1	2	3	4	5	6	7	8	9	10
Position of mark y (cm)	6.05	6.20	6.25	6.35	6.40	6.50	6.55	6.60	6.70	6.75

(a) Determine the equation of the best fitting straight line using:
  (i) Graphical method
  (ii) Method of sequential difference
  (iii) Method of extended difference
  (iv) Method of least squares

(b) Also, determine the internal estimate of uncertainty in the values of slope in each of the above-mentioned procedure for fitting the straight line relationship.

Solution

(i) The scatter diagram shown in Fig. 23.6 is obtained by plotting the points (1, 6.05), (2, 6.20), . . . , (10, 6.75).

A straight line which approximates the data (using personal judgement) is shown by the dashed line in the Fig. 23.5. This is, in fact, one of many possible lines which could have been constructed.

Choosing any two points (x_1, y_1) and (x_2, y_2) on the approximating straight line, the straight line relation can be determined from the equation

$$y - y_1 - \frac{y_2 - y_1}{x_2 - x_1}(x - x_1)$$

Two such points chosen in Fig. 23.5 are A (2, 6.15) and B (7, 6.55). Therefore, the required equation becomes

$$y - 6.15 = \frac{6.55 - 6.15}{7 - 2}(x - 2)$$

or

$$y = 0.08x + 5.59$$

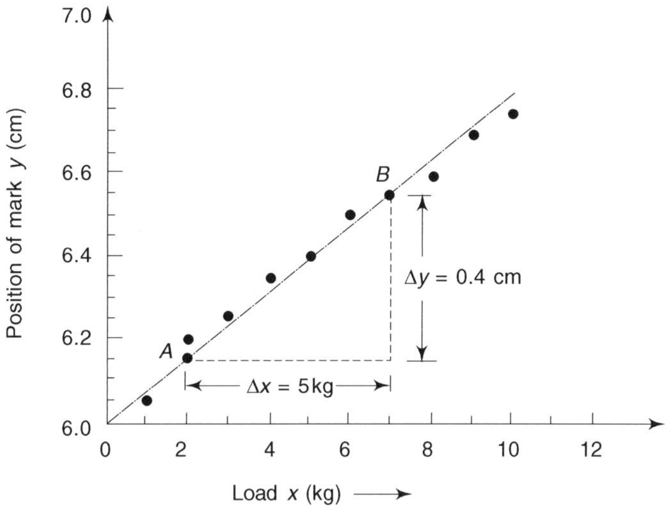

Fig. 23.5 *Determination of straight line relationship which approximates the data using graphical method.*

In this equation the slope $a_1 = 0.08$ and the intercept $a_0 = 5.99$.

As explained in Sec. 23.41, it is not possible to determine the internal estimate of uncertainty in the slope and the intercept values.

(ii) The calculation of slope using the method of sequential difference can be made from the following:

	Load x_i *(kg)*	*Position of mark* y_i *(cm)*	*Difference* $\Delta y_i = y_{n+1} - y_n$	*Difference* $\Delta x_i = x_{n+1} - x_n$	*Slope* $\Delta y_i/\Delta x_i$
	1	6.05			
			0.15	1	0.15
	2	6.20			
			0.05	1	0.05
	3	6.25			
			0.10	1	0.10
	4	6.35			
			0.05	1	0.05
	5	6.40			
			0.10	1	0.10
	6	6.50			
			0.05	1	0.05
	7	6.55			
			0.05	1	0.05
	8	6.60			
			0.10	1	0.10
	9	6.70			
			0.05	1	0.05
	10	6.75			
Total	55	64.35			0.70

From the above table we get,

$$\text{mean slope } \bar{a}_1' = 0.70/9 = 0.078$$

$$\bar{X} = 5.50$$

and $$\bar{Y} = 6.44$$

Now, $\bar{X}$ and $\bar{Y}$ are assumed to lie on the straight line $y = \bar{a}_0' + \bar{a}_1' x$. Since the value of $\bar{a}_1'$ is known, the value of mean intercept $\bar{a}_0'$ found by substituting the values of $\bar{X}$ and $\bar{Y}$ in the equation of the straight line becomes

$\bar{a}_0' = 6.01$

Thus, the equation of the straight line using the method of sequential differences becomes

$$y = 0.078x + 6.01$$

From the value, of slopes, we determine the standard deviation

$$\sigma_n = 0.034.$$

Now, the internal estimate of uncertainty in the slope $= \dfrac{0.034}{\sqrt{(9-1)}} = 0.012.$

Thus, the slope of the equation becomes

$$\bar{a}_1' = 0.078 \pm 0.012$$

(iii) The calculation of the slope using the method of extended differences can be made from the following:

	Load x_i *(kg)*	*Position of mark* y_I *(cm)*	*Difference* $\Delta y_i = (y_{n+5} - y_n)$	*Difference* $\Delta x_i = (x_{n+5} - x_n)$	*Slope* $\Delta y_i/\Delta x_i$
	1	6.05	0.45	5	0.09
	2	6.20	0.35	5	0.07
	3	6.25	0.35	5	0.07
	4	6.35	0.35	5	0.07
	5	6.40	0.35	5	0.07
	6	6.50			
	7	6.55			
	8	6.60			
	9	6.70			
	10	6.75			
Total	55	64.35			0.37

From the above table we get

mean slope $\bar{a}_1'' = 0.074$

Substituting the values of $\bar{X}$ and $\bar{Y}$ in the straight line equation

$$y = \bar{a}_0'' + \bar{a}_1'' x$$

$$\bar{a}_1'' = 6.03$$

Thus, the equation of the straight line using the method of sequential difference becomes

$$y = 0.074x + 6.03$$

From the values of slopes, we determine the standard deviation

$$\sigma_n = 0.008.$$

Now, the internal estimate of uncertainty in the slope $= \dfrac{0.008}{\sqrt{(5-1)}}$ or 0.004

Thus, the slope of the equation becomes

$$\bar{a}_1'' = 0.074 \pm 0.004$$

(iv) The calculation of slope using the method of least squares can be made from the following:

	Load x_i (kg)	Position of mark y_i (cm)	x_i^2	y_i^2	$x_i y_i$
	1	6.05	1	36.60	6.05
	2	6.20	4	38.44	12.40
	3	6.25	9	39.06	18.75
	4	6.35	16	40.32	25.40
	5	6.40	25	40.96	32.00
	6	6.50	36	42.25	39.00
	7	6.55	49	42.90	45.85
	8	6.60	64	43.56	52.80
	9	6.70	81	44.89	60.30
	10	6.75	100	45.56	67.50
Total	55	64.35	385	414.54	360.05

Using Eq. (23.22) we get,

$$\text{slope } a_1 = \frac{n\sum x_i y_i - \sum x_i \sum y_i}{n\sum x_i^2 - \left(\sum x_i\right)^2}$$

$$= \frac{(10)(360.05) - (55)(64.35)}{(10)(385) - (55)^2}$$

$$= \frac{61.25}{825} = 0.074$$

Further, using Eq. (23.21) we get,

$$\text{intercept } a_0 = \frac{\sum y_i \sum x_i^2 - \sum x_i \sum x_i y_i}{n\sum x_i^2 - \left(\sum x_i\right)^2}$$

$$= \frac{(64.35)(385) - (55)(360.05)}{(10)(385) - (55)^2}$$

$$= \frac{4972}{825} = 6.02$$

Thus, the least square regression line becomes

$$y = 0.074x + 6.02$$

Now, to determine the internal estimate of uncertainty in the slope and intercept values, we first determine the internal estimate of uncertainty in the values of y.

Using Eq. (23.30) we get,

$$\sigma_n^2(y) - \frac{1}{n}\left\{\sum y_i^2 - a_1 \sum x_i y_i - a_0 \sum y_i\right\}$$

$$= \frac{1}{10}\ \{414.54 - 0.074\ (360.05) - 6.02\ (64.35)\}$$

$$= \frac{1}{10} \{414.55 - 26.64 - 387.39\}$$

$$= 0.052$$

Now, the internal estimate of uncertainty in y values becomes

$$U_n(y) = \frac{\sigma_n(y)}{\sqrt{(10-2)}}$$

$$= \frac{\sqrt{0.052}}{\sqrt{8}} = 0.081$$

Using Eq. (23.35), the uncertainty in the slope values becomes

$$U_n(a_1) = \frac{\sqrt{n}}{\sqrt{\Delta}} U_n(y)$$

$$= \sqrt{\frac{10}{825}}\, 0.081 = 0.0088$$

Thus, the slope of the equation, using the method of least squares becomes

$$a_1 = 0.074 \pm 0.0088$$

Further, using Eq. (23.40) we get

$$U_n(a_0) = \frac{\left(\sum x_i^2\right)^{1/2}}{(\Delta)^{1/2}} U_n(y)$$

Substituting the values of $\sum x_i^2$, Δ and $U_n(y)$ we get

$$U_n(a_0) = \frac{\sqrt{385}}{\sqrt{825}} \times 0.081$$

$$= 0.055$$

Thus, the intercept of the equation, using the method of least squares becomes

$$a_0 = 6.02 \pm 0.055$$

Finally, the linear least square regression equation of the data with error analysis becomes

$$y = (0.074 \pm 0.0088)\, x + (6.02 \pm 0.055)$$

23.4 LEAST SQUARES EQUATIONS OF SECOND DEGREE AND HIGHER

Suppose our data points are not linear. We fit the curve by minimising the sum of error squares estimated from the equation of order n. Let us first consider the expansion to a second degree equation and from that proceed to a general equation for nth degree regression.

The general form of the second degree equation is

$$y = a_0 + a_1 x + a_2 x^2 \qquad (23.41)$$

Now, if the values of x are more accurate than y, then we estimate the values of ordinate y^* from the observed values of x; that is

$$y_i^* = a_0 + a_1 x_i + a_2 x_i^2 \qquad (23.42)$$

where y_i^* are the estimated values of the dependent variable y_i.

Now the sum of error squares S_e is given by:

$$S_e = \sum (y_i - a_0 - a_1x_i - a_2x_i^2)^2 \qquad (23.43)$$

For S_e to be minimum, we should have

$$\frac{\partial S_e}{\partial a_0} = 0, \quad \frac{\partial S_e}{\partial a_1} = 0 \quad \text{and} \quad \frac{\partial S_e}{\partial a_2} = 0$$

Consequently we have,

$$\sum (y_i - a_0 - a_1x_i - a_2x_i^2) = 0 \qquad (23.44)$$

$$\sum x_i(y_i - a_0a_1x_i - a_2x_i^2) = 0 \qquad (23.45)$$

and

$$\sum x_i^2 (y_i - a_0 - a_1x_i - a_2x_i^2) = 0 \qquad (23.46)$$

Equations (23.44) to (23.46) can be simplified and regrouped in the following form:

$$\sum y_i = a_0n + a_1 \sum x_i + a_2 \sum {}_i^2 \qquad (23.47)$$

$$\sum y_ix_i = a_0 \sum x_i + a_1 \sum x_i^2 + a_2 \sum x_i^2 \qquad (23.48)$$

$$\sum y_i x_i^2 = a_0 \sum x_i^2 + a_i \sum x_i^2 + a_2 \sum x_i^2 \qquad (23.49)$$

We then solve Eqs. (23.47) to (23.49) simultaneously to determine the values of a_0, a_1 and a_2.

It may be mentioned that Eqs. (23.47) to (23.49) are called the normal equations for the least square parabola. These equations can be easily remembered by observing that they can be obtained formally by multiplying Eq. (23.41) by 1, x, x^2, respectively and summing on both sides of the resulting equations. This technique can be extended to obtain the normal equations for least square cubic curves, least square quartic curves and, in general, for the polynomial of nth order.

Thus, the normal equations for the least square polynomial of nth order can be written as

$$\left.\begin{aligned} \sum y_ix_i^0 &= a_0 \sum x_i^2 + a_1 \sum x_i^2 + \ldots + a_n \sum x_i^2 \\ \sum y_ix_i &= a_0 \sum x_i + a_1 \sum x_i^2 + \ldots + a_n \sum x_i^2 \\ &\vdots \\ \sum y_ix_i^n &= a_0 \sum x_i^n + a_1 \sum x_i^{n+1} + \ldots + a_n \sum x_i^{2n} \end{aligned}\right\} \qquad (23.50)$$

Solving these $(n + 1)$ equations simultaneously (preferably using a digital computer) we get the values of a_0, a_1, . . ., a_n which gives the best least square polynomial equation for the given data.

Problem 23.2 *The total time required by a driver to bring a vehicle to stop from the moment he perceives the danger consists of: (i) reaction time (time elapsed between the recognition of danger and the application of brakes), and (ii) braking time (the time required for stopping the vehicle after the application of the brakes). The following table gives the stopping distance D (in m) during whole of this time starting from the instant the danger is perceived (i.e. reaction time plus the braking time) observed for a car travelling at V (m/s).*

Speed V (m/s)	15	20	25	30	35
Distance D (m)	30.7	42.8	57.4	70.5	91.3

By using the coding principle, transform the values of V into x = –2, –1, 0, 1 and 2 or otherwise, fit the least square parabola of the form $D = a_0 + a_1V + a_2V^2$.

Solution It may be noted that in this problem, there are odd number of data, the coding technique greatly simplifies the solution of the problem. Therefore, introducing

$$x_i = \frac{V_i - 25}{5}, \text{ we get}$$

x_i	D_i	$x_i D_i$	$x_i^2 D$
–2	30.7	–61.4	122.8
–1	42.8	–42.8	42.8
0	57.4	0.0	0.0
1	70.5	70.5	70.5
2	91.3	182.6	365.2
Total 0	292.7	148.9	601.3

$$\text{Also } \sum x_i^2 = 10, \quad \sum x_i^3 = 0 \text{ and } \sum x_i^4 = 34$$

Using the normal equation of the least square parabola [Eqs. 23.47 to 23.49] we get,

$$292.7 = 5a_0 + a_1(0) + a_2(10)$$

$$148.9 = a_0(0) + 10a_1 + 12(0)$$

and

$$601.3 = 10a_0 + a_1(0) + 34a_2$$

Solving the above equations we get,

$$a_0 = 56.27$$

$$a_1 = 14.89$$

$$a_2 = 1.14$$

Now, the least square parabolic equation in terms of x and D becomes

$$D = 56.27 + 14.89x + 1.14x^2$$

This equation in terms of original equation becomes

$$D = 56.27 + 14.89\left(\frac{V-25}{5}\right) + 1.14\,\frac{(V-25)^2}{5^2}$$

$$= (10.37 + 0.7V + 0.0456V^2) \text{ m}$$

Problem 23.3 *The removal of a drug from the circulatory system follows a law of the form $C = Ae^{-kt}$ where C is the concentration of the drug in the blood and t the time passed since the administration of drug.*

(a) Determine the values of constants A and k for the given data and hence write the equation connecting C and t assuming that the values of t are more accurate than C.

(b) Repeat the calculations of part (a) if the values of C are assumed to be more accurate than the values of t.

t (in h)	0.5	1.0	1.5	2.0	2.5	3.0
C (in mg/ml)	81.45	68.9	60.3	50.5	43.82	38.2

Solution

(a) Since $C = Ae^{-kt}$, we get,

$$\ln C = \ln A - kt$$

Let $\ln C = y$

and $\ln A = a_0$

$\therefore$ it can be written as

$$y = a_0 - kt$$

which is a linear equation.

Using the given data, we make the following table:

t_i	C_i	$y_i = ln\ C_i$	y_i^2	t_iy_i	t_i^2
0.5	81.45	4.40	19.36	2.20	0.25
1.0	68.9	4.23	17.89	4.23	1.0
1.5	60.3	4.10	16.81	6.15	2.25
2.0	50.5	3.92	15.37	7.82	4.0
2.5	48.32	3.78	14.29	9.46	6.25
3.0	38.20	3.64	13.25	10.92	9.0
Total 10.5		24.07	96.97	40.79	22.75

Using Eqs. (23.21) and (23.22) we get the least square regression values of slope and intercept when the values of t are more accurate than C.

$$-k = \frac{(6)(40.79)-(10.5)(24.07)}{(6)(22.75)-(10.5)^2} = -0.3046$$

or $k = 0.3046$

and
$$a_0 = \frac{(24.07)(22.75)-(10.5)(40.79)}{(6)(22.75)-(10.5)^2} = 4.5447$$

Now $\ln A = 4.5447$

$\therefore$ $A = e^{4.5447} = 94.132$

Hence, knowing the values of A and k, the relationship between C and t (when values of t are more accurate than C) becomes

$$C = 94.132e^{-0.3046t}$$

(b) When values of C are assumed more accurate than t then,

Equation $\ln C = \ln A - kt$ can be written as

$$t = \frac{\ln A}{k} - \frac{\ln C}{k}$$

Substituting $\ln A = a_0$ and $\ln C = y$ we get,

$$t = \frac{a_0}{k} - \frac{y}{k}$$

$$= b_0 + b_1 y$$

where
$$b_0 = \frac{a_0}{k} \quad \text{and} \quad b_1 = \frac{-1}{k}$$

Now, Using Eqs. (23.27) and (23.28) we obtain the values of slope and intercept as

$$b_1 = \frac{(6)(40.79) - (10.5)(24.07)}{(6)(96.97) - (24.07)^2}$$

$$= -3.257$$

$$b_0 = \frac{(10.5)(96.97) - (24.07)(40.79)}{(6)(96.97) - (24.07)^2}$$

$$= \frac{36.37}{2.455}$$

$$= 14.815$$

Now, $$b_1 = \frac{-1}{k}$$

or $$k = \frac{-1}{b_1} = \frac{1}{3.257} = 0.3071$$

Further, $$b_0 = \frac{a_0}{k}$$

or $$a_0 = b_0 k$$

$$= (14.815)\ (0.3071)$$

$$= 4.55 = \ln A$$

$\therefore$ $$A = e^{4.55} = 94.63$$

Hence, knowing the values of A and k, the relationship between C and t (when the values of C are more accurate than t) becomes

$$C = 94.63e^{-0.3071t}$$

Review Questions

23.1 Fill up the blanks in the following:

(i) The method of least squares minimises the ______ and ______ values of the variable which is comparatively less accurate.

(ii) Centroidal points namely $\overline{X}$ and $\overline{Y}$ ______ linear least square regression curve.

(iii) A parabolic curve, $y = a + bx + cx^2$ can be linearised by plotting ______ versus ______.

(iv) The normal equation for a parabolic curve passing through the origin (i.e. $y = ax^2$) is ______.

(v) The equation $y = ab^x + c$ represents a ______ (name the type of curve).

(vi) The expression $pq^{b^x} + d$ represents a ______ curve.

(vii) In linear least square regression, the term $\left\{n\sum x_i^2 - \left(\sum x_i\right)^2\right\} = 0$ only if ______.

(viii) A curve of the type $y = \sqrt{ax + b}$ can be linearised by plotting ______ versus ______.

(ix) The constants a and b for exponential data (i.e. $y = ae^{bx}$) can be easily determined by plotting the data on ______ graph paper.

(x) The uncertainty in the values of relatively inaccurate values in linear regression, say y, is given by the expression ______.

23.2 Indicate whether the following statements are true or false. If false, rewrite the correct statement.

(i) The method of extended differences is less accurate than the method of sequential differences for fitting a straight line through the data points.

(ii) The least square line need not always pass through the centroid of the data.

(iii) If y values of data are regressed over x values or vice versa, the resulting least square equation is always the same.

(iv) The method of sequential differences always takes into account all the data points while fitting a straight line through them.

(v) In a bivariate relationship, the method of least squares minimises the sum of error squares in both the variables.

(vi) The curve of the type, $y = (10)^x$ is a power law curve.

23.3 n pairs of observations $(x_1, y_1), (x_2, y_2), \ldots, (x_n, y_n)$ were seen to fit a curve $y = \beta x^2 + \gamma x^3$ (which passes through the origin). Determine the normal equations of the least square regression of y values on x values. Also determine the expressions for determining the values of β and γ.

23.4 Given the following data, would the method of sequential differences yield a good straight line fit. Give reasons to support your answer (do not determine the constants a_0 and a_1).

x_i	1	2	3	4	5	6	7	8
y_i	4	6	9	15	20	26	28	30

23.5 In an air-cooled engine cylinder simulation study, a circular cooling fin was heated in the central region and placed in an air stream. Ten thermocouples were placed at equal distances radially on the fin for temperature measurements. During a test run, the following data was obtained:

Radius R, (cm)	*Temperature T (°C)*	*Radius R (cm)*	*Temperature T (°C)*
2	75.5	12	62.1
4	73.1	14	59.4
6	70.5	16	57.2
8	67.7	18	54.8
10	64.2	20	52.2

Determine the coefficients for the curve of the form, $T = C_0 + C_1R$, using:

(i) graphical method,

(ii) method of sequential differences,

(iii) method of extended differences,

(iv) method of least squares.

23.6 Standard weights x_i are suspended on a spring and the corresponding lengths y_i are measured. It is known that for a linear spring the lengths should follows the linear relation:

$$y = \alpha + \beta x$$

(a) Determine, using the principle of least squares, the values of α and β, for the following set of data:

x_i	y_i
2	7.35
4	8.25
6	9.20
8	10.20
10	11.00
12	12.05

(b) Determine also the internal estimate of uncertainties in the values of α and β.

23.7 A set of machine bearings of a particular make were tested for wear at different operating temperatures controlled by an oil bath. The following test results were obtained:

Operating temperature x (in °C)	100	150	200	250	300	350	400
Amount of wear y (in mg/100 h of operation)	3.2	5.2	5.8	7.9	9.6	11.7	13.2

(a) Plot the points on a graph and verify that a linear relationship exists between x and y.
(b) Find the linear least square curve regressing y on x (i.e. assuming the temperature values given in the data are without error).
(c) Estimate the amount of wear at 325 and 0 °C.
(d) Comment on the estimated amount of wear given by the equation for an operating temperature of 0 °C.

23.8 An experiment has been carried out to investigate the temperature dependence of the resistance of a copper wire. The ideal variation is represented by:

$$R = R_0 (1 + \alpha T)$$

where R is the resistance at temperature T (°C), R_0 the resistance at 0°C and α the temperature coefficient of resistance. The following observations of R and T were obtained:

Temperature T (°C)	*Resistance* R (Ω)	*Temperature* T (°C)	*Resistance* R (Ω)
10	12.4	50	14.6
20	13.0	60	15.2
30	13.7	70	15.3
40	13.9	80	16.0

(a) Using the method of least squares, determine the value for the slope and intercept for the best fitting straight line assuming temperature values to be accurate.
(b) Hence evaluate the best value of α.
(c) Determine the standard deviation for the slope and intercept.
(d) Hence evaluate the standard deviation in α.

23.9 A study was made to determine the effectiveness of a drug in lowering the heart rate in adults. The observations were made on the reduction of heart rate (in beats per minute) following the administration of accurately measured, selected levels of dosages of the drug (in mg). The results were as follows:

Drug dosage x (in mg)	1.0	1.25	1.50	1.75	2.0	2.25	2.5	2.75	3.0
Reduction in heart beats per minute y	9	11	12	13	13	16	18	19	18

Determine the best fitting linear relationship of regressing y values on x values. Comment on the result of substituting $x = 0$ in the equation you have obtained.

23.10 The polytropic expansion/compression processes usually follow the thermodynamic relationship of the form $PV^n = C$, where n and C are constants. Using the least squares principle, determine from the following data:
(a) The values of n and C, (b) write the equation connecting P and V, and (c) estimate P when $V = 100.0$.

Volume V(in mm^3)	40.2	54.2	62.0	80.3	99.5	138.6	150.2	194.2
Pressure P(in N/mm^2)	89.2	60.8	48.5	33.5	25.1	19.2	14.5	10.2

23.11 Fit the curve $y = a/x^2$ to the following data by:
(a) Direct application of the method of least squares.
(b) Transformation of the equation into a linear relationship.

x_i	10	20	30	40	50	60
y_i	0.02245	0.00545	0.00295	0.00145	0.00095	0.00065

23.12 (a) The governing equation of a simple pendulum is:

$$T = 2\pi \sqrt{\frac{L}{g}}$$

What function of T and L would you plot to obtain a linear relationship between these variables?
(b) The following results of time periods were obtained for different lengths of a simple pendulum. Determine the value of g if the lengths of the pendulum were measured comparatively more accurately than that of time periods.

L (cm)	25.1	35.3	49.5	55.9	63.1	72.3
T(s)	1.0	1.2	1.4	1.5	1.6	1.7

(c) Would you obtain a significantly different value of g if T values are more accurate as compared to the measured values of L.

23.13 The following recordings of the number of bacteria, N, present in a culture at time t (h) were made.

t (in h)	0	2	4	6	8
N (bacterial count)	1000	1425	2000	2800	3600

(a) Plot the time of bacterial growth t against the bacterial count.
(b) Plot the time of bacterial growth t against logarithm of the bacterial count, on a separate graph.
(c) Which of these two do you consider closest to the linear relationship?
(d) Calculate the best straight line through the points you have chosen (either bacterial count or log of bacterial count) by the method of least squares.
(e) Estimate the number of bacteria at $t = 10$.

23.14 You are given n pairs of observations (x_i, y_i). Fit the least square curve $y = \ln(a + bx)$ in the given data by converting the given relation to a linear form:

$$e^y = a + bx$$

and using the data in the form of e^{yi} and x_i.

x_i	2	3	4	5	6	7
y_i	1.95	2.16	2.41	2.55	2.70	2.81

23.15 Find the normal equations for fitting least square curves of the following forms:

(a) $y = y_0 \exp(-h^2x^2)$; (b) $y = a + \frac{b}{x} + \frac{c}{x^2}$; (c) $y = \log(a + bx)$; (d) $y = ab^z + c$

Answers

23.1 (i) sum of square between measured, estimated; (ii) lies on the;

(iii) $\frac{y - y_1}{x - x_1}$, x; (iv) $\sum x_i^2 yi = a\sum x_i^4$; (v) modified exponential curve;

(vi) Modified Gompertz; (vii) all data points are same; (viii) y^2, x

(ix) semilog (x) $\sqrt{\frac{\sum y_i^2 - a_1 \sum x_i y_i - a_0 \sum yi}{(n)(n-2)}}$

23.2 (i) False; (ii) False; (iii) False; (iv) False; (v) False; (vi) False

23.3 (i) $\sum y_i x_i^2 = \beta \sum x_i^4 + \gamma \sum x_i^5$; (ii) $\sum y_i x_i^3 = \beta \sum x_i^5 + \gamma \sum x_i^6$

(iii) $\beta = \frac{(\sum yix_i^2)(\sum x_i^6) - (\sum yix_i^3)(\sum x_i^5)}{(\sum x_i^4)(\sum x_i^6) - (\sum x_i^5)^2}$; (iv) $\gamma = \frac{(\sum yix_i^3)(\sum x_i^4) - (\sum y_i x_i^2)(\sum x_i^5)}{(\sum x_i^4)(\sum x_i^6) - (\sum x_i^5)^2}$

23.4 Since the intervals of x values are uniform, therefore the slope in the method of sequential difference depends on the first and the last points of the data, and no weightage is given to the variation of the intermediate points. Therefore, the method of sequential difference would give erroranous results.

23.5 (ii) $T = 77.90 - 1.294\ R$; (iii) $T = 78.04 - 1.306\ R$; (iv) $T = 78.05 - 1.307\ R$

23.6 (i) $\alpha = 6.4$; (ii) $\beta = 0.4678$; (iii) $U(\alpha) = 0.016$; (iv) $U(\beta) = 0.0021$

23.7 (b) $y = 0.0334\ x - 0.27$

(c) 10.585 mg/100 h, –0.27 mg/100 h

(d) The wear at 0°C given by the equation is negative which is not possible in practice. Therefore, we can say that the linear relationship does not hold good for data points near 0 °C and hence the extrapolation of the given data near 0°C is not advisable.

23.8 (a) $R = 12.018 + 0.0499T$; (b) best value of $\alpha = 4.152 \times 10^{-3}$/°C;

(c) (i) σ(slope) 0.00955 Ω/°C, (ii) σ (intercept) = 0.4823 Ω;

(d) $\sigma(\alpha) = 7.95 \times 10^{-4}$/°C

23.9 (a) $y = 5.0x + 4.33$

(b) when $x = 0$, $y = 4.33$ i.e., there is a reduction in heart beats per minute when no drug is administered. This is obviously not possible. Hence, the extrapolation of linear least square regression is not valid near $x = 0$

23.10 (a) 1.347, 12822.11; (b) $PV^{1.347} = 12822.11$; (c) $P = 25.939$ N/mm^2

23.11 (a) $y = 2.211\left(\frac{1}{x^2}\right)$; (b) same as (a).

23.12 (a) The simple pendulum equation can be linearised by plotting $T = f\left(\sqrt{L}\right)$. If we substitute $T = y$ and $\sqrt{L} = x$, then the equation becomes $y_i = a_1 x_I$, where $a_1 = \dfrac{2\pi}{\sqrt{g}}$.

(b) By the method of least squares: $a_1 = \dfrac{\sum y_i \; x_i}{\sum x_i^2} = 2.00428$

which gives the value $g = 9.827$ m/s^2.

(c) When L values are regressed over T values, the value of g obtained is equal to 9.826 which is not significantly different in this particular data. It shows that errors in the measurement of T and L are of the same order.

23.13 (d) $N = 1024.65e^{0.1618t}$ (e) $[N]_{t=10\text{s}} = 5167$ (round figure).

23.14 $y = \ln (3.085 + 1.948x)$

23.15 (a) (i) $\sum y_i' = c'n - h^2 \sum x_i^2$ and

(ii) $\sum y_i' x_i^2 = c' \sum x_i^2 - h^2 \sum x_i'$, where $y' = \ln y$ and $c' = \ln y_0$

(b) (i) $y_i = an + \dfrac{b}{\sum x_i} + \dfrac{c}{\sum x_i^2}$

(ii) $\sum (y_i/x_i) = \dfrac{a}{\sum x_i} + \dfrac{b}{\sum x_i^2} + \dfrac{c}{\sum x_i^3}$ and

(iii) $\sum (y_i / x_i^2) = \dfrac{a}{\sum x_i^2} + \dfrac{b}{\sum x_i^3} + \dfrac{c}{\sum x_i^4}$

(c) (i) $\sum y_i' = a_n + b \sum x_i$ and

(ii) $\sum y_i' \, x_i = a \sum x_i + b \sum x_i^2$ where $y' = (10)^y$

(d) (i) $\sum y_i = a \sum \; + cn$ and

(ii) $\sum y_i x_i' = a \sum x_i'^2 + c \sum x_i'$; where $x' = (b)^x$

APPENDICES

Appendix

A-1

Fundamental and Derived Quantities in International System of Units

We come across two types of physical quantities, viz. fundamental quantities and derived quantities. The latter are functions of the fundamental quantities. The fundamental quantities can be further subdivided into the following two categories:

1. Primary fundamental quantities that are the most commonly occurring basic quantities, viz. mass, length and time.
2. Auxiliary fundamental quantities that occur in the thermal, electrical or illumination systems, viz. temperature, electric current and luminous intensity.

The standardised units of these quantities depend on the system of units used in practice. In the past, gravitational system of units, also known as cgs system of units defined the primary fundamental quantities, i.e. length, mass, and time as centimetre, gram and second. In fact, the name cgs units is a short form of centimetre, gram and second. This system had the advantage of the metric system but the unit quantities of mass and length were rather small. Therefore, the units of the derived quantities were also quite small. Further, fps units (short form of foot, pound and second) or English system of units was used in U.K., U.S.A. and other countries along with cgs system of units. The advantage in this system was that the unit quantities of mass and length namely, pound and foot were 'reasonable' quantities. However, the main disadvantage was that it did not conform to the metric system. In 1935, an Italian engineer Giorgi suggested a new system of units known as mksA system. In this the units of length, force (in place of mass), time and electric current were metre, kilogram force, second and ampere, respectively. This was adpoted by many countries. In 1960, at the international Conference of Weights and Measures, a new comprehensive system of units known as S.I. units (which is a short form the French word, *Systeme International d'Unites*) was evolved. This system of units has been adopted as legal in France and many other countries in the world. The advantages of the S.I. system are:

1. It is a metric system with base 10. Therefore, the decimal relationship between the units of same quantity makes it possible to express any small or large quantity in terms of power to the base 10.

2. It is more comprehensive because it defines the units of both primary fundamental as well as auxiliary fundamental quantities. In other words, it gives six basic units of mass, length and time as well as temperature, electric current and luminous intensity.
3. It is associated with coherent system of units. In other words, the product or quotient of any two base quantities results in a unit measured quantity. For example, unit length divided by unit time gives rise to a unit velocity.
4. In addition, it defines another base unit, the mole, which is the amount of a substance that equals in weight to its molecular weight in grams.
5. Further, two supplementary units of the plane angle (in radian) and solid angle (in steradian) are also defined in this system of units.
6. Lastly, this system has standardised the abbreviations of different units for complete uniformity and consistency.

Table A1.1 gives the symbols, dimensions and the units of the various fundamental, supplementary and the derived quantities in S.I. units.

Table A1.1 *Units, Dimensions and Symbols of Fundamental, Supplementary and Derived Physical Quantities*

Physical Quantity	*Symbol*	*Dimension*	*Unit*	*Unit symbol*
Fundamental				
Length	l	L	metre	m
Mass	m	M	kilogram	kg
Time	t	T	second	s
Temperature	T	θ	kelvin	K
Electric current	I	I	ampere	A
Luminous intensity			candela	Cd
Amount of substance			mole	mol
Supplementary				
Plane angle	α, β, γ	$[L]^\circ$	radian	rad
Solid angle	Ω	$[L^2]^\circ$	steratian	sr
Derived				
Density	ρ	$L^{-3} M$	kilogram per cubic metre	kg/m^3
Acceleration	a	LT^{-2}	metre per second squared	m/s^2
Force	F	LMT^{-2}	newton	N ($kg\text{-}m/s^2$)
Angular acceleration	α	T^{-2}	radian per second squared	rad/s^2
Energy, work	W	$L^2 MT^{-2}$	joule	J (N m)
Power	P	$L^2 MT^{-3}$	watt	W (J/s)
Pressure, stress	p	$L^{-1} MT^{-2}$	pascal	Pa (N/m^2)
Frequency	f	T^{-1}	hertz	Hz (1/s)
Electric current	I	I	ampere	A
Electric charge	Q	TI	coulomb	C (As)
Potential difference	V	$L^2 MT^{-3} I^{-1}$	volt	V (W/A)
Electric resistance	R	$L^2 MT^{-3} I^2$	ohm	Ω (V/A)
Electric capacitance	C	$L^{-2} M^{-1} T^4 I^2$	farad	F (A s/V)
Magnetic flux	ϕ	$L^2 MT^{-2} I^{-1}$	weber	Wb (V s)
Electric inductance	L	$L^2 MT^{-2} I^2$	henry	H (V s/A)
Magneitc flux density	B	$MT^{-2} I^{-1}$	tesla	T (Wb/m^2)
Luminous flux			lumen	1m (cd sr)
Luminance			candela per square metre	cd/m^2
Illumination			lux	lx (lm/m^2)

Appendix

A-2

Derivation of Solution for Step Response of Second-Order System

The final solution will be derived as below, for the case $\xi < 1$ or the underdamped case. Referring to Section 3.2.3 for the second-order system, the total solution is the sum of complementary solution of Eq. (3.63a) and particular solution of Eq. (3.63b). This gives

$$x_o(t) = \exp(-\xi\omega_n t)\left[c_1 \cos\omega_n \sqrt{1-\xi^2}\, t + c_2 \sin\omega_n \sqrt{1-\xi^2}\, t\right] + Kx_s \qquad \text{(A2.1)}$$

Using initial condition at

$$t = 0, \quad x_0(t) = 0$$

gives $\quad c_1 = -Kx_s$

and using initial conditions at

$$t = 0, \quad \frac{dx_o(t)}{dt} = 0, \text{ gives}$$

$$c_2 = -Kx_s \frac{\xi}{\sqrt{1-\xi^2}}$$

Substituting in Eq. (A.21) gives

$$x_0(t) = Kx_s\left[1 - e^{-\xi\omega_n t}\left(\cos\omega_n \sqrt{1-\xi^2}\, t + \frac{\xi}{\sqrt{1-\xi^2}} \sin\omega_n \sqrt{1-\xi^2}\, t\right)\right] \qquad \text{(A2.2)}$$

Equation (A2.2) represents an oscillatory decaying solution with frequency $\omega_n \sqrt{1-\xi^2} = \omega_d$, called damped frequency and amplitude decaying with time due to the term $e^{-\xi w_n t}$.

The above equation may also be written is an alternative form by substituting the coefficient of cos $\omega_d t$ and sine $\omega_d t$ terms in Eq. (A2.2) as

$$\frac{\xi}{\sqrt{1-\xi^2}} = R \cos \phi$$

$$1 = R \sin \phi$$

With the above, Eq. (A2.2) can be written as

$$x_o(t) = K_{x_S} \left[1 - \frac{e^{-\xi \omega_n t}}{\sqrt{1-\xi^2}} . \sin (\sqrt{1-\xi^2}\ \omega_n t + \phi) \right] \tag{A2.3}$$

where $$R = \frac{1}{\sqrt{1-\xi^2}}$$

and $$\phi = \sin^{-1} \sqrt{1-\xi^2}$$

Appendix

A-3

Auto-Correlation Functions of a Random Signal

The auto-correlation function of $x_i(t)$ is defined as:

$$R(\tau) = \lim_{T\to\infty} \frac{1}{2T} \int_{-T}^{T} x_i(t)\, x_i\,(t+\tau)\, \mathrm{d}t \tag{A3.1}$$

The function $R(\tau)$ is obtained by finding the values of the random signal at times t and $(t + \tau)$, τ being the time delay (Fig. A3.1). Its value is found for a certain time interval τ and the process is repeated for various values of τ. For a fast random signal, $R(\tau)$ versus τ curve shows a sharp peak, and for a slow random signal, the curve is flat, as shown in Fig. A3.2.

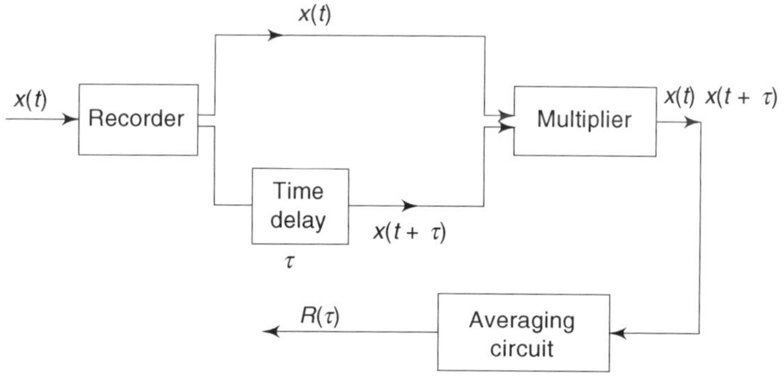

Fig. A3.1 *Determination of auto-correlation function R(τ)*

A frequency decomposition of $R(\tau)$ can be made as below by taking its Fourier transform $F(\omega)$

$$f(\omega) = \int_{-\infty}^{\infty} R(\tau)\, e^{-i\omega\tau}\, \mathrm{d}\tau \tag{A3.2}$$

where $f(\omega)$ is the Fourier transform of $R(\tau)$.

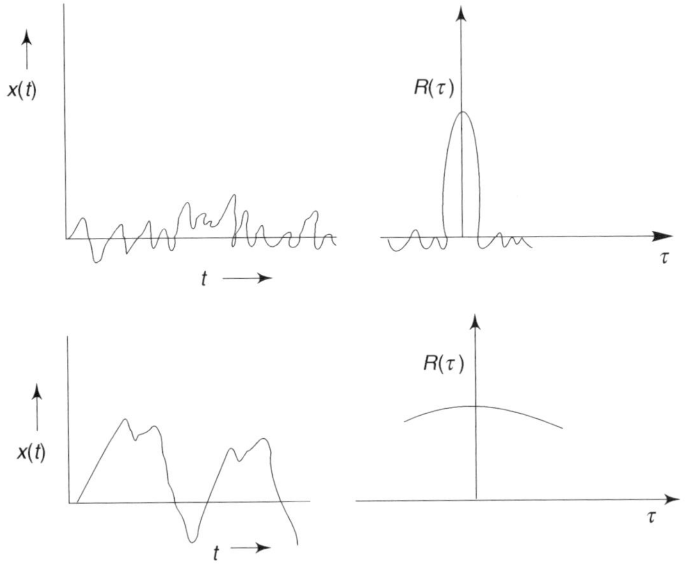

Fig. A3.2 *Auto-correlation function for various signals*

Its inverse transform is:

$$R(\tau) = \frac{1}{2\pi} \int_{-\infty}^{\infty} f(\omega) e^{i\omega\tau} \, d\omega \tag{A3.3}$$

For $\tau = 0$, Eq. (A3.3) is

$$R(0) = \frac{1}{2\pi} \int_{-\infty}^{\infty} f(\omega) \, d\omega \tag{A3.4}$$

Since $R(0)$ would be the mean square value of the random signal $\overline{x}_i^2$, we can write,

$$\overline{x}_i^2 = \frac{1}{2\pi} \int_{-\infty}^{\infty} f(\omega) d\omega \tag{A3.5}$$

(from Sec. 3.3.4)

Also

$$\overline{x}_i^2 = \int_{-\infty}^{\infty} s(\omega) \, d\omega \tag{A3.6}$$

$s(\omega)$ being the mean spectral density of the signal. Since $f(\omega)$ is the Fourier transform of the auto-correlation function $R(\tau)$, we can say from Eqs. (A3.5) and (A3.6) that the Fourier transform of auto-correlation function is proportional to the mean spectral density of the signal. Thus the relation between $R(\tau)$ and $s(\omega)$ can be noticed.

Appendix

A-4

Principal Strain and Stress Relations

The general relation for strain ε_θ and shear strain γ_θ on a plane at any angle θ from axis are:

$$\varepsilon_\theta = (\varepsilon_x + \varepsilon_y)/2 + (\varepsilon_x - \varepsilon_y)/2 \cos 2\theta + \gamma_{xy}/2 \sin 2\theta.$$

$$\gamma_\theta = (\varepsilon_x - \varepsilon_y) \sin 2\theta - \gamma_{xy} \cos 2\theta.$$

where ε_x and ε_y are direct strains in 'x' and 'y' directions respectively and γ_{xy} is shear strain.

Putting $d\varepsilon_\theta/d\theta = 0$, we get the direction of the principal strains i.e. angle θ_p with chosen x-axis, viz.

$$\tan 2\theta_p = \gamma_{xy}/(\varepsilon_x - \varepsilon_y) \qquad \text{(A4.1)}$$

Two angles θ_p, 90° apart may be obtained, one corresponding to maximum normal stress and the other corresponding to the minimum one.

$$\sin 2\theta_p = \pm \tan 2\theta_p/(1 + \tan^2 2\theta_p)^{1/2}$$

$$= \pm \gamma_{xy}/[(\varepsilon_x - \varepsilon_y)^2 + \sigma_{xy}^2]^{1/2}$$

and $$\cos 2\theta_p = \pm (\varepsilon_x - \varepsilon_y)/[(\varepsilon_x - \varepsilon_y)^2 + \sigma_{xy}]^{1/2}$$

Substituing the above in the equation for ε_θ, we get the maximum and minimum values of principal strains as below (using + and – signs, respectively, for maximum and minimum values)

$$\varepsilon_{max} = (\varepsilon_x + \varepsilon_y)/2 + 1/2[(\varepsilon_x - \varepsilon_y)^2 + \gamma_{xy}^2]^{1/2} \qquad \text{(A4.2)}$$

$$\varepsilon_{min} = (\varepsilon_x + \varepsilon y)/2 - 1/2[(\varepsilon_x - \varepsilon_y)^2 + \gamma_{xy}^2]^{1/2} \qquad \text{(A4.3)}$$

Further, the relations between principal stresses and strains for the plane stress case are:

$$\varepsilon_x = \sigma_x/E - \mu\, \sigma_y/E$$

$$\varepsilon_y = \sigma_y/E - \mu\, \sigma_x/E$$

where E = Young's modulus and

μ = Poisson's ratio

From the above, we get

$$\sigma_x = [E/(1 - \mu^2)] (\varepsilon_x + \mu\varepsilon_y)$$

$$\sigma_y = [E/(1 - \mu^2)] (\varepsilon_y + \mu\varepsilon_x)$$

Further, the expressions for principal stresses σ_{max} and σ_{min} are given as below, in terms of principal strains ε_{max} and ε_{min}.

$$\sigma_{max} = [E/(1 - \mu^2)] (\varepsilon_{max} + \mu\varepsilon_{min} \quad (A4.4)$$

$$\sigma_{min} = [E/(1 - \mu^2)] (\varepsilon_{min} + \mu\varepsilon_{max}) \quad (A4.5)$$

Further, it can be shown that the equation for the shear stress on any plane in terms of σ_x, σ_y and θ is

$$\tau_\theta = 1/2 (\sigma_x - \sigma_y) \sin 2\theta - \tau_{xy} \cos 2\theta$$

The angle at which shear stress is maximum can be found easily. The value of

$$\tau_{max} = 1/2[(\sigma_x - \sigma_y)^2 + (2\tau_{xy})^2]^{1/2}$$

Appendix

A-5

Statistical Properties of a Pair of Random Signals

Figure A5.1 shows a pair of signals, viz. force 'f' and response 'x', in time 't' and frequency domain 'ω'.

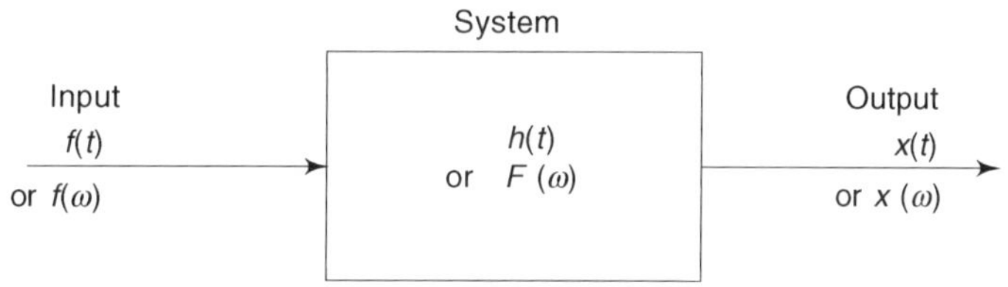

Fig. A5.1 *Transfer functions in time t and frequency ω domains*

$h(t)$ is called unit impulse forcing function in time domain 't' and $H(\omega)$ is called transfer function in frequency domain 'ω'.

The response to a unit impulse forcing function $h(t - \tau)$ is given by:

$$x(t) = \int_{-\infty}^{t} f(\tau)\, h\,(t - \tau)\, \mathrm{d}\tau.$$

where $\quad h\,(t - \tau) = 0 \quad \text{for} \quad t < \tau.$

and τ is the time at which impulse is applied.

Response $x(t)$ may be rewritten as:

$$x(t) = h(t) * f(t)$$

where the symbol * denotes the convolution operation. Taking the Fourier Transform of the above, it can be shown that:

$$x\,(\omega) = H\,(\omega)\, f\,(\omega),$$

which gives the Fourier transform $H\,(\omega)$ in the frequency domain. As mentioned in Sec. 15.7, the Fourier transforms can only be obtained for signals satisfying Dirichlet condition and the same cannot be obtained for random signals $x(t)$ and $f(t)$. However, the Fourier transforms can be obtained for cross-correlation and auto-correlation functions of the random signals which are seen to satisfy the dirichlet conditions.

Cross-correlation and Auto-correlation Functions Cross-correlation function relating to the above pair of random signals is:

$$R_{fx}(\tau) = \lim_{T \to \infty} \frac{1}{T} \int_{-\frac{T}{2}}^{\frac{T}{2}} f(t)\, x(t+\tau)\, dt$$

where 'γ' is the time delay and 'T' the time duration.
For auto-correlation function $R_{xx}(\tau)$ and $R_{ff}(\tau)$,

$$R_{xx}(\tau) = \lim_{T \to \infty} \frac{1}{T} \int_{-\frac{T}{2}}^{\frac{T}{2}} x(t)\, x(t+\tau)\, dt$$

A similar expression can be written for $R_{ff}(\tau)$.

Further, $$H(\omega)\, H^*(\omega) = \frac{x(\omega)\, x^*(\omega)}{f(\omega)\, f^*(\omega)} = \frac{S_{xx}(\omega)}{s_{ff}(\omega)}$$

or $$|H(\omega)|^2\, S_{ff}(\omega) = S_x\, x(\omega) \quad \text{(A5.1)}$$

where $S_{xx}(\omega)$ is called auto-spectral density function of $R_{xx}(\tau)$ and $S_{ff}(\omega)$ is auto-spectral density function of $R_{ff}(\tau)$.

or $$S_{xx}(\omega) = \int_{-\infty}^{\infty} R_{xx}(\tau) e^{-i\omega\tau}\, d\tau$$

$$S_{ff}(\omega) = \int_{-\infty}^{\infty} R_{ff}(\tau)\, e^{-i\omega\tau}\, d\tau.$$

The cross-spectral density function $S_{fx}(\omega)$ is given by the Fourier transforms of the corresponding correlation function viz. R_{fx}.

or $$S_{fx}(\omega) = \int_{-\infty}^{\infty} R_{fx}(\tau)\, e^{-i\omega\tau}\, d\tau$$

Further transfer function $H(\omega)$ may be written as

$$H(\omega) = \frac{S_x(\omega)}{S_f(\omega)} = \frac{S_x(\omega)\, S_x^*(\omega)}{S_f(\omega)\, S_x^*(\omega)}$$

$$= \frac{S_{xx}}{S_{fx}} = H_2(\omega) \quad \text{(A5.2)}$$

where S_x^* is complex conjugate of S_x.

Similarly $$H(\omega) = \frac{S_x(\omega)}{S_f(\omega)} \frac{S_f^*(\omega)}{S_f^*(\omega)} = \frac{S_{xf}}{S_{ff}} = H_1(\omega) \quad \text{(A5.3)}$$

The ratio of $H_1(\omega)$ and $H_2(\omega)$ is called γ^2, the coherence, which represents the quality of experimental analysis.

It will be shown below that $\gamma^2 < 1$ if noise is present in the system signals. This can be improved by averaging a large number of samples of signals.

Coherence In order to study the effect of noise at the output of a system shown in Fig. A5.2, it is seen that

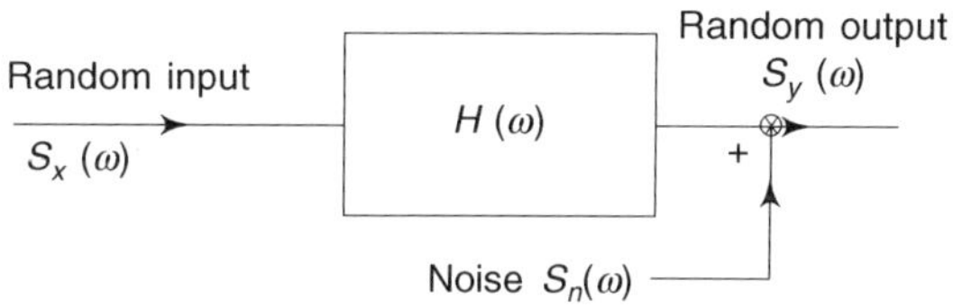

Fig. A5.2 *Random input–output system with noise*

$$S_{yy} = S_y\, S_y^* \tag{A5.4}$$

Now,
$$S_y = H\, S_x + S_n \tag{A5.5}$$
$$S_y^* = H^*\, S_x^* + S_n^* \tag{A5.6}$$

Substituting Eqs. (A5.5) and (A5.6) in Eq. (A5.4),

$$S_{yy} = H^*\, H\, S_x\, S_x^* + H^*\, S_x^*\, S_n + H\, S_x\, S_n^* + S_n\, S_n^* \tag{A5.7}$$

If x and n are not correlated, $S_x^*\, S_n$ and $S_x\, S_n^*$ disappear in Eq. (A4.4) on averagings, giving

$$S_{yy} = |H|^2\, S_{xx} + S_{nn} \tag{A5.8}$$

Substituting Eq. (A5.5) in (A5.6)

$$S_{yx} = (HS_x + S_n)\, S_x^*$$
$$= H\, S_{xx} + S_{nx}$$

Since
$$S_{nx} = 0,$$
$$S_{yx} = H\, S_{xx} \tag{A5.9}$$

or
$$|S_{yx}|^2 = |H|^2\, S_{xx}^2 \tag{A5.10}$$

Using Eqs. (A5.2) and (A5.3),

Coherence
$$\gamma^2 = \frac{H_1(\omega)}{H_2(\omega)}$$
$$= \frac{|S_{yx}|^2}{S_{xx}\, S_{yy}}$$

Using Eqs. (A5.8) and (A5.10), use get

$$\gamma^2 = \frac{|H|^2\, S_{xx}}{|H|\, S_{xx}^2 + S_{nn}} \tag{A5.11}$$

If noise = θ, $\gamma^2 = 1$, since $S_{nn} = \theta$. Thus $\gamma^2 < 1$ if noise is present.

Bibliography

Allocca, J.A. and Stuart Allen, *Transducers: Theory and Applications*, Reston Publishing Co., VA-1984.

Barney, G.C., *Intelligent Instrumentation*, Prentice-Hall of India Pvt. Ltd., New Delhi, 1988.

Beckwith, Thomas G., N. Buck Lewis and D. Marangoni Roy, *Mechanical Measurements*, 3rd Ed., Addison-Wesley Reading, Massachussetts, 1982.

Bentley, J.P., *Principles of Measurement Systems*, Pearson Education, New Delhi, 1995.

Bolton, W., *Mechatronics* 3rd Ed., Pearson Education, New Delhi, 2003.

Brignell, J. and White, N., *Intelligent Sensor Systems*, Institute of Physics Publishing, London, Revised Ed., 1996.

Collacott, R.A., *Mechanical Fault Diagnosis and Condition Monitoring*, Chapman and Hall, London, 1977.

Cromwell, Leslie, Weibell, F.J. and Pfeiffer E.A., *Biomedical Instrumentation and Measurements*, 2nd Ed., Prentice-Hall, N.J. 1991.

Dally, J.W. and W.F. Riley, *Experimental Stress Analysis*, 3rd Ed., McGraw-Hill, New York 1991.

Dally, J.W., William, R.F. and McConnell K.G., *Instrumentation for Engineering Measurements*, 2nd Ed., John Wiley and Sons, N.Y. 1993.

Doeblin, E.A. and Manik D.N., *Measurement Systems, Application and Design*, 5th Ed., Tata McGraw Hill Education Private Ltd., 2004.

Figliola, R.S. and Beasley, D.E. *Theory and Design for Mechanical Measurements*, John Wiley and Sons, N.Y. 1991.

Frank, R., *Understanding Smart Sensors*, Artech House Inc., U.S.A., 2000.

Helfrick, A.D. and Cooper, W.D., *Modern Electronic Instrumentation and Measurement Techniques*. Prentice-Hall of India, 1990.

Holman, J.P., *Experimental Methods for Engineers*, 7th Ed., Tata McGraw Hill Education Private Ltd., 2001.

Khandpur, R.S., *Handbook of Biomedical Instrumentation*, Tata McGraw Hill Education Pvt. Ltd., New Delhi, 1987.

Murty, D.V.S., *Transducers and Instrumentation*, 2nd Ed. Prentice-Hall of India Pvt. Ltd., New Delhi, 2008.

Nakra, B.C., *Theory and Applications of Automatic Controls*, New Age International (P) Ltd. New Delhi, 1998.

Nakra, B.C., Yadava, G.S. and Thuestad, L., *Vibration Measurement and Analysis*, National Productivity Council, New Delhi, 1989.

Nottingk, B.E. (Editor), "*Instrumentation Reference Book*, Butterworths, London, 2nd Ed. 1996.

Padmanabhan, T.R., *Industrial Instrumentation—Principles and Design*, Springer-Verlag, London, 2000.

Patranabis, D., *Sensors and Transducers*, Wheeler Publishing, New Delhi, 1997.

Rangan, C.S., G.R. Sarma, and V.S.V. Mani, *Instrumentation—Devices and Systems*, Tata McGraw Hill Education Private Ltd., New Delhi, 1997.

Raj, B, Jayakumar T. and Thavasimuthu M., *Practical Non-destructive Testing*, Narosa Publishing House, N. Delhi, 2nd Ed. 2002.

Index

Made in United States
Orlando, FL
04 November 2026

90204511R00363